Lecture Notes in Computer Science   11881

More information about this series at http://www.springer.com/series/7409

Reynold Cheng · Nikos Mamoulis ·
Yizhou Sun · Xin Huang (Eds.)

Web Information Systems Engineering – WISE 2019

20th International Conference
Hong Kong, China, January 19–22, 2020
Proceedings

 Springer

Editors
Reynold Cheng (iD)
University of Hong Kong
Hong Kong SAR, China

Nikos Mamoulis
University of Ioannina
Ioannina, Greece

Yizhou Sun
University of California
Los Angeles, CA, USA

Xin Huang (iD)
Hong Kong Baptist University
Kowloon Tong, Hong Kong SAR, China

ISSN 0302-9743   ISSN 1611-3349 (electronic)
Lecture Notes in Computer Science
ISBN 978-3-030-34222-7   ISBN 978-3-030-34223-4 (eBook)
https://doi.org/10.1007/978-3-030-34223-4

LNCS Sublibrary: SL3 – Information Systems and Applications, incl. Internet/Web, and HCI

Due to the problems/protests in Hong Kong, WISE 2019 has been postponed from November 26–30, 2019 until January 19–22, 2020.

This Springer imprint is published by the registered company Springer Nature Switzerland AG
The registered company address is: Gewerbestrasse 11, 6330 Cham, Switzerland

Preface

Welcome to the proceedings of the 20th International Conference on Web Information Systems Engineering (WISE 2019), held in Hong Kong, China, during November 26–30, 2019. This year marks the 20th anniversary of the conference, which was first held in Hong Kong in 2000. The series of WISE conferences aims to provide an international forum for researchers, professionals, and industrial practitioners to share their knowledge in the rapidly growing area of web technologies, methodologies, and applications. The first WISE event took place in Hong Kong, China (2000). Then the trip continued to Kyoto, Japan (2001); Singapore (2002); Rome, Italy (2003); Brisbane, Australia (2004); New York, USA (2005); Wuhan, China (2006); Nancy, France (2007); Auckland, New Zealand (2008); Poznan, Poland (2009); Hong Kong, China (2010); Sydney, Australia (2011); Paphos, Cyprus (2012); Nanjing, China (2013); Thessaloniki, Greece (2014); Miami, USA (2015); Shanghai, China (2016); Puschino, Russia (2017); Dubai, UAE (2018); and this year, WISE 2019 returns to Hong Kong, China, supported by the Hong Kong Polytechnic University and City University of Hong Kong.

A total of 211 research papers were submitted to the conference for consideration, and most of them were reviewed by three reviewers. Finally, 50 submissions were selected as regular papers (an acceptance rate of 23.7%). The research papers cover the areas of blockchain, deep learning, machine learning, recommender systems, data mining, web-based applications, graph learning, knowledge graphs, graph mining, text mining, and crowdsourcing.

In addition to regular research papers, the WISE 2019 program also featured the International Workshop on Web Information Systems in the Era of AI. This year's tutorial program included: (1) Knowledge Graph Data Management: Models, Methods, and Systems; (2) Intelligent Knowledge Lakes: The Age of Artificial Intelligence and Big Data; and (3) Local Differential Privacy: Tools, Challenges, and Opportunities.

The Organization Committee members of WISE 2019 included: the general co-chairs, Prof. Christian Jensen, Prof. Qing Li, and Prof. Tamer Ozsu; the program co-chairs, Prof. Reynold Cheng, Prof. Nikos Mamoulis, and Prof. Yizhou Sun; the workshop co-chairs, Prof. Leong Hou U and Prof. Jian Yang; the industry chair, Prof. Feifei Li; the tutorial and panel chairs, Prof. Kamal Karlapalem and Prof. Yunjun Gao; the demo chair, Dr. Yi Cai; the sponsor chair, Prof. Henry Chan; the finance chair, Prof. Howard Leung; the local arrangement co-chairs, Prof. Hong Va Leong, Prof. Man Lung Yiu, and Mr. Andrew Jiang; the publication chair, Dr. Xin Huang; the publicity co-chairs, Dr. Panagiotis Bouros, Dr. An Liu, and Dr. Wen Hua; and the WISE Steering Committee Representative, Prof. Yanchun Zhang.

We would like to sincerely thank our keynote speakers:

- Professor Wei Zhao, Chief Research Officer, American University of Sharjah, UAE; and
- Professor Mohamed Mokbel, University of Minnesota at Twin Cities, USA.

Special thanks are due to the members of the international Program Committee and the external reviewers for a rigorous and robust reviewing process. We are also grateful to the Hong Kong Polytechnic University, City University of Hong Kong, Springer Nature Switzerland AG, and the International WISE Society for supporting this conference. The WISE Organizing Committee is also grateful to the workshop organizers for their great efforts in helping promote web information system research to broader domains.

We expect that the ideas that have emerged in WISE 2019 will result in the development of further innovations for the benefit of scientific, industrial, and social communities.

November 2019   Christian Jensen
                                                          Qing Li
                                                          Tamer Ozsu
                                                          Reynold Cheng
                                                          Nikos Mamoulis
                                                          Yizhou Sun

The original version of cover and frontmatter was revised: The conference date was postponed by the organizers from November 26–30, 2019 to January 19–22, 2020.

Organization

General Co-chairs

Christian Jensen	Aalborg University, Denmark
Qing Li	The Hong Kong Polytechnic University, Hong Kong SAR, China
Tamer Ozsu	University of Waterloo, Canada

Program Co-chairs

Reynold Cheng	University of Hong Kong, Hong Kong SAR, China
Nikos Mamoulis	University of Ioannina, Greece
Yizhou Sun	University of California at Los Angeles, USA

Workshop Co-chairs

Leong Hou U.	University of Macau, Macau SAR, China
Jian Yang	Macquarie University, Australia

Industry Chair

Feifei Li	Alibaba, China

Tutorial and Panel Chair

Kamal Karlapalem	IIIT, India
Yunjun Gao	Zhejiang University, China

Demo Chair

Yi Cai	South China University of Technology, China

Sponsor Chair

Henry Chan	The Hong Kong Polytechnic University, Hong Kong SAR, China

Finance Chair

Howard Leung	City University of Hong Kong, Hong Kong SAR, China

Local Arrangement Co-chairs

Hong Va Leong   The Hong Kong Polytechnic University,
                             Hong Kong SAR, China
Man Lung Yiu   The Hong Kong Polytechnic University,
                             Hong Kong SAR, China
Andrew Jiang   Macao Convention & Exhibition Association,
                             Macau SAR, China

Publication Chair

Xin Huang   Hong Kong Baptist University, Hong Kong SAR,
                             China

Publicity Co-chairs

Panagiotis Bouros   Johannes Gutenberg University Mainz, Germany
An Liu   Soochow University, China
Wen Hua   Queensland University, Australia

WISE Steering Committee Representative

Yanchun Zhang   Victoria University, Australia

Program Committee

Karl Aberer   École Polytechnique Federale de Lausanne,
                             Switzerland
Marco Aiello   University of Stuttgart, Germany
Bernd Amann   Sorbonne University, France
Chutiporn Anutariya   Asian Institute of Technology, Thailand
Nikos Armenatzoglou   Amazon Web Services, USA
Devis Bianchini   University of Brescia, Italy
Klemens Böhm   Karlsruhe Institute of Technology, Germany
Xin Cao   The University of New South Wales, Australia
Barbara Catania   DIBRIS, University of Genoa, Italy
Bogdan Cautis   Télécom ParisTech, France
Tsz Nam Chan   The University of Hong Kong, Hong Kong SAR,
                             China
Kevin Chang   University of Illinois at Urbana-Champaign, USA
Wei Chen   Chinese Academy of Agricultural Sciences, China
Xiaojun Chen   Shenzhen University, China
Jiefeng Cheng   Tencents, China
Jinchuan Chen   Renmin University, China
Muhao Chen   University of California at Los Angeles, USA

Dickson K. W. Chiu	The University of Hong Kong, Hong Kong SAR, China
Byron Choi	Hong Kong Baptist University, Hong Kong SAR, China
Theodoros Chondrogiannis	University of Konstanz, Germany
Dario Colazzo	Paris Dauphine University, France
Alexandra Cristea	Durham University, UK
Valeria De Antonellis	University of Brescia, Italy
Anton Dignos	Free University of Bozen-Bolzano, Italy
Ha Loc Do	Alibaba, China
Lei Duan	Sichuan University, China
Georgios Fakas	Uppsala University, Sweden
Yixiang Fang	The University of New South Wales, Australia
Yunjun Gao	Zhejiang University, China
Claude Godart	University of Lorraine, France
Daniela Grigori	University Paris-Dauphine, France
Tobias Grubenmann	The University of Hong Kong, Hong Kong SAR, China
Xin Huang	Hong Kong Baptist University, Hong Kong SAR, China
Haibo Hu	The Hong Kong Polytechnic University, Hong Kong SAR, China
Jiafeng Hu	Googles Shanghai, China
Jyun-Yu Jiang	University of California at Los Angeles, USA
Panos Kalnis	King Abdullah University of Science and Technology, Saudi Arabia
Verena Kantere	University of Ottawa, Canada
Georgia Kapitsaki	University of Cyprus, Cyprus
Panagiotis Karras	Aarhus University, Denmark
Kyoung-Sook Kim	Tokyo Institute of Technology, Japan
Hong Va Leong	The Hong Kong Polytechnic University, Hong Kong SAR, China
John Liagouris	ETH Zurich, Switzerland
Kewen Liao	Charles Darwin University, Australia
Jianxin Li	Deakin University, Australia
Guoliang Li	Tsinghua University, China
An Liu	Soochow University, China
Yanhua Li	Worcester Polytechnic Institute, USA
Hui Li	Xiamen University, China
Guanfeng Liu	Macquarie University, Australia
Siqiang Luo	The University of Hong Kong, Hong Kong SAR, China
Jianming Lv	South China University of Technology, China

Fenglong Ma	Dalian University of Technology, China
Hui Ma	Victoria University of Wellington, New Zealand
Jiangang Ma	Federation University, Australia
Yun Ma	City University of Hong Kong, Hong Kong SAR, China
Abyayananda Maiti	Indian Institute of Technology Patna, India
Silviu Maniu	Université Paris-Sud, France
Sarana Nutanong	Vidyasirimedhi Institute of Science and Technology, Thailand
George Papastefanatos	Information Systems Management Institute, Athena Research Center, Greece
Kostas Patroumpas	Information Systems Management Institute, Athena Research Center, Greece
Mirjana Pavlovic	École Polytechnique Federale de Lausanne, Switzerland
Dimitris Plexousakis	Institute of Computer Science, FORTH, Greece
Nicoleta Preda	University Paris Saclay, France
Dimitris Sacharidis	Vienna University of Technology, Austria
Heiko Schuldt	University of Basel, Switzerland
Konstantinos Semertzidis	IBM Research, Ireland
Jingbo Shang	University of Illinois at Urbana-Champaign, USA
Caihua Shan	The University of Hong Kong, Hong Kong SAR, China
Chuan Shi	Beijing University of Posts and Telecommunications, China
Jieming Shi	National University of Singapore, Singapore
Mauro Sozio	Télécom ParisTech, France
Kostas Stefanidis	Tampere University, Finland
Stefan Tai	TU Berlin, Germany
Chaogang Tang	China University of Mining and Technology, China
Bo Tang	Southern University of Science and Technology, China
Dimitri Theodoratos	NJIT, USA
Panayiotis Tsaparas	University of Ioannina, Greece
Leong Hou U.	University of Macau, Macau SAR, China
Hongzhi Wang	Harbin Institute of Technology, China
Jin Wang	University of California at Los Angeles, USA
Lizhen Wang	Yunnan University, China
Xin Wang	Tianjin University, China
Tim Weninger	University of Notre Dame, USA
Shiting Wen	Ningbo Technological Institute of Zhejing University, China
Dingming Wu	Shenzhen University, China
Mingjun Xiao	University of Science and Technology of China, China
Xiaokui Xiao	National University of Singapore, Singapore

Xike Xie	University of Science and Technology of China, China
Zhenguo Yang	City University of Hong Kong, Hong Kong SAR, China
Carl Yang	University of Illinois at Urbana-Champaign, USA
Hongzhi Yin	The University of Queensland, Australia
Man Lung Yiu	The Hong Kong Polytechnic University, Hong Kong SAR, China
Sira Yongchareon	Auckland University of Technology, New Zealand
Demetrios Zeinalipour-Yazti	University of Cyprus, Cyprus
Detian Zhang	Soochow University, China
Jilian Zhang	Jinan University, China
Peixiang Zhao	Florida State University, USA
Yudian Zheng	Twitter, USA
Lihua Zhou	Yunnan University, China
Yi Zhuang	Zhejiang GongShang University, China
Andreas Zuefle	George Mason University, USA

Contents

Recommender Systems

Data Mining

Graph Learning

Knowledge Graphs

Graph Mining

Text Mining

BlockChain and Crowdsourcing

GroExpert: A Novel Group-Aware Experts Identification Approach in Crowdsourcing

Qianli Xing[1(✉)], Weiliang Zhao[1], Jian Yang[1], Jia Wu[1], Qi Wang[1], and Mei Wang[2]

[1] Department of Computing, Macquarie University, Sydney, Australia
qianli.xing@students.mq.edu.au,
{weiliang.zhao,jian.yang,jia.wu}@mq.edu.au, qi.wang20@students.mq.edu.au
[2] School of Computer Science and Technology, Donghua University, Shanghai, China
wangmei@dhu.edu.cn

Abstract. Measuring workers' abilities is a way to address the long standing problem of quality control in crowdsourcing. The approaches for measuring worker ability reported in recent work can be classified into two groups, i.e., *upper bound-based approaches* and *lower bound-based approaches*. Most of these works are based on two assumptions: (1) workers give their answers to a task independently and are not affected by other workers; (2) a worker's ability for a task is a fixed value. However realistically, a worker's ability should be evaluated as a relative value to those of others within a group. In this work, we propose an approach called GroExpert to identify experts based on their relative values in their working groups, which can be used as a basis for quality estimation in crowdsourcing. The proposed solution employs a fully connected neural network to implement the pairwise ranking method when identifying experts. Both workers' features and groups' features are considered in GroExpert. We conduct a set of experiments on three real-world datasets from the Amazon Mechanical Turk platform. The experimental results show that the proposed GroExpert approach outperforms the state-of-the-art in worker ability measurement.

Keywords: Crowdsourcing · Group-aware · Worker ability

1 Introduction

In recent years, crowdsourcing has proven to be an efficient and cost-effective way to handle tasks that are hard for computers to carry out [13], e.g., sentiment analysis, answer selecting and ranking, and natural language understanding. When crowdsourcing is used in these tasks, most of the existing crowdsourcing studies and online crowdsourcing platforms, such as Amazon Mechanical Turk[1], employ a

---

[1] https://www.mturk.com/.

© Springer Nature Switzerland AG 2019
R. Cheng et al. (Eds.): WISE 2019, LNCS 11881, pp. 3–17, 2019.
https://doi.org/10.1007/978-3-030-34223-4_1

redundancy-based strategy that assigns multiple relatively inexpensive workers to a task and aggregates the answers given by these workers to work out the correct answer [27]. However, the workers are normally with diverse skills, interests, personal objectives, and technological resources [4]. As a result, low-quality or even noisy answers can be introduced that can compromise the practicability of the applications built upon the crowdsourcing result [6,9,18]. Thus, a large number of works in the literature are dedicated to mitigating the impact of low-quality or noisy answers. The main objective of these works is to measure the ability level of workers that indicates the probability of workers to handle a task correctly in crowdsourcing. We refer to workers with high ability as experts.

The existing worker's ability measurement approaches can be classified into two groups, i.e., *upper bound-based approaches* and *lower bound-based approaches*. *Upper bound-based approaches* aim to improve the upper bound of answers quality. These approaches mainly focus on identifying high ability workers by considering their answering accuracy in the context of task features and worker features, such as task difficulty [17], task topics [7,17,22], task types [5,16,19,20], and worker demographics information [14]. The common drawback of these approaches is that they are usually designed for some specific types of tasks which may not suitable for others [23]. For example, the approaches designed for speech understanding tasks are not suitable for vision understanding. On the other hand, *lower bound-based approaches* try to guarantee the lower bound of answers quality in crowdsourcing. They try to identify and filter out the spammers and malicious workers who provide noisy answers [1,10]. All legitimate workers are equally treated in the label aggregation.

Most of the above works take the worker ability as an absolute value rather than a relative one. And these works have two main drawbacks. Firstly, these approaches aim to find high ability workers (experts) among all the workers. The long-tail phenomenon indicates that lots of tasks receive very few answers [11,27]. As a result, the experts identified using these approaches may not actually give answers to most of the tasks. In order to get the quality answer for a specific task, we need to identify those workers with relatively high ability values in answering this task among a group of workers. Secondly, these approaches only consider worker features and ignore group features. When employing redundancy-based methods, multiple workers who answered the same task virtually form a group. The ability of a worker should be evaluated as a relative value in the group of workers by considering both features of workers and groups. We will not consider task features in this work due to the constraint that the tasks do not have a rich set of features in these datasets we are experimenting.

In this work, we develop an approach called GroExpert that identifies experts in their working groups. We employ a fully connected neural network to implement the pairwise ranking method to assign an ability score to each worker. This score represents his/her expertise level comparing to other workers in the group. Both workers' features and groups' features are considered. We have the assumption that experts provide more accurate answers than non-experts, which adopted by most of the works in crowdsourcing.

The contributions of this work are summarized as follows:

- We propose a group-aware experts identification approach to evaluate the ability of workers in crowdsourcing environments. Our approach considers both worker features and worker group features. Worker features and worker group features are extracted from their historical information in the dataset. To the best of our knowledge, this is the first work to evaluate the worker ability as a relative value in crowdsourcing.
- A deep neural network is employed to implement the pairwise ranking method. Each worker is assigned an ability score which represents his/her expertise level comparing to other workers in the group. Based on these scores, experts in groups could be identified.
- Experiments against three real-world datasets from Amazon Mechanical Turk platform have been conducted. Experimental results show the improvement of the answer quality of the proposed approach compared with the existing state-of-the-art approaches.

The paper is organized as follows. In Sect. 2, we review the related worker ability approaches in the crowdsourcing environments. Section 3 provides the notations and problem specification for our approach. Section 4 proposes our group-aware approach for measuring worker ability. Section 5 presents the experimental results and discussions. Finally, we conclude this paper in Sect. 6.

2   Related Work

There has been quite a lot of work on evaluating worker ability in the research area of crowdsourcing. Existing approaches fall in the group of upper bound-based approaches or the group of the lower bound-based approaches.

2.1   Upper Bound-Based Approaches

The upper bound-based approaches mainly focus on identifying high ability workers by considering their answering accuracy in the context of task features and worker features. One of the important features of tasks is their difficulty. The work in [17] evaluated the worker ability which was directly affected by the task difficulty. [7,17] believed that workers have different ability degrees across different tasks. They proposed a framework to estimate the different ability degrees of workers and assigned tasks to them accordingly. [26] used the knowledge base, e.g., Wikipedia, to analyze the domains of tasks. By contrast, [22] classified different tasks into different latent topics, and evaluated the ability degrees of workers for different latent topics.

Different from the above studies, some works focused on dealing with different types of tasks. [5,16,20] focused on single-choice or decision making tasks. [19] focused on general tasks such as image description and language translation. They proposed an estimation method with workers referred to as evaluators who assessed answers given by other workers. They used the pairwise-based method to evaluate the ability of both workers and evaluators. [15] identified experts for knowledge intensive tasks. [14] identified experts by considering worker demographic information (major, education level, age, etc).

2.2 Lower Bound-Based Approaches

The lower Bound-based Approaches mainly focus on identifying and filtering out the spammers and malicious workers who provide noisy answers [1,10]. And these workers were excluded to get involved in any tasks. These workers are harmful to crowdsourcing platforms and would lead a crowdsourcing platform to a failure [8]. All the legitimate workers were equally treated in the label aggregation in these approaches.

In order to reduce the impact of spammers, traditional trust management techniques could be used in crowdsourcing but they seldom can be used directly [24]. [24] evaluated workers' trustworthiness by extending existing trust models. [21] enhanced their trust model with a fusion method. They fused untrustworthy answers from the crowd while simultaneously learning the trustworthiness of workers. [1] proposed a reputation system that considered the ability of requesters and the number of tasks completed. In this reputation system, they proposed a graph data model to analyze the relations between workers and requesters. Unlike most of the existing approaches, [10] considered a much broader class of adversarial workers. They assigned a trust score for every worker and filtered out spammers.

Most of the above works have two main drawbacks. Firstly, these approaches aim to find high ability workers (experts) among all workers, which can be inaccurate since these identified experts may not give answers to the interested tasks. Research shows that many tasks actually have received very few answers [11,27]. Secondly, these approaches only consider worker features and ignore group features. When employing redundancy-based methods, multiple workers who answered the same task virtually form a group, in which group dynamics can contribute to workers' performance. Therefore the ability of a worker should be evaluated as a relative value to other workers in the group.

The proposed solution belongs to the group of upper bound-based approaches. However different from the existing approaches, in this work, we propose a group-aware experts identification approach by evaluating the ability of a worker as a relative value to other workers in the group. Furthermore, both features of the workers and their groups are considered.

3   Notations and Problem Specification

In this section, we provide the notations in this work and describe the problem specification of measuring worker ability.

Notations: In a crowdsourcing platform, a common scenario is that a data requester has a set of workers to answer a set of tasks. A worker has a set of features. Normally, there are several to answer one task. Multiple workers groups are formed according to the tasks they have answered. Each group has its own features which are composed of features of workers in the group. A score function assigns each worker a value according to the features and group features of this worker. In this work, we have the following notations:

- $T = \{t_1, \cdots, t_n\}$ denotes the set of n tasks.
- $W = \{w_1, \cdots, w_m\}$ denotes the set of m workers.
- G_{t_i} denotes the set of workers who answered t_i.
- $WF_{w_j} = \{wf^1_{w_j}, \cdots, wf^p_{w_j}\}$, $wf^i_{w_j}$ $(i = 1, 2, ...p)$ is a single feature of worker w_j, p is the total number of features.
- $GF_{G_{t_i}} = \{WF_{w_j} | w_j \in G_{t_i}\}$ denotes the set of group features of G_{t_i}.
- $V^{G_{t_i}}_{w_j}$ is the feature vector for worker w_j in group G_{t_i} with all features in WF_{w_j} and GF_{g_i} as its components.
- $s^{t_i}_{w_j} = FK(V^{G_{t_i}}_{w_j})$, $FK(V^{G_{t_i}}_{w_j})$ is the score function with $V^{G_{t_i}}_{w_j}$ as its input.
- $S_{G_{t_i}} = \{s^{t_i}_{w_j} | w_j \in G_{t_i}\}$ is the set of scores for workers in G_{t_i}.

Problem Specification: Given a set of workers $W = \{w_1, \cdots, w_m\}$, a set of tasks $T = \{t_1, \cdots, t_n\}$. Task $t_i \in T$ is answered by u workers $(u = |G_{t_i}|)$; group G_{t_i} is for task t_i; worker $w_j \in G_{t_i}$ has the feature vector $V^{G_{t_i}}_{w_j}$. The $s^{t_i}_{w_j}$ is the ability score of worker w_j, $s^{t_i}_{w_j} = FK(V^{G_{t_i}}_{w_j})$. $G_{t_i} = \{s^{t_i}_{w_j} | w_j \in G_{t_i}\}$ is the set of ability scores of workers in the group G_{t_i}.

4 GroExpert Approach

In this work, we propose a group-aware experts identification approach to measure worker ability, referred to as GroExpert. In order to calculate the ability scores of workers, we employ a fully connected neural network to implement the pairwise ranking method. As shown in Fig. 1, the GroExpert approach can be divided into three phases, i.e., Phase 1: Pre-process datasets, Phase 2: Train the score function and Phase 3: Calculate the ability scores of workers. Based on the scores, we are able to identify experts in individual groups. In the following sections, we describe Phase 1, Phase 2 and Phase3. Based on the ability scores, we are able to identify the experts in individual groups.

4.1 Phase 1: Pre-process Datasets

In Phase 1, we pre-process worker answering records in datasets and construct feature vectors with features of individual workers and features of individual task groups. When the worker w_j answers the task t_i, we use WF_{w_j} and GF_{t_i} to construct the feature vector $V^{t_i}_{w_j}$. This procedure is shown in Fig. 2.

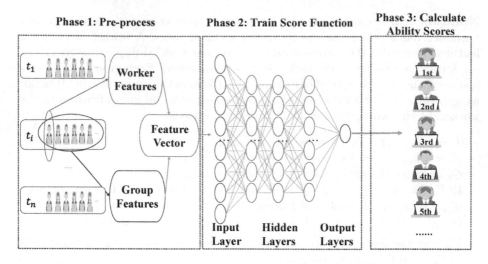

Fig. 1. GroExpert approach: Phase 1: Pre-process; Phase 2: Train score function; Phase 3: Calculate ability scores.

For the vector $V_{w_j}^{t_i}$, We use the a one-hot encoding-based method to represent it. As the vector $V_{w_j}^{G_{t_i}}$ contains all features in WF_{w_j} and GF_{g_i}, the dimensions of the $V_{w_j}^{G_{t_i}}$ is $p \times (m+1)$. We use the first p dimensions to represent the features of w_j, $p \times m$ dimensions to represent the features of workers in the task group G_{t_i}.

In this work, we focus on two features of the worker as **Reputation** and **Quantity Ratio**. These two features are commonly considered in the worker ability problem.

Reputation: Crowdsourcing platforms often establish a system to quantify the worker ability. For the worker w_j, we use R_{m_j} to represent this feature.

Quantity Ratio: The ratio between the number of tasks answered by a worker and the number of total tasks. For the worker w_j, we use Q_{m_j} to represent this feature. For worker w_j, $WF_{w_j} = \{Reputaion, QuantityRation\}$ and $p = 2$. Figure 3 shows the details. Here we suppose $m = 9$ and denote w_1, w_4 and w_7 with 1 (they have answered t_3 ($G_{t_3} = /w_1, w_4, w_7/$)), other workers with 0 (they have not answered). We also show the construction procedure of $V_{w_7}^{t_3}$.

4.2 Phase 2: Train the Score Function

We develop a fully connected deep neural network to learn the relations between the feature vectors and ability scores. The deep neural network is trained in the Feed-forward and the Back-propagation processes. Phase 2 in Fig. 1 illustrates the neural network architecture. The input vectors of our approach are $\{V_{w_j}^{t_i}\}$ (for $j = 1, 2, ..., m$; $i = 1, 2, ..., n$). We denote the output of hidden layers with

Feature Vector $V_{w_j}^{t_i}$
WF_{w_j}
$WF_{w_1}WF_{w_2}WF_{w_3}WF_{w_4}WF_{w_5}WF_{w_j}$
WF_{w_j}
$WF_{w_1}WF_{w_2}WF_{w_3}WF_{w_4}WF_{w_5}WF_{w_j}$
Worker Features
Group Features
t_i
...
worker W_j

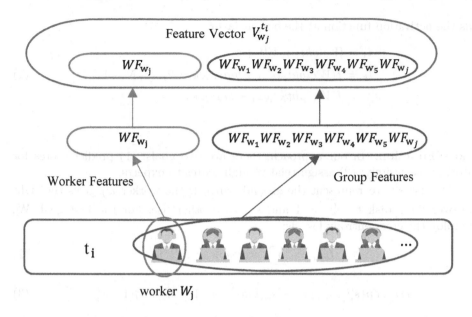

Fig. 2. Construct feature vectors for workers in groups

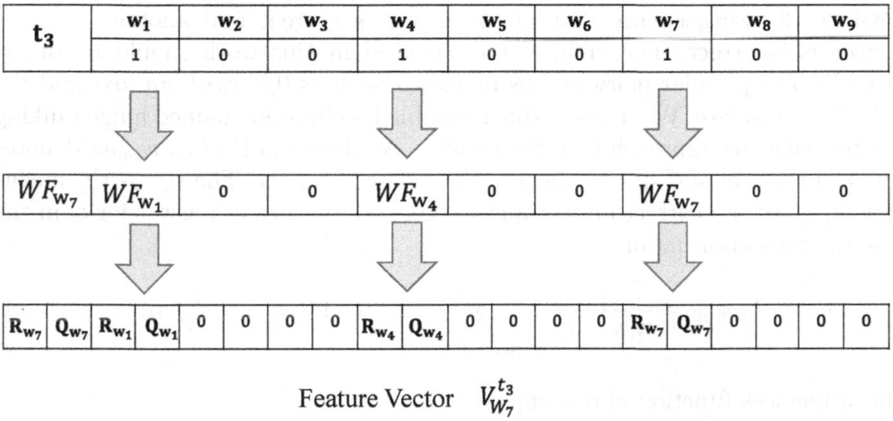

t_3
w_1 w_2 w_3 w_4 w_5 w_6 w_7 w_8 w_9
1 0 0 1 0 0 1 0 0
WF_{w_7} WF_{w_1} 0 0 WF_{w_4} 0 0 WF_{w_7} 0 0
R_{w_7} Q_{w_7} R_{w_1} Q_{w_1} 0 0 0 0 R_{w_4} Q_{w_4} 0 0 0 0 R_{w_7} Q_{w_7} 0 0 0 0
Feature Vector $V_{W_7}^{t_3}$

Fig. 3. Vector example of w_7 in group G_{t_3}

$l_k, k = 1, 2, ..., N-1$, the k_{th} weight with $Weight_k$, the bias term with $bias_k$, and the final output is score sets $s_{w_j}^{t_i}$ (for $j = 1, 2, ..., m$; $i = 1, 2, ..., n$). We use the $SELU^2$ as the activation functions at the hidden layers and use the $linear^2$

[2] https://keras.io/activations.

as the activation function at the output layer.

$$l_1 = Weight_1 \times V_{w_j}^{t_i} + b_1$$
$$l_k = f(Weight_k l_{k-1} + bias_k), k = 2, \ldots, N-1 \qquad (1)$$
$$s_{w_j}^{t_i} = f(Weight_N l_{N-1} + bias_N)$$

Loss Function. In our approach, there are two goals: (1) predict scores for workers accurately; (2) assign relative high scores to experts.

We use $s_{w_j}^{\bar{t_i}}$ to represent the ground truth. If the worker w_j gives the right answer for a task t_i, $s_{w_j}^{\bar{t_i}} = 1$ and $s_{w_j}^{\bar{t_i}} = 0$ otherwise. For the first goal, We employ the cross entropy loss as:

$$L_{acc} = \sum_{w_j \in W} crossen(s_{w_j}^{t_i}, s_{w_j}^{\bar{t_i}}) \qquad (2)$$

$$crossen(s_{w_j}^{t_i}, s_{w_j}^{\bar{t_i}}) = -(s_{w_j}^{\bar{t_i}} logs_{w_j}^{t_i} + (1 - s_{w_j}^{\bar{t_i}})log(1 - s_{w_j}^{t_i})), \qquad (3)$$

where the $s_{w_j}^{t_i}$ represents the predicted the ability score for the worker w_j in the group G_{t_i} and $s_{w_j}^{\bar{t_i}}$ represents the ground truth.

For the second goal, we compare the worker-pairs extracted in G_{t_i}. We only consider the worker-pairs that one's answer is correct and another worker's answer is incorrect. The architecture we used in this work should be differentiable. The popular pairwise loss functions such as 0/1 *ranking loss* and the NDCG are discrete. We choose a differentiable loss function named hinge ranking loss to make our approach be differentiable. As shown in Eq (4), w_a, w_b denotes a pair of workers and $s_{w_a}^{t_i}, s_{w_b}^{t_i}$ denote the scores of w_a, w_b. The $s_{w_a}^{t_i} \succ s_{w_b}^{t_i}$ means that w_a gives the correct answer and w_b gives the incorrect answer for t_i in the ground truth information.

$$L_{hinge} = \frac{2}{n(n-1)} \sum_{w_a, w_b : s_{w_a}^{t_i} \succ s_{w_b}^{t_i}} max(1 - s_{w_a}^{t_i} + s_{w_b}^{t_i}, 0) \qquad (4)$$

The whole loss function of our approach is:

$$L = L_{acc} + L_{hinge}. \qquad (5)$$

In this work, we utilize Stochastic Gradient Descent [25] to minimize L.

4.3   Phase 3: Calculate the Ability Score

The score function $FK(V_{w_j}^{G_{t_i}})$ is carried out by a fully connected neural network. After training the score function, we could measure the ability of the worker. For the new group G_{t_i}, we measure the ability of worker w_j by calculating the ability score $s_{w_j}^{t_i}$. Based on the scores, we are able to identify the experts in the group.

5 Experiments

We implement the experiments in Python on a server with CPU 3.6 GHZ and 100 GB memory. The neural network used in GroExpert (Phase 2) is implemented in Keras [3]. For all the experiments, we run them five times and calculate the average over these results.

5.1 Experimental Settings

Datasets. In the experiments, we use three real-world datasets collected from Amazon Mechanical Turk platform. These three datasets named Product, Emotions, and Face Sentiment. All of them are provided by [27]. We select these datasets among many public datasets[3] with two reasons. Firstly, these 3 datasets cover three common types of tasks. The tasks of Product are decision making tasks which ask workers to give a true or false answer to a specific task. Emotions belong to the type of numeric tasks. Each task is composed of a text and a range $[-100, 100]$. And workers are asked to estimate the degree of emotion by assigning an indication number. Face Sentiment belongs to the type of single-choice tasks, which workers are asked to select an option among four choices for a given task. Through the experiments on different types of tasks, we evaluate the effect of different task types on the performance of our approach. Secondly, the selected datasets have ground truth labeled. The details of these datasets are shown in Table 1 where we list statistics for each dataset: total number of tasks $|T|$, the total number of workers $|M|$ and the size of worker group $|g|$ which is the number of workers answering the same task.

Table 1. Statistics of the datasets

Dataset	$|T|$	$|W|$	$|g|$
Product	8315	176	3
Emotions	700	38	10
Face sentiment	572	27	9

Worker Features and Group Features. For each worker, we obtain worker reputation and quantity ratio as their features by exploiting the whole datasets. Due to the limited information in datasets, we take the answering accuracy of workers to represents their reputation. For the worker w_j, we use R_{m_j} to represent worker reputation, and we use Q_{m_j} to represent Quantity Ration, where $|T|$ denotes the number of all tasks, $|T_C|$ denotes the number of correctly answered tasks, and $|T_A|$ denotes the number of tasks answered by w_j. For the task t_i, we get group features after we got all workers features in this group.

---

[3] http://dbgroup.cs.tsinghua.edu.cn/ligl/crowddata/.

Then, we combine the worker feature and group feature to construct $V_{m_j}^{t_i}$ (see Figs. 2 and 3).

$$R_{w_j} = \frac{|T_C|}{|T|} \qquad (6)$$

$$Q_{w_j} = \frac{|T_A|}{|T|} \qquad (7)$$

Evaluation Metrics. The main objective of measuring the ability level of workers is to improve the answer quality collected from crowdsourcing platform. We use the metric named *Accuracy* in Eq. (8) to represent the quality of answers. We use $|T_C|$ to denote the number of correctly predicted tasks and use $|T_P|$ to denote the total number of predicting tasks.

There are several metrics that measure the ranking quality, such as Precision at position (P@n), Mean Average Precision (MAP) and Normalized Discounted Cumulative Gain (NDCG@n). However, these metrics mainly focus on the quality of the whole ranking of workers instead of the quality of answers.

$$Accuracy = \frac{|T_C|}{|T_P|} \qquad (8)$$

Comparison Methods. We compare our GroExpert method with the following four baseline methods:

- **Confusion Matrix (CM)**: this is the most widely used baseline method for worker ability modeling, which represents the probability distribution of workers who correctly answer the task [5,16]. A variety of worker features are incorporated to enhance the power of CM. We did not compare to these methods as we don't use the features they incorporate.
- **Basic Accuracy (Basic-Acc)**: this method models the worker ability according to their accuracy which is obtained from the ground truth. The method is extremely similar to the approval rate in Amazon Mechanical Turk.
- **RankSVM**: this linear method compares each pair of workers and transfers the ranking problem to the classification problem by leveraging the SVM classifier [12].
- **RankNET**: this is a pairwise-based ranking method which takes advantage of the neural network [2].

The confusion matrix and the basic accuracy methods are classical accuracy-based approaches, and RankSVM and RankNET are two state-of-the-art pairwise ranking approaches. We do not compare our method with the reputation-based methods because these methods focus on filtering out the bad workers while our method aims to find out the best workers.

Table 2. *Accuracy(%)* of GroExpert approach compared with baseline methods

Dataset	Product	Emotions	Face sentiment
Basic-Acc	63.12	22.05	59.39
CM	90.54	-	65.06
RankSVM	92.17	17.96	64.19
RankNET	92.64	20.020	65.06
GroExpert	**93.59**	**31.22**	**67.25**

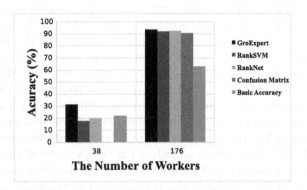

Fig. 4. Performance at different worker numbers

5.2 Performance Comparison

In this section, we compare our GroExpert method with 4 baseline methods and evaluate the performance. The results are summarized in Table 2. Our proposed GroExpert method performs the best against three different types of datasets. For simplicity, we select one expert in the group and take his/her answer as the predicted answer. Specifically, GroExpert method outperforms the 4 baseline methods in Product by an average of 8.9725%, ranging from 0.95% to 30.47%. In Emotions, our method outperforms the other 4 methods by an average of 11.15%, ranging from 9.17% to 13.26%. And in Face Sentiment, our method outperforms the 4 baseline methods by an average of 2.89%, ranging from 2.19% to 4.13%. RankNET and RankSVM perform similarly over all three datasets. By leveraging the power of the neural network, RankNET performs better than RankSVM. However, RankNET doesn't consider the group feature as we do. For accuracy-based methods, Confusion Matrix always performs better than the basic accuracy methods. However, the confusion matrix is not suitable for numeric tasks. As shown in Fig. 4, we improve the *Accuracy* a lot when the number of workers is small (we use 38 as an example). And when the number of workers is large (we use 176 as an example), we also achieve the best performance.

The *Accuracy* of our GroExpert varies in three types of tasks. It achieves the best performance in Product (93.59%), medium performance in Face Sentiment (67.25%) and worst performance in Emotions (31.22%). The reason is that the worker quality varies in these three types of tasks. The average worker accuracy can indicate worker quality, which is 78.57% in Product, 60.19% in Face Sentiment and 28.78% in Emotion. When we take all the workers in the dataset as a big group, the quality of this big group strongly affect the performance of our GroExpert.

5.3   Training Size

In order to measure the performance of our GroExpert approach, we conduct experiments on different training size over three datasets. The training size varies from {10%, 20%, 30%, 40%, 50%, 60%, 70%, 80%, 90%} of the total records in datasets and the comparison results are shown in Fig. 5 for Product, Emotion, and Face Sentiment, respectively. In Product, the *Accuracy* of all the methods are stable and are above 90% except the Basic Accuracy method. All the methods only need to rank 3 workers in Product, which may lead to high *Accuracy*. The *Accuracy* of GroExpert and RankNET fluctuates with the increase of the training size both in Emotion and Face Sentiment. Because the neural network used in these two models suffers data sparsity and overfitting problems. By contrast, the *Accuracy* of RankSVM increases slightly with the increase of the training size in Emotion. As can be seen in Fig. 5, for most of the training size, our method outperforms the baseline methods, this is because we leverage the neural network and take the group feature into consideration.

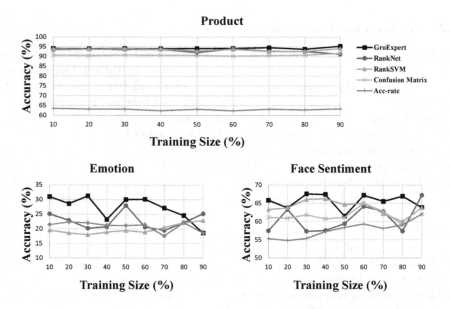

Fig. 5. Performance comparison for different methods

5.4 Sensitivity to Number of Hidden Layers and Number of Hidden Units

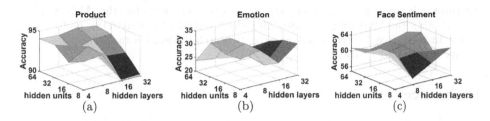

Fig. 6. Performance comparison for different hidden layers and hidden units

In this subsection, we discuss the influence of two important parameters as the number of hidden layers and the number of hidden units. The results over the three datasets are shown in Fig. 6(a), (b) and (c) respectively. To improve the performance of our method, a basic idea is to increase the number of hidden layers and hidden units. Therefore, we conduct experiments to investigate the influence of different hidden layers and hidden units. The number of hidden layers varies from 4 to 32 and the number of hidden units varies from 8 to 64.

For the dataset Product, the performance achieves the best (94.10%) when the number of hidden layers is 4 and the number of hidden units is 32. While the worst performance (90.21%) occurs when the number of hidden layers is 16 and the number of hidden units is 8. For the dataset Emotion, the performance achieves the best (33.06%) when the number of hidden layers is 8 and the number of hidden units is 16. While the worst performance (22.04%) occurs when the number of hidden layers is 32 and the number of hidden units is 32. For the dataset Face Sentiment, the performance achieves the best (63.88%) when the number of hidden layers is 4 and the number of hidden units is 8. While the worst performance (56.50%) occurs when the number of hidden layers is 8 and the number of hidden units is 8. The performance is not obviously improved as the increase in the number of hidden layers over all three datasets.

6 Conclusions

In this work, we propose an approach called GroExpert to identify experts in their working groups as a basis for future quality estimation in crowdsourcing. A fully connected neural network is employed to implement the pairwise ranking method when identifying experts. We take into consideration of both worker and group features. The experimental results against three real-world datasets show that our approach is effective and outperforms the existing baseline methods.

The proposed GroExpert approach provides a foundation for a wide range of solutions to problems in crowdsourcing environments, e.g., task assignment, truth inference, and reward mechanism development. Task requesters can assign tasks to workers with high ability scores, reward more to workers with higher ability scores, or have a higher trust degree on workers with higher ability scores.

Acknowledgements. This work was supported in part by the MQNS (No. 9201701203), the MQEPS (No. 96804590), the MQRSG (No. 95109718), and in part by the Investigative Analytics Collaborative Research Project between Macquarie University and Data61 CSIRO.

References

1. Allahbakhsh, M., Ignjatovic, A., Benatallah, B., Bertino, E., Foo, N., et al.: Reputation management in crowdsourcing systems. In: CollaborateCom, pp. 664–671. IEEE (2012)
2. Burges, C., et al.: Learning to rank using gradient descent. In: Proceedings of the 22nd International Conference on Machine Learning, pp. 89–96. ACM (2005)
3. Chollet, F., et al.: Keras (2015). https://keras.io
4. Daniel, F., Kucherbaev, P., Cappiello, C., Benatallah, B., Allahbakhsh, M.: Quality control in crowdsourcing: a survey of quality attributes, assessment techniques, and assurance actions. ACM Comput. Surv. (CSUR) **51**(1), 7 (2018)
5. Dawid, A.P., Skene, A.M.: Maximum likelihood estimation of observer error-rates using the EM algorithm. Appl. Stat. **28**, 20–28 (1979)
6. Donmez, P., Carbonell, J.G.: Proactive learning: cost-sensitive active learning with multiple imperfect oracles. In: Proceedings of the 17th ACM Conference on Information and Knowledge Management, pp. 619–628. ACM (2008)
7. Fan, J., Li, G., Ooi, B.C., Tan, K.l., Feng, J.: iCrowd: an adaptive crowdsourcing framework. In: Proceedings of the 2015 ACM SIGMOD, pp. 1015–1030. ACM (2015)
8. Hirth, M., Hoßfeld, T., Tran-Gia, P.: Cheat-detection mechanisms for crowdsourcing. University of Würzburg, Technical report, vol. 4 (2010)
9. Hu, H., Zheng, Y., Bao, Z., Li, G., Feng, J., Cheng, R.: Crowdsourced poi labelling: location-aware result inference and task assignment. In: 2016 IEEE 32nd International Conference on Data Engineering (ICDE), pp. 61–72. IEEE (2016)
10. Jagabathula, S., Subramanian, L., Venkataraman, A.: Reputation-based worker filtering in crowdsourcing. In: Advances in Neural Information Processing Systems, pp. 2492–2500 (2014)
11. Jain, A., Sarma, A.D., Parameswaran, A., Widom, J.: Understanding workers, developing effective tasks, and enhancing marketplace dynamics: a study of a large crowdsourcing marketplace. Proc. VLDB Endow. **10**(7), 829–840 (2017)
12. Joachims, T.: Optimizing search engines using clickthrough data. In: Proceedings of the Eighth ACM SIGKDD, pp. 133–142. ACM (2002)
13. Li, G., Wang, J., Zheng, Y., Franklin, M.J.: Crowdsourced data management: a survey. IEEE Trans. Knowl. Data Eng. **28**(9), 2296–2319 (2016)
14. Li, H., Zhao, B., Fuxman, A.: The wisdom of minority: discovering and targeting the right group of workers for crowdsourcing. In: Proceedings of the 23rd International Conference on World Wide Web, pp. 165–176. ACM (2014)
15. Li, J., Baba, Y., Kashima, H.: Hyper questions: unsupervised targeting of a few experts in crowdsourcing. In: Proceedings of the 2017 ACM on Conference on Information and Knowledge Management, pp. 1069–1078. ACM (2017)

16. Liu, Q., Peng, J., Ihler, A.T.: Variational inference for crowdsourcing. In: Advances in Neural Information Processing Systems, pp. 692–700 (2012)
17. Ma, F., et al.: Faitcrowd: fine grained truth discovery for crowdsourced data aggregation. In: Proceedings of the 21th ACM SIGKDD, pp. 745–754. ACM (2015)
18. Sheng, V.S., Provost, F., Ipeirotis, P.G.: Get another label? Improving data quality and data mining using multiple, noisy labelers. In: Proceedings of the 14th ACM SIGKDD, pp. 614–622. ACM (2008)
19. Sunahase, T., Baba, Y., Kashima, H.: Pairwise hits: quality estimation from pairwise comparisons in creator-evaluator crowdsourcing process. In: AAAI, pp. 977–984 (2017)
20. Venanzi, M., Guiver, J., Kazai, G., Kohli, P., Shokouhi, M.: Community-based Bayesian aggregation models for crowdsourcing. In: Proceedings of the 23rd WWW, pp. 155–164. ACM (2014)
21. Venanzi, M., Rogers, A., Jennings, N.R.: Trust-based fusion of untrustworthy information in crowdsourcing applications. In: Proceedings of the 2013 International Conference on Autonomous Agents and Multi-Agent Systems, pp. 829–836. International Foundation for Autonomous Agents and Multiagent Systems (2013)
22. Welinder, P., Branson, S., Perona, P., Belongie, S.J.: The multidimensional wisdom of crowds. In: Advances in Neural Information Processing Systems, pp. 2424–2432 (2010)
23. Yin, L., Han, J., Zhang, W., Yu, Y.: Aggregating crowd wisdoms with label-aware autoencoders. In: Proceedings of the 26th IJCAI, pp. 1325–1331. AAAI Press (2017)
24. Yu, H., Shen, Z., Miao, C., An, B.: Challenges and opportunities for trust management in crowdsourcing. In: IEEE/WIC/ACM, pp. 486–493. IEEE Computer Society (2012)
25. Zhang, T.: Solving large scale linear prediction problems using stochastic gradient descent algorithms. In: Proceedings of the Twenty-First International Conference on Machine Learning, p. 116. ACM (2004)
26. Zheng, Y., Li, G., Cheng, R.: DOCS: a domain-aware crowdsourcing system using knowledge bases. Proc. VLDB Endow. **10**(4), 361–372 (2016)
27. Zheng, Y., Li, G., Li, Y., Shan, C., Cheng, R.: Truth inference in crowdsourcing: is the problem solved? Proc. VLDB Endow. **10**(5), 541–552 (2017)

Detecting Fraudulent Accounts on Blockchain: A Supervised Approach

Michał Ostapowicz$^{(\boxtimes)}$ and Kamil Żbikowski$^{(\boxtimes)}$

Institute of Computer Science, Faculty of Electronics and Information Technology,
Warsaw University of Technology, ul. Nowowiejska 15/19, 00-665 Warsaw, Poland
michal.ostapowicz.stud@pw.edu.pl, kamil.zbikowski@ii.pw.edu.pl

Abstract. Applications of blockchain technologies got a lot of attention in recent years. They exceed beyond exchanging value and being a substitute for fiat money and traditional banking system. Nevertheless, being able to exchange value on a blockchain is at the core of the entire system and has to be reliable. Blockchains have built-in mechanisms that guarantee whole system's consistency and reliability. However, malicious actors can still try to steal money by applying well known techniques like malware software or fake emails. In this paper we apply supervised learning techniques to detect fraudulent accounts on Ethereum blockchain. We compare capabilities of Random Forests, Support Vector Machines and XGBoost classifiers to identify such accounts basing on a dataset of more than 300 thousands accounts. Results show that we are able to achieve recall and precision values allowing for the designed system to be applicable as an anti-fraud rule for digital wallets or currency exchanges. We also present sensitivity analysis to show how presented models depend on particular feature and how lack of some of them will affect the overall system performance.

Keywords: Blockchain · Anti-fraud · Supervised · Xgboost · Random forests · SVM · Ethereum

1 Introduction

Recent developments in digital currencies gave birth not only to a completely new way of exchanging value, but also to such areas like distributed trust management. Those advances may replace traditional notary services or payment processing companies in the near future [12]. Such advances are possible to achieve thanks to technology called blockchain that, in its basis, is as an immutable, distributed database. First public blockchain, called Bitcoin, was launched in 2009 and, not surprisingly, from the very beginning attracted fraudulent actors that tried to take advantage of other participants. These actors very often try to convince others to send them digital currency to their accounts by using different techniques like malware or fake emails. Due to the publicly available data, information about account once denoted as fraudulent can be shared and

© Springer Nature Switzerland AG 2019
R. Cheng et al. (Eds.): WISE 2019, LNCS 11881, pp. 18–31, 2019.
https://doi.org/10.1007/978-3-030-34223-4_2

available without limitations. Quite contrary to traditional financial systems, all the transfers to and from such account can be freely viewed and analyzed. The availability of this data gives us an opportunity to verify if there is a meaningful relation between operations done on the account and this account being fraudulent.

In this paper, we propose a novel approach for detecting fraudulent accounts on Ethereum network. Ethereum is a blockchain that has some significant improvements over Bitcoin [5]. Those improvements allow to write and execute contracts (called smart contracts) more easily. These contracts give an opportunity for many different actors to engage in complex agreements that are fully executable and can be verified with the use of the underlying protocol. More details on Ethereum can be found in [11].

In the first stage, we automatically gathered available data about accounts and transactions. Then, we created explanatory variables out of raw data. They represent aggregates and statistics computed over volumes and time. In the next stage, we tested three classifiers and compared their results in the context of possible applications. They can strongly depend on different use cases that may put more importance on precision than on recall or the other way round. The contribution of this study can be summarized as follows:

- We proposed a novel approach for identifying fraudulent accounts on Etherum blockchain that is easily transferable to other blockchains, like Bitcoin.
- We conducted a thorough analysis of three different machine learning algorithms for the task of classification accounts to "fraudulent" or "not fraudulent" class.
- We conducted a sensitivity analysis in order to verify how much we depend on particular explanatory variables. This is a test that allow us to address the potential problem of a look-ahead bias that may or may not exist within the data that we gathered.

2   Related Work

Detecting fraudulent activity in financial operations is a well known problem. Both researchers and practitioners put a lot of attention to developing new tools that would correctly identify new attack vectors. This is an endless battle in which both sides use their creativity and new technologies. A comprehensive survey on fraud detection techniques can be found in [8]. More recent surveys on fraud prevention systems and detecting financial fraud through data mining algorithms can be found in [1] and [2] respectively.

Quah and Sriganesh [10] used Self Organizing Maps (SOM) to detect credit card frauds. They took an approach that if a transaction is similar to all transactions in a set of genuine transactions, it is also considered genuine. On the other hand, if it looks like any of the transactions in a set of fraudulent, then it is also considered fraudulent. In addition to the basic task of clustering input data, Self Organizing Maps are also used to detect and extract hidden patterns. According to the authors, in real financial systems that verify each transaction on multiple layers, SOM may also serve as a filter for the layers following it. In the case described by the authors, SOM receives an input data vector consisting of client, account and transaction features.

In [6] authors used supervised learning methods to tackle similar problem. They used logistic regression, Support Vector Machine (SVM) and random forest. Apart from using typical transaction features as an algorithm's input (e.g. order value, type of items ordered, payment method), through abstraction and combination they engineered several new variables such as binary evaluated compliance of the country of the card transaction with the country to which the purchased items are to be delivered. Eventually, the authors used 71 features to describe each transaction. The best results were obtained using random forest method, which is why it was used in further analysis. As it turned out, despite quite good results in recognizing frauds, they were not good enough to fully automate verification of transactions.

In case of transfers done through blockchain transactions, fraud detection can be a more complicated task as most of the time we are not in possession of geographical and personal data of participants. Pham and Lee [9] in their article dealt with detecting frauds in the Bitcoin network. The network data was modeled as two graphs: a user graph and a transaction graph which were used to detect anomalies (e.g. fraudulent and suspicious users). They had information about 30 cases of theft in the Bitcoin network, which were later used to verify their results. In both graphs, each vertex was represented with 12 features, such as the input and output stage, the average time between transactions, the creation date and activity time. As the first step in the analysis they applied k-means algorithm to group all graph nodes. As the authors pointed out, this algorithm is not used to find anomalies, but it may be useful, because the points that diverge from the rest are expected to be found far from the centroids calculated with k-means algorithm. They wanted to investigate if anomalies in user graph, clearly refer to anomalies in the transaction graph, i.e. whether "suspicious" users were involved in "suspicious" transactions. To find anomalies in these groups authors used a method based on the Mahalanobis distance and Support Vector Machine (SVM). Suspected users and transactions indicated by both algorithms overlapped to a large degree. In both methods extreme values were indicated as suspicious, i.e. vertices with the largest or smallest degrees. That approach allowed to detect two authentic anomalies: one theft (detected by the Mahalanobis distance based method) and one loss caused by a corruption in a hashing function (detected by the SVM). These results do not seem to be statistically significant primarily due to a limited number of known thefts (or anomalies in general).

3   Methodology

3.1   Data Preparation

The data used in the analysis came from the Etherscan.io website, which is one of the most popular Ethereum blockchain browsers. It provides information about all transactions in the network, mined blocks and user accounts. Over 2 500 wallets were reported by the users as related to illegal activities and marked as "Hack/Phishing". Using the Etherscan API it was possible to download information about all transactions in which given wallet participated. Some of the wallets tagged as fraudulent had no transactions at all or were involved mostly in ERC20 token trade. They were not included in the dataset. After this correction we analyzed 2 200 wallets marked as involved in illegal activity. In addition to fraudulent transactions data, we also collected information about transactions from 349 999 randomly selected wallets out of the 65 564 460 existing (as of 28th May 2019) in the Ethereum network. They were not marked as suspicious and were considered non-fraudulent.

Based on the work of [9] we decided to create 13 explanatory variables concerning transaction data of each account. Explanatory variables are presented in Table 1. The dataset was divided into two parts: a training set with 281 760 samples and a validation set with 70 439 samples.

Table 1. Explanatory variables

Variable name	Variable description
IT	Amount of incoming transactions
OT	Amount of outgoing transactions
UIT	Amount of unique incoming transactions
UOT	Amount of unique outgoing transactions
AVIT	Average value of the incoming transaction
AVOT	Average value of the outgoing transaction
VIT	Total value of all incoming transactions
VOT	Total value of all outgoing transactions
ATIT	Average time between incoming transactions
ATOT	Average time between outgoing transactions
AGP	Average gas price
AGL	Average gas limit
DUR	Active duration (time in days since the first until the last transaction)

3.2 Experiment Setup

The prediction problem definition here is a classic example of a binary classification. We examined following classifiers: Random Forests, Support Vector Machines and XGBoost in order to determine their capabilities of making accurate predictions for a given dataset. Figure 1 presents data and system architecture for the conducted experiment. As a first step we downloaded data using the Etherscan API, which then was aggregated to create 13 variables presented in the Table 1. In the next step, using grid search with 10-fold cross-validation we tried to find set of parameters that could give the best results for the three supervised learning algorithms that we chose.

Data gathered from Etherscan did not allow to accurately determine the moment of marking particular account as a fraudulent one. It can be possible that certain aggregates that we use for training are biased and data used to compute them was gathered after the moment of marking a particular account as fraudulent. It is possible that some of the transactions can be a result of the public exposure of an account. This would not be a problem if were only interested in devising a method for simple classification of account. However, if we would like to use proposed method as an early warning system then we will have to take a moment of an exposure into consideration. We address this issue by conducting performance analysis after removing most important explanatory variables. As the final step we did a validation check on a part of a dataset that was not used for the training purposes. Result from this step were reported in the following sections.

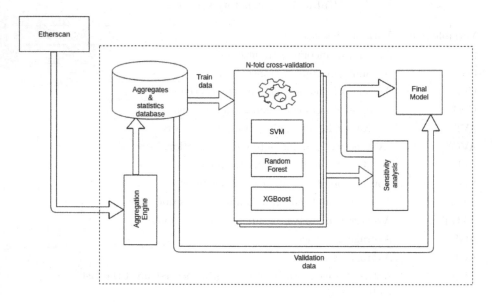

Fig. 1. System and data architecture for the conducted experiment

3.3 Prediction Models

The Support Vector Machine (SVM) classifier is a binary classifier algorithm that looks for an optimal hyperplane as a decision function in a high-dimensional space [3]. Having a training dataset $\{\mathbf{x}_k, y_k\} \in \mathbb{R}^n \times \{-1, 1\}$ where $\mathbf{x}_k$ are the training examples and y_k are the class labels at first we map $\mathbf{x}$ into a higher dimensional space via a function Φ, then computing a decision function in the form of:

$$f(x) = \langle \mathbf{w}, \Phi(\mathbf{x}) \rangle + b \tag{1}$$

by maximizing the distance between the set of points $\Phi(\mathbf{x}_k)$ to the hyperplane parameterized by $(\mathbf{w}, b)$. The class label of $\mathbf{x}$ is given by the sign of $f(\mathbf{x})$. The optimization problem for the SVM classifier with penalized misclassified examples can be written as:

$$\min_{\mathbf{w},\xi} \frac{1}{2} ||\mathbf{w}||^2 + \sum_{i=1}^{m} C\xi_i, \tag{2}$$

subject to:

$$y_i f(\mathbf{x}_i) \geq 1 - \xi_i, \tag{3}$$

With variables α_i defined such that:

$$\mathbf{w} = \sum_{i}^{m} \alpha_i y_i \mathbf{x_i}, \tag{4}$$

by solving for the Lagrangian dual of the problem 2, we obtain the simplified problem:

$$\max_{\alpha} Q(\alpha) = \sum_{i=1}^{m} \alpha_i - \frac{1}{2} \sum_{i=1}^{m} \sum_{j=1}^{m} \alpha_i \alpha_j y_i y_j \varphi(\mathbf{x_i}) \varphi(\mathbf{x_j}) \tag{5}$$

subject to:

$$\sum_{i=1}^{m} \alpha_i y_i = 0, \tag{6}$$

$$\alpha_i \geq 0. \tag{7}$$

Random Forest is a classifier consisting of a collection of tree-structured classifiers $\{h(\mathbf{x}, \Theta_k), k = 1, ...\}$ where $\{\Theta_k\}$ the are independent identically distributed random vectors and each tree casts a unit vote for the most popular class at input. [4] For each tree in the random forest new training set is generated, by drawing with replacement from the original training set. Tree is grown on the new training set using random feature selection at each node. The resulting trees are not pruned.

XGBoost is a scalable machine learning system for tree boosting proposed by Chen and Guestrin [7]. The impact of this system has been lately recognized in a number of machine learning and data mining challenges. For example, among the 29 challenge winning solutions published at Kaggle's blog during 2015, 17 solutions used XGBoost.

Considering training dataset $\{\mathbf{x}_k, y_k\} \in \mathbb{R}^n \times \{-1, 1\}$ where $\mathbf{x}_k$ are the training examples, y_k are the class labels and n is number of features, the output of model is voted or averaged by a collection F of k regression trees:

$$\hat{y}_i = \phi(\mathbf{x}_i) = \sum_{i=1}^{k} f_k(\mathbf{x}_i), f_k \in F \tag{8}$$

Each regression tree contains a continuous score on each of the leaves (w_i represents score on the i-th leaf). To learn the set of functions used in the model, the following objective needs to be minimized

$$L(\phi) = \sum_i l(\hat{y}_i, y_i) + \sum_k \Omega(f_k) \tag{9}$$

l is the training loss function which measures how well the model fits on training data. The second term Ω penalizes the complexity of the model and is defined as:

$$\Omega(f) = \gamma T + \frac{1}{2}\lambda\|w\|^2 \tag{10}$$

where the γ is the complexity of each leaf, T is the number of leaves in a decision tree and λ is a parameter to scale the penalty. If we apply the second-order Taylor expansion to the loss function and remove the constant terms we obtain the objective at the t-th iteration in the form of:

$$\tilde{L}^{(t)} = \sum_{i=1}^{n}[g_i f_t(\mathbf{x}_i) + \frac{1}{2}h_i f_t^2(\mathbf{x}_i)] + \Omega(f_t) \tag{11}$$

where g_i and h_i are respectively first and second derivative of the loss function.

4   Empirical Results

Our objective was to find a prediction model that could be used as a real-world fraud detection system. Due to the high class imbalance we decided to focus our assessment of a particular algorithm on analyzing recall and precision statistics. For different parameters configurations we obtained results with either high recall and low precision or low recall and high precision. The former one has an obvious advantage of capturing most of the frauds that were present in a dataset. On the other hand, it is completely useless for a real world applications in which all the alerts have to be manually analyzed by a human being.

As we included almost all of the fraudulent transaction and only minor sample of non-fraudulent, we had distribution in which probability of a random account being a fraudulent one was significantly higher than in the real-world. Because of that, we could not rely on precision statistic as it is vulnerable to this problem. Instead of using precision as a false alarm verification cost estimator we decided to use false positive rate. It fits our purpose since it does not depend on the total amount of frauds in the dataset.

4.1   Random Forest Results

For random forest we decided to tune number of variables randomly sampled as candidates at each split (mtry), minimum size of terminal nodes (min.node.size) and different cut-off probabilities i.e. probability above which sample is actually predicted as a non-fraud.

As we can see in Table 2 biggest impact on the results has threshold which determines final predicted class. Larger threshold causes less samples to be classified as non-fraud and therefore an increase of recall and at the same time increase in FPR which we would like to keep low.

Instead of choosing one configuration which would be a trade-off between recall and false positive rate, we decided to distinguish classifiers able to find as many actual fraudulent accounts as possible (maximizing recall) and a classifiers that make as few mistakes in predicting fraud class as possible (minimizing false positive rate). Validation results presented in Table 2, are similar to the ones we got with cross-validation and confirm, the best configurations are: Conf. 3 in terms of FPR and Conf. 19 in terms of recall. For chosen configurations of random forest we created confusion matrices (presented in Tables 3 and 4) that help to better analyze performance of this classifier on the dataset that is highly imbalanced.

Table 2. Validation results for random forests

	Configuration value			Cross-validation results [%]				
	mtry	min.node.size	Probability	Specificity	Recall	Precision	FPR	F1
Conf.1	3	1	0.5	99.97	24.36	83.33	0.03	37.7
Conf.2	6	1	0.5	99.96	25.52	80.29	0.04	38.73
Conf.3	**3**	**10**	**0.5**	**99.98**	**23.67**	**85.71**	**0.02**	**37.09**
Conf.4	6	10	0.5	99.97	24.59	83.46	0.03	37.99
Conf.5	3	1	0.65	99.93	30.16	72.63	0.07	42.62
Conf.6	6	1	0.65	99.92	32.02	70.41	0.08	44.02
Conf.7	3	10	0.65	99.94	30.16	76.47	0.06	43.26
Conf.8	6	10	0.65	99.93	32.02	72.63	0.07	44.44
Conf.9	3	1	0.8	99.79	42	55.35	0.21	47.76
Conf.10	6	1	0.8	99.73	44.08	50	0.27	46.86
Conf.11	3	10	0.8	99.81	41.76	57.32	0.19	48.32
Conf.12	6	10	0.8	99.75	44.32	52.47	0.25	48.05
Conf.13	3	1	0.9	99.31	54.06	32.5	0.69	40.59
Conf.14	6	1	0.9	99.19	54.52	29.3	0.81	38.12
Conf.15	3	10	0.9	99.34	54.52	33.76	0.66	41.7
Conf.16	6	10	0.9	99.24	55.22	30.95	0.76	39.67
Conf.17	3	1	0.99	90.67	83.53	5.22	9.33	9.83
Conf.18	6	1	0.99	90.79	83.06	5.26	9.21	9.89
Conf.19	**3**	**10**	**0.99**	**90.31**	**84.92**	**5.12**	**9.69**	**9.65**
Conf.20	6	10	0.99	90.63	83.29	5.19	9.37	9.77

Table 3. Confusion matrix for Conf. 3 random forest

Prediction	Actual value		Total
	Fraud	Non-fraud	
Fraud	102	17	119
Non-fraud	329	69991	70320
Total	431	70008	70439

Table 4. Confusion matrix for Conf. 19 random forest

Prediction	Actual value		Total
	Fraud	Non-fraud	
Fraud	366	6786	7152
Non-fraud	65	63222	63287
Total	431	70008	70439

Table 5. Validation results for SVM

	Configuration value		Cross-validation results [%]				
	Cost	Gamma	Specificity	Recall	Precision	FPR	F1
Conf 1.	**1**	**0.077**	**72.62**	**87.47**	**1.93**	**27.38**	**3.77**
Conf 2.	1	0.100	75.03	86.77	2.09	24.97	4.09
Conf 3.	1	0.500	79.52	84.69	2.48	20.48	4.82
Conf 4.	1	1.000	84.38	83.99	3.20	15.62	6.17
Conf 5.	1	2.000	85.84	82.60	3.47	14.16	6.65
Conf 6.	5	0.077	76.78	84.92	2.20	23.22	4.29
Conf 7.	5	0.100	77.72	84.92	2.29	22.28	4.47
Conf 8.	5	0.500	84.00	83.99	3.13	16.00	6.04
Conf 9.	5	1.000	85.60	83.76	3.46	14.40	6.64
Conf 10.	5	2.000	87.38	79.35	3.72	12.62	7.12
Conf 11.	10	0.077	77.64	84.69	2.28	22.36	4.44
Conf 12.	10	0.100	78.33	85.38	2.37	21.67	4.61
Conf 13.	10	0.500	85.07	83.29	3.32	14.93	6.39
Conf 14.	10	1.000	85.99	83.06	3.52	14.01	6.76
Conf 15.	10	2.000	88.00	76.80	3.79	12.00	7.23
Conf 16.	50	0.077	78.87	85.15	2.42	21.13	4.71
Conf 17.	50	0.100	79.44	85.15	2.49	20.56	4.83
Conf 18.	50	0.500	86.09	83.06	3.55	13.91	6.80
Conf 19.	50	1.000	87.35	80.05	3.75	12.65	7.17
Conf 20.	**50**	**2.000**	**89.41**	**75.87**	**4.23**	**10.59**	**8.01**

4.2 Support Vector Machine Results

For the purpose of training Support Vector Machines we chose the radial basis function as a kernel and additionally we increased cost of misclassifying samples to better address the problem of class imbalance in the dataset. The tuned parameters were: cost of constraints violation (cost) and kernel parameter gamma. As shown in Table 5 SVM achieved high recall, but with quite low precision for almost all configurations. If we only consider recall, Conf 1. was better than random forests' Conf 19. with significantly higher false positive rate. Actually, no set of parameters was able to get false positive rate lower than 10%. If we also had to choose configuration with the lowest FPR, Conf. 20 would be the best candidate.

4.3 XGBoost Results

In case of XGBoost we analyzed following hyperparameters in different configurations: maximum depth of a tree (max.depth), minimum sum of instance weight needed in a child (min.child.weight), subsample ratio of columns when constructing each tree (colsample) and, as in random forests, cut-off probability. As for the training itself, we set maximum number of iterations to 2000 with learning rate parameter set to 0.1 using early stop if error does not decrease in 100 consecutive iterations.

Even though we built classifiers for 240 combinations of hyperparameters we decided to present only 20 most interesting. In Table 6 Conf. 1 - Conf. 10 have the smallest false-positive rate and the other 10 configurations have significantly larger recall. Looking at the classification results we can draw a similar conclusion as in the case of random forest - cut-off probability is the most important

Table 6. Validation results for XGBoost

	Configuration value				Cross-validation results [%]				
	max.depth	colsample	min.c.w	prob.	Specificity	Recall	Precision	FPR	F1
Conf 1.	**6**	**0.25**	**1**	**0.50**	**99.95**	**31.32**	**78.03**	**0.05**	**44.70**
Conf 2.	9	0.25	1	0.50	99.95	30.16	78.30	0.05	43.55
Conf 3.	3	0.25	2	0.50	99.94	31.32	76.27	0.06	44.41
Conf 4.	3	0.25	1	0.50	99.95	32.02	78.41	0.05	45.47
Conf 5.	3	0.50	1	0.50	99.94	32.71	77.05	0.06	45.93
Conf 6.	6	0.50	1	0.50	99.94	32.02	76.24	0.06	45.10
Conf 7.	9	0.50	1	0.50	99.94	32.48	76.09	0.06	45.53
Conf 8.	3	0.75	1	0.50	99.95	33.18	79.89	0.05	46.89
Conf 9.	6	0.75	1	0.50	99.94	31.32	77.14	0.06	44.55
Conf 10.	9	0.75	1	0.50	99.94	32.71	77.05	0.06	45.93
Conf 11.	3	0.50	8	0.99	93.79	79.35	7.30	6.21	13.36
Conf 12.	3	0.25	8	0.99	93.58	80.51	7.16	6.42	13.16
Conf 13.	3	0.25	4	0.99	93.74	80.05	7.30	6.26	13.37
Conf 14.	3	0.50	4	0.99	93.99	79.58	7.54	6.01	13.77
Conf 15.	3	1.00	4	0.99	94.20	78.42	7.68	5.80	14.00
Conf 16.	**3**	**1.00**	**8**	**0.99**	**93.89**	**80.51**	**7.50**	**6.12**	**13.72**
Conf 17.	3	0.25	1	0.99	93.97	78.89	7.45	6.03	13.61
Conf 18.	3	0.75	8	0.99	93.83	79.58	7.35	6.17	13.46
Conf 19.	3	0.25	2	0.99	93.91	79.81	7.47	6.09	13.66
Conf 20.	3	0.75	4	0.99	94.05	80.05	7.65	5.95	13.96

parameter for the outcome. After examining the other parameters we were not able to clearly describe their exact impact for the results. As shown in Table 6 validation results confirmed, Conf. 1 and Conf. 16 being the best in their categories, but slightly worse than the best two random forest configurations.

4.4 Sensitivity Analysis

Decision to conduct sensitivity analysis was motivated by our inability to indicate the exact moment of the marking any particular account as fraudulent and thus aggregated transactions data might be contaminated with transactions that happened after an alert on Etherscan has been raised for a particular account. This may lead to look-ahead bias since we are using data that was unknown at the moment of detecting a fraudulent account. In our approach we investigated what impact on the quality of the classifiers excluding the most important and potentially biased variables might have.

Importance of considered variables is not as easily determined when using SVM as in random forest or XGBoost. Furthermore, none of the SVM results was as satisfactory (in terms of recall) as the best of random forests or XGBoost. These two observations led to omission of SVM in our sensitivity analysis.

Explanatory variables importances were calculated separately for each of the best configurations and are presented in the Fig. 2.

Considering random forests variable importance (sometimes called "gini importance") is defined as the total decrease in node impurity weighted by the probability of reaching that node averaged over all trees in the forest. Impurity is defined as:

$$G = \sum_{i=1}^{C} p(i) * (1 - p(i)) \tag{12}$$

with C being the number of classes and $p(i)$ being the probability of picking a datapoint with class i.

In case of XGBoost relative variable importance is measured as the Gain which is contribution of the corresponding feature to the model calculated by taking each feature's contribution for each tree in the model. If we define $G_j = \sum_{i \in I_j} g_i$ and $H_j = \sum_{i \in I_j} h_i$ (based on the Eq. 11) where I_j is the set of indices of data points assigned to the j-th leaf, we can express Gain as:

$$Gain = \frac{1}{2} \left[\frac{G_L^2}{H_L + \lambda} + \frac{G_R^2}{H_R + \lambda} + \frac{(G_L + G_R)^2}{H_L + H_R + \lambda} \right] - \gamma \tag{13}$$

This formula can be decomposed as (1) the score on the new left leaf (2) the score on the new right leaf (3) The score on the original leaf (4) regularization on the additional leaf.

As we can see in Fig. 2 the most important variables for each classifier are usually connected with the incoming and the least important with the outgoing transactions. The only variable that is either first or second in terms of importance for all three classifiers is average time between incoming transactions. For the XGBoost we decided to apply a minor change to the chosen configurations. Instead of stopping after having no decrease of error in 100 consecutive iterations, XGBoost would do 2000 iterations regardless of the results.

As presented in Tables 7 and 8 random forest turned out to be more resistant to cutting off important variables. Even though false positive rate for Conf. 19 is high, with 8 variables excluded we are still able to detect almost 70% of all frauds.

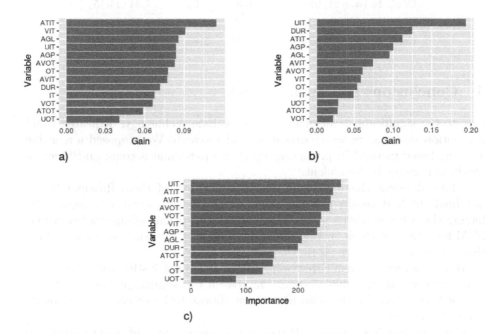

Fig. 2. Variable importance for: (a) XGBoost Conf. 1, (b) XGBoost Conf. 16, (c) Random Forest Conf. 3

Table 7. Validation results for random forests with n most important variables excluded

	Validation results [%]				
	Specificity	Recall	Precision	FPR	F1
Conf. 3 ($n = 2$)	99.98	15.55	81.71	0.02	26.12
Conf. 3 ($n = 4$)	99.98	14.62	84	0.02	24.90
Conf. 3 ($n = 8$)	99.98	7.66	71.74	0.02	13.84
Conf. 19 ($n = 2$)	89.52	82.37	4.62	10.48	8.74
Conf. 19 ($n = 4$)	89.38	81.67	4.52	10.62	8.57
Conf. 19 ($n = 8$)	88.66	68.91	3.60	11.34	6.86

Table 8. Validation results for XGBoost with n most important variables excluded

	Validation results [%]				
	Specificity	Recall	Precision	FPR	F1
Conf. 1 ($n = 2$)	99.95	26.68	75.16	0.05	39.38
Conf. 1 ($n = 4$)	99.95	17.63	68.46	0.05	28.04
Conf. 1 ($n = 8$)	99.98	2.78	54.55	0.02	5.30
Conf. 16 ($n = 2$)	92.66	76.33	6.02	7.34	11.15
Conf. 16 ($n = 4$)	90.69	71.46	4.51	9.31	8.49
Conf. 16 ($n = 8$)	87.03	62.41	2.88	12.97	5.50

5   Conclusions and Future Work

Due to the significant developments in blockchain technology, dedicated fraud prevention systems are an important area of research. We proposed a machine learning based method for predicting whether a particular account on Ethereum blockchain might be fraudulent.

Three different classifiers were analyzed and out of them Random Forest obtained the best results in terms of recall and false positive rate separately, having the other statistics at the reasonable level (in one of the configurations SVM had the best recall for the validation set but at the same time it had three times worse false positive rate).

Best recall for Random Forest was 84.92%. It did not justify using this model in any real-world anti-fraud system. The reason was significant amount of type I error being made by that classifier where almost 10% percent of all accounts would be alerted.

Configuration 3 for Random Forest that achieved 0.02% of false positive rate was still able to detect 23.67% of all frauds. This result can be perceived as a good candidate for an automated anti-fraud system. If we would like to deploy such a system on any cryptocurrency exchange or within cryptocurrency wallet we will mark as fraudulent one in five thousands accounts.

As for future work, we would like to obtain data from exchanges that will help determine whether proposed method can be applied in the current form or is needing further enhancements.

Conducted sensitivity analysis showed that proposed model are not too sensitive for particular explanatory variables but one of future research directions may include estimating exact moments of marking particular account as fraudulent. Then, we would not take a risk of our training set being vulnerable to look-ahead bias.

References

1. Abdallah, A., Maarof, M.A., Zainal, A.: Fraud detection system: a survey. J. Netw. Comput. Appl. **68**, 90–113 (2016)

2. Bhardwaj, A., Gupta, R.: Financial frauds: data mining based detection-a comprehensive survey. Int. J. Comput. Appl. **156**(10) (2016)
3. Boser, B.E., Guyon, I.M., Vapnik, V.N.: A training algorithm for optimal margin classifiers. In: Proceedings of the Fifth Annual Workshop on Computational Learning Theory, pp. 144–152. ACM (1992)
4. Breiman, L.: Random forests. Mach. Learn. **45**(1), 5–32 (2001)
5. Buterin, V., et al.: A next-generation smart contract and decentralized application platform. White paper (2014)
6. Carneiro, N., Figueira, G., Costa, M.: A data mining based system for credit-card fraud detection in e-tail. Decis. Support Syst. **95**, 91–101 (2017). https://doi.org/10.1016/j.dss.2017.01.002. http://www.sciencedirect.com/science/article/pii/S0167923617300027
7. Chen, T., Guestrin, C.: Xgboost: a scalable tree boosting system. CoRR abs/1603.02754 (2016). http://arxiv.org/abs/1603.02754
8. Kou, Y., Lu, C.T., Sirwongwattana, S., Huang, Y.P.: Survey of fraud detection techniques. In: IEEE International Conference on Networking, Sensing and Control, vol. 2, pp. 749–754. IEEE (2004)
9. Pham, T., Lee, S.: Anomaly detection in bitcoin network using unsupervised learning methods. CoRR abs/1611.03941 (2016). http://arxiv.org/abs/1611.03941
10. Quah, J.T., Sriganesh, M.: Real-time credit card fraud detection using computational intelligence. Expert Syst. Appl. **35**(4), 1721–1732 (2008). https://doi.org/10.1016/j.eswa.2007.08.093. http://www.sciencedirect.com/science/article/pii/S0957417407003995
11. Wood, G., et al.: Ethereum: a secure decentralised generalised transaction ledger. Ethereum Proj. Yellow Pap. **151**, 1–32 (2014)
12. Wörner, D., Von Bomhard, T., Schreier, Y.P., Bilgeri, D.: The bitcoin ecosystem: disruption beyond financial services? (2016)

Locking Mechanism for Concurrency Conflicts on Hyperledger Fabric

Lu Xu[1], Wei Chen[1,2], Zhixu Li[1], Jiajie Xu[1], An Liu[1], and Lei Zhao[1(✉)]

[1] School of Computer Science and Technology, Soochow University, Su Zhou, China
lxu7@stu.suda.edu.cn, {robertchen,zhixuli,xujj,anliu,zhaol}@suda.edu.cn
[2] Institute of Artificial Intelligence, Soochow University, Su Zhou, China

Abstract. Hyperledger Fabric is a popular permissioned blockchain platform and has great commercial application prospects. However, the limited transaction throughput of Hyperledger Fabric hampers its performance, especially when transactions with concurrency conflicts are initiated. In this paper, we focus on transactions with concurrency conflicts and propose a novel method LMLS, which contains the following two components, to optimize the performance of Hyperledger Fabric. Firstly, we design a locking mechanism to discovery conflicting transactions at the beginning of the transaction flow. Secondly, we optimize the ledger storage based on the locking mechanism, where the database indexes corresponding to conflicting transactions are changed and temporally stored in ledger to improve the processing efficiency. Extensive experiments conducted on three datasets demonstrate that the proposed novel methods can significantly increase transaction throughput in the case of concurrency conflicts, and maintain high efficiency in transactions without concurrency conflicts.

Keywords: Hyperledger Fabric · Concurrency · Locking mechanism

1 Introduction

Blockchain technologies have become popular these years and can be applied to different domains. Unlike a common database system, a Blockchain is a distributed, shared ledger system where the nodes do not fully trust each other. Each node holds the copy of the ledger which is represented as a chain of blocks, with each block being a sequence of transactions. With the characteristics of decentralization, distrust and tamper-proof, blockchain is adopted in a wise variety of industries. A number of blockchain platforms have been developed, including Bitcoin [16], Ethereum [3], Hyperledger Fabric [5] etc. Among them, Hyperledger Fabric is a representative blockchain platform and has attracted much attention due to the wide application range of it.

Hyperledger Fabric is a permissioned blockchain platform which is highly suitable for developing enterprise-class applications and has a modular design.

© Springer Nature Switzerland AG 2019
R. Cheng et al. (Eds.): WISE 2019, LNCS 11881, pp. 32–47, 2019.
https://doi.org/10.1007/978-3-030-34223-4_3

In Hyperledger Fabric, the identity of each participant is known and authenticated cryptographically. Different from many blockchains whose nodes are peer-to-peer, nodes in Hyperledger Fabric are of different types. The nodes in Hyperledger Fabric contain Client, Peer and Orderer, and each of them performs individual duty in the transaction flow. A transaction is initiated by Client and send to endorsing Peers. Endorsing Peers do endorsement and send response to Client, then Client broadcasts the transaction proposal and response to Orderer which orders them into blocks. The blocks containing some transactions are delivered to all Peers. At last, Peers update the ledger and the transaction flow finishes. In addition, Hyperledger Fabric has better scalability and security, and superior in performance [18] such as latency and throughput to other blockchain platforms. Hyperledger Fabric which our work focuses on is currently being used in many different applications such as Global Trade Digitization [23], SecureKey [8] and Everledger [4].

Hyperledger Fabric has received a lot of concerns, but has exposed many problems at the same time. The main problem is the performance of transaction processing, that is, blockchain system including Hyperledger Fabric can only handle a huge volume of transactions with a low throughput. Some papers analyze the performance of Hyperledger Fabric, Gupta et al. [14,15] present two models to optimize the temporal query performance of Hyperledger Fabric. Thakkar et al. [22] study the impact of various configuration parameters on the performance of Hyperledger Fabric. Gorenflo et al. [13] improve the throughput of Hyperledger Fabric by reducing computation and I/O overhead during the transaction flow. Although these studies have made great contributions, their proposed methods cannot be directly used to tackle the following task, i.e., multiple operations updating the same data in the ledger simultaneously. This is because approaches developed in existing work can only conduct the operations having no conflicting transactions. Unfortunately, this problem, which is called concurrency conflicts, is ubiquitous in Hyperledger Fabric where the data is distributedly stored. We define the concurrency conflict in Hyperledger Fabric as multiple proposals updating the same data in the ledger simultaneously. Since a transaction passes through multiple nodes and the transaction flow is relatively complicated, transactions with concurrency conflicts are discovered in the final step, which leads to the inefficient processing of transactions in Hyperledger Fabric.

To address above mentioned problem, we propose a novel method LMLS. Firstly, a locking mechanism is proposed to discovery conflicting transactions at the beginning of the transaction flow. For example, there are two transactions that are transferred to the same account at the same time. Since the previous transaction first updated the account data, the conflict of data inconsistency occurred in the latter transaction, which caused the transfer to fail. If there are multiple times of the above transactions, the processing efficiency will be low. The locking mechanism can prevent some conflicting transactions from occupying resources of the nodes. We use redis [7] to implement the locking mechanism which mainly contains locking and unlocking. When a transaction request is initiated, it is first checked to ensure if its corresponding key is locked, thereby determining whether the transaction is a conflicting transaction. Moreover, a listener is used to control the lock and unlock operations. Secondly, based on the locking mechanism, database indexes corresponding to conflicting transactions are changed and temporally stored to improve processing efficiency. In Hyperledger Fabric, the data is stored as a key-value pair $\langle k, v \rangle$. We transform the index of the data corresponding to the conflict transaction from k to (k, d), where d is a unique identifier of a transaction and (k, d) is the composite key generated by k and d. This allows conflicting transactions who share the same key not to fail. That is to say, based on LMLS, we can address concurrency conflicts in Hyperledger Fabric. To sum up, the contributions of this paper are as follows.

- To the best of our knowledge, we are the first to improve the performance of Hyperledger Fabric in transaction processing by considering concurrency conflicts.
- To tackle the issue of concurrency conflicts, we design a novel method LMLS which contains Locking Mechanism and Ledger Storage.
- The experimental results show that our method can significantly increase transaction throughput in the case of concurrency conflicts and maintain high efficiency in transactions without concurrency conflicts.

The rest of the paper is organized as follows: We present the related work in Sect. 2 and formulate the problem in Sect. 3. Section 4 gives a brief introduction of Hyperledger Fabric architecture. In Sect. 5 we propose LMLS method to improve the performance of Hyperledger Fabric with concurrency conflicts. In Sect. 6, experiments are conducted to validate the effectiveness of the proposed method. Finally, we conclude this paper in Sect. 7.

2   Related Work

Efficient handling of concurrency conflicts is a hot research topic in distributed database, and conflicting transactions are also existing in Hyperledger Fabric which is a distributed system. Hyperledger Fabric is a recent system that is still undergoing rapid development. Hence, there is relatively little work on the performance analysis of the system or suggestions for architectural improvements. Next, we will introduce the recent work related to this research.

Analyzing Blockchain Performance. Blockchain performance analysis is an emerging area. Recently the BLOCKBENCH system [12] benchmarked the popular blockchain implementations - Hyperledger Fabric, Ethereum and Parity [6] against a set of database workloads. Similar efforts include - benchmarking Hyperledger Fabric and Ethereum against transactional workloads [18]. They find that Hyperledger Fabric outperforms Ethereum in all metrics. Our paper focuses on improve the performance of Hyperledger Fabric.

Analyzing Hyperledger Fabric Performance. Some studies have also looked at performance studies of Hyperledger Fabric, and analyzed the performance from multiple perspectives. For example, Nasir *et al.* [17] compare the performance of Hyperledger Fabric 0.6 and 1.0 which find that the 1.0 version outperforms the 0.6 version. Baliga *et al.* [10] show that application-level parameters such as the read-write set size of the transaction and chaincode as well as event payload sizes significantly impact transaction latency.

Optimizing Transaction Processing Performance. Many studies have proposed the optimization of the performance for processing transactions in Hyperledger Fabric. In recent work, Thakkar *et al.* [22] study the impact of various configuration parameters on the performance of Hyperledger Fabric. They identify some major performance bottlenecks and provide some optimizations such as MSP cache, parallel VSCC validation. Gupta *et al.* [14,15] present two models to optimize the temporal query performance of Hyperledger Fabric. Gorenflo *et al.* [13] improve the throughput of Hyperledger Fabric by reducing computation and I/O overhead during the transaction flow. Sharma *et al.* [20] study the use of database techniques to reorder transaction to remove serialization conflicts and abort transactions which have no chance to commit early to improve the performance of Hyperledger Fabric.

Optimizing Other Aspects of Performance. In addition, Some papers have optimized the performance of other aspects of Hyperledger Fabric, i.e., channel, oderer component. As known to all, Hyperledger Fabric's orderer component can be a bottleneck so Sousa *et al.* [21] study the use of the well-known BFT-SMART [11] implementation as a part of Hyperledger Fabric to improve it. Androulaki *et al.* [9] study the use of channels for scaling Fabric. However, this work does not present a performance evaluation to quantitatively establish the benefits from their approach. Raman *et al.* [19] study the use of lossy compression to reduce the communication cost of sharing state between Fabric endorsers and committers. However, their approach is only applicable to scenarios which are insensitive to lossy compression, which is not the general case for blockchain-based applications.

However, only few studies have looked at the issues concurrency conflicts on blockchain. Thus, to improve the performance of Hyperledger Fabric, we focuses on concurrency conflicts of transactions on this platform.

3 Problem Definition

3.1 The Problem of Concurrency Conflicts in Hyperledger Fabric

Although Hyperledger Fabric has a higher transaction throughput than other permissioned blockchain systems and some papers have studied its transaction performance, they almost assume that multiple requests do not modify the same data in the ledger at the same time. However, when multiple requests want to modify the same data simultaneously, Hyperledger Fabric will process one of the requests and successfully modify the value, and the rest will return

"MVCC_READ_CONFLICT" errors, which cannot be successfully updated. In detail, according to the transaction flow of Hyperledger Fabric, both requests should be sent to Peers for endorsement, and the results of endorsement will be sent to Orderer. Orderer packages and sorts the transaction proposals and responses, then send them to all Peers for final validation. In the process of validation, Peers need to ensure that the current state of the ledger is consistent with the state of the ledger in which the transaction is generated. When multiple requests are initiated at the same time, one of the requests update the value of the data first, causing errors in the remaining requests when the requests verify consistency and returning failures. Such concurrency conflicts result in lower efficiency in processing transactions.

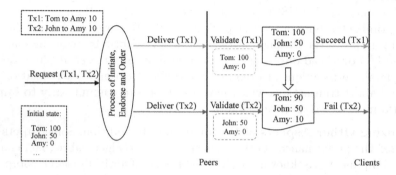

Fig. 1. The instance for conflicting transactions (Tx1 represents Tom transfers $10 to Amy and Tx2 represents John transfers $10 to Amy).

3.2   The Instance for Conflicting Transactions

Specifically, as shown in Fig. 1, there are three people Tom, John and Amy. In the initial state, the account balance of Tom, John and Amy is $100, $50, and $0. At some point Tom and John simultaneously transfer $10 to Amy, that is, there are two requests to update Amy's account balance at the same time. Here, they are initiated almost simultaneously, through endorsement by Peers, ordering by Orderers. Then, the two transactions are packed into the block and successively delivered to Peers for verification. It should be noted that the transactions in the block contain much information, one of them is the status of the ledger when the transaction is initiated (here, the status of the ledger is the initial state shown in Fig. 1). Without loss of generality, we assume that Tx1 arrives earlier, and Peers compare the local ledger with the initial state in Tx1 (the values corresponding to Tom are both 100 and to Amy are both 0) finding that they are consistent. Therefore, the balance of Tom is successfully updated to $90 and the balance of Amy is successfully updated to $10. However, at this time, Tx2 is delivered to Peers, and repeating the above comparison, Peers find it is not consistent with the current value of the local ledger (the value corresponding to Amy in local ledger is 10 while the value in Tx2 is 0). Thus, the request of Tx2 is failed to update the ledger and it should be initiated again.

Problem Formalization. Given a set of transactions with concurrency conflicts in Hyperledger Fabric, a novel method LMLS is designed to tackle the problem, where a locking mechannism and the optimization of ledger storage are developed.

4 The Hyperledger Fabric Architecture

4.1 Nodes in Hyperledger Fabric

Nodes are the communication entities of the blockchain. Different from many blockchains whose nodes are peer-to-peer, nodes in Hyperledger Fabric play different roles in the network. There are three types of nodes shown in Fig. 2:

Client. A Client represents an entity operated by the end user. A Client submits transaction proposal to the Endorser Peer and broadcasts proposal and response to Orderer.

Peer. A Peer is mainly responsible for reading and writing the ledger by executing chaincode. All Peers are committing peers (Committers) responsible for maintaining the state and the ledger. Peers can additionally take up a special role of an endorsing peer (Endorser). The endorsing peer is a dynamic role, and Peer is the endorsement node only when the application initiates a transaction endorsement request to it, otherwise it is a normal committing peer.

Orderer. A number of Orderers make up ordering service. Since the Hyperledger Fabric is a distributed system, a ledger is stored on each node. When each node wants to modify the state of the ledger, there must be a mechanism to ensure the consistency of all these operations, which is the orderer service. Orderers are responsible for ordering the unpackaged transactions into blocks.

4.2 Transaction Flow

Figure 2 depicts the transaction flow which involves 5 steps. This flow assumes that the application user has registered and enrolled with the organization's certificate authority (CA). The transaction flow is as follows:

(1) Initiating Transaction. Client using Fabric SDK constructs a transaction proposal and sends the proposal which is signed with credentials to one or more endorsement Peers simultaneously.

(2) Endorsement. First, the endorsing Peers verify the signature (using MSP). Second, the endorsing Peers take the transaction proposal arguments as inputs and execute the chaincode against the current state database to produce transaction results including a response, read set and write set. Third, the results,

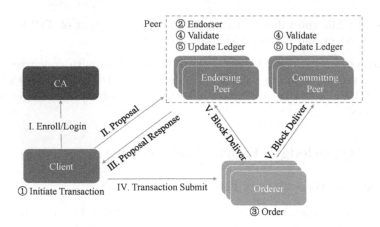

Fig. 2. The transaction flow of Hyperledger Fabric.

along with the endorsing Peer's signature and a YES/NO endorsement state-ment are passed back as a proposal response to Client. Client will collect enough proposal responses from Peers and verify if the result are same.

(3) Ordering. Client broadcasts the transaction proposal and response within a transaction message to the Orderer. The Orderer orders them chronologically by channel, and creates blocks of transactions per channel.

(4) Validation. The blocks containing some transactions are delivered to all Peers. Peers need to verify the signature by Orderer and need to do VSCC validation. A VSCC validation will check if the endorsement policy is satisfied, if not, the transaction will be marked invalid.

(5) Ledger Updated. Each Peer appends the block to the local ledger, and for each valid transaction the write sets are committed to the state-db which stores the current state of all keys.

5   Proposed Method LMLS

In order to solve the concurrency conflict problems in Hyperledger Fabric, we propose the following novel method LMLS to optimize the transaction flow to increase efficiency. Firstly, a locking mechanism is proposed so that conflicting transactions can be discovered at the beginning of the transaction flow. Secondly, based on the lock mechanism, we add a database index for conflicting transac-tions and change the storage way of conflicting transactions, so that they can be temporarily stored in the database. The above methods can effectively improve the performance of Hyperledger Fabric with concurrency conflicts.

5.1   Locking Mechanism

By analyzing the existing problems of Hyperledger Fabric, the main reason for the inefficiency is that invalid transactions (which ultimately failed to successfully update the ledger) are found to be invalid after almost completing the whole transaction flow. Therefore, we consider adding a locking mechanism at the beginning of the transaction process. The locking mechanism can prevent some of the conflicting transactions from occupying resources of the nodes, so that some invalid transactions can be found in the early stage of the transaction flow, thereby improving efficiency.

Implementation of the Locking Mechanism. In this paper, we use redis [7] to implement the locking mechanism. Redis is essentially a database of key-value types. Due to the advantages of redis in performance and concurrency, the use of redis scenarios is mostly a highly concurrent scenario. The idea of implementation is not complicated. In general, we can be divided into two steps: locking and unlocking. First introduce the process of locking, the distinguished name of a task in the request as a key to the redis. If there is a request with the same distinguished name arriving, try to insert it into redis. If it can be successfully inserted, return *True*, that is, it is successfully locked and will get a lock identifier. Otherwise, return *False*, that is, the other request with the same distinguished name is operating, and the lock fails. The process of unlocking is relatively simple. The lock identifier is passed as a parameter to check whether the lock exists. If it exists, the lock identifier can be deleted from the redis.

Listener. To determine when to unlock, we used a listener which can be used to know when the transaction was successfully written to the blockchain. Because of knowing that the transaction has been written to the block, the identifier can be unlocked. In this paper, we use Hyperledger Fabric officially provided listening interface ChannelEventHub [2]. Transaction processing in Hyperledger Fabric is a long operation. As a result the applications must design their handling of the transaction lifecycle in an asynchronous fashion. We mainly use registerTx-Event interface to listen the transaction flow. When a transaction is initiated, a transaction listener is registered and returns a specific sequence number as the identifier. When the transaction is written to the blockchain, it will be listened to by the listener, and the listener will call the function to unlock the lock identifier corresponding to the transaction.

5.2   Optimization of Ledger Storage

Although the lock mechanism can cause invalid transactions to be discovered earlier, users need to re-initiate these transactions which does not improve the user experience. When multiple conflicting transactions are initiated simultaneously, there will still be only one transaction that can be successfully updated to the blockchain ledger and the other transactions need to be initiated again. Therefore, based on the locking mechanism, we improve the storage of the blockchain ledger and transform the database indexes to avoid concurrency conflicts.

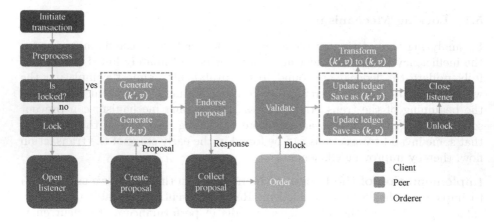

Fig. 3. The complete transaction flow with LMLS.

In Hyperledger Fabric, the data in ledger is stored in key-value pair. For a key k, the latest pair is called the current state of the key k which is stored in state-db, while all the pairs including the latest pair form the historical states of key k which is stored in history-db. Obviously, the collection of current states for all keys is termed as state-db, and the collection of historical states is termed as history-db. In this paper, all the changes transactions initiated are in the current state, so we only pay attention to state-db.

Usually, we modify the data in state-db by initiating a proposal. In this paper, we assume that each time a proposal is initiated, only one data in state-db is modified, that is, a transaction T generates a proposal P, which corresponds to a key-value pair $\langle k, v \rangle$ in state-db. If two transactions T_i and T_j are initiated at the same time, two proposals P_i and P_j will be generated, corresponding to the key-value pairs $\langle k_i, v_i \rangle$ and $\langle k_j, v_j \rangle$ in the state-db. If $k_i = k_j$, this is the case of concurrency conflicts. In order to effectively avoid conflicts and enable both proposals to be successfully executed, we transform the database indexes of state-db. Specifically, for conflicting transactions, we transformed $\langle k, v \rangle$ to $\langle (k, d), v \rangle$ where (k, d) is the composite key generated by k and d, and d is a transaction id for transaction T, which is a unique identifier that is randomly generated. For transactions T_i and T_j, without losing generality, we assume that T_i is processed before T_j, then we transform $\langle k_j, v_j \rangle$ to $\langle k'_j, v_j \rangle$ where k'_j represents the composite key (k_j, d_j). Thus, k_i and k'_j are not equal and both transactions T_i and T_j can update the ledger avoiding concurrency conflicts.

5.3   Steps of LMLS

Combining the ledger storage improvements with locking mechanism, the steps of LMLS are shown in Fig. 3, which can be divided into the following steps.

I. A user initiates a transaction, and Client pre-processes the transaction, including obtaining the key k of the data that the transaction wants to update. Client checks if k is locked. If it is, directly turn to III, otherwise, turn to II.

II. Lock k and get a lock identifier l.

III. Client opens the listener, generates the corresponding transaction proposal, and sends the proposal to Peers.

IV(i). If k obtains the corresponding lock identifier l, Peers generate the key-value pair $\langle k^{'}, v \rangle$ according to the transaction id, and endorse to simulate the execution of smart contracts.

IV(ii). If k does not obtain l, Peers generate the key-value pair $\langle k, v \rangle$, and endorse to simulate the execution of smart contracts.

V. Peers return the endorsement result to Client, and Client sends the proposal and result to Orderer which order and package them to new block. Orderer send the packaged block to Peers, and Peers perform the final verification.

VI(i). If k obtains the corresponding lock identifier l, Peers save $\langle k^{'}, v \rangle$ into state-db to update ledger. Client listens to the operation and closes the listener.

VI(ii). If k does not obtain l, Peers save $\langle k, v \rangle$ into state-db to update the ledger. Client listens to the operation, then it unlocks the lock identifier l corresponding to k first and closes the listener.

VII. After all the above steps are finished, $\langle k^{'}, v \rangle$ will merge with $\langle k, v \rangle$ by chaincode safely and the former will be deleted.

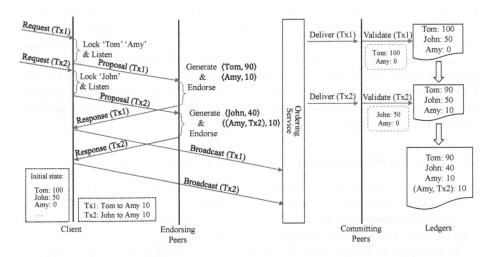

Fig. 4. The example for LMLS to process conflicting transactions (Tx1 represents Tom transfers \$10 to Amy and Tx2 represents John transfers \$10 to Amy).

5.4   Examples for LMLS

Continue the example in Sect. 3, we assume that Tx1 in Fig. 4 arrives earlier, then Client locks two keys ('Tom' and 'Amy') in Tx1 and starts listening. Subsequently, the request of Tx2 is initiated, at this time Client only locks the key 'John' , then starts listening. When the above two proposals are sent to Peers, Peers generate the corresponding key-value pair respectively and endorse them. The difference is that for Tx1, two key-value pairs $\langle Tom, 90\rangle$ and $\langle Amy, 10\rangle$ are generated, but for Tx2, a key-value pair $\langle John, 40\rangle$ and a composite index-key-value pair $\langle (Amy, Tx2), 10\rangle$ are generated. Then, the two transactions are ordered and delivered to Peers where validation need to be done. In this example, Peers first validate Tx1. They compare the local ledger with the initial state in Tx1 (the values corresponding to Tom are both 100 and to Amy are both 0) finding that it is consistent. Therefore, the balance of Tom is successfully updated to $90 and the balance of Amy is successfully updated to $10. Next, Peers validate Tx2, since it has be known as a conflicting transaction in the previous process where the value corresponding to Amy is being operated by another request, a composite key-value pair $\langle (Amy, Tx2), 10\rangle$ will be added to the ledger instead of $\langle Amy, 20\rangle$. In addition, the balance of John will be successfully updated to $40. As shown in Fig. 4, there are two indexes related to Amy in the final ledger where the sum of them is 20. When we request to query Amy's balance, it will return 20 instead of 10.

6   Experiments and Analysis

6.1   Experiment Setup

Since there are many concurrencies in the trading scenario, we implement a concurrency scenario, which can be used for trading, with a chaincode [1]. Our chaincode enables users to register their accounts, deposit, withdraw and transfer and check balances. In this paper, we mainly simulated saving money with concurrency. We use Fabric release v1.2, single peer setup running on a Lenovo T430 machine with 8 GB RAM, dual core Intel i5 processor. We use a single peer but we keep the consensus mechanism turned on. We use all default configuration settings to run our experiments.

6.2   Compared Methods and Metrics for Experiments

We compare the performance of our method LMLS with the original Hyperledger Fabric system. Although existing methods [13, 22] also work on the performance of transaction processing, their results are not comparable here, as their methods only work for transactions without concurrency conflicts.

In this paper, we compare the performance of LMLS and Fabric with following metrics: (1) Total time - the time cost to process all transactions. (2) Success rate - the ratio of transactions successfully written to the ledger to all transactions. (3) Throughput - the amount of transactions successfully written into the ledger per unit time.

6.3   Datasets

We carry out experiments with three synthetically generated datasets. We implement a data generator to generate sets of transactions. In each transaction $\{username, operation, amount\}$, operation denotes the type of the transaction, such as deposit and withdrawal. The generated datasets are as follows.

- **DS1:** In this dataset, the accounts for all transactions are the same, that is, each transaction deposits for the same account. The number of transactions is 10K.
- **DS2:** In this dataset, the accounts for all transactions are not necessarily the same. The number of transactions and accounts are 10K and 1000.
- **DS3:** In this dataset, the accounts for all transactions are different, that is, each transaction deposits for different accounts. Therefore, there is no concurrency conflict in this dataset. The number of transactions is 1K.

6.4   Experiment Results

Experiment for DS1. First, we do experiment in DS1 which the accounts for all transactions are the same. We change the transaction arrival rate, which is the average of transactions initiated per second, from 10 tps to 200 tps. Four groups of experiment are tested which with different transaction volume N of

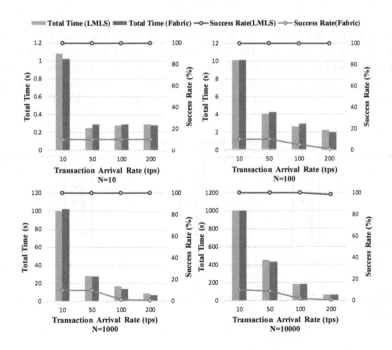

Fig. 5. Time cost and success rate of LMLS and Fabric at different transaction arrival rates in DS1 (N denotes the transaction volume).

10, 100, 1K and 10K. We test time cost and success rate, and the results can be seen from Fig. 5.

As can be seen from Fig. 5, on the one hand, regardless of the transactions volume, the total time of the two methods is similar, which shows that LMLS does not reduce the efficiency of the system in processing transactions. On the one hand, LMLS obviously has a higher success rate and the success rate can reach 100% no matter how high transaction arrival rate is. However, except in the case of a transaction volume of 10, with the increase of the transaction arrival rate, the success rates of Fabric have decreased significantly. Especially when the transaction arrival rate rises to 200 tps, the success rate is close to 0%. This shows that LMLS can successfully handle almost all transactions in the case of high concurrency conflicts, while Fabric cannot. Thus, LMLS is more suitable for scenarios with concurrency conflicts and the efficiency is obviously better than Fabric.

Experiment for DS2. To further validate the performance of our methods, we do experiment in DS2 which the accounts for all transactions are not necessarily the same. In the experiment, we initiate multiple transactions with concurrency conflicts and these transactions will modify different accounts. We define the average transaction number per user as n, the computation method as follows:

$$n = \frac{\sum_{i=1}^{u} a_i}{u} \tag{1}$$

where u denotes the number of accounts modified in the experiment and a_i denotes the number of times the i-th account modified. For convenience, the number of times of each account modified is the same in our experiment, that is, $\forall i, j \in \{1, 2, ..., u\}$, $a_i = a_j$, thus, $n = a_i$ ($i = 1, 2, ..., u$). We test the throughput for two methods varying n and the results can be seen from Fig. 6.

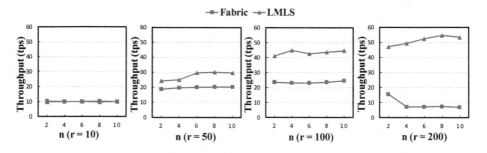

Fig. 6. Throughput of LMLS and Fabric in DS2 varying n (r denotes the transaction arrival rate).

In Fig. 6, n denotes the average transaction number per user and r denotes the transaction arrival rate. First, we can see that with the increase of n, the gap of throughput between Fabric and LMLS is increased. The results illustrate that LMLS is more suitable for scenarios with multiple concurrency conflicts. Second, with the increase of n, the throughput of LMLS is generally on the rise, while Fabric's unchanged, moreover, when $r = 200$, its throughput drops significantly. This shows that the efficiency of Fabric is greatly reduced in high-concurrency scenarios, but LMLS not. Third, horizontally comparing the four line charts, we can see that, as r increases, the throughput of LMLS is also increasing, which illustrates LMLS also performs well in the case of high concurrency conflicts. The experimental results show that our method is significantly more efficient than fabric in complex trading scenarios involving concurrency conflicts.

Experiment for DS3. In order to verify the efficiency of our method in transactions without concurrent conflicts, we do experiment in DS3 which the accounts for all transactions are different, that is, no concurrency conflict in this dataset. As shown in Fig. 7, we test the throughput of two methods at different arrival rates from 1 tps to 200 tps. We can see that as the transaction arrival rate increases, the throughput of Fabric as well as LMLS is increasing. In the absence of concurrency conflicts, LMLS performance is similar to Fabric. Although we add a lock mechanism, our efficiency has not decreased. The experimental results show that our methods are applicable regardless of whether there are scenarios with concurrency conflicts or without.

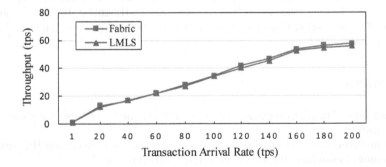

Fig. 7. Throughput of LMLS and Fabric at different transaction arrival rates in DS3.

The Cost Analysis. On the one hand, we consider the time overhead. LMLS compared to Fabric have the cost of lock-mechanism construction time. However, compared to the time it takes for the system to process the transactions, the time to build a lock is negligible, as the experimental results show. In addition, although LMLS changes the database indexing method, it does not increase the time overhead of storage.

On the other hand, we analyze the storage cost of two methods. LMLS builds composite key-value pairs for each transaction with concurrency conflicts, so the number of key-value pairs on state-db increase. However, we eventually merge the composite pairs with the original pairs. Therefore, in general, storage cost has not increased. Moreover, in Hyperledger Fabric, regardless of whether the transaction is valid, it will be stored in the block if it has been sorted by Orderer. Therefore, in the case of transactions with concurrency conflicts, Fabric will package a large number of invalid transactions into blocks. In contrast, LMLS can reduce the cost of block storage.

7   Conclusion

In this paper, we focus on optimize the performance of Hyperledger Fabric by improving the handling efficiency of transactions with concurrency conflicts. We propose a novel method LMLS to optimize the performance of Hyperledger Fabric. Firstly, we design a locking mechanism to discovery conflicting transactions at the beginning of the transaction flow. Secondly, we optimize the ledger storage based on the locking mechanism, where the database indexes corresponding to conflicting transactions are changed and temporally stored in ledger. To validate the performance of the proposed solutions, extensive experiments are conducted and results demonstrate that our method outperforms the original method.

Acknowledgements. This work was supported by the National Natural Science Foundation of China (Grant No. 61572335, 61572336, 61902270), and the Major Program of Natural Science Foundation, Educational Commission of Jiangsu Province, China (Grant No. 19KJA610002), and the Natural Science Foundation, Educational Commission of Jiangsu Province, China (Grant No. 19KJB520052, 19KJB520050), and Collaborative Innovation Center of Novel Software Technology and Industrialization, Jiangsu, China.

References

1. Chaincodes. http://hyperledger-fabric.readthedocs.io/en/release-1.2/chaincode4n oah.html
2. ChannelEventHub. https://fabric-sdk-node.github.io/ChannelEventHub.html
3. Ethereum blockchain app platform. https://ethereum.org/
4. Everledger: A digital global ledger. https://www.everledger.io/
5. Hyperledger fabric. https://www.hyperledger.org/projects/fabric
6. Parity. https://www.parity.io/
7. Redis. https://redis.io/
8. Securekey: Building trusted identity networks. https://securekey.com/
9. Androulaki, E., Cachin, C., De Caro, A., Kokoris-Kogias, E.: Channels: horizontal scaling and confidentiality on permissioned blockchains. In: Lopez, J., Zhou, J., Soriano, M. (eds.) ESORICS 2018. LNCS, vol. 11098, pp. 111–131. Springer, Cham (2018). https://doi.org/10.1007/978-3-319-99073-6_6
10. Baliga, A., Solanki, N., Verekar, S., Pednekar, A., Kamat, P., Chatterjee, S.: Performance characterization of hyperledger fabric. In: CVCBT, pp. 65–74 (2018)
11. Bessani, A.N., Sousa, J., Alchieri, E.A.P.: State machine replication for the masses with BFT-SMART. In: DSN, pp. 355–362 (2014)

12. Dinh, T.T.A., Wang, J., Chen, G., Liu, R., Ooi, B.C., Tan, K.: BLOCKBENCH: a framework for analyzing private blockchains. In: Salihoglu, S., Zhou, W., Chirkova, R., Yang, J., Suciu, D. (eds.) SIGMOD, pp. 1085–1100 (2017)
13. Gorenflo, C., Lee, S., Golab, L., Keshav, S.: Fastfabric: scaling hyperledger fabric to 20,000 transactions per second. CoRR abs/1901.00910 (2019)
14. Gupta, H., Hans, S., Aggarwal, K., Mehta, S., Chatterjee, B., Jayachandran, P.: Efficiently processing temporal queries on hyperledger fabric. In: ICDE, pp. 1489–1494 (2018)
15. Gupta, H., Hans, S., Mehta, S., Jayachandran, P.: On building efficient temporal indexes on hyperledger fabric. In: CLOUD, pp. 294–301 (2018)
16. Nakamoto, S.: Bitcoin: a peer-to-peer electronic cash system (2008)
17. Nasir, Q., Qasse, I.A., Talib, M.A., Nassif, A.B.: Performance analysis of hyperledger fabric platforms. Secur. Commun. Netw. **2018**, 1–14 (2018)
18. Pongnumkul, S., Siripanpornchana, C., Thajchayapong, S.: Performance analysis of private blockchain platforms in varying workloads. In: ICCCN, pp. 1–6 (2017)
19. Raman, R.K., et al.: Trusted multi-party computation and verifiable simulations: a scalable blockchain approach. CoRR abs/1809.08438 (2018)
20. Sharma, A., Schuhknecht, F.M., Agrawal, D., Dittrich, J.: How to databasify a blockchain: the case of hyperledger fabric. CoRR abs/1810.13177 (2018)
21. Sousa, J., Bessani, A., Vukolic, M.: A byzantine fault-tolerant ordering service for the hyperledger fabric blockchain platform. In: DSN, pp. 51–58 (2018)
22. Thakkar, P., Nathan, S., Viswanathan, B.: Performance benchmarking and optimizing hyperledger fabric blockchain platform. In: MASCOTS, pp. 264–276 (2018)
23. White, M.: Digitizing global trade with Maersk and IBM. https://www.ibm.com/blogs/blockchain/2018/01/digitizing-global-trade-maersk-ibm/

Handling Conditional Queries
on Hyperledger Fabric Efficiently

Tianlu Yan[1], Wei Chen[1,2], Pengpeng Zhao[1], Zhixu Li[1], An Liu[1],
and Lei Zhao[1(✉)]

[1] School of Computer Science and Technology, Soochow University, Su Zhou, China
tlyan@stu.suda.edu.cn,
{robertchen,ppzhao,zhixuli,anliu,zhaol}@suda.edu.cn
[2] Institute of Artificial Intelligence, Soochow University, Su Zhou, China

Abstract. As a popular consortium blockchain platform, Hyperledger Fabric has received increasing attention recently. When conducting queries that meet some specific conditions on such platform, we need to search ledger data which usually has multiple attributes. Although efficiently handling conditional queries can be leveraged to support various use-cases, it presents significant challenges as data on Hyperledger Fabric is organized on file-system and exposed via limited API. To tackle the problem, we propose the following novel methods in this paper. In the first one, we use all conditions of the query to create composite keys before executing it. To further improve the performance of conditional queries on Fabric, we build an index called AUP in the second method, where we also study the update of AUP during transactions. The extensive experiments conducted on the real-world dataset demonstrate that the proposed methods can achieve high performance in terms of efficiency and memory cost.

Keywords: Hyperledger Fabric · Ledger data · Conditional queries

1 Introduction

In recent years, blockchain technologies have attracted wide attention and been used in many real applications. This is because they get rid of the centralized storage and can guarantee the data security. A blockchain is a shared, distributed ledger that records transactions between different nodes in a verifiable and permanent way where nodes do not trust each other [18]. Each node in the blockchain network holds the same ledger which contains multiple blocks. A block usually has a list of transactions and encloses the hash of its immediate previous block, where transaction data can be saved in a ledger only after it has passed a series of validations. Note that, blockchain network can be divided into three categories, namely private network, public network and consortium network. In a public network, anyone can join the network to perform transactions. In a private network, there are only a limited range of participating nodes; the

© Springer Nature Switzerland AG 2019
R. Cheng et al. (Eds.): WISE 2019, LNCS 11881, pp. 48–62, 2019.
https://doi.org/10.1007/978-3-030-34223-4_4

access of data has strict rights management, and only participants have the write permission. The consortium chain is available for participants of a specific group. It internally specifies multiple pre-selected nodes as billers, and the generation of each block is determined by all pre-selected nodes. The consortium network is suitable for enterprise applications, each node in the network can be owned by different organizations, and enterprises can integrate the values of multiple systems without having to bring in a trusted third-party.

Hyperledger Fabric [4] is an enterprise-grade and open-source consortium blockchain platform. Like many other blockchain systems (e.g., Ethereum [3], Parity [6]), it divides data into two states: current and historical states. Data is ingested on this system in form of key-value pairs. For a given key, the latest pair is called current state and others are called historical states. Two typical databases in the system are StateDB and HistoryDB. StateDB includes the collection of current states for all keys. HistoryDB includes the collection of historical states for all keys and can be used to quickly locate the position of data in ledger. The historical data is distributed across a large number of blocks on file-system, which leads to the low efficiency of a query with multiple conditions (We refer to it as conditional query in this work). This is because, given a key, the Hyperledger Fabric will return all the historical data of it, based on which we can get the results meeting the given conditions, during an API call.

Obviously, an efficient method is necessary to conduct the conditional query in aforementioned case. Note that, although existing studies have made great contributions in blockchain query [12,13,20,21], the two main techniques, granular access control and indexes constructed based on StateDB, proposed by them can not be directly used to efficiently handle conditional queries on blockchain. This is because, on one hand, nodes are authorized to join in the Hyperledger Fabric network, then there is no need to create additional granular access control for it; on the other hand, it is time consuming to query the whole ledger data before updating the index. Assuming that a user executes a conditional query containing multiple conditions, the conventional query methods need to return all data meeting the first condition and then filter the data according to other conditions, which leads to large time cost. Additionally, conventional methods usually bring a lot of data redundancy, which is demonstrated in Sect. 6. Having observed these weaknesses, we propose the following novel methods, i.e., CCK and AIM. In the first one, we create a composite key for the given query based on the associated conditions of it. Then, we use the composite key to create a new key-value pair before executing data insertion, which can avoid the filtration of historical data. In the second one, to solve the data redundancy problem of the first method, we build an index called AUP for HistoryDB based on LevelDB [5], and the value of each key in AUP consists of corresponding keys of current states.

Considering a use-case, an author α publishes a publication p in a venue v, a key-value pair $<\alpha, (v, o)>$ is inserted into Hyperledger Fabric ledger, and o denotes the other information of the publication, such as title, time and URL. We are interested in querying all publications that are published in the venue

v and belong to the author α. In the first method CCK, we use α and v to create composite key (α, v). By this way, we convert the above key-value pair to $<(\alpha, v), o>$. Based on this method, the processing of filtering publications that belong to α but are not published in v can be avoided. However, we need to create multiple key-value pairs for the publication with multiple authors in this method, which leads to the problem of data redundancy. To solve it, in the second method AIM, we build AUP to record all authors having relationships with the publication to be stored. The key-value pairs in AUP are in the form of $<(\alpha, v), \varepsilon(S_\alpha)>$, where S_α represents all authors of the publication p, and $\varepsilon(S_\alpha)$ denotes all authors that have co-authored with α in history. While inserting a new key-value pair $<(S_\alpha, v), o>$ into blockchain, it inserts $<S_\alpha, "">$ into HistoryDB firstly, and then create $<(\alpha, v), \varepsilon(S_\alpha)>$ in AUP for each author in S_α.

In this study, we have designed novel methods to conduct conditional queries on Hyperledger Fabric with high performance. To sum up, we make the following contributions.

- We are the first to study the problem of efficiently handling conditional queries on Hyperledger Fabric.
- To avoid the process of filtering candidates, we propose the method CCK. To tackle the data redundancy problem brought by CCK, we build an index AUP in the second method AIM.
- We conduct extensive experiments on DBLP, and the results demonstrate that the proposed approaches can achieve high performance in terms of efficiency and memory cost.

The rest of this paper is organized as follows. In Sect. 2, we brefily view existing work related to the research of blockchain. Section 3 presents the background of Hyperledger Fabric. In Sect. 4, we formulate the problem and present notations used in this work. We introduce the proposed methods in Sect. 5 and report the experimental results in Sect. 6. This paper is concluded in Sect. 7.

2 Related Work

Though blockchain analysis is an emerging area, it has received significant attention and a lot of studies have been made on it. These studies are mainly divided into two categories: security and performance. In terms of security, [14] makes a survey of blockchain security issues and challenges, [15] discusses the applicability of blockchain to intrusion detection, and identifies open challenges. There is also a lot of work focused on the performance of blockchain, including [11,17,18]. They mainly concentrate on realizing higher throughputs and lower latencies by using different consensus algorithms, encryption methods. In [8], authors analyze how fundamental and circumstantial bottlenecks in Bitcoin [1] limit the ability of its current peer-to-peer overlay network to support substantially higher throughputs and lower latencies.

2.1   Performance Modeling of Blockchain Networks

The authors of [19] contrast PoW-based blockchains to those BFT-based state machine replication and discuss proposals to overcome scalability limits and outline key outstanding open problems in the quest for the "ultimate" blockchain fabric(s). In [10], they first describe BLOCKBENCH, which is the first evaluation framework for analyzing private blockchains and serves as a fair means of comparison for different platforms and enables deeper understanding of different system design choices, and then they use BLOCKBENCH to conduct comprehensive evaluation of three major private blockchains: Ethereum, Parity and Hyperledger Fabric. They measure the overall performance of the platforms and draw conclusions across the three platforms. [9] is similar to [10], they discuss several research directions for bringing blockchain performance closer to the realm of databases. Zheng et al. [22] provide an overview of blockchain architecture firstly and compare some typical consensus algorithms used in different blockchains.

2.2   Performance Evaluation of Hyperledger Fabric

In existing work, [7] introduces the design and the architecture of Hyperledger Fabric, and presents the performance of a single Bitcoin like crypto currency application on Fabric, called Fabcoin, which uses CLI command to emulate client instead of using a SDK. [12,13,20,21] pay more attention to how to efficiently handle queries in the blockchain platform. [20,21] handle the problem of flexible queries by using granular access control, both of them improve performance by changing encryption methods. [12,13] are the most similar work to our queries, they both propose two method to processe temporal queries on Fabric.

In spite of the great contributions made by the aforementioned studies, none of them consider conditional queries on Fabric. To tackle the problem, we propose two methods in this paper, i.e., composite key based method CCK and AUP index based method AIM, and details are presented in Sect. 5.

3   Background

A Hyperledger Fabric network contains peer nodes, ordering service nodes and clients. A peer node in the network of Fabric is divided into an endorsing node or a committing node. The endorsing node executes the chaincode (a.k.a. smart contract [16]) logic to endorse a transaction, but the committing node does not has the chaincode logic. Although they are different in this point, both of them maintain the ledger in a file system. An ordering service node participates in the consensus protocol and the process of block generation. The client can initiate a transaction proposal to invoke a chaincode function, which can perform read and write operations on shared ledger data by defined ledger APIs. Further, the transaction flow in Hyperledger Fabric consists of 4 phases, (1) Endorsement Phase - simulating the transaction on endorser nodes and collecting the state changes; (2) Orderering Phase - ordering transactions through a consesus proto-col; (3) Validation Phase - verifying the block signature and all transactions in a block; and (4) Commitment Phase - committing valid transaction data to the ledger.

3.1   Data Storage Structure

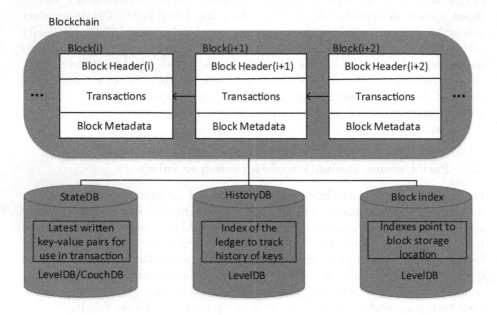

Fig. 1. The structure of data storage in a single-chain.

In Fabric, all valid transactions are stored in blocks, and all blocks are stored in the file system. A simple structure of single-chain data storage is presented in Fig. 1. It contains StateDB, HistroyDB and block index. The StateDB stores the current state of each key and supports LevelDB and CouchDB [2]. The HistoryDB stores the historical state of each key. It records the change of each key in StateDB and is implemented by LevelDB. In fact, it does not store the real value of each key and can be used to quickly locate the position of transaction in the block. Hyperledger Fabric provides a variety of block indexing methods. The content of the block index is the file location pointer, which consists of three parts: the file number, the offset within the file, and the number of bytes occupied by the block. The block index can be used to quickly find the position of blocks.

If we want to add a new state or change the current state of a key, we need to initiate a transaction proposal, executing which successfully, a new key-value pair will be added to a block. The value of the key in StateDB is changed, but the previous key-value pair is still stored in the ledger if it had the value of the key before. Additionally, a new key-value pair will also be inserted into the HistoryDB.

3.2 Accessing Historical States

Hyperledger Fabric provides specific APIs, such as **GHFK** and **CK**, which are used in our proposed methods CCK and AIM.

GetHistoryForKey(k) (GHFK [13]): This is an API provided by Hyperledger Fabric to access the historical states. For a given key k, this call returns all the past states of key k in the history.

CreateCompositeKey(ob, ks) (CK): This is an API provided by Fabric to combine the given attributes ks and object type ob to form a composite key, which can be used as a key to access historical states.

Specifically, when initiating a transaction proposal to get historical states of a given key k, we need to execute a GHFK call. During the execution of the GHFK call, it retrieves all keys in HistoryDB and each key is start with k firstly. Then it analyses all these keys to get the list of block numbers and transaction numbers. Next, it queries the block index to get the location of blocks and then deserializes all blocks to access transaction data according to transaction numbers. Finally, it extracts out all the values. That is to say, the GHFK call needs to retrieve the historical data from multiple blocks and returns an iterator in the end. The more values accessed through this iterator, the larger the number of blocks that need to be deserialized.

4   Problem Statement

In this section, we present all the notations used throughout the paper in Table 1, and then we formulate the problem.

In Fabric, handling conditional queries requires to deserialize blocks that satisfy all query conditions. For example, in DBLP, given an author α and a venue v, when we want to get all publications that belong to the author α and published in venue v, we need to deserialize blocks that satisfy these two conditions: (1) the block contains a transaction which ingests a key-value pair with key equals to α; (2) this pair describes a publication which is published in the given venue v.

Currently, abovementioned conditional query is time-consuming on Fabric, as such operation is not directly supported by Fabric. If intending to query all publications that meet those conditions, we firstly need to query all publications belong to the given author. During this process, we need to deserialize multiple blocks. Then we still need to filter publications according to the venue. Therefore, some deserialized blocks are useless. Larger the number we need to filter, more time the operation will spend. Besides, if we create a key-value pair for each author of a publication, it will lead to a large number of redundancy, since a publication usually has multiple authors and Fabric does not provide any

Table 1. Definitions of notations.

Notation	Definition
α	An author of a publication in DBLP
o	The other information of a publication in DBLP
S_α	All authors of one publication of α, $S_\alpha = \{\alpha_1, \alpha_2, ..., \alpha_n\}$
$\varepsilon(S_\alpha)$	All authors of all publications of α
v	The venue of a publication in DBLP
K	The result of creating composite key by calling CK
V	The value of a key in AUP
S_K	The collection of the results of executing CK
$\varepsilon(\alpha)$	The set of publications belonging to author α
$\varepsilon(\alpha, v)$	The set of publications belonging to α and published in v
$\varepsilon(S_\alpha, v)$	The set of publications belonging to S_α and published in v

indexing capability on the data in HistoryDB. Due to the redundancy, it takes a lot of time to ingest the publication on the ledger. However, if we don't create the key-value for each author, we can not get all information of the publication with multiple authors, when we only know an author.

Problem Formulation. Given a query, which contains multiple conditions, our goal is to obtain values that satisfy all conditions by conducting the query with the proposed methods on Fabric.

5   Proposed Methods

In this section, we present three methods to execute conditional queries and describe problems encountered during execution. The second method CCK is designed based on composite keys to avoid filtration process and the third method AIM can reduce redundancy by creating index. In order to better explain the proposed methods, we discuss the details of them based on DBLP.

5.1   Baseline Method

In this subsection, we present our baseline method for executing conditional queries on Fabric.

For each publication in $\varepsilon(S_\alpha, v)$, when we want to insert it into ledger, firstly, we need to obtain all authors in $S_\alpha = \{\alpha_1, \alpha_2, ..., \alpha_n\}$, and then the client initiates n transaction proposals to save this publication. Given a query associated with an author α and a venue v, to search all publications belonging to α and published in venue v, we firstly executes a GHFK call, then obtain the set $\varepsilon(\alpha)$. Next, we still need to remove all publications that are not published in venue v from $\varepsilon(\alpha)$. Finally, the remained publications in $\varepsilon(\alpha)$ are the results of the query. Note that, with the increase of the number of publications that are not published in the venue v, more publications should be removed, which leads to a lot of time cost.

5.2 Composite Key Based Method CCK

To address the problem of the baseline method, we design a novel method CCK, the details of which are discussed as follows, based on composite key.

For each publication in $\varepsilon(S_\alpha, v)$, we use each author in $S_\alpha = \{\alpha_1, \alpha_2, ..., \alpha_n\}$ and a venue v to create a composite key K by calling CK firstly. Then, we create n composite keys for this publication and invoke n transactions to save it. Based on these composite keys, we can conduct the following query. For example, given a query q associated with an author α and a venue v, with the goal of obtaining all publications belonging to α and published in venue v. We firstly use the author α and venue v to create a composite key K, then execute a GHFK call, during which the key K will be compared with all composite keys generated in CCK. Note that, each GHFK call precisely accesses those blocks that contain corresponding publications belonging to α and published in venue v, on ledger. Finally, we can directly get all publications $\varepsilon(\alpha, v)$.

Compared with the baseline method, CCK is more efficient to query all publications that satisfy all conditions, since the filtration process has been avoided. However, in CCK, the number of transactions to be invoked should equal to the number of authors in a publication. That is to say, we have to save the same publication multiple times, which results in massive redundancy.

5.3 AUP Index Based Method AIM

In this part, we build an index AUP to solve the problem of redundancy. For each publication in $\varepsilon(S_\alpha, v)$, we use each author α in $S_\alpha = \{\alpha_1, \alpha_2, ..., \alpha_n\}$ and venue v to create composite key K by calling CK. As an author may publish multiple publications in a same venue, the value V of each K is also a composite key, we create it with the Algorithm 1. The composite key consists of corresponding keys of current states. We create a key-value pair for each author α in $S_\alpha = \{\alpha_1, \alpha_2, ..., \alpha_n\}$ and insert it into AUP. Although a publication may belong to multiple authors, we only need to save the same publication one times, and then we invoke a transaction to save the publication. We do not use the CK as the first parameter is meaningless in this method. In Algorithm 1, we use a separator '#' to split each key. For example, we add '#' between $key1$ and $key2$, and the final result is in the form of $key1$ '#' $key2$. The reason for choosing '#' as a separator is: there is no '#' in the names of author and venue. By this way, we can separate keys accurately.

Considering a example, when a new publication data need to be saved to ledger, we first use each author in S_α and v to create composite key K by calling CK, then initiate a transaction proposal to commit data to ledger. Next, we query the AUP to get the value V of key K. If V is empty, we use all authors of the publication as a key to create composite key by Algorithm 1, which used as

---

Algorithm 1. Creating a Composite Key

Input: keys of current states(key_1,key_2,...,key_i)($0 \leqslant i \leqslant n$)
Output: composite key: V
1 Receive keys;
2 namespace ← '#';
3 **for** $i=0$ to n **do**
4 | **if** key_i *does not contain* '#' **then**
5 | | V ← V+namespace+key_i;
6 | **end**
7 **end**
8 Return V;

Algorithm 2. Splitting Multiple Values

Input: V(a value in AUP) and L(the length of the value)
Output: S_K: a collection of the splited keys
1   Receive value;
2   namespace ← '#';
3   index ← 0;
4   **for** $i=0$ to L **do**
5 | **if** *Value[i]==namsepace* **then**
6 | | components ← append(components,Value[index:i]);
7 | | index ← $i+1$;
8 | **end**
9 | append the components into S_K;
10 **end**
11 Return S_K;

---

the value of K. Otherwise, we split the value V with the Algorithm 2, where we still use '#' as separator and get the collection S_K. If the new key is different from any key in S_K, we append the new key to S_K, and then we use S_K to create new composite key NV by Algorithm 1 and put this new key-value pair $<K,NV>$ into AUP. Otherwise, we don't need to do anything. Finally, When the transaction is completed sucessfully, the publication data is saved to ledger.

The process of a conditional query is shown in Algorithm 3 explicitly. Firstly, we use the author α and venue v to create composite key K by calling CK. Then we use K to query AUP and get the value V. Next, we need to separate V and get the collection of keys S_K. Finally, we execute a group of GHFK calls based on the keys in S_K and get the collection $\varepsilon(\alpha, v)$, which is the result that we want to get.

During the design of AUP, we use Mutex in the Go language. Mutex is a commonly used method to control shared resource access, which ensures that only one goroutine can access shared resources at the same time. For example, if we use 4000 goroutines to execute transactions, after one goroutine queries the AUP to get the value of a the given key, another goroutine updates the value of the key, which will make the value obtained by the previous goroutine incorrect. Then, the incorrect value will lead to the loss of data. To solve the problem, we use Mutex to create the index AUP. When a goroutine writes to the AUP, other goroutines need to wait until the previous goroutine has finished writing.

Algorithm 3. Process of a Conditional Query

Input: an author(α)and the venue(v)
Output: $\varepsilon(\alpha,v)$:All publications belong to α and published in v
1 Receive α and v;
2 $K \leftarrow$ use α and v to create composite key by calling CK;
3 $V \leftarrow$ query AUP with K;
4 $S_K \leftarrow$ split V with Algorithm 2;
5 $L \leftarrow$ get the length of keys;
6 **for** $i=0$ to L **do**
7 $\quad$ call GHFK with the i-th key in S_K;
8 $\quad$ append the result of GHFK to $\varepsilon(\alpha,v)$;
9 **end**
10 Return $\varepsilon(\alpha,v)$;

---

6 Experiment

6.1 Fabric Instance

We use Hyperledger Fabric v1.3 and the implemented network consists of a single organization. The organization contains three nodes, a CA node, an endorsing node and an ordering service node with one public channel avaliable for communication. The endorsing node is configured to use CouchDB as the StateDB. We use Fabric SDK to emulate clients and run the entire system by using docker containers on a server. The server is equipped with 24 Intel(R) Xeon(R) CPU E5-2630 v2 processors at 2.60 GHz, for a total 256 GB of RAM. We keep all nodes turned on and use all default configuration settings to run our experiments.

6.2 System Workload

We carry out our experiment evaluation using DBLP. The total number of publications in DBLP is 4146645. As each publication in DBLP usually has multiple authors, we create a record with the same publication for those authors respectively. Finally, the total number of records is 12508891, in which 8362245 records are redundant. The total number of different authors publishing publications in different venues is 7843756. We divide all these data into 7 groups according to the ratio $r(r=j/i$, i represents the number of publications belong to α, j represents the number of publications belong to α and published in v). Groups are shown in Table 2. In this paper, we measure the performance of methods using the following metrics - (1) Query execution times - time taken to execute the conditional query. (2) Insertion times - time taken to insert data into Fabric ledger. (3) Memory cost - memory size occupied by all data.

Table 2. All groups and the number of members of each group.

Group	1	2	3	4	5	6	7
$r(\%)$	100 - 18	18 - 15	15 - 12	12 - 9	9 - 6	6 - 3	3 - 0
Total number	3218585	274878	421421	512382	612054	1034199	1770226

6.3 Experimental Evaluation

Table 3 shows the performance of three methods: baseline, CCK and AIM. We randomly select 1000 records from each group to execute 1000 queries at a time, which we execute 1000 times and take the average query time as the result. The query time is calculated from the time when the query transaction proposal is initiated until the response information is received.

Table 3. Query time of each method.

Group	Query time of baseline	Query time of CCK	Query time of AIM
1	29.03(s)	12.77(s)	9.57(s)
2	50.14(s)	12.31(s)	9.52(s)
3	55.44(s)	11.73(s)	8.51(s)
4	70.84(s)	11.70(s)	8.27(s)
5	80.92(s)	11.46(s)	7.93(s)
6	109.20(s)	10.85(s)	7.29(s)
7	174.74(s)	9.87(s)	6.07(s)

6.4 Time Cost of Baseline

As we can see from the Table 3, with the ratio r decreases, the baseline method takes more time. This is because as the ratio r decreases, the author we used to query has more publications. When we want to get all the publications that meet the conditions, we need to call the GHFK. The Fabric firstly queries the HistoryDB to get all keys that satisfy the conditions. The key in HistoryDB consists of the key of a current data, block number and transaction number. Then it uses block numbers to query block index to get all blocks and deserializes the content of these blocks. Next, it uses transaction numbers to get transactions and extracts out the values inserted. Finally, the GHFK call returns an iterator and we get values from the iterator. The more values are accessed through this iterator, the more blocks are deserialized. Therefore, given an author, the more publications belong to the authors, the more blocks need to be deserialized, the more time we will take to execute query transaction.

Consider the query in baseline method, it needs to get all blocks that contain publications belong to author α. It hence deserializes all these blocks and need to remove publications that are not published in venue v. As the number of publications that are not published in venue v increases, it needs to deserialize more and more blocks and removes more and more publications that do not satisfy the conditions. The bottleneck of the first method is that to retrieve publications belong to author α and published in venue v, we need to deserialize all blocks containing publications belongs to author α. Larger the number of publications that are not published in venue v, worse is hence the performance of baseline method.

6.5   Time Cost of CCK

The third column of Tabel 3 presents the performance of CCK. When we execute queries in group 1, CCK takes 12.77s which takes 16.26s less time than the baseline method. When we execute queries in group 3, CCK takes 11.73s which takes 43.71s less time than baseline. As the ratio decreases, the performance of CCK method becomes better. This is because with the decrease of ratio, the number of publications belong to the author α and published in venue v becomes smaller, and the number of blocks that we need to deserialize also becomes smaller. We are able to achieve this improvement by using CCK because we exactly know which block contains publications belong to author α and published in venue v. That is to say, we just need to get blocks that contain publications belong to author α and published in venue v. This effect becomes more severe, when we execute queries in group 7. Considering the case when an author has total x publications, in which y publications published in venue v and the data of each publication is stored in different blocks. When we execute queries with the baseline method, we need to deserialize x blocks and remove $x - y(\mathrm{x} \geq \mathrm{y})$ publications from the result. However, if we use CCK, we only need to deserialize y blocks. The larger $x - y$, the higher the performance of CCK. This is equivalent to the smaller ratio, the higher the performance of CCK. The time-cost by using CCK is much smaller than that by using the baseline method.

6.6   Time Cost of AIM

We next analyze the time-cost of using AIM to execute conditional queries, it is not much different from CCK. This is because in the AIM, we also create composite key and we exactly know which block contains the data that meets our conditions. So the number of block we need to deserialize is same. However, CCK has a big problem, it brings a lot of redundancy. We need to use each author in a publication and the venue of the publication to create composite key (α, v), and we need to take (α, v) and the other information o as a key-value pair to insert into the ledger. So if a publication has n $(n \geq 1)$ authors, it will generate n key-value pairs and wherein $n - 1$ are duplicates, which lead to the size of ledger created by using CCK is bigger than the ledger created by using AIM and the cardinality of the ledger data that performs conditional queries becomes larger. That is the reason why AIM is a little better than CCK in query performance. Besides, the redundancy causes us to spend a lot of time inserting these key-value pairs into ledger. In our experiment, we ingest a publication in one transaction. So the total number of transaction is 12508891 by using CCK and baseline methods, and we execute these transactions with 4000 goroutines. Both baseline method and CCK method cost more than 13 h to finish these transactions. However, when we use AIM method, the total number of transaction is 4146646 and it costs 5 h 29 m to finish these transaction. By using AIM method, we save more than 2 times time, which we can see from Table 4. We build the index during the process of a transaction. In fact, the data is continuously streaming in. If we do not build the index during the process of a transaction, when we execute queries, we may can not get the new data immediately because it has not yet been saved to the index. Beside, if we do not use this method, when we want to construct index, we will need to querying ledger before, which will cost a lot of time.

6.7 Memory Cost of the Three Methos

In addition, by constructing the index AUP, we also save data storage space. Specifically, let us use $|P|$ and $|I|$ to denote the average size of a transaction data in block and the key-value pair in AUP ($|P| > |I|$) respectively. In baseline method and CCK, the total size of all data is $12508891|P|$. In AIM, the total size of all data is $4146646|P| + 2234392|I|$. The difference between these two values is $8362245|P| - 2234392|I|$, and $8362245|P| - 2234392|I| > 0$. Therefore, AIM saves more data storage space than baseline method and CCK.

Table 4. The data insertion time of different methods.

Methods	Baseline	CCK	AIM
Transaction number	12508891	12508891	4146646
Data insertion time	13 h 8 m	13 h 12 m	5 h 29 m

6.8 Analysis

From the above three methods, we can see that the AIM has the best performance. It solves the problem of redundancy, improves the efficiency of queries and data insertion. Then, we get two conclusions. Firstly, when we execute conditional queries, and the key which we want to use has a large number of unrelated values need to be removed, the best method is to use all conditions to create a composite key. Then we can use this composite key to execute queries, which can help deserialize a small number of blocks and directly find blocks containing values that we want to get without the process of filtration. Secondly, when multiple keys have a same value, we can create index to reduce the time of data insertion and reduce redundancy. Just like the use-case in our experiment, multiple authors have a same publication, we reduce the time of inserting the publication into the ledger by creating an index AUP. By combining the method of creating composite key and building index, the performance of both queries and inserting data have a significant improvement.

In addition, methods presented in this paper can also be generalized to other conditional queries. For example, we can use the proposed methods to get a medical history of a patient in a certain department in the medical field.

7 Conclusion and Future Work

In this paper, we present three methods to handle conditional queries on Hyperledger Fabric. We use the first method as our baseline, both CCK and AIM easily outperform the baseline. We benchmark these three methods and we also conduct a comprehensive study to understand and analyse the conditional queries performance on Hyperledger Fabric by creating composite keys and building an index. Besides, the process of building index is included in an transaction. Not only does it saves more time during the process of insertion data, but also we can get data in a timely manner.

In our future work, we can further improve the performance of conditional queries in Hyperledger Fabric by using different methods of creating composite key and building index. As the static structure of LevelDB consists of six main parts and keys with the same prefix are adjacent in the file.

Acknowledgements. This work was supported by the National Natural Science Foundation of China (Grant No. 61572335, 61572336, 61902270), and the Major Program of Natural Science Foundation, Educational Commission of Jiangsu Province, China (Grant No. 19KJA610002), and the Natural Science Foundation, Educational Commission of Jiangsu Province, China (Grant No. 19KJB520052, 19KJB520050), and Collaborative Innovation Center of Novel Software Technology and Industrialization, Jiangsu, China.

References

1. Bitcoin. https://bitcoin.org/en/getting-started/. Accessed 10 June 2019
2. Couchdb. https://couchdb.apache.org/. Accessed 10 June 2019
3. Ethereum. https://www.ethereum.org/. Accessed 10 June 2019
4. Hyperledger fabric. https://www.hyperledger.org/projects/fabric. Accessed 10 June 2019
5. LevelDB. https://github.com/syndtr/goleveldb/. Accessed 10 June 2019
6. Parity. https://www.parity.io/. Accessed 10 June 2019
7. Androulaki, E., et al.: Hyperledger fabric: a distributed operating system for permissioned blockchains. In: Proceedings of the Thirteenth EuroSys Conference, p. 30. ACM (2018)
8. Croman, K., et al.: On scaling decentralized blockchains. In: Clark, J., Meiklejohn, S., Ryan, P.Y.A., Wallach, D., Brenner, M., Rohloff, K. (eds.) FC 2016. LNCS, vol. 9604, pp. 106–125. Springer, Heidelberg (2016). https://doi.org/10.1007/978-3-662-53357-4_8

9. Dinh, T.T.A., Liu, R., Zhang, M., Chen, G., Ooi, B.C., Wang, J.: Untangling blockchain: a data processing view of blockchain systems. IEEE Trans. Knowl. Data Eng. **30**(7), 1366–1385 (2018)
10. Dinh, T.T.A., Wang, J., Chen, G., Liu, R., Ooi, B.C., Tan, K.L.: Blockbench: a framework for analyzing private blockchains. In: Proceedings of the 2017 ACM International Conference on Management of Data, pp. 1085–1100. ACM (2017)
11. Gervais, A., Karame, G.O., Wüst, K., Glykantzis, V., Ritzdorf, H., Capkun, S.: On the security and performance of proof of work blockchains. In: Proceedings of the 2016 ACM SIGSAC Conference on Computer and Communications Security, pp. 3–16. ACM (2016)
12. Gupta, H., Hans, S., Aggarwal, K., Mehta, S., Chatterjee, B., Jayachandran, P.: Efficiently processing temporal queries on hyperledger fabric. In: 2018 IEEE 34th International Conference on Data Engineering (ICDE), pp. 1489–1494. IEEE (2018)
13. Gupta, H., Hans, S., Mehta, S., Jayachandran, P.: On building efficient temporal indexes on hyperledger fabric. In: 2018 IEEE 11th International Conference on Cloud Computing (CLOUD), pp. 294–301. IEEE (2018)
14. Lin, I.C., Liao, T.C.: A survey of blockchain security issues and challenges. IJ Netw. Secur. **19**(5), 653–659 (2017)
15. Meng, W., Tischhauser, E.W., Wang, Q., Wang, Y., Han, J.: When intrusion detection meets blockchain technology: a review. IEEE Access **6**, 10179–10188 (2018)
16. Omohundro, S.: Cryptocurrencies, smart contracts, and artificial intelligence. AI Matters **1**(2), 19–21 (2014)
17. Pongnumkul, S., Siripanpornchana, C., Thajchayapong, S.: Performance analysis of private blockchain platforms in varying workloads. In: 2017 26th International Conference on Computer Communication and Networks (ICCCN), pp. 1–6. IEEE (2017)
18. Thakkar, P., Nathan, S., Viswanathan, B.: Performance benchmarking and optimizing hyperledger fabric blockchain platform. In: 2018 IEEE 26th International Symposium on Modeling, Analysis, and Simulation of Computer and Telecommunication Systems (MASCOTS), pp. 264–276. IEEE (2018)
19. Vukolić, M.: The quest for scalable blockchain fabric: proof-of-work vs. BFT replication. In: Camenisch, J., Kesdoğan, D. (eds.) iNetSec 2015. LNCS, vol. 9591, pp. 112–125. Springer, Cham (2016). https://doi.org/10.1007/978-3-319-39028-4_9
20. Zhang, X., Poslad, S.: Blockchain support for flexible queries with granular access control to electronic medical records (EMR). In: 2018 IEEE International Conference on Communications (ICC), pp. 1–6. IEEE (2018)
21. Zhang, X., Poslad, S., Ma, Z.: Block-based access control for blockchain-based electronic medical records (EMRs) query in ehealth. In: 2018 IEEE Global Communications Conference (GLOBECOM), pp. 1–7. IEEE (2018)
22. Zheng, Z., Xie, S., Dai, H., Chen, X., Wang, H.: An overview of blockchain technology: architecture, consensus, and future trends. In: 2017 IEEE International Congress on Big Data (BigData Congress), pp. 557–564. IEEE (2017)

Machine Learning

Learning to Fuse Multiple Semantic Aspects from Rich Texts for Stock Price Prediction

Ning Tang[1], Yanyan Shen[1(✉)], and Junjie Yao[2]

[1] Department of Computer Science and Engineering,
Shanghai Jiao Tong University, Shanghai, China
{tnbaby,shenyy}@sjtu.edu.cn
[2] School of Computer Science and Software Engineering,
East China Normal University, Shanghai, China
junjie.yao@sei.ecnu.edu.cn

Abstract. Stock price prediction is challenging due to the non-stationary fluctuation of stock price, which can be influenced by the stochastic trading behaviors in the market. In recent years, researchers have focused on exploiting massive text data such news and tweets to predict stock price, achieving promising outcomes. Existing methods typically compress each text into a fixed-length representation vector, whereas rich texts may involve multiple semantic aspect-level information that has different effects on the future stock price. In this paper, we propose a novel Multi-head Attention Fusion Network (MAFN) to exploit aspect-level semantic information from texts to enhance prediction performance. MAFN employs the encoder-decoder framework, where the encoder adopts the multi-head attention mechanism to automatically learn the aspect-level text representations via different attention heads. Furthermore, we subtly fuse the learned representations by discarding the dross and selecting the essential. The decoder generates stock price sequence by incorporating textual information and historical price dynamically via the hierarchical attention. Experimental results on real data sets show the superior performance of MAFN against several strong baselines as well as the effectiveness of exploiting and fusing fine-grained aspect-level textual information for stock price prediction.

Keywords: Stock price prediction · Multi-head attention · Encoder-decoder

1 Introduction

The stock market is one of the largest financial markets in the world, attracting millions of investors for stock trading, who aim to buy stocks at low price and sell them at high price to maximize gains. Stock price prediction, trying to determine the future selling price of a company stock, is indisputably important

© Springer Nature Switzerland AG 2019
R. Cheng et al. (Eds.): WISE 2019, LNCS 11881, pp. 65–81, 2019.
https://doi.org/10.1007/978-3-030-34223-4_5

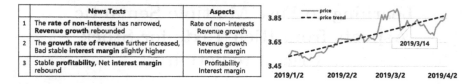

	News Texts	Aspects
1	The **rate of non-interests** has narrowed, **Revenue growth** rebounded	Rate of non-interests Revenue growth
2	The **growth rate of revenue** further increased, Bad stable **interest margin** slightly higher	Revenue growth Interest margin
3	Stable **profitability**, Net **interest margin** rebound	Profitability Interest margin

Fig. 1. Three pieces of news in 2019-3-14 and the stock price for the Bank of China. Each news contains multiple semantic aspects, and they share some aspects such as revenue growth and interest margin.

to help investors make good investment decisions [17]. However, predicting stock price accurately is challenging due to the non-stationary fluctuation of the stock price time series [1], which is typically influenced by the highly stochastic trading behaviors from numerous stock investors in the market. Fortunately, the increasing amount of text data accessible from social media, such as tweets and news, has become an important source of information that sheds light on the trend of future stock price [7, 10, 13]. This paper focuses on exploiting text information towards more accurate stock price prediction.

Existing literature have applied various Natural Language Processing (NLP) techniques to incorporate massive text data for enhancing stock price prediction, which can be generally categorized into two groups: feature-based methods [21] and neural network-based methods [10, 23]. The former category involves hand-crafted feature extraction from texts, such as sentiment words [21] and structural events [6, 7]. These approaches have two major drawbacks: (i) the identification of discriminative features requires financial knowledge from domain experts, and the set of useful features often evolves over time due to the high dynamics of the stock market; (ii) the proposed prediction models can hardly capture deep feature interactions due to the limited model capacity, thus resulting in unsatisfactory performance. The second category introduces neural network models to automatically learn features and their interactions for stock price prediction. Hu et al. [10] proposed a hybrid hierarchical attention network to learn a weight for each text indicating its quality and trustworthiness for stock price prediction. Xu et al. [23] designed variational architecture with RNN network to capture the stochastic latent factors of stock market in a generative manner.

While deep neural networks are effective in exploiting latent semantics from text data, they typically compress all the textual information into a fixed-length representation vector and fail to distinguish fine-grained aspects involved in the texts. To be specific, text data such as news tend to contain multiple semantic aspects about a company and these aspects can have diverse sentiment signals that affect future stock price in different ways. For instance, consider the three pieces of news for the Bank of China and its stock price within the same time period in Fig. 1. There are mainly four aspects involved in the news, namely rate of non-interest, revenue growth, profitability and interest margin. Each news has commented on more than one aspects for the Bank of China. We notice that the comments on revenue growth are quite positive while the comments on interest margin are negative. In practice, revenue growth is directly related to the company's income and is more accurate to reflect the true situation of the company compared with interest margin. Therefore, the positive comments on revenue growth is more significant, which is consistent with the growing trend of the stock price. The above example highlights the importance of (i) identifying multiple semantic aspects from massive texts and (ii) subtly fusing their influence on stock price for future prediction. Intuitively, the ignorance of these two factors would impede the development of an accurate predictor. However, to the best of our knowledge, none of the existing methods have exploited fine-grained aspect-level information from texts for stock price prediction.

To address the problem of multi-aspect identification and fusion, we propose a Multi-head Attention Fusion Network (MAFN) for stock price prediction based on text data. At a high level, MAFN employs the encoder-decoder framework. During the encoding stage, we first supply the embeddings of text words into a bidirectional recurrent neural network to form the primary representation of each text. We then apply the Multi-head Attention mechanism [22] to extract aspect-level information from the text representation automatically. The key idea is to project the primary representation into different latent semantic subspaces via learnable projection matrices, where each semantic subspace implicitly corresponds to a semantic aspect. A novel attempt of our approach is to enforce the attention mechanism [15] to dynamically fuse multiple aspect-level text representations based on the computed attention scores during each encoding step. The decoder is implemented by an LSTM that absorbs both historical stock price sequence and fused textual information to produce the final prediction result. To be specific, we treat the fused text representations from previous time steps as the contexts and introduce a new attentive read layer that discriminates the importance of these contexts and generates an aggregated context vector at each decoding step. The resultant context vector and the stock price will be combined and fed to the decoder together. In addition, we introduce an attentive prediction layer to attend to different hidden states of the decoder for final prediction.

The contributions of this paper can be summarized as follows:

- We propose to exploit fine-grained aspect-level information from massive texts to enhance the performance of stock price prediction, and develop an end-to-end neural network model, named Multi-head Attention Fusion Network (MAFN). MAFN follows the encoder-decoder framework that encodes aspect-level text information and decodes the stock price sequence effectively.
- We introduce a multi-head attentive fusion layer in the encoder to extract different aspects from texts automatically and distinguish the importance of different aspects via the attention mechanism. We then fuse aspect-level text representations based on the computed attention weights.
- We design a hierarchical attention layer in the decoder to dynamically absorb the fused text representations and stock price from previous time steps for future stock price prediction.
- We conduct extensive experiments on two real datasets: A share and NAS-DAQ stock markets. The results show that (i) on average MAFN achieves 10%

and 3% improvement in RMSE and MAE respectively, compared with various baseline methods; (ii) multi-head attentive fusion layer effectively extracts aspects from texts and fuses these aspects according to their importance.

The remainder of this paper is organized as follows. We review the related works on stock price prediction in Sect. 2. We provide the definitions and problem in Sect. 3. We present our proposed method in Sect. 4. The experimental results are described in Sect. 5, and we conclude the paper in Sect. 6.

2   Related Work

The increasing amount of text data contains rich information that indicates the status of the listed companies and affects their stock price implicitly. Various approaches have been proposed to enhance stock price prediction performance based on massive text data, which in general, can be classified into two categories: feature-based methods and neural network-based methods.

Feature-based methods typically require time-consuming feature engineering over texts. The features such as the number of sentiment words/phases and events are extracted to develop a predictor for future stock price. Li et al. [13] proposed a statistical model to detect financial sentiment words and studied their influence on future stock price. In order to consider the structural information among words, Ding et al. [6,7] proposed to extract structured events from the news and combined them with the knowledge graph to enrich the text representations. Si et al. [18] built a social network of stocks from tweets and based on the mood of stocks from its neighbors for stock prediction. However, requiring financial knowledge from experts and lack of deep feature interactions limit the model capacity and can not coordinate the rapid evolution of stock market (Table 1).

Table 1. Comparison of different models

Category	Model	Features	Literature
Feature-based	SVR	Sentiment words	Li et al. [13]
	CNN	Structural events	Ding et al. [6]
	CNN	Knowledge graph	Ding et al. [7]
	VAR	Social relation	Si et al. [18]
Neural network-based	RNN	Text content	Chung et al. [5]
	RNN with DFT	Price	Zhang et al. [25]
	HAN	News content	Hu et al. [10]
	VAE	Tweets and price	Xu et al. [23]

To address the aforementioned limitations of the feature-based methods, another line of researches developed neural network models to learn text information for stock price prediction. For instance, recurrent neural networks (RNNs) and its variants are widely applied to model the sequential information from texts to enhance prediction accuracy [5]. Zhang et al. [25] adopted Recurrent Neural Network to capture long term pattern and short term pattern of stock price via Discrete Fourier Transform. Hu et al. [10] proposed a hybrid hierarchical attention network to learn different weights for the texts and constructed an attentive recurrent layer to predict the stock price trend. In contrast to the above discriminative models, Xu et al. [23] introduced a generative model based on the variational architecture to utilize textual features and historical stock price for future price prediction.

3 Preliminaries

3.1 Definitions and Problem

Definition 1 (Stock Price Data $\mathcal{D}_S$). *Let S denote the set of stocks in the market. The stock price data $\mathcal{D}_S = \{D_s\}_{s\in S}$ records price for all the stocks in S over T time intervals. In this paper, we consider closing price, opening price, lowest price and highest price as stock price information, of any stock obtained in trading days. We have $D_s = \{d_t^s\}_{t=1}^T$, where $d_t^s = [y_{c,t}^s, y_{o,t}^s, y_{l,t}^s, y_{h,t}^s]$ means above stock price of stock $s \in S$ during the t-th trading day, respectively.*

Definition 2 (Text Data $\mathcal{M}_S$). *The text data refers to the textual information that describes the companies of the stocks, such as news and tweets. Given a set S of stocks, the text data $\mathcal{M}_S = \{M_s\}_{s\in S}$ contains texts related to the stocks in S. Specifically, we denote by $M_s = \{m_t^s\}_{t=1}^T$ all the text data for stock s over T trading days. m_t^s can include multiple texts, represented as $m_t^s = \{n_{t,i}^s\}_{i=1}^l$, where l is the number of relevant texts. Each text $n_{t,i}^s \in m_t^s$ is a sequence of words, represented as $\{w_{t,i,j}^s\}_{j=1}^p$, where p is the number of the involved words.*

In this paper, we aim to exploit the rich text information for predicting future stock closing price, which can be formulated as the following regression problem.

Definition 3 (Stock Price Prediction Problem). *Let S be the set of stocks. Given stock price data $\mathcal{D}_S = \{D_s\}_{s\in S}$ and text data $\mathcal{M}_S = \{M_s\}_{s\in S}$ obtained from the previous T trading days, we aim to predict the stock closing price for any stock in S during the next trading day, i.e., $y_{c,T+1}^s$, for any $s \in S$.*

3.2 Long Short-Term Memory

Before delving into the details of our proposed model, we first review an important variant of recurrent neural networks (RNNs): Long Short-Term Memory (LSTM), which is the basis for the following sections. LSTM [9] have been used for various sequence modeling tasks, such as language translation [20], trajectory prediction [2] and video classification [24]. Different from the conventional RNNs that suffer from the gradient vanishing and exploding problems, LSTM enforces the gating mechanism to control information flow in the network. Specifically, the regular LSTM involves a memory cell $\mathbf{c}$ with the cell state $\mathbf{h}$ and three different gates: input gate $\mathbf{i}$, forget gate $\mathbf{f}$, output gate $\mathbf{o}$. During the time step t, LSTM absorbs the input $\mathbf{x}_t$ and computes the updated memory cell $\mathbf{c}_t$ and cell state $\mathbf{h}_t$ using the following equations:

$$
\begin{aligned}
\mathbf{i}_t &= \sigma(\mathbf{W}_{ix}\mathbf{x}_t + \mathbf{W}_{ih}\mathbf{h}_{t-1} + \mathbf{W}_{ic} \odot \mathbf{c}_{t-1} + \mathbf{b}_i) \\
\mathbf{f}_t &= \sigma(\mathbf{W}_{fx}\mathbf{x}_t + \mathbf{W}_{fh}\mathbf{h}_{t-1} + \mathbf{W}_{fc} \odot \mathbf{c}_{t-1} + \mathbf{b}_f) \\
\mathbf{c}_t &= \mathbf{f}_t \odot \mathbf{c}_{t-1} + \mathbf{i}^t tanh(\mathbf{W}_{cx}\mathbf{x}_t + \mathbf{W}_{ch}\mathbf{h}_{t-1} + \mathbf{b}_c) \\
\mathbf{o}_t &= \sigma(\mathbf{W}_{ox}\mathbf{x}_t + \mathbf{W}_{oh}\mathbf{h}_{t-1} + \mathbf{W}_{oc} \odot \mathbf{c}_t + \mathbf{b}_o) \\
\mathbf{h}_t &= \mathbf{o}_t \odot tanh(\mathbf{c}_t)
\end{aligned}
\tag{1}
$$

where σ is the sigmoid function, $\odot$ is the Hadamard product, $\mathbf{W}$ and $\mathbf{b}$ are the weight matrices and biases to be learned, respectively.

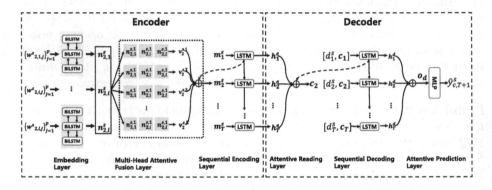

Fig. 2. The overview of the proposed model MAFN

4   MAFN: Fusing Aspect-Level Textual Information for Stock Price Prediction

Figure 2 depicts the overview of our proposed stock price prediction model called Multi-head Attention Fusion Network (MAFN), which follows the general encoder-decoder framework [4,15]. Consider a stock s in S, the encoder in MAFN takes the text data $\{m_t^s\}$ as input and learns to extract and fuse fine-grained aspect-level textual information automatically. The decoder in MAFN incorporates both textual information learned from the encoder and the historical stock price data $\{d_t^s\}$, and finally produces the stock closing price in the next trading day. As the text data and stock price are sequences of size T, we adopt LSTM as the basic structure for both encoder and decoder. We next elaborate the details of the proposed MAFN.

4.1 Encoding with Multi-Head Attention

The *encoder* aims to capture the semantic information of text data $\{m_t^s\}_{t=1}^T$ by exploiting multiple aspect-level representations, which consists of three layers: *embedding layer, multi-head attentive fusion layer* and *sequential encoding layer*. Recall that $m_t^s = \{n_{t,i}^s\}_{i=1}^l$ contains l texts for stock s at t-th day. The embedding layer embeds $n_{t,i}^s$ into a low-dimensional latent vector $\mathbf{n}_{t,i}^s$. The multi-head attentive fusion layer transforms $\mathbf{n}_{t,i}^s$ into K different semantic subspaces $\{\mathbf{n}_{t,i}^{s,k}\}_{k=1}^K$ via projection matrices, and each semantic subspace is associated with a particular latent aspect-level representation. It then fuses multiple aspect-level representations for all the texts in m_t^s using the attention mechanism, and obtains a unified vector $\mathbf{m}_t^s$ that fuses the aspect-level semantics of m_t^s. Finally, the sequential encoding layer feeds a sequence of the encoded text representations $\{\mathbf{m}_t^s\}_{t=1}^T$ into an LSTM and obtains the corresponding hidden states $\{\mathbf{h}_1^e, \cdots, \mathbf{h}_T^e\}$.

Embedding Layer. Given a text set $m_t^s = \{n_{t,i}^s\}_{i=1}^l$ for stock s at the t-th trading day, the *embedding layer* aims to learn a latent vector representation $\mathbf{n}_{t,i}^s$ for each text $n_{t,i}^s \in m_t^s$. Since each text composes a sequence of words, we first use the pre-trained Word2vec [12] as the initial word embeddings, i.e., $\{\mathbf{w}_{t,i,j}^s\}_{j=1}^p$. We then apply a bi-directional LSTM (Bi-LSTM) to explore the latent semantics of the embed word sequence in two directions, forward and backward, as follows:

$$\overrightarrow{\mathbf{h}}_{t,i,j}^s = \overrightarrow{\mathrm{LSTM}}(\mathbf{w}_{t,i,j}^s, \overrightarrow{\mathbf{h}}_{t,i,j-1}^s) \tag{2}$$

$$\overleftarrow{\mathbf{h}}_{t,i,j}^s = \overleftarrow{\mathrm{LSTM}}(\mathbf{w}_{t,i,j}^s, \overleftarrow{\mathbf{h}}_{t,i,j-1}^s) \tag{3}$$

$$\mathbf{e}_{t,i,j}^s = (\overrightarrow{\mathbf{h}}_{t,i,j}^s + \overleftarrow{\mathbf{h}}_{t,i,j}^s)/2 \tag{4}$$

where $j \in [1, p]$. $\overrightarrow{\mathrm{LSTM}}$ and $\overleftarrow{\mathrm{LSTM}}$ are respectively the forward and backward LSTMs based on Eq. (1). The two hidden states $\overrightarrow{\mathbf{h}}_{t,i,j}^s$ and $\overleftarrow{\mathbf{h}}_{t,i,j}^s$ in each step j are averaged to acquire the context-aware word embedding $\mathbf{e}_{t,i,j}^s$. Then the final text representation $\mathbf{n}_{t,i}^s$ is computed by averaging over all the context-aware word embeddings as $\mathbf{n}_{t,i}^s = \frac{1}{p}\sum_{j=1}^p \mathbf{e}_{t,i,j}^s$.

Multi-Head Attentive Fusion Layer. Given a set of text representations $\{\mathbf{n}_{t,i}^s\}_{i=1}^l$, we first aim to exploit multiple aspect-level representations for each text automatically. Our insight is that (i) the text representations produced by the embedding layer are expected to focus on the complete space of the latent semantics, whereas a transformed subspace is useful to reflect the semantics for a specific aspect; and (ii) each piece of the aspect-level information has its distinct contribution to the future stock closing price as illustrated in Fig. 1. Following our insight, we propose a multi-head attentive fusion layer to allow the encoder to attend to multiple aspect-level semantics of the text data. The multi-head attention mechanism has been widely applied in sequence modeling tasks [22], where each head corresponds to a particular semantic subspace of sentences. In our context, we project the text representation $\mathbf{n}_{t,i}^s$ into different subspaces via K learnable projection matrices $\{\mathbf{W}^k\}_{k=1}^K$ as follows:

$$\mathbf{n}_{t,i}^{s,k} = \mathbf{W}^k \mathbf{n}_{t,i}^s, \quad k \in [1, K] \tag{5}$$

where $\mathbf{n}_{t,i}^{s,k}$ encodes the information from the perspective of the k-th aspect.

We find that in practice the number of texts in the t-th trading day varies over different stocks and there exists redundant textual information for the same stock. To keep the most significant information in each semantic subspace, we use max pooling from every dimension over l projected text vectors as $\mathbf{v}_t^{s,k} = \mathrm{MaxPooling}(\mathbf{n}_{t,1}^{s,k}, \cdots, \mathbf{n}_{t,l}^{s,k})$, where $\mathbf{v}_t^{s,k}$ retains the textual information for aspect k over all the texts in m_t^s.

To fuse the aspect-level textual information, we use the typical attention mechanism to compute a probability distribution over all the aspects and distinguish the importance of aspect-level semantics on future stock closing price. Specifically, we perform Luong attention [15] over the K latent vectors $\{\mathbf{v}_t^{s,k}\}_{k=1}^K$ and compute the attention score for each vector as follows:

$$\alpha_t^k = \frac{\exp(\mathbf{v}_t^{s,k}\mathbf{W}\mathbf{h}_{t-1}^e)}{\sum_{k=1}^K \exp(\mathbf{v}_t^{s,k}\mathbf{W}\mathbf{h}_{t-1}^e)} \tag{6}$$

where $k \in [1, K]$ and $\mathbf{W}$ is the weight matrix to be learned. $\mathbf{h}_{t-1}^e$ is the hidden state of the LSTM encoder in the previous trading day which is computed using the sequential encoding layer described below. It is worth mentioning that incorporating $\mathbf{h}_{t-1}^e$ into the computation of the attention scores allows the model to filter insignificant or noisy textual information that is inconsistent with the general short-term textual information.

We then compute the final text representation $\mathbf{m}_t^s$ during the t-th trading day by fusing all the aspect-level information using $\{\alpha_t^k\}_{k=1}^K$. Formally, we perform weighted sum over $\{\mathbf{v}_t^{s,k}\}_{k=1}^K$ as $\mathbf{m}_t^s = \sum_{k=1}^K \alpha_t^k \mathbf{v}_t^{s,k}$.

We summarize the key advantages of our multi-head attentive fusion layer. First, the identification of latent aspect subspace is achieved via the multi-head attention mechanism automatically, and the number of aspects can be acquired via cross-validation. Second, we apply max pooling to address the problem of different text numbers per day and is effective to eliminate duplicated textual information. Finally, we take the temporal tendency of the textual information into account to fuse the aspect-level information dynamically.

Sequential Encoding Layer. This layer is to capture the temporal dependency among the fused aspect-level information over T time steps. To be specific, we employ the LSTM to implement the sequential encoding layer that absorbs the sequence $\{\mathbf{m}_t^s\}_{t=1}^T$ and outputs the corresponding encoder hidden states $\{\mathbf{h}_t^e\}_{t=1}^T$ for modeling the deep semantics of the fused text representations. Formally, we have:

$$\mathbf{h}_t^e = \mathrm{LSTM}(\mathbf{m}_t^s, \mathbf{h}_{t-1}^e), \quad t \in [1, T] \tag{7}$$

4.2   Decoding with Hierarchical Attention

The *decoder* aims to predict future stock closing price $\hat{y}^s_{c,T+1}$ based on the learned textual information from the encoder $\{\mathbf{h}^e_1, \cdots, \mathbf{h}^e_T\}$ and the historical stock price $\{d^s_t\}^T_{t=1}$. It consists of three layers: *attentive reading layer, sequential decoding layer* and *attentive prediction layer*. The attentive reading layer aggregates the hidden states from the encoder using the traditional attention mechanism for each decoding step. The resultant vector at step $t \in [1, T]$ can be considered as the context to predict the stock closing price of the next trading day. This follows our observations that not all the fused textual information from previous days are equally important and the importance may change dynamically over time. The sequential decoding layer takes the context vectors $\{\mathbf{c}_t\}^T_{t=1}$ and the stock price $\{d^s_t\}^T_{t=1}$ over T steps as input and generates the decoder hidden states $\{\mathbf{h}^d_1, \cdots, \mathbf{h}^d_T\}$. In the attentive prediction layer, we employ the temporal attention to identify the importance of each decoder hidden state and feed the aggregated state to a fully connected layer for producing the stock closing price in day $T + 1$.

Attentive Reading Layer. At each decoding step $t \in [1, T]$, the attentive reading layer takes as input the encoder hidden states $\{\mathbf{h}^e_1, \cdots, \mathbf{h}^e_T\}$ and the previous hidden state $\mathbf{h}^d_{t-1}$ produced by the sequential decoding layer below. We first calculate the attention weight for each encoder hidden state $\mathbf{h}^e_{t'}$ as follows:

$$\beta_{t,t'} = \frac{\exp(\mathbf{h}^e_{t'}\mathbf{W}\mathbf{h}^d_{t-1})}{\sum_{t'=1}^T \exp(\mathbf{h}^e_{t'}\mathbf{W}\mathbf{h}^d_{t-1})} \tag{8}$$

where $\mathbf{W}$ is the weight matrix to be learned. We then combine all the encoder hidden states and obtain a context vector $\mathbf{c}_t$ for the current decoding step t as $\mathbf{c}_t = \sum_{t'=1}^T \beta_{t,t'}\mathbf{h}^e_{t'}$. Note that $\mathbf{c}_t$ changes as per decoding step due to the dynamics of attention weights.

Sequential Decoding Layer. This layer adopts an LSTM to absorb both the context vectors and stock price to extract features in a sequential manner. Specifically, during each decoding step t, we concatenate the context vector $\mathbf{c}_t$ and the stock price d^s_t, and then update the hidden state $\mathbf{h}^d_t$ of the decoder LSTM using the following equation:

$$\mathbf{h}^d_t = \text{LSTM}([\mathbf{c}_t, d^s_t], \mathbf{h}^d_{t-1}) \tag{9}$$

Attentive Prediction Layer. Our goal is to predict the stock closing price in day $T + 1$. We observe that a simple aggregation over all the decoder hidden states for final prediction may compromise the prediction accuracy because the features in different time steps can have different effects on the future stock closing price. This inspires us to employ the temporal attention to discriminate the importance of temporal features for final prediction. In this layer, we compute the temporal attention for each decoder hidden state $\mathbf{h}_t^d$ as follows:

$$\gamma_t = \frac{\exp(\mathbf{h}_t^d \mathbf{W} \mathbf{h}_T^d)}{\sum_{i=1}^{T} \exp(\mathbf{h}_i^d \mathbf{W} \mathbf{h}_T^d)} \tag{10}$$

We then compute the weighted sum over all the decoder hidden states using $\{\gamma_t\}_{t=1}^{T}$ and feed the result into a fully connected layer for producing $\hat{y}_{c,T+1}^s$. Formally, we have:

$$\mathbf{o}^d = \sum_{t=1}^{T} \gamma_t \mathbf{h}_t^d \tag{11}$$

$$\hat{y}_{c,T+1}^s = \mathbf{w}_f \mathbf{o}^d + b \tag{12}$$

where $\mathbf{w}_f$ and b are respectively the weight vector and bias to be learned.

4.3   Learning and Optimization

In essence, stock price prediction can be formulated as a regression problem. Compared with predicting an up/down label, we argue that predicting the exact stock closing price is more accurate and valuable. Hence, we use the following objective function for our model:

$$L = \sum_{\mathbf{x} \in \mathcal{S}} (\hat{y}_c(\mathbf{x}) - y_c(\mathbf{x}))^2 \tag{13}$$

where $\mathcal{S}$ denotes the training set in which each training example $\mathbf{x}$ includes the historical stock price and the associated texts over continuous T trading days. $y_c(\mathbf{x})$ and $\hat{y}_c(\mathbf{x})$ denote the actual and the predicted stock closing price in day $T+1$, respectively.

To prevent overfitting, we use dropout [19] and L_2 regularization techniques. Specifically, we enforce dropout at the output layer of every time step in both encoder and decoder, and add the L_2 regularization term over all model parameters to the loss function. The final objective function is:

$$L = \sum_{\mathbf{x} \in \mathcal{S}} (\hat{y}_c(\mathbf{x}) - y_c(\mathbf{x}))^2 + \lambda ||\mathbf{W}||^2 \tag{14}$$

where $\mathbf{W}$ denotes all the model parameters and λ is a hyperparameter to control the regularization strength. Instead of using stochastic gradient descent (SGD), we apply Adam optimizer [11] to minimize the above loss function, which is more appropriate for non-stationary objectives and dynamically tunes the learning rate to faster converge.

5   Experiments

5.1   Experiment Settings

Datasets. We use two real datasets to evaluate the performance of our proposed method.

- **News&Price** contains 131 stocks from 10 different fields and over 30,000 financial news collected from Sina[1]. The whole dataset starts from 2014/01/01 to 2018/01/01, which is split into training set (2014/01/01–2017/01/01), validation set (2017/01/01–2017/06/01) and test set (2017/06/01–2018/01/01).
- **Tweets&Price** is published in [23], which contains 88 stocks from 9 different fields and tweets collected from twitter[2]. The whole dataset starts from 2014/01/01 to 2016/01/01, which is split into training set (2014/01/01–2015/08/01), validation set (2015/08/01–2015/10/01) and test set (2015/10/01–2016/01/01).

Settings and Compared Methods. As the scale of different stocks varies sharply, we normalize the price of each stock into [0, 1] with its min and max price. We adopt pretrained word embeddings for the news and tweets: 300-dimension Chinese Financial News word vectors [12] for news and 50-dimension Glove Twitter word vectors [16] for tweets. We use the validation set to early stop the training process and perform grid search the best hyperparameter values.

We compare our proposed MAFN with the following methods:

- **Average**: a naive predictor using the mean of historical price data as the prediction for future price. We use 10 as the size of time window for getting historical price.
- **Random Forest** [14]: a classical regression tree method using stock price.
- **XGBoost** [3]: a highly effective and scalable tree boosting method using stock price.
- **SVR**: a classical extended regression model based on support vector machine with stock price.
- **Attentive LSTM** [8]: attention-based LSTM model for stock price prediction using stock price.
- **HAN** [10]: a strong classification model for stock price prediction, extracting features from text with hierarchical attention.
- **StockNet** [23]: a state-of-the-art generative model for stock movement classification, which uses both price and text data.

Metrics. We use RMSE $= \sqrt{\frac{\sum_{i=1}^{N}(\hat{y}_{c,i}-y_{c,i})^2}{N}}$ and MAE $= \frac{\sum_{i=1}^{N}|\hat{y}_{c,i}-y_{c,i}|}{N}$ as the metrics for regression methods. N is the total number of test instances. $y_{c,i}$ and $\hat{y}_{c,i}$ are the actual and predicted closing price, respectively.

---

[1] finance.sina.com.cn.

[2] twitter.com.

Table 2. Comparison results for regression methods.

Dataset	News&Price		Tweets&Price	
Metric	MAE ($\times 10^{-2}$)	RMSE ($\times 10^{-2}$)	MAE ($\times 10^{-2}$)	RMSE ($\times 10^{-2}$)
Average	2.807	6.291	3.881	6.791
Random forest	3.046	3.942	3.078	3.935
XGBoost	3.202	3.882	2.853	3.854
SVR	1.745	2.439	2.481	3.181
Attentive LSTM	1.952	2.846	2.282	2.979
MAFN	**1.524**	**2.181**	**2.223**	**2.889**

To compare with classification methods, we obtain the same labels as in [23]. Specifically, we convert our predicted price into up and down labels using the rule: $l_{T+1} = \mathbf{1}\{\hat{y}_{c,T+1} > y_{c,T}\}$, where $\mathbf{1}$ is the indicator function.

Following the previous work [23], we adopt two metrics for classification methods: ACC $= \frac{|\{(x,l_T) \in Test\ Set | \hat{l}_T = l_T\}|}{N}$ and MCC $= \frac{tp \times tn - fp \times fn}{\sqrt{(tp+fp)(tp+fn)(tn+fp)(tn+fn)}}$, where $\hat{l}_T$ is the predicted label, and tp, tn, fp, fn are true positive, true negative, false positive and false negative computed from confusion matrix, respectively.

5.2   Comparison Results

We first compare MAFN with five regression methods: Average, Random Forest, XGBoost, SVR and Attentive LSTM. As shown in Table 2, SVR and Attentive LSTM perform best among all the baselines over News&Price and Tweets&Price datasets, respectively. MAFN achieves 1.524×10^{-2} MAE and 2.181×10^{-2} RMSE on News&Price, outperforming SVR by 12.7% and 10.6% respectively. As for Tweets&Price, MAFN achieves 2.6% and 3% improvement in MAE and RMSE compared with Attentive LSTM. Attentive LSTM performs better than Random Forest and XGBoost on both datasets, and is slightly worse than SVR on News&Price, showing the advantages of neural networks over traditional machine learning methods. Furthermore, MAFN performs better than Attentive LSTM and SVR on both datasets, indicating the effectiveness of exploiting fine-grained aspect-level textual information for stock price prediction.

Table 3. Comparison results for classification methods. The results labeled with * are copied from [23] directly following the same setting.

Dataset	News&Price		Tweets&Price	
Metric	ACC/%	MCC	ACC/%	MCC
HAN	51.09	0.0071	57.64*	0.0518*
StockNet	52.20	0.0171	**58.23***	0.0808*
Attentive LSTM	52.89	0.0436	56.45	0.1244
MAFN	**53.61**	**0.0561**	57.36	**0.1530**

Table 4. Ablation test results. The values in brackets are the performance loss against MAFN.

Dataset	News&Price		Tweets&Price	
Metric	MAE($\times 10^{-2}$)	RMSE($\times 10^{-2}$)	MAE($\times 10^{-2}$)	RMSE($\times 10^{-2}$)
MAFN (-HA-MA)	2.634 (+1.11)	3.526 (+1.34)	2.654 (+0.43)	3.521 (+0.63)
MAFN (-MA)	2.763 (+1.24)	3.734 (+1.55)	2.804 (+0.58)	3.531 (+0.64)
MAFN (-HA)	2.186 (+0.66)	2.961 (+0.78)	2.381 (+0.16)	2.991 (+0.10)
MAFN	**1.524**	**2.181**	**2.223**	**2.889**

Next, we compare MAFN with two classification methods: HAN and Stock-Net. From Table 3, we can see that MAFN performs the best on two datasets in terms of MCC, which achieves over 20% improvement compared with HAN and StockNet. On the News&Price dataset, MAFN achieves the highest accuracy 53.61%. HAN has the lowest accuracy and MCC than all the other models. One possible reason is that HAN only uses news information, while historical stock price is critical for predicting future value. Furthermore, Attentive LSTM which only uses price information can beat StockNet which utilizes both price and news, while MAFN consistently outperforms Attentive LSTM on both datasets. The advantages of MAFN could be explained in two aspects: (i) MAFN leverages useful text data while Attentive LSTM does not; (ii) MAFN adopts multi-head attention to precisely extract aspect-level textual information and employs the hierarchical attention to select relevant textual features, while StockNet performs a simple concatenation over price data and text representations.

5.3 Ablation Tests

We conduct ablation studies to illustrate the rationality and effectiveness of the proposed model architecture. We remove some parts from the complete model architecture to identify the most crucial component as follows:

- MAFN (-MA): uses the encoder-decoder framework where the encoder computes the average of text representations without multi-head attention.
- MAFN (-HA): uses the encode-decoder framework with multi-head attention where the decoder does not use the hierarchal attention.
- MAFN (-HA-MA): uses the encoder-decoder framework without multi-head attention in encoder and hierarchical attention in decoder.

As shown in Table 4, removing either hierarchical attention or multi-head attention would lead to performance loss. MAFN (-MA) achieves the worse performance among all the counterpart models. This indicates that multi-head attention contributes most to the performance gain compared with other components. Meanwhile, MAFN (-HA-MA) performs slightly better than MAFN (-MA), but is still much worse than the complete model. This implies that text data may not always be informative for stock price prediction, and the inclusion

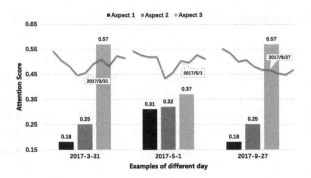

Fig. 3. Visualization of multi-head attention.

of noisy textual features may hurt the prediction performance. As MAFN (-HA) is the best model compared with other counterparts in MAFN, the lower performance loss verifies the effectiveness of hierarchical attention in selecting useful text information in temporal dimension.

5.4   Visualization of Multi-Head Attention Mechanism

To better illustrate the effectiveness of multi-head attention, we visualize the aspect attention scores as well as the true price trend of the New Hope Group in Fig. 3. Table 5 shows some example news in three days which have the corresponding aspect attention distributions in Fig. 3. From the results, we have the following observations: First, our model can automatically extract different aspects from text data. On 2017-03-31 and 2017-09-27, the texts contain similar aspects about transformation and chairman, and hence we obtain similar attention distributions. On 2017-5-1, we observe completely different aspects about integration and pass rate, and the corresponding attention distribution is different than the other two days. Second, our model can dynamically assign different importance weights to different aspects. The predictions on 2017-3-31 and 2017-9-27 are different: up and down, respectively. The only difference is the sentiment about transformation: positive on 2017-3-31 and negative on 2017-9-27. In terms of the attention scores in Fig. 3, both 2017-3-31 and 2017-9-27 have the highest attention score in aspect 3 whereas their prediction results are different. Though we cannot figure out the exact semantics of each aspect, it is reasonable to conjecture that aspect 3 is about the transformation plan for the New Hope Group.

Table 5. Different daily news examples.

Date	News
2017-3-31	Liu Yonghao, Chairman of New Hope Group, talks about the joint venture of enterprises in the economic transition period
	Liu Chang, chairman of the new hope, talked about the transformation effect last year, and the profit-increasing pig strategy
2017-5-1	New hope dairy industry talks about China's quality, and seeks integration
	In the first quarter, the pass rate of dairy products was 100%, and the milk enterprises opened low-temperature fresh cards
2017-9-27	New hope Liu Yonghao hopes to launch a competition for outstanding entrepreneurs
	The new hope transformation plan is far from successful, with 30 billion yuan in revenue for 340 million net profit

6 Conclusion

This paper aims to incorporate rich text information to enhance stock price prediction accuracy. Our key insight is to exploit multiple aspect-level information from texts and subtly fuse these fine-grained information according to their contributions on future stock closing price. We propose a Multi-head Attention Fusion Network (MAFN) that follows the general encoder-decoder framework to encode aspect-level textual information and decode stock closing price sequence accordingly. MAFN applies the multi-head attention mechanism to learn aspect-level text representations via different attention heads and distinguish the importance of each representation dynamically. The decoder absorbs the fused text representations and historical stock price by evaluating the discriminative effects of features via the hierarchical attention. The experimental results show that our model outperforms various baseline methods and can effectively utilize aspect-level information for accurate stock price prediction.

Acknowledgements. This work is supported by the National Key Research and Development Program of China (No. 2018YFC0831604). Yanyan Shen is also supported by NSFC (No. 61602297). Junjie Yao is supported by NSFC 61502169, U1509219 and SHEITC.

References

1. Adam, K., Marcet, A., Nicolini, J.P.: Stock market volatility and learning. J. Finan. **71**(1), 33–82 (2016)
2. Alahi, A., Goel, K., Ramanathan, V., Robicquet, A., Fei-Fei, L., Savarese, S.: Social LSTM: human trajectory prediction in crowded spaces. In: Proceedings of the IEEE Conference on Computer Vision and Pattern Recognition, pp. 961–971 (2016)

3. Chen, T., Guestrin, C.: XGBoost: a scalable tree boosting system. In: Proceedings of the 22nd ACM SIGKDD International Conference on Knowledge Discovery and Data Mining, pp. 785–794. ACM (2016)
4. Cho, K., et al.: Learning phrase representations using RNN encoder-decoder for statistical machine translation. arXiv preprint arXiv:1406.1078 (2014)
5. Chung, J., Ahn, S., Bengio, Y.: Hierarchical multiscale recurrent neural networks. In: International Conference on Learning Representations (2017)
6. Ding, X., Zhang, Y., Liu, T., Duan, J.: Using structured events to predict stock price movement: an empirical investigation. In: Proceedings of the 2014 Conference on Empirical Methods in Natural Language Processing (EMNLP), pp. 1415–1425 (2014)
7. Ding, X., Zhang, Y., Liu, T., Duan, J.: Deep learning for event-driven stock prediction. In: Twenty-Fourth International Joint Conference on Artificial Intelligence (2015)
8. Feng, F., Chen, H., He, X., Ding, J., Sun, M., Chua, T.S.: Improving stock movement prediction with adversarial training. arXiv preprint arXiv:1810.09936 (2018)
9. Hochreiter, S., Schmidhuber, J.: Long short-term memory. Neural Comput. **9**(8), 1735–1780 (1997)
10. Hu, Z., Liu, W., Bian, J., Liu, X., Liu, T.Y.: Listening to chaotic whispers: a deep learning framework for news-oriented stock trend prediction. In: Proceedings of the Eleventh ACM International Conference on Web Search and Data Mining, pp. 261–269. ACM (2018)
11. Kingma, D.P., Ba, J.: Adam: a method for stochastic optimization. arXiv preprint arXiv:1412.6980 (2014)
12. Le, Q.V., Mikolov, T.: Distributed representations of sentences and documents. In: International Conference on Machine Learning, pp. 1188–1196 (2014)
13. Li, Q., Wang, T., Li, P., Liu, L., Gong, Q., Chen, Y.: The effect of news and public mood on stock movements. Inf. Sci. **278**, 826–840 (2014)
14. Liaw, A., Wiener, M., et al.: Classification and regression by randomforest. R News **2**(3), 18–22 (2002)
15. Luong, M.T., Pham, H., Manning, C.D.: Effective approaches to attention-based neural machine translation. arXiv preprint arXiv:1508.04025 (2015)
16. Pennington, J., Socher, R., Manning, C.: Glove: global vectors for word representation. In: Proceedings of the 2014 Conference on Empirical Methods in Natural Language Processing (EMNLP), pp. 1532–1543 (2014)
17. Preethi, G., Santhi, B.: Stock market forecasting techniques: a survey. J. Theor. Appl. Inf. Technol. **46**(1), 24–30 (2012)
18. Si, J., Mukherjee, A., Liu, B., Pan, S.J., Li, Q., Li, H.: Exploiting social relations and sentiment for stock prediction. In: Proceedings of the 2014 Conference on Empirical Methods in Natural Language Processing (EMNLP), pp. 1139–1145 (2014)
19. Srivastava, N., Hinton, G., Krizhevsky, A., Sutskever, I., Salakhutdinov, R.: Dropout: a simple way to prevent neural networks from overfitting. J. Mach. Learn. Res. **15**(1), 1929–1958 (2014)
20. Sundermeyer, M., Schlüter, R., Ney, H.: LSTM neural networks for language modeling. In: Thirteenth Annual Conference of the International Speech Communication Association (2012)
21. Tabari, N., Seyeditabari, A., Peddi, T., Hadzikadic, M., Zadrozny, W.: A comparison of neural network methods for accurate sentiment analysis of stock market tweets. In: Alzate, C., et al. (eds.) MIDAS/PAP -2018. LNCS (LNAI), vol. 11054, pp. 51–65. Springer, Cham (2019). https://doi.org/10.1007/978-3-030-13463-1_4

22. Tao, C., Gao, S., Shang, M., Wu, W., Zhao, D., Yan, R.: Get the point of my utterance! learning towards effective responses with multi-head attention mechanism. In: IJCAI, pp. 4418–4424 (2018)
23. Xu, Y., Cohen, S.B.: Stock movement prediction from tweets and historical prices. In: Proceedings of the 56th Annual Meeting of the Association for Computational Linguistics (vol. 1: Long Papers), pp. 1970–1979 (2018)
24. Yue-Hei Ng, J., Hausknecht, M., Vijayanarasimhan, S., Vinyals, O., Monga, R., Toderici, G.: Beyond short snippets: deep networks for video classification. In: Proceedings of the IEEE Conference on Computer Vision and Pattern Recognition, pp. 4694–4702 (2015)
25. Zhang, L., Aggarwal, C.C., Qi, G.: Stock price prediction via discovering multi-frequency trading patterns, pp. 2141–2149 (2017)

Transfer Learning via Feature Selection Based Nonnegative Matrix Factorization

Thirunavukarasu Balasubramaniam[1]([⊠]), Richi Nayak[1], and Chau Yuen[2]

[1] Queensland University of Technology, Brisbane, Australia
{thirunavukarasu.balas,r.nayak}@qut.edu.au
[2] Singapore University of Technology and Design, Singapore, Singapore

Abstract. Transfer learning has been successfully used in recommender systems to deal with the data sparsity problem. Existing techniques assume that the source and target domains share the same feature space. This paper proposes a new direction in transfer learning where the source and target domains can have different feature space. The proposed technique, Feature Selection based Nonnegative Matrix Factorization (FSNMF), selects the useful features that can minimize the cost function of the target domain. The features of the source domain are learned using NMF and their importance is measured using the gradient principle. Experiments with real-world datasets show the effectiveness of FSNMF in comparison to state-of-the-art relevant transfer learning techniques.

Keywords: Transfer learning · Feature selection · Nonnegative matrix factorization · Recommender systems

1 Introduction

With the massive amounts of data generated by Web 3.0 applications such as Amazon, Netflix, Facebook, and Twitter, it is essential that users have quick access to the information of their interest. A recommender system has become an integral component of Information Filtering systems that recommends "selected" items to users based on their previous and their "alike" users' activities [1,10]. The dramatic growth of the internet population introduces new challenges to recommender systems. Data sparsity is a key problem caused by the fewer interactions of users with items [28]. For instance, only 10% of Twitter users create about a total of 80% of contents on the network [23]. Around 80% of Amazon users review only a single item [29]. Identifying users similarity or association, a fundamental requirement of recommender systems, becomes a significantly challenging task under these circumstances.

Fortunately, these web applications collect both explicit and implicit user data that include auxiliary information like reviews and social trust along with user ratings [27]. Researchers have used the auxiliary information in collaborative filtering to improve the recommender system's performance [15]. The auxiliary

R. Cheng et al. (Eds.): WISE 2019, LNCS 11881, pp. 82–97, 2019.
https://doi.org/10.1007/978-3-030-34223-4_6

information can provide more features in addition to the features derived by the explicit (sparse) feedback data and identify the similar users more effectively.

Transfer learning has been used to utilize auxiliary information in the collaborative filtering based recommender systems [17]. Transfer learning is a knowledge transfer technique where the knowledge is transferred from one domain (i.e., source domain) to another domain (i.e., target domain) [16]. The performance is improved because the knowledge from the source domain helps to improve the learning ability of the target domain.

A fundamental requirement of transfer learning techniques is transfer the required knowledge from the source to target domain effectively [16]. Usually, the knowledge transfer occurs based on the assumption that both source and target domains are strongly related according to the feature space that they share [18]. However, in the existing methods, the features learned from the source domain do not reflect the features of the target domain if they are not closely related. Using irrelevant features from the source domain can degrade the transfer learning performance in the target domain [4,25,33].

This paper focuses on transferring relevant knowledge from the source domain that can help to minimize the cost function in the target domain. As the knowledge is transferred via the learned features, the feature selection technique based on NMF, called as Feature Selection based Nonnegative Matrix Factorization (FSNMF), is presented. We propose to calculate the feature importance for each source feature using the gradient principle that measures how much a source feature can minimize the cost function of the target domain. Based on the calculated feature importance, a subset of source features are selected to transfer knowledge to the target domain. We propose to use NMF based feature selection as factorization based methods have shown superior performance in handling sparse data when compared to other collaborative filtering techniques [12]. Moreover, the input features generated from matrix factorization have shown efficient recommendation generation in emerging deep learning techniques [8].

The selective transfer learning is an emerging research [25,30]. The differences between the proposed FSNMF and the existing selective transfer learning are two-fold. Firstly, the existing selective transfer learning process focuses on selected instances from multiple source domains that have higher transferability. However, FSNMF selects a subset of features from the source domain. Moreover, selecting the source instances is a pre-processing task in existing methods, while it is a learning task in FSNMF. Secondly, unlike other methods where an appropriate source domain is selected from the multiple source domains, FSNMF focuses on relevant knowledge transfer from a single domain.

The contributions of this paper are two-fold. Firstly, to our best of knowledge, this is the first work that studies feature selection for relevant knowledge transfer from the source domain to the target domain. Secondly, the proposed gradient principle based feature importance measure determines how much a source feature can minimize the cost function of the target domain.

2 Related Work

Transfer Learning for Recommendation Systems: Data sparsity, due to missing user ratings, is a significant challenge in recommender systems. Transfer learning has emerged as a successful method to utilise auxiliary information that can reduce the effect of sparsity [16,17,27].

Earlier attempts used transfer learning for recommender systems by only using the closely related source knowledge to learn the target domain [18,19]. In [18], the rating information is converted into likes and dislikes to form binary auxiliary information. This auxiliary information acts as the source domain that adds like/dislike knowledge to the rating data (target domain). Similarly, in [22], the features are learned from the labelled demographic data (i.e., source domain) and transferred to the unlabelled data (i.e., target domain). In both the cases, the source and target domain are derived from the same information and satisfies the assumption of sharing common feature space. However, if the source and target domains are completely different, these methods are reported to fail or even lead to negative transfer [4,16,25].

There exist a handful of methods that focus on transferring knowledge from distinct source domain. [31] proposed a technique to transfer knowledge from the source domain with review text to the rating target domain. Recently, the knowledge graph associated with movies is used as the source domain to recommend movies to users in the target domain [20]. Tags associated with items have also been used as auxiliary information to generate movie recommendations in the target domain [6,7]. User-to-user social trust is another significant auxiliary information that has been used previously [11]. Though the source domain is distinct from the target domain, these methods transfer knowledge based on the common assumption that both these domains share the same feature space and transfer the entire knowledge between two sources.

However, not all the source features are relevant to the target domain, regardless the extent of the similarity between two domains. To improve the effectiveness of transfer learning, we propose a method that transfers only the relevant features from the source domain to the target domain.

Transfer Learning Techniques: Transfer learning techniques can be classified as parameter-based [32], instance-based [5] and factorization-based [18,20]. Parameter-based approaches using Support Vector Machine algorithm regularize the parameters in the target domain based on the parameters learned in the source domain [32]. In instance-based approaches [5], the useful source domain instances are selected to boost the target domain learner. Factorization-based approaches using NMF have been studied in recent years due to their superiority in handling sparse data sets [18,20]. They have been found flexible to adapt various constraints within two domains [2,14].

We summarise the existing factorization based methods in three categories as shown in Table 1. Coordinate System Transfer (CST) [18] and TagMatrix [6] use the principle coordinates as the bridge between the source and target domains.

The source domain features are used as the initialization for the target domain during factorization. Due to the additional bridging matrix, these matrix tri-factorization methods become computationally expensive. Moreover, TagMatrix [6] is applicable only when tag information is available as the source domain.

Table 1. Summary of the existing work

Category	Method	Auxiliary information
Matrix tri-factorization	CST [18]	Any
	TagMatrix [6]	Tags
Regularization	Trust propagation [11]	Social trust
	Tag-inferred [7]	Tags
	GR [14]	Any
Matrix co-factorization	SFS [20]	Any
	Collective factorization [19]	Any

On the other hand, methods such as Trust Propagation [11], Tag-Inferred [7], and Graph Co-Regularization (GR) [14] adapt a regularization technique for transfer learning where the target features are learned with an additional constraint. For example, the features of target domain such as social trust and tags are regularized in [11] and [7] respectively. However, these methods are not generic and cannot be applied to other datasets with different auxiliary information. Unlike other techniques, GR imposes regularization to preserve the geometrical structure of the source domain in the target domain.

Methods such as Shared Feature Space (SFS) [20] and Collective Factorization [19], learn the source and target domains together with a shared feature space. These types of methods are applicable to any auxiliary information as long as two domains share common features.

All these categories of methods have a common assumption that the source and target domains are highly similar. They do not consider a situation where not all the source features are essential in learning the target domain. This paper proposes a new direction in transfer learning by transferring only a subset of relevant features and proposes a novel concept and method of feature selection technique based transfer learning via NMF.

3 Feature Selective Nonnegative Matrix Factorization (FSNMF) Transfer Learning

3.1 Problem Definition

Let the source and target domains be denoted by S and T respectively. The sparse data from T is modeled as a matrix, $\mathbf{B} \in \mathbb{R}^{N \times M}$ with missing values. S can have multiple information, but we are only interested in the information that shares dimension N or/and M of $\mathbf{B}$. Based on which dimension is shared between the source and target domain, $\mathcal{S}$ can be represented in two different ways. The components of source domain $\mathcal{S}$ sharing the knowledge on dimension N with $\mathcal{T}$ is denoted as $\mathbf{A}^{(1)} \in \mathbb{R}^{N \times T}$ and sharing the knowledge on dimension M with $\mathcal{T}$ is denoted as $\mathbf{A}^{(2)} \in \mathbb{R}^{M \times T}$. T indicates the un-shared dimension of $\mathcal{S}$ that differentiates the source domain from the target domain.

Definition 1 (Transfer Learning): Transfer learning uses the knowledge (features) from the source domain $\mathcal{S}$ to help improve the learning capability of the target task in the target domain $\mathcal{T}$ where $\mathcal{S} \neq \mathcal{T}$ but $\mathcal{S}$ and $\mathcal{T}$ are closely related. The relevant source knowledge that should be transferred to the target domain is learned by the feature selective factorization process.

Consider a recommendation problem where the task is to recommend items to users based on the user-item interactions recording the user feedback (i.e., rating) on each item. This (highly-sparse) user-item interaction is the target domain $\mathcal{T}$ that can be represented as a rating matrix $\mathbf{B} \in \mathbb{R}^{N \times M}$ where N is the number of items and M is the number of users. According to previous studies [18], a rating higher than 3 is considered as like/true/1 and less than 3 is considered as dislike/false/0, and is fed as matrix value in $\mathbf{B}$. There exist some auxiliary information such as users' written reviews on items and users' social trust that can be considered as the source domain $\mathcal{S}$. The review information can be represented as the document-term matrix weighted using a weighted scheme denoted as $\mathbf{A}^{(1)} \in \mathbb{R}^{N \times T}$ for the common N items in $\mathcal{T}$ where T is the number of terms. The social trust information can be represented as the (binary) user-user matrix as $\mathbf{A}^{(2)} \in \mathbb{R}^{M \times M}$ for the common M users in $\mathcal{T}$. The relevant knowledge from $\mathbf{A}^{(1)}$ and $\mathbf{A}^{(2)}$ can be transferred to predict missing values in $\mathbf{B}$. The predicted missing values assist in providing accurate top-K item recommendations to the users.

3.2 Proposed Feature Selective Nonnegative Matrix Factorization

Figure 1 shows the overall process of FSNMF. The first task is to factorize the source matrix (e.g., $\mathbf{A}^{(1)}$) into two low-ranked factor matrices, $\mathbf{W_S}$ and $\mathbf{H_S}$, with NMF. The NMF optimization process of the source matrix $\mathbf{A}^{(1)}$ can be represented as [13],

$$f(\mathbf{W_S}, \mathbf{H_S}) = \left\| \mathbf{A}^{(1)} - \mathbf{W_S} \mathbf{H_S'} \right\| \tag{1}$$

where $\|.\|$ indicates the frobenius norm and $\mathbf{H_S'}$ indicates the transpose matrix of $\mathbf{H_S}$. Eq. (1) is a non-convex optimization problem. The well known Alternating Least Square (ALS) algorithm is commonly used where each factor matrix is updated alternatively using the following update rules [24].

$$\mathbf{W_S} = \mathbf{W_S} \frac{\mathbf{A}^{(1)} \mathbf{H_S'}}{\mathbf{W_S} \mathbf{H_S} \mathbf{H_S'}} \tag{2}$$

$$\mathbf{H_S} = \mathbf{H_S} \frac{\mathbf{W_S'} \mathbf{A}^{(1)}}{\mathbf{W_S'} \mathbf{W_S} \mathbf{H_S}} \tag{3}$$

Similarly, the NMF optimization of the target matrix $\mathbf{B}$ is formulated as,

$$f(\mathbf{W_T}, \mathbf{H_T}) = \|\mathbf{B} - \mathbf{W_T} \mathbf{H_T'}\| \tag{4}$$

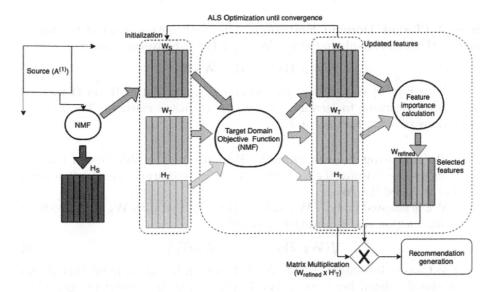

Fig. 1. The FSNMF process

Traditional transfer learning techniques assume that $\mathcal{S}$ and $\mathcal{T}$ share the same features (ie. $\mathbf{W_T} \leftarrow \mathbf{W_S}$). Therefore, the factorization process to solve Eq. (4) is initialized with $\mathbf{W_S}$ instead of $\mathbf{W_T}$. Unlike traditional techniques that either depend on the direct adaptation of source features [18] or impose the target features to be similar to source features [14], we propose a novel feature selective technique which carefully selects features that can effectively minimize the target objective function Eq. (4). The proposed transfer learning via feature selection consists of three steps.

1. $\mathbf{W_T}$ is randomly initialized to minimize the objective function Eq. (4). This random initialization learns only the target domain ($\mathcal{T}$) features as represented by $\mathbf{B}$. This will generate the first set of R features.
2. We also adapt the features learned from the source domain $\mathbf{W_S}$ as the second initialization to minimize the objective function Eq. (4). This is based on the assumption that $\mathbf{W_T} \leftarrow \mathbf{W_S}$. This will generate the second set of R features.
3. From $2R$ features generated from $\mathbf{W_T} \in \mathbb{R}^{N \times R} \cap \mathbf{W_S} \in \mathbb{R}^{N \times R}$, top-$R$ features are selected using the gradient principle based feature selection technique [9]. Gradient principle enables to use the gradient values of all the elements in a factor matrix to measure how much each element in the factor matrix can minimize the objective function. We propose to use this as the feature importance value and select relevant features to learn the target domain.

It should be noted that the top-R features are selected from both domains ($\mathbf{W_T}$ and $\mathbf{W_S}$) that are found useful to minimize the target domain's objective function. We conjecture that these relevant features help to learn the target domain to recommend top-K items to the users. We explain each step in detail.

Step 1 (Target Domain Features): The NMF optimization of the target matrix $\mathbf{B}$ with randomly initialized $\mathbf{W_T}$ and $\mathbf{H_T}$ is represented as,

$$f(\mathbf{W_T}, \mathbf{H_T}) = \|\mathbf{B} - \mathbf{W_T}\mathbf{H'_T}\| \tag{5}$$

Solving Eq. (5) is equivalent to performing traditional NMF on $\mathbf{B}$ without any transfer learning to identify two low-ranked factor matrices. The features learned are only the features of $\mathcal{T}$.

Step 2 (Adapting Source Domain Features): The objective of transfer learning is to use features of $\mathcal{S}$ in $\mathcal{T}$. We explain the scenario where the dimension N of $\mathbf{B}$ is shared (i.e., $\mathbf{W_S}$).

If $\mathbf{W_S}$ is adapted from $\mathcal{S}$, $\mathbf{W_T}$ in Eq. (5) is replaced with $\mathbf{W_S}$. The modified objective function can be formulated as [20],

$$f(\mathbf{W_S}, \mathbf{H_T}) = \|\mathbf{B} - \mathbf{W_S}\mathbf{H'_T}\| \tag{6}$$

The transfer learning using Eq. (6) is based on the assumption that $\mathcal{S}$ and $\mathcal{T}$ are closely related hence using $\mathbf{W_S}$ in target domain is effective only when $\mathbf{W_S}$ shares common features with $\mathbf{W_T}$. Using this technique when the source and target domain are distinct can cause negative transfer, as feeding irrelevant features from source domain to the target domain changes the original features of the target domain. Therefore, an effective approach will be to transfer only the relevant features from the source domain to the target domain. Moreover, the negative transfer can be avoided if the source features that can minimize the cost of target domain are carefully selected.

We propose to solve both Eqs. (5) and (6) independently to calculate $\mathbf{W_T}$ and $\mathbf{W_S}$ respectively. With both calculated matrices, we carefully select features those are important to minimize the target domain objective function Eq. (4). We first solve Eq. (5) using ALS [24] to calculate $\mathbf{W_T}$ as follows,

$$\mathbf{W_T} = \mathbf{W_T}\frac{\mathbf{B}\mathbf{H'_T}}{\mathbf{W_T}\mathbf{H_T}\mathbf{H'_T}} \tag{7}$$

Similarly, Eq. (6) is solved to calculate $\mathbf{W_S}$ as,

$$\mathbf{W_S} = \mathbf{W_S}\frac{\mathbf{B}\mathbf{H'_T}}{\mathbf{W_S}\mathbf{H_T}\mathbf{H'_T}} \tag{8}$$

Both Eqs. (7) and (8) are solved by fixing $\mathbf{H_T}$ to the randomly initialized value. This ensures that $\mathbf{H_T}$ is shared during the iterative computation of $\mathbf{W_T}$ and $\mathbf{W_S}$.

If the size of $\mathbf{W_T}$ is $\mathbb{R}^{N \times R}$ and $\mathbf{W_S}$ is $\mathbb{R}^{N \times R}$, we have a total of $2R$ features for N items. We now propose to measure the feature importance of all the $2R$ features and select top-R features among them.

Step 3 (Feature Importance Calculation): We propose a feature selection technique based on gradient principle to select R relevant features from $2R$ features. (Note: Each column of the matrix represents a feature.) Since the factor matrices are updated using the gradient value, it is effective and efficient to calculate the feature importance using gradient principle. The gradient principle technique [9] measures how much each element (ie. cell) in a factor matrix can minimize the objective function and it is denoted as element importance $\mathbf{E_S}$ and $\mathbf{E_T}$ for source and target factor matrix respectively. Therefore, for each column of $\mathbf{E_S}$ and $\mathbf{E_T}$, the sum of elements in that column measures the feature importance.

For simplicity let us represent,

$$\mathbf{G_T} = \mathbf{G_S} = \mathbf{BH'_T} \tag{9}$$

$$\mathbf{Q_T} = \mathbf{W_T H_T H'_T} \tag{10}$$

$$\mathbf{Q_S} = \mathbf{W_S H_T H'_T} \tag{11}$$

With the pre-computed $\mathbf{G_T}$ and $\mathbf{Q_T}$, the element importance matrix $\mathbf{E_T} \in \mathbb{R}^{N \times R}$ of $\mathbf{W_T}$ is calculated using the gradient principle [9] as,

$$\mathbf{E_T} = -(\mathbf{G_T} * \mathbf{W_T}) - 0.5 * (\mathbf{Q_T} * \mathbf{W_T} * \mathbf{W_T}) \tag{12}$$

Equation (12) is the difference between the target domain cost (Eq. (5)) before and after updating the factor matrix. It measures how much the target domain cost (Eq. (5)) can be minimized by updating each element in $\mathbf{W_T}$.

The column-wise sum of elements in $\mathbf{E_T}$ gives the feature importance $\mathbf{v_t} \in \mathbb{R}^{1 \times R}$ and is defined as,

$$\mathbf{v_t(r)} = \sum_{n=1}^{N} \mathbf{E_T}(n, r) \tag{13}$$

where $\mathbf{v_t(r)}$ indicates the feature importance value of rth target feature.

Similarly, with the pre-computed $\mathbf{G_S}$ and $\mathbf{Q_S}$, the element importance matrix $\mathbf{E_S} \in \mathbb{R}^{N \times R}$ of $\mathbf{W_S}$ is calculated using the gradient principle as,

$$\mathbf{E_S} = -(\mathbf{G_S} * \mathbf{W_S}) - 0.5 * (\mathbf{Q_S} * \mathbf{W_S} * \mathbf{W_S}) \tag{14}$$

Equation (14) measures how much the target domain cost (Eq. (6)) can be minimized by updating each element in $\mathbf{W_S}$.

The column-wise sum of elements in $\mathbf{E_S}$ gives the feature importance $\mathbf{v_s} \in \mathbb{R}^{1 \times R}$ and is defined as,

$$\mathbf{v_s(r)} = \sum_{n=1}^{N} \mathbf{E_S}(n, r) \tag{15}$$

where $\mathbf{v_s(r)}$ indicates the feature importance value of rth source feature.

Now, for $2R$ features from $\mathbf{W_T}$ and $\mathbf{W_S}$, we have the feature importance value calculated as $\mathbf{v_t}$ and $\mathbf{v_s}$ using Eqs. (13) and (15) respectively. The $2R$

features are sorted based on the feature importance value and the top-R features are selected. The selected features are represented as $\mathbf{W_{refined}} \in \mathbb{R}^{N \times R}$. Selecting only the source domain features may lead to negative transfer learning whereas selecting only the target domain features will lose additional knowledge from the source domain. Therefore, this paper selects a subset of features from both the source and target domains to effectively optimize the target domain objective function.

$\mathbf{W_T}$ and $\mathbf{W_S}$ are concatenated together into $\mathbf{W_{all}} \in \mathbb{R}^{N \times 2R}$ as,

$$\mathbf{W_{all}} = (\mathbf{W_T}|\mathbf{W_S}) \tag{16}$$

where $|$ indicates the matrix horizontal concatenation.

Now the features of $\mathbf{W_{all}}$ is sorted in descending order based on $\mathbf{v_t}$ and $\mathbf{v_s}$, and it is defined as,

$$\mathbf{W_{sorted}} = sort(\mathbf{W_{all}}) \tag{17}$$

The top-R features from $\mathbf{W_{sorted}} \in \mathbb{R}^{N \times 2R}$ is selected as,

$$\mathbf{W_{refined}} = \sigma_{\leq R}(\mathbf{W_{sorted}}) \tag{18}$$

where $\sigma_{\leq R}$ indicates that top-R features are selected.

Note that the learning process of $\mathbf{H_T}$ is same for both Eqs. (5) and (6). With the selected features $\mathbf{W_{refined}}$, the update rule becomes,

$$\mathbf{H_T} = \mathbf{H_T} \frac{\mathbf{W'_{refined}}\mathbf{B}}{\mathbf{W'_{refined}}\mathbf{W_{refined}}\mathbf{H_T}} \tag{19}$$

Since there is no transfer occurring on dimension M of $\mathbf{B}$ using $\mathbf{H_T}$, there is no requirement of selecting features for $\mathbf{H_T}$. So the overall update process is in the order of $(\mathbf{W_T}, \mathbf{W_S}, \mathbf{W_{refined}}, \mathbf{H_T})$. This is repeated cyclic until convergence or stopped. Algorithm 1 details the process.

Algorithm 1. Feature Selection based Nonnegative Matrix Factorization (FSNMF)

Input: Target matrix $\mathbf{B} \in \mathbb{R}^{N \times M}$; Randomly initialized factor matrices $\mathbf{W_T} \in \mathbb{R}^{N \times R}$, $\mathbf{H_T} \in \mathbb{R}^{M \times R}$; Source features shared on dimension N of $\mathbf{B}$, $\mathbf{W_S} \in \mathbb{R}^{N \times R}$; Rank R; $\mathbf{v_t} = \varnothing$; $\mathbf{v_s} = \varnothing$; maxiters.
Compute: $\mathbf{G_T}$, $\mathbf{G_S}$, $\mathbf{Q_T}$ and $\mathbf{Q_S}$ using Eqs. (9), (10) and (11);
Output: Learned factor matrices $\mathbf{W_{refined}}$, and $\mathbf{H_T}$
for *maxiters* **do**
    Update $\mathbf{W_T}$ via ALS update rule Eq. (7).
    Update $\mathbf{W_S}$ via ALS update rule Eq. (8).
    for $r = 1 : R$ **do**
        | Calculate feature importance ($\mathbf{v_t(r)}$ and $\mathbf{v_s(r)}$) by Eqs. (13) and (15).
    end
    Select top-R features ($\mathbf{W_{refined}}$) using Eq. (18).
    Update $\mathbf{H_T}$ using Eq. (19).
end

4   Experiments and Results

Experiments are conducted to answer the following questions.

Q1. What is the effect of sparsity in top-K item recommendation generation and how FSNMF helps to overcome sparsity problem?

Q2. How the (source and target) domain dissimilarity affects the transfer learning capability?

Q3. What is the accuracy and runtime performance of FSNMF and other benchmark methods in top-K item recommendation?

Datasets: Two real-world datasets (Table 2) have been used in experiments. The Amazon[1] review datasets consist of users and their feedbacks as ratings about the corresponding products/items. The datasets also have users' written reviews for the items. We combine all the reviews from the users for an item creating one review document for one item. The TF-IDF weighting scheme [21] is used to represent item-review matrix. The rating matrix (i.e., target domain) and item-review matrix (i.e., source domain) shares common items (ie. $\mathbf{W_T} \leftarrow \mathbf{W_S}$). The Epinions[2] dataset consists of users and their feedbacks as ratings about the products/items. It also consists of user reviews for the items. Similar to Amazon dataset, we create the item-review matrix. It also consists of who-trust-whom social trust information which is represented as a (binary) user-user matrix where 1 indicates users trust each other. The rating matrix (i.e., target domain) and user-user matrix (i.e., source domain) shares common users (ie. $\mathbf{H_T} \leftarrow \mathbf{H_S}$).

Table 2. Statistics of the datasets. D1: Amazon instant video (ratings + review); D2: Amazon musical instrument (ratings + review); D3: Epinions (ratings + review); D4: Epinions (ratings + social trust)

	D1	D2	D3	D4	
# of items (N)	1138	900	5000	9948	$\mathbf{B} \in \mathbb{R}^{N \times M}$
# of users (M)	4717	1429	7946	7900	
# of ratings	23013	10261	11679	215294	
Density	0.0042%	0.0063%	0.0002%	0.0022%	
% of Positive ratings	82.47%	87.92%	74.31%	71.07%	
Auxiliary (item reviews)	✓	✓	✓	✗	$\mathbf{A}^{(1)} \in \mathbb{R}^{N \times T}$
# of terms (T)	38769	17290	37964	-	
Avg. # of terms per item	34.06	19.21	07.59		
Auxiliary (user social trust)	✗	✗	✗	✓	$\mathbf{A}^{(2)} \in \mathbb{R}^{M \times M}$

Table 2 details the statistics of four datasets. In D1, the auxiliary information is additional written reviews along with ratings on 1138 items. Out of 23013

---

[1] http://jmcauley.ucsd.edu/data/amazon/.
[2] http://www.trustlet.org/downloaded_epinions.html.

ratings, 82.47% are positive ratings. It also reflects that nearly 82% of written reviews are positive opinions. Hence, the auxiliary domain and target domain share a highly similar context. Similar to D1, D2 also shares a highly similar context among the source and target domain. Comparatively, the similarity between the source and target domains for D3 falls below 75% and the auxiliary domain of D4 is different from the source domain. Therefore, we conjecture that it is more challenging to transfer the knowledge from the source domain to the target domain for D3 and D4.

We partition the dataset into 80% training and 20% test set to evaluate the performance of FSNMF with all benchmarks.

Evaluation Metrics: The objective is to recommend items to the users, hence the evaluation metrics are described in the context of top-K recommendations. Five evaluation metrics are used.

$$Precision\ at\ K\ (P@K) := \frac{|\{relevant\ items\}| \cap |\{retrieved\ items\}|}{|\{retrieved\ items\}|} \quad (20)$$

$$Recall\ at\ K\ (R@K) := \frac{|\{relevant\ items\}| \cap |\{retrieved\ items\}|}{|\{relevant\ items\}|} \quad (21)$$

$$F1\ score\ at\ K\ (F1@K) := 2\left(\frac{Precision \times Recall}{Precision + Recall}\right) \quad (22)$$

$$Root\ Means\ Square\ Error\ (RMSE) := \sqrt{\sum_{(u,i)}(b_{ui} - \hat{b}_{ui})^2/z} \quad (23)$$

$$Mean\ Absolute\ Error\ (MAE) := \sum_{(u,i)}(b_{ui} - \hat{b}_{ui})/z \quad (24)$$

where b_{ui} is the original ratings and $\hat{b}_{ui}$ is the predicted ratings. z indicates the total number of ratings.

Experimental Setup and Benchmarks: All the experiments were executed on Intel (R) Xeon (R) CPU E5-2665 0 @ 2.40 GHz model with 16 GB RAM. The source of FSNMF is made available for academic research purpose[3]. We choose a latest method from each category that can be applied with any auxiliary information, as discussed in Table 1.

- Traditional NMF [3] with the multiplicative updating (ALS) algorithm (i.e., without transfer learning) is used to show the performance improvement achieved with FSNMF that adds transfer learning in NMF.
- Coordinate System Transfer (CST) [18] is a Matrix Tri-Factorization based transfer learning method. The user/item knowledge from the source domain is used as the initialization to learn the target domain through the principle coordinates (latent features).

---

[3] https://github.com/thirubs/FSNMF.

- Graph Co-Regularization (GR) [14] is a regularization based transfer learning approach. GR learns the source and target domain together where the target domain features are learned with an additional graph regularization.
- Shared Feature Space (SFS) [20] is a Matrix Co-Factorization based technique. The source and target domain features are learned together with a shared latent space. For example, in D4, the user factor is shared with the user-item rating matrix and user-user social trust matrix.

4.1   Results

Table 3 and Figure 2 show the top-K recommendation generation performance of FSNMF and other benchmarks. Overall the FSNMF provides the best performance compared to the other benchmarks.

Effect of Sparsity and Domain Similarity: All transfer learning techniques achieve better performance over the NMF without transfer learning. The benchmarked transfer learning techniques achieve 1–3% improvement in $F1$ score for D1 whereas FSNMF achieves up to 11% higher $F1$ score. For D2, FSNMF achieves up to 42% improved performance while other benchmarks achieve only 22–25% improvement. Reasons of better performance of transfer learning on D2 as compared to D1 are as follows. Firstly, the target rating matrix of D1 is more sparser than D2. Secondly, the percentage of positive ratings is 5% higher for D2 when compared to D1. Since most of the reviews and ratings are positive in D2, the source domain and target domain are highly similar and shares more common features. Therefore, domain similarity have a positive influence in learning the target domain. The target domain of D3 is highly spare when compared to D1 and D2. Due to this reason, the benchmark transfer learning, GR shows nearly no improvement. This clearly shows the incapability of GR to deal with sparse datasets. On the other hand, FSNMF shows up to 6% improved performance.

Effect of Domain Dissimilarity: Dataset D4 is different from the rest of the datasets as social trust is used instead of reviews. The reviews and ratings are closely related (highly similar) as one's reviews reflect the ratings, whereas the social trust is entirely a different context. The benchmark techniques show up to 11% higher performance in comparison to NMF, while FSNMF shows nearly 27% higher performance. Due to the additional diagonal matrix required for CST, it ran out of time (o.o.t) for bigger datasets like D4. This shows the superiority of FSNMF in transferring knowledge from two different domains.

Accuracy and Runtime Performance: Tables 4 and 5 show the accuracy performance and running time respectively. FSNMF shows, on average, 2% improved accuracy against the NMF. Though FSNMF is 2.5 *times* slower than NMF, it is 2 to 22.6 *times* faster than other transfer learning techniques. This shows that FSNMF is an accurate and efficient transfer learning technique.

Table 3. top-K recommendation performance of each method on all datasets

D1	$P@1$	$P@5$	$P@10$	$P@50$	$P@100$	$R@1$	$R@5$	$R@10$	$R@50$	$R@100$
NMF [3]	0.335	0.251	0.231	0.221	0.221	0.065	0.155	0.192	0.220	0.221
CST [18]	**0.519**	0.292	0.251	0.236	0.236	**0.165**	0.233	0.236	0.236	0.236
GR [14]	0.425	0.297	0.270	0.257	0.257	0.105	0.201	0.232	0.257	0.257
SFS [20]	0.371	0.277	0.256	0.247	0.247	0.081	0.179	0.217	0.246	0.247
FSNMF	0.446	**0.362**	**0.343**	**0.331**	**0.336**	0.123	**0.273**	**0.314**	**0.336**	**0.336**
D2	$P@1$	$P@5$	$P@10$	$P@50$	$P@100$	$R@1$	$R@5$	$R@10$	$R@50$	$R@100$
NMF [3]	0.348	0.213	0.198	0.194	0.194	0.145	0.189	0.193	0.194	0.194
CST [18]	**0.650**	0.460	0.439	0.435	0.435	0.324	0.432	0.435	0.435	0.435
GR [14]	0.550	0.447	0.444	0.444	0.444	0.266	0.413	0.436	0.444	0.444
SFS [20]	0.521	0.406	0.403	0.403	0.403	0.244	0.372	0.395	0.402	0.402
FSNMF	0.637	**0.612**	**0.611**	**0.610**	**0.610**	**0.345**	**0.577**	**0.603**	**0.610**	**0.610**
D3	$P@1$	$P@5$	$P@10$	$P@50$	$P@100$	$R@1$	$R@5$	$R@10$	$R@50$	$R@100$
NMF [3]	0.242	0.228	0.227	0.227	0.227	0.215	0.227	0.227	0.227	0.227
CST [18]	0.266	0.244	0.244	0.244	0.244	0.239	0.244	0.244	0.244	0.244
GR [14]	0.242	0.229	0.229	0.229	0.229	0.215	0.229	0.229	0.229	0.229
SFS [20]	0.272	0.252	0.252	0.252	0.252	0.247	0.252	0.252	0l252	0.252
FSNMF	**0.297**	**0.281**	**0.280**	**0.280**	**0.280**	**0.283**	**0.280**	**0.280**	**0.280**	**0.280**
D4	$P@1$	$P@5$	$P@10$	$P@50$	$P@100$	$R@1$	$R@5$	$R@10$	$R@50$	$R@100$
NMF [3]	0.406	0.243	0.189	0.153	0.150	0.064	0.125	0.141	0.153	0.150
CST [18]	o.o.t	o.o.t	o.o.t	o.o.t	o.o.t	o.o.t	o.o.t	o.o.t	o.o.t	o.o.t
GR [14]	0.42	0.369	0.216	0.182	0.181	0.070	0.1433	0.165	0.180	0.181
SFS [20]	0.521	0.348	0.295	0.265	0.266	0.103	0.205	0.237	0.264	0.266
FSNMF	**0.607**	**0.475**	**0.437**	**0.422**	**0.422**	**0.127**	**0.304**	**0.371**	**0.421**	**0.422**

Table 4. $RMSE$ and MAE

Method	$RMSE$					MAE				
	D1	D2	D3	D4	Avg	D1	D2	D3	D4	Avg
NMF [3]	0.988	0.989	0.997	0.994	0.992	0.987	0.989	0.997	0.994	0.992
CST [18]	0.998	0.994	0.997	-	0.996	0.998	0.993	0.997	-	0.996
GR [14]	**0.962**	0.969	0.999	0.992	0.981	0.988	0.966	0.999	0.991	0.986
SFS [20]	0.988	0.975	**0.995**	0.983	0.985	0.988	0.972	**0.995**	0.982	0.984
FSNMF	0.973	**0.944**	0.998	**0.981**	**0.974**	**0.972**	**0.939**	0.997	**0.980**	**0.972**

Sensitivity of the Parameter R (Rank): The selection of R in factorization, to choose the number of features in the original data, is an NP-hard problem [26]. Based on the experiments with FSNMF, a rank ≥ 100 shows minimal change in the RMSE. Hence the rank of the factorization process is set to 100.

Table 5. Running time (in seconds) of each dataset and method

Without transfer learning	Method	D1	D2	D3	D4	Avg.
	NMF [3]	32.10	01.01	01.85	26.60	15.39
With transfer learning	CST [18]	1658.81	881.97	912.32	o.o.t	863.26
	GR [14]	110.78	03.01	664.74	168.89	236.66
	SFS [20]	134.75	03.06	131.84	39.70	78.07
	FSNMF	**57.39**	**02.23**	**54.73**	**35.38**	**38.09**

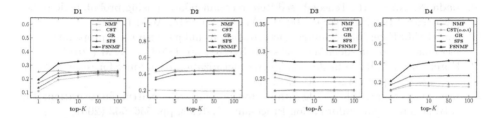

Fig. 2. The $F1@K$ performance of all the methods for item recommendations

5 Conclusion

In this paper, we propose a novel Feature Selection based Nonnegative Matrix Factorization (FSNMF) technique for transfer learning, which carefully selects features to transfer from source domain to a target domain. In particular, the gradient principle based feature importance calculation measures how much a source feature can minimize the objective function of a target domain. Therefore, the features selected help to learn the target domain more effectively by transferring only the relevant knowledge from the source domain. We believe that FSNMF opens a new door to improve transfer learning by selective features. Many research issues such as feature selective transfer learning for multiple source domains can be further examined.

References

1. Aggarwal, C.C., et al.: Recommender Systems. Springer, Cham (2016). https://doi.org/10.1007/978-3-319-29659-3
2. Balasubramaniam, T., Nayak, R., Yuen, C.: Understanding urban spatio-temporal usage patterns using matrix tensor factorization. In: ICDMW, pp. 1497–1498. IEEE (2018)
3. Berry, M.W., Browne, M., Langville, A.N., Pauca, V.P., Plemmons, R.J.: Algorithms and applications for approximate nonnegative matrix factorization. Comput. Stat. Data Anal. **52**(1), 155–173 (2007)
4. Cao, Z., Long, M., Wang, J., Jordan, M.I.: Partial transfer learning with selective adversarial networks. In: CVPR, pp. 2724–2732. IEEE (2018)
5. Dai, W., Yang, Q., Xue, G.R., Yu, Y.: Boosting for transfer learning. In: ICML, pp. 193–200. ACM (2007)

6. Fang, Z., Gao, S., Li, B., Li, J., Liao, J.: Cross-domain recommendation via tag matrix transfer. In: ICDMW, pp. 1235–1240. IEEE (2015)
7. Hao, P., Zhang, G., Martinez, L., Lu, J.: Regularizing knowledge transfer in recommendation with tag-inferred correlation. IEEE Trans. Cybern. **49**(1), 83–96 (2017)
8. He, X., Liao, L., Zhang, H., Nie, L., Hu, X., Chua, T.S.: Neural collaborative filtering. In: Proceedings of the 26th International Conference on World Wide Web, pp. 173–182. International World Wide Web Conferences Steering Committee (2017)
9. Hsieh, C.J., Dhillon, I.S.: Fast coordinate descent methods with variable selection for non-negative matrix factorization. In: SIGKDD, pp. 1064–1072. ACM (2011)
10. Ifada, N., Nayak, R.: Tensor-based item recommendation using probabilistic ranking in social tagging systems. In: WWW, pp. 805–810. ACM (2014)
11. Jamali, M., Ester, M.: A matrix factorization technique with trust propagation for recommendation in social networks. In: RecSys, pp. 135–142. ACM (2010)
12. Koren, Y., Bell, R., Volinsky, C.: Matrix factorization techniques for recommender systems. Computer **42**(8), 30–37 (2009)
13. Lee, D.D., Seung, H.S.: Algorithms for non-negative matrix factorization. In: Advances in Neural Information Processing Systems, pp. 556–562 (2001)
14. Long, M., Wang, J., Ding, G., Shen, D., Yang, Q.: Transfer learning with graph co-regularization. IEEE TKDE **26**(7), 1805–1818 (2014)
15. McAuley, J., Targett, C., Shi, Q., Van Den Hengel, A.: Image-based recommendations on styles and substitutes. In: SIGIR, pp. 43–52. ACM (2015)
16. Pan, S.J., Yang, Q.: A survey on transfer learning. IEEE TKDE **22**(10), 1345–1359 (2009)
17. Pan, W.: A survey of transfer learning for collaborative recommendation with auxiliary data. Neurocomputing **177**, 447–453 (2016)
18. Pan, W., Xiang, E.W., Liu, N.N., Yang, Q.: Transfer learning in collaborative filtering for sparsity reduction. In: AAAI (2010)
19. Pan, W., Yang, Q.: Transfer learning in heterogeneous collaborative filtering domains. Artif. Intell. **197**, 39–55 (2013)
20. Piao, G., Breslin, J.G.: Transfer learning for item recommendations and knowledge graph completion in item related domains via a co-factorization model. In: Gangemi, A., et al. (eds.) ESWC 2018. LNCS, vol. 10843, pp. 496–511. Springer, Cham (2018). https://doi.org/10.1007/978-3-319-93417-4_32
21. Salton, G., Buckley, C.: Term-weighting approaches in automatic text retrieval. Inf. Process. Manag. **24**(5), 513–523 (1988)
22. Shang, J., Sun, M., Collins-Thompson, K.: Demographic inference via knowledge transfer in cross-domain recommender systems. In: ICDM. IEEE (2018)
23. Southern, M.: 10% of Twitter users are creating 80% of tweets (2019). https://www.searchenginejournal.com/10-of-twitter-users-are-creating-80-of-tweets/305101/
24. Takane, Y., Young, F.W., De Leeuw, J.: Nonmetric individual differences multidimensional scaling: an alternating least squares method with optimal scaling features. Psychometrika **42**(1), 7–67 (1977)
25. Tan, B., Zhang, Y., Pan, S.J., Yang, Q.: Distant domain transfer learning. In: AAAI (2017)
26. Vavasis, S.A.: On the complexity of nonnegative matrix factorization. SIAM J. Optim. **20**(3), 1364–1377 (2009)
27. Verbert, K., et al.: Context-aware recommender systems for learning: a survey and future challenges. IEEE Trans. Learn. Technol. **5**(4), 318–335 (2012)

28. Wibowo, A.T.: Generating pseudotransactions for improving sparse matrix factorization. In: RecSys, pp. 439–442. ACM (2016)
29. Woolf, M.: A statistical analysis of 1.2 million Amazon reviews (2014). https://minimaxir.com/2014/06/reviewing-reviews/
30. Xiao, L., Min, Z., Yongfeng, Z., Yiqun, L., Shaoping, M.: Learning and transferring social and item visibilities for personalized recommendation. In: CIKM. ACM (2017)
31. Xin, X., Liu, Z., Lin, C.Y., Huang, H., Wei, X., Guo, P.: Cross-domain collaborative filtering with review text. In: IJCAI (2015)
32. Yang, J., Yan, R., Hauptmann, A.G.: Adapting SVM classifiers to data with shifted distributions. In: ICDMW, pp. 69–76. IEEE (2007)
33. Ying, W., Zhang, Y., Huang, J., Yang, Q.: Transfer learning via learning to transfer. In: ICML, pp. 5072–5081 (2018)

Learning Restricted Deterministic Regular Expressions with Counting

Xiaofan Wang[1,2] and Haiming Chen[1(✉)]

[1] State Key Laboratory of Computer Science, Institute of Software,
Chinese Academy of Sciences, Beijing 100190, China
{wangxf,chm}@ios.ac.cn
[2] University of Chinese Academy of Sciences, Beijing, China

Abstract. Regular expressions are widely used in various fields. Learning regular expressions from sequence data is still a popular topic. Since many XML documents are not accompanied by a schema, or a valid schema, learning regular expressions from XML documents becomes an essential work. In this paper, we propose a restricted subclass of single-occurrence regular expressions with counting (RCsores) and give a learning algorithm of RCsores. First, we learn a single-occurrence regular expressions (SORE). Then, we construct an equivalent *countable finite automaton* (CFA). Next, the CFA runs on the given finite sample to obtain an updated CFA, which contains counting operators occurring in an RCsore. Finally we transform the updated CFA to an RCsore. Moreover, our algorithm can ensure the result is a *minimal* generalization (such generalization is called *descriptive*) of the given finite sample.

Keywords: Schema inference · Regular expressions · Counting · Descriptive generalization

1 Introduction

Regular expression are widely used in information extraction, network security, database management, programming languages, etc. Nowadays, mining potential knowledge from sequence data has become a common task in many research areas and application scenarios [9,20,24,27]. The technologies of learning regular expressions have also obtained more and more attention and development. For example, many XML documents are not accompanied by a schema, or a valid schema [1,4,5,23], learning regular expressions from XML documents will facilitate the diverse applications of XML Schema, such as data processing, automatic data integration, and static analysis of transformations [10,21,22]. In this paper, we focus on learning regular expressions from XML documents.

For any given positive data, Gold specified that the class of regular expressions cannot be learned [15]. Even Bex et al. claimed that the class of

Work supported by National Natural Science Foundation of China under Grant Nos. 61872339, 61472405.

R. Cheng et al. (Eds.): WISE 2019, LNCS 11881, pp. 98–114, 2019.
https://doi.org/10.1007/978-3-030-34223-4_7

deterministic regular expressions cannot be learned [3]. Therefore, there are many works focusing on learning subclasses of deterministic regular expressions [2,3,6,7,11,12]. Deterministic regular expressions [8] require that each symbol in the input word can be unambiguously matched to a position in the regular expression without looking ahead in the word. Single-occurrence regular expressions (SOREs) [6,7] are classic subclass of deterministic regular expressions (standard). However, SOREs do not support counting, which is an extension of standard regular expressions used in XML Schema [14,16–19,25,26]. Then, we propose a restricted subclass of single-occurrence regular expressions with counting (RCsores). Our experiments (see Table 3) showed that the proportion of RCsores is 89.45% for 425,275 regular expressions extracted from XSD files, which were grabbed from Open Geospatial Consortium (OGC) XML Schema repository[1]. I.e., the majority of schemas in above real-world XSD files use RCsores. Therefore, it is necessary to study a learning algorithm for RCsore. Compared with Gold-style learning [15], the descriptive generalization [12,13] does not require to learn an exact representation of the target language, but can lead to a compact and powerful model [13]. Thus, our learning algorithm is based on the descriptive generalization [12,13].

For learning algorithms of SOREs, Bex et al. [7] proposed RWR and RWR_ℓ^2 [7]. Freydenberger et al. [12] presented the learning algorithm $Soa2Sore$ [12]. Additionally, [7] (resp. [12]) mentioned the future work, which is that SOREs extended with counting can be learnt by an additional post-processing step following the algorithm RWR (resp. $Soa2Sore$). However, the additional post-processing may result in the problem of overgeneralization [25]. For solving this problem, Wang et al. [25] proposed the class ECsores (see Definition 2), and the corresponding learning algorithm $InfECsore$ [25]. However, although the ECsore learnt by $InfECsore$ is descriptive of any given finite sample, the recall of $InfEC-sore$ is lower[2]. Additionally, every possibly repeated subexpression of the ECsore can be extended with counting, then the algorithm $InfECsore$ needs plenty of accurate counting such that it is not efficient to process larger samples. Wang et al. [26] also proposed a subclass cSOREs, which are a subclass of ECsore, and the corresponding learning algorithm $InfcSORE$ [26], but the learnt cSORE is not descriptive of any given finite sample[3]. Therefore, we propose a new subclass RCsore and the corresponding method for learning RCsore. Although RCsores are also subclass of ECsores, for any given finite sample, our algorithm not only can ensure the learnt RCsore is descriptive of the given finite sample (w.r.t. the class of RCsores), but also can ensure that the recall for the expression derived by our algorithm can be higher than that for the expression learnt by

---

[1] http://schemas.opengis.net/.

[2] For instance, the original expression in XSD can be denoted by $r_0 = (a|b)^{[1,6]}$, given sample $\{ba, aa, abaa, aabaa\}$, the ECsore learnt by $InfECsore$ is $r_1 = (b?a^{[1,2]})^{[1,2]}$. However, the learnt RCsore can be $r_2 = (b?a)^{[1,4]}$. Let $S_1 = \{s|s \in \mathcal{L}(r_0), s \in \mathcal{L}(r_1)\}$ and $S_2 = \{s|s \in \mathcal{L}(r_0), s \in \mathcal{L}(r_2)\}$. Then, $|S_1| = 14$ and $|S_2| = 25$. Thus, $\frac{|S_1|}{|\mathcal{L}(r_0)|} < \frac{|S_2|}{|\mathcal{L}(r_0)|}$.

[3] Let $S = \{b, abd, ad, cddcdd\}$, the cSORE learnt by $InfcSORE$ is $r_3 = ((a?b?|c)d?)^{[1,4]}$, however, there is a cSORE $r_4 = (a?b?|c?(d^{[1,2]})?)^{[1,2]}$ such that $\mathcal{L}(r_3) \supset \mathcal{L}(r_4) \supseteq S$.

InfECsore. Moreover, for a smaller sample, the learnt RCsore has better generalization ability (higher precision and recall) than the learnt ECsore. And the
learning algorithm of RCsore is more efficient than that of ECsore for processing
larger samples.

The main contributions of this paper are as follows.

- We infer a SORE and construct an equivalent countable finite automaton
  (CFA) [25].
- The CFA runs on the given finite sample to obtain an updated CFA, which
  has updated the counting operators that will occur in an RCsore.
- We convert the updated CFA to an RCsore and prove that the generated
  RCsore is descriptive of any given finite language.

The paper is structured as follows. Section 2 gives the basic definitions.
Section 3 presents the learning algorithm of the RCsore, and proves the RCsore
generated by our algorithm is descriptive of any given finite language. Section 4
presents experiments. Section 5 concludes the paper.

2   Preliminaries

2.1   Regular Expression with Counting

Let Σ be a finite alphabet of symbols. $\mathcal{R}_c$ is a set (non-empty) of regular
expressions with counting over Σ. ε, $a \in \Sigma$ are regular expressions in $\mathcal{R}_c$.
For regular expressions $r_1, r_2 \in \mathcal{R}_c$, the disjunction $(r_1|r_2)$, the concatenate
$(r_1 \cdot r_2)$, the Kleene-star r_1^*, and counting (*numerical occurrence constraints* [14])
$r_1^{[m,n]}$ are also regular expressions in $\mathcal{R}_c$. $m \in \mathbb{N}$, $n \in \mathbb{N}_{/1}$, $\mathbb{N} = \{1, 2, 3, \cdots\}$,
$\mathbb{N}_{/1} = \{2, 3, 4, ...\} \cup \{+\infty\}$, and $m \leq n$. For a regular expression $r \in \mathcal{R}_c$,
$\mathcal{L}(r^{[m,n]}) = \{w_1 \cdots w_i | w_1, \cdots, w_i \in \mathcal{L}(r), m \leq i \leq n\}$. Note that r^+, $r?$, and r^*
are used as abbreviations of $r^{[1,+\infty]}$, $r|\varepsilon$, and $r^{[1,+\infty]}|\varepsilon$, respectively. Usually, we
omit concatenation operators in examples. $|r|$ denotes the length of r, which is
the number of symbols and operators occurring in r plus the sizes of the binary
representations of the integers [14]. For a finite sample S, $|S|$ denotes the number
of strings in S. $\varnothing$ denotes the empty set. For space consideration, all omitted
proofs can be found at http://github.com/GraceFun/InfRCsore.

2.2   SORE, ECsore and RCsore

SORE is defined as follows.

Definition 1 (SORE [6,7]). *Let Σ be a finite alphabet. A single-occurrence
regular expression (SORE) is a standard regular expression over Σ in which
every terminal symbol occurs at most once.*

Example 1. $(ab)^+$ is a SORE, while $(ab)^+a$ is not.

Definition 2 (ECsore [25]). *Let Σ be a finite alphabet. An ECsore is a regular expression with counting over Σ in which every terminal symbol occurs at most once. For a regular expression r, an ECsore forbids immediately nested counters, expressions of form $(r?)?$ and $(r?)^{[m,n]}$.*

ECsore does not use the Kleene-star and the iteration operations. And ECsores are deterministic by definition.

Definition 3 (RCsore). *Let Σ be a finite alphabet. An RCsore is an ECsore over Σ. For regular expressions r_1, r_2 and r_3, an RCsore forbids expressions of form $(r_1 r_2 r_3)^{[m_1,n_1]}$ where $\varepsilon \in \mathcal{L}(r_1)$, $\varepsilon \in \mathcal{L}(r_3)$ and $r_2 \in \{e^{[m_2,n_2]}, e?\}$ for regular expression e ($\varepsilon \notin \mathcal{L}(e)$).*

According to the definition, RCsores are a subclass of ECsores. ECsores are deterministic regular expressions, so are the RCsores.

Example 2. $(a|b^{[1,2]})^{[3,4]}(c?d)^{[1,+\infty]}$, $(a^{[3,4]}b)^{[1,2]}$, and $((a?b?|c)(d^{[2,3]})?)^{[1,2]}$ are RCsores, also ECsores, while $a?b^+a$ is not a SORE, therefore neither an RCsore nor an ECsore. However, the expressions $(a?b^{[1,2]}c?)^{[1,2]}$ and $(a?b?c?)^{[1,2]}$ are ECsores, not RCsores. $(a^{[1,2]})^{[1,2]}$, $((a^{[1,2]})?)^{[1,2]}$ and $((a^{[1,2]})?)?$ are forbidden.

2.3 Descriptivity

We give the notion of descriptive expressions and automata.

Definition 4 (Descriptivity [12]). *Let $\mathcal{D}$ be a class of regular expressions or finite automata over some alphabet Σ. A $\delta \in \mathcal{D}$ is called $\mathcal{D}$-descriptive of a non-empty language $S \subseteq \Sigma^*$ if $\mathcal{L}(\delta) \supseteq S$, and there is no $\gamma \in \mathcal{D}$ such that $\mathcal{L}(\delta) \supset \mathcal{L}(\gamma) \supseteq S$.*

If a class $\mathcal{D}$ is clear from the context, we simply write *descriptive* instead of $\mathcal{D}$-descriptive.

Proposition 1. *Let Σ be a finite alphabet. There exists an RCsore-descriptive RCsore r for every language $L \subseteq \Sigma^*$.*

2.4 Countable Finite Automaton

Definition 5 (Countable Finite Automaton [25]). *A Countable Finite Automaton (CFA) is a tuple $(Q, Q_c, \Sigma, \mathcal{C}, q_0, q_f, \Phi, \mathsf{U}, \mathsf{L})$. The members of the tuple are described as follows:*

- *Σ is a finite and non-empty alphabet.*
- *q_0 and q_f : q_0 is the initial state, q_f is the unique final state.*
- *Q is a finite set of states. $Q = \Sigma \cup \{q_0, q_f\} \cup \{+_i\}_{i \in \mathbb{N}}$.*
- *$Q_c \subset Q$ is a finite set of counter states. Counter state is a state q ($q \in \Sigma$) that can directly transit to itself, or a state $+_i$. For each subexpression (excluding single symbol $a \in \Sigma$) under the iteration operator, we associate a unique counter state $+_i$ to count the minimum and maximum number of repetitions of the subexpression, respectively.*

- C is finite set of counter variables that are used for counting the number of repetitions of the subexpressions under the iteration operators. $C = \{c_q | q \in Q_c\}$, for each counter state q, we also associate a counter variable c_q.
- $U = \{u(q) | q \in Q_c\}$, $L = \{l(q) | q \in Q_c\}$. For each subexpression under the iteration operator, we associate a unique counter state q such that $l(q)$ and $u(q)$ are the minimum and maximum number of repetitions of the subexpression, respectively.
- Φ maps each state $q \in Q$ to a set of tuples consisting of a state $p \in Q$ and two update instructions. $\Phi: Q \mapsto \wp(Q \times ((L \times U \mapsto (\mathbf{Min}(L \times C), \mathbf{Max}(U \times C))) \cup \{\emptyset\}) \times ((C \mapsto \{\mathbf{res}, \mathbf{inc}\}) \cup \{\emptyset\}))$. ($\emptyset$ denotes empty instruction.)

Definition 6 (Transition Function of a CFA [25]). *The transition function δ of a CFA $(Q, Q_c, \Sigma, C, q_0, q_f, \Phi, U, L)$ is defined for any configuration (q, γ, θ) and the letter $y \in \Sigma \cup \{\dashv\}$*

(1) $y \in \Sigma$: $\delta((q, \gamma, \theta), y) = \{(z, f_\alpha(\gamma, \theta), g_\beta(\theta)) | (z, \alpha, \beta) \in \Phi(q) \wedge (z = y \vee ((y, \alpha, \beta) \notin \Phi(q) \wedge z \in \{+_i\}_{i \in \mathbb{N}}))\}$.

(2) $y = \dashv$: $\delta((q, \gamma, \theta), \dashv) = \{(z, f_\alpha(\gamma, \theta), g_\beta(\theta)) | (z, \alpha, \beta) \in \Phi(q) \wedge (z = q_f \vee z \in \{+_i\}_{i \in \mathbb{N}})\}$.

3   Inference of RCsores

Our learning algorithm works in the following steps.

(1) We infer a SORE for a given finite sample. (2) A CFA is equivalently transformed from the SORE obtained from (1). (3) The CFA transformed from step (2) runs on the same finite sample used in step (1) to obtain an updated CFA, which has updated the counting operators that will occur in an RCsore. (4) We convert the updated CFA in step (3) to an RCsore.

Algorithm 1. *InfRCsore*

Input: a finite sample S;
Output: an RCsore-descriptive RCsore;
1: A SORE $r_s = InfSore(SOA(S))$;
2: CFA $A = ConsCFA(r_s)$;
3: CFA $A' = Counting(A, S)$;
4: $r = GenRCsore(A')$;
5: **return** r;

Algorithm 1 is the framework of our learning algorithm. Algorithm *SOA* [12] constructs the single-occurrence automaton (SOA) [7,12] for the given finite sample S. Algorithm *InfSore* is described in Sect. 3.1, algorithm *ConsCFA* is given in Sect. 3.2, algorithm *Counting* is showed in [25], algorithm *GenRCsore* is presented in Sect. 3.4.

3.1   Inferring Standard Deterministic Regular Expression: SORE

The problem of learning SORE was solved by Bex et al. and Freydenberger et al. Bex et al. proposed the learning algorithm RWR [7] and its variants. Freydenberger et al. [12] proved the results of RWR with its variants are not descriptive of any given finite sample, and then presented the learning algorithm *Soa2Sore* [12]. However, the SORE learnt by *Soa2Sore* is descriptive of the language, which is the set of the strings accepted by the SOA that is built for the given finite sample [12]. Despite of that, we still can infer a SORE such that an RCsore, which is descriptive of the given finite sample, can be derived from the obtained SORE.

Algorithm 2 learns a SORE from the given finite sample. First, a SORE is inferred by $Soa2Sore$. Then, the SORE is converted to a normal form (SORE). Theorem 1 demonstrates that the normal form is more approximate to the given finite sample than the SORE learnt by $Soa2Sore$.

Algorithm 2. *InfSore*

Input: a finite sample S;
Output: a SORE r_s;
1: A SORE $r_0 = Soa2Sore(SOA(S))$;
2: Let $r_{f_1} = (r_1? \cdots r_k?)^+$ $(k \geq 2)$;//r_i $(1 \leq i \leq k)$ is a regular expression
3: Let $r_{f_2} = (r_1| \cdots |r_k)^+$ where $r_i \in \{e_i^+, e_i\}$ $(k \geq i \geq 1)$;//e_i is a regular expression
4: Let $r_{f_3} = (r_1 r_2^+ r_3)^+$ where $\varepsilon \in \mathcal{L}(r_1)$ and $\varepsilon \in \mathcal{L}(r_3)$;
5: **if** Case (1): r_0 contains the expression of the form r_{f_1} **then**
6:    for all expressions of form r_{f_1}: r_{f_1} is converted to $r'_{f_1} = (r_1| \cdots |r_k)^+$;
7: **if** Case (2): r_0 contains the expression of the form r_{f_2}, where $r_i = e_i$ **then**
8:    for all expressions of form r_{f_2}: r_{f_2} is converted to $r'_{f_2} = (e_1^+| \cdots |e_k^+)^+$;
9: **if** Case (3): r_0 contains the expression of the form r_{f_3} **then**
10:    for all expressions of form r_{f_3}: r_{f_3} is converted to $r'_{f_3} = (r_1 r_2 r_3)^+$;
11: Let $r_s = r_0$; **return** r_s;

In Algorithm 2, if the SORE r_0 does not contain any one expression of the forms r_{f_1}, r_{f_2} and r_{f_3} (which are specified in lines 4, 2 and 3, respectively), then *InfSore* directly outputs r_0, i.e., $r_s = r_0$. Note that, except for case (1) (in line 5), other cases are equivalent conversions for r_0. The conversion in case (2) (in line 7) is mainly used to easily construct a CFA in the next section and track as many subexpressions as possible (which can be repeated) in a SORE. For processing r_0 to a normal form r_s, it takes $\mathcal{O}(|r_0|)$ time. Let the built SOA in line 1 contain n_s nodes and t_s transitions. $Soa2Sore$ takes $\mathcal{O}(n_s t_s)$ time to infer a SORE. Thus, the time complexity of algorithm *InfSore* is $\mathcal{O}(n_s t_s)$ $(n_s t_s > |r_0|)$.

Example 3. For sample $S = \{a, acc, acbb, bab\}$, the result of algorithm $Soa2Sore$ is $r_0 = ((a(c^+)?)|b)^+$. Let the SORE $r_s := InfSore(SOA(S))$, then the SORE $r_s = ((a(c^+)?)^+|b^+)^+$.

Theorem 1. *For any given finite sample S, let $r_0 = Soa2Sore(SOA(S))$, and let $r_s := InfSore(SOA(S))$, then $\mathcal{L}(r_0) \supseteq \mathcal{L}(r_s) \supseteq S$.*

According to Theorem 1, $\mathcal{L}(r_s)$ is more approximate to the given finite sample than $\mathcal{L}(r_0)$. Therefore, we can obtain a descriptive RCsore, which is extended from the expression of form r_s.

3.2   Translating SORE to CFA

To avoid plenty of accurate counting in a CFA, the CFA should be constructed from a specific structure, instead of being learnt from a given finite sample [25]. Therefore, in this section, we present how to translate a SORE to a CFA. First, we construct the state-transition diagram of a CFA by traversing the syntax tree of the SORE, which is obtained from Sect. 3.1. Then, the detailed descriptions of the CFA are similar with that described in [25]. Theorem 2 shows that an equivalent CFA can be transformed from an RCsore.

Algorithm 3 first constructs the state-transition diagram of a CFA by using Algorithm 4, then presents the detailed descriptions of the CFA. The state-transition diagram of a CFA is a finite directed graph, denoted by G. Algorithm 4 constructs a directed graph G by traversing a syntax tree. The entire process is similar to the preorder traversal of the binary tree. For a syntax tree T, $T.L$ and $T.R$ denote the left subtree and the right subtree of T, respectively. For a graph G, $G. \prec (v)$ denotes the set of all immediate predecessors of v in G, $G. \succ (v)$ denotes the set of all immediate successors of v in G. Some subroutines in Algorithm 4 are as follows.

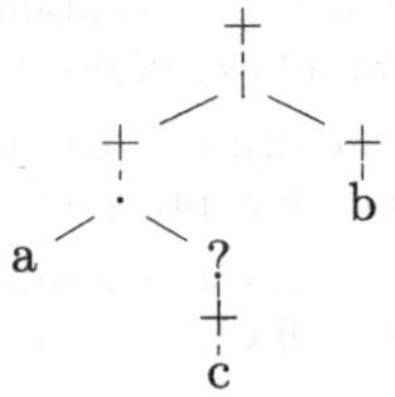

Fig. 1. The syntax tree of expression $((a(c^+)?)^+|b^+)^+$.

$Conn_G(t, G_1, G_2)$. According to label t, a new graph G is constructed by connecting graphs G_1 and G_2. If $t =$ '·', then add edges $\{(v_1, v_2)|v_1 \in G_1. \prec (q_f), v_2 \in G_2. \succ (q_0)\}$; remove nodes $G_1.q_f$, $G_2.q_0$ and their associated edges; let

---

Algorithm 3. *ConsCFA*

Input: a syntax tree T;
Output: a CFA $\mathcal{A}$;
1: $G = Cons_G(T)$;
2: CFA $\mathcal{A} = (Q, Q_c, \Sigma, \mathcal{C}, G.q_0, G.q_f, \Phi(\mathcal{R}), \mathsf{U}, \mathsf{L})$;
3: **return** $\mathcal{A}$;

$G.q_0 = G_1.q_0$. If $t =$ '|', then add new nodes q_0, q_f; add edges $\{(q_0, v_1)|v_1 \in G_1. \succ (q_0) \cup G_2. \succ (q_0)\}$, and $\{(v_2, q_f)|v_2 \in G_1. \prec (q_f) \cup G_2. \prec (q_f)\}$; remove nodes $G_1.q_0$, $G_1.q_f$, $G_2.q_0$, $G_2.q_f$ and their associated edges; let $G.q_0 = q_0$.

$Add^+(G, +_i)$. G is a graph, and $+_i$ (a counter state in CFA) is a node. Add^+ adds node $+_i$ (initially, $i = 1$) into the graph G. Add new node q_f; let $\mathcal{R}_{+_i} = \{v | v \in G. \succ (q_0)\}$; add edges $\{(+_i, v_1) | v_1 \in G. \succ (q_0)\}$; add edges $\{(v_2, +_i) | v_2 \in G. \prec (q_f)\}$; remove node $G.q_f$ and its associated edges; add edge $(+_i, q_f)$. The set of $\mathcal{R}_{+_i}$ is established to specify the transition entrances for state $+_i$ to count the minimum and maximum number of repetitions of the subexpression under the iteration operator. Each $\mathcal{R}_{+_i}$ is a global variable. Let $\mathcal{R} = \{\mathcal{R}_{+_i}\}_{i \in \mathbb{N}}$.

In Algorithm 3, after the state-transition diagram G

Algorithm 4. $Cons_G$

Input: a syntax tree T;
Output: a directed graph $G(V, E)$;
1: if $T = \emptyset$ return $\emptyset$;
2: if $T.label \in \Sigma$ then
3:     Add new nodes $q = T.label$, q_0, and q_f;
4:     return $G(\{q_0, q, q_f\}, \{(q_0, q), (q, q_f)\})$;
5: if $T.label = '.'$ then
6:     $G_1 = Cons_G(T.L)$; $G_2 = Cons_G(T.R)$;
7:     return $Conn_G(T.label, G_1, G_2)$;
8: if $T.label \in \{+, ?\}$ then
9:     $G = Cons_G(T.L)$;
10:     if $T.label = '+'$ then
11:         if $T.L.label \in \Sigma$ then
12:             add edge $(G.T.L.label, G.T.L.label)$;
13:         else $G = Add^+(G, +_i)$; inc i;
14:     if $T.label = '?'$ then
15:         add edge $(G.q_0, G.q_f)$;
16:     return G
17: if $T.label = '|'$ then
18:     $G_1 = Cons_G(T.L)$; $G_2 = Cons_G(T.R)$;
19:     $G = Conn_G(T.label, G_1, G_2)$;
20:     return G;

---

of a CFA is constructed, the CFA $\mathcal{A}$ is then obtained. In line 2, [25] shows the detailed descriptions of the CFA $\mathcal{A}$. Note that, $\Phi(\mathcal{R})$ denotes that $\mathcal{R}$ is a parameter in Φ.

For any SORE r obtained in Sect. 3.1, the time complexity of constructing the corresponding syntax tree is $\mathcal{O}(|r|)$, and the preorder traversal of the syntax tree used to construct the state-transition diagram of a CFA also requires $\mathcal{O}(|r|)$ time. Therefore, the time complexity of constructing a CFA is $\mathcal{O}(|r|)$.

Example 4. For the expression $((a(c^+)?)^+ | b^+)^+$, the syntax tree can be seen in Fig. 1. The corresponding state-transition diagram can be seen in Fig. 2(a).

Theorem 2. *For any given SORE r, there is a CFA $\mathcal{A}$ such that $\mathcal{L}(\mathcal{A}) = \mathcal{L}(r)$.*

3.3   Counting with CFA

The constructed CFA in Sect. 3.2 runs on the given finite sample, which is the same set of strings used to generate the SORE in Sect. 3.1. The CFA counts the minimum and maximum number of repetitions of the subexpressions under

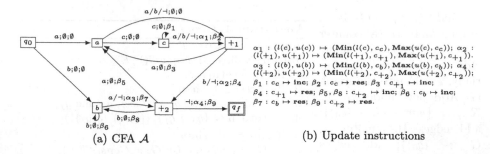

(a) CFA $\mathcal{A}$ (b) Update instructions

Fig. 2. (a) is the CFA $\mathcal{A}$ for regular language $\mathcal{L}(((a(c^+)?)^+|b^+)^+)$. The label of the transition edge is $(y;\alpha_i;\beta_j)$ $(i,j \in \mathbb{N})$, y $(y \in \Sigma \cup \{\dashv\})$ is a current letter; (b) specifies that, α_i is an update instruction for the lower bound and upper bound variables, and β_j is an update instruction for the counter variable.

the iteration operators. Counting rules are given by transition functions of the CFA. We use the algorithm *Counting* proposed in [25] to run the CFA. Let $\mathcal{A}$ denote the constructed CFA and S denote the given finite sample. After the CFA $\mathcal{A}$ recognized the sample S, let $\mathcal{A}'$ denote the CFA $\mathcal{A}$ which has updated the the minimum and maximum number of repetitions of the subexpressions under the iteration operators. Let $\mathcal{A}' = Counting(\mathcal{A}, S)$, and $\mathsf{C} = \{(l(q), u(q))|l(q) = \mathcal{A}'.\mathsf{L}.l(q), u(q) = \mathcal{A}'.\mathsf{U}.u(q), q \in \mathcal{A}'.Q_c\}$. The elements in C are counting operators, which will be introduced into an RCsore. The time complexity of *Counting* is $\mathcal{O}(N\overline{L})$ time, where $N = |S|$ and $\overline{L}$ is the average length of the strings in S [25].

Example 5. For the sample $S = \{a, acc, acbb, bab\}$, $r_s = ((a(c^+)?)^+|b^+)^+$ is the SORE obtained from Sect. 3.1, the CFA $\mathcal{A}$ showed in Fig. 2 runs on the sample S. Then, the tuples in C are listed as follows: $(l(c), u(c)) = (1, 2)$, $(l(b), u(b)) = (1, 1)^4$, $(l(+_1), u(+_1)) = (1, 1)$, $(l(+_2), u(+_2)) = (1, 3)$. $l(+_1)$ and $u(+_1)$ (resp. $l(+_2)$ and $u(+_2)$) are the minimum and maximum number of repetitions of the subexpression $(a(c^+)?)$ (resp. $(a(c^+)?|b)$), respectively. Note that the minimum numbers of repetitions of symbol c are both 0 in strings a and bab. In Sect. 3.4, we will convert expression $c^{[1,2]}$ to $(c^{[1,2]})?$.

3.4   Generating RCsore

In this section, we transform the updated CFA $\mathcal{A}'$ obtained in Sect. 3.3 to an RCsore. Since the algorithm *GenECsore* can convert a CFA to an descriptive ECsore (w.r.t. the class of ECsores). We still can use the algorithm *GenECsore*

---

[4] Note that, the CFA $\mathcal{A}$ runs on S, the direct counting result for b is $(l(b), u(b)) = (1, 2)$. However, $(l(b), u(b))$ is subsequently updated by *Counting* that b can be repeated by using the counting operator $[l(+_2), u(+_2)] = [1, 3]$.

to derive an RCsore, the constructed CFA in this paper is equivalent to an RCsore, not an equivalent representation of an ECsore. Then, for an updated CFA $\mathcal{A}'$, the algorithm *GenECsore* can convert the CFA $\mathcal{A}'$ to an descriptive RCsore (w.r.t. the class of RCsores).

Algorithm 5 converts the updated CFA to an RCsore. Theorem 3 demonstrates the finally obtained RCsore is descriptive of any given finite sample. Assume that the updated CFA contains n_c nodes and t_c transitions. *GenECsore* takes $\mathcal{O}(n_c t_c)$ time to infer an ECsore [25]. Then, the time complexity of generating RCsore is $\mathcal{O}(n_c t_c)$.

Algorithm 5. *GenRCsore*

Input: the updated CFA $\mathcal{A}'$

Output: an RCsore r;

1: $r = GenECsore(\mathcal{A}')$;

2: **return** r;

Example 6. The tuples in C obtained from algorithm *Counting* are as follows. $(l(c), u(c)) = (1, 2)$, $(l(b), u(b)) = (1, 1)$, $(l(+_1), u(+_1)) = (1, 1)$ and $(l(+_2), u(+_2)) = (1, 3)$. For the updated CFA $\mathcal{A}'$, the generated RCsore is $((a(c^{[1,2]})?)|b)^{[1,3]}$.

Theorem 3. *For any given finite language S, let $r :=InfRCsore(S)$, the time complexity of algorithm InfRCsore is $\mathcal{O}(n_c t_c + N\overline{L})$ and r is an RCsore-descriptive RCsore for S.*

Let $\mathcal{A}_c$ and $\mathcal{A}_g$ denote the CFAs constructed in this paper and in literature [25], respectively. Assume that the CFA $\mathcal{A}_g$ contains n_g nodes and t_g transitions. The time complexity of *InfECsore* is $\mathcal{O}(n_g t_g + N\overline{L})$ [25]. $\mathcal{A}_c$ and $\mathcal{A}_g$ are equivalent representations of RCsore and ECsore, respectively. The CFA $\mathcal{A}_g$ can contain more nodes labeled $+_i$ ($i \in \mathbb{N}$) than the CFA $\mathcal{A}_c$. And the transitions in $\mathcal{A}_g$ can be also more than that in $\mathcal{A}_c$. Thus, $n_c t_c \leq n_g t_g$.

4 Experiments

In this section, we validate our algorithm on real-world XML data and generated XML data. We also provide evaluations of our algorithm in terms of generalization ability and time performance.

4.1 Data and Experiments

Table 3 demonstrates the practicability of RCsores, then we evaluate our algorithm on XML data. We obtained XML documents (*dblp-2018-04-01.xml*) conforming to DTD from DBLP Computer Science Bibliography corpus[5], from which we extracted the elements: inproc(eedings), article, phdth(esis), incolle(ction), and procee(dings). We obtained XML documents conforming to

---

[5] http://dblp.org/xml/release/.

XSD form Mondial corpus[6], from which the elements count(ry), provin (ce) and city are extracted. In order to validate on diverse XSDs, a number of real-world XSDs listed in Table 2 are searched from Google. However, we do not find the corresponding XML data, so we randomly generated them by using ToXgene[7]. The samples employed in the experiments are available at http://github.com/GraceFun/InfRCsore.

Table 1 lists the results of the learning algorithms *Soa2Sore*, *InfECsore* and *InfRCsore* on real-world XML data. Note that, based on descriptive general-ization, *Soa2Sore* is the first algorithm being used to infer a SORE [12], and *InfECsore* is the algorithm being applied to learn a most practical subclass of deterministic regular expressions with counting: ECsore [25]. For each of the elements inproc(eedings), article and procee(dings), the corresponding expression learnt by *InfRCsore* is not only more precise than the corresponding expres-sion in original DTD, but also more precise than the corresponding expression computed by *Soa2Sore*. Also, the result of *InfRCsore* is more general than the result of *InfECsore*, such that the learnt RCsore covers more XML data satis-fying the corresponding original DTD than the learnt ECsore. For phdth(esis) and incolle(ction), the learnt RCsores are identical to the corresponding expres-sions computed by *InfECsore*. For each of elements count(ry), provin(ce) and city, the result of *InfRCsore* and the result of *InfECsore* are the same, and the corresponding RCsore and ECsore both are more precise than the corresponding expression generated by *Soa2Sore* and the corresponding expression in original XSD.

Table 2 lists a number of the expressions extracted from real-world XSDs and the results of the learning algorithms *Soa2Sore*, *InfECsore* and *InfRCsore* on generated XML data. For ep1, the learnt RCsore is identical to the learnt ECsore, they both indicate that more symbols or subexpressions can have numerical occurrence constraints, but are allowed to occur more times by the nested coun-ters. For ep2, the learnt RCsore is identical to the learnt ECsore, they both are identical to the corresponding original XSD. This implies the original XSDs such as shown by ep2 could be precisely learnt by *InfRCsore* and *InfECsore*. For ep3 and ep4, although the learnt RCsores forbid the expressions learnt by *InfECsore*, which are more precise than the corresponding original XSD, even are identical to the corresponding original XSD for ep3, the learnt RCsores are more general than the learnt ECsores. Especially, for ep4, the learnt RCsore covers more XML data satisfying the corresponding original XSD than the learnt ECsore. For ep5, the learnt RCsore has the same higher nesting depth of counting operators with the learnt ECsore.

---

[6] http://www.dbis.informatik.uni-goettingen.de/Mondial/#XML.
[7] http://www.cs.toronto.edu/tox/toxgene/.

Table 1. Results of *Soa2Sore*, *InfECsore* and *InfRCsore* on real-world XML data. The left column gives element names, sample size for *Soa2Sore*, *InfECsore* and *InfRCsore*, respectively. The right column lists original DTD/XSD, the results of *Soa2Sore*, the results of *InfECsore* and the results of *InfRCsore*, respectively.

Element **Sample size**	Original segment of DTD/XSD Result of *Soa2Sore* Result of *InfECsore* Result of *InfRCsore*
inproc. 2153167 2153167 2153167	$(a\|b\|\cdots\|v)^*$ $(b^*(ck?)?(r\|a\|m)?(o\|(dj?)\|f\|n\|q\|e\|l)^*)^+$ $(b?(ck?)?((r\|a^{[1,45]}\|m^{[1,3]})^{[1,3]})?((o^{[1,87]}\|(dj?)\|f\|n\|q\|e\|l)^{[1,6]})?)^{[1,5]}$ $((b\|(ck?)\|r\|a^{[1,34]}\|m^{[1,3]}\|o^{[1,51]}\|(dj?)\|f\|n^{[1,2]}\|q\|e\|l\|)^{[5,11]})?$
article 1796920 1796920 1796920	$(a\|b\|\cdots\|v)^*$ $(b^*(((a^*(c\|e)?)\|m\|n\|q)(((j\|((f\|r)d?)\|h\|i)k?)\|p\|l)^*o^*)^+)$ $((b^{[1,5]})?((((a^{[1,69]})?(c\|e)?)^{[1,2]}\|m\|n\|q)^{[1,3]}((((j\|((f\|r)d?)\|h\|i)k?)^{[1,3]}\|p\|l)^{[1,3]})?$ $(o^{[1,116]})?)^{[1,3]})$ $(((b^{[1,5]})?(a^{[1,69]}\|c\|e\|q\|m^{[1,2]}\|n\|((j\|((f\|r)d?)\|h\|i)k?)^{[1,4]}\|p\|l\|o^{[1,116]})^{[1,9]})?$
phdth. 64943 64943 64943	$(a\|b\|\cdots\|v)^*$ $(a^*c(((((p\|(fk?)\|u)t?j?)\|e)(i\|l\|m\|s)^*)^+q?)$ $((a^{[1,3]})?c((e\|((u\|(fk?)\|p)t?j?))((s^{[1,3]}\|m^{[1,5]}\|l\|i)^{[1,3]})?)^{[1,5]}q?)$ $((a^{[1,3]})?c((e\|((u\|(fk?)\|p)t?j?))((s^{[1,3]}\|m^{[1,5]}\|l\|i)^{[1,3]})?)^{[1,5]}q?)$
procee. 58959 58959 58959	$(a\|b\|\cdots\|v)^*$ $(((a?(b\|c))^+h?)?(i\|s\|d)?(j\|q\|l\|(fr?)\|t\|e\|(pg?)\|m)^*)^*$ $((((a?(b\|c))^{[1,32]}h?)?((i\|s^{[1,2]}\|d)^{[1,2]})?((j\|q\|l\|(fr?)\|t\|e\|(pg?)\|m^{[1,3]})^{[1,5]})?)^{[1,4]})?$ $(((a?(b\|c))^{[1,32]}h?)\|i\|s^{[1,2]}\|d\|j\|q\|l\|(fr?)\|t\|e\|(pg?)^{[1,2]}\|(m^{[1,2]})^{[3,9]})?$
incolle. 46750 46750 46750	$(a\|b\|\cdots\|v)^*$ $(a^*c((d(j\|p)?)\|f\|r\|(ev?)\|l\|m)^*(o^+\|n\|q)?)$ $((a^{[1,49]})?c(((d(j\|p)?)\|f\|r\|(ev?)\|l\|m)^{[3,6]})?(o^{[2,104]}\|n\|q)?)$ $((a^{[1,49]})?c(((d(j\|p)?)\|f\|r\|(ev?)\|l\|m)^{[3,6]})?(o^{[2,104]}\|n\|q)?)$
count. 244 244 244	$(a^+b?c^*d?\cdots k?(l?\|m?)n?o^+p^*\cdots s^*(t^*\|u^*))$ $(ab?c^+(de?)?(f(g(hi)?)?j?k?)?(m?\|l)n?o^+p^*\cdots t^*u^+)$ $(ab?c^{[1,25]}(de?)?(f(g(hi)?)?j?k?)?(l\|m?)n?o^{[1,2]}(p^{[1,12]})?(q^{[1,8]})?(r^{[1,8]})?(s^{[1,16]})?$ $(t^{[1,2]})?u^{[1,306]})$ $(ab?c^{[1,25]}(de?)?(f(g(hi)?)?j?k?)?(l\|m?)n?o^{[1,2]}(p^{[1,12]})?(q^{[1,8]})?(r^{[1,8]})?(s^{[1,16]})?$ $(t^{[1,2]})?u^{[1,306]})$
provin. 1443 1443 1443	$(a^+b?c?d^*e^*)$ $(a^+b?c?d^*e^*)$ $(a^{[1,4]}b?c?(d^{[1,6]})?(e^{[1,5]})?)$ $(a^{[1,4]}b?c?(d^{[1,6]})?(e^{[1,5]})?)$
city 3383 3383 3383	$(a^+b?c?d?e?f^*g^*h^*)$ $(a^+b?(cde?)?f^*g^*h^*)$ $(a^{[1,5]}b?(cde?)?(f^{[1,10]})?(g^{[1,4]})?(h^{[1,3]})?)$ $(a^{[1,5]}b?(cde?)?(f^{[1,10]})?(g^{[1,4]})?(h^{[1,3]})?)$

Table 2. Results of *Soa2Sore*, *InfECsore* and *InfRCsore* on generated XML data.

Element	Original segment of XSD
	Result of *Soa2Sore*
Sample	Result of *InfECsore*
size	Result of *InfRCsore*
ep1	$((a|b|c|d|e|f)^{[1,10]})?$
941	$(a|b|c|d|e|f)^{+}$
941	$(a^{[1,3]}|b^{[1,4]}|c^{[1,3]}|d^{[1,4]}|$ $e^{[1,3]}|f^{[1,4]})^{[2,6]}$
941	$(a^{[1,3]}|b^{[1,4]}|c^{[1,3]}|d^{[1,4]}|$ $e^{[1,3]}|f^{[1,4]})^{[2,6]}$
ep2	$(a^{[10,20]}|b^{[30,40]})^{[3,5]}$
188	$(a|b)^{+}$
188	$(a^{[10,20]}|b^{[30,40]})^{[3,5]}$
188	$(a^{[10,20]}|b^{[30,40]})^{[3,5]}$
ep3	$(((a|b)?c?(d|e)?)^{[2,48]})$
988	$((a|b)?(d|e)?c?)^{+}$
988	$(((a|b)?(d|e)?c?)^{[2,48]})$
988	$(a|b|c|d|e)^{[6,45]}$
ep4	$(a?b?c?def?g?h?)^{[1,1000]}$
500	$(a?b?c?def?g?h?)^{+}$
500	$(a?b?c?(de)^{[1,10]}f?g?h?)^{[1,100]}$
500	$(a?b?c?def?g?h?)^{[1,597]}$
ep5	*None*
48	$(a|(b(c|d)^{+}))^{+}$
48	$((b(d^{[1,2]}|c^{[1,2]})^{[1,8]})^{[1,2]}|a^{[1,3]})^{[1,9]}$
48	$((b(d^{[1,2]}|c^{[1,2]})^{[1,8]})^{[1,2]}|a^{[1,3]})^{[1,9]}$

Table 3. Proportions of SOREs, ECsores, and RCsores.

Subclasses	% of XSDs
SOREs	80.74
ECsore	93.53
RCsore	89.45

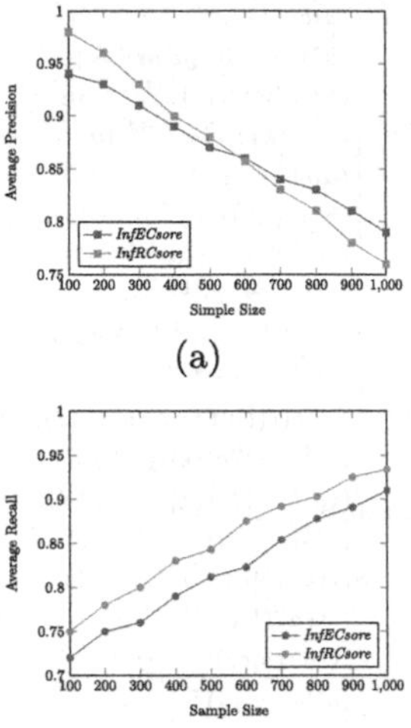

(a)

(b)

Fig. 3. (a) is average precision as a function of the sample size for each of *InfECsore* and *InfRCsore*. (b) is average recall as a function of the sample size for each of *InfECsore* and *InfRCsore*.

4.2 Performance

Generalization Abilities. Since the corresponding results of the algorithms *InfECsore* and *InfRCsore* have different generalization abilities for the same sample (such as ep3 and ep4 showed in Table 2), we evaluate the algorithms *InfECsore* and *InfRCsore* by computing the precision and recall. We specify that, the learnt expression with higher precision and recall has better generalization ability. The average precision and average recall, which are as functions of sample size, respectively, are the average values over 1000 expressions.

We randomly extracted the 1000 expressions from XSDs, which were grabbed from OGC XML Schema repository[8]. Each one of the 1000 expressions contains the counters, where the upper bounds are less than 100. To learn each extracted expression e_0, we randomly generated corresponding XML data by using ToX-gene, the samples are extracted from the XML data, each sample size is that

---

[8] http://schemas.opengis.net/.

listed in Fig. 3. And we define precision (p) and recall (r). Let positive sample (S_+) be the set of the all strings accepted by e_0, and let negative sample (S_-) be the set of the all strings not accepted by e_0. Let e_1 be the expression derived by *InfECsore* or *InfRCsore*. A true positive sample (S_{tp}) is the set of the strings, which are in S_+ and accepted by e_1. While a false negative sample (S_{fn}) is the set of the strings, which are in S_+ and rejected by e_1. Similarly, a false positive sample (S_{fp}) is the set of the strings, which are in S_- and accepted by e_1. While a true negative sample (S_{tn}) is the set of the strings, which are in S_- and rejected by e_1. Then, let $p = \frac{|S_{tp}|}{|S_{tp}|+|S_{fp}|}$ and $r = \frac{|S_{tp}|}{|S_{tp}|+|S_{fn}|}$. Note that, for an RCsore, we can construct an equivalent counter automata [14]. The constructed counter automata can decide whether the samples S_+ and S_- can be recognized or not, then we can obtain $|S_{tp}|$, $|S_{fp}|$ and $|S_{fn}|$.

As the sample size increases, compared with the results of *InfECsore*, the plots in Fig. 3(a) demonstrate that the precision for the expression learnt by *InfRCsore* is higher for a smaller sample, but is lower for a larger sample. However, the plots in Fig. 3(b) illustrate that, for any given sample, the recall for the expression learnt by *InfRCsore* is higher than that for the expression derived by *InfECsore*. The reason is that, for the same sample, the learnt RCsore can have more constrains than the learnt ECsore such that some subexpressions without counting operators. This will reduce that the learnt RCsore is expressive enough to cover more XML data. In summary, *InfRCsore* has better generalization ability for a smaller sample.

Time Performance. Although Theorem 3 implies that, for learning a RCsore, the algorithm *InfRCsore* can be faster than the algorithm *InfECsore*, the quantitative analyses of time performance about the algorithms *InfRCsore* and *InfECsore* should be given. Then, we present the evaluation about running time in different size of samples and different size of alphabets. Our experiments were conducted on a ThinkCentre M8600t-D065 with an Intel core i7-6700 CPU (3.4GHz) and 8G memory. And all codes were written in C++.

Table 4(a) shows the average running times in seconds for *InfRCsore* and *InfECsore* as a function of sample size, respectively. Table 4(b) shows the average running times in seconds for *InfRCsore* and *InfECsore* as a function of alphabet size, respectively. We still randomly extracted expressions from XSDs according to the above mentioned method. 1000 expressions of alphabet size 15 are chosen that, to learn each one of them, we randomly generated corresponding XML data by using ToXgene, the samples are extracted from the XML data, each sample size is that listed in Table 4(a). The running times listed in Table 4(b) are averaged over 1000 expressions of that sample size. Another 1000 expressions with distinct alphabet size listed in Table 4(b) are chosen that, to learn each one of them, we also randomly generated corresponding XML data by using ToXgene, the samples are extracted from the XML data, but the corresponding sample size is 1000. The running times listed in Table 4(a) are averaged over 1000 expressions of that alphabet size.

The running times of *InfRCsore* as compared with that of *InfECsore* are reported in Table 4(a). They show that *InfRCsore* is more efficient than *InfEC-sore* on large samples. However, Table 4(b) illustrates that the speed of *InfRCsore* varies widely when the alphabet size is over 20. Thus, the time performances of *InfRCsore* and *InfECsore* demonstrate that the algorithm *InfRCsore* is more efficient for processing large data sets.

Table 4. (a) and (b) are average running times in seconds for *InfRCsore* and *InfECsore* as the functions of sample size and alphabet size, respectively.

(a)

sample size	time(s) ($|\Sigma| = 15$)	
	InfRCsore	*InfECsore*
100	0.044	0.043
1000	0.052	0.079
10000	0.142	0.394
100000	0.989	3.488
1000000	11.389	21.22

(b)

alphabet size	time(s) ($|S| = 1000$)	
	InfRCsore	*InfECsore*
5	0.049	0.071
10	0.054	0.075
20	0.067	0.141
50	0.631	0.280
100	1.711	1.269

5   Conclusion

This paper proposed a restricted subclass of deterministic regular expressions with counting: RCsores and the corresponding learning algorithm. The main steps include learning a SORE, constructing an equivalent CFA, running the CFA to obtain an updated CFA, and converting the updated CFA to an RCsore. Compared with previous work, for any given finite language, our algorithm not only can learn a descriptive RCsore, which has higher recall for any sample, but also has better generalization ability for smaller sample, and is more efficient for processing larger sample. A future work is extending the SORE with counting, interleaving, and unorder concatenation, studying the practical issues and the learning algorithms.

References

1. Barbosa, D., Mignet, L., Veltri, P.: Studying the XML Web: gathering statistics from an XML sample. World Wide Web **9**(2), 187–212 (2006)
2. Bex, G.J., Gelade, W., Neven, F., Vansummeren, S.: Learning deterministic regular expressions for the inference of schemas from XML data. In: Proceedings of the 17th International Conference on World Wide Web, pp. 825–834. ACM (2008)
3. Bex, G.J., Gelade, W., Neven, F., Vansummeren, S.: Learning deterministic regular expressions for the inference of schemas from XML data. ACM Trans. Web **4**(4), 1–32 (2010)
4. Bex, G.J., Martens, W., Neven, F., Schwentick, T.: Expressiveness of XSDs: from practice to theory, there and back again. In: Proceedings of the 14th International Conference on World Wide Web, pp. 712–721. ACM (2005)
5. Bex, G.J., Neven, F., Van den Bussche, J.: DTDs versus XML Schema: a practical study. In: Proceedings of the 7th International Workshop on the Web and Databases: Colocated with ACM SIGMOD/PODS 2004, pp. 79–84. ACM (2004)

6. Bex, G.J., Neven, F., Schwentick, T., Tuyls, K.: Inference of concise DTDs from XML data. In: International Conference on Very Large Data Bases, Seoul, Korea, pp. 115–126, September 2006
7. Bex, G.J., Neven, F., Schwentick, T., Vansummeren, S.: Inference of concise regular expressions and DTDs. ACM Trans. Database Syst. **35**(2), 1–47 (2010)
8. Brüggemann-Klein, A., Wood, D.: One-unambiguous regular languages. Inf. Comput. **142**(2), 182–206 (1998)
9. Bui, D.D.A., Zeng-Treitler, Q.: Learning regular expressions for clinical text classification. J. Am. Med. Inform. Assoc. **21**(5), 850–857 (2014)
10. Che, D., Aberer, K., Özsu, M.T.: Query optimization in XML structured-document databases. VLDB J. **15**(3), 263–289 (2006)
11. Freydenberger, D.D., Kötzing, T.: Fast learning of restricted regular expressions and DTDs. In: Proceedings of the 16th International Conference on Database Theory, pp. 45–56. ACM (2013)
12. Freydenberger, D.D., Kötzing, T.: Fast learning of restricted regular expressions and DTDs. Theory Comput. Syst. **57**(4), 1114–1158 (2015)
13. Freydenberger, D.D., Reidenbach, D.: Inferring descriptive generalisations of formal languages. J. Comput. Syst. Sci. **79**(5), 622–639 (2013)
14. Gelade, W., Gyssens, M., Martens, W.: Regular expressions with counting: weak versus strong determinism. SIAM J. Comput. **41**(1), 160–190 (2012)
15. Gold, E.M.: Language identification in the limit. Inf. Control **10**(5), 447–474 (1967)
16. Hovland, D.: Regular expressions with numerical constraints and automata with counters. In: Leucker, M., Morgan, C. (eds.) ICTAC 2009. LNCS, vol. 5684, pp. 231–245. Springer, Heidelberg (2009). https://doi.org/10.1007/978-3-642-03466-4_15
17. Kilpeläinen, P., Tuhkanen, R.: Towards efficient implementation of XML Schema content models. In: Proceedings of the 2004 ACM Symposium on Document Engineering, pp. 239–241. ACM (2004)
18. Kilpeläinen, P., Tuhkanen, R.: One-unambiguity of regular expressions with numeric occurrence indicators. Inf. Comput. **205**(6), 890–916 (2007)
19. Latte, M., Niewerth, M.: Definability by weakly deterministic regular expressions with counters is decidable. In: Italiano, G.F., Pighizzini, G., Sannella, D.T. (eds.) MFCS 2015. LNCS, vol. 9234, pp. 369–381. Springer, Heidelberg (2015). https://doi.org/10.1007/978-3-662-48057-1_29
20. Lee, M., So, S., Oh, H.: Synthesizing regular expressions from examples for introductory automata assignments. In: ACM SIGPLAN Notices, vol. 52, pp. 70–80. ACM (2016)
21. Manolescu, I., Florescu, D., Kossmann, D.: Answering XML queries on heterogeneous data sources. In: VLDB, vol. 1, pp. 241–250 (2001)
22. Martens, W., Neven, F.: Typechecking top-down uniform unranked tree transducers. In: Calvanese, D., Lenzerini, M., Motwani, R. (eds.) ICDT 2003. LNCS, vol. 2572, pp. 64–78. Springer, Heidelberg (2003). https://doi.org/10.1007/3-540-36285-1_5
23. Mignet, L., Barbosa, D., Veltri, P.: The XML Web: a first study. In: Proceedings of the 12th International Conference on World Wide Web, pp. 500–510. ACM (2003)
24. Moreo, A., Eisman, E.M., Castro, J.L., Zurita, J.M.: Learning regular expressions to template-based FAQ retrieval systems. Knowl.-Based Syst. **53**, 108–128 (2013)
25. Wang, X., Chen, H.: Inferring deterministic regular expression with counting. In: Trujillo, J.C., et al. (eds.) ER 2018. LNCS, vol. 11157, pp. 184–199. Springer, Cham (2018). https://doi.org/10.1007/978-3-030-00847-5_15

26. Wang, X., Chen, H.: Learning a subclass of deterministic regular expression with counting. In: Douligeris, C., Karagiannis, D., Apostolou, D. (eds.) KSEM 2019. LNCS (LNAI), vol. 11775, pp. 341–348. Springer, Cham (2019). https://doi.org/10.1007/978-3-030-29551-6_29
27. Xie, Y., Yu, F., Achan, K., Panigrahy, R., Hulten, G., Osipkov, I.: Spamming botnets: signatures and characteristics. ACM SIGCOMM Comput. Commun. Rev. **38**(4), 171–182 (2008)

Generating Adversarial Examples
by Adversarial Networks
for Semi-supervised Learning

Yun Ma[1($\boxtimes$)], Xudong Mao[2], Yangbin Chen[1], and Qing Li[2]

[1] Department of Computer Science, City University of Hong Kong,
Hong Kong, China
mayun371@gmail.com, robinchen2-c@my.cityu.edu.hk
[2] Department of Computing, The Hong Kong Polytechnic University,
Hong Kong, China
xudong.xdmao@gmail.com, qing-prof.li@polyu.edu.hk

Abstract. Semi-Supervised Learning (SSL) has exhibited strong effectiveness in boosting the performance of classification models with the aid of a large amount of unlabeled data. Recently, regularizing the classifier with the help of adversarial examples has proven effective for semi-supervised learning. Existing methods hypothesize that the adversarial examples are based on the pixel-wise perturbation of the original samples. However, other types of adversarial examples (e.g., with spatial transformation) should also be useful for improving the robustness of the classifier. In this paper, we propose a new generalized framework based on adversarial networks, which is able to generate various types of adversarial examples. Our model consists of two modules which are trained in an adversarial process: a generator mapping the original samples to adversarial examples which can fool the classifier, and a classifier that tries to classify the original samples and the adversarial examples consistently. We evaluate our model on several datasets, and the experimental results show that our model outperforms the state-of-the-art methods for semi-supervised learning. The experiments also demonstrate that our model can generate adversarial examples with various types of perturbation such as local spatial transformation, color transformation, and pixel-wise perturbation. Moreover, our model is also applicable to supervised learning, performing as a regularization term to improve the generalization performance of the classifier.

Keywords: Semi-supervised learning · Adversarial networks · Adversarial examples

1 Introduction

Deep learning has launched a profound reformation in both supervised learning [12,13] and semi-supervised learning [23,25]. Basically, semi-supervised learning aims to improve the generalization performance of the classification models

Y. Ma and X. Mao—Contributed equally to this work.

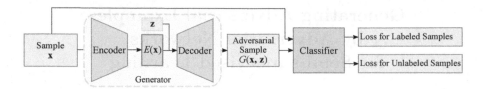

Fig. 1. Overview of the proposed framework.

based on a large amount of unlabeled data but a small amount of labeled data. Recently, there have been proposed two classes of promising techniques: (1) deep generative models based methods [3,15] and (2) perturbation based methods [17,24]. Deep generative models based methods benefit from the effectiveness of unsupervised learning models, capturing the underlying data distribution by training on the unlabeled data. On the other hand, the idea of perturbation based methods is to smooth the model prediction by forcing the classifier to output consistent results between the original samples and the corresponding perturbed variants.

Recently, virtual adversarial training (VAT) [20], a different perturbation based method, has been proposed and achieved great success in semi-supervised learning. Unlike previous methods [17,24] that apply random perturbation to the input data, VAT imposes the perturbation in the most adversarial direction. In particular, VAT first generates adversarial examples by perturbing the original samples to most greatly deviate the output distribution of the classifier, and then enforces the classifier to minimize the distributional divergence between prediction on the original samples and the corresponding adversarial examples. This improves the robustness of the classifier against the adversarial examples of both labeled and unlabeled data.

However, the objective function of VAT limits itself to only utilizing the adversarial examples with pixel-wise perturbation. Indeed, other types of adversarial examples (e.g., spatial transformation) should also be useful to improve the smoothness of the output distribution, since convolutional neural networks have been proven to be very sensitive to the spatial transformation [2,29].

To overcome this limitation, in this paper we propose a generalized framework of using adversarial examples for semi-supervised learning based on adversarial networks. Unlike VAT which limits the adversarial examples as the type of pixel-value based perturbation, our model is able to learn various adversarial examples with different types of perturbation. Inspired by the generative adversarial networks (GAN) [10], our model consists of two players, a generator and a classifier, which are trained in an adversarial manner: the generator aims to generate adversarial examples to fool the classifier, while the target of the classifier is to classify the original samples and the adversarial examples consistently. This framework is applicable to unlabeled data by making the classifier minimize the distance between the output distribution on the original samples and their corresponding adversarial examples. Minimizing this distance based on unlabeled data improves the generalization performance of the classifier for semi-supervised learning.

As Fig. 1 shows, the generator is designed as an encoder-decoder architecture, and the adversarial examples originate from the encoded latent space instead of the input space, allowing the model to learn different transformations in the latent space. Moreover, we concatenate a noise vector with the encoded vector, making it possible to output multiple adversarial examples for each input sample.

We present comprehensive experiments to show that our model outperforms VAT for both semi-supervised and supervised learning. The experimental results also demonstrate that our model is able to create adversarial examples with various types of perturbation such as local spatial transformation, color transformation, and pixel-wise perturbation. These adversarial examples are able to regularize the model from more diverse directions.

Our contributions in this paper can be summarized as follows:

- We propose a new semi-supervised learning framework by adversarially training a classifier and an adversarial example generator.
- We design an encoder-decoder architecture for the generator, allowing the model to learn various types of perturbation such as local spatial transformation and color transformation. These types of adversarial examples can further regularize the model for semi-supervised learning.
- We evaluate the proposed model on both synthetic data and three benchmark datasets, and the experimental results demonstrate that our model outperforms the state-of-the-art methods for both semi-supervised and supervised learning.

2 Related Work

2.1 Adversarial Examples

Neural network based machine learning models have been discovered to be vulnerable to adversarial examples [27]. By adding a carefully designed slight perturbation to the original sample, adversarial examples can mislead the model to make significantly different decisions. Several works have been dedicated to explaining the cause of adversarial examples. Initially, Szegedy et al. [27] conjectured the existence of adversarial examples is due to the high nonlinearity of the models. Later, Goodfellow et al. [11] introduced the linear hypothesis stating that the linear nature in high-dimensional spaces is the reason for the model's vulnerability to adversarial perturbations. Cisse et al. [5] stated that the robustness of models to adversarial examples is highly related with the Lipschitz constant of the networks. Gilmer et al. [9] argued the model behavior on adversarial examples is a natural result in the high-dimensional data manifold.

Some works are investigating how to craft adversarial examples to effectively attack classification models. Szegedy et al. [27] created an adversarial example for a given input by iteratively optimizing the objective function, i.e., minimizing the probability of correct class for the norm-constrained adversarial example, using the L-BFGS algorithm. Goodfellow et al. [11] introduced the fast gradient sign method based on their proposed linear hypothesis. The adversarial transformation network [4] generated targeted adversarial examples with a generator architecture whose objective is to minimize the L2 loss between the adversarial examples and the original samples. Xiao et al. [28] adopted a GAN to generate the adversarial perturbations.

In this paper, we employ the adversarial examples to facilitate semi-supervised learning, instead of pursuing the highest attack success rate and best perceptual similarity as in above works on adversarial example generation.

2.2   Deep Models for Semi-supervised Learning

Generative Models Based Methods. Recently, GAN [10] has achieved great success in generative models due to its capability of generating high-quality images [1,19,22], and has shown the effectiveness of applying to semi-supervised learning. Existing methods [8,25,26] adversarially optimized a generator and a classifier. In particular, the generator tries to generate realistic samples, while the classifier tries to correctly predict labels for true data and discriminate the true data from the generated ones. Kumar et al. [16] further penalized the variation of the classifier along the tangent directions around the real samples. However, Li et al. [18] found that combining the twos roles, classification and discrimination, may cause some incompatible problems. Thus they proposed to decouple the two roles to two independent networks to achieve better optimization of the classifier for semi-supervised learning. Moreover, Dai et al. [7] theoretically showed that good semi-supervised learning requires a bad generator, and proposed the complement generator, which generated samples in low-density regions, to further improve the generalization ability of the classifier.

Perturbation Based Methods. Perturbation based methods facilitate semi-supervised learning by encouraging the model to learn a decision boundary smooth enough in the local neighborhood of each data sample. Sajjadi et al. [24] proposed to minimize the difference between the network outputs on multiple passes of the same input sample, with each pass associated with a random transformation and perturbation. Laine et al. [17] proposed a similar Π model by enforcing the classifier to have consistent predictions for two input realizations with different random perturbation, and a temporal ensembling model pushing the current prediction to approach predictions from history. However, such isotropic smoothing over random perturbations is found to be insufficient to defend the perturbations in the adversarial direction [11,27]. Focusing on the adversarial examples, Goodfellow et al. [11] proposed the adversarial training, aiming at improving the model robustness by teaching the classifier to correctly classify both original training samples and their adversarial examples. Miyato et al. [20] further proposed VAT to extend the concept on unlabeled data. In particular, VAT smooths the output distribution (as in [17,24]) of the classifier by enforcing it to output similar distribution for pairs of the original sample and its corresponding virtual adversarial sample.

Our proposed model is related to both generative and perturbation based methods. Specifically, unlike other generative models trying to synthesize samples from underlying data distribution, the generator in our framework is to create the adversarial examples. Then the adversarial examples are utilized to regularize the smoothness of the classifier together with the original samples.

3 Method

In this section, we first present the problem definition, and then introduce some background of the current pixel-value based perturbation methods for semi-supervised learning. Lastly, we present our approach and discuss the advantages of our approach.

3.1 Problem Definition

We start with defining a set of notations. Let $\mathcal{D}_l = \{(x_i^l, y_i^l) | i = 1, \ldots, N_l\}$ and $\mathcal{D}_u = \{x_i^u | i = 1, \ldots, N_u\}$ be respectively a labeled dataset and an unlabeled dataset, where x_i denotes an input vector, and y_i denotes an output label. Our objective is to learn a classifier C based on $\mathcal{D}_l$ and $\mathcal{D}_u$, and we use $C(\cdot | x)$ to represent the output distribution of the classifier conditional on the input.

3.2 Current Pixel-Value Based Perturbation for SSL

Current state-of-the-art perturbation based method for semi-supervised learning is VAT [20]. VAT is based on the adversarial training model proposed in [11], which adds the pixel-wise perturbation in the most anisotropic direction that causes misclassification for the labeled data. VAT extends this to the unlabeled data by defining the current inferred labels as the virtual labels for the unlabeled data. Then VAT enforces the classifier to output similar distribution for pairs of the original sample and the corresponding adversarial sample. This can smooth the output distribution of the classifier, which in turn improves the robustness of the classifier against the adversarial examples of both labeled and unlabeled data.

The objective function of VAT is defined to minimize

$$\mathcal{L}_{\mathrm{nll}}(C) + \beta \mathcal{L}_{\mathrm{vat}}(C) \tag{1}$$

with

$$\begin{aligned}
\mathcal{L}_{\mathrm{nll}}(C) &= \mathbb{E}_{(x,y) \sim \mathcal{D}_l} \left[-\log C(y|x) \right], \\
\mathcal{L}_{\mathrm{vat}}(C) &= \mathbb{E}_{x \sim \mathcal{D}_l \cup \mathcal{D}_u} \left[D[C(\cdot|x), C(\cdot|x + r_{\mathrm{vadv}})] \right], \\
r_{\mathrm{vadv}} &= \underset{\|r\|_2 \leq \epsilon}{\arg\max}\, D\left[C(\cdot|x), C(\cdot|x + r) \right],
\end{aligned} \tag{2}$$

where $\mathcal{L}_{\mathrm{nll}}(C)$ is the typical negative log-likelihood loss for the labeded data, $\mathcal{L}_{\mathrm{vat}}(C)$ is the regularization term to penalize the inconsistency between the model predictions on x and its adversarial example $x + r_{\text{vadv}}$, β is a hyper-parameter trading off the two loss terms, $\epsilon > 0$ is the norm constraint for the adversarial perturbation r_{vadv}, and $D[p, q]$ is a non-negative function that measures the divergence between two distributions p and q.

From Eq. 2, we can see that r_{vadv} directly modifies pixel values. However, this will limit VAT to adversarial examples with pixel-wise perturbation. We argue that other types of adversarial examples (e.g., spatial transformation) should also be useful to improve the robustness of the classifier, since convolutional neural networks have been proved to be very sensitive to global transformations [2] or local transformations [29].

3.3   Our Approach

As Fig. 1 shows, our model consists of two players: a generator and a classifier. The original sample x is first mapped by an encoder-decoder architectured generator G to the adversarial sample $G(x, z)$, where z is a noise vector added to the encoded vector $E(x)$, making the generator able to generate multiple adversarial examples for each x. The generator and the classifier are trained in an adversarial process [10]: the generator tries to generate adversarial examples to fool the classifier into making different decisions with those on the original samples, while the target of the classifier is to classify the original samples and adversarial examples consistently. As a result, in addition to the basic negative log-likelihood loss for the labeled data, our model involves an adversarial loss and a reconstruction loss, as the following describes, to achieve the regularization with the generated adversarial examples.

Adversarial Loss. The adversarial loss, which is the most critical one in our model, defines the minimax game between the generator and the classifier. For both labeled and unlabeled data, the loss is defined based on some divergence methods D to measure the difference between the output distribution of the classifier C on the original samples and their corresponding adversarial examples: $D[C(\cdot|x), C(\cdot|G(x, z))]$. In particular, we enforce the classifier to minimize this divergence to achieve smooth output distribution in the local neighborhood around each training sample, while enforcing the generator to maximize this divergence to generate effective adversarial examples. Thus the adversarial loss can be formulated as

$$\min_{C} \max_{G} \mathcal{L}_{\text{adv}}(G, C) = \mathbb{E}_{x \sim \mathcal{D}_l \cup \mathcal{D}_u} \left[D[C(\cdot|x), C(\cdot|G(x, z))] \right], \tag{3}$$

where $D[p, q]$ is a non-negative function that measures the divergence between two distributions p and q. The adversarial loss does not require ground-truth labels of training samples, making it applicable to semi-supervised learning as well as supervised learning. One alternative choice is that the objective of the generator for labeled data can be formulated as maximizing the negative log-likelihood loss, however, we find this strategy does not provide further improvement on the performance in our preliminary experiments.

Algorithm 1. Minibatch stochastic gradient descent training of our model.

for number of training iterations do
- Sample minibatches of labeled samples from $\mathcal{D}_l$ and unlabeled samples from $\mathcal{D}_u$.
- Sample minibatches of noise samples from noise prior $p(z)$.
- Update the classifier by descending its stochastic gradient according to Equation 5.
- Update the generator by ascending its stochastic gradient according to Equation 5.

end for

Reconstruction Loss. Recall that the adversarial examples refer to slightly perturbed variants of original input samples. Therefore, the generated adversarial example is required to look similar to its corresponding original sample. We employ a reconstruction loss to impose such similarity. Formally, we can define the reconstruction loss as

$$\min_G \mathcal{L}_{\text{reconst}}(G) = \mathbb{E}_{x \sim \mathcal{D}_l \cup \mathcal{D}_u}[\|x - G(x, z)\|_2^2]. \tag{4}$$

Full Objective. Based on the adversarial loss and the reconstruction loss, we define the full objective of our model as

$$\min_C \max_G \mathcal{L}(G, C) = \mathcal{L}_{\text{nll}}(C) + \alpha \mathcal{L}_{\text{adv}}(G, C) - \lambda \mathcal{L}_{\text{reconst}}(G), \tag{5}$$

where $\mathcal{L}_{\text{nll}}(C) = \mathbb{E}_{(x,y) \sim \mathcal{D}_l}[-\log C(y|x)]$ is the typical negative log-likelihood loss for the labeled data. α and λ are used to control the weights of the adversarial loss and the reconstruction loss. In fact, λ is similar to the hyperparameter ϵ in VAT which constrains the norm magnitude of the perturbation. The training procedure of our model is illustrated in Algorithm 1.

Latent Space Based Adversarial Example Generation. To generate adversarial examples beyond the pixel-wise perturbation, we adopt an encoder-decoder architecture for the generator. In particular, we generate the adversarial examples from the encoded latent space, instead of the input space directly. This makes it possible to learn high-level transformations such as spatial transformation and color transformation. We achieve this by differentiating the responsibilities between the encoder and the decoder: the encoder is only responsible for mapping the samples to a latent space and the decoder is responsible for reconstructing the samples and producing the adversarial examples. Therefore, we associate the encoder only with $\mathcal{L}_{\text{reconst}}(G)$, while associate the decoder with the full loss of G. Moreover, we concatenate a noise vector z with the encoded vector $E(x)$, which makes the generator able to output multiple adversarial examples for each input sample. In our experiments (see Sect. 4.4), we show that our model is able to generate adversarial examples with spatial transformation, color transformation, or pixel-wise perturbation. Such diverse types of adversarial examples are expected to provide better regularization power.

Table 1. The network architectures of the adversarial example generator and classifier used on synthetic data. "fc." stands for "fully connected".

Generator	Classifier
2D input	
Encoder: fc. 10 tanh	fc. 100 ReLU
Decoder: fc. 2	fc. 50 ReLU
2D output	fc. 10 Softmax

4   Experiments

In this section, we first validate the proposed model on synthetic data for semi-supervised learning, and then compare our model with various strong baselines on three benchmark datasets, i.e., MNIST, SVHN, and CIFAR-10, for both semi-supervised and supervised learning. Finally, we visualize some adversarial examples generated by our model, which exhibit more diversities and thus explains the superiority of our model in semi-supervised learning.

4.1   Implementation Details

We implement our model with PyTorch [21]. The network architectures of the generator and the classifier used for the synthetic data and the three benchmark datasets are shown in Tables 1, 2, and 3. We adopt Adam optimizer [14] to update the model parameters. The batch sizes of labeled and unlabeled data are set to 100 and 300 respectively for MNIST, 32 and 128 for SVHN and CIFAR-10, and 8 and 500 for synthetic datasets. For the divergence measurement D in Eq. 3, we adopt L2 distance for MNIST and SVHN, and Kullback-Leibler divergence for CIFAR-10 and synthetic datasets. The weight for the adversarial loss α is set to 1.0 for synthetic datasets, MNIST and SVHN, and 2.0 for CIFAR-10. The weight for the reconstruction loss λ is set to 1.0 for synthetic datasets, 0.01 for MNIST, and 0.02 for SVHN and CIFAR-10. The learning rate is set to 0.001 for synthetic datasets, MNIST and SVHN, and 0.003 for CIFAR-10.

4.2   Semi-supervised Learning on Synthetic Data

We first evaluate our proposed model on two synthetic datasets to provide an intuitive explanation of our model. The synthetic data are based on two circles with two different radiuses, 0.2 and 0.5. We created two synthetic datasets by applying zero-mean Gaussian noise with two different standard deviations $\sigma = 0.01$ and $\sigma = 0.06$, as shown in the first column of Fig. 2. Each dataset contains 8 labeled data points and 1500 unlabeled data points. We abbreviate the two datasets as Circles (0.01) and Circles (0.06).

For both our model and VAT, we adopt identical network architecture for the classifier, which consists of three fully connected layers. Moreover, the generator of our model is a neural network with two fully-connected layers. Details of the network architectures are shown in Table 1. We fixed the weights for the

Table 2. The network architectures of the classifiers used on the MNIST, SVHN, and CIFAR-10 datasets. "fc." stands for "fully connected".

MNIST	SVHN	CIFAR-10
28×28 Gray Image	32×32 RGB Image	32×32 RGB Image
	3×3 conv. 128 lReLU $(\alpha) = 0.1$	
	3×3 conv. 128 lReLU $(\alpha) = 0.1$	
	3×3 conv. 128 lReLU $(\alpha) = 0.1$	
	2×2 stride 2 max-pool, dropout 0.5	
fc. 1200 ReLU	3×3 conv. 256 lReLU $(\alpha) = 0.1$	
fc. 600 ReLU	3×3 conv. 256 lReLU $(\alpha) = 0.1$	
fc. 300 ReLU	3×3 conv. 256 lReLU $(\alpha) = 0.1$	
fc. 150 ReLU	2×2 stride 2 max-pool, dropout 0.5	
	3×3 conv. 512 lReLU $(\alpha) = 0.1$	
	1×1 conv. 256 lReLU $(\alpha) = 0.1$	
	1×1 conv. 128 lReLU $(\alpha) = 0.1$	
	global average pool	
	fc. 10 Softmax	

Table 3. The network architectures of the adversarial example generators used on the MNIST, SVHN, and CIFAR-10 datasets. "fc." stands for "fully connected".

MNIST	SVHN	CIFAR-10
28×28 Gray Image	32×32 RGB Image	32×32 RGB Image
Encoder		
		3×3 conv. 16 ReLU
		3×3 stride 2 conv. 32 ReLU
fc. 1000 ReLU	4×4 stride 2 conv. 64 ReLU	3×3 stride 2 conv. 64 ReLU
fc. 500 ReLU	4×4 stride 2 conv. 128 ReLU	
	3×3 conv. 128 ReLU	Residual block with
		3×3 conv. 64 ReLU
		3×3 conv. 64
Decoder		
		Residual block with
		3×3 conv. 64 ReLU
fc. 500 ReLU	3×3 conv. 128 ReLU	3×3 conv. 64
fc. 1000 ReLU	4×4 stride 2 deconv. 64 ReLU	
fc. 784 tanh	4×4 stride 2 deconv. 3 tanh	3×3 stride 2 deconv. 32 ReLU
		3×3 stride 2 deconv. 16 ReLU
		3×3 conv. 3 tanh
28×28 Gray Image	32×32 RGB Image	32×32 RGB Image

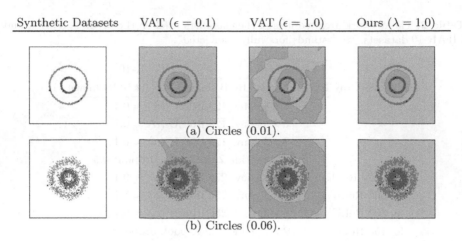

Synthetic Datasets   VAT ($\epsilon = 0.1$)   VAT ($\epsilon = 1.0$)   Ours ($\lambda = 1.0$)

(a) Circles (0.01).

(b) Circles (0.06).

Fig. 2. Comparison between VAT and our model on the Circles (0.01) and Circles (0.06) datasets. Green and orange points denote unlabeled data from two different classes, and labeled data are marked with black crosses. The first column shows the datasets, and the last three columns show the decision boundaries learned by different models for the two datasets. (Color figure online)

Table 4. Test performance of semi-supervised learning methods on Circles (0.01) and Circles (0.06).

Models	Test error rate (%)	
	Circles (0.01)	Circles (0.06)
VAT ($\epsilon = 0.1$)	**0.00** (± 0.00)	24.61 (± 4.58)
VAT ($\epsilon = 1.0$)	4.59 (± 5.64)	5.10 (± 4.28)
Ours ($\lambda = 1.0$)	**0.00** (± 0.00)	**4.75** (± 5.31)

regularization terms of VAT and our model (i.e., β in Eq. 1 and α in Eq. 5) as 1.0, and searched the optimal hyper-parameter ϵ in VAT and λ in our model for each dataset.

The results are shown in Fig. 2, and we have the following three major observations. First, VAT ($\epsilon = 0.1$) and VAT ($\epsilon = 1.0$) learn the decision boundary correctly on Circles (0.01) and Circles (0.06), respectively. However, VAT ($\epsilon = 0.1$) fails to learn on Circles (0.06), and VAT ($\epsilon = 1.0$) fails to learn on Circles (0.01). Second, our model is able to learn the decision boundary correctly on both datasets with the same $\lambda = 1.0$. This demonstrates the robustness of our model in hyper-parameters. Third, the decision boundary learned by VAT ($\epsilon = 1.0$) on Circles (0.06) contains a green area outside the red circle area. This is not reasonable although this decision boundary can classify the samples correctly. The reason for this phenomenon is that: The adversarial examples of the inner circle points will reach the outer green area when ϵ is large, and VAT

Table 5. Test performance of semi-supervised learning methods on MNIST, SVHN, and CIFAR-10, with 1000, 1000, and 4000 labeled data samples respectively. No data augmentation is utilized. Our results are averaged over 5 runs. "-" means the result is not reported by the corresponding paper.

Models	Test error rate (%)		
	MNIST	SVHN	CIFAR-10
TSVM [6]	5.38	-	-
Pseudo Ensembles Agreement [3]	2.87	-	-
Deep Generative Model [15]	2.40 (±0.02)	-	-
Ladder Networks [23]	**0.84** (±0.08)	-	20.4 (±0.47)
CatGAN [26]	1.73 (±0.18)	-	19.58 (±0.58)
ALI [8]	-	7.42 (±0.65)	17.99 (±1.62)
Improved GAN [25]	-	8.11 (±1.3)	18.63 (±2.32)
Triple GAN [18]	-	5.77 (±0.17)	16.99 (±0.36)
Π model [17]	-	5.43 (±0.25)	16.55 (±0.29)
FM-GAN+Jacob.-reg+Tangents [16]	-	4.39 (±1.2)	16.20 (±1.6)
GoodSSLwithBadGAN [7]	-	4.25 (±0.03)	14.41 (±0.03)
VAT [20]	1.27 (±0.11)	4.28 (±0.10)	13.15 (±0.21)
Our Model	1.17 (±0.10)	**3.93** (±0.07)	**12.97** (±0.10)

will classify this area as the same type of the inner circle points. In contrast, our model is able to learn the red area correctly.

To further verify the advantage of our model against VAT, we run 10 times for each model and report the average test error rate in Table 4. We can observe that VAT classify the samples correctly on Circles (0.01) and Circles (0.06) with $\epsilon = 0.1$ and $\epsilon = 1.0$, respectively, but they fail to learn on the other dataset. On the contrary, our model is able to classify the samples correctly on both datasets with the same $\lambda = 1.0$.

4.3 Semi-supervised and Supervised Learning on MNIST, SVHN, and CIFAR-10

We evaluate our model on three widely used benchmark datasets: MNIST, SVHN, and CIFAR-10. While our model is proposed for semi-supervised learning, it can be seamlessly applied to the supervised learning task. Therefore, we validate its performance on both semi-supervised and supervised learning on all three datasets. The network architecture of the classifier is the same as VAT [20] for fair comparison. Specifically, the classifier for MNIST is a neural network with five fully-connected layers. For SVHN and CIFAR-10, we follow the network architecture used in [17,18,20].

Table 6. Test performance of supervised learning methods on MNIST, SVHN, and CIFAR-10. Data augmentation has been applied on SVHN and CIFAR-10. Our results are averaged over 5 runs. "-" means the result is not reported by the corresponding paper.

Models	Test error rate (%)		
	MNIST	SVHN	CIFAR-10
Superivised-only	1.09 ($\pm$0.02)	2.79 ($\pm$0.08)	6.58 ($\pm$0.10)
Ladder Networks [23]	**0.57** ($\pm$0.02)	-	-
Π model [17]	-	2.54 ($\pm$0.04)	5.56 ($\pm$0.10)
Temporal Ensembling [17]	-	2.74 ($\pm$0.06)	5.60 ($\pm$0.10)
Adversarial Training [11]	0.78	-	-
RPT [20]	0.84 ($\pm$0.03)	-	6.30 ($\pm$0.04)
VAT [20]	0.64 ($\pm$0.05)	-	5.81 ($\pm$0.02)
Our Model	0.61 ($\pm$0.04)	**2.49** ($\pm$0.06)	**5.51** ($\pm$0.02)

Semi-supervised Learning. We randomly select 1000, 1000, and 4000 labeled data samples for MNIST, SVHN, and CIFAR-10 from the full training data samples (which include 60,000, 73,257, and 50,000 training samples for MNIST, SVHN, and CIFAR-10) and use the rest as unlabeled data. Following VAT [20], we also adopt an additional conditional entropy loss $\mathcal{L}_{ent}$:

$$\mathcal{L}_{ent}(C) = \mathbb{E}_{x \sim \mathcal{D}_l \cup \mathcal{D}_u} \Big[- \sum_y C(y|x) \log C(y|x) \Big], \qquad (6)$$

which has been proven helpful for semi-supervised learning.

Table 5 shows the test performance of our model against state-of-the-art methods on the three datasets. The most related work to our model is VAT [20] which also utilizes the adversarial examples for model regularization, and the other baselines are based on random perturbations or generative models. From Table 5, we have the following two major observations. First, our model outperforms VAT for all the three datasets. Second, our model also outperforms all the random perturbation based and generative model based methods, except Ladder Networks [23] on MNIST, which we conjecture is because the skip connections in Ladder Networks [23] can best fit the MNIST dataset, making the Ladder Networks the strongest method for MNIST.

Supervised Learning. Following [20], we apply data augmentation for SVHN using random translation and for CIFAR-10 using random translation and horizontal flip. The results for supervised learning are shown in Table 6. Similar to the semi-supervised learning, our model also outperforms all the baseline methods except Ladder Networks [23] on MNIST, which demonstrate that our model can also benefit supervised learning by regularizing the classifier with the generated adversarial examples.

Original Adversarial Original Adversarial

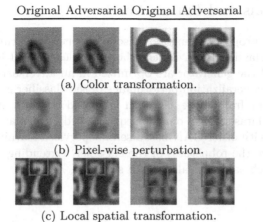

(a) Color transformation.

(b) Pixel-wise perturbation.

(c) Local spatial transformation.

Fig. 3. Selected types of perturbation on SVHN learned by our model.

Original Adversarial Original Adversarial

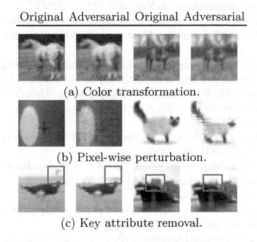

(a) Color transformation.

(b) Pixel-wise perturbation.

(c) Key attribute removal.

Fig. 4. Selected types of perturbation on CIFAR-10 learned by our model.

4.4   Visualization of Generated Adversarial Examples

As stated in the previous sections, our model is not restricted to the pixel-value based perturbation. Unlike VAT [20], we decouple the adversarial example generation from the classifier with our generator. In Figs. 3 and 4, we show some types of adversarial examples on SVHN and CIFAR-10 generated by our model in the learning process. We can observe that our model is able to generate various types of perturbation such as color transformation, local spatial transformation, as well as the pixel-wise perturbation. Our adversarial example types are more diverse than those in VAT (see Fig. 5 in [20]), which implies better regularization and explains our advantage over VAT in semi-supervised and supervised learning.

5   Conclusions and Future Work

In this paper, we propose a generalized semi-supervised learning framework by regularizing the classifier with adversarial examples yielded from a generator. The classifier and the generator are optimized in an adversarial manner, at the end of which the generalization performance of the classifier gets improved. The effectiveness of our framework has been validated on both synthetic datasets and benchmark datasets. For our subsequent study, we plan to integrate our proposed model with other GAN-based semi-supervised learning methods, so as to further improve the robustness of the classifier. Extending our framework to other domains such as text is also worth studying.

References

1. Arjovsky, M., Chintala, S., Bottou, L.: Wasserstein GAN. arXiv:1701.07875 (2017)
2. Azulay, A., Weiss, Y.: Why do deep convolutional networks generalize so poorly to small image transformations? arXiv:1805.12177 (2018)
3. Bachman, P., Alsharif, O., Precup, D.: Learning with pseudo-ensembles. In: Proceedings of the Neural Information Processing Systems (NeurIPS), pp. 3365–3373 (2014)
4. Baluja, S., Fischer, I.: Learning to attack: adversarial transformation networks. In: Association for the Advancement of Artificial Intelligence (AAAI) (2018)
5. Cisse, M., Bojanowski, P., Grave, E., Dauphin, Y., Usunier, N.: Parseval networks: improving robustness to adversarial examples. In: International Conference on Machine Learning (ICML) (2017)
6. Collobert, R., Sinz, F.H., Weston, J., Bottou, L.: Large scale transductive SVMs. J. Mach. Learn. Res. (JMLR) (2006)
7. Dai, Z., Yang, Z., Yang, F., Cohen, W.W., Salakhutdinov, R.: Good semi-supervised learning that requires a bad GAN. In: Advances in Neural Information Processing Systems (NeurIPS) (2017)
8. Dumoulin, V., et al.: Adversarially learned inference. In: International Conference on Learning Representations (ICLR) (2017)
9. Gilmer, J., et al.: Adversarial spheres. arXiv:1801.02774 (2018)
10. Goodfellow, I., et al.: Generative adversarial nets. In: Advances in Neural Information Processing Systems (NeurIPS), pp. 2672–2680 (2014)
11. Goodfellow, I.J., Shlens, J., Szegedy, C.: Explaining and harnessing adversarial examples. In: International Conference on Learning Representations (ICLR) (2015)
12. He, K., Gkioxari, G., Dollár, P., Girshick, R.: Mask R-CNN. In: International Conference on Computer Vision (ICCV) (2017)
13. He, K., Zhang, X., Ren, S., Sun, J.: Deep residual learning for image recognition. In: Computer Vision and Pattern Recognition (CVPR) (2016)
14. Kingma, D.P., Ba, J.: Adam: a method for stochastic optimization. In: Proceedings of the International Conference on Learning Representations (ICLR) (2015)
15. Kingma, D.P., Rezende, D.J., Mohamed, S., Welling, M.: Semi-supervised learning with deep generative models. In: Advances in Neural Information Processing Systems (NeurIPS) (2014)
16. Kumar, A., Sattigeri, P., Fletcher, T.: Semi-supervised learning with GANs: manifold invariance with improved inference. In: Proceedings of the Neural Information Processing Systems (NeurIPS), pp. 5534–5544 (2017)

17. Laine, S., Aila, T.: Temporal ensembling for semi-supervised learning. In: International Conference on Learning Representations (ICLR) (2017)
18. Li, C., Xu, K., Zhu, J., Zhang, B.: Triple generative adversarial nets. In: Advances in Neural Information Processing Systems (NeurIPS) (2017)
19. Mao, X., Li, Q., Xie, H., Lau, R.Y., Wang, Z., Smolley, S.P.: Least squares generative adversarial networks. In: International Conference on Computer Vision (ICCV) (2017)
20. Miyato, T., Maeda, S.I., Koyama, M., Ishii, S.: Virtual adversarial training: a regularization method for supervised and semi-supervised learning. IEEE Trans. Pattern Anal. Mach. Intell. (2018)
21. Paszke, A., et al.: Automatic differentiation in PyTorch. In: Proceedings of the Advances in Neural Information Processing Systems (NeurIPS) Workshop (2017)
22. Radford, A., Metz, L., Chintala, S.: Unsupervised representation learning with deep convolutional generative adversarial networks. arXiv:1511.06434 (2015)
23. Rasmus, A., Valpola, H., Honkala, M., Berglund, M., Raiko, T.: Semi-supervised learning with ladder networks. In: Advances in Neural Information Processing Systems (NeurIPS) (2015)
24. Sajjadi, M., Javanmardi, M., Tasdizen, T.: Regularization with stochastic transformations and perturbations for deep semi-supervised learning. In: Advances in Neural Information Processing Systems (NeurIPS) (2016)
25. Salimans, T., et al.: Improved techniques for training GANs. In: Advances in Neural Information Processing Systems (NeurIPS), pp. 2226–2234 (2016)
26. Springenberg, J.T.: Unsupervised and semi-supervised learning with categorical generative adversarial networks. In: International Conference on Learning Representations (ICLR) (2016)
27. Szegedy, C., et al.: Intriguing properties of neural networks. In: International Conference on Learning Representations (ICLR) (2014)
28. Xiao, C., Li, B., Zhu, J.Y., He, W., Liu, M., Song, D.: Generating adversarial examples with adversarial networks. In: International Joint Conference on Artificial Intelligence (IJCAI) (2018)
29. Xiao, C., Zhu, J.Y., Li, B., He, W., Liu, M., Song, D.: Spatially transformed adversarial examples. In: International Conference on Learning Representations (ICLR) (2018)

Deep Learning

Dual Path Convolutional Neural Network for Student Performance Prediction

Yuling Ma[1,2], Jian Zong[1], Chaoran Cui[3(✉)], Chunyun Zhang[3], Qizheng Yang[1], and Yilong Yin[1(✉)]

[1] School of Software, Shandong University, 250100 Jinan, China
mayuling@mail.sdu.edu.cn, ylyin@sdu.edu.cn
[2] School of Information Engineering, Shandong Yingcai College, 250104 Jinan, China
[3] School of Computer Science and Technology,
Shandong University of Finance and Economics, 250014 Jinan, China
crcui@sdufe.edu.cn

Abstract. Student performance prediction is of great importance to many educational domains, such as academic early warning and personalized teaching, and has drawn numerous research attention in recent decades. Most of the previous studies are based on students' historical course grades, demographical data, in-class study performance, and online activities from e-learning platforms, e.g., Massive Open Online Courses (MOOCs). Thanks to the widely used of campus smartcard, it supplies an opportunity to predict students' academic performance with their off-line behavioral data. In this study, we seek to capture three student behavioral characters, including duration, variation and periodicity, and predict students' performance based on the three types of information. However, it is highly challenging to extract efficient features manually from the huge amount of raw smartcard records. Besides, it is not trivial to construct a good predictive model for some majors with limited student samples. To address the above issues, we develop a novel end-to-end deep learning method and propose Dual Path Convolutional Neural Network (DPCNN) for student performance prediction. Moreover, we introduce multi-task learning to our method and predict the performance of students from different majors in a unified framework. Experimental results demonstrate the superiority of our approach over the state-of-the-art methods.

Keywords: Student performance prediction · Campus behavior · Convolutional Neural Networks (CNN) · Multi-task learning

1 Introduction

As one of the most popular topics in educational data mining, student performance prediction plays a crucial role in many educational domains, e.g., student academic early warning and personalized teaching [1–3]. For example, based on the results of a predictive model, the instructor can provide personalized

© Springer Nature Switzerland AG 2019
R. Cheng et al. (Eds.): WISE 2019, LNCS 11881, pp. 133–146, 2019.
https://doi.org/10.1007/978-3-030-34223-4_9

intervention and guidance to improve student learning, especially for those low-performance students [2]. In recent decades, extensive research effort has been devoted to student performance prediction [4]. Owing to the convenience of collecting data, a large portion of studies focus on e-learning platforms, e.g., MOOCs [5,6], and predict students' performance based on online study activity logs. However, these data concerned with online activities is hardly captured in off-line learning scenarios. With students' historical course grades, demographical data, and their study records on target course (i.e., the course to be predicted), the other series of researches construct predictive models by direct use of part or whole of the aforementioned data [2,3,7]. However, these studies generally suffered from limited efficient features/predictors.

Thanks to the development of information technology, campus smartcards are widely used in colleges, which record about students' campus activities in an unobtrusive way. It supplies an opportunity to predict students' academic performance from the new perspective of campus behaviors. Recent studies illustrate that such real-time digital records generally can reveal some behavioral factors correlated with student academic performance [8,9]. Intuitively, a certain swiping card behavior of a student may reflect an incident that happened to him/her, e.g., swiping smartcards in the library, campus supermarket and dormitory generally means studying, shopping and relaxing, respectively. Each implicit incident may be correlated with students' academic performance, e.g., if a student spends long time in library, there is a very high probability that he/she is a diligent student and will achieve good academic performance.

Motivated by the aforementioned analysis, in this paper, we aim to model three behavioral characters, including duration, variation and periodicity, and construct predictive models for students' performance with the three types of information. Besides, considering inconsistent course settings across majors, we view constructing predictive model for different majors as different tasks. However, it is highly challenging to construct good predictive models owing to the two following issues:

(1) Traditional handcrafted features are highly dependent upon human experts and domain knowledge, and it is thus highly challenging to extract efficient features manually from huge amount of raw smartcard records.
(2) The number of students varies from major to major. For the majors with limited student samples, it is not trivial challenging to train a good predictive model.

To address the above issues, in this study, we exploit a novel end-to-end deep learning approach to predict student academic performance. Recent results indicate that the implicit features extracted from the Convolutional Neural Networks (CNN) are very efficient [10], and it has been empirically illustrated that CNN has powerful ability to hierarchically capture the spatial structural information [11]. Benefiting from these findings, we employ CNN to learn features from the raw smartcard records, and propose a Dual Path CNN method, called DPCNN, to model the aforementioned three behavioral characters. Specifically, we represent students as tensors by direct use of raw records, each dimension of which denote time, location and date of swiping card behaviors, respectively. Then two types of filters are designed according to the size of student tensor, and utilized in the dual path network to model duration and variation, respectively. By taking the date axis as the depth of convolutions, periodicity can be modeled. Besides, given limited student samples in some majors, we introduce multi-task learning [12] to our method. Through the shared convolutional layers followed by task-specific fully-connected networks, predictive models for different majors can be trained in a unified framework. In this way, the problem of data scarcity can be alleviated. Our contributions are four-fold:

- Instead of extracting features manually, we exploit end-to-end learning style to predict students' performance. To the best of our knowledge, such a deep learning approach for student performance prediction in traditional teaching scenarios has not been previously reported.
- We propose a dual path CNN method, which is comprised of dual path convolutions followed by three-layer fully-connected networks, and three aforementioned behavioral characters can be modeled based on well-designed filters.
- We construct predictive models for different majors simultaneously following the idea of multi-task learning. Benefiting from relatedness between majors, the problem of data scarcity can be effectively alleviated.
- Experimental results demonstrate the superiority of our approach over the state-of-the-art methods.

In the following, we will briefly review related works, then the proposed method DPCNN is detailed in Sect. 3. We report experimental results and analysis in Sect. 4, followed by the conclusion and future work in Sect. 5.

2   Related Work

As one of the most important research branches of educational data mining, there has been a large body of work on student performance prediction in recent decades. Owing to the convenience of collecting data, many efforts have been devoted to predicting performance based on online activity logs from e-learning platforms, including MOOCs [5,6,13–17], Intelligent Tutoring Systems (ITS) [18–20], Learning Management Systems (LMSs) [1,21–24], Hellenic Open University (HOU) [25,26], and other platforms [27–31]. For example, Ren et al. predicted grades using data from MOOC server logs, such as the average number of daily study sessions, total video viewing time, number of videos a student watches, and number of quizzes [6]. Macfadyen et al. developed predictive models of student final grades based on LMS tracking data, including the number of discussion messages posted, number of mail messages sent, and number of assessments completed [23]. Zafra et al. predicted students' performance (i.e., pass or fail) with the information about quizzes, assignments and forums stored in Moodle, which is a free learning management system [24]. As can be seen, the above studies for e-learning platforms have mainly relied on the data about students' online activities, which is hardly accessed in off-line study scenarios. Another line of studies utilized students' demographical data, in-class study performance, and their past course grades to construct predictive models for student performance [2,3,7,32–36]. To name a few, Huang et al. predicted students' final grades for a course based on scores in three mid-term exams and grades in four pre-requisite courses [2]. Meier et al. predicted students' final grades based on the performance assessments on homework assignments, mid-term exam, course project, and final exam [3]. Ma et al. predicted students' performance prior to a course's commencement with their historical course grades as well as course description [7]. Marbouti et al. utilized the in-class performance factors, including grades for attendance, quizzes, and weekly homework, to predict at-risk students [32]. However, these researches generally suffer from limited efficient features.

Recently, there is a growing trend to predict students' performance based on their behavioral data, which is instantly recorded in campus smartcards [8,9]. In [8], the authors extracted two high-level behavioral characters, including orderliness and diligence, to predict students' GPA ranking. In [9], besides orderliness and diligence, two more factors, i.e., sleeping pattern and friend factors, were extracted to construct predictive models. However, these features are extracted in a manual way, which are highly dependent upon human ingenuity and prior knowledge.

3   Framework

In this section, we first represent student samples as tensors based on their smartcard records, then the DPCNN framework is proposed, followed by multi-task learning and the implementation Details.

3.1   Student Representation

In our dataset, smartcard records cover the period from September 01, 2013 to August 31, 2015 (i.e., 730 days totally), and the time of swiping card in each day varies from 6am to 12 pm (i.e., 18 h totally), which may occur at 12 campus places, e.g., the library, canteen and dormitory. As aforementioned, instead of extracting features manually, we seek to learn representations for student samples with deep learning methods. Therefore, we denote a student sample as a tensor $X \in R^{t \times l \times d}$ by using the raw records directly. Here, t denotes the number of time intervals that a day is split into, l denotes the number of campus places where swiping card behaviors may occur, and d denotes the number of days during the period covered by smartcard records. If a student X has a record of swiping smartcard at the jth campus place in the ith time interval of the kth day, X_{ijk} equals 1, otherwise it equals 0. As can be seen easily, in this study, the value of l and d is 12 and 730, respectively. Additionally, we divide the swiping card period in a day into 18 time bins, each of which spans 1 h. Thus, t is equal to 18. With tensors, we can analysis students' behavioral data from multiple views, i.e., temporal dimension, spatial dimension and periodic dimension.

3.2   Dual Path CNN

As aforementioned, CNN has powerful ability to hierarchically capture the spatial structural information. We thus choose CNN as the backbone to construct our framework. In this part, we first analyze how to adopt convolutional operations to model the behavioral characters, and then details the proposed method.

Given a student tensor $X \in R^{t \times l \times d}$, we attempt to employ filters of different size to model different behavioral characters. They are as follows: (1) filters of size $\alpha \times l$ with taking the date axis as the depth of convolutions. Here, l equals to the width of the tensor, which denotes the number of campus places (i.e., 12 in this work), and $\alpha \leq t$ is a hyperparameter, which means how many time intervals can be observed per convolution. Trough convolutional operations upon the tensor X along the axis of time intervals, swiping smartcard behaviors at different time in a day can be observed, and the changing patterns of these behaviors in temporal domain can be captured. In this way, the behavioral character *duration* can be modeled; (2) filters of size $t \times 1$ with taking the date axis as the depth of convolutions. Here, t is consistent with that in $X \in R^{t \times l \times d}$, both of which equal the number of time intervals a day split into (i.e., 18 in this study), and the width of such filters is set to be one. The reason is that we consider only one campus place per convolutional operation. Similarly, with the proceeding of convolutions along the axis of location, swiping card behaviors at different places can be observed, and the changing patterns of these behaviors in spatial domain can be captured. In this way, the behavioral character *variation* can be modeled. Additionally, the depth of both the above-mentioned convolutions are the date axis, and thus periodicity of campus behaviors can be modeled.

Based on the above analysis, we propose a novel Dual Path Convolutional Neural Network, called DPCNN, to model the aforementioned high-level behavioral characters. Figure 1 presents the architecture of DPCNN, which is comprised of dual path convolutional layers followed by a three-layer fully-connected neural network. The aforementioned two types of filters are exploited as a start in dual path convolutions, i.e., filters of size $\alpha \times l$ for the top path (i.e., Path1) to model duration, and filters of size $t \times 1$ for the bottom path (i.e., Path2) to model variation. Here, l and t is equal to 12 and 18, respectively, as aforementioned. The hyperparameter α is empirically set to be 3 based on the data used in our study. In order to fetch more information, more filters of size $\alpha \times l$ / $t \times 1$ can be adopted in the Path1/Path2 as the first-level convolutional layer. Besides, it was empirically observed that layerwise stacking of convolutions often yielded better representations [37]. Thus more levels of convolutional layers are employed in the dual path structure. Due to limited storage, the first level of convolutions in dual path both exploit 64 filters, which is followed by three more levels of convolutions with 128, 256, 512 filters, respectively. Finally, the output features generated from path1 and features from path2 are both rearranged to be feature

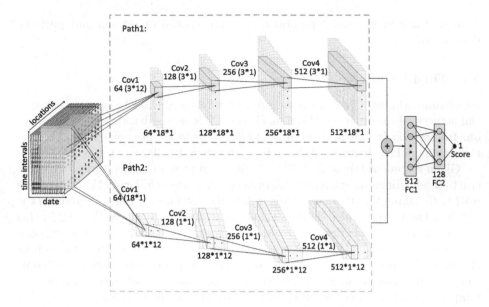

Fig. 1. DPCNN architecture. Cov: Convolution; ⊕: Concatenation operation; FC: Fully-connected

vectors, which are concatenated and then fed into a three-layer fully-connected network to make predictions.

In our study, the labels are the ranking of students' accumulated Grade Point Average (GPA), and thus we follow the idea of learning-to-rank to train our model. Formally, we denote each pairwise comparison by a triple $(X_i; X_j; y)$, where $X_i, X_j \in R^{t \times l \times d}$ are two student samples, and $y \in \{+1, -1\}$ is a label. $y = +1$ denotes that the former student (i.e., X_i) is ahead of the latter X_j's ranking, and $y = -1$ means the reverse. We denote the dataset consisted of n pairwise comparisons as $D = \left\{ \left(X_i^k, X_j^k, y^k \right) \right\}_{k=1}^n$. The goal of our task is to learn a mapping function $f(X) \rightarrow R$ that can give the predictive real value for each student sample. The desired mapping function is obtained by minimizing the hinge loss function as follows:

$$L = \sum_{(X_i, X_j, y) \in D} max(0, \ y(f(X_j) - f(X_i)) + 1) \tag{1}$$

3.3 Multi-task Learning

The data used in this study are behavioral records of 8199 undergraduate students from 19 majors, and the label of a student is his or her ranking based on students' GPA in his or her major. Due to inconsistent course settings across majors, it may be irrational to construct a common predictive model with mixing data of different majors brutally. Therefore, we view constructing models for different majors as different tasks. However, the number of students varies across majors from 100 to 600. For the majors with limited students, it is highly challenging to train a good predictive model. Multi-task learning is an empirically good solution, which can train classifiers for multiple related tasks simultaneously [12]. Though we view constructing classifiers for different majors as different tasks, these tasks may be correlated owing to the similarity of courses and teaching styles from similar majors, such as computer science and electronic engineering [9]. We thus introduce multi-task learning to our framework. Hard parameter sharing is the most commonly used approach to multi-task learning in neural networks, which generally applied by sharing the hidden layers between all tasks, while keeping several task-specific output layers [38]. Motivated by this, we let all the tasks share representation learning layers (i.e., the dual path convolutional layers of DPCNN) while remaining the final three-layer fully-connected networks task-specific. Extension experiments illustrate the appealing effectiveness of multi-task learning, which is reported later.

3.4 Implementation Details

The python libraries, including "torch" and "torchvision", are used to build our network. As aforementioned, in our study, student samples are denoted by tensors of size $18 \times 12 \times 730$. Thus, filters of size $\alpha \times 12$ and 18×1 are utilized to model duration and variation, respectively. The hyperparameter α is empirically set to be 3. Additionally, we take the axis of date as the depth of convolutions to model periodicity. With the limited storage, we utilize 64 filters to fetch more information in the first-level convolutions, and three more levels of convolutions are exploited to yield a better representation, with 128, 256 and 512 filters, respectively. More specifically, in DPCNN, the top path (i.e., Path1) starts with a 64 filters of size 3×12 convolutional layer, followed by three levels of convolutional layers with 128, 256 and 512 filters of size 3×1, respectively. Similarly, the Path2 starts with a 64 filters of size 18×1 convolutional layer, followed by three levels of convolutional layers with 128, 256 and 512 filters of size 1×1, respectively. In our framework, the output of the fully connected layer is a single real value corresponding to the predictive score of a student. The stride of convolutions is 1. The network can be effectively optimized with the Adam method.

4 Experiments

In this section, we first details the data used in our study, and then introduce two performance measures. Finally, the proposed framework DPCNN are compared with the state-of-the-art methods, and the effects of dual path structure and multi-task learning are studied.

4.1 Data Description

The data used in this paper has been publicly accessed on a national undergraduate big data competition platform. It consists of 13,700,000 smartcard records of 8199 undergraduate students from 19 majors, which cover the period from September 01, 2013 to August 31, 2015. These data records a large volume of students' campus behaviors, such as paying for meals, entering/exiting the dormitories, and entering the library. These behaviors may occur at 12 different campus places, including cafeterias, campus supermarket, library, dormitory, laundry room, campus bathroom, boiled water room, printing center, academic affairs office, school bus, campus hospital, and card center. Besides, academic performance data is also supplied, which denotes a student's GPA ranking in his or her major. Following the idea of learn-to-rank, the input samples are pairs of student. In the implements, due to limited computational resources, we randomly select approximately 200,000 pairs of student. 70% and the remained 30% are utilized to train and test our model, respectively.

4.2   Evaluation Metrics

Since we follow the idea of learning-to-rank to train our model, we exploit Spearman's rank correlation coefficient [39], which is one of most important ranking-based methods, to evaluate the performance of the proposed method. Spearman coefficient can measure the correlation between the predicted rank and the actual rank, which can be defined as

$$\rho = 1 - \frac{6 \sum_{i=1}^{m} (\hat{r}(X_i) - r(X_i))^2}{m(m^2 - 1)} \tag{2}$$

where m is the number of students under consideration, $\hat{r}(X_i)$ and $r(X_i)$ are the predicted rank and the actual rank of the student X_i, respectively. The higher the Spearman coefficient, the better the prediction performance.

Besides Spearman coefficient, we also care about accuracy of our model. Given a data set $D = \{(X_i^k, X_j^k, y^k)\}_{k=1}^{n}$, the accuracy is defined as below.

$$acc = \frac{1}{n} \sum_{k=1}^{n} \mathbb{I}(\hat{y}^k = y^k) \tag{3}$$

Here for predicate π, $\mathbb{I}(\pi)$ equals 1 if π holds and 0 otherwise, $\hat{y}^k$ and y^k are the output label and the actual label, respectively.

4.3   Performance Comparison with State-of-the-Art Methods

In this part, to demonstrate the effectiveness of the proposed network, we compare the DPCNN model against the following two state-of-the-art methods for student performance prediction.

- **RankNet**, a well-known supervised learning to rank algorithm, was utilized to predict the ranking of students with two high-level behavioral characters including orderliness and diligence [8]. The two features were extracted from smartcard records in a manual way. Specifically, the authors calculated orderliness based on two behaviours: taking showers in dormitories and having meals in cafeterias, and roughly estimated diligence based on two other behaviours: entering/exiting the library and fetching water in teaching buildings. More details can be found in [8].

- **MTLTR-APP** is a multi-task predictive framework based on a learning-to-rank algorithm proposed in [9]. This method took both the difference of majors and the difference of semesters into account, and considered constructing predictive models for students' performance in different semesters as different tasks, even if the students came from the same major. MTLTR-APP can thus capture inter-semester correlation, inter-major correlation with constraints upon model parameters. Three handcrafted behavior features (i.e., orderliness, diligence, and sleep pattern) as well as student similarity, were employed to predict student performance.

Table 1. Comparison of DPCNN with the state-of-the-art methods

Methods	acc (Accuracy)	ρ (Spearman coefficient)
RankNet	0.5980	0.2800
MTLTR-APP	0.6012	0.2905
DPCNN	**0.7658**	**0.6964**

We implement the two above methods on our dataset, to demonstrate the effectiveness of the proposed method. It is necessary to mention that in our dataset, the label information is the ranking based on students' accumulated GPA rather than GPA in each semester. Inter-semester correlation is thus discarded when we implement the MTLTR-APP method. Table 1 shows the performance of DPCNN and the two above methods. It can be observed that the proposed DPCNN has the highest accuracy as well as Spearman coefficient. Compared with the RankNet method, the proposed DPCNN further improves the accuracy and Spearman coefficient by an absolute value 16.78% and 41.64%, respectively. Likewise, DPCNN also makes large improvements against MTLTR-APP, i.e. 16.46% on accuracy and 40.59% on Spearman coefficient. The better results demonstrate the proposed dual path network is capable of learning better representation from huge amount of raw smartcard records, compared with handcrafted features used in the RankNet and MTLTR-APP approaches.

4.4 Effect of Dual Path Structure

In order to demonstrate the appealing effectiveness of the dual path architecture, we intentionally design two kinds of networks, which merely owns the top path (i.e., the bottom path is discarded) and the bottom path (i.e., the top path is discarded), respectively. For convenience, we denote them as Single-Path1 network and Single-Path2 network, respectively. The results are reported in Table 2.

Table 2. Comparison of DPCNN with single path networks

Methods	acc (Accuracy)	ρ (Spearman coefficient)
Single-Path1 network	0.7540	0.6353
Single-Path2 network	0.7326	0.5990
DPCNN	**0.7658**	**0.6964**

As can be seen from Tabel 2, the DPCNN is obviously superior to both Single-Path1 network and Single-Path2 network on the two evaluation metrics. In particularly, the DPCNN obtains the Spearman coefficient of 0.6964%, which makes large improvements, i.e. 6.11% compared with Single-Path1 network and 9.74% compared with Single-Path2 network. The reason may be that Single-Path1 network merely utilize filters of size 3×12, and it thus only can capture the two behavioral characters including duration and periodicity, i.e., variation is missing. Likewise, Single-Path2 network merely utilize filters of size 18×1, and it thus can capture variation and periodicity, but lose duration. Benefitting from the dual path structure, DPCNN can capture all of the three behavioral characters, and thus make a better prediction.

4.5 Contribution of Multi-task Learning

As aforementioned, we take student performance prediction for different majors as different tasks due to the consistent course settings across majors. To alleviate the issue of data scarcity, we follow the idea of multi-task learning and construct predictive models for multiple majors in a unified framework, i.e., a novel deep network of sharing the convolutional layers (i.e., representation learning layers) while keeping the final three-layer fully-connected networks (i.e., output layers) task-specific. In this part, we care about the benefits from multi-task learning. To this end, we predict student performance with the two following single-task methods. The results are reported in Table 3.

- **Single-task network** constructs predictive models for each major separately without considering the relatedness between tasks. Specifically, we divide the data set into 19 subsets corresponding to 19 majors. Each task owns the whole network DPCNN (i.e., without sharing convolutions), and predictive model for each major is trained one by one. In this method, the average performance is reported.
- **Mixed-data method** views constructing predictive models for all majors as a whole task. In other words, the whole DPCNN framework are shared, and it trains a common model for all majors through mixing data of different majors brutally, i.e., the whole DPCNN framework are shared, without task-specific layers.

As can be seen from Table 3, the DPCNN achieves obviously better performance compared with Single-task network as well as the Mixed-data method.

Table 3. Comparison of DPCNN with single-task methods

Methods	acc (Accuracy)	ρ (Spearman coefficient)
Single-task network	0.7507	0.5889
Mixed-data method	0.7259	0.6100
DPCNN	**0.7658**	**0.6964**

Specifically, first, Mixed-data method obtains the worst performance on accuracy, and thus illustrate that it is irrational to train a common model for different majors. Second, the performance of DPCNN makes a great improvement compared with that of Single-task network, i.e., 1.51% for accuracy and 10.75% for Spearman coefficient. The reason may be that DPCNN constructs different predictive models for different majors in a unified framework, and the relatedness between tasks can be implicitly exploited. Third, intuitively, higher accuracy is generally accompanied by a higher Spearman coefficient. Surprisingly, when we compare Single-task network with Mixed-data method, we find that the Single-task network is superior to the Mixed-method on accuracy, while inferior to Mixed-method on the Spearman coefficient. The result may be that in the implements, we sample some pairs of student from each major rather than utilizing all the pairs to construct predictive models.

5   Conclusion and Future Work

In this paper, we predict academic performance based on a large-scale students' behavioral data. Instead of using handcrafted features, we exploit end-to-end deep learning method. To model the three behavioral characters including duration, variation and periodicity, we propose dual path convolutional neural networks. Through dual path convolutions upon student samples, which are represented as tensors, duration and variation can be modeled, respectively. Besides, by taking the date dimension of tensors as the depth of convolutional operations, periodicity can be modeled. Then we introduce multi-tasking learning into our framework, and let multiple tasks share the common convolutional layers while remaining the final three-layer fully-connected networks task-specific. By comparing with two baselines, we show the effectiveness of our proposed DPCNN for predicting academic performance. Moreover, extension experiments illustrate the effectiveness of both dual path structure and multi-task learning.

Though the proposed approach DPCNN can achieve a better presentation automatically as well as better performance, it fails to show the relationship between campus behaviors and academic performance. Thus more deep learning methods, e.g., attention model, can be explored and exploited for our future study. Besides, it should be noted that there exist many other factors affecting student performance, such as psychological status, in-class study behaviors, and historical course grades. Thus, it is also highly appealing to consider more factors to predict student performance in the future.

Acknowledgements. This work was supported by National Natural Science Foundation of China (Grant 61701281, 61573219, 61703234, and 61876098), Shandong Provincial Natural Science Foundation (Grant ZR2017QF009, Grant ZR2016FM34), Shandong Science and Technology Development Plan (Grant J18KA375), Shandong Province Higher Educational Science and Technology Program (Grant J17KA065), and the Fostering Project of Dominant Discipline and Talent Team of Shandong Province Higher Education Institutions.

References

1. Rianne, C., Chris, S., Ad, K., Uwe, M.: Predicting student performance from LMS data: a comparison of 17 blended courses using Moodle LMS. IEEE Trans. Learn. Technol. **10**(1), 17–29 (2017)
2. Huang, S., Fang, N.: Predicting student academic performance in an engineering dynamics course: a comparison of four types of predictive mathematical models. Comput. Educ. **61**, 133–145 (2013)
3. Meier, Y., Xu, J., Atan, O., Schaar, M.V.D.: Predicting grades. IEEE Trans. Signal Proces. **64**(4), 959–972 (2016)
4. Romero, C., Ventura, S.: Educational data mining: a review of the state of the art. IEEE Trans. Syst. Man. Cybern. C **40**(6), 601–618 (2010)
5. Qiujie, L., Rachel, B.: The different relationships between engagement and outcomes across participant subgroups in massive open online courses. Comput. Educ. **127**, 41–65 (2018)
6. Ren Z., Rangwala H., Johri A.: Predicting performance on MOOC assessments using multi-regression models. arXiv preprint arXiv:1605.02269 (2016)
7. Ma, Y.L., Cui, C.R., Nie, X.S., et al.: Pre-course student performance prediction with multi-instance multi-label learning. Sci. China Inf. Sci. **62**(2), 200–205 (2019)
8. Cao, Y., Gao, J., Lian, D., et al.: Orderliness predicts academic performance: behavioural analysis on campus lifestyle. J. Roy. Soc. Interface **15**(146) (2018)
9. Yao, H., Lian, D., Cao, Y., et al.: Predicting academic performance for college students: a campus behavior perspective. ACM Trans. Intel. Syst. Tec. **10**(3), 1–21 (2019)
10. Razavian, A.S., Azizpour, H., Sullivan, J., et al.: CNN features off-the-shelf: an astounding baseline for recognition. arXiv preprint arXiv:1403.6382 (2014)
11. Zhang J., Zheng Y., Qi D.: Deep spatio-temporal residual networks for citywide crowd flows prediction. In: 31st AAAI Proceedings on Artificial Intelligence, pp. 1655–1661. AAAI, San Francisco (2017)
12. Zhang Y., Yang Q.: A survey on multi-task learning. arXiv preprint arXiv:1707.08114 (2017)
13. Wang, F., Chen, L.: A nonlinear state space model for identifying at-risk students in open online courses. In: 9th International Proceedings on Educational Data Mining, Raleigh, NC, USA, pp. 527–532 (2016)
14. Li, W., Gao, M., Li, H., Xiong, Q.Y., et al.: Dropout prediction in MOOCs using behavior features and multi-view semi-supervised learning. In: International Proceedings on Neural Networks, pp. 3130–3137. IEEE, Vancouver (2016)
15. He, J.Z., Bailey, J., Rubinstein, B., Zhang, R.: Identifying at-risk students in massive open online courses. In: 29th AAAI Proceedings on Artificial Intelligence, pp. 1749–1755. AAAI, Austin (2015)

16. Mi, F., Dit-Yan, Y.: Temporal models for predicting student dropout in massive open online courses. In: 2015 IEEE International Proceedings on Data Mining Workshop, pp. 256–263. IEEE, Atlantic City (2015)
17. Kim, B.H., Vizitei, E., Ganapathi, V.: GritNet: student performance prediction with deep learning. In: 11st International Proceedings on Educational Data Mining, Buffalo, NY, USA, pp. 625–629 (2018)
18. Trivedi, S., Pardos, Z.A., Heffernan, N.T.: Clustering students to generate an ensemble to improve standard test score predictions. In: International Conference on Artificial Intelligence in Education, Christchurch, New Zealand, pp. 377–384 (2011)
19. Thai-Nghe, N., Schmidt-Thieme, L.: Multi-relational factorization models for student modeling in intelligent tutoring systems. In: 7th International Conference on Knowledge and Systems Engineering. IEEE, Chongqing (2015)
20. Suleyman, C., Luo, S., Yan, P.X., Ron, T.: Probabilistic latent class models for predicting student performance. In: International Conference on Information and Knowledge Management, pp. 1513–1516. ACM, San Francisco (2013)
21. Er, E.: Identifying at-risk students using machine learning techniques: a case study with is 100. Int. J. Mach. Learn. Comput. **2**(4), 476–480 (2012)
22. Hu, Y.H., Lo, C.L., Shih, S.P.: Developing early warning systems to predict students online learning performance. Comput. Hum. Behav. **36**, 469–478 (2014)
23. Macfadyen, L.P., Dawson, S.: Mining lms data to develop an early warning system for educators: a proof of concept. Comput. Educ. **54**(2), 588–599 (2010)
24. Zafra, A., Romero, C., Ventura, S.: Multiple instance learning for classifying students in learning management systems. Expert Syst. Appl. **38**(12), 15020–15031 (2011)
25. Kotsiantis, S.B., Pierrakeas, C.J., Pintelas, P.E.: Preventing student dropout in distance learning using machine learning techniques. Appl. Artif. Intell. **18**(5), 411–426 (2004)
26. Xenos, M.: Prediction and assessment of student behaviour in open and distance education in computers using bayesian networks. Comput. Educ. **43**(4), 345–359 (2004)
27. Wang, A.Y., Newlin, M.H., Tucker, T.L.: A discourse analysis of online classroom chats: predictors of cyber-student performance. Teach. Psychol. **28**(3), 222–226 (2001)
28. Wang, A.Y., Newlin, M.H.: Predictors of performance in the virtual classroom: identifying and helping at-risk cyber-students. J. High Educ. **29**(10), 21–25 (2002)
29. Essa, A., Ayad, H.: Student success system: risk analytics and data visualization using ensembles of predictive models. In: 2nd Proceedings on Learning Analytics and Knowledge, Vancouver BC, Canada, pp. 158–161 (2012)
30. Lopez, M.I., Luna, J.M., Romero, C., Ventura, S.: Classification via clustering for predicting final marks based on student participation in forums. JEDM **4** (2012)
31. Wu, R.Z., Liu, Q., Liu, Y.P., et al.: Cognitive modelling for predicting examinee performance. In: 24th Proceedings of the International Joint Conference on Artificial Intelligence, pp. 1017–1024. AAAI Press, Buenos Aires (2015)
32. Marbouti, F., Diefes-Dux, H.A., Madhavan, K.: Models for early prediction of at-risk students in a course using standards-based grading. Comput. Educ. **103**, 1–15 (2016)
33. Kimberly, E.A., Matthew, D.P.: Course signals at purdue: using learning analytics to increase student success. In: 2nd Proceedings on Learning Analytics and Knowledge, pp. 267–270. ACM, Vancouver (2012)

34. Ashay, T., Shajith, I., Bikram, S., et al.: Predicting student risks through longitudinal analysis. In: 20th Proceedings of International Conference on Knowledge Discovery and Data Mining, pp. 1544–1552. ACM, New York (2014)
35. Gedeon, T.D., Turner, S.: Explaining student grades predicted by a neural network. In: Proceedings of International Joint Conference on Neural Networks, Nagoya, pp. 609–612 (2002)
36. Acharya, A., Sinha, D.: Early prediction of students performance using machine learning techniques. Int. J. Comput. Appl. **107**(1), 37–43 (2014)
37. Bengio, Y., Courville, A., Vincent, P.: Representation learning: a review and new perspectives. IEEE Trans. Pattern Anal. **35**(8), 1798–1828 (2013)
38. Ruder S.: An overview of multi-task learning in deep neural networks. arXiv preprint arXiv:1706.05098v1 (2017)
39. Spearman, C.: The proof and measurement of association between two things. Am. J. Psychol. **100**(3/4), 441–471 (1987)

A Case Based Deep Neural Network Interpretability Framework and Its User Study

Rimmal Nadeem[1], Huijun Wu[1,2], Hye-young Paik[1,2(✉)], and Chen Wang[2]

[1] School of Computer Science and Engineering, UNSW, Sydney, Australia
rimmal.nadeem@student.unsw.edu.au, huijun.wu@data61.csiro.au,
h.paik@unsw.edu.au
[2] Data61, CSIRO, Sydney, Australia
chen.wang@csiro.au

Abstract. Despite its popularity, the decision making process of a Deep Neural Network (DNN) model is opaque to users, making it difficult to understand the behaviour of the model. We present the design of a Web-based DNN interpretability framework which is based on the core notions in case-based reasoning approaches where exemplars (e.g., data points considered similar to a chosen data point) are utilised to help achieve effective interpretation. We demonstrate the framework via a Web based tool called Deep Explorer (*DeX*) and present the results of user acceptance studies. Our studies showed the effectiveness of the tool in gaining a better understanding of the decision making process of a DNN model as well as the efficacy of the case-based approach in improving DNN interpretability.

Keywords: Deep neural network interpretability · Visualisation · Decision boundaries · Interpretable machine learning

1 Introduction

Despite having the capability to outperform humans in many tasks, the inner workings of DNN models often lack transparency and interpretability, leading to a black box like behaviour [1]. For a user, it is difficult to understand the path DNN models take to come to a decision. An interesting question that remains unanswered is: What training data influenced the result?

A common approach to examine the behaviour of a DNN model is by a trial-and-error based method where users would guess what types of training data or features might be added or tweaked to affect the model. This approach tends to be ad-hoc and time consuming. It would thus be useful to have an assisting tool which help users to see and understand the decision making process of a DNN in an intuitive manner.

In this work, we propose *DeX*, which is an interactive Web tool that helps users better understand the decision making process of a DNN. The tool visualises the similarity of data points in a DNN model transformed space for users

© Springer Nature Switzerland AG 2019
R. Cheng et al. (Eds.): WISE 2019, LNCS 11881, pp. 147–161, 2019.
https://doi.org/10.1007/978-3-030-34223-4_10

to infer the decision criteria of the model. The visualisation elements of *DeX* include (i) numerical values representing the probability distribution of these data points among labels, (ii) clusters of data points in the DNN transformed space, (iii) k-nearest neighbours of a specified data point, and (iv) boundaries characterized by the model. *DeX* visualises a model's training, test points and decision boundaries, allowing users to carry out an interactive exploration of different models utilizing the dataset, as well as a visual comparison of two different models. Using this tool, users can also spot potential weaknesses of models. We have conducted user studies of the tool to demonstrate the effectiveness of the proposed system.[1]

2   Related Work

There is an immense amount of related work relating to the domain of interpretability of Deep Learning Models. We have selected the following topic areas as closely related work.

Prototype Finding and Criticisms: A case-based reasoning approach [2] aims to find examples or prototypes of the solutions of similar past problems that could help solve new problems. In the context of deep learning interpretation, this approach aims to find training data points that are close to other data points within their own classes and far away from those in different classes. Prototype methods involves presenting a minimal subset of "representative" samples from a data set that can serve as a condensed view of the data set. For example, prototype finding is solved as a set cover optimization problem by Bien et al. [3].

Another similar approach is criticisms, where data points that do not quite fit the model are identified and referred to as *criticism samples*. Together with prototypes, criticism can help humans build a better mental model of the complex data space. For example, Kim et al. [4] make use of Maximum Mean Discrepancy-critic which intends to find outliers in a class (referred as criticisms) that differ the most to other data points belonging to the same class.

However, we argue that criticisms and prototypes alone cannot give users sufficient information about why a model makes a certain classification decision on a data point. We can give more contextual information to the users in terms of presenting other similar data points in the neighbouring classes and characterising the differences between classes.

Visualization: Various visualisation techniques are used to explain the decision making process of a deep learning model. The works in [5–7] present visualisation techniques that aim to identify the most important training data that led to certain model predictions (e.g., highlighting different areas of input images that could help reveal the fact that the model factor in the local structure of an image rather than the general scene). Yosinki et al. [8] also introduced a visualisation tool that shows the activations formed on each layer of a trained deep learning

---

[1] A video presentation of the system is available from: https://youtu.be/E87X9U53 sXg.

model as it processes input data. Looking at the evolution of live activations during training helps shape valuable insights about how the model works.

In our system, we allow for a visual comparison between two different models and their decision boundaries resulting from their training data. The tool also allows easy identification of model weaknesses such as training and testing errors. Subsequently, it will work towards providing answers to some of the frequently asked questions users normally have in regards to the black-box workings of DNN. We apply the work of Wu et al. [9] to improve the interpretability of the DNNs through visualising the training & test data along with the model's decision boundaries, which will enable users to have a significantly better understanding of the model predictions and their decision-making process.

Interpretable Decision Boundaries: In this work, we make use of the previous work done by Wu et al. [9] a method to enhance users' understanding of a shared DNN model. Given a training dataset and a model, the algorithm proposed in this approach selects a small set of training data that *best characterizes* the model's decision boundaries. These data points are meant to give useful information to the users to infer how a model prediction is made in relation to them. They employed a max-margin based approach to select the most representative training data that largely contributed to the forming of the decision boundaries of a DNN model referred as interpretable decision boundaries. These training data points are organized via an Explicable Boundary Tree (EB-tree) based on the distances in the DNN transformed space.

3   Case-Based Interpretability Framework

The proposed framework is comprised of two parts: core interpretability processing layer and Web-browser based visualisation and interactivity layer. Figure 1 illustrates the *DeX* architecture.

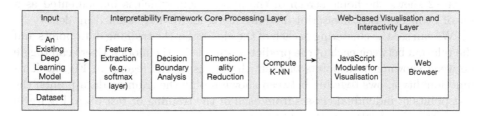

Fig. 1. An overview of the framework

3.1   The Core Processing Layer

To start, a dataset[2] is processed by a chosen DNN model and a Tensorflow module extracts the features - the data representation in the hidden layers and

---

[2] We used MNIST [10] in our case study implementation.

the softmax layer of the model - from the model transformed space. The features are in high dimension which is not ideal for visualisation. We use the following three techniques to aid the visualisation of the features:

- t-SNE [11] for dimensionality reduction: the t-SNE algorithm is known to preserve the local distances of the high-dimensional data. Using t-SNE, we present the model's representational data points in clusters, showing the most similar data points as learned by the model. This forms the basis of our case-based approach to interpretation where we enable users to *view examples of similar or dissimilar cases* to infer the positioning of a new data point within the model,
- pre-calculation of K Nearest Neighbours for each data point: K-NN searches the data for the K instances that most resemble the point of interest. These K instances can essentially be seen as the points the model looked at in order to place/classify the point of interest, highlighting a model's decision process. To obtain the nearest neighbours for the models at hand, we use the FALCONN [12] library for Python which is based on Locality-Sensitive Hashing and a Euclidean distance nearest neighbour search,
- the decision boundary analysis: as mentioned before, we utilised the algorithm proposed in [9] for this task. The analysis selects the most representative training data that largely contributed to the forming of the decision boundaries of the DNN model. The results of this analysis is used by the visualisation layer as another way to inspect the model's behaviour. The boundary inspection can help users better understand the classification results, as the nodes in the boundary show examples of data points that the model considered most ambiguous in making decision. It can highlight weaknesses of the model.

3.2   Web-Based Visualisation and Interactivity

Figure 2 shows the home screen of the application, which is implemented as a set of browser-based Javascript modules using an MNIST trained model.

We can see the clustering of each of the classes for the training data points and the relation between each of the predictions made by the model. Users can toggle between showing data points of their choice, using the legend in Fig. 3a. This functionality allows toggling between showing only the training data points, the test points or individual classes from each of the training or test data. Figure 3b shows the visualisation with only testing points.

Upon clicking on a data point, the data point's k-NNs are highlighted where k is set to 5 by default (Figs. 4a and b). The raw image of the data point clicked on is also displayed in a larger format on the top left corner of the visualisation with a button to 'Explore k-NN'. With this feature users can visually see the relative distance of the points in a 2D mapping gained from the higher dimensional space. Hovering over each of the data points on the plot shows a tool-tip with the raw image corresponding to the data point (Fig. 4b). This is to fulfil the main aim of interpretability as the users will find it useful to know what each of the data points actually looks like. This makes it more intuitive to the user

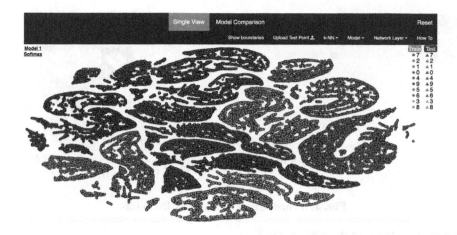

Fig. 2. Home screen of the application

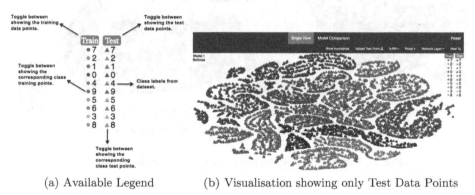

(a) Available Legend   (b) Visualisation showing only Test Data Points

Fig. 3. Available visualisation controls

which features of the image may have lead to the classification decision, whether it be mis-classified or correct.

The application allows spotting of two kinds of errors, training and test errors. Training error refers to the incorrect predictions a model makes on training data. This training data has been utilised to train the model and this essentially does not imply that the model once trained will have 100% accurate performance when connected back on the training data itself. In Fig. 5a, we can see two blue training points predicted wrongly into the grey cluster. This itself will reveal to the user a weakness in their model and using the feature of hovering over the data points to look at the image associated with the points, the user can work towards improving their model.

The second is a test error, which is the error received whilst running the trained model on the test data. This data is used to reveal the true accuracy of the model. In Fig. 5b, we can observe an orange test point mis-classified into

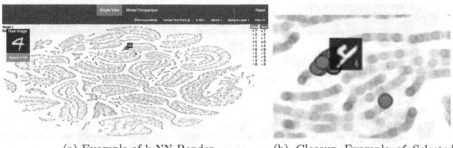

(a) Example of k-NN Render   (b) Closeup Example of Selected
                                    Data Point's k-NN Render

Fig. 4. Visualisation showing k-NN points

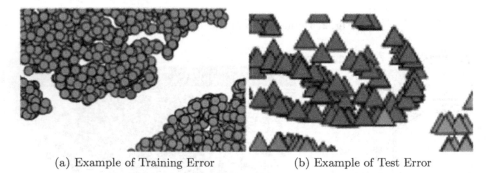

(a) Example of Training Error   (b) Example of Test Error

Fig. 5. Visualisation highlighting errors (Color figure online)

the blue class cluster highlighting a test error. Once again, the easy spotting of different colours with one cluster reveals a weakness in the model.

Another feature available is the exploration of the decision boundaries of the model (Fig. 6a). The decision boundaries presented is the region of the model in which the output label of a classifier is ambiguous. The decision boundaries should help users better understand the similarity or dissimilarity of features embodied in the boundary and in the test data. Observing if the nodes in the decision boundaries are clearly separated or not can reveal the weakness of a model. To improve their model users can use the data in the congested areas where nodes are found close together as a seed to fine tune the model.

Seeing the k-NN and the corresponding raw images, users can more easily infer how the model classified the data point based on their similarities. In Fig. 6b, we see an interesting example of the k-NN render of a data point. The data point selected with true label 0 (red cluster) is predicted incorrectly to be of class 4 (purple cluster). Looking at the raw image of this point, a human can interpret this number as either a 4, a 9 (brown cluster) or a 0. We can see this is exactly what the model thought as well as the k-NN appear in 3 of the clusters and hence a weak point can be pinpointed in the model in this area.

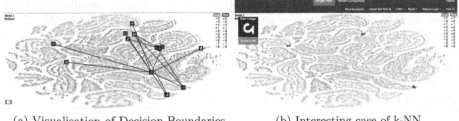

(a) Visualisation of Decision Boundaries   (b) Interesting case of k-NN

Fig. 6. Other visualisation effects (Color figure online)

When selecting a data point, users can further inspect the data point's k-NN by clicking on the 'Explore k-NN' button to see more detailed display of the raw images and labels corresponding to the k-NN. This will be useful in the case when a user wants to better interpret the relationships and similarities within decisions made by the model.

We also implemented a feature of uploading a test data point to show its k-NN in the boundaries. This is similar to an instant visualisation of exactly the path the model takes in making a decision depending on input data. Currently the test points allowed to be uploaded are from the existing test dataset only.

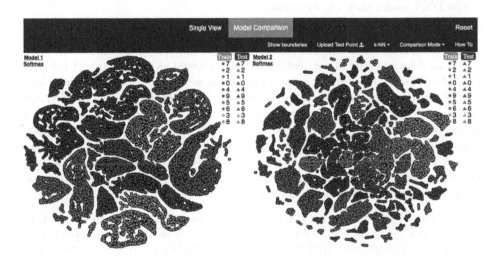

Fig. 7. A model comparison view

The model comparison feature (Fig. 7) allows the rendering of a side-by-side comparison of two different models. Users have a variety of options available to choose from in the comparison mode (permutation of different models and layers). All the features mentioned before are also available in this visualisation mode. Whatever action a user performs on one of the models will render the same action to be performed on the other: An interactive visualisation that allows users to carry out and interpret model comparison more efficiently. This makes evaluation of models much easier by dealing with behaviour rather than numerical values. *DeX* also have various other features including the option to choose the value of k for the k-NN render, zooming in and out of the visualisation, a help menu, ability to view different layers, all of which have been aimed at mitigating the black-box phenomenon of DNN.

4   Experiments

We have conducted user acceptance studies using the case study implementation with the MNIST dataset. We describe the testing setup and summary result, followed by detailed descriptions of the findings by user tasks/goals.

Setup: We invited 13 participants having basic understanding of neural networks and machine learning for a user acceptance study. The participants who took part had a wide range of experience levels with neural networks and similar themes, which was shown to affect the results of this user study. The study was carried out at 3 different locations, all on displays of 1920×1200 resolution. We aimed to answer the following two research questions in this study:

1. Determine if the *DeX* application allows users to gain a better insight about the DNN model they are exploring.
2. Evaluate how efficient the user interface of *DeX* is.

We designed 8 tasks in total. The testing is concluded by a few follow up questions. The first part of completing the task allowed us to gain insight into the usability of *DeX*, whereas the second part of answering the follow up questions highlighted the usefulness of *DeX* in improving the interpretability of DNN. For each user, 5 min were allocated to introduce them to the aim of the research project and the *DeX* application. 30 min were allocated to complete the tasks, including the answering of the follow up questions. At the end of each user study session, participants were given an online post-questionnaire to complete in their own time. Some participants exceeded the 30 min time due to curiosity of some of the features i.e. they asked many questions in between.

Result Summary: We recorded the time took by each participant in completing each of the task. Average task completion times (in seconds) are reported in Table 1. The participants took longest to carry out task 4 (finding the training/test errors) while users struggled most with task 6 covering the decision boundaries.

4.1   Detailed Results

Exploring k-NN (Tasks 2–3): In these tasks, participants were asked to use the application to explore the k-NN of a random point of their choice from the presented model and then later change the number of k-NN being displayed.

Table 1. Tasks and result summary for user acceptance testing

Task No./Description	Results	Time
1. Which one is a better model? (Using numerical statistics)	13/13 people correctly guessed the better model	N/A
2. Find and explore the k-NN of a random point	13/13 people found k-NN helpful for understanding model decisions	54.9
3. Change the number of k-NN being explored	13/13 people correctly completed the task	16.7
4. Find a train and test error	13/13 people found it helpful to discover the error points using the visualisation in understanding model behaviour	72.3
5. Upload a test data point	12/13 people correctly completed the task	37.8
6. Explore the decision boundaries	12/13 people found the decision boundaries to be insightful in discovering model behaviour	71.2
7. Look at a different layer	12/13 people found it useful having the option to look at different layers	22.2
8. Compare models using Model Comparison feature	13/13 people preferred this comparison method over numerical comparison	61.5

Whilst the participants carried out the required tasks, we were able to determine the usability of the feature and were able to gain knowledge of the practicality of this method through our follow up questions.

The users were required to carry out these tasks by using the mouse pointer to click over a data point in order to highlight the required neighbours. The user's first interaction with the application consisted of hovering over the data points to explore the tool tips containing their corresponding images. Many participants thought they had to use their own knowledge to calculate the k-NN of a point. Once they were guided that the application would do this automatically and that they can utilize the help menu, they proceeded to figure out how to display the k-NN. Some of the participants did not require any assistance to complete these tasks. Moreover, a few of the users found that the 'Explore k-NN' was not obvious as it did not looked like a click-able button. So when asked to look at the k-NN in a different view, the participants spent plenty of time looking around the user interface and using the help menu. Other than this when asked to change the number of neighbors being shown, 100% of the participants succeeded to do so without any assistance. All participants found the result of the k-NN to be similar to what they expected and liked the ability to see the two different views.

In the follow up questions, when the participants were asked whether the ability to view the k-NN of a chosen point is helpful in understanding the behavior of the model, all of the replies were in favour (Table 1). This accentuated the practicality and convenience of using k-NN as a method to improve the interpretability of Deep Learning Models. Participants highlighted that outliers would be easy to spot and interesting to explore the similar cases. A point was also raised about how this methodology would allow understanding complex models more easily as the k-NN represent what the model deems most 'similar' when looking at data points. A few participants highlighted the problems with the black-box behaviour of a model and how this representation gave them a better insight into the model's 'thinking'.

Spotting of Errors (Task 4): Before beginning these tasks to spot the errors, participants asked for an explanation for what train and test errors were. Some even went to the help menu to find this information. This highlighted that we may need to include this type of information in the how-to and help menu. A few participants complained that it was hard to see the test points in combination with the train points, since there were so many. Participants did not figure out the functionality of the toggle buttons as toggles existed between the test and train data points. A few users also struggled to find the errors since there existed none for train but some for test.

However, after understanding the definitions of training/test errors, some commented that it was very useful that the visualisation in *DeX* basically highlighted the errors for them, that it was easy to spot the errors due to the colour differences in clusters in comparison to just reading the accuracy values.

Another approach participants took in identifying errors was to look at the k-NN and search for mislabeled points in the 'Explore k-NN'. Furthermore they hovered over the data points in the visualisation to see the corresponding images in the clusters. When designing the tool, we did not anticipate that these methods could be used to spot errors, so it was very insightful in observing participants doing so.

Also, using the same approach, two participants discovered human errors which consisted of training data being labeled incorrectly: a 5 was found in the 3 cluster because it was labeled as a 3. This was not the models fault and highlighted how human error can affect a models behavior. In fact, we were not aware of this human error ourselves until it was pointed out by the participants during the study. This underlined the effectiveness of visualising the dataset along with similar images for each data point.

Participants suggested that we provide additional information such as the true vs. predicted labels in order to make training or test errors more explicit. Another suggestion was made to consider colour-blind users and how they may not be able to spot errors in the current colour scheme involving red and green.

During the testing for this feature, we also found a bug in the toggles and how the colours did not match up to the toggle actions.

100% of the participants highlighted that they would rather use *DeX* to discover errors existing in their models (as shown in Table 1), since the errors were easy to spot and complimented by the k-NN feature. Feedback was received regarding the k-NN feature in *DeX* revealing more information about what data points the model considered similar and discover the reason why the error may be labeled as is. The metric results in Table 1 and feedback received emphasized the effectiveness of the t-SNE clustering and test error spotting mechanism as well as re-validated the feasibility of using k-NN as methods to improve the interpretability of deep learning models.

Uploading a Test Point (Task 5): Participants found this technique very useful, as it would allow for fast revealing of the result of model classification for a test point. All participants agreed that the tool would be very useful if new unseen test points could also be classified using the same method of uploading a test point. Participants revealed that they would use the tool to see if test points are correctly placed and further be able to explore the point of interest's neighbours. Many appreciated the fact that this visualisation allows doing such exploration, as the conventional way of feeding the point to a classification work-flow, e.g., Tensorflow, gives a black-box result and does not reveal much about the decision of the model. Users liked the feature of being able to quickly figure out if the model classified a funny looking digit wrong before technically improving the model to classify the digits correctly. The result and feedback confirmed the efficiency of the upload test point feature within the *DeX* application.

One user suggestion was to change the name of "upload test point" to 'check test point' to better describe what it does currently as no new test points can be currently uploaded for evaluation.

Decision Boundaries (Task 6): Participants found this feature to be the most confusing. This was because when asked to look at the decision boundaries they went in with the pre-conceived idea that the term 'boundaries' referred to the edges of each of the class clusters. Furthermore, when the interpretable boundaries were displayed they could not spot the information icon to find out what boundaries were actually telling them. Once aided to look at the help information they were able to understand what was happening. The feedback given was that having to look at the 'boundaries' as such was confusing due to previous knowledge of boundaries being the gap between two differing clusters. A few participants were not able to grasp the meaning of the line connecting two nodes. Since it was called decision boundaries, they thought that the line outlined a boundary where in fact the line was just a connection between two nodes which the model finds most ambiguous. The participants suggested that the lines should not be called 'decision boundaries' but something along the lines of an 'ambiguity tree'.

After reading the explanation through the information icon, participants understood what the decision boundaries were actually showing: Many saying that it seems to tell you a big picture/overview of how the model is performing and how to improve the model accordingly. After viewing the boundaries, some even went on to identify a few cluster areas where improvement can be made by adding more data points. Overall, 12/13 participants found the decision boundaries insightful in discovering the model at hand (Table 1) which signified the effectives of using decision boundaries as a method to better interpret deep learning models.

Viewing Different Layers of a Model (Task 7): Users found the option to view different layers highly beneficial. After being explained the functionality of different layers in a model, even the participants with little knowledge of the purpose of different layers found the visualisation to be highly useful. The feature of viewing different layers was highlighted to explain the incremental improvement nature of a DNN model in terms of generating the final results, which the users found insightful. This showed the practicality and benefits of allowing exploration of different layers as a feature in *DeX*. This was further supported by the number of people who found it a useful option to understand the model (12/13 metric, in Table 1).

Model Comparison (Task 1 and 8): One of our main goals was to test the advantage of comparing models visually, using the comparison feature we offered in *DeX*, over a conventional method such as reading the numerical accuracy of the models. For this, we set the participants to carry out a comparison of two models using only the numerical accuracy of the models vs. using the *DeX*'s model comparison feature. All of the participants answered correctly as to which model was better using both methods. The participants revealed their thinking process whilst answering our follow up questions. In the numerical accuracy method, the participants explained they chose the model with the higher number as being better of the two. In the visual method using *DeX*, users stated that the visual component allowed them to explore the model better and grasp better information about it, which is what lacked in the numerical comparison method. Various participants also expressed that using the visualisation in combination with the numerical details of the model will make a very powerful comparison technique. *DeX* indeed offers this, the numerical details were omitted during the testing in order to draw an effective comparison between comparing models using numerical analysis vs. comparing models using our visualisation methods.

When using the visualisation for model comparison, the expected path taken was to explore the models using all the features evaluated prior to this feature i.e. k-NN, decision boundaries, finding errors etc. However most participants based their judgement on the visualisation and visibility of clustering and errors only. Most of the users who had basic yet limited knowledge of neural networks were able to intuitively arrive at the decision of which model was better due to the "neatness" of the clustering. Three participants also went on to use the decision boundaries due to the understanding they gained through the information they gained from Task 7. These observations brought attention to the most useful features (t-SNE visualisation and decision boundaries) of *DeX* that were used to understand a model in order to make a comparison.

There were also some cases where users did not notice the model comparison tab in the navigation bar and proceeded to compare models by switching between viewing the different models from the model dropdown menu. At the end of the model comparison when the user had come to a decision, they highlighted how it was a bit annoying having to switch between two of the models to compare details. When the facilitator pointed out they could carry out the model comparison side by side, the participants were intrigued and appreciative of the feature. Although already having arrived at a decision for the better model, they used the Model Comparison feature to re-explore both models side-by-side and complimented on the practicality of such a feature, further validating the utility of the Model Comparison feature.

Table 2. Post task questionnaire. All the answers were given in a 5-point scale of agreement scores (1: strongly disagree, 2: disagree, 3: neither agree nor disagree, 4: agree, 5: strongly agree). The '+' next to the average score indicates the higher the score the better whereas the '*' indicates the lower the score the better.

Statement	Avg.	Std. Dev.
1 - I found the system unnecessarily complex	3.22*	1.13
2 - I found the system very cumbersome to use	1.90*	0.99
3 - I think that I would need the support of a technical person to be able to use this system	1.78*	0.63
4 - I needed to learn a lot of things before I could get going with this system	2.00*	0.94
5 - I found there was too much inconsistency in this system	1.89*	0.74
6 - I think that I would like to use this system frequently	4.00+	0.67
7 - I thought the system was easy to use	3.89+	0.87
8 - I found the various functions in this system were well integrated	3.80+	0.31
9 - I would imagine that most people would learn to use this system very quickly	4.11+	0.87
10 - I felt very confident using the system	3.70+	0.82
11 - I would use this system to evaluate models of my own	4.67+	0.67
12 - The user interface is a great tool to understand DNN models better	4.22+	0.92

4.2   Post-questionnaire

The result of the post-task questionnaire is presented in Table 2. The goal of these questions was to get the users experience and opinion during the user study and capture any insights we may have missed. The first 10 questions are selected from the Software Usability Scale (SUS) questionnaire [13]. These results of the participant's answers to the questions demonstrates the effectiveness and usefulness of the proposed system for better interpreting of deep learning models and a better technique to carry out model comparison. For the first 4 statements the lower is the score the better it is and for the remainder the higher the better. Out of the results shown in Table 2, 11/12 statements achieved desirable scores. Statement 1 received an average of 3.22 which was borderline, whilst we were looking for a lower score. This response can be linked to the observation that users found the system a little difficult to learn at the beginning due to not being able to navigate the help menu.

4.3 General Discussion

Although a pre-requisite of participating in the user study was to have a basic understanding of neural networks, machine learning and clustering, we had a wide range of knowledge levels within our participants. This is because we asked the participants to self-rate their proficiency and experience on the topics. Interestingly, participants doing research in the domain of NN asked many questions while performing the tasks resulting in increase in their total completion time. Experienced participants were positively surprised by the immediate effects of the visualisation and being able to find out vital information about a model.

On the other hand, many of the users with limited knowledge of neural networks also asked questions i.e. what the data set was, what the visualisation was showing, while reading the help information spread throughout the application. After learning about what was going on, they were also positive about the tool. All participants on average spent the least time on tasks 3 and 7 because these involved selecting menu items only.

The participants also had some concerns regarding the loss of information that may be caused due to the dimensionality reduction as they were not familiar with what t-SNE does, once explained they were content with the work. Many users used the phrase 'like de-bugging the DNN model' to describe their experience with the tool, after exploring the various aspects of the model using the application.

Users had the general feedback of the placement and visibility of action buttons being improved: clickable buttons and menus were most times unnoticed by the users despite the clear headings. For some of the tests carried out, the interface took some time to load features, which was affected by the long-distance between the hosting server and a client machine. During the other half of the tests we were situated at the same location as the server and the interface did not show lagging. Users who experienced the slow interface suggested implementing a loading feedback mechanism so users are not confused as to thinking the interface is not working. Another feedback provided was to have a short video introducing the functionality of the system along with the help menu for a fast learning of all of the features.

The visualisation was highly appreciated by all of the users. There was a highly positive reflection on being able to identify various errors; weaknesses and ways a model could be improved. Most users commented on being visual learners so being able to see the clustering of a model was highly effective for them. Features that require improvement included speed/responsiveness, feedback and placement of some action items (mentioned above).

5   Conclusion

DNN interpretability is becoming an increasingly important concept in deep learning research. In this paper, we presented the design and implementation of a Web-based DNN interpretability framework which takes a case-based reasoning approach where similar data points and decision boundaries are utilised to help achieve intuitive interpretation. Our user acceptance studies showed that the tool is effective in gaining a better understanding of the inner workings of a DNN model.

Acknowledgements. The authors thank all participants who took part in the application user study.

References

1. Sturm, I., Lapuschkin, S., Samek, W., Müller, K.R.: Interpretable deep neural networks for single-trial EEG classification. J. Neurosci. Methods **274**, 141–145 (2016)
2. Aamodt, A., Plaza, E.: Case-based reasoning: foundational issues, methodological variations, and system approaches. AI Commun. **7**(1), 39–59 (1994)
3. Bien, J., Tibshirani, R.: Prototype selection for interpretable classification. Ann. Appl. Stat., 2403–2424 (2011)
4. Kim, B., Khanna, R., Koyejo, O.O.: Examples are not enough, learn to criticize! Criticism for interpretability. In: Advances in Neural Information Processing Systems, pp. 2280–2288 (2016)
5. Zeiler, M.D., Fergus, R.: Visualizing and understanding convolutional networks. In: Fleet, D., Pajdla, T., Schiele, B., Tuytelaars, T. (eds.) ECCV 2014. LNCS, vol. 8689, pp. 818–833. Springer, Cham (2014). https://doi.org/10.1007/978-3-319-10590-1_53
6. Fong, R.C., Vedaldi, A.: Interpretable explanations of black boxes by meaningful perturbation. In: IEEE Conference on Computer Vision, pp. 3449–3457 (2017)
7. Koh, P.W., Liang, P.: Understanding black-box predictions via influence functions. In: The 34th International Conference on Machine Learning, pp. 1885–1894 (2017)
8. Yosinski, J., Clune, J., Nguyen, A., Fuchs, T., Lipson, H.: Understanding neural networks through deep visualization. arXiv preprint arXiv:1506.06579 (2015)
9. Wu, H., Wang, C., Yin, J., Lu, K., Zhu, L.: Sharing deep neural network models with interpretation. In: Conference on World Wide Web, pp. 177–186 (2018)
10. Deng, L.: The mnist database of handwritten digit images for machine learning research [best of the web]. IEEE Signal Process. Mag. **29**(6), 141–142 (2012)
11. Maaten, L.V.D., Hinton, G.: Visualizing data using t-SNE. J. Mach. Learn. Res. **9**, 2579–2605 (2008)
12. Andoni, A., Indyk, P., Laarhoven, T., Razenshteyn, I., Schmidt, L.: Practical and optimal LSH for angular distance. In: Advances in Neural Information Processing Systems, pp. 1225–1233 (2015)
13. Brooke, J.: Sus-a quick and dirty usability scale. In: Usability Evaluation in Industry, vol. 189, no. 194, pp. 4–7 (1996)

Personalized Book Recommendation Based on a Deep Learning Model and Metadata

Yiu-Kai Ng$^{(\boxtimes)}$ and Urim Jung

Computer Science Department, Brigham Young University, Provo, UT 84602, USA
ng@compsci.byu.edu, urimjung1@gmail.com

Abstract. Reading books is one of the widely-adopted methods to obtain knowledge. Through reading books, one can obtain life-long knowledge and maintain them. Additionally, if multiple sources of information can be obtained from various books, then obtaining relevant books is desirable. This can be done by book recommendation. There are, however, a number of challenges in designing a book recommender system. One of the challenges is to suggest relevant books to users without accessing their actual content. Unlike websites or blogs, where the crawler can simply scrape the content and index the websites for web search, book contents cannot be accessed easily due to copyright laws. Because of this problem, we have considered using data such as book records, which contains various metadata of a book, including book description and headings. In this paper, we propose an elegant and simple solution to the book recommendation problem using a deep learning model and various metadata that can infer the content and the quality of books without utilizing the actual content. Metadata, which include Library Congress Subject Heading (LCSH), book description, user ratings and reviews, which are widely available on the Internet. Using these metadata are relatively simple compared to approaches adopted by existing book recommender systems, yet they provide essential and useful information of books.

Keywords: Book recommendation · Deep learning · Metadata

1 Introduction

Reading books enhances our understanding on the content covered in a book and offer us an opportunity to learn new knowledge. According to [8], many of the college students believe reading book is directly linked to academic success in college. For people who are not in college, reading books helps them learn throughout their lives.

Instead of accessing the content of a book using its hard-copy archived in a library or made available in a book store, electronic copies became available online through online services such as Google Books or Amazon. In addition,

© Springer Nature Switzerland AG 2019
R. Cheng et al. (Eds.): WISE 2019, LNCS 11881, pp. 162–178, 2019.
https://doi.org/10.1007/978-3-030-34223-4_11

book reviews and ratings can be downloaded by customers and users so that they can filter sub-standard books and make the best choice. Several book recommendation systems have been developed to recommend relevant books to users [7] based on machine learning algorithms or other techniques such as data mining. However, these algorithms require accessing the actual book content which is not widely available due to the copyright law. Instead, we propose an elegant and effective solution to the problem by using metadata associated with books. Metadata are useful, since they offer useful information of the corresponding books. We consider book descriptions, LCSH, user ratings, and reviews to rank books.

Our book recommender is designed for solving the *information overload* problem while minimizing the *time* and *efforts* imposed on readers in discovering unknown, but suitable, books for pleasure reading or knowledge acquisition. Our recommender first identifies a set of *candidate books*, among the ones archived at a website, with topics related to a number of books preferred by the user U. Our recommender is a self-reliant recommender which, unlike others, does not rely on personal tags nor access logs to make book recommendation. It is unique, since it explicitly determines categories of books that match the one preferred by users using a deep learning algorithm, besides considering the subject headings, user ratings, content descriptions, and sentiment on books that are available online.

Our proposed solution provides book stores and libraries diverse and effective book recommendation. In addition, the users can have a satisfying experience with the book recommendation system in terms of saving time and efforts in searching for relevant and interested books to read. Furthermore, our book recommender system is significantly differed from existing approaches, since we do not consider any data mining technique. By simply aggregating the information provided by metadata of books, we effectively recommend books that are relevant to the user's information needs.

2   Related Work

A number of book recommenders [6,15] have been proposed in the past. Amazon's recommender [6] suggests books based on the purchase patterns of its users. Yang et al. [15] analyze users' access logs to infer their preferences and apply the collaborative filtering (CF) strategy, along with a ranking method, to make book suggestions. Givon and Lavrenko [4] combine the CF strategy and social tags to capture the content of books for recommendation. Similar to the recommenders in [4,15], the book recommender in [12] adopts the standard user-based CF framework and incorporates semantic knowledge in the form of a domain ontology to determine the users' topics of interest. The recommenders in [4,12,15] overcome the problem that arises due to the lack of initial information to perform the recommendation task, i.e., the cold-start problem. However, the authors of [4,15] rely on user access logs and social tags, respectively to recommend books, which may not be publicly available and are not required by our recommender. Furthermore, the recommender in [12] is based on the existence of a book ontology, which can be labor-intensive and time-consuming to construct [2].

Zhu and Wang [17] adopt relational data mining algorithm for recommending books. They apply the Apriori data mining algorithm to eliminate mismatched book records and effectively perform data mining using optimization. This approach reduces the amount of book data to be considered. Mooney and Roy [7] apply the contend-based book recommendation approach to obtain the descriptions of books and develop a machine learning algorithm to categorize the text. After categorizing the text, they utilize user profile and use the Bayesian learning algorithm to find the appropriate book for the specific user. Sohail et al. [14] solve the book recommendation problem by constructing an opinion-mining algorithm which relies on the reviews written by users to extract the users' opinions on books for making recommendation. All of these approaches are significantly differed from ours, since the latter simply relies on topic analysis and matadata of books in making book recommendation to its users.

3 Our Book Recommender System

We first utilize a deep neural network model to classify a book B given by a user U who also provides a number of preferred books in a profile. Based on the category of B, we filter books in a collection that are in the same category as B, called *candidate books* CB. Hereafter, we consider different *features* (presented in Sects. 3.2 to 3.5) of books in CB to rank them (in Sect. 3.6) accordingly.

3.1 The Recurrent Neural Network (RNN) Model

We employ a recurrent neural network (RNN) as our classifier, since RNNs produce robust models for classification. Similar to other deep neural networks, RNNs are both trained (optimized) by the backpropagation of error and comprised of a series of layers.

- An *input* layer is a vector or matrix representation of the data to be modeled.
- A few *hidden*, or *latent*, layers of activation nodes, sometimes referred to as "neurons", are included. Each of the hidden layer is designed to map its previous layer to a higher-order (and often higher-dimensional) representation of the features which aims to be more useful in modeling the output than the original features.
- An *output* layer produces the desired output for classification or regression tasks.

The output is produced by propagating numeric values forward. The network is trained by backpropagating the *error*[1] from the output layer backwards. Unlike other network structures, a RNN takes into account the *ordering* of tokens within sequences, rather than simply accounting for the existence of certain values or combinations of values in that sequence. For example, the terms 'car' and 'repair' may appear in a sentence, but the sentiment of that sentence depends on whether

---

[1] An error is the relative divergence of the produced output from the ground truth.

or not they appear adjacent to each other and in that order. For complex textual tasks such as this example, RNNs tend to outperform bag-of-words models which are unable to capture important *recurrent patterns* that occur within sentences.

RNNs achieve the recurrent pattern matching through its *recurrent layer(s)*. A recurrent layer is one which contains a single recurrent unit through which each value of the input vector or matrix passes. The recurrent unit maintains a *state* which can be thought of as a "memory". As each value in the input iteratively passes through the unit at time step t, the unit updates its state h_t based on a function of that input value x_t and its own previous state h_{t-1} as $h_t = f(h_{t-1}, x_t)$, where f is any non-linear activation.

Recurrent layers are designed to "remember" the most important features in sequenced data no matter if the feature appears towards the beginning of the sequence or the end. In fact, one widely-used implementation of a recurrent unit is thus named "Long-Short Term Memory", or LSTM. The designed RNN accurately classifies our data set of books solely based on their sequential text properties.

Table 1. Dimensions and number of parameters of layers in the RNN

Layer	Output dimensions	Total parameters	Trainable parameters
Input	72	0	0
Embedding	72 × 300	1,950,000	0
Bi-directional GRU	72 × 128	140,160	140,160
Global Max Pooling (1D)	128	0	0
Dropout 1	128	0	0
Dense Hidden	64	8,256	8,256
Dropout 2	64	0	0
Dense Output	31	845	845
Total		**2,099,279**	**149,261**

Feature Representation. To utilize a RNN, we need to provide the network with sequential data as input and a corresponding ground-truth value as its target output. Each data entry has to first be transformed in order to be fed into the RNN. Attributes of book entries were manipulated as follows:

Label. The label consists of the category of a book, each of which is the top 31 categories pre-defined by Thriftbooks[2]. Since RNN cannot accept strings as an output target, each unique category string is assigned a unique integer value, which is transformed into a one-hot encoding[3] to be used later as the network's prediction target.

---

[2] https://www.thriftbooks.com/sitemap/.

[3] A one-hot encoding of an integer value i among n unique values is a binarized representation of that integer as an n-dimensional vector of all zeros except the ith element, which is a one.

Features. Features are extracted from the data set S as the *brief description* of a book, which is called a *sentence* of an entry, and is accessible from the book-affiliated websites such as Amazon[4]. Words in a brief description are transformed into *sequences*, or ordered lists of tokens, i.e., unigrams and special characters such as punctuation marks. Each sequence is padded with an appropriate number of null tokens such that each sequence was of uniform length. We have considered only the first 72 tokens in each sentence when representing the features, since over 90% of sentences in S contain 72 or fewer tokens. We considered the 6,500 most commonly-occurring tokens in S.

Text. While extracting features, we have chosen not to remove stopwords, since we prefer not to lose any important semantic meaning, e.g., 'not', within term sequences nor punctuation, since many abstracts include mathematical symbols, e.g., '|', which especially correlate to certain categories. We did, however, convert all of the text in a sentence to lowercase because the particular *word embedding* which we used did not contain cased characters.

Network Structure. We first discuss our RNN used for classifying book categories. Table 1 summaries different layers, their dimensions, and their parameters in our RNN.

The Embedding Layer. A design goal of our neural network is to capture relatedness between different English words (or tokens) with similar semantic meanings. For example, the phrase "he said" has a similar semantic meaning to the phrases "he says" or "she said". Our neural network begins with an embedding layer whose function is to learn a *word embedding* for the tokens in the vocabulary of our dataset. A word embedding maps tokens to respective n-dimensional real-valued vectors. Similarities in semantic meanings between different tokens ought to be captured in the word embedding by corresponding vectors which are also similar either by Euclidean distance, or by cosine similarity, or both. For example, the n-dimensional vector for 'he' may be similar to the vector for 'she' by cosine similarity, or the vector for 'says' may be close in Euclidean space to the vector for 'said'.

The *embedding layer* contains 1,950,000 parameters, since there are 6,500 vectors, one for each token in the vocabulary, and each vector comes with 300 dimensions, and all of which could be trained. Due to the large amount of time it would take to properly train the word embedding from scratch, we have performed two different tasks: (i) we have loaded into the embedding layer as weights an uncased, 300-dimensional word embedding, GloVe, which has been pre-trained on documents on the Web, and (ii) we have decided to freeze, i.e., not train, the embedding layer at all. The pre-trained vectors from GloVe sufficiently capture semantic similarity between different tokens for our task and they are not required to be further optimized. Since the embedding layer was not trained, it simply served to transform the input tokens into a 300-dimensional

---

[4] www.amazon.com.

space. Therefore, instead of the 72-element vector which we started with, the embedding layer outputs a 72×300 real-valued matrix.

The Bi-directional GRU Layer. Following the embedding layer in our network is one type of recurrent layer – a bi-directional GRU, or Gated Recurrent Unit, layer. A GRU is a current state-of-the-art recurrent unit which is able to 'remember' important patterns within sequences and 'forget' the unimportant ones.

This layer effectively 'reads' the text, or 'learns' higher-order properties within a sentence, based on certain ordered sequences of tokens. The number of trainable parameters in a single GRU layer is $3 \times (n^2 + n(m + 1))$, where n is the output dimension, or the number of time steps through which the input values pass, and m is the input dimension. In our case, $n = 64$, since we have chosen to pass each input through 64 time steps, and $m = 300$ which is the dimensionality of each word vector in the embedding space. Since our layer is bi-directional, the number of trainable parameters is twice that of a single layer, i.e., $2 \times 3 \times (64^2 + 64 \times 301) = 140{,}160$, the greatest number of trainable parameters in our network.

The recurrent layer outputs a 72×128 matrix, where 72 represents the number of tokens in a sequence, and 128 denotes the respective output values of the GRU after each of 64 time steps in 2 directions.

The Global Max-Pooling Layer (1D). At this point in the network, it is necessary to reduce the matrix output from the GRU layer to a more manageable vector which we eventually use to classify the token sequence into one of the 31 categories. In order to reduce the dimensionality of the output, we pass the matrix through a *global max-pooling* layer. This layer simply returns as output the maximum value of each column in the matrix. Max-pooling is one of several pooling functions, besides sum- or average-pooling, used to reduce the dimensionality of its input. Since pooling is a computable function, not a learnable one, this layer cannot be optimized and contains no trainable parameters. The output of the max-pooling layer is a 128-dimensional vector.

The Dropout Layer 1. Our model includes at this point a dropout layer. Dropout, a common technique used in deep neural networks which helps to prevent a model from overfitting, occurs when the output of a percentage of nodes in a layer are suppressed. The nodes which are chosen to be dropped out are probabilistically determined at each pass of data through the network. Since dropout does not change the dimensions of the input, this layer in our network also outputs a 128-dimensional vector.

The Dense Hidden Layer. Our RNN model includes a dense, or fully-connected, layer. A *dense layer* is typical of nearly all neural networks and is used for discovering hidden, or latent, features from the previous layers. It transforms a vector x with N elements into a vector y with M inputs by multiplying x by a $M \times N$ weight matrix W. Throughout training, weights are optimized via backpropagation.

The Dropout Layer 2. Before classification, our RNN model includes another dropout layer to again avoid overfitting to the training sequences.

The Dense Output Layer. At last, our RNN model includes a final dense layer which outputs 31 distinct values, each value corresponding to the relative probability of the input belonging to one of the 31 unique categories. Each instance is classified according to the category corresponding to the highest of the 31 output values.

3.2   LCSH

The Library of Congress provides a unique tag, known as Library of Congress Subject Heading, denoted LCSH, for each book prior to its publication. Unlike social media, where users can create a tag to a post suitable to their taste, Library of Congress maintains standardized tags, which come from a controlled vocabulary, from where a subject heading is constructed [3]. Based on this fact, we can effectively measure the closeness of two books in terms of their subject areas by applying our *word correlation factor (WCF)* to compute the *similarity* between their corresponding tags, which consists of a sequence of keywords, in LCSH.

The word-correlation factor between keywords i and j, denoted $Sim(i, j)$, is pre-computed using 880,000 documents in the Wikipedia collection (wikipedia. org/)[5] based on their *frequency of co-occurrence* and *relative distance* in each Wikipedia document.

$$Sim(i, j) = \frac{\sum_{w_i \in V(i)} \sum_{w_j \in V(j)} \frac{1}{d(w_i, w_j)+1}}{|V(i)| \times |V(j)|} \tag{1}$$

where $d(w_i, w_j)$ is the *distance* between words w_i and w_j in any Wikipedia document D, $V(i)$ ($V(j)$, respectively) is the set of *stem* variations of i (j, respectively) in D, and $|V(i)| \times |V(j)|$ is the *normalization* factor.

Although WordNet[6] provides synonyms, hypernyms, holonyms, and antonyms for a given word, there is no partial degree of similarity measures (closeness), i.e., weights, assigned to any pair of words. For this reason, word-correlation factors are more sophisticated in measuring word similarity than word pairs in WordNet.

The word correlation factor of keywords w_1 and w_2 is assigned a value between 0 and 1, such that '1' denotes an *exact* match and '0' denotes *total dissimilarity* between w_1 and w_2. Note that even for highly similar, non-identical words, they are on the order of 5×10^{-4} or less. For example, the degree of similarity between "tire" and "wheel" is 3.1×10^{-6}, which can be treated as 0.00031% *similar* and 99.99% *dissimilar*. As we prefer to ascertain how likely the words are on a scale of 0% to 100% in sharing the same semantic meaning,

---

[5] Words within the Wikipedia documents were *stemmed* and *stopwords* were removed.
[6] wordnet.princeton.edu/.

we further *scale* the word-correlation factors. Since correlation factors of non-identical word pairs are less than 5×10^{-4} and word pairs with correlation factors below 1×10^{-7} do not carry much weight in the similarity measure, we use a logarithmic scale, i.e., *ScaledSim*, which assigns words w_1 and w_2 the similarity value V of 1.0 if they are *identical*, 0 if $V < 1 \times 10^{-7}$, and a value between 0 and 1 if $1 \times 10^{-7} \leq V \leq 5 \times 10^{-4}$, which is formally defined as

$$ScaledSim(w_1, w_2) = \begin{cases} 1 & \text{if } w_1 = w_2 \\ Max(0, 1 - \dfrac{ln(\frac{5 \times 10^{-4}}{Sim(w_1, w_2)})}{ln(\frac{5 \times 10^{-4}}{1 \times 10^{-7}})}) & \text{Otherwise} \end{cases} \quad (2)$$

where $Sim(w_1, w_2)$ is the *word-correlation factor* of w_1 and w_2 defined in Eq. 1. We computes the *degree of similarity* of any two LCSHs L and C using

$$LimSim(L, C) = \frac{\sum_{i=1}^{m} Min(1, \sum_{j=1}^{n} ScaledSim(i, j))}{m} \quad (3)$$

where m and n denote the number of keywords in the LCSHs L and C, respectively, i and j are the keywords in L and C, respectively, and $ScaledSim(i, j)$ is as defined in Eq. 2.

Using the *LimSim* function, instead of simply adding the *ScaledSim* value of each keyword in L with respect to each keyword in C, we *restrict* the highest possible sentence-similarity value between L and C to 1, which is the value for *exact* matches. By imposing this constraint, we ensure that if L contains a keyword K that is (i) an *exact* match of a keyword in C, and (ii) similar to (some of) the other words in C, then the degree of similarity of L with respect to C cannot be significantly impacted/affected by K to ensure a balanced similarity measure of L with respect to C.

3.3   User Ratings

Making recommendations for users based on their past behaviors is crucial and is in essence learning hidden factors which drive users' decision-making process, and rating prediction is such an approach. In this paper, we apply rating prediction for making book recommendations. The higher a predicted rating on a book B for user U using the ratings of books previously viewed by U is, the more likely B is appealed to U. To reduce the problem of finding a user's decision latent-factor model to finding the set of users who make similar decisions, matrix factorization (MF) is a sophisticated rating prediction approach to use such a decision latent-factor model.

To predict unknown ratings on books, a recommender is given a $m \times n$ sparse matrix of known user-book ratings. Singular value decomposition (SVD) [5] can be employed to deduce each user and book latent-factor vectors by factoring out the user and book latent-factor matrices from the user-book rating matrix. Traditional SVD, however, requires the given matrix to be dense. Assuming that all the missing entries are either zero or averages of other entries and applying classical SVD to fill the matrix is going to result in intolerable inaccuracy in the predictions. To handle the sparseness problem, we apply the Funk SVD Learning Algorithm, which is the current state-of-the-art SVD algorithm popularized by Simon Funk in solving the Netflix 100M rating problem. The basic idea is to employ techniques of gradient descent to iterate through the set of known ratings to minimize the squared error of the predicted rating. This iterative process involves the following steps: (i) before the training starts, a predicted rating was guessed to be the average book rating plus the user offset, (ii) for each given user-book rating, the prediction in the previous iteration is updated in the opposite direction of the gradient, and (iii) step (ii) was repeated until prediction error converges to zero.

3.4   User Reviews

In addition to user ratings, we consider common user reviews on books, which can be used for measuring the overall sentiment [12] towards books, to determine the most desirable books to be recommended. Quite often a user writes a user review on a book without providing a rating, and vice versa. Given that user ratings offer only an absolute value without any additional information on a book, while the user reviews contribute additional sentiments to the book. For example, assume that a user gives the same ratings on two different books. Based on the ratings we have to assume that the two books are equally good or equally bad. However, suppose the user makes the comment "Decently written" on the first book, and "Decently written, but I liked the concept" on the second book. With the additional comments, we can claim that the second book is more desirable than the first, since positive sentiment is made towards the second book. For this reason, users' reviews can be used as a supplement to the users' ratings to make suitable book recommendations to users. Sentiment book reviews can easily be found through multiple book websites.

In order to apply users' book reviews in our recommender system, we first determine the polarity of each word w in each review r of a book BK such that w is positive (negative, respectively) if its positive (negative, respectively) SentiWordNet[7] (sentiwordnet.isti.cnr.it) score is higher than its negative (positive, respectively) counterpart. We calculate the overall sentiment score of the reviews made on BK, denoted $StiS(BK)$, by subtracting the sum of its negative words' scores from the sum of its positive words' scores, which reflects the overall sentiment orientation, i.e., positive, negative, or neutral, of the reviews on BK. As the length of the comments on BK can significantly affect the overall sentiment on BK, i.e., the longer each review is, the more sentiment words are in the review, and thus the higher (lower, respectively) its sentiment score is, we normalize the sentiment score of BK by dividing the sum of the SentiWordNet scores of the words in the reviews with the number of sentiment words in the reviews on BK, which yields

---

[7] SentiWordNet, a lexical resource for opinion mining, assigns to each word in WordNet three sentiment scores: positivity, objectivity (i.e., neutral), and negativity. A SentiWordNet score is bounded between -1 and 1, inclusively.

Table 2. TF-IDF weighting scheme used in the enhanced cosine similarity measure in Eq. 6

Condition	Weight assignment
$B_i \in B$ and $P_{B_i} \in P_B$	$V_{B_i} = tf_{B_i,B} \times idf_{B_i}$ and $V_{P_{B_i}} = tf_{P_{B_i},P_B} \times idf_{P_{B_i}}$
$B_i \in B$ and $P_{B_i} \notin P_B$	$V_{B_i} = tf_{B_i,B} \times idf_{B_i}$ and $V_{P_{B_i}} = \frac{\sum_{c \in HS_{B_i}} tf_{c,P_B} \times idf_c}{|HS_{B_i}|}$
$B_i \notin B$ and $P_{B_i} \in P_B$	$V_{B_i} = \frac{\sum_{c \in HS_{P_{B_i}}} tf_{c,B} \times idf_c}{|HS_{P_{B_i}}|}$ and $V_{P_{B_i}} = tf_{P_{B_i},P_B} \times idf_{P_{B_i}}$

$$StiS(BK) = \sum_{i=1}^{n} \frac{\sum_{j=1}^{m} SentiWordNet(Word_{i,j})}{|Rev_i|} \quad (4)$$

where n is the number of reviews on BK, m is the number of words in the kth $(1 \leq k \leq n)$ review on BK, $Word_{i,j}$ $(1 \leq i \leq n, 1 \leq j \leq m)$ is the jth word in the ith review, and $|Rev_i|$ is the number of words in the ith review of BK.

As the highest (lowest, respectively) SentiWordNet score of any word is 1 (-1, respectively), $LS < StiS(BK) \leq HS$, where $-0.9 \leq HS \leq 1$, $-1 \leq LS \leq 0.9$, and $HS - LS = 0.1$. $StiS(BK)$ is further scaled so that its value, denoted $StiS_{Scaled}(BK)$, is bounded between 0 and 1, since a *negative* $StiS(BK)$ value can be returned if the overall sentiment of BK leans towards the negative region. Equation 5 assigns the normalized value to $StiS(BK)$.

$$StiS_{Scaled}(BK) = CL(StiS(BK)) + \frac{0.9 - FL(StiS(BK))}{2}$$

$$CL(StiS(BK)) = \frac{\lceil StiS(BK) \times 10 \rceil}{10}, FL(StiS(BK)) = \frac{\lfloor StiS(BK) \times 10 \rfloor}{10} \quad (5)$$

3.5 Content Similarity Measure

We depend on the user profile P of a user U[8], which is a set of books preferred by U, to infer U's interests/preferences. To determine the degree to which the content of a candidate book B in appeals to U, we compute the *content similarity* between B and each book P_B in P, denoted $CSim(B, P)$ as defined in Eq. 6, using a "bag-of-words" representation on the *brief descriptions* of B and P_B obtained from book-affiliated websites, such as Amazon[9]. To compute $CSim(B, P)$, we employ an enhanced version of the *cosine similarity measure*, which relaxes the exact-matching constraint imposed by the cosine measure and explores words in the description of B that are *analogous to*, besides the *same as*, words in the description of P_B.

$$CSim(B, P) = \max_{P_B \in P} \frac{\sum_{i=1}^{n} VB_i \times VP_{B_i}}{\sqrt{\sum_{i=1}^{n} VB_i^2} \times \sqrt{\sum_{i=1}^{n} VP_{B_i}^2}} \quad (6)$$

---

[8] If a user does not offer a user profile P, then we simply treat the book provided by the user as the only book in P.

[9] www.amazon.com.

where B and P_B are represented as n-dimensional vectors $VB = <VB_1, \ldots,$ $VB_n>$ and $VP_B = <VP_{B_1}, \ldots, VP_{B_n}>$, respectively, n is the number of distinct words in the descriptions of B and P_B, and VB_i (VP_{B_i}, respectively), which is the *weight* assigned to word B_i (P_{B_i}, respectively), is calculated as shown in the equations in Table 2.

HS_w in Table 2 is the set of words that are *highly similar* to, but not the *same* as, a given word w in the description of a book Bk, which is either B or P_B, $|HS_w|$ is the size of HS_w, $tf_{w,Bk} = \frac{f_{w,Bk}}{\sum_{w \in Bk} f_{w,Bk}}$ is the normalized *term frequency* of w in Bk, and $idf_w = log\frac{N}{n_w}$ is the *inverse document frequency* for w in the collection of books N archived at a social bookmarking site, where n_w is the number of books in N that include w in their descriptions. Relying on the *tf-idf* weighting scheme, we prioritize discriminating words that capture the content of its respective book.

The *max* function in Eq. 6 emulates the "most pleasure" strategy (commonly applied in game theory and group profiling [10]). Applying this strategy, we select the *highest* possible score among the ones computed for each P_B in P and B. The *larger* the number of exact-matched or highly-similar words in the descriptions of both B and P_B is, the *more likely* B is a relevant recommendation for U, and guarantees that B is highly similar to at least one of the books of interest to U. We adopt the cosine measure (in Eq. 6), which has been effectively applied to determine the degree of resemblance between any two items in content-based recommenders.

3.6   Combining Ratings

Based on computed scores of *LCSH*, *user ratings*, *user reviews*, and *content similarity measure* for each candidate book B, we apply the *Borda Count voting scheme* [1] to determine the *ranking* score for B. The Borda Count voting scheme is a positional-scoring procedure such that given k (≥ 1) candidates, each voter casts a vote for each candidate according to his/her preference. A candidate that is given a first-place vote receives k-1 points, a second-ranked candidate k-2 points, and so on up till the last candidate, who is awarded no points. Hereafter, the points assigned to each candidate across all the voters are added up and the candidate with the *most* points *wins*.

We employ the Borda Count strategy to generate a single ranking score for B, denoted $Borda(B)$, that regards all the features scores of B as equally important in determining the degree to which a user is interested in B. Using Eq. 7, we assign (i) $k = |CandBks|$, which is the number of candidate books selected for a user U, and (ii) $C = 4$, which is the number of voters, i.e., the four ranked lists of the four features. Candidate books with the top-10 Borda scores are recommended to U.

$$Borda(B) = \sum_{c=1}^{C} (k - S_c^B) \tag{7}$$

where S_c^B is the position on the ranking of B based on the cth ranked list to be fused.

We adopt Borda, since its combination algorithm is *simple* and *efficient*, which requires neither training nor compatible relevance scores that may not be available [1], and its performance is competitive with other existing aggregation strategies [1].

4   Experimental Results

In this section, we evaluate our recommender and compare its performance with others.

4.1   Datasets

We have chosen a number of book records included in the Book-Crossing dataset to conduct the performance evaluation of our recommender[10]. The book-crossing dataset was collected by Cai-Nicolas Ziegler [18] in 2004 with data extracted from BookCrossing.com. It includes 278,858 users who provide, on the scale of 1 to 10, 1,149,780 ratings on 271,379 books. Each book record includes a user_ID, the ISBN of a book, and the rating provided by the user (identified by user_ID) on the book. We used Amazon.com AWS advisement API to verify that the ISBNs from the book-crossing dataset are valid. The 271,379 books in the Book-Crossing dataset is denoted as BKC_DS.

4.2   Accuracy of Our RNN Classifier

Using a 80/10/10% training/validation/test split of the data as mentioned in Sect. 4.1, we achieved 73% classification accuracy on book test data. The accuracy could not be higher likely because of the high amounts of overlap between distinct keywords in the brief description of books with different categories, such as "Deep Learning Computing" and "Theory of computation". With 73% accuracy, we still successfully classify 3 out of 4 articles, which is way above the baseline "best-guesser" classifier. Other bag-of-words modeling techniques with which we have experimented, i.e., logistic regression, SVM, and Multinomial Naïve Bayes [10], showed lower results.

4.3   Evaluation Using Individual Versus Combined Features

In order to justify the necessity of employing all of the four features adopted by our recommender for identifying and ranking appealing books for a user, we have conducted an empirical study which analyzes the capability of each individual feature in making useful book recommendations and compares its performance

---

[10] Other datasets can be considered as long as they contain user_IDs, book ISBNs, and ratings.

with employing all the features. As shown in Fig. 1, our book recommender that consider all the features significantly outperforms each of the individual features in terms of obtaining the lowest prediction error rates among all the features and thus in making useful suggestions to its users based on the rating prediction errors. The combined feature model achieves the highest prediction accuracy, which is less than *half* a rating (out of 10) away from the actual rating. The results clearly indicate that we take the advantage of the individual strength of each feature and greatly improves its effectiveness and the ranking of its suggested books. The overall prediction error of using all the features is 0.41 (see Fig. 1), is a statistically significant improvement ($p < 0.01$) over the prediction error achieved by any individual feature based on the Wilcoxon signed-ranked test.

4.4 Comparing Book Recommendation Systems

In this section, we compared our recommender with exiting book recommenders that achieve high accuracy in recommendations on books based on their respective model.

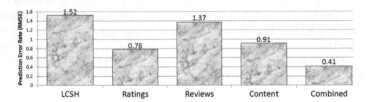

Fig. 1. Prediction error rates of the individual features and the combined prediction model

- **MF.** Yu et al. [16] and Singh et al. [13] predict ratings on books and movies based on matrix factorization (MF), which can be adopted for solving large-scale collaborative filtering problems. Yu et al. develop a non-parametric matrix factorization (NPMF) method, which exploits data sparsity effectively and achieves predicted rankings on items comparable to or even superior than the performance of the state-of-the-art low-rank matrix factorization methods. Singh et al. introduce a collective matrix factorization (CMF) approach based on relational learning, which predicts user ratings on items based on the items' genres and role players, which are treated as unknown values of a relation between entities of a certain item using a given database of entities and observed relations among entities. Singh et al. propose different stochastic optimization methods to handle and work efficiently on large and sparse data sets with relational schemes. They have demonstrated that their model is practical to process relational domains with hundreds of thousands of entities.
- **ML.** Besides the matrix factorization methods, probabilistic frameworks have been introduced for rating predictions. Shi et al. [11] propose a joint matrix

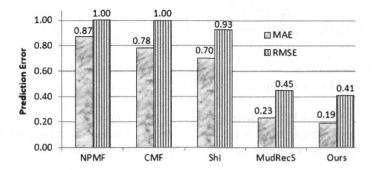

Fig. 2. The MAE and RMSE scores for various book recommendation systems based on BKC_DS, the BookCrossing dataset

factorization model for making context-aware item recommendations.[11] Similar to ours, the matrix factorization model developed by Shi et al. relies not only on factorizing the user-item rating matrix but also considers contextual information of items. The model is capable of learning from user-item matrix, as in conventional collaborative filtering model, and simultaneously uses contextual information during the recommendation process. However, a significant difference between Shi et al.'s matrix factorization model and ours is that the contextual information of the former is based on mood, whereas ours makes recommendations according to the contextual information on books.

– MudRecS [9] makes recommendations on books, movies, music, and paintings similar in content to other books, movies, music, and paintings, respectively that a MudRecS user is interested in. MudRecS does not rely on users' access patterns/histories, connection information extracted from social networking sites, collaborated filtering methods, or user personal attributes (such as gender and age) to perform the recommendation task. It simply considers the users' ratings, genres, role players (authors or artists), and reviews of different multimedia items. MudRecS predicts the *ratings* of multimedia items that match the interests of a user to make recommendations.

Figure 2 shows the Mean Absolute Error and RMSE scores of our and other recommender systems on the BKC_DS dataset. *Root Mean Square Error* (RMSE) and *Mean Absolute Error* (MAE) are two performance metrics widely-used for evaluating rating predictions on multimedia data. Both RMSE and MAE measure the *average magnitude* of *error*, i.e., the average prediction error, on incorrectly assigned ratings. The error values computed by RMSE are squared before they are summed and averaged, which yield a relatively *high* weight to errors of *large* magnitude, whereas MAE is a *linear* score, i.e., the absolute values of individual differences in incorrect assignments are weighted equally in the average.

---

[11] The system was originally designed to predict ratings on *movies* but was implemented by [9] for additional comparisons on *books* as well.

$$RMSE = \sqrt{\frac{\sum_{i=1}^{n}(f(x_i) - y_i)^2}{n}}, MAE = \frac{1}{n}\sum_{i=1}^{n}|f(x_i) - y_i| \qquad (8)$$

where n is the total number of items with ratings to be evaluated, $f(x_i)$ is the rating predicted by a system on item x_i ($1 \leq i \leq n$), and y_i is an expert-assigned rating to x_i.

As the MAE and RMSE scores shown in Fig. 2, our book recommender significantly outperforms other book recommender systems on rating predictions of the respective books based on the Wilcoxon Signed-Ranks Test ($p \leq 0.05$).

4.5   Human Assessment on Our Recommender

We further evaluated our recommender to determine whether its suggestions are perceived as preferable by ordinary users, which offers another perspective on the performance of the recommender. The additional evaluation is based on real users' assessments of the recommender which goes beyond the offline performance analysis conducted and presented in previous subsections. To accomplish this task, we conducted a user study using Amazon's Mechanical Turk (MT)[12], a "marketplace for work that requires human intelligence", which allows individuals or businesses to programmatically access thousands of diverse, on-demand workers and has been used to collect user feedback for multiple information retrieval tasks.

Table 3. Sampled books and their corresponding subject area employed in the user study conducted using Mechanical Turk

Book title	Subject area
The Autobiography of Benjamin Franklin	History
Fast Food Nation: The Dark Side of . . .	Cooking
The 7 Habits of Highly Effective Teens	Parenting
Think and Grow Rich: The Landmark . . .	Business
Code Complete	Computer & Tech
Healthy Sleep Habits, Happy Child	Medical
Scary Stories to Tell in the Dark	Horror

In the user study, we used a set of 100 randomly-sampled books with diverse subject areas. (A number of sampled books used in this study and their corresponding subject areas is shown in Table 3.) We created a HIT (Human Intelligent Task) on MT so that for each sampled book, each appraiser was presented with a list of *five* ranked recommended books suggested by our recommender,

---

[12] https://www.mturk.com/mturk/welcome.

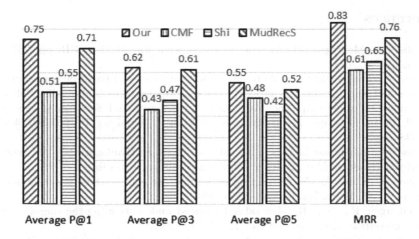

Fig. 3. Evaluation using Mechanical Turks for our book recommender

CMF, Shi, and MudRecS, respectively and asked to select the ones that are relevant to the sampled book. The user study was conducted between March 12 and March 23, 2019 on MT. Altogether, there were 715 responses among the HITs used in the study. Based on the corresponding set of responses provided by MT appraisers, we have verified that users tend to favor our recommended books for a given book. (See Fig. 3 for the results of the empirical study.)

We evaluated and compared the performance of our recommender with CMF, Shi, and MudRecS based on average P@1 (Precision at rank position 1), P@3, and P@5, and MRR (Mean Reciprocal Rank). These values are easy to compute to produce a single performance value and is readily understandable. Figure 3 shows the performance ratios computed using MT appraisers, which indicates that highly-ranked books recommended by us were treated as relevant by the MT appraisers, and the results are statistically significant ($p < 0.03$).

5 Conclusions

Reading books can enrich one's life with knowledge and deep understanding of various topics, and over the years the book industry has become an influential global consumer market. According to Statista[13], approximately 74% of the population in the U.S.A. consumed at least one book and books published in the higher education market generated nearly 4 billion US dollars in the year of 2017. With the huge amount of books available these days, various book recommendation systems have been proposed to meet user's book searching needs. Unlike many of the existing book recommender systems, our proposed book recommender simply relies solely on a deep learning model and book metadata to make personalized book recommendations. The empirical study demonstrates that our recommender outperforms well-known book recommenders.

---

[13] https://www.statista.com/topics/1177/book-market/.

References

1. Aslam, J., Montague, M.: Models for metasearch. In: ACM SIGIR, pp. 267–276 (1997)
2. Ding, Z.: The development of ontology information system based on Bayesian network and learning. In: Jin, D., Lin, S. (eds.) Advances in Multimedia, Software Engineering and Computing. Advances in Intelligent and Soft Computing, vol. 129, pp. 401–406. Springer, Heidelberg (2012). https://doi.org/10.1007/978-3-642-25986-9_62
3. Elmer, E.: Rasmuson Library. University of Alaska Fairbanks, Library of Congress Subject Headings (2014). https://library.uaf.edu/ls101-lc-subject
4. Givon, S., Lavrenko, V.: Predicting social-tags for cold start book recommendations. In: ACM RecSys. pp. 333–336 (2009)
5. Kleibergen, F., Paap, R.: Generalized reduced rank tests using the singular value decomposition. Econometrics **133**(1), 97–126 (2006)
6. Linden, G., Smith, B., York, J.: Amazon.com recommendations: item-to-item collaborative filtering. Internet Comput. **7**(1), 76–80 (2003)
7. Mooney, R., Roy, L.: Content-based book recommending using learning for text categorization. In: ACM DL 2000, pp. 195–204 (2000)
8. Owusu-Acheaw, M., Larson, A.: Reading habits among students and its effect on academic performance: a study of students of Koforidua Polytechnic. LPP **1130** (2014)
9. Qumsiyeh, R., Ng, Y.: Predicting the ratings of multimedia items for making personalized recommendations. In: SIGIR, pp. 475–484 (2012)
10. Ricci, F., Rokach, L., Shapira, B., Kantor, P.: Recommender System. Handbook. Springer, Cham (2011). https://doi.org/10.1007/978-3-319-29659-3
11. Shi, Y., Larson, M., Hanjalic, A.: Mining mood-specific movie similarity with matrix factorization for context-aware recommendation. In: Workshop on CARS, pp. 34–40 (2010)
12. Siersdorfer, S., Chelaru, S., Nejdl, W., Pedro, J.: How useful are your comments: analyzing and predicting YouTube comments and comment ratings. In: WWW, pp. 891–990 (2010)
13. Singh, A., Gordon, G.: Relational learning via collective matrix factorization. In: ACM SIGKDD, pp. 650–658 (2008)
14. Sohail, S., Siddiqui, J., Ali, R.: Book recommendation system using opinion mining technique. In: IEEE ICACCI, pp. 1609–1614 (2013)
15. Yang, C., Wei, B., Wu, J., Zhang, Y., Zhang, L.: CARES: a ranking-oriented CADAL recommender system. In: JCDL, pp. 203–212 (2009)
16. Yu, K., Zhu, S., Lafferty, J., Gong, Y.: Fast nonparametric matrix factorization for large-scale collaborative filtering. In: ACM SIGIR, pp. 211–218 (2009)
17. Zhu, Z., Wang, J.: Book recommendation service by improved association rule mining algorithm. In: ICMLC, pp. 3864–3869 (2007)
18. Ziegler, C., McNee, S., Konstan, J., Lausen, G.: Improving recommendation lists through topic diversification. In: WWW, pp. 22–32 (2005)

DINRec: Deep Interest Network Based API Recommendation Approach for Mashup Creation

Yong Xiao[1,2], Jianxun Liu[1,2(✉)], Rong Hu[1,2], Buqing Cao[1,2],
and Yingcheng Cao[1,2]

[1] Key Laboratory of Knowledge Processing and Networked Manufacturing,
Hunan University of Science and Technology, Xiangtan 411201, Hunan, China
ljx529@gmail.com
[2] School of Computer Science and Engineering, Hunan University of Science
and Technology, Xiangtan 411201, Hunan, China

Abstract. Recommending appropriate APIs for Mashup creation has become a challenge as the number of APIs from different sources grows fast. In order to understand the relationships among multiple ecosystem APIs, most existing API recommendation methods focus on semantic similarity relationships but underutilize the composition and cooperation relationships between APIs, which may lead to low recommendation precision. In view of this problem, a Deep Interest Network based API Recommendation approach (DINRec) for Mashup development is proposed in this paper. In this approach, APIs are chosen incrementally for compositing into a Mashup and in that process the embedding vector of the Mashup's existing composition features will be updated adaptively by using Deep Interest Network. Moreover, a Doc2simu model is used to help training industrial deep networks with relatively small amounts of dataset. Finally, some experiments on real-world dataset are implemented to verify the efficiency of our proposed approach.

Keywords: Deep Interest Network · Doc2simu model · API recommendation · Mashup

1 Introduction

Mashup technique has got a far-reaching impact in recent years which provides a flexible way for fulfilling dynamic and customized Web service developer requirements and tackles the functional limitations of individual Application programming interfaces (APIs). However, as the number of APIs grows rapidly, how to recommend appropriate ones for Mashup creation to satisfy users' requirements becomes a challenge. For example, as of May 16, 2019, the dominant website ProgrammableWeb has published 21,552 web APIs under 484 categories. If a developer wants to build a Mashup related with messaging, ProgrammableWeb search engine will return a list containing 1,576 Web APIs. It is a difficult task to go through these lists of results and select the desired APIs.

© Springer Nature Switzerland AG 2019
R. Cheng et al. (Eds.): WISE 2019, LNCS 11881, pp. 179–193, 2019.
https://doi.org/10.1007/978-3-030-34223-4_12

Some existing methods focus on keyword or semantic matching while others are based on Quality of Service (QoS) prediction [1] for service recommendation. However, keyword-based matching is usually imprecise while semantic-based matching is expensive to construct in practice. In addition, QoS is unstable and lagging that may affect the precision of real-time prediction. In view of these shortcomings, in recent years, some machine learning techniques such as Latent Dirichlet Allocation (LDA) [2, 3] or Relational Topic Model (RTM) [4, 5] were used to learn topic from the services' descriptions or the users' requirements. In some other studies, Word2vec and Doc2vec [6] were used to extract deep semantic features between words and vectorize descriptions of API and Mashups.

In this paper, we also apply machine learning techniques to recommend APIs to Mashup, while considering the cooperation and composition relationships between APIs simultaneously. Generally, the function of a Mashup is implemented by several APIs, or all APIs realize the complex function of the Mashup. Therefore, the prior chosen APIs may affect the selection strategy of the subsequent APIs while creating a Mashup. For example, a Mashup BBC Browser described as "Maps channel program information to relevant Twitter account" contains three APIs, i.e., BBC Nitro, Twitter and Facebook. Suppose that BBC Nitro API and Twitter API have been chosen for creating this Mashup, then the probability of Facebook API being recommended to the Mashup would increase. The first reason is that an API which is category-similar to the prior selected APIs would not likely to be recommended to the same Mashup. Second, it is more reasonable to recommend an API that can complete the function of the Mashup that have not been completed by the prior selected APIs. For these reasons, Facebook API will be recommended to BBC Browser Mashup according to the prior selected BBC Nitro API and Twitter API.

To realize this conception, we proposed a Deep Interest Network (DIN) [7] based API recommendation approach, called DINRec. By introducing a local activation unit, DINRec adaptively learn the representation vector of composition and cooperation relationships between selected APIs and candidate API, moreover, this representation vector varies over different candidate APIs, which makes it possible to update embedding vector of the prior selected APIs when gradually add APIs into the Mashup. Through this mechanism, the prior selected APIs with higher cooperation to the candidate API will get higher activated weights before they get into the multilayer perceptron. We conduct some experimental studies to gain insight about this phenomenon. The final results show that DINRec perfectly integrates the functional semantics and composition relationship of Mashups.

A recent check of ProgrammableWeb.com's statistics shows that the number of APIs used by Mashups only covers a quarter of the amounts of total APIs, and most Mashups only contain less than 3 Web APIs. However, training industrial deep networks with few features is prone to over-fitting. To solve this problem, in this paper, we present a Doc2simu model to train the text vectors of all Web APIs, and choose the ones with high similarities as the extended dataset help to support effective training in industrial networks. The process of the method proposed in this paper is illustrated in Fig. 1. The contributions of this paper are summarized as follows:

(1) We propose a recommendation method based on Deep interest Network (DIN-Rec), which can improve the expressive ability of feature model and better capture the diversity characteristics related to functional semantics and composition relationships of Mashups.
(2) We present a Doc2simu model to help training industrial deep networks by extending dataset based on the Doc2vec model and cosine similarity.
(3) We conduct experiments on a real-world dataset crawled from ProgrammableWeb to evaluate the effectiveness of DINRec.

The rest of this paper is organized as follows. Section 2 presents the process and structure of DINRec. Section 3 discusses and analyzes the experimental results and variable parameters. Section 4 describes the related works and Section 5 draws a conclusion of the paper.

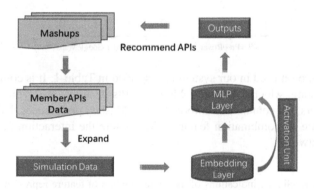

Fig. 1. The framework of DINRec.

2 Process of DINRec

2.1 Feature Representation

Feature Description. Describing the features of Mashups and their member APIs is the fundamental task to obtain the functional semantics and composition relationships of Mashups. Formally, the feature of a Mashup is defined as follow:

Definition 1 (Feature of a Mashup). The feature of a Mashup can be defined as a tuple $F = (F^M, F^A)$. . In this tuple, $F^M = (N^M, T^M, C^M, D^M)$, where N^M is the name of the Mashup, T^M is the tags of the Mashup, C^M is the category of the Mashup and D^M is the description text of the Mashup; $F^A = \{F_i^A | 0 \leq i \leq N_i\}$ (N^A, T^A, C^A, D^A), where $N^A = \{n_{A,i} | 1 \leq i \leq n\}$, $T^A = \{t_{A,i} | 1 \leq i \leq n\}$, $C^A = \{c_{A,i} | 1 \leq i \leq n\}$ and $D^A = \{d_{A,i} | 1 \leq i \leq n\}$, $n_{A,i}, t_{A,i}, c_{A,i}, d_{A,i}$ represent the i-th API's name, tag, category and description text feature, respectively, n is the number of member APIs for each Mashups.

Feature Representation. As the above shows, features in our recommendation tasks is mostly in a multi-group categorial form, based on this, we use one-hot encoding to represent features in this paper. One-hot encoding is simple to compute and understand, and employed frequently when it is necessary to represent a categorical variable in a neural network, which is normally transformed into high-dimensional sparse binary features [8, 9]. Mathematically, encoding vector of i-th feature group is formularized as $t_i \in R^{K_i}$. K_i denotes the dimensionality of feature group i, which means feature group i contains K_i unique APIs. $t_i[j]$ is the j-th element of t_i and $t_i[j] \in \{0, 1\}$, $\sum_{j=1}^{K_i} t_i[j] = k$. Vector t_i with $k = 1$ refers to one-hot encoding and $k > 1$ refers to multi-hot encoding. Then one instance can be represent as $x = [t_1^T, t_2^T, \ldots t_M^T]^T$ in a group-wise manner, where M is number of feature groups, $\sum_{i=1}^{M} K_i = K$, K is dimensionality of the entire feature space. In this way, the aforementioned instance with one-hot encoding and multi-hot encoding of features are illustrated as:

$$\underbrace{[0, \ldots, 1, \ldots, 0]}_{N^M = \text{PropRover}} \quad \underbrace{[0, \ldots, 1, \ldots, 1, \ldots, 0]}_{N^A = \{\text{FeedBurner,GoogleMaps}\}}$$

The whole feature set used in our system is described in Table 1. It is composed of four categories, among which Mashup's MemberAPIs features are typically multi-hot encoding vectors and contain rich information of Mashup preferences. Note that in our setting, there are no combination features. We capture the interaction of features with deep neural network.

Table 1. Simple indications of the representation of feature representation.

Category	Feature Group GROUP	Dimension	Type	Ids per Instance
Mashup Features	Name	$\sim 10^4$	One-hot	1
	Tag	$\sim 10^3$	One-hot	1
	Cate	$\sim 10^3$	One-hot	1
Mashups' MemberAPIs features	APIs_Names	$\sim 10^2$	Multi-hot	$\sim 10^2$
	APIs_Tags	~ 10	Multi-hot	~ 10
	APIs_Cates	~ 10	Multi-hot	~ 10
Candidate API features	API_Name	$\sim 10^4$	One-hot	1
	API_Tag	$\sim 10^3$	One-hot	1
	API_Cate	$\sim 10^3$	One-hot	1
Content features	Map	$\sim 10^2$	One-hot	1
	Game	$\sim 10^2$	One-hot	1

2.2 Deep Interest Network

In this section, we will introduce the framework of Deep Interest Network (DIN). The architecture of it can be illustrated in the Fig. 2, which consists of several parts:

Embedding Layer. As the inputs are high dimensional binary vectors, embedding layer is used to transform them into low dimensional dense representations. For the i-th feature group of t_i, let $W^i = \left[w_1^i, \ldots, w_j^i, \ldots, w_{1K_t}^i \right] \in R^{D \times K_t}$ represent the i-th embedding dictionary, where $w_j^i \in R^D$ is an embedding vector with dimensionality of D. Embedding operation follows the table lookup mechanism, as illustrated in Fig. 2.

- If t_i is one-hot vector with j-th element $t_i[j] = 1$, the embedded representation of t_i is a single embedding vector $\text{t}_i = w_j^i$.
- If t_i is multi-hot vector with $t_i[j] = 1$ for $j \in \{i_1, i_2, \ldots, i_k\}$, the embedded representation of t_i is a list of embedding vectors: $\{e_{i_1}, e_{i_2}, \ldots e_{i_k}\} = \left\{ w_{i_1}^i, w_{i_2}^i, \ldots w_{i_k}^i \right\}$.

Pooling Layer and Concat Layer. Notice that different Mashups have different numbers of APIs. So that the number of non-zero values for multi-hot behavioral feature vector t_i varies across instances, causing the lengths of the corresponding list of embedding vectors to be variable. As fully connected networks can only handle fixed-length inputs, it is a common practice [8, 10] to transform the list of embedding vectors via a pooling layer to get a fixed-length vector:

$$e_i = pooling(e_{i_1}, e_{i_2}, \ldots e_{i_k}) \tag{1}$$

average pooling, which apply element-wise sum/average operations to the list of embedding vectors. Both embedding and pooling layers operate in a group-wise manner, mapping the original sparse features into multiple fixed length representation vectors. Then all the vectors are concatenated together to obtain the overall representation vector for the instance.

Activation Unit. From the above steps, we obtain a fixed-length representation vector of Mashup composition by pooling all the embedding vectors over the Mashup composition feature group, as Eq. (1). This representation vector stays the same for a given Mashup, in regardless of what candidate APIs are. In order to solve this problem, DIN pay attention to the representation of locally activated intentions to recommend APIs for Mashup. Instead of expressing all Mashup's diverse composition with the same vector, DIN adaptively calculate the representation vector of Mashup's composition by taking into consideration the relevance of existing composition to recommend candidate APIs for Mashup. And this representation vector varies over different candidate APIs, so that we can select novel APIs for Mashup incrementally based on the composition relationship information.

From the Fig. 2, we can observe that DIN introduces a novel designed local activation unit. Specifically, activation units are applied on the Mashup composition features, which performs as a weighted sum pooling to adaptively calculate Mashup representation v_U given a candidate API A, as shown in Eq. (2):

$$v_U(A) = f(v_A, e_1, e_2, \ldots, e_H) = \sum_{j=1}^{H} a(e_j, v_A)e_j = \sum_{j=1}^{H} w_j e_j \tag{2}$$

where $\{e_1, e_2, \ldots, e_H\}$ is the list of embedding vectors of composition of Mashup U with length of H, v_A is the embedding vector of API A. In this way, $v_U(A)$ varies over

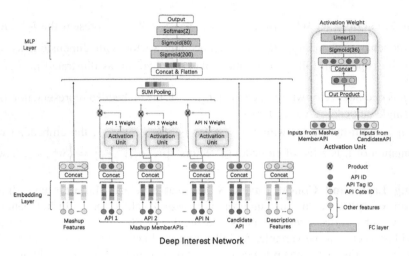

Fig. 2. DIN model structure.

different APIs. a($\cdot$) is a feed-forward network with output as the activation weight, as illustrated in Fig. 2. Apart from the two input embedding vectors, a($\cdot$) adds the out product of them to feed into the subsequent network, which is an explicit knowledge to help relevance modeling. Local activation unit of Eq. (2) shares similar ideas with attention methods which are developed in NMT task [11]. However different from traditional method, the constraint of $\sum_i w_i = 1$ is relaxed in Eq. (2), aiming to reserve the intensity of Mashup composition. That is, normalization with softmax on the output of a($\cdot$) is abandoned. Instead, value of $\sum_i w_i$ is treated as an approximation of the intensity of activated Mashup composition to some degree. For example, if a Google Maps API has been chosen for creating a Mashup of travel class. Given two candidate APIs of Bing Maps and World Weather Online, World Weather Online may get larger value of v_U (higher intensity of preference) than Bing Maps, because it complements the function of this Mashup and avoids choosing category-similar APIs for the Mashup. Traditional attention methods lose the resolution on the numerical scale of v_U by normalizing of the output of a($\cdot$). We have tried LSTM to model Mashup's invoked APIs dataset in the sequential manner. But it shows no improvement, we leave it for future research.

MLP. Given the concatenated dense representation vector, fully connected layers are used to learn the combination of features automatically. Recently developed methods [8, 12, 13] focus on designing structures of MLP for better information extraction.

Loss. The objective function used in base model is the negative log-likelihood function defined as:

$$L = -\frac{1}{N} \sum_{(x,y) \in S} (y \log p(x) + (1 - y) \log(1 - p(x))) \tag{3}$$

where S is the training set of size N, with x as the input of the network and $y \in \{0, 1\}$ as the real label, $p(x)$ is the output of the network after the softmax layer, representing the predicted probability of sample x being recommended.

3   Experiment

3.1   Dataset Description and Doc2simu Preprocess

Dataset Description. To evaluate the performance of different APIs recommendation methods, we crawled 6415 real Mashups which invoke 1595 APIs from the ProgrammableWeb site and the overall statistics of our datasets is show in Table 2. For each Mashups or APIs, we firstly obtained their descriptive text and then performed a preprocessing process to get their standard description information. Figure 3 presents the statistics of APIs distribution in Mashups on the crawled dataset. From the Fig. 3, we can see that, 53.1%/25.1%/10.4% Mashups respectively invoke 1/2/3 APIs. Totally, more than 99% Mashups invoke 1–10 APIs. Therefore, we report experiment results obtained by recommending 1 to 10 APIs for target Mashup in this section.

Table 2. Statistic of our ProgrammableWeb dataset.

Projects	Mashup	API
Number of entities	6415	1595
Number of categories	375	127
Number of tags	996	964

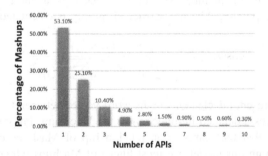

Fig. 3. Web APIs distribution of Mashups in the crawled dataset.

Doc2simu Preprocess. As the Fig. 3 shows, the dataset mentioned above is relatively small and most Mashups invoked only a small amount APIs which to a great extent will cause over-fitting in industrial depth network, and it may not be tolerated for our recommendation system. To remit this problem, we set up a Doc2simu model to expand our dataset. We followed several steps to clean and preprocess them.

First, we retrieved the three sections for each invoked APIs, including the name, primary category, primary tag and description. Then we processed the description document of all APIs by using tokenizer and stemming, meanwhile we removed those illegal characters including digits and special characters (e.g., &, % and $, etc.), and removed general or stop words in the end. The rest of the words were validated using dictionary.

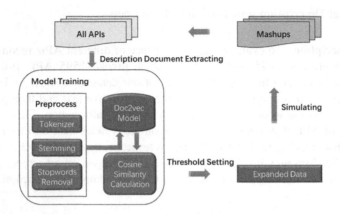

Fig. 4. Doc2simu preprocess.

Next, we put the ultima corresponding description document of each API into the Doc2vec [14] model for training and get the corresponding word vector, then the cosine similarity between each API and all other APIs word vectors is calculated, and the semantic similarity matrix is obtained. Finally, we select APIs, whose cosine similarity is more than 0.88 with the member APIs of each Mashups, as the extended simulation dataset of the corresponding Mashup. Cosine similarity is calculated as follows:

$$\cos(x, y) = \frac{\sum_{i=1}^{n}(x_i \times y_i)}{\sqrt{\sum_{i=1}^{n}(x_i)^2} \times \sqrt{\sum_{i=1}^{n}(y_i)^2}} \tag{4}$$

where x_i and y_i represent the elements in the word vector between two different APIs x and y. Figure 4 gives a detailed introduction to the Doc2simu Preprocess. In addition, in order to ensure that the composition relationship of Mashups can be taken into account when training our model, we must filter out Mashups which contain three or more APIs.

After the above treatment, our dataset has become rich, and with more than 3 MemberAPIs for each Mashups. Features include API_id, cate_id, Mashup's invoked APIs_id_list and cate_id_list. Let all MemberAPIs of a Mashup be $(b_1, b_2, \ldots, b_k, \ldots, b_n)$, the task is to predict the $(k+1)$-th MemberAPIs by making use of the first k MemberAPIs. Training dataset is generated with k = 1, 2, ..., n-2 for each Mashups. In the test dataset, we predict the last one given the first n-1 MemberAPIs. For all models, we use SGD as the optimizer with exponential decay, in which learning rate starts at 1 and decay rate is set to 0.1. The activation function is set to be sigmoid function.

3.2 Metrics

In recommendation field, Area Under Receiver Operator Characteristic Curve (**AUC**) is a widely used metric [15]. It measures the goodness of order by ranking all the APIs with recommendation, including intra-Mashups and inter-Mashups orders. A variation of Mashups weighted AUC is introduced in [16, 17] which measures the goodness of intra- Mashups order by averaging AUC over Mashups. We adapt this metric in our experiments. For simplicity, we still refer it as AUC. It is calculated as follows:

$$AUC = \frac{\sum_{i=1}^{n} \#impression_i \times AUC_i}{\sum_{i=1}^{n} \#impression_i} \tag{5}$$

where n is the number of Mashups, $\#impression_i$ and AUC_i are the number of impressions and AUC corresponding to the i-th Mashup.

Besides, we introduce Average Precision (**AP**) to evaluate the performance of all methods. AP is calculated as the area under the precision-recall curve. AP considers two measurements (i.e., precision and recall) simultaneously. It has been widely used in Information Retrieval [18] and Computer Vision [19]. Hence AP is defined as:

$$AP = \frac{1}{2} \sum_i (Pre(i) + Pre(i-1)) \times (Re(i) - Re(i-1)) \tag{6}$$

where $Pre(i)$ and $Re(i)$ are the precision and recall at the i-th threshold, respectively. Larger AUC and AP values indicate better performance.

3.3 Performance Comparison

We compare DINRec with the following strong baselines that are designed for Service Recommendation:

- LR [20]. Logistic regression (LR) is a widely used shallow model before deep networks for recommendation task. We implement it as a weak baseline.
- NMF (nonnegative matrix factorization) [21]. This approach employs matrix factorization to user-item matrix with a constraint that the factorized matrix is positive.
- FM [19]. This approach is the traditional factorization machine. It concatenates user id and item id as sparsity feature, and learns the interactions between users and items to complete the user-item matrix.
- Wide&Deep [8]. In real industrial applications, Wide&Deep model has been widely accepted. It consists of two parts: (i) wide model, which handles the manually designed cross product features, (ii) deep model, which automatically extracts nonlinear relations among features. Wide&Deep needs expertise feature engineering on the input of the "wide" module. We follow the practice in [13] to take cross-product of Mashups composition and candidates as wide inputs. For example, in our dataset, it refers to the cross-product of Mashup rated APIs and candidate APIs.
- DeepFM [13]. This approach combines the power of factorization machines for recommendation and deep learning for feature learning in a new neural network architecture.

In this part, we conduct experiments by randomly removing a part of Mashup-API pairs from the Mashup-API interaction matrix to make the matrix with different densities (i.e., 10% to 90%) in diverse models. For example, 10% denotes that we remove 90% entries on the Mashup-API matrix. And then we set the 10% entries as training set, the remaining 90% entries as testing set [22]. The results of performance comparison on our dataset is shown in Table 3. Obviously, all the deep networks beat LR model significantly, which indeed demonstrates the power of deep learning. DeepFM with specially designed structures preforms better than Wide&Deep. The performance of FM is better than NMF due to learning the interactions between Mashups and APIs. In addition, the performance of DeepFM is better than FM due to applying deep neural network to learn the high-dimensional interactions between Mashups and APIs. DINRec performs best among all the competitors. We owe this to the design of local activation unit structure in DIN. DIN pays attentions to the locally related Mashup composition relationship by soft-searching for parts of Mashup invoked APIs that are relevant to candidate API. With this mechanism, DIN obtains an adaptively varying representation of Mashup composition relationship, greatly improving the expressive ability of model compared with other deep networks. The table only presents the results of training data sparsity that is 10%, 20%, 80% and 90%, all results and impact of training data sparsity will be described and discussed in the next subsection.

3.4 Impact of Training Dataset Sparsity

The training dataset sparsity is an important factor to impact recommendation performance. It represents how much information considered by Mashup on APIs we can utilize. To study impact of training data density, we set it from 10% to 90% with a step value of 10%. From Fig. 5, it can be found that our DINRec model achieves the best performance under all training data density. The performance of FM based approaches (e.g., FM and DeepFM) is better than LR and NMF. The performance of Wide&Deep is only worse than DINRec and DeepFM. Moreover, the AUC and AP values grow up with the increasing of training dataset sparsity. It is reasonable because when there is more training dataset, there is more information between Mashups and APIs will be collected, which is benefit for improving the recommendation accuracy.

3.5 Impact of Cosine Similarity Setting

In this subsection, we performed an empirical study on the effect of different cosine similarity setting of DINRec on the results. As mentioned in Section 1, it is necessary to find a suitable cosine similarity threshold when we extend our datasets by using Doc2simu model. To investigate the effect of this threshold, we performed an experiment under different setting of the threshold including 0.80, 0.82, 0.84, 0.86, 0.88, 0.90, 0.92, 0.94, 0.96. The results are listed in Fig. 6. It can be observed that when increasing similarity setting value from 0.80 to 0.96, the AUC and AP value show an upward trend at first, and then a downward trend. Obviously, the best performance of AUC and AP value is achieved when the value is set as 0.88. This phenomenon demonstrates that larger similarity setting values bring better recommendation results. But why the AUC and AP values drop when the cosine similarity value is greater than

Table 3. Performance comparison.

Methods	Sparsity of training dataset							
	Training dataset = 10%		Training dataset = 20%		Training dataset = 80%		Training dataset = 90%	
	AUC	AP	AUC	AP	AUC	AP	AUC	AP
LR	0.6302	0.5353	0.6383	0.5364	0.6331	0.5293	0.6354	0.5345
NMF	0.6552	0.5291	0.6953	0.5358	0.7724	0.5814	0.7921	0.5867
FM	0.7814	0.6742	0.7987	0.6811	0.8184	0.6994	0.8219	0.7067
Wide&Deep	0.8166	0.7198	0.8251	0.7282	0.8479	0.7473	0.8581	0.7526
DeepFM	0.8403	0.7317	0.8488	0.7402	0.8773	0.7677	0.8831	0.7754
DINRec	**0.8573**	**0.7686**	**0.8652**	**0.7714**	**0.8934**	**0.7934**	**0.9052**	**0.8016**

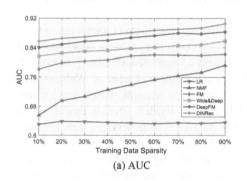
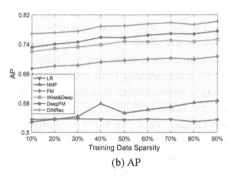

(a) AUC   (b) AP

Fig. 5. Impact of training dataset sparsity.

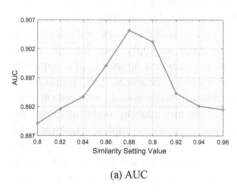
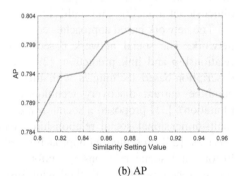

(a) AUC   (b) AP

Fig. 6. Impact of cosine similarity setting.

0.88 is a question that worth thinking about. Indeed, in theory, the higher the similarity the higher the final result should be better, but it can' be ignored that the corresponding amount of training dataset will be less. In the depth learning model, the amount of training dataset is proportional to the good results. For example, when the similarity is set to 0.86, there are 199314 training datasets, but when the similarity is set to 0.96, the training dataset is only 43832.

4 Related Works

Web API recommendation technique plays an important role in service-oriented computing and effectively improves the quality of service discovery [23]. A number of research works have been done on Web API recommendation. They are mainly classified into three types: Collaborative filtering-based methods, Semantic based methods, and network based methods.

Collaborative filtering-based methods make use of user activities and past interactions to learn preferences and generate recommendations. [24] incorporated functional interest, QoS preference and diversity feature to recommend top-N diversified Web services to users. [25] proposed a collaborative filtering approach to predict missing QoS based on the information of similar Web users and services. [26] incorporate user, topic, and service-related latent factors into service discovery and recommendation.

The semantic based approaches aimed at finding the highest matching degree services via semantic similarity computation. [1] proposed a semantic content-based recommendation approach by analyzing the context of intended service. [26] considered simultaneously both rating dataset and semantic content dataset of Web services using a probabilistic generative model. [25] proposed a semantic-based service discovery framework, consisting of user model, context model, service model and a service discovery process. The similarity usually calculates from services' functionality description with some topic model, such as LDA topic model. [27] presented a recommendation system to design Mashup applications, relying on the multi-dimensional information, such as similar Mashups, similar Web APIs, cooccurrence and popularity of Web APIs. [28] advanced the current state of the art for Web API search and ranking from mashups developers' point of view, by addressing two key issues: multi-dimensional modeling and multi-dimensional framework for selection.

The network based approaches consist of two parts: social network and information network. The social network based approaches tend to apply user interest, social relationship and link prediction. [29] proposed a combined approach that improves description-based techniques with these social ranking measures. [30] proposed to combine current discovery techniques (exploration) with social information (exploitation). [31] proposed a social-aware service recommendation model by exploring multi-dimensional social relationships among potential users, topics, Mashups, and services. [27] presented an approach based on user interest from their Mashup usage history and social relationships information. [32] proposed a social network-based service recommendation method with trust enhancement by employing matrix factorization and random walk algorithm. The information network based approaches mainly employ different kinds of information and multiple semantic meanings of meta paths to recommend service. [33] proposed an efficient consistent regularization framework to enhance Mashup discovery by leveraging HIN between Mashups and their components. [34] proposed to recommend services for Mashup creation by exploiting different types of relationships in service related HIN. Inspired by the above approaches and in view of their shortcomings, we propose a novel recommendation approach that integrates Mashup functional semantics to composition structure approach.

5 Conclusion and Future Work

This paper introduces an effective service recommendation approach for Mashup creation based on DIN. Excessive reliance on functional semantic information in previous research work is a bottleneck for capturing the diversity of Mashups composition relationship. To improve the expressive ability of the traditional models, a novel approach named DINRec is proposed to activate related Mashup composition relationships and obtain an adaptive representation vector for prior selected APIs which varies over different candidate APIs. Besides, a novel technique is introduced to help training industrial deep networks with small-scale dataset and further improve the performance of DINRec. Our method was examined on extended ProgrammableWeb dataset. The results demonstrate that our method outperforms several state-of-the art methods. In future work, neoteric activation unit and textual features of APIs under our framework deserves further investigation.

Acknowledgment. This work was supported in part by the National Natural Science Foundation of China under Grant 61572187, Grant 61872139, Grant 61772193 and Grant 61702181, in part by the Natural Science Foundation of Hunan Province under Grant 2017JJ2098, Grant 2018JJ2136, Grant 2018JJ3190 and Grant 2018JJ2139, and in part by the Educational Commission of Hunan Province of China under Grant 17C0642.

References

1. Zhong, Y., Fan, Y., Huang, K., Tan, W., Zhang, J.: Time-aware service recommendation for mashup creation. IEEE Trans. Serv. Comput. **8**(3), 356–368 (2014)
2. Cao, B., Liu, X.F., Liu, J., Tang, M.: Domain-aware Mashup service clustering based on LDA topic model from multiple data sources. Inf. Softw. Technol. **90**, 40–54 (2017)
3. Bai, B., Fan, Y., Tan, W., Zhang, J.: SR-LDA: Mining effective representations for generating service ecosystem knowledge maps. In: 2017 IEEE International Conference on Services Computing (SCC), pp. 124–131 (2017)
4. Cao, B., Liu, J., Wen, Y., Li, H., Xiao, Q., Chen, J.: QoS-aware service recommendation based on relational topic model and factorization machines for IoT Mashup applications. J. Parallel Distrib. Comput. (2018)
5. Li, C., Zhang, R., Huai, J., Sun, H.: A novel approach for API recommendation in mashup development. In: 2014 IEEE International Conference on Web Services, pp. 289–296 (2014)
6. Chen, Q., Sokolova, M.: Word2Vec and Doc2Vec in unsupervised sentiment analysis of clinical discharge summaries. arXiv preprint arXiv:1805.00352 (2018)
7. Zhou, G., et al.: Deep interest network for click-through rate prediction. In: Proceedings of the 24th ACM SIGKDD International Conference on Knowledge Discovery & Data Mining, pp. 1059–1068, July 2018
8. Cheng, H.T., et al.: Wide & deep learning for recommender systems. In: Proceedings of the 1st Workshop on Deep Learning for Recommender Systems (ACM), pp. 7–10 (2016)
9. Shan, Y., Hoens, T.R., Jiao, J., Wang, H., Yu, D., Mao, J.C.: Deep crossing: web-scale modeling without manually crafted combinatorial features. In: Proceedings of the 22nd ACM SIGKDD International Conference on Knowledge Discovery and Data Mining, pp. 255–262 (2016)

10. Covington, P., Adams, J., Sargin, E.: Deep neural networks for youtube recommendations. In: Proceedings of the 10th ACM Conference on Recommender Systems, pp. 191–198 (2016)
11. Bahdanau, D., Cho, K., Bengio, Y.: Neural machine translation by jointly learning to align and translate. arXiv preprint arXiv:1409.0473 (2014)
12. Qu, Y., et al.: Product-based neural networks for user response prediction. In: 2016 IEEE 16th International Conference on Data Mining (ICDM), pp. 1149–1154 (2016)
13. Guo, H., Tang, R., Ye, Y., Li, Z., He, X.: DeepFM: a factorization-machine based neural network for CTR prediction. arXiv preprint arXiv:1703.04247 (2017)
14. Le, Q., Mikolov, T.: Distributed representations of sentences and documents. In: Proceedings of the 31st International Conference on Machine Learning, pp. 1188–1196 (2014)
15. Fawcett, T.: An introduction to ROC analysis. Pattern Recogn. Lett. 27(8), 861–887 (2006)
16. Zhu, H., et al.: Optimized cost per click in taobao display advertising. In: Proceedings of the 23rd ACM SIGKDD International Conference on Knowledge Discovery and Data Mining, pp. 2191–2200 (2017)
17. He, R., McAuley, J.: Ups and downs: modeling the visual evolution of fashion trends with one-class collaborative filtering. In: Proceedings of the 25th International Conference on World Wide Web, International World Wide Web Conferences Steering Committee, pp. 507–517 (2016)
18. Yue, Y., Finley, T., Radlinski, F., Joachims, T.: A support vector method for optimizing average precision. In: Proceedings of the 30th Annual International ACM SIGIR Conference on Research and Development in Information Retrieval, pp. 271–278 (2007)
19. Rendle, S.: Factorization machines. In: 2010 IEEE International Conference on Data Mining, pp. 995–1000 (2010)
20. Hosmer Jr., D.W., Lemeshow, S., Sturdivant, R.X.: Applied Logistic Regression, vol. 398. Wiley, Hoboken (2013)
21. Lee, D.D., Seung, H.S.: Algorithms for non-negative matrix factorization. In: Advances in Neural Information Processing Systems, pp. 556–562 (2001)
22. Xie, F., Chen, L., Ye, Y., Zheng, Z., Lin, X.: Factorization machine based service recommendation on heterogeneous information networks. In: 2018 IEEE International Conference on Web Services (ICWS), pp. 115–122 (2018)
23. Xia, B., Fan, Y., Tan, W., Huang, K., Zhang, J., Wu, C.: Category-aware API clustering and distributed recommendation for automatic Mashup creation. IEEE Trans. Serv. Comput. 8(5), 674–687 (2014)
24. Klusch, M., Fries, B., Sycara, K.: Automated semantic web service discovery with OWLS-MX. In: Proceedings of the Fifth International Joint Conference on Autonomous Agents and Multiagent Systems, pp. 915–922 (2006)
25. Xu, S.Y., Raahemi, B.: A semantic-based service discovery framework for collaborative environments. Int. J. Simul. Modelling 15(1), 83–96 (2016)
26. Yao, L., Sheng, Q.Z., Ngu, A.H., Yu, J., Segev, A.: Unified collaborative and content-based web service recommendation. IEEE Trans. Serv. Comput. 8(3), 453–466 (2014)
27. Cao, B., Liu, J., Tang, M., Zheng, Z., Wang, G.: Mashup service recommendation based on user interest and social network. In: 2013 IEEE 20th International Conference on Web Services, pp. 99–106 (2013)
28. Bianchini, D., De Antonellis, V., Melchiori, M.: WISeR: a multi-dimensional framework for searching and ranking web APIs. TWEB 11(3), 19:1–19:32 (2017)
29. Tapia, B., Torres, R., Astudillo, H.: Simplifying Mashup component selection with a combined similarity- and social-based technique. In: Proceedings of the 5th International Workshop on Web APIs and Service Mashups, p. 8 (2011)

30. Torres, R., Tapia, B., Astudillo, H.: Improving Web API discovery by leveraging social information. In: Proceedings of the IEEE International Conference on Web Services, pp. 744–745 (2011)
31. Xu, W., Cao, J., Hu, L., Wang, J., Li, M.: A social-aware service recommendation approach for Mashup creation. In: 2013 IEEE 20th International Conference on Web Services, pp. 107–114 (2013)
32. Deng, S., Huang, L., Xu, G.: Social network-based service recommendation with trust enhancement. Expert Syst. Appl. **41**(18), 8075–8084 (2014)
33. Wan, Y., Chen, L., Yu, Q., Liang, T., Wu, J.: Incorporating heterogeneous information for Mashup discovery with consistent regularization. In: Pacific-Asia Conference on Knowledge Discovery and Data Mining, pp. 436–448 (2016)
34. Liang, T., Chen, L., Wu, J., Dong, H., Bouguettaya, A.: Meta-path based service recommendation in heterogeneous information networks. In: Sheng, Q.Z., Stroulia, E., Tata, S., Bhiri, S. (eds.) ICSOC 2016. LNCS, vol. 9936, pp. 371–386. Springer, Cham (2016). https://doi.org/10.1007/978-3-319-46295-0_23

Recommender Systems

Co-purchaser Recommendation Based on Network Embedding

Jihong Chen[1], Wei Chen[1,2], Jinjing Huang[1], Jinhua Fang[1], Zhixu Li[1], An Liu[1], and Lei Zhao[1(✉)]

[1] School of Computer Science and Technology, Soochow University, Suzhou, China
jhchen@stu.suda.edu.cn,
{robertchen,huangjj,jhfang,zhixuli,anliu,zhaol}@suda.edu.cn
[2] Institute of Artificial Intelligence, Soochow University, Suzhou, China

Abstract. Although recommending co-purchasers for a target buyer on the group buying is an interesting problem, existing studies haven't paid attention to this topic. Different from the collaborator recommendation that only considers users with high similarity to the target user, co-purchaser recommendation takes both users with high and weak similarity into account, and the recommendation results can achieve high recall and diversity. However, the task turns out to be a challenging problem since it is hard to make a precise recommendation for buyers with weak similarity. To address the problem, we propose the following two methods. In the first one, we directly impose a penalty to the weakly similar co-purchasers in the embedding space. To further improve the recommendation performance, in the second one, we smoothly increase the co-occurrence probability of the weakly similar co-purchasers by truncated bias walk. Our experimental results on real datasets show that the proposed methods, particularly the latter, can effectively complete the co-purchaser recommendation and has a high recommendation performance.

Keywords: Group buying · Collaborator recommendation · Network embedding · Truncated walk

1 Introduction

Co-purchase, also known as group buying, in which people with the same merchandise interests form a group and conduct the purchase together to achieve discounts [1]. In recent years, we have witnessed the prosperity of it in some online shopping services (e.g., Taobao[1], Groupon[2], and PDD[3]) benefit from the advanced electronic payment technology and convenient express service.

In real applications, the co-purchase usually includes the following steps: Firstly, merchants promise to offer products or services with discount on the

---

[1] https://www.taobao.com/.
[2] https://www.groupon.com/.
[3] https://www.pinduoduo.com/.

© Springer Nature Switzerland AG 2019
R. Cheng et al. (Eds.): WISE 2019, LNCS 11881, pp. 197–211, 2019.
https://doi.org/10.1007/978-3-030-34223-4_13

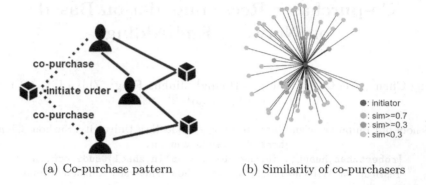

(a) Co-purchase pattern   (b) Similarity of co-purchasers

Fig. 1. A toy example of copurchase: (a) simple co-purchase pattern; (b) similarity of co-purchasers in one co-purchase on the TaoBao. The distance between nodes ∝ their shortest path and *sim* represents cosine similarity of their shopping history.

condition that a certain number of customers would make the purchase; Then, an initiator manually invite friends, followers, and like-minded people to participate in the purchase; Finally, co-purchasers accept the invitation and benefit from lower price which is unavailable to the individual buyer [2].

Most group recommendation algorithms generate an item suggestion for a group [3]. However, as the above discussion shows, the co-purchase is a distinctive group activity, in which members of the group are uncertain until the purchase order is submitted, that is to say, we need to pay more attention to who is the appropriate co-purchaser for an initiator, rather than focusing on what is the right item for a group like the traditional group recommendation. How to choose the appropriate co-purchasers? There are two methods for the problem: the manual invitation of the initiator and the automatic recommendation of the recommendation system. Although the former is a classic solution commonly used by industry [2], it still has many flaws, such as inefficiency, insufficient demand for co-purchaser, and the limited quantity of participants. Compared with the first method, the latter one is a more promising method, as the superiority of the recommendation system has been proven in many other areas [4].

If we consider a co-purchase transaction as a collaboration between the initiator and the co-purchaser, then we can solve the co-purchaser problem build upon the positive experiences of previous collaborator recommendation tasks. Much literature has been published [5–7] on collaborator recommendation systems as well as their real-world applications, such as co-author recommendation in the academic social network [5,6,8], developer recommendation in the open source community [7], and co-star recommendation in the film industry [9].

In the above collaborator recommendation tasks, finding robustly similar users for the target user is a core task, for example, in an academic network, people tend to repetitively collaborate with fellow researchers with close researcher topics [6,8]. It is also a classic idea in many recommendation algorithms, such as the typical user-based collaborative filtering approach distinguish the target users interests and preferences by aggregating the highest similar users [4]. However, the co-purchaser recommendation is a special scenario, in which not all co-purchasers have high similarity with the initiator. As shown in Fig. 1b, a large number of weakly similar users also participated in the co-purchase transaction, but they are usually not noticed by existing recommendation methods.

The co-purchaser recommendation is a challenging task, since the identification of potential co-purchasers from the weakly similar users is not easy. To tackle the problem, two embedding strategies are proposed, which capture weakly similar co-purchasers from different perspectives. In the first one, we propose a multi-layered learning architecture with PathSim [5] diffusion, namely PathSim Diffused Structural Deep Network Embedding (PDSDNE), which connects weakly similar users by PathSim and directly impose a penalty to the mapping error of the weakly similar users. Obviously, it is a forthright strategy that is beneficial to the weakly similar user, but it will inevitably damage the original network structure. In the second one, we devise a co-occurrence model based on truncated walking paths, namely co-purchasers to vectors (cop2vec). More specifically, cop2vec can smoothly improve the co-occurrence probability of the weakly similar co-purchasers by truncated bias walk, and thus learn a more reasonable representation for co-purchasers. In this way, not only those co-purchasers who are highly similar to the initiator are close to the initiator, but also the potential co-purchasers with weakly similar to the initiator.

The contributions of our paper are summarized as follows:

- To the best of our knowledge, this is the first work that shows how to recommend co-purchasers in group buying. This is an important subject because co-purchaser recommendation has been proved to be more effective than the handcrafted invitation.
- We propose PDSDNE and cop2vec, two efficient co-purchaser recommendation strategies, which effectively perceive weakly similar co-purchasers.
- Through extensive experiments, demonstrates the efficacy and scalability of the presented methods in the co-purchaser recommendation task.

The rest of the paper is organized as follows. Section 2 presents related work. Section 3 introduces the embedding methods (PDSDNE and cop2vec) with the details of how to capture the weakly similar users for the co-purchaser recommendation. Section 4 describes the experimental setup and presents qualitative and quantitative results. Section 5 gives the conclusion with future work.

2 Related Work

2.1 Similarity Search

Similarity search is a basic operation in collaborative filtering, and it can be directly used as a simple strategy to find a collaborator [5,11]. When the input is in the form of a scoring matrix, the common similarity search approach including Cosine Similarity and Pearson Correlation Coefficient [11]. In the network analysis task [12], a large number of similarity search methods with different definitions of similarity have been proposed such as Common Neighbors, Jaccard Index, and Adamic-Adar Index, In addition to the above mentioned local-based function, Sun et al. [5] proposed a path-based similarity measure to suit peer objects.

2.2   Network Embedding

The low-dimensional representation learning of recommendation objects is a classic approach to the recommendation system [8,13,14], for example, one of the most efficient and best used recommend methods is matrix factorization in which users and items are represented in a low-dimensional latent factors space [4]. Network embedding aims at learning low-dimensional vectors for the vertices of a network [10,15,16], such that the proximities among the original network are preserved in the low-dimensional space.

Recent progress in neural embedding methods for linguistic tasks has dramatically advanced state-of-the-art Natural Language Processing (NLP) capabilities. These methods attempt to map words and phrases to a low dimensional vector space that captures semantic relations between words [17]. Specifically, Skip-gram with Negative Sampling (SGNS), also known as word2vec, set new records in various NLP tasks. Inspired by it, DeepWalk [15] is proposed as a method for learning the latent representations of the nodes of a social network. The method aims to transplant the word-context concept in documents into networks, and combines truncated random walk with Skip-gram model to achieve this. We can utilize the model to learn the low-dimensional and distributed embedding of nodes as it facilitates the preservation of its structural context— local neighborhoods—in the original network. On this basis, WALKLETS [18] and node2vec [19] further extend DeepWalk utilize high-order proximities and bias walk. LINE [16] is recently proposed embedding approach for large scale networks. By design, LINE learns two representations separately, one preserving first-order proximity and the other preserving second-order proximity. Then, Wang et al. extended the method using a deep autoencoder [20].

3   Co-purchaser Recommendation

3.1   Formalizations

In this section, we first introduce the concept of interaction networks, and then give a formal definition of the co-purchase recommendation problem.

Interaction Networks. An interaction network is defined as a graph $G = (V, E)$, where V and E represent the node set and the edge set. For example, one can represent the interaction network in Fig. 1a with buyers and products as nodes, wherein edges indicate the interactions, such as the purchase (buyer to product) and the trust (buyer to buyer). In order to ensure data consistency, the edges are unweighted.

Co-purchaser Recommendation. Network embedding is to learn a low-dimensional vector for each node. Let U be the vector set of all buyers and let P be the vector set of all possible products. On the basis, we define the co-purchaser recommendation goal as follows: Given a buyer i and a product j, we can now assign a co-purchase score S to each buyer c, it can be written as

$$S(c, (i, j)) = U_c \cdot U_i{}^T + U_c \cdot P_j{}^T \tag{1}$$

3.2 Multi-layered Learning Architecture with PathSim

In this section, we first define the notation of PathSim diffusion. Then we introduce the multi-layered learning model of PDSDNE. At last we present some discussions and analysis on the model.

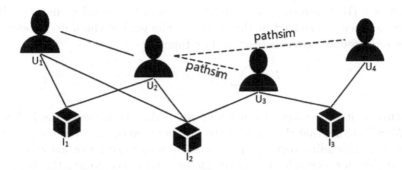

Fig. 2. The PathSim diffused network (The initiator links to the weakly similar users by PathSim).

Given a network $G = (V, E)$, we can obtain its adjacency matrix $S \in \mathbb{R}^{V \times V}$. we have $s_{ij} = 1$ if there exists a link between i and j, and $s_{ij} = 0$ otherwise. for each row $s_i = \{s_{ii}\}_{j-1}^{n}$. In reality, the observed links only account for a small portion. There exist many co-purchasers who have some connectivity with the initiator but no direct links, especially weakly similar co-purchasers. we define the PathSim diffused matrix $P \in \mathbb{R}^{V \times V}$ by extending PathSim measurement proposed in [5] as follows:

$$p_{ij} = \begin{cases} s_{ij} & \text{if } s_{ij} \neq 0 \\ \frac{2 \times path(i,j)}{path(i,i) + path(j,j)} & \text{if the shortest length between } i \text{ and } j < R \\ 0 & \text{otherwise} \end{cases} \tag{2}$$

Where $path(i, j)$ is the number of paths between i and j, $path(i, i)$ is the number of paths between i and i, $path(j, j)$ is the number of paths between j and j. Notice the length of all paths is the shortest length of the path between i and j. Where R is the range of the PathSim diffusion. In theory, the score of the p can measure the connectivity between vertexes and normalized by the visibility of vertexes. As shown in Fig. 2, there is 1 path between u_2 and u_4, 1 path between u_4 and u_4 and 3 paths between u_2 and u_2. We can calculate that the score of the p is 0.5.

Intuitively, if two vertexes share many common neighbors, they tend to be similar. As shown in Fig. 2, u_1 has the same shopping history as u_2, so they are similar and can be purchasing together. to model the neighbor structure, also known as the second-order proximity, autoencoders have emerged as one of the commonly used building blocks [20,21]. An autoencoder performs two actions, i.e. the encoder and decoder. The encoder consists of multiple non-linear functions $f(\cdot) = f_{\theta_k}(\cdots f_{\theta_1}(\cdot))$ that map the input data to the representation space. The decoder also consists of multiple functions $g(\cdot) = g_{\hat{\theta}_1}(\cdots g_{\hat{\theta}_k}(\cdot))$ mapping the representations in representation space to reconstruction space. Let us assume that $f_{\theta_1}(x) = \sigma(W_1 x + b_1)$ and $g_{\hat{\theta}_1}(x) = \sigma(\hat{W}_1 x + \hat{b}_1)$, where σ is the activation function, $\theta = (W, b)$ are the parameters involved in the encoder, and $\hat{\theta} = (\hat{W}, \hat{b})$ are the parameters involved in the decoder. The goal of the autoencoder is to minimize the following reconstruction loss function

$$L_n = \sum_i \|s_i - g(f(s_i))\|_2^2 \tag{3}$$

Naturally, it is necessary for network embedding to preserve the link structure. We wish to see that the stronger the link between the two vertexes, the more similar their embedding vectors. Many classical recommendation algorithms have the objective, for example, in matrix factorization techniques, the higher the user's rating of the item, the more overlapping their latent vectors. In addition, by adding the penalty of PathSim score, these weakly similar co-purchasers will be close to the initiator in the embedding space. The loss function for this goal is defined as follows:

$$L_l = \sum_{i,j} p_{i,j} \|f(s_i) - f(s_j)\|_2^2 \tag{4}$$

To preserve both neighborhood structure and link structure, we jointly minimize the objective function by combining Eqs. 4 and 3:

$$L = L_l + \alpha L_n \tag{5}$$

As shown in previous works [20], we use stochastic gradient descent (SGD) to optimize the model. The key step is to calculate the partial derivative of the parameters $\{\theta, \hat{\theta}\}$. Ultimately, the embedding vectors can be computed by the encoder. However, While the PathSim diffusion can be beneficial to weakly similar co-purchasers. It can also damage the original network structure and bring some negative impact on the general reconstruction of the network. We want to use a smoother way to perceive weakly similar co-purchasers and minimize the impact on the basic network features.

3.3   Co-occurrence Model Based on Truncated Walk

For the consideration of being self-contained, we briefly review the key idea of the co-occurrence model. The co-occurrence model is first used for linguistic tasks, and attempt to map words to a low dimensional vector space that captures semantic relations between words. Specifically, the SGNS model aims to maximize the co-occurrence probability among the words that appear within a window. Inspired by it, DeepWalk [15] is proposed as a method for learning the latent representations of the nodes of a social network. The method samples a set of paths from the input graph using the truncated random walk. Each path sampled from the graph corresponds to a sentence from the corpus, where a node corresponds to a word. Given a path consisting of nodes $w1 - wk$, The co-occurrence model objective is to maximize the following term:

$$\frac{1}{K}\sum_{i=1}^{K}\sum_{-c<j<c}\log P(w_{i+j}|w_i) \tag{6}$$

Where c is the context window size. Applying negative sampling [17], P is defined as:

$$P(w_{i+j}|w_i) = \sigma(\mathbf{u_i^T u_j}) + \sum_{t\in NS}\sigma(-\mathbf{u_i^T u_t}) \tag{7}$$

Where $\sigma(x) = 1/(1 + \exp(-x))$, and NS is the negative samples for w_i.

By applying the co-occurrence model in formula 6, Frequently co-occurring nodes in a path share similar neighborhoods (In this section, the definition of neighborhoods is slightly different from PDSDNE, it usually refers to the window in paths, not just the one-hop neighbors in networks.) and get similar embedding [13,15]. For example (see Fig. 3a), u_1 may co-occur with u_2 most frequently, we will naturally recommend u_1 as a co-purchaser to u_2. However, it is still challenging to recommend weakly similar co-purchaser like u_3. Unfortunately, a large number of co-purchaser have distributed the long tail of similarity, and they are difficult to be perceived by the existing recommendation approaches.

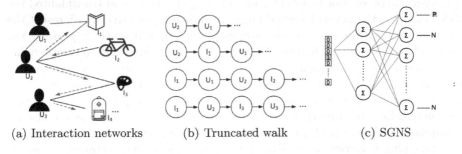

(a) Interaction networks   (b) Truncated walk   (c) SGNS

Fig. 3. Overview of co-occurrence model

To address the weakly similar co-purchaser problem, We propose a novel neighborhoods sampling strategy that is beneficial to the weakly similar co-purchasers, Which can smoothly improve the co-occurrence probability of the weakly similar co-purchasers by truncated bias walk.

General Neighborhoods Sampling Strategy. Network embedding methods based on the SGNS architecture reconstructs network features by learning the notion of neighborhoods. We first briefly introduce the general neighborhoods sampling strategy—truncated random walk, formally, a random walk begins at the source node s and gets a node sequence of fixed length le, let n_i denote the ith node in the sequence, starting with $n_0 = s$. The node n_i is generated by the following distribution.

$$P(n_i = v | n_{i-1} = u, i < le) = \begin{cases} \frac{\pi_{uv}}{\sum_{x \in \Gamma(u)} \pi_{ux}} & \text{if } v \in \Gamma(u) \\ 0 & \text{otherwise} \end{cases} \tag{8}$$

where $\Gamma(u)$ is the one-hop neighbors of node u, and π_{uv} is the unnormalized transition probability between nodes u and v (e.g., the edge weights w_{ux}).

However, the simple way not allow us to account for the network structure and guide our search procedure to explore different types of network neighborhoods. Additionally, the farther nodes are difficult to capture, and may not even be touched in the finite number of the truncated walk. As shown in Fig. 1a, consider a truncated random walk arrived at the purchaser node u_2, after which the walk will have multiple paths to reach another purchaser node u_1, that is, u_2 will frequently coexist with u_1 in the node sequence generated by walks, and finally the SGNS model maps two nodes that frequently coexist into two close feature vectors. In contrast, there are rare opportunities to travel from u_2 to u_3, that is, u_2 will rarely coexist with u_3, and finally the SGNS model maps two nodes that rarely coexist into two irrelevant feature vectors.

Biased Neighborhoods Sampling Strategy. Prior studies have found the equivalence between word-context and node-neighborhood and transplanted the SGNS model to the network embedding. The daily corpus can only represent the common word feature, likewise, the truncated random walk con only preserve the basic and general network feature. We want to get more information that benefits weakly similar users. For example, a student may face the following scenarios on the group buying platform: he may co-purchase with his classmates, which is very intuitive because they are robustly similar; he may also co-purchase with buyers of a safety helmet because they have a consistent need for helmets; he might even co-purchase with buyers of a rucksack, however, there are incongruities between their shopping behavior. In reality, the last scenario is very common. There is no aligned preference between co-purchasers, just an intersection under a large category (e.g., outdoor activities).

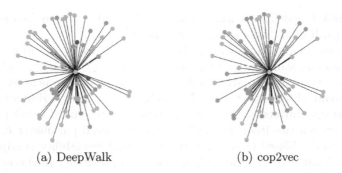

(a) DeepWalk   (b) cop2vec

Fig. 4. The number of times co-occurrence of co-purchaser in one co-purchase on the TaoBao (•: initiator, •: co-occurrence> 100, •: co-occurrence> 50, •: co-occurrence> 10, •: co-occurrence<= 10). (Color figure online)

Building on the above observations, we design a flexible neighborhood sampling strategy which allows us to perceive the weakly related nodes effectively and sensitively. We achieve this by developing a flexible bias walk procedure that can explore farther neighborhoods with co-purchase tendencies. For example, a bias walk that just traversed edge (t, u) and now resides at node u. The walk now needs to decide on the next step so it evaluates the transition probabilities π_{ux} on edges (u, x) leading from u. We set the unnormalized transition probability to $\pi_{ux} = \alpha_{pkl}(t, u, x) \cdot w_{ux}$, where

$$
\alpha_{pkl}(t, u, x) = \begin{cases} p & \text{if} \quad t == x \\ k \cdot \text{sim}(t, x) & \text{if} \quad t \in I \quad \text{and} \quad x \in I \\ \frac{lw}{1 + |(w_t - w_x|} & \text{if} \quad x \in \Gamma(t) \\ 1 & \text{otherwise} \end{cases}
\tag{9}
$$

In the equation, p, k, and l are the preset biased parameters that control the tendency of truncated walks. w_t is the purchase edge associated with t. U and I are the users set and the items set. $\text{sim}(t, x)$ denotes the approximate index between item nodes t and x. We simply set the approximate index to $\text{sim}(t, x) = (\Gamma(i) \cap \Gamma(j))/(\Gamma(i) \cup \Gamma(j))$, although we can calculate a more accurate approximate index using side information attached to products.

Parameter p controls the likelihood of immediately revisiting a node in the walk. If we set a value greater than 1, it would lead the walk to explore the nodes that have already visited, and this would keep the walk "local" close to the starting node [19]. Setting it to a low value ensure the walk spreads out at a faster rate and avoids "bigram" redundancy in node sequences.

Parameter k is the key to ensure that the initiator node was able to perceive the weakly similar co-purchaser. Setting k to a high value, the walking strategy encourages the walk to diffuse along the chains (the red line in Fig. 3a) that are composed of related goods. These chains are like backbones in the network, by approaching the chains, the paths generated by walks makes more meaningful when the walk is moving far away. That is, the farther co-purchasers attached to the chain will more likely coexist with the initiator.

Going back to Fig. 3a, Buyer u_2 bought a product i_2 in online shopping, consider a random walk that just traversed edge (i_2, u_2) and now resides at node u_2. There are several alternative nodes (i_3, u_1, i_2, i_1) on the next step. At this point, we could observe that i_3 (safety helmet) has a high similarity with i_2 (bicycle) because buyers of the two commodities are almost overlapping. The similarity between two items is amplified by the biased parameter k, and then propagated to the biased factor, and the transition probability is adjusted to a larger value. That is to say, the walk has the high possibility to choose i_3 on the next step, and the walking path is like a backbone of the interaction network. Finally, the purchasers of i_3, such as u_3, will appear in the walking path and form a co-occurrence with u_1 and u_2. SGNS model will capture the phenomenon of co-occurrence and map it to the embedding space.

Parameter l allows us to adjust the stay rate of the walk. If two buyers have a consistent preference for one item or two items get a consistent rating by one buyer, the item or buyer have a higher value of the stay. The higher the numeric of the parameter, the larger the influence of the stay rate, and vice versa.

By adjusting the biased parameters, the biased strategy of walking can flexibly explore the neighborhoods of nodes in interaction networks. In particular, the parameters allow our walk procedure to generate more meaningful co-occurrence paths for the co-purchaser recommendation. A toy example is shown in Fig. 4, the weakly similar co-purchasers (cyan nodes in Fig. 4a) get a higher number of times co-occurrence. As discussed in the formula 6, the weakly similar co-purchasers will gain better embedding vectors because they have higher relevance to the initiator. In addition, the biased walk is a smooth strategy and does not damage the original network information. That is, the original structure of networks and the adaptability to weakly similar co-purchasers both can be taken into account.

4   Experiments

In this section, we conduct various experiments to demonstrate the effectiveness of our proposed methods. First, we describe three real-world datasets on online shopping and visualize the embedding of a small number of purchasers. Secondly, we evaluate the methods by the top-k purchaser recommendation task. Finally, we report the co-purchaser detection experimental results on multiple online shopping datasets and present the influence of biased parameters.

4.1   Datasets

We design experiments on three widely adopted online shopping datasets, including Epinions, Amazon Electrol and TaoBao IJCAI16. Note, Amazon and TaoBao are processed into 5-core subsets, which all users and items have at least five records. Additionally, to enhance the diversity of truncated walk, we add a trust edge between two buyers when they have multiple co-purchase. Table 1 shows some statistics about datasets.

Table 1. Statistics of datasets

	Buyer nodes	Product nodes	Purchase edges	Trust edges
Epinions	40163	61273	664823	269649
Amazon	158694	61848	1612208	574868
TaoBao	35295	17617	250489	71244

4.2 Visualization of the Embeddings

In this part, we visualize the embeddings of buyers learned by PDSDNE and cop2vec. We compared two classic embedding methods like SVD and Deep-Walk. The results are shown in Fig. 5 where the buyers of the same item were highlighted with the same color. While the PDSDNE can be effective for the 2D embedding, it can also present a sparse form. Network embedding methods based on the truncated walk has a natural advantage in dealing with this problem, DeepWalk can map purchasers of the same item more closely. On that basis, cop2vec can further compact these nodes that are mapped to remote locations due to the weak similarity.

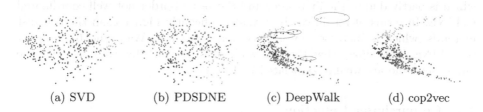

(a) SVD   (b) PDSDNE   (c) DeepWalk   (d) cop2vec

Fig. 5. Visualization of purchaser (Color figure online)

4.3 Top-k Purchaser Recommendation

Although purchaser recommendation is not common on many e-commerce platforms, it is a critical part of group buying because we need to decide whom to recommend products to. To split the test set, We randomly selected 20% items from the TaoBao dataset and removed their 50% purchaser. After the model training, we choose Top-k close purchasers for the item in the embedding space, which are considered to be the most likely buyers to purchase the item. In order to comprehensively evaluate the effectiveness of the recommendation, we not only employ two state-of-the-art embedding models as baselines, including LINE [16] and DeepWalk [15]. But also fully compared the two proposed methods.

For a fair comparison, we use a 128 dimensions vector to denote a node in all methods. In LINE, as suggested in [16], the representation is directly concatenated by first-order (dimension 64) and second-order (dimension 64). In addition, we still use the same parameters for the truncated walk. The number

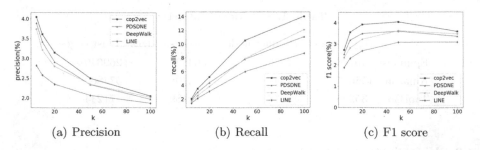

Fig. 6. The performance evaluation of Top-k purchaser recommendation.

of walks per node is 50, and the walk length is 30. The context window is 8, and the size of negative samples is 5. In PDSDNE, the structure of two-layer encoder are 1000 and 128, and this is also the case with the decoder. In cop2vec, the biased parameters are tuned to be optimal.

As shown in Fig. 6, the walk-based network embedding method (DeepWalk and cop2vec) outperforms the proximity-based method LINE. When k is taken as 50, the performance of cop2vec is better. Compared with PDSDNE, precision is improved by 6%. Compared with LINE, the precision is improved by 23%, which is partly due to the first-order and the second-order not well coordinated in LINE. In terms of recall, cop2vec was significantly higher than the contrast methods, which is increased by 41% compared to DeepWalk and 67% higher than LINE. This shows that the bias walk strategy can effectively perceive the purchasers who are weakly associated with items.

4.4 Co-purchaser Detection

In this section, we evaluate our proposed method on the co-purchaser recommendation task. Given a purchase initiator and his order for a certain item, we want to select the possible co-purchaser candidates. Note that current group buying platforms encourage buyers to sign in using social accounts, that is, we can give priority to recommending co-purchaser from a group of social accounts, rather than recommending co-purchaser from the whole buyers.

We choose 20% of the items from datasets and remove their n purchaser as a true buyer set. Additionally, we add $(n-1)/2$ unpurchased users as a false buyer set for each selected item. Where n is 50% of an item's total number of buyers. Select two buyers from the true buyers set, one as the initiator and one as the co-purchaser to form a positive sample, and finally generate $n(n-1)/2$ positive samples. Select a buyer from true buyers set and false buyers set respectively, one as the initiator and one as the co-purchaser to form a negative sample, and finally generate $n(n-1)/2$ negative samples. We use AUC (Area Under Curve) score to evaluate co-purchase intentions of positive and negative samples, where the co-purchase intention can be represented by the Hadamard product of the embedding vectors.

Table 2. AUC (Area Under Curve) scores for co-purchaser prediction

	SVD	CN	LINE	SDNE	DeepWalk	node2vec	PDSDNE	cop2vec
Epinions	0.789	0.708	0.765	0.744	0.802	0.806	0.815	**0.832**
Amazon	0.620	0.549	0.546	0.589	0.656	0.655	0.582	**0.687**
TaoBao	0.703	0.645	0.668	0.681	0.683	0.696	0.717	**0.764**

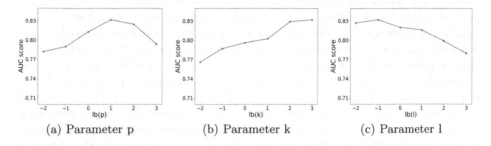

(a) Parameter p   (b) Parameter k   (c) Parameter l

Fig. 7. Parameter sensitivity

We conduct experiments on three different scale datasets and compare them with two traditional methods including SVD (Singular Value Decomposition), CN (Common Neighbors), and not just network embedding methods. The performances on three datasets are summarized as Table 2. We observe that cop2vec is consistently better than all the comparison methods.

On Epinions dataset, the performance of co-purchaser recommendations is the best, and we attribute this to a large number of real trust edges on the dataset. The AUC score of PDSDNE is 9% higher than SDNE, 6% higher than LINE, and 2% lower than cop2vec. On Amazon dataset, the walk-based network embedding method is significantly higher than other types of methods, and the worst-performing DeepWalk is also 20% better than the proximity-based embedding method LINE. We can see that the performance of cop2vec gain is more significant on Taobao dataset, the AUC score is 7% higher than PDSDNE, 11% higher than DeepWalk, and 19% higher than Common Neighbors.

4.5 Parameter Sensitivity

We investigate the parameter sensitivity in this section. Specifically, we mainly evaluate how the different choices of biased parameters affect the results of the co-purchaser recommendation. We report AUC score on the Epinions in Fig. 7. Intuitively, we can see that the performance raises when the value of parameter p increase, as shown in [19], a high p ensures that the walk does not go too far from the start node. We also observe that performance tends to saturate once the biased parameter k reaches around 8. Interestingly, we keep the parameter l at a small figure and get a good performance. This experiment suggests that we don't need to pay too much attention to the "closed-loop" structure in the co-purchaser recommendation task.

5   Conclusions

As an emerging online shopping form, group buying has been restricted by the co-purchaser recommendation problem. Both the handcrafted invitation and the classic collaborative filtering do not solve the problem well. In this paper, we present network embedding based methods to address the co-purchaser recommendation challenge. To cope with the problem that traditional algorithms are desensitized to the weakly similar nodes, we propose two novel co-purchaser recommendation method, namely PDSDNE and cop2vec, particularly the latter, which effectively perceive weakly similar nodes and maintain the original network information. Experiments on real-world datasets verify the effectiveness of our proposed approaches. For future work, incorporating side information such as stores, product categories, and attributes of buyers constitutes a heterogeneous network with more diversity of the bias walk, which may further improve the co-purchaser recommendation performance.

Acknowledgment. This work was supported by the National Natural Science Foundation of China (Grant No. 61572335, 61572336, 61902270), and the Major Program of Natural Science Foundation, Educational Commission of Jiangsu Province, China (Grant No. 19KJA610002), and the Natural Science Foundation, Educational Commission of Jiangsu Province, China (Grant No. 19KJB520052, 19KJB520050), and Collaborative Innovation Center of Novel Software Technology and Industrialization, Jiangsu, China.

References

1. Liu, H., Wang, W., Liu, D., Wang, H., Du, N.: HappyGo: a field trial of local group buying. In: ACM 2012 Conference on Computer Supported Cooperative Work, pp. 505–508. ACM, Seattle (2012)
2. Prospectus of Pinduoduo (A famous group buying platform in China). https://sec.gov/Archives/edgar/data/1737806/000104746918004833/a2235994zf-1.htm. Accessed 20 Jan 2019
3. Qin, D., Zhou, X., Chen, L.: Dynamic Connection-based Social Group Recommendation. IEEE Trans. Knowl. Data Eng. (2018)
4. Yue, S., Martha, L., Alan, H.: Collaborative filtering beyond the user-item matrix: a survey of the state of the art and future challenges. ACM Comput. Surv. **47**(1), 1–45 (2014)
5. Sun, Y., Han, J., Yan, X., Yu, P.S., Wu, T.: PathSim: meta path-based top-k similarity search in heterogeneous information networks. Proc. VLDB Endowment **4**(11), 992–1003 (2011)
6. Gustavo A.P., Hector G.C., Francisco J.C., Lucia R.: Recommending intra-institutional scientific collaboration through coauthorship network visualization. In: Proceedings of the 2013 Workshop on Computational Scientometrics: Theory & Applications, pp. 7–12. ACM, San Francisco (2013)

7. Chen, X.: Study on cooperator recommendation of virtual collaborative community. J. Softw. **8**(11), 2908–2916 (2013)
8. Chen, T., Sun, Y.: Task-guided and path-augmented heterogeneous network embedding for author identification. In: Proceedings of the Tenth ACM International Conference on Web Search and Data Mining, pp. 295–304. ACM, Cambridge (2017)
9. Guo, Z., Li, H.: Link prediction of actor cooperation lationship in heterogeneous information network. Comput. Eng. **43**(1), 219–225 (2017)
10. Cui, P., Wang, X., Pei, J., Zhu, W.: A survey on network embedding. IEEE Trans. Knowl. Data Eng. **31**(5), 833–852 (2019)
11. McLaughlin, M.R., Herlocker, J.L.: A collaborative filtering algorithm and evaluation metric that accurately model the user experience. In: Proceedings of the 27th Annual International ACM SIGIR Conference on Research and Development in Information Retrieval, pp. 329–336. ACM, Sheffield (2004)
12. Víctor, M., Fernando, B., Juan-Carlos, C.: A survey of link prediction in complex networks. ACM Comput. Surv. **49**(4), 1–33 (2016)
13. Wang, J., Huang, P., Zhao, H., Zhang, Z., Zhao, B., Lee, D.L.: Billion-scale commodity embedding for E-commerce recommendation in Alibaba. In: Proceedings of the 24th ACM SIGKDD International Conference on Knowledge Discovery, pp. 839–848 (2018)
14. Wen, Y., Guo, L., Chen, Z., Ma, J.: Network embedding based recommendation method in social networks. In: Companion of the Web Conference 2018, pp. 11–12 (2018)
15. Perozzi, B., Al-Rfou, R., Skiena, S.: DeepWalk: online learning of social representations. In: Proceedings of the 20th ACM SIGKDD International Conference on Knowledge Discovery and Data Mining, pp. 701–710. ACM, New York (2014)
16. Tang, J., Qu, M., Wang, M., Zhang, M., Yan, J., Mei, Q.: LINE: large-scale information network embedding. In: Proceedings of the 24th International Conference on World Wide Web, Florence, Italy, pp. 1067–1077 (2015)
17. Mikolov, T., Sutskever, I., Chen, K., Corrado, G.S., Dean, J.: Distributed representations of words and phrases and their compositionality, In: Advances in Neural Information Processing Systems, pp. 3111–3119 (2013)
18. Perozzi, B., Kulkarni, V., Chen, H., Skiena, S.: Don't walk, skip!: Online learning of multi-scale network embeddings. In: Proceedings of the 2017 IEEE/ACM International Conference on Advances in Social Networks Analysis and Mining, New York, USA, pp. 258–265 (2017)
19. Grover, A., Leskovec, J.: node2vec: scalable feature learning for networks, In: Proceedings of the 22nd ACM SIGKDD International Conference on Knowledge Discovery and Data Mining, pp. 855–864. ACM, San Francisco (2016)
20. Wang, D., Cui, P., Zhu, W.: Structural deep network embedding. In: Proceedings of the 22nd ACM SIGKDD International Conference on Knowledge Discovery and Data Mining, pp. 1225–1234. ACM, San Francisco (2016)
21. Cao, S., Lu, W., Xu, Q.: Deep neural networks for learning graph representations. In: Thirtieth AAAI Conference on Artificial Intelligence, Arizona, February, pp. 1145–1152 (2016)

Community-Based Recommendations on Twitter: Avoiding the Filter Bubble

Quentin Grossetti[1], Cédric du Mouza[1], and Nicolas Travers[1,2(✉)]

[1] CEDRIC Lab, CNAM Paris, Paris, France
{quentin.grossetti,cedricdu.mouza}@cnam.fr
[2] Research Center, Léonard de Vinci Pôle Universitaire,
Paris La Défense, Paris, France
nicolas.travers@devinci.fr

Abstract. Due to their success, social network platforms are considered today as a major communication mean. In order to increase user engagement, they rely on recommender systems to personalize individual experience by filtering messages according to user interest and/or neighborhood. However some recent results exhibit that this personalization of content might increase the *echo chamber* effect and create *filter bubbles*. These filter bubbles restrain the diversity of opinions regarding the recommended content. In this paper, we first realize a thorough study of communities on a large Twitter dataset to quantify how recommender systems affect users' behavior and create filter bubbles. Then we propose the *Community Aware Model* (CAM) to counter the impact of different recommender systems on information consumption. Our results show that *filter bubbles* concern up to 10% of users and our model based on similarities between communities enhance recommender systems.

Keywords: Twitter · Communities · Filter bubble · Recommender system

1 Introduction

Social networking has become a major way to share and discover information on the Internet. Users generally connect since they know each other in real life or share a common interest. Since received content from the flow is related to people with whom they are connected to, users may consequently find their opinions constantly echoed back which creates an *echo chamber* [8], that may skew their point of view. Moreover, it has been theorized that this phenomenon is reinforced by recommender systems [18] massively used to enhance users' engagement by personalizing individual experience. Consequently, they tend to focus on highly relevant messages mainly based on users' neighborhood and/or interests. Recently critics argued that such systems are impoverishing user opportunities to be displayed to diversified information, so called the *"filter bubble"*.

The link between Recommender Systems (RS) and filter bubbles is not clearly characterized in literature and we particularly target this issue in this paper. So

© Springer Nature Switzerland AG 2019
R. Cheng et al. (Eds.): WISE 2019, LNCS 11881, pp. 212–227, 2019.
https://doi.org/10.1007/978-3-030-34223-4_14

we first extract communities with a traditional community detection algorithm in a real `Twitter` dataset. This algorithm which relies mainly on topological properties (so not topic-centric) will group people who are close and strongly connected in the network, because they know each other, are geographically close and/or share common interests. Then we perform analysis to detect *filter bubbles* by measuring how often messages leave their community of origin and we try to understand how RS focus on content originated from a reduced number of communities. To achieve this we propose to characterize users by a community profile based on their interactions with communities through messages provenance. Then we show that recommendations provided by RS may differ from users' community profile and generate a *filter bubble* for some users. Therefore, we advocate the fact that *filter bubbles* can be characterized by topology-based communities, further works on opinion mining are out of the scope of this article.

Our second objective is to tackle this *filter bubble* effect for these users through a re-ranking of their recommendations to be more respectful of their community profile. Our proposal can be deployed on top of any RS without modifying its implementation. We show that our solution significantly improve the quality of recommendations by matching more closely users' community profile and by reducing the *filter bubble* effect at a limited computation cost.

In a nutshell, the main contributions of the paper are:

1. A community analysis to study how information is propagated through communities to characterize *echo chambers*,
2. A measure and an analysis of the *filter bubble* effect from respectively a community and a user's point of view,
3. A novel re-ranking strategy that relies on users' community profile and the community network to reduce the *filter bubble* effect.

2   Related Work

Most popular social network platforms such as *Facebook, Baidu, Twitter* or *Instagram* gather millions of users. To help them find relevant content, these platforms largely rely on RS. Recently, some works have shown that these platforms have to face two simultaneous effects that affect user points of view. First, the *"echo chamber"* phenomenon means that some users tend to consume only information from the same ideological alignment. This leads to biased opinions. The second effect, due to the personalization of content from recommender systems, traps users in a *"filter bubble"* as described by Eli Pariser [18].

Studies on *"echo chambers"* were initially conducted in social sciences to investigate how people tend to bind with similar people, creating communities and having difficulties to access opposite view points. It has been partially described and analyzed in [7,16]. They conclude that people tend to choose news articles from sources aligned with their political opinions. [3] also shows that people tend to connect to each other on social platforms following an homophily behavior, so to bind with similar people. A large study [5] focusing on filter bubbles and echo chambers states that this phenomenon is not limited to the

Table 1. Main features of the Twitter dataset

# nodes	2,2M	# edges	325.5M	# tweets	3,002M
Avg. path length	3.7	Avg. out-deg	57.8	Max out-deg	349K
Diameter	15	Avg. in-deg	69.4	Max in-deg	185K

digital era since social media users only mimic traditional offline reading habits. In short, *echo chambers* is a natural phenomenon which has existed for a long time before Facebook and the echo chamber on social networks is due to this real-life behavior, homophily, which is only replicated on social platforms.

While it is commonly admitted that *echo chambers* exist, there is no indisputable evidence of the existence of *"Filter Bubbles"*. Indeed, it is unclear whether recommender system algorithms amplify the echo chamber phenomenon or not. Some studies have tried to quantify this phenomenon. [5] studied web-browsing habits of 50,000 US-located people. To our knowledge, this is the largest study on *filter bubbles* and *echo chambers* phenomena. They observed a counterintuitive behavior: users with the highest "ideological segregation" rely more on recommender systems to find new information but also are more exposed to opposite perspectives. Thus, people using recommendation systems (RS) are the ones seeing more different points of view. Another study [17] related to movie recommendations made by the *GroupLens* team has a similar conclusion. This work on *filter bubbles* asserts that RS actually lower the chances of being trapped into a *filter bubble*. *Facebook* also conducted a similar study [1] on their algorithm which is used to filter the feed of users. They conclude that it only decreases by 1% the chances of seeing posts corresponding to opposing views.

Models aiming at bursting an *echo chamber* to create more "peaceful" debates on a specific topic, such as gun control or Obamacare, have been presented in [6]. In this work the authors propose to add edges between people having opposite views in order to reduce controversy in the network. [12] proposes a model where the user gives a specific point of view in order to see how recommendation change based on this new perspective. A similar idea is developed in [8]. However, these solutions are difficult to deploy in practice because they rely on the will of the users to change their viewpoints. Our approach largely differs from existing work since it is, to the best of our knowledge, the first approach to use communities as a tool to observe *echo chambers* and *filter bubbles* effects, and to propose a re-ranking strategy of the recommendation to reduce the filter bubble phenomenon.

3   Community Analysis

In order to estimate the importance of the *filter bubbles* and *echo chambers'* phenomena induced by recommender systems' usage, we first extract communities from the social graph with the traditional Louvain method. Then we try to have a better understanding of the communities these algorithms produce and we study the behavior of users regarding the community they belong to.

3.1 Twitter Dataset

We present here the main characteristics of our Twitter dataset introduced in *[anonymous]*. It is based on a connected component extracted from the graph made provided by Kwak et al. [14] which has been updated since 2017 thanks to the Twitter API [1]. We collected the incoming edges (followers), out-coming edges (followees) and all the tweets published by the associated accounts. Observe that due to the API limit we only retrieved the last 3,200 messages for each node. Table 1 summarizes the main features of the dataset.

We can notice that, with more than 2 million users and 3 billion messages, we have a mean number of 1,375 published tweets per user. We detect that around 12% of these tweets, so on average 150 tweets per user, correspond to a retweet action. Our analysis also exhibits that 92% of the tweets are never retweeted. It means that recommendations mainly focus on a small part of the messages. As shown in [11] users tend to have more similar profiles with users within a 2-hop distance in the graph (called homophily [13]). This homophily has an impact on information propagation: people close to each other in the network tend to have a higher number of retweets in common.

3.2 Communities' Detection

To characterize the *echo chamber* phenomenon and the information propagation between users, we identify and study communities in our dataset. Scalable community detection algorithms are proposed in literature, like Infomap, Louvain and Label Propagation. Note that these methods only use the network topology and not topics, user profiles or exchanged content to extract communities. Moreover they associate users to a single community. The Louvain algorithm we have adopted is tailored for directed graphs [4,15]. It consequently suits to the Twitter network. It maximizes the modularity of clusters inside the graph that will produce denser components (*i.e.,* maximizing the number of connection triplets). However note that we also performed similar work for Infomap and got very similar results. To explain the filter bubble effect, we try to understand the rationale for the formation of a community. We first label the communities according to their main feature(s). Remember that a user belongs to a single "community" according to the considered community-detection algorithms, and that these communities are built by considering only the topology of the underlying social graph. We focus on the 105 more representative communities, *i.e.,* those with more than 100 users identified by the Louvain method. To determine the labels, we adopt the following three-step process:

(1) Most followed users inside each community are selected (most central users),
(2) We find most frequent terms occurring in the tweets of these users and we check important features from their profiles like *age, location, language, etc.,*
(3) Based on these two kinds of information we provide the most representative tag for each entity.

---

[1] https://dev.twitter.com/rest/public.

Some improvements may be considered like performing named-entity extraction rather than only relying on term frequencies, for instance. However it turned out that our basic strategy provides good labeling since users who have strong common interests, such as "Sports" for instance, are highly connected and form a community we effectively tagged as "Sports".

3.3 The Community Network

We exploit here the detected communities to enlighten the echo chamber effect. The objective is to quantify how information spreads outside the community to which it has been attached to. We first link a tweet to a community, then we find out how many communities it reaches. This quantification could be seen as a *propagation measure* inside the social network. This allows us to study the presence of echo chambers at both users and communities level.

Community Membership of a Message. To track messages "activity" we need to identify the way to attach messages to a community. Two options can mainly achieve this: a message belongs to the community from which it occurs first or to the community in which it obtained most likes/retweet.

It appears that 90% of retweeted messages obtain a high popularity in the community from which it comes from. The remaining 10% belong to small communities and naturally become famous when they reach larger communities. In the following we decide to identify the message community membership based on the community where it was written initially in order to emphasize the influence of small communities on bigger ones.

Correlation Between Popularity and Spread. Now we have communities and messages, we can measure the popularity of messages and how they propagate throughout the *community network*.

Figure 1 shows the distribution of retweeted messages with respect to the number of reached communities. We can see that 80% of retweeted messages reach at most 2 communities, and among them, half remains internal to the community they belong to. This distribution is characterized by a *Power Law*: $Cx^{-\alpha}$ (with $C = 200$, $\alpha = 2.2$ and $x_{min} > 1$ for probabilities). As expected C is really high stating that the probability that a tweet remains in a community is high. According to α, this classical value (typically between 2 and 3) indicates that communities have far connections between each other. This experiment confirms the fact that most of the messages are rarely retweeted while few very popular messages reach high numbers of communities. It underlines the existence of an echo-chamber effect inside communities.

According to this analysis, we conclude that most of the tweets hardly ever leave their community, especially if they are not popular.

4   Filtering Bubble

The objective is to analyze how recommender systems create or reinforce the *echo chamber* phenomenon at community and user levels. We study the filter bubble

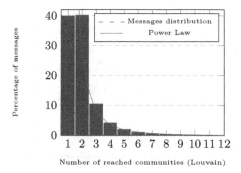

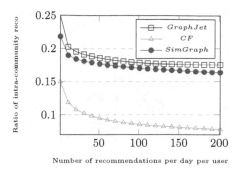

Fig. 1. Msg. wrt. # reached communities

Fig. 2. Ratio of intra-community reco

effect with three different recommendation systems: *GraphJet* [19] proposed by *Twitter*, Collaborative Filtering [2] (called *CF*) and *SimGraph* [11]. To achieve this, we consider recommendations produced for samples of 25 users randomly extracted from each community obtained by Louvain.

4.1   Community-Level Approach

A global approach to quantify the *filter bubble* effect is to compute the proportion of intra-community recommendations. When the proportion of users' recommendations belongs to its own community is too high (*intra-community recommendations*, opposite of the diversity), it implies that a *filter bubble* effect could lead to the reinforcement (or apparition) of an *echo chamber* effect.

In Fig. 2 we plot the ratio of intra-community recommendations regarding the number of recommendations proposed per day (for each user). We find out that *GraphJet* tends to propose less "diverse" recommendations than *CF* with on average 23% of intra-community recommendations. This could be explained by the random walk-based algorithm behind *GraphJet* that would give more opportunities to recommend messages in the neighborhood, which corroborates conclusions of Fig. 1. At the opposite *CF* computes similarities between users from the whole graph independently from the topology and tends to provide more diversified recommendations than other solutions, in terms of community provenance. *SimGraph* results are between *CF* and *GraphJet* since it mixes both topology and similarity (*i.e.,* homophily).

We also notice that independently of the number of recommendations proposed, the diversity is constant after 20 recommendations. Consequently, in the following we fix the recommendation number to 20 per day. As expected, in Fig. 2 *filter bubbles* aren't visible due to average values over every user.

To study the filter bubble at community scale, we display in Fig. 3 the ratio of intra-community recommendations per community along with their size for the CF recommendation algorithm. Community labels come from Sect. 3.2. Due to space limitations, we do not display Figures for the other algorithms but they behave similarly. We observe that for all recommendation algorithms, there is a

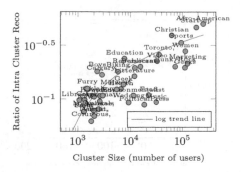

Fig. 3. CF intra-Recommendations

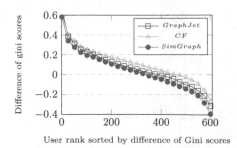

Fig. 4. Diff. of Gini coefficients between likes and recos

logarithmic correlation between community size and intra-cluster recommendations. The rationale is that the bigger a community is, the higher the chances are for its users to receive a recommendation from this community. However this experiment reveals that a global approach isn't sufficient to exhibit a particular community being concerned by a filter bubble.

4.2 Local Approach

Since we cannot detect filter bubbles with a global approach at community-level, we attempt to see whether this phenomenon can be observed at user-level. Therefore, we analyze communities' diversity for which recommended tweets are issued from. For this, we apply for each user the *Gini* coefficient [9] on the aggregate number of received recommendations per community. The *Gini* coefficient measures the ratio of inequalities within a set of values, *i.e.,* its diversity.

Users with high Gini scores seem to be trapped into a filter bubble. It is due to the RS which provides recommendations issued from few different communities. However after analyzing their profiles, we observe that these users have in fact a very specific usage of the platform (*e.g.,* football player's account only interact with sports messages). Therefore the RS by recommending only sports messages just follow the usage of the user maintaining the *echo-chamber* effect.

Consequently we believe that we must consider users' profile in the platform to determine if they are in a bubble or not. We thus consider the difference between user's interactions and RS recommendations. We propose to show this effect by computing the difference between the Gini coefficient of users' profile (list of effectively "*liked*" communities) and the one from the recommender system (list of "*recommended*" communities). Results are plotted in Fig. 4. High values mean that the recommendations are too diversified compared to the real user behavior while low values lead to a *bubble effect* with fewer communities concerned by recommendations compared to the real user behavior.

We see that 30% of the users are faced with less diversified recommendations than their own profile. This effect is mainly due to a frequent behavior of the user who "likes" many messages from a particular community and less frequently from "random" ones. However, recommender systems focus mostly on this main community and provide recommendations mainly issued from this community.

5   CAM - A Community-Aware Model

Thanks to this preliminary but essential study, we are now able to detect a filter-bubble effect on users' community profile with topology-based communities. We propose in the following our Community-Aware Model whose objective is to reduce the filter-bubble impact. It can be deployed on top of a RS and it enhances it with a new scoring function which permits re-ranking the recommendations. Observe that our approach is consequently independent of the choice of the RS and may be consequently deployed in any existing social network platform.

5.1   Community Profiles

So consider a user u and a social network where n communities were detected by a community detection algorithm. Let $\overrightarrow{Pu}$ be the user's u community profile represented as a normalized vector: $\overrightarrow{Pu} = (pc_1, pc_2, \ldots, pc_n)$ where pc_i denotes the rate of messages from the community c_i among all the messages he liked.

Suppose that a recommender system RS produces a list of recommendations $LReco_u$ for the user u from which only the top-k items are extracted and presented to u. The main idea is to re-rank $LReco_u$ by considering, for each message, its community of origin. The end goal consists in finding a top-k which corresponds more precisely to the user community profile $\overrightarrow{Pu}$.

Note that naive models which attempt to pick up the required number of messages from $LReco_u$ in each community of $\overrightarrow{Pu}$ wouldn't be successful. Indeed, due to too low recommendation scores or to a period where the corresponding community is less active, some communities from a profile $\overrightarrow{Pu}$ are not present (or insufficiently present) in $LReco_u$. Besides, with such naive approaches, a message with a high recommendation score which is not issued from a community appearing in $\overrightarrow{Pu}$ will also be discarded, even if the community is topologically and/or thematically closed to, which contributes to the filter bubble effect.

Since our community analysis reveals that some communities are thematically very close to, we propose that our re-ranking model takes into account this similarity and consequently modifies the scores produced by RS even for messages from communities which are not in $\overrightarrow{Pu}$.

Our model relies on the impact of items on communities called $\overrightarrow{V_U}$ and the user's profile $\overrightarrow{Pu}$. It tries to minimize the distance between $\overrightarrow{V_U}$ and $\overrightarrow{Pu}$.

5.2   Community Similarity Score

We first need to determine a measure of similarity between communities which takes into account (1) topology, (2) semantic information and (3) flows of information between these communities. We propose the following similarity measure to estimate how similar two communities can be.

Definition 1. *(Community Similarity Score) The asymmetric similarity measure between a community c_i and c_j is estimated as follows:*

$$sim(c_i, c_j) = \alpha\ Links(c_i \to c_j) + \beta\ Sem(c_i, c_j) + \gamma\ Flow(c_i \to c_j) \qquad (1)$$

where Links is the ratio of the number of links from c_i which are directed to c_j among its outgoing links, Sem represents the similarity (see Sect. 6.1) between the main topics of c_i and c_j, and Flow corresponds to the link importance which relies on the proportion of circulating tweets (retweets) from c_i to c_j. α, β and γ are constants which can be tuned according to the behavior of the underlying RS (see Sect. 6.2) in order to target relevance and/or filter bubbles.

Based on this similarity measure we can build the Community Similarities Matrix $(CSM = (sim_{ij})_{1 \leq i,j \leq n})$. Observe that this matrix is not symmetric since we consider links' direction and information propagation (flow).

5.3   Community-Aware Recommendations

We consider that each item I is associated to a community score vector $\overrightarrow{I}$ which captures how this item is thematically and topologically close to each community. To compute the vector $\overrightarrow{I}$ of an item I we rely on the community-similarity matrix CSM. So $\overrightarrow{I}$ corresponds to the community similarities from column c_i of the CSM matrix to which community I is associated to.

Our model intends to propose a set of recommendations U, selected from the recommendation list $LReco_u$ produced by RS, with a community score vector $\overrightarrow{V_U}$ which matches as much as possible the user profile $\overrightarrow{Pu}$. The community score vector $\overrightarrow{V_U}$ of a set of recommendations U is the aggregation of different normalized community score vectors of each item in U: $\overrightarrow{V_U} = \sum_{I \in U} \overrightarrow{I} / \| \sum_{I \in U} \overrightarrow{I} \|$.

Finding the set of recommended items U whose community profile $\overrightarrow{V_U}$ matches as much as possible the profile $\overrightarrow{Pu}$ can be modeled as a distance minimization problem between $\overrightarrow{V_U}$ and $\overrightarrow{Pu}$:

$$\begin{cases} U = argmax_{LReco_u} |\overrightarrow{Pu} - \overrightarrow{V_U}| \\ |U| = k \end{cases} \qquad (2)$$

However, determining the new recommendations based only on the distance with the user profile, regardless of the importance of the recommended content, may lead to recommend content of lower interest for the user. So another objective for our approach consists in the following maximization problem:

$$\begin{cases} U = argmax_{LReco_u} \sum_{I \in U} recom(u, I) \\ |U| = k \end{cases} \qquad (3)$$

where $recom(u, I)$ denotes the score of item I for user u provided by the RS.

Consequently the objective of our re-ranking algorithm is expressed as a multi-objective optimization problem determined by both Eqs. 2 and 3.

5.4   Avoiding the Filter Bubble

A traditional strategy to determine a solution to a multi-objective optimization problem is scalarization where no solution satisfies both objectives. Scalarizing is an a priori method, which transforms the multi-objective optimization problem into a single-objective optimization problem.

Table 2. Recommendation scores for Joe

Ranking	Origin community	Score
1	A	0.8
2	A	0.7
3	A	0.5
4	B	0.4
5	C	0.1

Table 3. Scores for all 3-item combinations

3-item set and their score			
{1,2,3}	0.81	**{1,2,4}**	**0.40**
{1,2,5}	0.57	{1,3,4}	0.41
{1,3,5}	0.54	{1,4,5}	0.48
{2,3,4}	0.42	{2,3,5}	0.47
{2,4,5}	0.47	{3,4,5}	0.43

To achieve this transformation, we propose to integrate the recommendation score when estimating the community score vector $\overrightarrow{I}$. Since our objective is to get a high global recommendation score, we attempt to discard first from our recommendation set, items with a low recommendation score.

Thus, we adopt for our community score vector $\overrightarrow{I}$ this new definition:

$$\overrightarrow{I} = \frac{1}{recom(u,I)} \times \overrightarrow{CSM(c_i)} \tag{4}$$

With this new definition, an item with a low recommendation score will significantly increase the different components of its community score vector. This item will have a high impact on $\overrightarrow{V_U}$ and increase the profile distance. Thus, this item is more likely to be replaced by another one in the final item set.

Example 1. Consider a user Joe to whom a recommender system proposes a list of recommendations $Reco_{Joe}$. Assume for this example that we limit the recommendations to the top-3 scores, so Joe receives the three recommendations originated from the community A. We suppose that there are only 3 communities and that there exists no similarity between them. Therefore CSM is the identity matrix. We assume that Joe interacts equally with these three communities; therefore his profile is: $\overrightarrow{P_{Joe}} = (0.33, 0.33, 0.33)$.

To re-rank the items by considering the user profile and the relevance of the messages, we compute the distance from Eq. 2 with the community score vector computed with Eq. 4.

We display in Table 3 the $distance = |\overrightarrow{P_{Joe}} - \overrightarrow{V_{Joe}}|$ for the different 3-item combinations. For our example, we see that the score for the best combination is 0.40 and corresponds to {1, 2, 4}. We see in Table 2 that these items have a high recommendation score, and this set better matches the user profile.

To determine the top-k recommendation set, we theoretically need to compute all the combinations of k items from U extracted from $LReco_u$ which consists of N items has a complexity of $\binom{N}{k}$. We escape the exponential complexity by adopting an interchange algorithm. So we initialize the recommendation set with the k top-rated items. Then we check for each of the $N - k$ remaining items if we can reduce the distance with the user profile by replacing one recommendation by this item. This algorithm has a N^2 complexity.

6   Experiments

We first detail the experimental protocol we adopted to measure the benefits of our re-ranking model CAM for both the relevance of recommendations and the number of users suffering from the filter bubble effect. We also study the impact of the different parameters in our model, *i.e.,* semantic similarities, the flow and the topological similarity, on the overall results. Our experiments reveal that our model can be tailored for different RS to provide significant gains. For all our experiments, we use the `Twitter` dataset described in Sect. 3.

6.1   Settings

To measure the filter bubble effect, we use the *Gini* coefficient (see Sect. 4.2) for a sample of users from our dataset. More precisely we measure the difference between the user's Gini coefficient computed for his community profile and the one computed for the community distribution of the recommendations. We consider that a user is affected by a filter bubble if this difference is lower than a given threshold of -0.2 (bottom right of Fig. 4). This -0.2 corresponds to the inflection point observed in Fig. 4 which characterizes 10% of the users.

We select the largest communities, with at least 1,000 users, from the USA found by the Louvain clustering method. They represent 38 communities. For each of these communities, we randomly extract 16 users, leading to 608 selected users. This choice of 16 users corresponds to the maximum number of users for the smallest community that retweeted at least twice, therefore users that can be targeted by a RS to give sufficient messages to re-rank. We chose to balance all the communities by an equal number of users.

Then we select messages' retweets which were retweeted at least twice. This constitutes a set of $132, 389, 409$ sharing actions, timely ordered. We split the set in two: the first 90% of actions (the oldest retweets) compose the training set and the last 10% the test set. While the former set is used to train the three methods, the latter one allows checking the recommendations with real retweets. Note that the test set captures 66 days of retweets from users in our dataset.

Then for each recommender system we compare *CF*, *GraphJet* and *SimGraph*, we observe the recommendations computed during the test set with and without applying our CAM algorithm. To estimate the CAM re-ranking score we determine its three components *Links*, *Semantic* and *Flow* as follows:

- The number of directed edges between communities is used to compute the *Links* weight, capturing the topological proximity between communities.
- In order to compute the *Semantic* similarity we rely on `Word2Vec` trained on Google News data [10]. We extract most frequent words from communities and combined them to create a vector thanks to `Word2Vec`. This method allows us to compute semantic distance between communities.
- We measure the *Flow* weight from the network of communities based on the proportion of tweets that circulates between corresponding communities through retweets (flow proximity).

We consider that a message is a *hit* if it is recommended to a user based on the training set, and we detect that it leads indeed to an interaction (retweet/like) in the test set. This prediction task can be seen as a relevance measure.

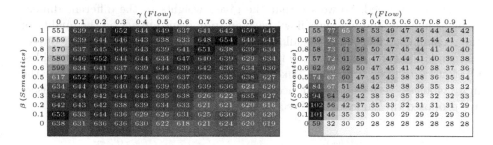

Fig. 5. *hits* and users suffering from filter bubble for *GraphJet* w.r.t. γ and β

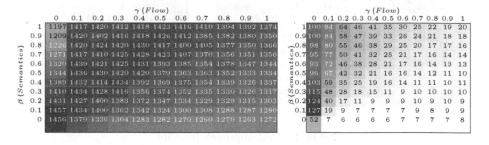

Fig. 6. *hits* and users suffering from filter bubble for CF model w.r.t. γ and β

6.2   Studying Weights' Impact

The re-ranking algorithm relies on the similarities between communities (Eq. 1). Since similarity scores depend on α, β and γ, we perform experiments to study the impact of each weight on the re-ranking quality. Each weight is bounded between 0 and 1 and we adopt a 0.1 padding for our experiments providing 11 different values for every weight which leads to $11^3 = 1,331$ different weights configurations. For space reason, we displayed only results with α set to 0.5 which showed a lower impact than β and γ that are considered here.

In Fig. 5 we plot the results for *GraphJet* from `Twitter`. The left table shows the number of accurate predictions (*hits*) made by the system w.r.t. β and γ weights. The right table represents the number of users among our 608 selected users who suffer from a bubble effect (those with a Gini difference lower than the -0.2 threshold). Results for respectively the collaborative system (*CF*) and for *SimGraph* are presented respectively in Figs. 6 and 7. We first observe a high variability of results depending on our parameters. The number of accurate recommendations - *hits* - ranges from 492 to 653 for `GraphJet` for instance, so a 32% difference. We notice the same order of variability for both *CF* and *SimGraph*. The variability is even more important for the number of users facing a filter bubble. For instance, we see that 6 users at a minimum are concerned by a filter bubble for the *CF* model while in the worst configuration there are 128 users concerned. So we see that the given weights to the different dimensions (semantics and messages flow) have an important impact on the quality of the recommendations and they allow us to efficiently boost the relevance of recommendations or to decrease the number of users suffering from filter bubbles. Obviously, the two scores are linked: the more we narrow users' interests the more chances we have to make accurate recommendations but the more users are proposed the same kinds of recommendations.

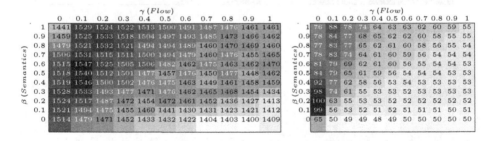

Fig. 7. *hits* and users suffering from filter bubble for *SimGraph* w.r.t. γ and β

Table 4. CAM approach benefits

	Initial		Best configuration	
	Hits	Filter Bubble	Hits	Filter Bubble
GraphJet	552	5.4%	630 (+14%)	4.6% (−15%)
CF	1,400	2.7%	1,348 (−3%)	0.9% (−64%)
SimGraph	1,468	10.0%	1,491 (+2%)	7.0% (−23%)

Overall we observe that the three RS tested show similar key trends when changing α, β and γ. Reducing weight β (the semantics similarity between communities) allows tweets whose topic is different from the users niche interest to be more likely recommended and therefore lowers the number of users suffering from the bubble effect. On the other hand, lowering this weight also induces that some relevant items will not be recommended to the user. The γ weight (the flow between communities) has an opposite effect. Higher γ values tend to recommend items from different communities with which the user is used to interacting but also to decrease at the same time the number of hits. Finally, we observe that α has a similar impact β on the results but with a lower amplitude.

Our experiments show that there does not exist a configuration where both the relevance of recommendations and the number of users in a filter bubble are optimized. So the different weights in our CAM model may be tuned according to the objectives of the recommender system. Thanks to our experiments, we can also determine for each recommender system the configurations which minimize the number of users suffering from the filter bubble effect (see below).

6.3   Gains Achieved with the CAM Approach

Our next experiment aims at illustrating the gain that we achieve by deploying the CAM model on top of existing recommender systems. So for each recommender system, we select the best weight setting to minimize the number of users affected by filter bubbles according to our observations in Figs. 5, 6 and 7.

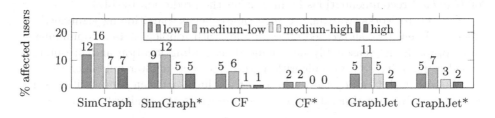

Fig. 8. Filter bubble users w.r.t. their activity without or with* CAM

We present in Table 4 the percentage of users facing a filter bubble with and without re-ranking the recommendations for the different recommender systems, along with the total number of *hits* we get. We observe that our re-ranking model successfully decreases the number of users affected by the filter bubble by 15% for $GraphJet$, 64% for CF and 23% for $SimGraph$. Additionally, by matching more to the user's profile we also improve the relevance of recommendations and boost the number of accurate predictions especially for $Graphjet$. Only for CF, removing 64% of filter bubble effects on affected users slightly decreases their relevance: -3% of hits.

Our model seems to remove users more successfully from a filter bubble for the CF model. This could be explained by choice possibilities of the re-ranking step. Sometimes, $GraphJet$ and $SimGraph$ hardly produce recommendations far in the social network, narrowing the possibilities of re-ranking while CF could compute a large list of recommendations for all users [11].

6.4   Users Activity and Filter Bubbles

In Fig. 8, we investigate the link between users' activity, *i.e.*, number of messages they interacted with, and the filter bubble. Users are assigned to a category (*i.e.*, low, medium-low, medium-high, high) according to the number of interactions they made on the platform. For each of these categories, we plot the percentage of users affected by a filter bubble. We observe that most users concerned by this phenomenon have a low activity. Users with low or medium-low activities correspond to more than 70% of affected users.

Due to fewer interactions, recommender systems focus on the known interest of these users. Therefore, this limits the scope of possible recommendations. Consequently, using *CAM* allows highlighting items that were poorly considered by the underlying recommender system, and impact those users much more.

7   Conclusion

In this paper we have presented a thorough study on information flow on *Twitter* and we showed that the filter bubble phenomenon only concerns a minority of users. We proposed the CAM approach which relies on similarities between communities to re-rank lists of recommendations in order to weaken the filter bubble effect for these users. Moreover our approach is able to boost the accuracy of *GraphJet* recommendations by increasing the prediction by 14%.

For future works, we want to investigate better partitioning strategies for *Twitter*. Even if we showcased filter bubbles with topology-based communities, our approach can reasonably be enhanced by finding location and/or opinion-based community detection algorithms to better detect filter bubble effects.

We also wish to study the evolution of the links between communities, since retweets evolve over time, it will have an impact on the similarity measure.

References

1. Bakshy, E., Messing, S., Adamic, L.A.: Exposure to ideologically diverse news and opinion on facebook. Science **348**(6239), 1130–1132 (2015)
2. Breese, J.S., Heckerman, D., Kadie, C.: Empirical analysis of predictive algorithms for collaborative filtering. In: UAI 2018, pp. 43–52 (1998)
3. Colleoni, E., Rozza, A., Arvidsson, A.: Echo chamber or public sphere? Predicting political orientation and measuring political homophily in twitter using big data. J. Commun. **64**(2), 317–332 (2014)
4. Dugué, N., Labatut, V., Perez, A.: A community role approach to assess social capitalists visibility in the Twitter network. SNAM **5**(1), 26 (2015)
5. Flaxman, S., Goel, S., Rao, J.M.: Filter bubbles, echo chambers, and online news consumption. Public Opin. Q. **80**(S1), 298–320 (2016)
6. Garimella, K., De Francisci Morales, G., Gionis, A., Mathioudakis, M.: Reducing controversy by connecting opposing views. In: WSDM, pp. 81–90 (2017)
7. Garrett, R.K.: Echo chambers online?: Politically motivated selective exposure among internet news users. JCC **14**(2), 265–285 (2009)

8. Gillani, N., Yuan, A., Saveski, M., Vosoughi, S., Roy, D.: Me, my echo chamber, and I: introspection on social media polarization. CoRR abs/1803.01731 (2018)
9. Gini, C.: Variabilità e mutabilità. Libreria Eredi Virgilio Veschi (1912)
10. Google: Word2vec (2013). https://code.google.com/archive/p/word2vec/
11. Grossetti, Q., Constantin, C., du Mouza, C., Travers, N.: An homophily-based approach for fast post recommendation in microblogging systems. In: Proceedings International Conference on Extending Database Technology (EDBT), Austria, pp. 1–12 (2018)
12. Kamishima, T., Akaho, S., Asoh, H., Sakuma, J.: Enhancement of the neutrality in recommendation. In: Decisions@ RecSys, pp. 8–14 (2012)
13. Hyung Kang, J., Lerman, K.: Using Lists to measure homophily on Twitter. In: AAAI, pp. 26–32 (2012)
14. Kwak, H., Lee, C., Park, H., Moon, S.B.: What is twitter, a social network or a news media? In: WWW, pp. 591–600 (2010)
15. Leicht, E.A., Newman, M.E.J.: Community structure in directed networks. Phys. Rev. Lett. **100**, 118–122 (2008)
16. Munson, S.A., Resnick, P.: Presenting diverse political opinions: how and how much. In: Human Factors in Computing Systems, pp. 1457–1466. ACM (2010)
17. Nguyen, T.T., Hui, P., Harper, F.M., Terveen, L.G., Konstan, J.A.: Exploring the filter bubble: the effect of using recommender systems on content diversity. In: WWW, pp. 677–686 (2014)
18. Pariser, E.: Beware online "filter bubbles" (2011). https://www.ted.com/talks/eli_pariser_beware_online_filter_bubbles
19. Sharma, A., Jiang, J., Bommannavar, P., Larson, B., Lin, J.: GraphJet: real-time content recommendations at Twitter. PVLDB **9**(13), 1281–1292 (2016)

Memory-Augmented Attention Network for Sequential Recommendation

Cheng Hu$^{(\boxtimes)}$, Peijian He, Chaofeng Sha, and Junyu Niu

School of Computer Science, Shanghai Key Laboratory of Intelligence Processing,
Fudan University, Shanghai 200433, China
{chenghu17,pjhe18,cfsha,jyniu}@fudan.edu.cn

Abstract. An increased interest in sequential recommendation has been observed in recent years. Many models have been proposed to leverage the sequential user-item interaction data, which includes those based on Markov Chain or recurrent neural networks. Most of these models are designed for the scenario where each historical record composed of single item. However, the records could be a subset of items (or session) such as music playlists and baskets in e-commerce applications. How to leverage the session structure to improve the effectiveness of the recommendation system is a challenge. To this end, we propose a MEmory-augmented Attention Network for Sequential recommendation (MEANS), to effectively recommend next items given the sequential session data. The most recent sessions are stored into external memory after a max-pooling operation. The long-term user preference are learned through an attention network which is stacked on the memory layer. Finally, the mixture of long-term and short-term preference is fed into the prediction layer to make recommendations. Extensive experiments on four real datasets show that MEANS outperforms various state-of-the-art sequential recommendation models.

Keywords: Sequential recommendation · Memory network · Attention mechanism

1 Introduction

With the growth of user-item interaction in various applications such as e-commerce Web sites or media streaming, the recommendation systems are deployed to support users in finding interested items. In traditional recommendation systems, the user-item interactions are feed into models such as nearest neighborhood based or matrix factorization. In recent years, much work have explored the temporal or sequential information in user historic records.

There are two lines of research in recommendation systems to leverage the temporal information, i.e., sequential recommendation and session-based recommendation. In session-based recommendation, the goal is to recommend the next item v_{t+1} given the prefix of current session $S = \{v_1, \cdots, v_t\}$. Most recent

© Springer Nature Switzerland AG 2019
R. Cheng et al. (Eds.): WISE 2019, LNCS 11881, pp. 228–242, 2019.
https://doi.org/10.1007/978-3-030-34223-4_15

work employ Markov chain or recurrent neural networks to learn the representation of the context of session which is feed into a prediction module [12]. In sequential recommendation, many methods based on neural networks are also proposed given an ordered or timestamped list of past user actions as an input [13]. The empirical results show that modeling long- and short- term user preference simultaneously can improve the performance of sequential recommendation models [26].

In some applications, the record at time slice t could be a session or transaction of actions or items [26]. Tailored to this sequential recommendation with session data, Ying et al. [26] recently proposed a model named SHAN which integrates a hierarchical attention network to recommend the next-items. However, we believe that the user's behavior history has not be exploited thoroughly by existing models. Consider a toy session such as {flour, fat, egg, sugar} which could be purchased for a cake. The local patterns within this session could be ignored when we treat each item separately. Therefore, the high-level information in the sessions should be exploited to capture the user preference. To this end, we propose a sequential recommendation model that leverages the users' most recent sessions, which can capture the hidden patterns within the sessions and the correlation between sessions. First we use a memory module to store the recent sessions after a max-pooling operation. We then design an attention network to learn user's long-term preference which is combined with the short-term preference to represent an user. Finally the mixture of long- and short-term preference is feed into a prediction layer to recommend the next-items users interest of. The model's parameters are estimated with Bayesian Personalized Ranking criterion. The effectiveness of proposed method are validated through extensive evaluation on four datasets.

The main contributions of the paper are summarized as follows:

1. We propose a novel framework named MEmory Augmented Network for Sequential recommendation (MEANS) that integrates the memory network and attention network.
2. We employ max-pooling operations to capture the local pattern within sessions, a memory module to maintain the recent session information, and an attention network to learn the importance of each session when learning the users' long-term preferences.
3. We conduct extensive experiments on four real datasets. The experimental results demonstrate the superiority of our proposed model compared to the state-of-the-arts.

2   Related Work

After several decades of development, many effective recommendation algorithms have emerged. The association rule method directly mines the patterns existing in user-item interaction [1]. The item-based collaborative filtering algorithm analyzes the user-item matrix to identify relationships between different items [4,19]. Later, a large number of model-based methods were proposed, which greatly affected the field of recommendation systems. Including PMF [17], MF [9], Restricted boltzmann machines [18], Bayesian methods [15].

In addition to the above state-of-the-art methods, there are a lot of related work. According to the application scenario of our proposed model, it is mainly related to two major methods:

2.1   Sequential Recommendation

It has been widely concerned that the user's long-term behavior is mined to accurately obtain user preferences. Pattern recognition, as the most basic data mining algorithm, also applies to problems in the background, such as [25]. Markov chains also have powerful capabilities in stochastic process modeling, such as [2,16]. Further, since the rise of the deep learning, the neural network structure is also used to model the user-item interaction. The RNNs model is specifically designed for sequential task. DREAM [27] uses the RNNs model to correlate all of the user's basket records. HGRU [14] relays and evolves latent hidden states of the RNNs across user sessions. [24] is based on long-short-term memory (LSTM) autoregressive models that dynamically capture user and item characteristics. Personal representation is contributed to a new type of Gated Recurrent Unit (GRU) to effectively produce personalized recommendations [5]. The time interval is also used as an input to the recurrent unit to accurately measure the relationship between continuous behavior of the user [28]. Multi-task learning is also used in sequential methods, such as [10], which combines the RNNs and MF. It has achieved good results.

Although deep learning is powerful enough, the requirements for resources are higher than the general method. Therefore, some methods simplify user continuous behavior modeling in order to achieve trade-off between model efficiency and complexity. For example, RUM [3] uses memory network to record the user's recent behaviors, combined with the attention mechanism to obtain user preferences. SHAN [26] divides user behavior into long-term and short-term, and then uses attention mechanism to get the final behavioral representation.

2.2   Session-Based Recommendation

In addition to modeling all historical behaviors of users, focusing on the last session to obtain user preferences has also become a research direction in the field of recommendation systems. In this respect, the neural network model also plays an important role. There are a lot of methods which are based on the GRU to learn the internal relationship of the session [6,8,21]. [20] is based on the RNNs model, supplementing the session representation with event information. NARM [11] uses GRU for global encoding and local encoding, respectively. The local encoding combines the attention mechanism to obtain the user purpose feature. Of course, the CNN-based model has also been tried in session-based recommendation [22]. In order to simplify the model complexity, HRM [23] directly uses the pooling operation to extract the session representation. Similar methods include [7].

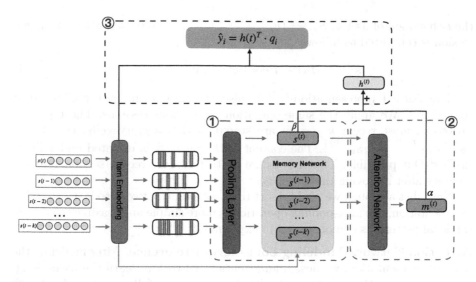

Fig. 1. The architecture of MEANS.

3   Proposed Method

3.1   Notations and Problem Formulation

In this paper, we study the next-item prediction problem following the problem formulation of [26]. We denote the user set as $U = \{u_1, u_2, \cdots, u_n\}$ and the item set as $V = \{v_1, v_2, \cdots, v_m\}$. For each user u, let $H_u(T) = \{S_u^{(1)}, \cdots, S_u^{(T)}\}$ denote her previous T sessions where each session is a subset of itemset V, i.e., $S_u^{(t)} \subseteq V$. For any time slice t, the user u's short-term preference is reflected by session $S_u^{(t)}$. And the long-term (before time t) is reflected by $H_u(t-1)$. Given the historical behavior of all users, we aim to recommend the next item users will consume.

3.2   Recommendation Framework: MEANS

Our proposed recommendation model consists of three components: (1) a memory module that transforms each of recent k sessions into a session vector then store them in the memory M, (2) an attention network that helps to generate the long-term user's preference representation, (3) a predictor layer that ranks items for the next-item recommendation. The architecture of MEANS is shown in Fig. 1.

Memory Module: Let $\boldsymbol{v}_j \in R^d$ be the embedding vector of item v_j which is the basic unit in MEANS. At session level, we aim to capture the local pattern within each session. To this end, we employ a max-pooling operation to extract the most salient features from every item-embedding dimension. Formally, given the t-th session of user u, $S_u^{(t)} = \{v_1^{(t)}, v_2^{(t)}, ..., v_n^{(t)}\}$, the representation of current session is computed as follows:

$$s_u^{(t)} = Max - Pooling(v_1^{(t)}, v_2^{(t)}, ..., v_n^{(t)}). \tag{1}$$

The learned representation $s_u^{(t)}$ is also the short-term user preference at time slice t. We apply the same operation to previous sessions. The representations of most recent k session are stored in the memory cells, i.e., $M_u = \{s_u^{(t-1)}, s_u^{(t-2)}, \cdots, s_u^{(t-k)}\}$. The content of the memory is updated under First-In-First-Out principle. That means we write the most recent session into the memory after removing the oldest one.

Note that although we could resort to an attention network to learn a session representation, the max-pooling operation is more simple and feasible to capture the local pattern as demonstrated in experiments.

Attention Network Modeling Long-Term Preference. After modeling the session representation, we design an attention network to learn the users' long-term preference. Here we firstly feed k sessions into a fully connected network followed by a rectifier linear unit transformation to get the importance score of each session. Then the scores are normalized through a softmax operation. Formally, the attention coefficient of each session stored in the memory is computed as follows:

$$h^{(i)} = ReLU(\boldsymbol{W}s_u^{(t-i)} + \boldsymbol{b}) \tag{2}$$

$$\alpha_i = \frac{\exp(s_u^{(t)} \cdot h^{(i)})}{\sum_{j=1}^{k} \exp(s_u^{(t)} \cdot h^{(j)})} \tag{3}$$

where $ReLU(x) = \max(0, x)$, $\boldsymbol{W} \in R^{d \times d}$ and $\boldsymbol{b} \in R^d$ are model's parameters.

Then the long-term representation of user u before time slice t is modeled as a weighted sum of the most recent k session which is stored in the memory. Formally, this representation can be formulated as follows:

$$\boldsymbol{m}_u^{(t)} = \sum_{i=1}^{k} \alpha_i s_u^{(t-i)}, \tag{4}$$

where the attention coefficient α_i could reflect the importance of s_u^{t-i} on further prediction. When both the long- and short- term preferences are ready, we arrive at the user's representation at time slice t as follows:

$$h_u^{(t)} = \alpha \boldsymbol{m}_u^{(t)} + \beta s_u^{(t)} \tag{5}$$

We tune the weighting parameter α and β to balance the influence of long-term and short-term preference, which will be detailed in the experiments.

Prediction Layer: In the prediction layer, we employ the inner product between the representations of user and item as the ranking function:

$$\hat{x}_{uv} = h_u^{(t)} \cdot \boldsymbol{v} \tag{6}$$

3.3   Model Training

The goal of our model is to recommend top-k items which users may interest. Therefore we resort to Bayesian Personalized Ranking criterion to ranking items pair-wisely. Given the training set $D_S = \{(u, v_i, v_j) : u \in U, v_i \in S_u^{(t)}, v_j \in V \setminus S_u^{(t)}\}$, the generic optimization criterion of BPR (BPR-OPT) [15] is defined as follows:

$$L_{BPR} = \sum_{(u,v_i,v_j) \in D_S} \ln \sigma(\hat{x}_{uv_i} - \hat{x}_{uv_j}) - \lambda_\Theta ||\Theta||^2, \tag{7}$$

where λ_Θ are model specific regularization parameters. The parameters $\Theta = \{V, W, b\}$ are estimated through stochastic gradient ascent. The tuning of hyper-parameters are detailed in the experiments.

4   Experiments

To evaluate the effectiveness of our proposed method, we conduct extensive experiments on four real datasets and compare against state-of-the-art session-based and sequential recommendation methods.

4.1   Experimental Setting

Datasets and Data Preparation. We conduct experiments on the following four datasets:

- **Gowalla**[1] contains the point-of-interest information from users.
- **TallM**[2] records the user's consumption and browsing behavior during the user's shopping process.
- **MovieLens**[3] dataset is collected from the Movielens web site, which contains the user's viewing status and user ratings.
- **Tafeng**[4] is a grocery shopping dataset, and collects users' transaction data. It covers products from furniture, food to office supplies.

Similar to [26], we evaluate comparison methods on data from recent years. We perform the following preprocessing on the four data sets: user actions in one day are treated as a session. All singleton sessions, cold-start users and items are removed. In the three data sets Gowalla, TallM, Movielens, the items that occur less than 20 times are regarded as cold-start items, and the users who have less than 5 session are treated as cold-start user. For Tafeng dataset, the items which occur less than 10 users are regarded as cold-start, and the users have less three sessions are treated as cold-start. After that, we randomly select 20% of sessions in the last several months as test set, and the rest are used for validation. We also randomly sample one item from the sessions as the next item to be predicted. After the above processing, the statistics of all data sets are summarized in the Table 1.

---

[1] http://www.yongliu.org/datasets.
[2] https://tianchi.aliyun.com/dataset/dataDetail?dataId=53.
[3] http://grouplens.org/datasets/movielens/20m/.
[4] http://www.bigdatalab.ac.cn/benchmark/bm/dd?data=Ta-Feng.

Table 1. Statistics of the evaluation datasets.

Datasets	#Users	#Items	#Sessions	Avg.s size	Avg.s per user
Gowalla	14470	11862	124218	2.9	8.5
TallM	17617	48804	116879	3.3	6.6
MovieLens	1418	4456	25717	11.2	18.1
Tafeng	12095	11024	75417	7.3	6.2

Baselines. To validate the performance of the proposed method, we compare with the following baselines:

- POP: This method counts the number of occurrences of items in all user interactions and recommended items based on their popularity.
- BPR [15]: Bayesian Personalized Ranking is a state-of-the-art recommendation framework which exploits the user-item interaction data. We adopt BPR-MF in our comparisons.
- HRM [23]: HRM utilizes pooling operations to obtain sequential feature from the baskets, which is pooled with the user's representation before making prediction. Here we use HRM with double max-pooling operations because it achieves the best results as shown in [23].
- GRU4Rec [6]: This is a state-of-the-art session-based recommendation model. The method uses the GRU to model the context of sessions.
- NARM [11]: This is a session-based recommendation model which integrates recurrent neural network and attention mechanism.
- HGRU [14]: The is a personalized session-based recommendation model which learns session-level and user-level representations. The combination of the two representation is passed to the next session as the initialization state and treated as the user's next representation.
- RUM [3]: This is a state-of-the-art sequential recommendation model based on memory network. RUM dynamically updates user memory with the most recent item, and use an attention mechanism to weight the recent items.
- SHAN [26]: This is a state-of-the-art sequential recommendation model for session data. SHAN employs a hierarchical attention network to learn long- and short-term preference which is combined as a hybrid user representation.

We also include a variant of our MEANS, named MEANS-, which uses the current session to recommend the next-item, i.e. set $\alpha = 0$.

Evaluation Metrics. We follow the evaluation protocol in [26] and adopt two common evaluation metrics:

$P@K$: In the scenario of predicting the next item, $P@K$ score is often used to measure the accuracy of the prediction. The value of $P@K$ indicates the proportion of correct predictions in the top-K recommendation list:

$$P@K = \frac{n_{hit}}{N}$$

where N is the number of test data, n_{hit} denotes the number of correct prediction. A hit occurs when the target item lies in the top-K recommendation list.

$MRR@K$: The average of the reciprocal ranks of the target items in the top-K recommendation list. The reciprocal rank is set to zero if the rank is larger than K. Formally, for a given test T, the $MMR@K$ is defined as follows:

$$MRR@K = \frac{1}{N} \sum_{v \in T} \frac{1}{Rank(v)}$$

The $MRR@K$ is a normalized score with range of $[0, 1]$. As the value increases, it indicates that the target item is ranked higher in the prediction list, which indicates better performance in the recommendation system.

Parameter Settings. We use the grid search to tune the hyper-parameter on all data sets, and choose the best setting by early stopping based on the $P@50$ score on the validation set. In the grid search, the hyper-parameter ranges as follows: embedding dimension size d in $\{20, 50, 100\}$, number of recent sessions k in $\{3, 5, 8\}$, learning rate η of stochastic gradient search in $\{0.001, 0.005, 0.01, 0.1, 1\}$, regularization parameter λ in $\{0.001, 0.01, 0.1, 1\}$. According to the empirical performance, we use the following hyper-parameters setting for all the tests on four datasets: $\{d : 50, k : 3, \eta : 0.01, \lambda : 0.01\}$. All items' embedding are initialized randomly with Normal distribution $N(0, 0.01)$. The weight parameters in attention network are initialized by sampling from the uniform distribution $U(-\sqrt{\frac{3}{d}}, \sqrt{\frac{3}{d}})$.

4.2 Impact of Hyper-parameters

First we study the impact of hyper-parameters on MEANS, which include the weights of long- and short- term preferences, and embedding dimensions. Due to the space limit, here we report the performance in terms of $P@K$ when varying the weights of long- and short- term preferences (α and β).

The importance of short-term preference on prediction performance has been validated by the pilot experiments with MEANS-, so we fix the weight of short-term preference (β) to 1 in most cases. Figure 2 shows the change in $P@K$ of MEANS on the four datasets as the weights of long-term (α) varies. From the figure, we can see that MEANS performs best on most datasets (especially on MoviveLens and Tafeng) when $\alpha = 0.2$, which confirm the effectiveness of taking advantage of the long-term preference. Therefore, we empirically fix α to 0.2 in the following experiments. Note that we also include the case when $\beta = 0$ in the figure. The results confirm the significance of short-term preference on next-item prediction.

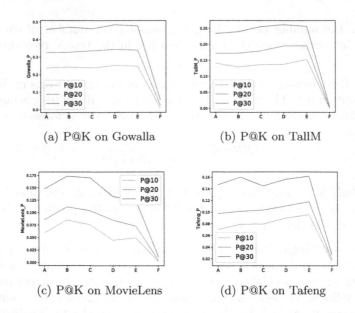

(a) P@K on Gowalla   (b) P@K on TallM

(c) P@K on MovieLens   (d) P@K on Tafeng

Fig. 2. The impact of different weight pairs of long- and short-term preference on each dataset. Since the results in terms of $MRR@K$ are similar to $P@K$, we only show the results of $P@K$. Here we adopt six weight pairs of long- and short-term. These weight pair settings correspond to the points: $A(0.0, 1.0)$, $B(0.2, 1.0)$, $C(0.5, 1.0)$, $D(0.8, 1.0)$, $E(1.0, 1.0)$, $F(1.0, 0.0)$.

We also analyze the impact of embedding size (d) on the performance of MEANS. We vary d in $\{20, 50, 100\}$. From the results in Fig. 3, we can see that MEANS performs best on Gowalla and Tafeng datasets when the embedding dimension is set to 50. Although results on Tall and MovieLens are best when $d = 100$, we keep $d = 50$ in the following experiments to tradeoff the prediction performance and computational efficiency.

4.3   Overall Performance Comparison

In this section, we compare the performance of different methods on four datasets. Table 2 summarizes the best performance obtained on embedding size 50. From the table, we can find that:

(1) Our proposed method achieves the best or competitive performance in terms of both metrics on Gowalla, MovieLens, and Tafeng datasets, significantly outperforms the state-of-the-art methods based on recurrent neural networks and attention networks. (On average, the relative improvement over the best baseline is 15.30% in terms of $P@K$). Although MEANS lose on TallM dataset in term of $MRR@50$, it also performs best under $P@50$. The results demonstrate that the effectiveness of MEANS, which captures local patterns through max-pooling operation on sessions and dynamically updates

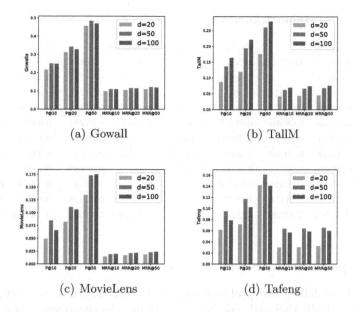

Fig. 3. The impact of embedding size (d) on the performance of MEANS.

the user long-term preference through the memory module and attention network. The out-performance of both MEANS and MEANS- indicates that the short-term preference hidden in the current session can be captured by max-pooling operation. That simple operation is suitable for the next-item prediction tasks.

(2) SHAN outperforms other methods in the most cases. It indicates that the hierarchical attention network in SHAN may capture the high-level nonlinear information. However, the local pattern may not be captured by distributing the attention weights on item-level as demonstrated in the following subsection.

(3) Both RUM and MEANS employ memory network to model the long-term preferences. However, RUM operates on item level in memory module, which can not capture the high-level information in the session and limit the capacity of the followed attention network. The inferiority of HRM could be due to that it just leverages the short-term preferences.

(4) The inferiority of GRU4Rec and HGRU in most cases suggests that the recurrent neural network may not be powerful as in natural language processing. The user preferences or pattern within sessions or between sessions could not be modeled recurrently.

Table 2. Overall performance on four data sets

Dataset	Gowalla		TallM		MovieLens		Tafeng	
Metric	P@50	MRR@50	P@50	MRR@50	P@50	MRR@50	P@50	MRR@50
POP	0.00067	–	0.00076	–	0.02295	0.00755	0.00474	0.00054
BPR	0.00101	0.0001	0.00152	–	0.01408	0.00117	0.00444	0.00038
HRM	0.42136	0.09794	0.08333	0.01981	0.09692	0.01318	0.16156	0.04249
GRU4Rec	0.22546	0.08986	0.19281	**0.0753**	0.08986	0.01294	0.03072	0.01071
NARM	0.37371	0.0891	0.20959	0.03685	0.05165	0.00503	0.13891	0.03841
HGRU	0.37626	0.10152	0.2609	0.06752	0.04639	0.0093	0.035	0.00988
RUM	0.41598	0.02781	0.0405	0.00378	0.08174	0.00733	0.15088	0.0647
SHAN	0.40909	0.10401	0.20768	0.04115	0.11893	0.01688	**0.1618**	**0.06504**
MEANS-	0.45857	0.11156	0.23412	0.06694	0.14797	0.02028	0.14695	0.04423
MEANS	**0.48541**	**0.12232**	**0.26164**	0.06952	**0.17339**	**0.0237**	0.16172	0.06592

4.4   Compares MEANS with SHAN

Recall that we follow the problem formulation in [26], now we compare the performance of MEANS with SHAN to investigate the effectiveness of our model designs.

From the results show in Fig. 4, we can see that MEANS consistently outperform SHAN. This may to due to that the local patterns in sessions are captured by our max-pooling operation, and the long-term preference is dynamically estimated more well by the integration of memory module and attention network in MEANS. However, the whole history before current session is treated as a whole session in SHAN. The long-term preference hidden in such long behavior sequences may not be captured by the attention network used in SHAN.

4.5   Performance over Sessions with Different Length

Recall that our model characterizes each session with a max-pooling operation to capture the local pattern which might be complicated when the length of session is long enough. Now we investigate the performance of our model over sessions with different length. The sessions are classified into short or long with respect to the mean length of sessions in each dataset. The percentages of each class in four datasets are shown in the Table 3.

Table 3. Statistics of session length

Dataset	Short sessions	Long sessions
Gowalla	82.77% (length $\leq$ 3)	17.23%
TallM	90.02% (length $\leq$ 3)	9.98%
MovieLens	76.49% (length $\leq$ 11)	23.51%
Tafeng	85.51% (length $\leq$ 7)	14.49%

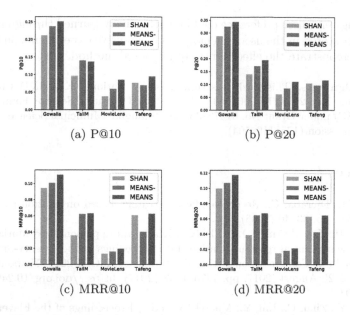

(a) P@10   (b) P@20

(c) MRR@10   (d) MRR@20

Fig. 4. Our method compares with SHAN on four data sets.

Table 4 shows the results of MEANS over four datasets. From the table, we find that: The performance of MEANS over short sessions are better than long ones, except under $P@50$ on the TallM dataset. Note that the performance over short sessions is still close to the overall performance. We leave improving prediction performance over long sessions on other datasets as future work.

Table 4. Performance over short sessions and long sessions

Dataset	Short sessions		Long sessions	
	P@50	MRR@50	P@50	MRR@50
Gowalla	0.48115	0.12291	0.39501	0.10577
TallM	0.23992	0.06877	0.33716	0.05588
MovieLens	0.1902	0.01798	0.05333	0.00207
Tafeng	0.18533	0.06312	0.11313	0.02776

5   Conclusion

In this paper, we propose a sequential recommendation model, named MEANS, which integrates max-pooling operation, memory network, and attention network. The high level information could be captured by the pooling operation, and the long-term user preference could be attentively learned through accessing the contents stored in the users' memory. The extensive experiments on four real datasets demonstrate the effectiveness of proposed method.

Acknowledgments. This work was supported by National Key R&D Program of China (Grant No. 2018YFB0904503), the National Natural Science Foundation of China (NFSC) under Grant No. 61572135, and Shanghai Municipal Science and Technology Commission (17511108504).

References

1. Adda, M., Djeraba, C.: Recommendation strategy based on relation rule mining. In: ITWP, pp. 33–40 (2005)
2. Cai, C., He, R., McAuley, J.J.: SPMC: socially-aware personalized Markov chains for sparse sequential recommendation. In: Proceedings of the Twenty-Sixth International Joint Conference on Artificial Intelligence, IJCAI 2017, Melbourne, Australia, 19–25 August, 2017, pp. 1476–1482 (2017). https://doi.org/10.24963/ijcai.2017/204
3. Chang, Y., Zhai, C., Liu, Y., Maarek, Y. (eds.) Proceedings of the Eleventh ACM International Conference on Web Search and Data Mining, WSDM 2018, Marina Del Rey, CA, USA, 5–9 February 2018. ACM (2018). https://doi.org/10.1145/3159652
4. Deshpande, M., Karypis, G.: Item-based top-N recommendation algorithms. ACM Trans. Inf. Syst. **22**(1), 143–177 (2004). https://doi.org/10.1145/963770.963776
5. Donkers, T., Loepp, B., Ziegler, J.: Sequential user-based recurrent neural network recommendations. In: Proceedings of the Eleventh ACM Conference on Recommender Systems, RecSys 2017, Como, Italy, 27–31 August 2017, pp. 152–160 (2017). https://doi.org/10.1145/3109859.3109877
6. Hidasi, B., Karatzoglou, A., Baltrunas, L., Tikk, D.: Session-based recommendations with recurrent neural networks. In: 4th International Conference on Learning Representations, ICLR 2016, San Juan, Puerto Rico, 2–4 May 2016, Conference Track Proceedings (2016)
7. Hu, L., Cao, L., Wang, S., Xu, G., Cao, J., Gu, Z.: Diversifying personalized recommendation with user-session context. In: Proceedings of the Twenty-Sixth International Joint Conference on Artificial Intelligence, IJCAI 2017, Melbourne, Australia, 19–25 August 2017, pp. 1858–1864 (2017). https://doi.org/10.24963/ijcai.2017/258
8. Jannach, D., Ludewig, M.: When recurrent neural networks meet the neighborhood for session-based recommendation. In: Proceedings of the Eleventh ACM Conference on Recommender Systems, RecSys 2017, Como, Italy, 27–31 August 2017, pp. 306–310 (2017). https://doi.org/10.1145/3109859.3109872
9. Koren, Y., Bell, R.M., Volinsky, C.: Matrix factorization techniques for recommender systems. IEEE Comput. **42**(8), 30–37 (2009). https://doi.org/10.1109/MC.2009.263
10. Lang, J. (ed.): Proceedings of the Twenty-Seventh International Joint Conference on Artificial Intelligence, IJCAI 2018, 13–19 July 2018, Stockholm, Sweden. ijcai.org (2018)

11. Li, J., Ren, P., Chen, Z., Ren, Z., Lian, T., Ma, J.: Neural attentive session-based recommendation. In: Proceedings of the 2017 ACM on Conference on Information and Knowledge Management, CIKM 2017, Singapore, 06–10 November 2017, pp. 1419–1428 (2017). https://doi.org/10.1145/3132847.3132926
12. Ludewig, M., Jannach, D.: Evaluation of session-based recommendation algorithms. User Model. User-Adapt. Interact. **28**(4–5), 331–390 (2018). https://doi.org/10.1007/s11257-018-9209-6
13. Quadrana, M., Cremonesi, P., Jannach, D.: Sequence-aware recommender systems. ACM Comput. Surv. **51**(4), 66:1–66:36 (2018). https://doi.org/10.1145/3190616
14. Quadrana, M., Karatzoglou, A., Hidasi, B., Cremonesi, P.: Personalizing session-based recommendations with hierarchical recurrent neural networks. In: Proceedings of the Eleventh ACM Conference on Recommender Systems, RecSys 2017, Como, Italy, 27–31 August 2017, pp. 130–137 (2017). https://doi.org/10.1145/3109859.3109896
15. Rendle, S., Freudenthaler, C., Gantner, Z., Schmidt-Thieme, L.: BPR: bayesian personalized ranking from implicit feedback. In: UAI 2009, Proceedings of the Twenty-Fifth Conference on Uncertainty in Artificial Intelligence, Montreal, QC, Canada, 18–21 June 2009, pp. 452–461 (2009)
16. Rendle, S., Freudenthaler, C., Schmidt-Thieme, L.: Factorizing personalized Markov chains for next-basket recommendation. In: Proceedings of the 19th International Conference on World Wide Web, WWW 2010, Raleigh, North Carolina, USA, 26–30 April 2010, pp. 811–820 (2010). https://doi.org/10.1145/1772690.1772773
17. Salakhutdinov, R., Mnih, A.: Probabilistic matrix factorization. In: Advances in Neural Information Processing Systems 20, Proceedings of the Twenty-First Annual Conference on Neural Information Processing Systems, Vancouver, British Columbia, Canada, 3–6 December 2007, pp. 1257–1264 (2007)
18. Salakhutdinov, R., Mnih, A., Hinton, G.E.: Restricted Boltzmann machines for collaborative filtering. In: Machine Learning, Proceedings of the Twenty-Fourth International Conference (ICML 2007), Corvallis, Oregon, USA, 20–24 June 2007, pp. 791–798 (2007). https://doi.org/10.1145/1273496.1273596
19. Sarwar, B.M., Karypis, G., Konstan, J.A., Riedl, J.: Item-based collaborative filtering recommendation algorithms. In: Proceedings of the Tenth International World Wide Web Conference, WWW 10, Hong Kong, China, 1–5 May 2001, pp. 285–295 (2001). https://doi.org/10.1145/371920.372071
20. Sen, S., Geyer, W., Freyne, J., Castells, P. (eds.): Proceedings of the 10th ACM Conference on Recommender Systems, Boston, MA, USA, 15–19 September 2016. ACM (2016)
21. Tan, Y.K., Xu, X., Liu, Y.: Improved recurrent neural networks for session-based recommendations. In: Proceedings of the 1st Workshop on Deep Learning for Recommender Systems, DLRS@RecSys 2016, Boston, MA, USA, 15 September 2016, pp. 17–22 (2016). https://doi.org/10.1145/2988450.2988452
22. Tuan, T.X., Phuong, T.M.: 3d convolutional networks for session-based recommendation with content features. In: Proceedings of the Eleventh ACM Conference on Recommender Systems, RecSys 2017, Como, Italy, 27–31 August 2017, pp. 138–146 (2017). https://doi.org/10.1145/3109859.3109900
23. Wang, P., Guo, J., Lan, Y., Xu, J., Wan, S., Cheng, X.: Learning hierarchical representation model for next basket recommendation. In: SIGIR, pp. 403–412 (2015). https://doi.org/10.1145/2766462.2767694

24. Wu, C., Ahmed, A., Beutel, A., Smola, A.J., Jing, H.: Recurrent recommender networks. In: Proceedings of the Tenth ACM International Conference on Web Search and Data Mining, WSDM 2017, Cambridge, United Kingdom, 6–10 February 2017, pp. 495–503 (2017). https://doi.org/10.1145/3018661.3018689

25. Yap, G.-E., Li, X.-L., Yu, P.S.: Effective next-items recommendation via personalized sequential pattern mining. In: Lee, S., Peng, Z., Zhou, X., Moon, Y.-S., Unland, R., Yoo, J. (eds.) DASFAA 2012. LNCS, vol. 7239, pp. 48–64. Springer, Heidelberg (2012). https://doi.org/10.1007/978-3-642-29035-0_4

26. Ying, H., et al.: Sequential recommender system based on hierarchical attention networks. In: Proceedings of the Twenty-Seventh International Joint Conference on Artificial Intelligence, IJCAI 2018, 13–19 July 2018, Stockholm, Sweden, pp. 3926–3932 (2018). https://doi.org/10.24963/ijcai.2018/546

27. Yu, F., Liu, Q., Wu, S., Wang, L., Tan, T.: A dynamic recurrent model for next basket recommendation. In: Proceedings of the 39th International ACM SIGIR conference on Research and Development in Information Retrieval, SIGIR 2016, Pisa, Italy, 17–21 July 2016, pp. 729–732 (2016). https://doi.org/10.1145/2911451.2914683

28. Zhu, Y., et al.: What to do next: modeling user behaviors by time-LSTM. In: Proceedings of the Twenty-Sixth International Joint Conference on Artificial Intelligence, IJCAI 2017, Melbourne, Australia, 19–25 August 2017, pp. 3602–3608 (2017). https://doi.org/10.24963/ijcai.2017/504

Multi-head Attentive Social Recommendation

Xu Luo[✉], Chaofeng Sha, Zijing Tan, and Junyu Niu

School of Computer Science, Shanghai Key Laboratory of Intelligence Processing,
Fudan University, Shanghai 200433, China
{xuluo17,cfsha,zjtan,jyniu}@fudan.edu.cn

Abstract. Recently social relationship among users has been exploited to improve the recommendation performance. The intuition behind most of these work is social homophily such that users are more similar to their neighbors. Attention mechanism or attention network from deep learning has been a popular component employed by recommendation models. However, how to attentively learn the influence between users remains pretty much open in the existing social recommendation models. In this paper, we propose a social recommendation model MAS, Multi-head Attentive Social Recommendation. The key to MAS is a multi-head attention network which can distinguish the impact of users' friends when predicting users' preference on different items. When compared to the state-of-the-art baseline methods on three real-world datasets, our method achieves the best performance.

Keywords: Collaborative filtering · Social recommendation · Attention network

1 Introduction

Recommender systems are pervasive in various web and e-commerce applications. The traditional approaches to personalized recommendation, such as neighborhood models and latent factor models, rely on the users' explicit or implicit feedbacks on the items. Through training models on the historical data, the recommender systems can predict the unknown ratings or preferences of users on items. However, there are abundant link structures between users in social media or other web applications that can potentially sway user's own opinion, because people may choose to change their own preferences to match others' responses. Many recent studies have shown that it is beneficial to take those social structures into account when building a recommender system [6,8,10,18], particularly in the setting when users have diverse interests or characteristics. Most social recommendation models rely on explicit feedback [14,26]. The social structure is exploited to regularize the user preference vectors when modeling the user-item interactions [17]. Recently, several social recommendation models have been developed for implicit feedback data, such as SBPR [31] and SPMC

© Springer Nature Switzerland AG 2019
R. Cheng et al. (Eds.): WISE 2019, LNCS 11881, pp. 243–258, 2019.
https://doi.org/10.1007/978-3-030-34223-4_16

[2]. SBPR uses an observation that *users tend to assign higher ranks to items that their friends prefer*. The impact of different friends on a user is treated as uniform. However, as noted in [2] that *different friends could have a different amount of impact on a user*. In SPMC, the impact of friends on a user is modeled as the closeness between them. However, the impact of friends on a user is static with respect to items. In this paper, we further assume that when predicting the preference of a user u towards different items, the impact of friends on u also depends on the items. For example, when user u considers to choose a Romance movie, she would be more likely influenced by some friends. In another time, she might be influenced by other friends when choosing a Thriller movie.

Based on this assumption, we propose a new social recommendation model, MAS (Multi-head Attentive Social recommendation) that explicitly exploits social impact of friends on users rating behaviors. Specifically, we first embed users and items into low-dimensional space through utilizing the user-item interaction. Then we enhance the user representations with a weighted sum of social representations of neighbors, which are learned through explicitly modeling the social relationship. The social representations are used as context information in multi-head attention network, which can stably estimate different influence of neighbors for users with respect to different items. We finally employ Bayesian Personalized Ranking framework to estimate the model's parameters. The advantages of our new approach over the existing ones are confirmed by extensive experiments on three real-world datasets. Our contributions can be summarized as follows:

1. We design a multi-head attention network to learn the diverse social impact of different friends on a user with respect to each item.
2. We propose a new social recommendation model which incorporates the user-item interaction and social structure into a unified model. The learned preference representation and social representation are combined to represent users.
3. We conduct extensive experiments on three real-world datasets, which include performance comparisons with state-of-the-art methods. The experimental results demonstrate the effectiveness of our proposed approach, even for cold-start users.

2   Related Work

In recent years, the social link between users is used to improve the prediction accuracy of the recommender system [6,18]. In [10], the authors propose Trustwalker which is based on a random walk model combining the trust-based and the collaborative filtering approach for recommendation. In [16], the authors propose to fuse the users' tastes and their trusted friends' favors together. Based on this intuition, they propose social trust ensemble to represent the formulation of the social trust restrictions on the recommender systems. Social regularization has been introduced in [17] to constrain matrix factorization objective functions.

In [30], a circle-based recommendation system is developed through inferring category-specific social trust circles from available rating data combined with social network data. The out-performance of using explicit social information is also demonstrated in the experiments of [15].

In [13], the authors propose to combine contextual information and social information to improve quality of recommendations. The matrix factorization is applied on the sub-matrix which is partitioned based on various context, and the social information is incorporated into a social regularization term. In [22], local and global social relations are exploited for recommendation. In their framework, the user preferences of two socially connected users are correlated locally and the global user reputation score is used to weight the importance of their ratings. Motivated by the heuristic that individuals will affect each other during the process of reviewing, a truster and trustee model is proposed in [29] to map users into the same latent feature spaces but with different implications that can explicitly describe the feedback how users affect or follow the opinions of others. They synthesized the two models to one fusing model simultaneously fitting available ratings and trust ties. TrustSVD proposed in [8] extends SVD++ [11] with social trust information. The trust matrix is decomposed into trust-feature matrix and trustee-feature matrix. However, the above models are developed for explicit feedback data.

A few recent works exploit tie strength or user dependencies for building social recommendation systems. The effects of distinguishing strong and weak ties in social recommendation are studied in [27]. They use Jaccard similarity to approximate the tie strength and incorporate the distinction of strong and weak ties into the Bayesian Personalized Ranking method. In their following work [26], the personalized preference of strong and weak ties are learned and incorporated into probabilistic matrix factorization for social recommendation. Assuming users' latent features follow a matrix variate normal distribution, the learnt positive and negative dependencies between users are incorporated into the probabilistic relational matrix factorization model proposed in [14]. In SREPS [12], the rating data, consumption data and social structure are incorporated to learn latent vectors in the essential preference space. In SERec model [25], the social information is used as social regularization and social boosting. Chaney et al. [3] propose social Poisson factorization, a Bayesian model that incorporates social network information.

Attention network has been successfully applied to tackle problems such as machine translation [1], and recommendation systems [28]. The advantages of recommendation models based on attention mechanism over traditional weighted contribution perspective [3] are demonstrated in recent work such as ACF [4] and NAIS [9]. The authors of [20] propose Attentive Recurrent Social Recommendation model that users' temporal complex dynamic interests and the static interests are mixed to model users preferences over time. The temporal social influence over time is learned through a temporal attention network. However, the performance of Static Attentive Social Recurrent Recommendation is inferior to ARSR and our model, as demonstrated in experiments of [20] and ours.

3 Proposed Method

In this section, we begin by setting up the notations before delving our proposed method MAS (Multi-head Attentive Social Recommendation). First we present the component modeling user-item interaction. We then design and elaborate a social attention network which is used to learn the social impact. Furthermore, we also extend our basic model with multi-head attention mechanism. Finally, the MAS model is developed after the introduction of the component modeling social structure.

3.1 Notations and Problem Formulation

We denote the user set as $U = \{u_1, u_2, \cdots, u_n\}$ and the item set as $I = \{i_1, i_2, \cdots, i_m\}$. In the recommendation systems that exploit social structure, we are given a social network $G = (U, E)$, where E is the set of social links between users, in addition to a rating matrix $R = [x_{ui}]_{n \times m}$. In this paper, we consider social recommendation with implicit feedback data where the user-item interaction $x_{ui} \in \{0, 1\}$ denotes the feedback of user u on item i. We denote R_u and N_u as the positive and negative feedback of user u, respectively. The user-user interactions describe the social connections between users, such as truster/trustee relations or friendships. We denote the friends of user u by $F(u)$.

Our goal is to recommend a list of items that a user may interested in by exploiting both the user-item interactions and social relationship between users.

3.2 Modeling User Feedback

As we aim to recommend top-N items to users, we employ Bayesian Personalized Ranking (BPR) as our basic learning model due to its effectiveness in exploiting the user-item implicit feedback. The basic assumption of individual pair-wise preference over two items used in BPR can be represented as follows:

$$i >_u j, i \in R_u, j \in N_u,$$

where the partial ranking relationship $i >_u j$ means that a user u is likely prefer a positive item $i \in R_u$ to a negative item $j \in N_u$ [19]. Given the training set $D_S = \{(u, i, j) : u \in U, i \in R_u, j \in N_u\}$, the generic optimization criterion of BPR (BPR-OPT) is defined as follows:

$$L_I = \sum_{(u,i,j) \in D_S} \ln \sigma(\hat{x}_{ui} - \hat{x}_{uj}) - \lambda_\Theta ||\Theta||^2, \tag{1}$$

where σ is the sigmoid function $\sigma(z) = \frac{1}{1+e^{-z}}$, and λ_Θ are model specific regularization parameters. The preference of the user u towards the item i, $\hat{x}_{ui}$ can be estimated through a ranking model.

In this paper, we adopt the BPR-MF framework [19] where the user-item interaction $\hat{x}_{ui}$ is estimated as follows:

$$\hat{x}_{ui} = q_i^T p_u \tag{2}$$

where $p_u \in \mathbb{R}^d$ and $q_i \in \mathbb{R}^d$ are latent preference vector of user u and item i, respectively.

Inspired by the work of node or social embedding [21], we introduce a social representation $c_u \in \mathbb{R}^d$ to the user u. Then we could enhance the representation of user u as follows:

$$p_u + \sum_{v \in F(u)} \alpha_{uv} c_v \qquad (3)$$

where α_{uv} measures the closeness between user u and v, or social impact of user v on user u.

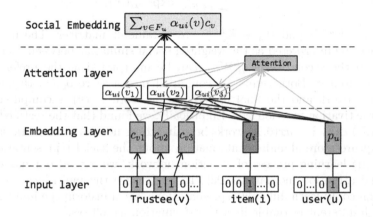

Fig. 1. An illustration of one head social attention network

It is essentially the inherent user interest p_u adjusted by a weighted sum of her friends' social preference vectors. The weight α_{uv} could be set as $\alpha_{uv} = 1/|F(u)|$ or $\alpha_{uv} = 1/\sqrt{|F(u)|}$. However, the amount of impact of different friends on user u is overlooked in this simplest setting. Another choice of α_{uv} is the closeness between user u and user v as proposed in SMPC [2]. However, as mentioned in Introduction section, our rating or buying behavior might be influenced by different friends on different items. Based on this assumption, we propose the following estimator $\hat{x}_{ui}$:

$$\hat{x}_{ui} = q_i^T \left(p_u + \sum_{v \in F(u)} \alpha_{ui}(v) c_v \right), \qquad (4)$$

where $\alpha_{ui}(v)$ denotes the amount of impact of user v on user u for item i. It can be seen that this formulation subsumes previous ones. We design an social attention network to learn the weight $\alpha_{ui}(v)$ in the next section.

3.3   Multi-head Social Attention

The goal of the social attention is to select friends who have large amount impact on user when predicting the behavior of a user on different items. We can derive a normalized score with a softmax function if we follow traditional settings of neural attention network [1]. Figure 1 illustrates the architecture of designed social attention network. Given the user's preference vector p_u, the friend's social representation vector c_v and the item representation vector q_i, we use a two-layer network to learn the attention weight $\alpha_{ui}(v)$ as follows:

$$o_{ui}(v) = p_u^T W_p c_v + q_i^T W_q c_v \tag{5}$$
$$\alpha_{ui}(v) = softmax(o_{ui}(v))$$
$$= \frac{\exp(o_{ui}(v))}{\sum_{v' \in F(u)} \exp(o_{ui}(v'))} \tag{6}$$

where $W_p \in \mathbb{R}^{d \times d_a}$ and $W_q \in \mathbb{R}^{d \times d_a}$ are parameter matrices. The first term $p_u^T W_p c_v$ can be interpreted as the weighted social similarity between user u and user v, and the second term $q_i^T W_q c_v$ can be interpreted as the preference of friend u to item i. Both terms are combined into a score $o_{ui}(v)$ through the attention network and the final representation score $o_{ui}(v)$ is computed by a nonlinear activation function $\phi(\cdot)$. Empirically, we found that the rectified linear unit (ReLU) $\phi(x) = max(0, x)$ works best due to its non-saturating gradient [7].

To capture more abundant information about the social representation and stabilize the learning process of attention network, we extend our model with multi-head attention mechanism [23]. To make sure the multi-head attention mechanism could learn different representations, we transform the input feature c_v through different learnable liner transformation as follows:

$$c_v^k = c_v W_c^k \tag{7}$$

The multi-head attention network has multi-parameters W^k and c^k. These independent attention mechanisms execute the transformation and output the following weight $\alpha_{ui}^k(v))$:

$$o_{ui}^k(v) = \phi(p_u^T W_p^k c_v^k + q_i^T W_q^k c_v^k) \tag{8}$$
$$\alpha_{ui}^k(v) = softmax(o_{ui}^k(v)) \tag{9}$$

After that, we employ averaging strategy to combine the multi-head attention outputs as follows:

$$\alpha_{ui}(v) = \frac{1}{K} \sum_{k=1}^{K} \alpha_{ui}^k(v) \tag{10}$$

3.4 Modeling Social Structure

In this section, we introduce the component modeling social structure which bridges the latent preference p_u of user u and social embedding c_u. We achieve this goal through another instantiation of BPR. Given the social network or graph $G = (U, E)$, the social proximity between a friend v and user u is estimated as $\hat{y}_{uv} = p_u^T c_v$. Borrowing the partial ranking assumption of BPR, we assume that $\hat{y}_{uv} \geq \hat{y}_{uv_n}$ for $v \in F(u)$ and $v_n \in U \backslash F(u)$. That means the social proximity between neighbors should be larger than strangers. This yields the following objective function:

$$\mathcal{L}_G = \sum_{(u,v,v_n) \in D_G} \ln \sigma(\hat{y}_{uvv_n}) \tag{11}$$

where $\hat{y}_{uvv_n} = \hat{y}_{uv} - \hat{y}_{uv_n}$ and $D_G = \{(u,v,v_n) : u \in U, v \in F(u), v_n \in U \backslash F(u)\}$. From the perspective of recommender system, if two users are strongly connected, it is very likely that they have similar preference and the friend should have more impact on him.

3.5   The Unified MAS Model

In this section, we aim to develop a unified MAS model. By incorporating the objective functions for modeling the user feedback and social structure respectively as Eqs. (1) and (11), we arrive at the unified objective function for our MAS model as follows:

$$\mathcal{L}_{MAS} = \mathcal{L}_I + \gamma \mathcal{L}_G - \lambda \mathcal{L}_r \tag{12}$$

where γ and λ are parameters to trade-off the contribution of corresponding terms, and $\mathcal{L}_r$ is the regularization term to avoid overfitting which is defined as follows:

$$\mathcal{L}_r = \frac{1}{2}(\sum_u ||p_u||^2| + \sum_v ||c_v||^2 + \sum_i ||q_i||^2 + ||W^k||).$$

The model parameters $\{p_u, c_v, q_i, w_j\}$ are estimated by maximizing the above regularized objective function through stochastic gradient ascent. W^k is a collection of attention network parameters W_c^k, W_p^k and W_q^k.

Learning Details. We describe some learning details which are essential to re-implement our method.

Sampling Methods. Since we use the pairwise learning method BPR to optimize model parameters, the sampled training instances have great impact on the recommendation performance. Therefore, we adopt two sampling strategies for the interaction component and the social component. For sampling an instance in the interaction component, we uniformly sample a negative item j from N_u to build a training triple (u,i,j) for user u and positive item i. For sampling an instance in the social component which is also a triple (u,v,v_n), we use the negative sampling to draw the negative user v_n according to the noise distribution $P_n(u)$. We set $P_n(u) \propto d_u^{3/4}$ as used in [21], where d_u is the out-degree of user u.

Pre-training. In our model, we integrate the attention network to better model user preference, but this could lead to hard fitting because of the non-linearity of neural network and sensitivity to initialization. On the other hand, we only consider user's neighbor relations during modeling social structure. In reality, users occur in similar contexts also tend to have similar preference. However, training complex social model and user preference simultaneously converges slowly and traps to local minimums. Besides, our aim is to train the recommendation model rather than to devise sophisticated model for social embedding. Therefore, we consider involving the pre-trained social embedding as prior knowledge. LINE [21] shows good performance for capture social structures for social embedding. We adopt the social embedding learned by LINE to initialize that of MAS which greatly facilitates the learning of the attention network and achieves better performance.

Training Process. We summarize the training process in Algorithm 1. To be specific, lines 1–3 initialize all embedding vectors and attention parameters; lines 6 performs uniform sampling; line 7 employs SGA to learn the embeddings and parameters in the interaction component. lines 10 performs negative sampling; line 11 employs SGA to learn the embeddings in the social component.

Algorithm 1. Optimization Algorithm of MAS Model

Input: interaction data R, social graph G,
      negative feedback of user u N_u, noise distribution of negative user v_n $P_n(u)$
Output: p_u, c_v, q_i, W^k
1: Initialize embedding vectors p_u, q_i
2: Initialize embedding vector c_v using pre-trained social embedding
3: Initialize attention parameters W^k
4: **while** *not converged* **do**
5:    **for** $(u, i) \in$ interaction data R **do**
6:        Uniformly sample a negative item j from N_u, build a training triple (u, i, j);
7:        update p_u, c_v, q_i, q_j, W^k by $\frac{\partial \mathcal{L}_I}{\partial p_u}, \frac{\partial \mathcal{L}_I}{\partial c_v}, \frac{\partial \mathcal{L}_I}{\partial q_i}, \frac{\partial \mathcal{L}_I}{\partial q_j}, \frac{\partial \mathcal{L}_I}{\partial W^k}$;
8:    **end for**
9:    **for** $(u, v) \in$ social graph G **do**
10:       Use negative sampling to draw a negative user v_n from $P_n(u)$, build a training triple (u, v, v_n);
11:       update p_u, c_v, c_{v_n} by $\frac{\partial \mathcal{L}_G}{\partial p_u}, \frac{\partial \mathcal{L}_G}{\partial c_v}, \frac{\partial \mathcal{L}_G}{\partial c_{v_n}}$;
12:    **end for**
13: **end while**

---

4 Experiments

In this section, we evaluate the effectiveness of the proposed MAS framework using three real-world datasets.

4.1   Experimental Settings

Datasets. Three publicly available datasets are used in our experiments, whose statistics are shown in Table 1.

- *Ciao*[1]. This dataset contains ratings on DVDs and social network which is crawled from a DVD community website in December, 2013.
- *Epinions*[2]. Epinions is an online consumer review website. The dataset is collected from this website and also contains ratings and social network.
- *Foursquare*[3]. This dataset is collected from the popular location-based social network Foursquare with the ratings of venues and social relations.

All three datasets consist of both ratings and directed social links. We removed those items that have less than five ratings. Since feedbacks in these datasets are all explicit ratings from 1 to 5, we treat the items rated higher than 2 as positive feedback Following [27], we randomly choose 70% of each user's positive items as the training set, 20% are held out as the validation set and the rest are used as the test set. The best parameters are determined according to the performance on validation set. Since a few top-ranked items can only be noticed by users [5], we use the top-K metrics like Precision, Recall, MAP, and NCDG to measure recommendation performance.

Table 1. Statistics of the datasets.

	Ciao	Epinions	Foursquare
#Users	7,260	27,488	51,055
#Items	10,812	22,565	43,593
#Observed feedback	139,257	436,883	201,298
#Social relations	110,709	392,662	269,144
#Average feedback	13.4	15.9	3.9
#Avg. social relations	15.2	14.3	5.3

Baselines. To validate the performance of our approach, we compare with the following baselines which aim to recommend with implicit feedback data.

- *BPR*. This is a recommendation method tailored to implicit feedback data combined with Matrix Factorization model [19].
- *SBPR*. This is a state-of-the-art recommendation model that benefits from modeling social relations. Introduced by [31], the model is based on the assumption that users and their social circles should have similar tastes/preferences towards items.

---

[1] https://www.cse.msu.edu/~tangjili/trust.html.
[2] http://www.trustlet.org/epinions.html.
[3] https://archive.org/details/201309_foursquare_dataset_umn.

- SPF^4. This is a probabilistic model that incorporates social network information into Poisson factorization [3].
- $SARSE$. Static Attentive Social Recurrent Recommendation is one of variants of Attentive Recurrent Social Recommendation model recently proposed in [20]. In SARSE, only the social information is leveraged to learn users' stationary interests.

For convenience, we denote the model with averaging social embeddings as "ABPR". In this paper, we focus on leveraging the social information to improve the recommendation performance. To be fair, we present the comparison results with SARSE, the static variant of DARSE.

Parameter Settings. For all models, we randomly initialize the parameters with Gaussian distribution, where the mean and standard deviation are 0 and 0.001, respectively. All models are trained with stochastic gradient descent with a batch size of 1024 at each iteration and the learning rate is 0.001 according to the turning process. We set the regularization parameter λ to 0.01. The parameters of our attention network are initialized with Glorot normal initializer. The latent factor number of attention network is 16 and the number of multi-head is 3. The weight of social component γ is 0.8. All these hyper-parameters are tuned on the validation set.

4.2 Performance Comparison

We first compare the performance of our models with other methods under different metrics. Then we investigate the performance comparison on cold-start users. Finally, we investigate the performance over users with different degrees.

Overall Comparison. Table 2 shows the performance comparison of four metrics with respect to the different recommendation methods on three datasets, where the embedding size includes 16 and 32. We have the following findings:

- As shown in Table 2, our proposed model MAS achieves competitive results than the other methods. The relative improvements over the best baselines are 24.1%, 8.7% and 3.3% in terms of Recall@10 on three datasets, respectively. This may due to that MAS could capture the different impact of user's friends effectively. In addition, the performance of MAS is better than SARSE which also adopts the attention mechanism, possibly because MAS assigns more appropriate social weights which is related to each item through the multi-head attention network. The improvements against SPF also verify the benefits of multi-head mechanism.
- The proposed ABPR is a simplified version of MAS which only aggregates the friends' embeddings averagely. Even so, ABPR still outperforms SBPR. In particular, in terms of Recall@10, the relative improvement over SBPR is

---

[4] https://github.com/ajbc/spf.

about 6% on average. While SBPR indirectly exploits the social connections, it only utilizes ratings information of user's friends. ABPR incorporates the interaction component and the social component together effectively which leads to better performance.

- All models taking social connections into account perform better than BPR on all datasets, which indicates that integrating social information can effectively improve performance when this information could be collected. We also conduct experiments when the embedding size sets to 32. Our model still outperforms all baselines on all datasets in terms of different evaluation metrics. Meanwhile, the performance of all models improves, when the latent embedding size increases from 16 to 32. Figure 2 shows the performance on three datasets in terms of Recall@K when varying K.

Comparison on Cold-Start Users. We further investigate the performance of different methods on cold-start users who consume less than 5 items. As shown in Table 3, our approach MAS significantly outperforms competitors which confirms the advantages of two components' incorporation and attention network again. The results indicate that MAS is effective for not only all users, but also cold-start users.

Comparison in Degrees. We then analyze the performance comparison on users with different degrees. Here, the degree of a user is the number of his friends. After removing isolated users, we divide the users into four groups (namely, 1–5, 6–15, 16–25 and 25+) in terms of their degrees. Then, we compare the models in terms of Recall@100. As can be seen from Fig. 3, our approach MAS performs the best on different groups in most cases. However, when the degree is quite large, MAS loses on Epinions and Foursquare. By examining the datasets, we find that there are only dozens of users in these groups which may causes the instability. However, MAS is more stable than ABPR on three datasets, which indicates the incorporation of attention mechanism brings more superiority.

Effect of Pre-training. To enhance our model, we also use a popular embedding method LINE [21] to initialize the social embedding which is able to capture the first-order and second-order proximity. From the Table 4, we can find that the performance of MAS with pre-trained social embeddings is improved significantly.

4.3   Attention Analysis

We also investigate the multi-head attention weights to show that our model is more explainable. To demonstrate this, we investigate the multi-head attention weights of a sampled user on target item shown in Table 5. Since ABPR aggregates the social embeddings averagely, it uniformly assigns weights to the four friends. In this case, the target item #885 and #5020 are sampled from

Table 2. Overall performance comparison with different dimension: $d = 16$ and $d = 32$. (The best scores are in bold.)

All	Metrics	$d = 16$					
		BPR	SBPR	SPF	SARSE	ABPR	MAS
Ciao	Pre@10	0.011771	0.011969	0.013795	0.012607	0.012739	**0.015578**
	Recall@5	0.025704	0.026720	0.029682	0.027483	0.025891	**0.035738**
	Recall@10	0.041705	0.041305	0.046343	0.044330	0.043497	**0.057512**
	MAP@100	0.006192	0.006749	0.007937	0.006807	0.006772	**0.009036**
	NDCG@100	0.071647	0.078791	0.100342	0.079136	0.078630	**0.112618**
Epinions	Pre@10	0.005430	0.005154	0.007215	0.005417	0.005799	**0.007672**
	Recall@5	0.015882	0.015016	0.018453	0.014979	0.016074	**0.019077**
	Recall@10	0.023777	0.022744	0.028749	0.023394	0.025221	**0.031253**
	MAP@100	0.003125	0.002999	0.004710	0.003567	0.003932	**0.005008**
	NDCG@100	0.047220	0.046970	0.068366	0.054218	0.058598	**0.071660**
Foursquare	Pre@10	0.015491	0.015914	0.014764	0.016774	0.016388	**0.017316**
	Recall@5	0.080396	0.083234	0.071552	0.084722	0.083648	**0.091292**
	Recall@10	0.128577	0.132335	0.123059	0.139419	0.136544	**0.144006**
	MAP@100	0.008165	0.008512	0.008788	0.008891	0.008744	**0.009211**
	NDCG@100	0.116142	0.121366	0.127044	0.125835	0.123654	**0.131144**
		$d = 32$					
Ciao	Pre@10	0.012541	0.012607	0.015050	0.012541	0.013553	**0.015512**
	Recall@5	0.026552	0.027740	0.032062	0.026710	0.028773	**0.037294**
	Recall@10	0.043851	0.044100	0.050020	0.043781	0.047155	**0.056443**
	MAP@100	0.006607	0.006953	0.008704	0.006883	0.007270	**0.009183**
	NDCG@100	0.077390	0.081609	0.106309	0.079903	0.085926	**0.113812**
Epinions	Pre@10	0.005593	0.005442	0.007591	0.005862	0.005906	**0.007697**
	Recall@5	0.015597	0.015222	**0.019498**	0.016257	0.015751	0.019203
	Recall@10	0.024583	0.024490	0.030298	0.025993	0.025438	**0.031148**
	MAP@100	0.003159	0.003089	0.004996	0.003865	0.003924	**0.005154**
	NDCG@100	0.047384	0.047907	0.072082	0.057266	0.058611	**0.073327**
Foursquare	Pre@10	0.016210	0.016485	0.014882	0.016374	0.016529	**0.017316**
	Recall@5	0.082646	0.083160	0.073936	0.083592	0.083660	**0.091742**
	Recall@10	0.135407	0.137932	0.123535	0.136601	0.137330	**0.144124**
	MAP@100	0.008513	0.008680	0.008825	0.008688	0.008777	**0.009222**
	NDCG@100	0.121073	0.123171	0.128066	0.122608	0.124192	**0.131713**

the test set and should be ranked higher. From the Table 5, we can see that MAS predicts a larger score on the target items successfully. As a comparison, ABPR predicts a relatively smaller score. We have the following findings: on the one hand, the attention weights can represent the diverse impact of friends. MAS assigns a lower weight on the friend #779 and a larger weight on the other friends to the item #885. On the other hand, the attention weights of friends may vary when the user consumes different items. MAS assigns a lower weight on the friend #3250 to the item #5020 which is different from the item #885. The reason may because that the preferred items of friends and the relationship between them have considerable influence on the user's decision.

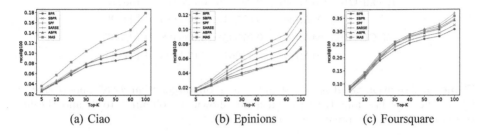

(a) Ciao   (b) Epinions   (c) Foursquare

Fig. 2. Performance under different values of K (measured by Recall@100).

Table 3. Cold-start performance comparison (embedding size 16).

Cold start	Metrics	BPR	SBPR	SPF	SARSE	ABPR	MAS
Ciao	Pre@10	0.004634	0.004802	0.005897	0.005055	0.005055	**0.006740**
	Recall@10	0.036226	0.038051	0.042825	0.038894	0.038192	**0.050688**
	MAP@100	0.002294	0.002776	0.003360	0.002796	0.002834	**0.004066**
	NDCG@100	0.034076	0.042236	0.054717	0.042990	0.042903	**0.064374**
Epinions	Pre@10	0.002818	0.002860	0.003674	0.002672	0.003027	**0.003674**
	Recall@10	0.021380	0.021791	0.027413	0.020566	0.022800	**0.027642**
	MAP@100	0.001596	0.001644	0.002307	0.001736	0.001924	**0.002328**
	NDCG@100	0.026728	0.027941	0.038933	0.030064	0.033406	**0.039222**
Foursquare	Pre@10	0.016550	0.017048	0.015555	0.017827	0.017438	**0.018479**
	Recall@10	0.141429	0.145835	0.134005	0.152445	0.149381	**0.157862**
	MAP@100	0.008799	0.009192	0.009179	0.009555	0.009388	**0.009885**
	NDCG@100	0.126055	0.131999	0.133405	0.135933	0.133425	**0.141354**

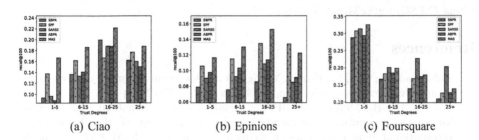

(a) Ciao   (b) Epinions   (c) Foursquare

Fig. 3. Performance comparison on users with different degrees (measured by Recall@100).

Table 4. Performance of MAS methods with (w/) and without (w/o) pre-trained social embeddings (embedding size 16).

	Metrics	Ciao	Epinions	Foursquare
Recall@10	MAS w/o	0.043786	0.025871	0.138270
	MAS w	0.057512	0.031185	0.144006
NDCG@100	MAS w/o	0.079351	0.058428	0.122587
	MAS w	0.112618	0.071693	0.131144

Table 5. Attention weights of a sampled user on target item #885 and #5020 on Ciao. The user has four trustees which are shown in column 1 to 4, and the last column represents the prediction score.

Friend ID	#779	#3250	#3901	#4207	Score	Friend ID	#779	#3250	#3901	#4207	Score
ABPR	0.250	0.250	0.250	0.250	0.026	ABPR	0.250	0.250	0.250	0.250	0.001
	0.205	0.165	**0.482**	0.146			0.275	0.201	**0.332**	0.190	
MAS	0.199	**0.427**	0.186	0.186	3.27	MAS	**0.291**	0.255	0.230	0.223	1.86
	0.034	0.213	0.251	**0.501**			0.177	0.292	0.198	**0.330**	

5   Conclusion and Future Work

In this paper, we propose a new social recommendation model, MAS, to exploit the implicit feedback data and social structure. We design a social attention network to learn the impact of users' friends when predicting users' preference on different items. The experiments on three real-world datasets demonstrate the effectiveness of our approach. In the future, we plan to extend our social recommendation model through utilizing graph attention networks proposed in [24].

Acknowledgments. This work was supported by National Key R&D Program of China (Grant No. 2018YFB0904503) and the National Natural Science Foundation of China (NFSC) under Grant No. 61572135.

References

1. Bahdanau, D., Cho, K., Bengio, Y.: Neural machine translation by jointly learning to align and translate. arXiv preprint arXiv:1409.0473 (2014)
2. Cai, C., He, R., McAuley, J.: SPMC: socially-aware personalized Markov chains for sparse sequential recommendation. In: Proceedings of the Twenty-Sixth International Joint Conference on Artificial Intelligence, IJCAI 2017, Melbourne, Australia, 19–25 August 2017, pp. 1476–1482 (2017)
3. Chaney, A.J., Blei, D.M., Eliassi-Rad, T.: A probabilistic model for using social networks in personalized item recommendation. In: RecSys, pp. 43–50 (2015)
4. Chen, J., Zhang, H., He, X., Nie, L., Liu, W., Chua, T.: Attentive collaborative filtering: Multimedia recommendation with item- and component-level attention. In: SIGIR, pp. 335–344 (2017)

5. Chen, L., Pu, P.: Users' eye gaze pattern in organization-based recommender interfaces. In: Proceedings of the 16th International Conference on Intelligent User Interfaces, pp. 311–314. ACM (2011)
6. Golbeck, J.: Generating predictive movie recommendations from trust in social networks. In: Proceedings of the Trust Management, 4th International Conference, iTrust 2006, Pisa, Italy, 16–19 May 2006, pp. 93–104 (2006)
7. Goodfellow, I., Bengio, Y., Courville, A., Bengio, Y.: Deep Learning, vol. 1. MIT Press, Cambridge (2016)
8. Guo, G., Zhang, J., Yorke-Smith, N.: TrustSVD: collaborative filtering with both the explicit and implicit influence of user trust and of item ratings. In: Proceedings of the Twenty-Ninth AAAI Conference on Artificial Intelligence, Austin, Texas, USA, 25–30 January 2015, pp. 123–129 (2015)
9. He, X., He, Z., Song, J., Liu, Z., Jiang, Y., Chua, T.: NAIS: neural attentive item similarity model for recommendation. IEEE Trans. Knowl. Data Eng. **30**(12), 2354–2366 (2018)
10. Jamali, M., Ester, M.: TrustWalker: a random walk model for combining trust-based and item-based recommendation. In: KDD, pp. 397–406 (2009)
11. Koren, Y.: Factorization meets the neighborhood: a multifaceted collaborative filtering model. In: KDD, pp. 426–434 (2008)
12. Liu, C., Zhou, C., Wu, J., Hu, Y., Guo, L.: Social recommendation with an essential preference space. In: Proceedings of the Thirty-Second AAAI Conference on Artificial Intelligence, New Orleans, Louisiana, USA, 2–7 February 2018 (2018)
13. Liu, X., Aberer, K.: Soco: a social network aided context-aware recommender system. In: 22nd International World Wide Web Conference, WWW 2013, Rio de Janeiro, Brazil, 13–17 May 2013, pp. 781–802 (2013)
14. Liu, Y., Zhao, P., Liu, X., Wu, M., Duan, L., Li, X.L.: Learning user dependencies for recommendation. In: Proceedings of the Twenty-Sixth International Joint Conference on Artificial Intelligence (IJCAI 2017), pp. 2379–2385 (2017)
15. Ma, H.: An experimental study on implicit social recommendation. In: SIGIR, pp. 73–82 (2013)
16. Ma, H., King, I., Lyu, M.R.: Learning to recommend with social trust ensemble. In: SIGIR, pp. 203–210 (2009)
17. Ma, H., Zhou, D., Liu, C., Lyu, M.R., King, I.: Recommender systems with social regularization. In: WSDM, pp. 287–296 (2011)
18. Massa, P., Avesani, P.: Trust-aware recommender systems. In: Proceedings of the 2007 ACM Conference on Recommender Systems, RecSys 2007, Minneapolis, MN, USA, 19–20 October 2007, pp. 17–24 (2007)
19. Rendle, S., Freudenthaler, C., Gantner, Z., Schmidt-Thieme, L.: BPR: Bayesian personalized ranking from implicit feedback. In: UAI, pp. 452–461 (2009)
20. Sun, P., Wu, L., Wang, M.: Attentive recurrent social recommendation. In: SIGIR, pp. 185–194 (2018)
21. Tang, J., Qu, M., Wang, M., Zhang, M., Yan, J., Mei, Q.: Line: Large-scale information network embedding. In: Proceedings of the 24th International Conference on World Wide Web, pp. 1067–1077. International World Wide Web Conferences Steering Committee (2015)
22. Tang, J., Hu, X., Gao, H., Liu, H.: Exploiting local and global social context for recommendation. In: IJCAI 2013, Proceedings of the 23rd International Joint Conference on Artificial Intelligence, Beijing, China, August 3–9, 2013, pp. 2712–2718 (2013)
23. Vaswani, A., et al.: Attention is all you need. In: NIPS, pp. 5998–6008 (2017)

24. Velickovic, P., Cucurull, G., Casanova, A., Romero, A., Liò, P., Bengio, Y.: Graph attention networks. In: ICLR (2018)
25. Wang, M., Zheng, X., Yang, Y., Zhang, K.: Collaborative filtering with social exposure: a modular approach to social recommendation. In: Proceedings of the Thirty-Second AAAI Conference on Artificial Intelligence, New Orleans, Louisiana, USA, 2–7 February 2018 (2018)
26. Wang, X., Hoi, S.C.H., Ester, M., Bu, J., Chen, C.: Learning personalized preference of strong and weak ties for social recommendation. In: Proceedings of the 26th International Conference on World Wide Web, WWW 2017, Perth, Australia, 3–7 April 2017, pp. 1601–1610 (2017)
27. Wang, X., Lu, W., Ester, M., Wang, C., Chen, C.: Social recommendation with strong and weak ties. In: CIKM, pp. 5–14 (2016)
28. Xiao, J., Ye, H., He, X., Zhang, H., Wu, F., Chua, T.: Attentional factorization machines: Learning the weight of feature interactions via attention networks. In: Proceedings of the Twenty-Sixth International Joint Conference on Artificial Intelligence, IJCAI 2017, Melbourne, Australia, 19–25 August 2017, pp. 3119–3125 (2017)
29. Yang, B., Lei, Y., Liu, D., Liu, J.: Social collaborative filtering by trust. In: IJCAI 2013, Proceedings of the 23rd International Joint Conference on Artificial Intelligence, Beijing, China, 3–9 August 2013, pp. 2747–2753 (2013)
30. Yang, X., Steck, H., Liu, Y.: Circle-based recommendation in online social networks. In: KDD, pp. 1267–1275 (2012)
31. Zhao, T., McAuley, J.J., King, I.: Leveraging social connections to improve personalized ranking for collaborative filtering. In: CIKM, pp. 261–270 (2014)

CPL: A Combined Framework of Pointwise Prediction and Learning to Rank for top-N Recommendations with Implicit Feedback

Nengjun Zhu and Jian Cao[✉]

Shanghai Institute for Advanced Communication and Data Science,
Department of Computer Science and Engineering, Shanghai Jiao Tong University,
Shanghai, China
{zhu_nj,cao-jian}@sjtu.edu.cn

Abstract. Pointwise prediction and Learning to Rank (L2R) are both widely used in recommender systems. Currently, these two types of approaches are often considered independently, and most existing efforts utilize them separately. Unfortunately, pointwise prediction tends to overfit the training data while L2R is more prone to higher variance, and both of them suffer one-class problems using implicit feedback. Therefore, we propose a new framework called CPL, where pointwise prediction and L2R are inherently combined to discriminate user preferences on unobserved items, to improve the performance of top-N recommendations. To verify the effectiveness of CPL, an instantiation of CPL, which is named CPLmg, is introduced. CPLmg is based on two components, i.e., FSLIM (Factorized Sparse LInear Method) and GAPfm (Graded Average Precision factor model), to perform pointwise prediction and L2R, respectively. The low-rank users' and item's latent factor matrices act as a bridge between FSLIM and GAPfm. Moreover, FSLIM dynamically rates an unobserved item for a user based on its similarity with observed items. These pseudo ratings are further utilized with a confidence score to rank items in GAPfm. Extensive experiments on two datasets show that CPLmg significantly outperforms the baselines.

Keywords: Recommender system · Implicit feedback · Collaborative filtering · Learning to Rank · Metrics optimization

1 Introduction

Recommender systems (RSs) have been widely adopted by many online services, since they are able to solve the information overload problem as well as facilitate interaction between users and systems. Most RSs infer users' interests through users' historical behaviors, either represented in explicit form or implicit form. Explicit feedback such as rating, which is given by users, can indicate a

© Springer Nature Switzerland AG 2019
R. Cheng et al. (Eds.): WISE 2019, LNCS 11881, pp. 259–273, 2019.
https://doi.org/10.1007/978-3-030-34223-4_17

user's interest in a particular product. However, this explicit feedback is not always available in practical systems. On the contrary, implicit feedback, such as users' browsing history, or even mouse movements, can be easily obtained from the system and do not burden the users. This information can also reflect users' preferences, although in an indirect way. Consequently, recommendation approaches based on implicit feedback are becoming more widely used [1].

Pointwise prediction and learning to rank (L2R) are two representative genres of the approaches for RSs. Pointwise prediction tries to estimate the value of an item to a user based on historical data with the aim of minimizing prediction errors. It is straightforward and effective when users' historical data is organized in rating forms. In domains where only implicit feedback is available, there are also two definitional levels, i.e., 1 for observed examples and 0 for missing ones, which can reflect the connection strength between a user and an item to some degree [2]. However, pointwise prediction models easily lead to large bias, i.e., overfitting of training data, since they are confined to being finely tuned to each value of individual examples, even these examples are noises. On the other hand, L2R methods explore the preferential relations among multiple items, i.e., the relation that a user prefers item i over item j, and consider the entire ranking list as a target for optimization. In contrast to pointwise prediction, L2R methods may cause high variances since they are not sensitive to small changes in the estimated value of each individual examples unless these examples are compared to the other ones. To balance variance and bias, existing approaches usually add regularization terms to target functions.

In this paper, we explore a new framework, CPL (a Combined framework of Pointwise prediction and L2R), which tries to balance bias and variance not only by regularization but also based on the inherent features of pointwise prediction and L2R. The ultimate goal of CPL is to find a balance between predicting an accurate value for each example (which is the goal of pointwise methods) and keeping the correct preferential relations between items (which is the goal of L2R methods). To verify the effectiveness of CPL, we choose SLIM [9] which is one of the pointwise prediction methods, and GAPfm [14] which is one of the L2R methods, as the two components to implement CPL. SLIM and GAPfm both have been demonstrated to have a stronger performance than other state-of-the-art approaches to top-N recommendations. SLIM utilizes the intuition of item-based K-nearest neighborhood (ItemKNN) collaborative filtering and makes use of the learning process of matrix factorization (MF) techniques to estimate the coefficients between every two items. The estimated coefficients are analogous to item similarities in the traditional ItemKNN method, but they are learned from observed data instead of being calculated based on items' attribute vectors. GAPfm, which addresses the top-N recommendation problem in domains with grade relevance data, takes the Graded Average Precision (GAP) metric as the extreme optimization objective function. However, we would like to utilize the highly discriminative trait of GAP to dynamically mine potential positive examples and to avoid the trap of suppression of preferences for items about which the user is unaware [13, 17].

To combine SLIM and GAPfm in a better way, we first revise their original versions. Then, we combine the new versions into CPLmg, which is an implementation of CPL. Specifically, we improve the SLIM to be a more general factor model, namely FSLIM. FSLIM inherits all the desirable characteristics of SLIM and the difference is that FSLIM constructs a dense representation both for users and items, which can improve the recommendation performance [6]. Then, the low-rank users' and item's latent factor matrices act as a bridge between FSLIM and GAPfm, so that the learned dense representation can be transferred to each other. Moreover, FSLIM dynamically rates unobserved items for a user based on the learned item similarities. These pseudo ratings are further utilized in GAPfm, and the confidence score of a pseudo rating to be a threshold, which separates unknown items to a positive example or to a negative example, is also updated dynamically in every training round. Thus, the combination of FSLIM and GAPfm results in the considerably improved learning accuracy of GPLmg.

The main contributions of this paper are as follows: (1) We introduce a new framework CPL to combine pointwise prediction and L2R methods to address the top-N recommendation problem. (2) We propose an implementation of CPLmg for CPL. In CPLmg, we combine the FSLIM and GAPfm models. FSLIM is extended from SLIM. Moreover, strategies are designed to better integrate FSLIM and GAPfm. (3) Extensive experiments, which show that CPLmg outperforms other baselines on various evaluation metrics, are conducted.

2 Related Work

Our proposed model, which is based on a combination of pointwise prediction and learning to ranking, addresses the top-N recommendation problem with implicit feedback. Therefore, it is related to state-of-the-art top-N recommendation technologies, including matrix factorization (MF) methods and L2R approaches.

MF is one of the most popular model-based collaborative filtering (CF) methods. It learns latent factor representations with respect to users and items, and models user preferences as the dot product of latent factor vectors. SLIM [9] is a particular case of MF. It directly learns a similarity matrix from the data and thus becomes a novel learning model. To address the quadratic computation problem of SLIM, a factorized similarity model FISM [5] is proposed. FISM factorizes the similarity matrix into two low-rank matrices. However, both SLIM and FISM do not produce a user-specific latent factor matrix. Thus, LRec [13], which is interpreted as a linear classification model for each user, is proposed to overcome this limitation. Currently, some work has explored the combination of MF and deep learning for recommendations. For instance, NeuMF [4] is a neural network-based CF method, and it is essentially a fusion of generalized matrix factorization and multi-layer perceptron.

L2R becomes a hot research area, since it directly models partial ordering relations between items, which happens to be consistent with top-N recommendation tasks. One key element of L2R methods is the objective measures, defined as either ranking error functions or optimization metrics. Thus, based on different objective measures, many L2R methods have been proposed. BPR [11] maximizes AUC metrics by utilizing the partial order relations between items. xCLiMF [15] is an L2R method based on expected reciprocal rank (ERR). Moreover, TFMAP [16] optimizes MAP metric directly. To alleviate the overfitting problem of L2R, GTRM [17] optimizes the group-oriented mean average precision (GMAP) which considers the similarities between items, and PRIGP [10] integrates item-based pairwise preferences and item group-based pairwise preferences into the framework based on BPR-OPT derived from BPR.

However, the above-mentioned approaches only utilize regularization terms to balance bias and variance, and none of them combine pointwise prediction and L2R for top-N recommendations.

3 Preliminaries

3.1 Definitions and Notations

Assume that the implicit feedback data is from M users' behaviors on N items, and we use the symbol u to index a user, the symbol i and j to index items, and the symbol k to index a latent factor. The set of all users and items are represented by $\mathcal{U} = \{1, 2, \ldots, u, \ldots, M\}$ and $\mathcal{I} = \{1, 2, \ldots, i, \ldots, N\}$, respectively. The matrices $\mathbf{P} \in \mathbb{R}^{M \times K}$ and $\mathbf{Q} \in \mathbb{R}^{N \times K}$ are latent factor matrices related to users and items, respectively. The entire set of users' historical feedback such as purchases/clicking are represented by a user-item interaction matrix $\mathbf{A} \in \mathbb{R}^{M \times N}$, in which each entry is represented by $\mathbf{A}_{ui} \in \{0, 1\}$, where $\mathbf{A}_{ui} = 1$ means user u has at some point interacted with item i (observed items), otherwise the entry is marked as 0 (unobserved items).

In the rest of the paper, vectors and matrices are both denoted by upper bold symbols, where the symbol with no subscript represents the matrix itself. The symbol with one subscript (e.g., $\mathbf{P}_u$) represents a vector extracted from its matrix by the row/column subscript index, and the symbol with two subscripts (e.g., $\mathbf{P}_{uk}$) represents the entry. A predicted value is denoted by the symbol with a wide tilde head (e.g., $\widetilde{\mathbf{A}}_{ui}$). Unless stated differently, all vectors are column vectors by default, but the vectors with the transposed subscript $^\top$ are row vectors (e.g., $\mathbf{P}_u^\top$ denotes the u-th row of $\mathbf{P}$).

3.2 SLIM

A parse linear method SLIM [9] has demonstrated very good performance for top-N recommendations. Different from traditional similarity models that calculate similarities based on attributes according to certain criteria, SLIM learns the item similarities from the data directly. That is, SLIM estimates a sparse aggregation coefficient matrix $\mathbf{W} \in \mathbb{R}^{N \times N}$, in which each entry $\mathbf{W}_{ij}$ can be viewed as the similarity between items i and j. Then the recommendation score from user u to an unobserved item i is computed as a sparse aggregation of all the observed items of the user, as follows:

$$\widetilde{\mathbf{A}}_{ui} = \mathbf{A}_u^\top \mathbf{W}_i \tag{1}$$

where $\mathbf{A}_u^\top$ is the row vector extracted from $\mathbf{A}$ by the row/user index u, and $\mathbf{W}_i$ is a column vector, which represents the i-th column vector of matrix $\mathbf{W}$. Then, SLIM estimates/learns the $\mathbf{W}$ by solving the following optimization problem:

$$\underset{\mathbf{W}}{\text{minimize}} \quad \frac{1}{2}||\mathbf{A} - \mathbf{A}\mathbf{W}||_F^2 + \frac{\beta}{2}||\mathbf{W}||_F^2 + \lambda||\mathbf{W}||_1$$

$$\text{subject to} \quad \mathbf{W} \geq 0$$

$$\text{diag}(\mathbf{W}) = 0 \tag{2}$$

Here, $||\mathbf{W}||_1$ is the entry-wise ℓ_1-norm of $\mathbf{W}$ which encourages sparsity, and $||\bullet||_F$ is the matrix Frobenius norm. The constraint $\text{diag}(\mathbf{W}) = 0$ prevents learned item similarities from being affected by the item itself. As for the nonnegativity constraint, [7] showed that it could be ignored without affecting performance.

3.3 GAPfm

GAPfm [14] is a listwise L2R method which directly optimizes a smoothed approximation of GAP metric [12]. GAP generalizes average precision (AP) to the case of multi-graded relevance, and inherits the most important properties of AP metric to guarantee that mistakes in recommended items at the top of the list carry a higher penalty than mistakes at the bottom of the list. The definition of GAP is as follows:

$$GAP_u = \frac{1}{Z_u} \sum_{i=1}^{N} \frac{I_{ui}}{R_{ui}} \sum_{j=1}^{N} I_{uj} \mathbb{I}(R_{uj} \leq R_{ui})$$

$$(\mathbb{I}(y_{ui} < y_{uj}) \sum_{l=1}^{y_{ui}} \delta_l + \mathbb{I}(y_{uj} \leq y_{ui}) \sum_{l=1}^{y_{uj}} \delta_l) \tag{3}$$

where R_{ui} denotes the ranked position of item i for user u, e.g., $R_{ui} = 1$ denotes the item is ranked in the first/highest position. $I_{ui} = 1$ $(I_{ui} \in \{0,1\})$ denotes the item is a positive example, otherwise it is a negative/missing example. y_{ui} denotes the grade of item i to user u. $\mathbb{I}(x)$ is a binary indicator function, i.e., it is equal to 1 if x is true, otherwise 0. $Z_u = \sum_{l=1}^{y_{max}} n_{ul} \sum_{c=1}^{l} \delta_c$ is a constant normalizing coefficient for user u, where n_{ul} denotes the number of items rated with grade l by user u, and δ_l denotes the thresholding probability that the user sets as a threshold of relevance at grade l, i.e., regarding items with grades equal or larger than l as relevant ones, and the others as irrelevant ones.

$$\delta_l = \begin{cases} \dfrac{2^l - 1}{2y_{max}}, & y_{max} > 1 \\ 1, & y_{max} = 1 \end{cases} \tag{4}$$

where $[1, y_{max}]$ is the scale of ratings. Then, with a small manipulation, Eq. (3) can be smoothed to be an optimization objective function with respect to the learned parameters, i.e., $\mathbf{P}$ and $\mathbf{Q}$, the details are given in the following sections.

4 Proposed Methodology

In this section, we introduce two components of CPL: (1) Factorized SLIM (FSLIM), and (2) GAPfm with sampling strategy. Then, we show in detail how to combine FSLIM and GAPfm to implement CPL.

4.1 Factorized SLIM (FSLIM)

Our proposed Factorized SLIM is a new version of SLIM that incorporates ideas from traditional matrix factorization (MF) methods and similarity approaches. We still define the recommendation score from user u to an unobserved item i as a sparse aggregation of the scores of all observed items by the user. However, the score of each item is no longer a defined value, i.e., 1 and 0, but is calculated as the dot product of the item's latent factor vector and the user's latent factor vector, as shown in Eq. (5). The dense representations of users and items introduce more information capabilities.

$$\widetilde{\mathbf{A}}_{ui} = \mathbf{P}_u^\top \sum_{j \in \mathcal{N}(i) \cap \mathcal{O}(u)} \mathbf{Q}_j \mathbf{W}_{ji} \tag{5}$$

where $\mathbf{W} \in \mathbb{R}^{N \times N}$ is a sparse aggregation coefficient matrix such as that in SLIM, and $\mathcal{O}(u)$ is the set of all observed items of user u. To speed up the learning process, $\mathcal{N}(i)$ representing the set of near neighborhoods of item i, is added to select items in $\mathcal{O}(u)$. This operation can be viewed as feature selection [9]. We utilize the cosine similarity, which is calculated based on co-click/co-visitation behaviors to items by users, to retrieve $iknn$ near neighborhoods of item i, i.e., $|\mathcal{N}(i)| = iknn$. Finally, taking into account all users, the loss function is defined as follows:

$$\mathcal{L}_F = \frac{1}{2} \sum_{u=1}^M \sum_{i=1}^N \|\mathbf{A}_{ui} - g(\mathbf{P}_u^\top \sum_{j \in \mathcal{N}(i) \cap \mathcal{O}(u)} \mathbf{Q}_j \mathbf{W}_{ij})\|_F^2$$
$$+ \frac{\beta_1}{2}\|\mathbf{P}\|_F^2 + \frac{\beta_2}{2}\|\mathbf{Q}\|_F^2 + \frac{\beta_3}{2}\|\mathbf{W}\|_F^2 + \lambda\|\mathbf{W}\|_1 \tag{6}$$

where $g(x) = 1/(1+e^x)$ is a sigmoid function, which is a common choice for one-class recommendation. We add the constraint $\text{diag}(\mathbf{W}) = 0$ to prevent learned item similarities from being affected by the item itself, and drop the nonnegativity constraint, i.e., $\mathbf{W} \geq 0$, compared to SLIM as the reason we aforementioned.

Stochastic gradient decent technology (SGD) is used to solve this optimization problem, and the gradients of the parameters are listed as follows:

$$\frac{\partial \mathcal{L}_F}{\partial \mathbf{P}_u} = -(\mathbf{A}_{ui} - g(\widetilde{\mathbf{A}}_{ui}))g'(\widetilde{\mathbf{A}}_{ui}) \sum_{j \in \mathcal{N}(i) \cap \mathcal{O}(u)} \mathbf{Q}_j \mathbf{W}_{ji} + \beta_1 \mathbf{P}_u \tag{7}$$

$$\frac{\partial \mathcal{L}_F}{\partial \mathbf{Q}_i} = - \sum_{j \in \mathcal{N}(i) \cap \mathcal{O}(u)} (\mathbf{A}_{uj} - g(\widetilde{\mathbf{A}}_{uj}))g'(\widetilde{\mathbf{A}}_{uj})\mathbf{P}_u \mathbf{W}_{ij} + \beta_2 \mathbf{Q}_i \tag{8}$$

$$\frac{\partial \mathcal{L}_F}{\partial \mathbf{W}_{ij}} = -(\mathbf{A}_{ui} - g(\widetilde{\mathbf{A}}_{ui}))g'(\widetilde{\mathbf{A}}_{ui})\mathbf{P}_u^\top \mathbf{Q}_j + \beta_3 \mathbf{W}_{ij} \pm \lambda \tag{9}$$

where $g'(x) = g(x)/(1 - g(x))$ is the derivative of function $g(x)$. Then, with a learning step size η_1, the parameters are updated using SGD.

4.2   GAPfm with Sampling Strategy

The work about GAPfm in [14] mainly focuses on graded relevance domains, such as rating data, and takes GAP as the objective metric in learning to rank. However, in domains with binary relevance data, we would still like to take full advantage of high informativeness and discriminative power of GAP to dynamically mine potential preferred items and to avoid the trap of the suppression of preferences for items about which the user is unaware. That is, we utilize the sparse aggregation coefficient matrix (the item similarity matrix) learned from FSLIM to estimate the pseudo rating of each item for each user, which can be demonstrated as:

$$y_{ui} = \begin{cases} g\Big(\mathbf{P}_u^\top \sum_{j \in \mathcal{N}(i) \cap \mathcal{O}(u)} \mathbf{Q}_j \mathbf{W}_{ji}\Big), & i \notin \mathcal{O}(u) \\ 1, & i \in \mathcal{O}(u) \end{cases} \tag{10}$$

It is likely that some values will be prefill into matrix $\mathbf{A}$. The closer the value of y_{ui} to 1, the more likely item i is a potential preferred item for user u, since the value of y_{ui} depicts the relationship between item i and the user. Thus, according to the value of y_{ui}, we select the top pn unobserved items as potential preferred items for user u, and record the indexes of all these items and already observed items into a set $\mathcal{O}'(u)$ as well as record the pseudo ratings (the values of y_{ui}) of all items in $\mathcal{O}'(u)$ into a set $Y(u)$. Then, we change the thresholding probability $\delta_u(y)$ as a confidence score of that pseudo rating $y \in Y(u)$ being the threshold value for user u, i.e., regarding items with a pseudo rating equal or larger than y as potential preferred ones, and the others as not preferred ones, as follows:

$$\delta_u(y) = \frac{exp(y)}{\sum_{t \in Y(u)} exp(t)} \tag{11}$$

The larger the value of $\delta_u(y)$ or of y, the more credible the result of this division. Then, we update the formulation of GAP in Eq. (3) as follows:

$$GAP_u = \frac{1}{Z_u} \sum_{i=1}^{N} \frac{I_{ui}}{R_{ui}} \sum_{j=1}^{N} S_{uij} I_{uj} \mathbb{I}(R_{uj} \le R_{ui}) \tag{12}$$

$$Z_u = \sum_{t \in Y(u)} n_{ut} \sum_{l \in Y(u) \& l \le t} \delta_u(l)$$

$$S_{uij} = \mathbb{I}(y_{ui} < y_{uj}) \sum_{t \in Y(u) \& t \le y_{ui}} \delta_u(t) + \mathbb{I}(y_{uj} \le y_{ui}) \sum_{t \in Y(u) \& t \le y_{uj}} \delta_u(t)$$

where R_{ui} denotes the ranked position of item i for user u, I_{ui} indicates whether the index of the item is in $\mathcal{O}'(u)$, and $\mathbb{I}(x)$ is a binary indicator function, such as those in GAPfm. Z_u is a constant normalizing coefficient for user u, where n_{ut} denotes the number of items rated with pseudo rating t to user u, S_{uij} is a intermediate variable whose value is related to the sorted list.

Then we use $g(x)$ function and parameters $\mathbf{P}, \mathbf{Q}$ to estimate the term of $\frac{1}{R_{ui}} \approx g(f_{ui})$ and $\mathbb{I}(R_{uj} \leq R_{ui}) \approx g(f_{uj} - f_{ui})$, where $f_{ui} = \mathbf{P}_u^\top \mathbf{Q}_i$, in Eq. (12) to get a smoothed version of GAP as follows:

$$GAP_u \approx \frac{1}{Z_u} \sum_{i=1}^{N} I_{ui} g(f_{ui}) \sum_{j=1}^{N} S_{uij} I_{uj} g(f_{uj} - f_{ui}) \tag{13}$$

Then, taking into account all users and adding two Frobenius norms $\|\mathbf{P}\|_F$ and $\|\mathbf{Q}\|_F$ as well as parameters β_4 and β_5 to control the magnitude of regularization, the final objective function of GAPfm is shown as follows:

$$\mathcal{L}_G = \sum_{u=1}^{M} \sum_{i=1}^{N} I_{ui} g(f_{ui}) \sum_{j=1}^{N} S_{uij} I_{uj} g(f_{u(j-i)}) - \frac{\beta_4}{2} \|\mathbf{P}\|_F^2 - \frac{\beta_5}{2} \|\mathbf{Q}\|_F^2 \tag{14}$$

Note that, Eq. (14) has dropped the coefficient $1/M$ and $1/Z_u$ since they are independent of the latent factors and have no influence on the optimization procedure. Now, we use the stochastic gradient ascent (SGA) to solve this optimization problem, and the gradients of the parameters are as follows:

$$\frac{\partial \mathcal{L}_G}{\partial \mathbf{P}_u} = \sum_{i=1}^{N} I_{ui} \Big(g'(f_{ui}) \sum_{j=1}^{N} I_{uj} S_{uij} g(f_{u(j-i)}) \cdot \mathbf{Q}_i$$

$$+ g(f_{ui}) \sum_{j=1}^{N} I_{uj} S_{uij} g'(f_{u(j-i)}) \cdot (\mathbf{Q}_j - \mathbf{Q}_i) \Big) - \lambda \mathbf{P}_u \tag{15}$$

$$\frac{\partial \mathcal{L}_G}{\partial \mathbf{Q}_i} = I_{ui} \Big(g'(f_{ui}) \sum_{j=1}^{N} I_{uj} S_{uij} g(f_{u(j-i)}) + \sum_{j=1}^{N} I_{uj}$$

$$[S_{uji} g(f_{uj}) - S_{uij} g(f_{ui})] g'(f_{u(j-i)}) \Big) \mathbf{P}_u - \lambda \mathbf{Q}_i \tag{16}$$

Then, we update the parameters in GAPfm using SGA with a learning rate η_2.

4.3   CPLmg Recommendation Model

Now, we introduce how to combine FSLIM and GAPfm under MF framework to implement CPLmg, so that they can mutually reinforce each other and can better learn from complex user-item interactions. We propose to train FSLIM and GAPfm using a multi-task learning approach [8] where the latent factor

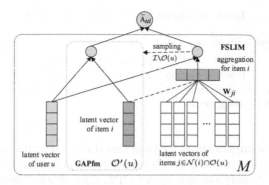

Fig. 1. The framework of CPLmg

matrices $\mathbf{P}$ and $\mathbf{Q}$ are shared underlying variables, as shown in Fig. 1. In particular, the matrices $\mathbf{P}$ and $\mathbf{Q}$ are jointly updated by FSLIM and GAPfm in each co-training round by modeling the task of pointwise methods and the task of L2R methods. The additional item similarity matrix $\mathbf{W}$ further helps GAPfm mine potential preferred items, which allows information transfer between two tasks [3]. Furthermore, the trade-off controlling parameters in CPLmg are the learning rate parameters, i.e., η_1 and η_2, since the relationship between the values of η_1 and η_2 determines the impact of each component on the model learning process. CPLmg is trained until both FSLIM and GAPfm are converged or until reaching the maximal number of iteration.

4.4   Time Complexity

The time complexity of CPLmg comprises two parts which accumulate linearly, i.e., the time cost of FSLIM and GAPfm. Thus, CPLmg finally takes $O(M|\bar{I}|^3 K + M|\bar{J}|(|\bar{I}| + \ln(|\bar{J}|))$ time to update the parameters in each iteration.

4.5   Recommendation

At the prediction phase, we measure the final preference score of unobserved items to each user as follows:

$$\widetilde{\mathbf{A}}_{ui} = \mathbf{P}_u^\top \sum_{j \in \mathcal{N}(i) \cap \mathcal{O}(u) \cup \{i\}} \mathbf{Q}_j \mathbf{W}'_{ji} \tag{17}$$

where $\mathbf{W}' = \mathbf{W} + w * \mathbf{I}$, and $\mathbf{I} \in \mathbb{R}^{N \times N}$ is an identity matrix, and w is a weight parameter of the combination of prediction functions in FSLIM, i.e., $\widetilde{\mathbf{A}}_{ui} = \mathbf{P}_u^\top \sum_{j \in \mathcal{N}(i) \cap \mathcal{O}(u)} \mathbf{Q}_j \mathbf{W}_{ji}$ and in GAPfm, i.e., $\widetilde{\mathbf{A}}_{ui} = \mathbf{P}_u^\top \mathbf{Q}_i$, respectively. The value of w is related to the trade-off controlling parameters of the framework CPL, and we set the value of the ratio of $\frac{\eta_2}{\eta_1}$ to w. The items from the set $\mathcal{I}\backslash\mathcal{O}(u)$ with the largest prediction values based on Eq. (17) are recommended to the user.

Table 1. The datasets used in evaluation

Dataset	#users	#items	#trans	Density	Ratings	Threshold[a]
AppData	20,467	40,259	1,022,339 (installations)	0.124%	–	–
ML100K	943	1,682	100,000 (ratings)	6.30%	1–5	1

[a]The user-item pairs with ratings equal or greater than *threshold* are positive examples, the others are the missing ones.

So far, we have implemented CPLmg based on two components: FSLIM and GAPfm. A question then arises: why does the combination of these methods promote the top-N performance under CPL framework. There are four possible reasons: (1) The learning directions of FSLIM and GAPfm are generally consistent when learning the model parameters $\mathbf{P}$ and $\mathbf{Q}$ through the gradient approach. Both of them tend to increase the value of a dot product with respect to positive examples. (2) The sampling procedure based on the item similarity matrix $\mathbf{W}$ brings more information from FSLIM to GAPfm to make GAPfm more informative. (3) The multi-task learning approach allows information transfer between the two tasks. (4) CPLmg balances the variance and bias of the model, as previously discussed. Thus, our proposed CPLmg approach can yield better performance for top-N recommendations, which is demonstrated by the following experiments.

5   Experimental Results

5.1   Datasets and Settings

Our experiments are based on two datasets, AppData and MovieLens-100K (ML100K). The characteristics of these two datasets are shown in Table 1.

The dataset AppData is from users' log files where the users' interactive behaviors with mobile applications are recorded for six months. Since we are more concerned about which applications the user will install on their smartphone, we only keep already installed mobile applications for users. Then, each observed user-item pair represents one record of the user installing the application.

The ML100K is a public dataset and it is organized in rating forms. However, since we only discuss the one-class recommendation problem in this paper, the ratings are converted to the appropriate binary form based on the threshold h, i.e., the user-item pairs with ratings higher than h are positive examples, the others are the missing ones. To simplify the analysis, we give the best value of the threshold in our experiments, i.e., $h = 1$. In the experiment, mobile applications and movies are the items to be recommended.

We randomly select records from the users' historical data to keep a certain number of observed items for each user as the test data, and set the rest of the records as the training set. For example, "Given 10" denotes that for each user, we randomly select ten observed items as unknowns in the training set, but as positive examples in the test set. Then, we measure the performance over these positive examples in the test data.

Precision is a widely used evaluation metric in RSs. It reflects the ratio of relevant items in the ranked list given a truncated position. In the case of top-N RSs, MAP (mean average precision) and MRR (mean reciprocal rank) are more practical, as they are position-related metrics. To better verify the properties of the model, we apply all these three metrics to evaluate the performance of the new and compared methods in this paper.

Table 2. Performance comparison based on the top-5 recommendation items

Method	AppData/Given 3				AppData/Given 10				
	Params[1]		Precision	MRR	MAP	Params		Precision	MRR	MAP
iPOP	–	–	0.0989	0.2874	0.1098	–	–	0.2830	0.6303	0.2264
ItemKNN	50	–	0.1126	0.3164	0.1290	50	–	0.3601	0.6089	0.2971
FISMauc	0.8	$1e^{-5}$	0.1002	0.2895	0.1108	0.9	$1e^{-5}$	0.2989	0.6338	0.2395
GAPfm	0.01	–	0.1186	0.3136	0.1348	0.01	–	0.3862	0.7053	0.3132
SLIM	0.1	0.5	0.1563	0.3948	0.1759	0.1	0.5	0.4017	0.7070	0.3177
FSLIM	0.06	0.14	0.1601	0.4158	0.1801	0.04	0.12	0.4072	0.7164	0.3173
NeuMF	10	–	0.1698	0.4209	0.1894	10	–	0.4098	0.7203	0.3184
CPLmg	245	22	**0.1799**	**0.4377**	**0.2022**	245	22	**0.4208**	**0.7345**	**0.3305**

Method	ML100K/Given 3				ML100K/Given 10				
	Params		Precision	MRR	MAP	Params		Precision	MRR	MAP
iPOP	–	–	0.0417	0.1153	0.0415	–	–	0.1324	0.3001	0.0823
ItemKNN	50	–	0.0697	0.1855	0.0696	50	–	0.1769	0.3885	0.1152
FISMauc	0.7	$5e^{-6}$	0.0491	0.1119	0.0413	0.6	$1e^{-6}$	0.1342	0.2948	0.0808
GAPfm	0.05	–	0.0923	0.2497	0.0987	0.05	–	0.1820	0.3998	0.1475
SLIM	0.2	0.6	0.0982	0.2519	0.1009	0.1	0.5	0.2232	0.4722	0.1537
FSLIM	0.002	0.005	0.1034	0.2590	0.1113	0.001	0.005	0.2398	0.4795	0.1599
NeuMF	8	–	0.1078	0.2601	0.1132	8	–	0.2399	0.4819	0.1614
CPLmg	350	35	**0.1194**	**0.2708**	**0.1298**	350	35	**0.2483**	**0.4916**	**0.1712**

5.2 Experimental Comparisons with Previous Models

We compare our methods CPLmg and FSLIM with six baselines as follows: (1) **iPOP** recommends a certain number of the most popular items from the training set to all users. (2) **ItemKNN** is a traditional item-based collaborative filtering method using Jaccard similarity. (3) **FISMauc** [5] considers ranking errors based on loss function and obtains better performance than FISMrmse, which considers the pointwise squared error loss function. Therefore, we do not further report on the performance of FISMrmse. (4) **NeuMF** [4], which is a state-of-the-art method using neural network-based collaborative filtering (NCF) framework. (5) **SLIM** and (6) **GAPfm** are related to two components of CPLmg, respectively. For each model, the parameters were empirically tuned to their optimal values in the experiments and they were recorded in Table 2, i.e., for ItemKNN, they are the number of neighbors; for FISMauc, they are the user-specific parameter α and the learning rate; for GAPfm, they are the regularization parameters; for SLIM and FSLIM, they are both the ℓ_1-norm and ℓ_2-norm regularization parameter; for NeuMF, it is the number of negative samples; for CPLmg, they are the number of near neighborhoods $iknn$ and the number of candidate potential positive examples pn.

Since the size of the recommendation window is limited in practice, we measure all the performance values in the experiments which are reported in this subsection based on the top-5 recommendation, and the results for top-10 recommendations are shown in the next subsection.

Table 2 shows that CPLmg achieves the best performance than the baselines according to all three metrics. Then, it is NeuMF, which is the state-of-the-art method using implicit feedback. It proves that the proposed CPLmg is highly competitive for top-N recommendation tasks for reasons previously discussed, and also proves that the combination of FSLIM and GAPfm is effective since the performance of FSLIM and GAPfm is not good as NeuMF before the combination. We can also observe that the performance of FSLIM is better than SLIM. This indicates that dense representations of the user and item matrix can better model the users' preferences. The results also show SLIM and GAPfm outperform the remaining methods, i.e., iPOP, ItemKNN, and FISMauc. This observation provides empirical evidence that SLIM and GAPfm approaches are more effective for top-N recommendations. This is one reason why we choose SLIM and GAPfm as the two components in our new framework. Furthermore, it can be noted that the performance of all methods is better when the number of given items increases from 3 to 10. The reason for this lies in the fact that more preferred items in the test data can better reveal the preferences of users and more preferred items in the test means a higher chance of ranking potential preferred items in the top positions.

5.3   Analysis of CPLmg Components

In this section, we describe the experiments conducted to explore the influence of the main parameters on CPLmg, i.e., the number of neighborhoods when conducting feature selection for FSLIM, the size of the candidate potential preferred items in GAPfm sampling process, and the learning rates. It is worth pointing out when we change the settings of one of these parameters, the others are set to their optimal values, e.g., $iknn = 245, pn = 22, \eta_1 = 5 \times 10^{-2}, \eta_2 = 10^{-4}$ for App-Data. All experiment results given in this section are under the condition, i.e., "Given 10". Due to the space limitation and without loss of generality, we only report the parameter influences for the top-10 recommendations on AppData. Similar results were observed on ML100K data.

iknn. We first conducted an experiment to investigate the influence of the number of near neighborhoods $iknn$. The results are shown in Fig. 2(a)–(c). We can observe that the values of all three metrics including precision, MRR, and MAP significantly increase at the beginning, then after the turning points, i.e., 240, all values decline. This proves the effectiveness of the feature selection algorithm and it might have an optimal value of $iknn$. The value of $iknn$ is not the bigger the better, since too many similar items may blur the preference information to

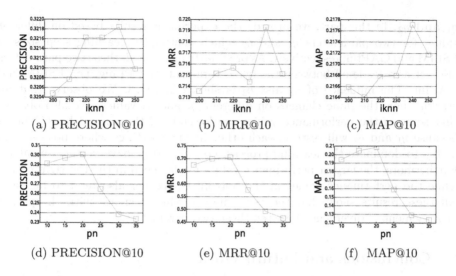

(a) PRECISION@10 (b) MRR@10 (c) MAP@10

(d) PRECISION@10 (e) MRR@10 (f) MAP@10

Fig. 2. Results on different parameters *iknn* and *pn* for top-10 recommendations

be learned for a user. Note that the *iknn* is not the final number of neighborhoods to be considered since $|\mathcal{N}(i) \cap \mathcal{O}(u)| \leq iknn, |\mathcal{O}(u)|$, and usually $|\mathcal{O}(u)|$ might be small in practice, e.g., the average number of installed applications over users in AppData is smaller than 50.

pn. The value of *pn* controls the number of candidate potential preferred items in the sampling process of GAPfm. The influence of *pn* on the recommendation performance is shown in Fig. 2(d)–(f). We can observe that precision, MRR, and MAP performance can be improved by properly increasing *pn*. However, when the increasement is over a turning point, i.e., 20, the performance starts to decline sharply. The reason for this is that a larger value of *pn* also introduces more false preferred items. This observation proves that it is critical to properly take into account missing values within the model in domains with binary implicit feedback, since the selected missing values can alleviate the overfitting risk.

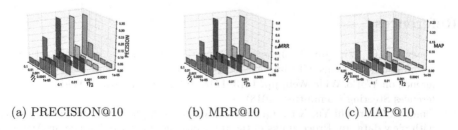

(a) PRECISION@10 (b) MRR@10 (c) MAP@10

Fig. 3. Results on different parameters η_1 and η_2 for top-10 recommendations

η_1 **and** η_2. In this part, we provide the experiment results based on different values for two parameters η_1 and η_2 which control the learning step sizes of FSLIM and GAPfm, respectively. As previously mentioned, these two parameters also act as a trade-off between the two components of CPLmg, i.e., FSLIM and GAPfm. The influence of η_1 and η_2 is shown in Fig. 3. We observe that all criteria show the same changes on different η_1 and η_2 values. We also observe that some of the performance values are lower than the normal level. This is because η_1 and η_2 will restrain each other in some settings where both η_1 and η_2 try to dominate the learning process, i.e., η_1 and η_2 have very close values. Furthermore, the largest performance values are fastened in the top right corner while the performance values in the bottom left corner also tend to increase. All these results show that the performance values increase with the proper increasement of divergence between these two parameters.

6   Conclusions and Future Work

In this paper, we proposed a new framework, CPL, where pointwise prediction and L2R are inherently combined to discriminate user preferences on unobserved items and to improve the performance of top-N recommendations. Moreover, to verify the effectiveness of CPL, we implement CPLmg which takes FSLIM and GAPfm as its two components, where FSLIM is a variant of SLIM by infusing dense representations. The components reinforce each other through information interchange based on the dense representations and aggregation coefficients. The final experiments prove that CPLmg is effective and outperforms the others on various evaluation metrics. There are some potential research topics for future study. Firstly, the combination approach between two components of CPLmg can be extended. We would like to explore a more complex combination. For instance, we can fuse two components based on the neural network framework motivated by NCF [4]. Secondly but not lastly, other models of pointwise prediction and L2R methods can be tried in the framework.

Acknowledgement. This work is partially supported by National Key Research and Development Plan (No. 2018YFB1003800).

References

1. Anyosa, S.C., Vinagre, J., Jorge, A.M.: Incremental matrix co-factorization for recommender systems with implicit feedback. In: Proceedings of the WWW Conference on World Wide Web, pp. 1413–1418. International World Wide Web Conferences Steering Committee (2018)
2. Ding, J., Guanghui Yu, X.H., Quan, Y.: Improving implicit recommender systems with view data. In: Proceedings of the International Joint Conference on Artificial Intelligence, pp. 3343–3349. AAAI Press (2018)
3. Dong, D., Zheng, X., Zhang, R., Wang, Y.: Recurrent collaborative filtering for unifying general and sequential recommender. In: Proceedings of the International Joint Conference on Artificial Intelligence, pp. 3350–3356. AAAI Press (2018)

4. He, X., Liao, L., Zhang, H., Nie, L., Hu, X., Chua, T.S.: Neural collaborative filtering. In: Proceedings of the WWW Conference on the World Wide Web, pp. 173–182. International World Wide Web Conferences Steering Committee (2017)
5. Kabbur, S., Ning, X., Karypis, G.: Fism: factored item similarity models for top-n recommender systems. In: Proceedings of the ACM SIGKDD International Conference on Knowledge Discovery and Data Mining, pp. 659–667. ACM (2013)
6. Larraín, S., Parra, D., Soto, A.: Towards improving top-n recommendation by generalization of slim. In: RecSys Posters (2015)
7. Levy, M., Jack, K.: Efficient top-n recommendation by linear regression. In: RecSys Large Scale Recommender Systems Workshop (2013)
8. Ma, J., Zhao, Z., Yi, X., Chen, J., Hong, L., Chi, E.H.: Modeling task relationships in multi-task learning with multi-gate mixture-of-experts. In: Proceedings of the 24th ACM SIGKDD International Conference on Knowledge Discovery and Data Mining, pp. 1930–1939. ACM (2018)
9. Ning, X., Karypis, G.: Slim: sparse linear methods for top-n recommender systems. In: Proceedings of the International Conference on Data Mining (ICDM), pp. 497–506. IEEE (2011)
10. Qiu, S., Cheng, J., Yuan, T., Leng, C., Lu, H.: Item group based pairwise preference learning for personalized ranking. In: Proceedings of the International ACM SIGIR Conference on Research and Development in Information Retrieval, pp. 1219–1222. ACM (2014)
11. Rendle, S., Freudenthaler, C., Gantner, Z., Schmidt-Thieme, L.: BPR: Bayesian personalized ranking from implicit feedback. In: Proceedings of the Conference on Uncertainty in Artificial Intelligence, pp. 452–461. AUAI Press (2009)
12. Robertson, S.E., Kanoulas, E., Yilmaz, E.: Extending average precision to graded relevance judgments. In: Proceedings of the International ACM SIGIR Conference on Research and Development in Information Retrieval, pp. 603–610. ACM (2010)
13. Sedhain, S., Menon, A.K., Sanner, S., Braziunas, D.: On the effectiveness of linear models for one-class collaborative filtering. In: Proceedings of the AAAI Conference on Artificial Intelligence, pp. 229–235. AAAI Press (2016)
14. Shi, Y., Karatzoglou, A., Baltrunas, L., Larson, M., Hanjalic, A.: GAPFM: optimal top-n recommendations for graded relevance domains. In: Proceedings of the ACM International Conference on Conference on Information and Knowledge Management, pp. 2261–2266. ACM (2013)
15. Shi, Y., Karatzoglou, A., Baltrunas, L., Larson, M., Hanjalic, A.: xCLiMF: optimizing expected reciprocal rank for data with multiple levels of relevance. In: Proceedings of the ACM Conference on Recommender Systems, pp. 431–434. ACM (2013)
16. Shi, Y., Karatzoglou, A., Baltrunas, L., Larson, M., Hanjalic, A., Oliver, N.: TFMAP: optimizing map for top-n context-aware recommendation. In: Proceedings of the International ACM SIGIR Conference on Research and Development in Information Retrieval, pp. 155–164. ACM (2012)
17. Zhu, N., Cao, J.: GTRM: a top-N recommendation model for smartphone applications. In: Proceedings of the IEEE International Conference on Web Services (ICWS), pp. 309–316. IEEE (2017)

Data Mining

RTIM: A Real-Time Influence Maximization Strategy

David Dupuis[1,2], Cédric du Mouza[3], Nicolas Travers[1,3(✉)],
and Gaël Chareyron[1]

[1] Léonard de Vinci Pôle Universitaire, Research Center, Paris La Défense, France
{david.dupuis,nicolas.travers,gael.chareyron}@devinci.fr
[2] Kwanko, Paris, France
[3] CEDRIC Lab, CNAM, Paris, France
prenom.nom@cnam.fr

Abstract. Influence Maximization (IM) consists in finding in a network the top-k influencers who will maximize the diffusion of information. However, the exponential growth of online advertisement is due to Real-Time Bidding (RTB) which targets users on webpages. It requires complex ad placement decisions in real-time to face a high-speed stream of users. In order to stay relevant, the IM problem should be updated to answer RTB needs. While traditional IM generates a static set of influencers, they do not fit with an RTB environment which requires dynamic influence targeting. This paper proposes RTIM, the first IM algorithm capable of targeting users in a RTB environment. We also analyze influence scores of users in several social networks and provide a thorough experimental process to compare static versus dynamic IM solutions.

Keywords: Real-Time Bidding · Influence Maximization · Social network

1 Introduction

Since Kempe et al. [16], Influence Maximization (IM) is a well studied maximum coverage problem which consists in finding the smallest subset of individuals in a social network who will maximize information diffusion through social influence. In this paper, we are interested in enhancing IM methods with Real-Time Bidding (RTB) constraints. We consider that a user can be influenced because he saw an ad, interacted with it, or purchased the product. To be more relevant, IM algorithms need to take into consideration time and targeting requirements of RTB.

However, online advertising revenue outpaced all other advertising strategies thanks to the advent of RTB [24,27] and Social Network Services (SNS). RTB is an online auction system which allows online advertisers to bid in real-time for ad locations on a webpage in less than $100\,ms$ [27]. RTB ad targeting is initially based on the content of the web page and users' consumer profile, but fails to

© Springer Nature Switzerland AG 2019
R. Cheng et al. (Eds.): WISE 2019, LNCS 11881, pp. 277–292, 2019.
https://doi.org/10.1007/978-3-030-34223-4_18

rely on the social value of each customer as suggested in [9]. As far as we know no RTB algorithms attempt to find an IM solution to improve bidding decisions. They only consider influence as a parameter and not as a propagation value on a network. It is important to note that there is potential here for IM to integrate full bidding, however this approach is left for future work.

Existing IM algorithms propose various optimization technics to statistically choose a seed set of users that maximizes influence. However, as far as we know, no existing IM algorithm can work within a real-time bidding environment and satisfy it's requirements. Indeed, whereas existing algorithms take hours to find seed sets up to 200 seeds in a large social network [1], they do not scale up or take into account real time streams of users.

This article targets the issue with the following constraints: (a) only an online user can be targeted, (b) deciding whether to target a user must be done in under 100ms, (c) the propagation influence score relies on a social network containing millions of users and relationships, (d) thousands of users must be targeted in real-time while maximizing scores of large seed sets.

Therefore, to target influential users in a RTB environment it is necessary to develop an IM algorithm capable of deciding in real-time which users are worth targeting. To achieve this we propose the RTIM approach which stands for Real-Time Influence Maximization.

Our main contributions are as follow:

- We propose an elegant approach for real-time influence maximization focusing on the stream of online users,
- We provide a deep analysis of users' influence scores for various social network datasets in order to showcase users' behavior in IM,
- We set up a thorough experimental setting for RTIM and IMM models on different social networks.

In this article, we first explain in Sect. 2 the state of the art on IM. Section 3 explains our two stages RTIM approach and Sect. 4 gives the RTIM model and algorithms. In order to understand the impact of our model, we propose an analysis of influence in different datasets in Sect. 5. This leads to the experimental process in Sect. 6 with a dynamic stream evaluation. Finally, we conclude and give some perspectives in Sect. 7.

2   IM State of the Art

Influence Maximization takes place in a social network graph $\mathcal{G} = (V, E)$ where V is the set of vertices (users), E the set of directed edges (influence relationships). In this graph $\mathcal{G}$, a user is **activated** if he has successfully been influenced by a neighbor and therefore influences his own outgoing neighbors. A **targeted user** is a user for whom a piece of information is shown to be propagated.

The goal of IM is to produce a **seed set** $\mathcal{S}$ of targeted users which maximizes its influence on $\mathcal{G}$. The optimal seed set, or the final result is defined as $\mathcal{S}^*$.

2.1 Propagation Models

Kempe et al. [16] propose two common propagation models: *Independent Cascade* (IC) and *Linear Threshold* (LT). *IC* considers that each user can be influenced by a neighbor independently of any of his other neighbors. *LT* considers that a user is activated if the sum of successful influence probabilities from his neighbors is greater than his activation threshold.

Under the *IC* model, time unfolds in discrete steps. At any time-step, each newly activated node $u_i \in V_a, \forall i \in V$ gets one independent attempt to activate each of its outgoing neighbors $v_j \in Out(u_i), \forall j \in V\{i\}$ with a probability $p(u,v) = e_{ij}$. In other words, e_{ij} denotes the probability of u_i influencing v_i.

As explained in [12] there is a real challenge in acquiring real-world data to build datasets containing accurate influence probabilities. Common practice is to use theoretical edge weight models. For *IC*, the `Constant` model is where each weight e_{ij} is given a constant probability $[3,8,10,11,13,16]$. Some define $p \in [0.01, 0.1]$ $[4,23]$. The `Tri-Valency` model is where the weights are randomly chosen from a list of probabilities such as $\{0.001, 0.01, 0.1\}$ $[3,6,15]$. Finally, the `Weighted Cascade` (WC) model is where $e_{ij} = \frac{1}{|In(v_j)|}$ where $In(v_j)$ is the number of neighbors that influence u $[3,4,6\text{–}8,10,11,16,19,25,26]$. Under WC, all neighbors that influence u_i do so with the same probability. Therefore, it is easier to influence a user with a low in-degree.

For *LT*, the general edge weight rule is that the sum of the weights which must equal 1. Therefore, WC applies to *LT*. Additional alternative models can be found in [21] where an extensive IM state-of-the art is done.

IC is very useful to model information diffusion when a single exposition to a piece of information from one source is enough to influence an individual. *LT* doesn't change the fundamental approach of our algorithm, it should not be difficult to extend to. For these reasons, we limit our approach to *IC*. In addition, we define the edge weights using the *WC* model, because we believe it better corresponds to the diversity of influence between individuals in a real-world social network.

2.2 Properties

Kempe et al. [16] prove that influence maximization is NP-Hard under both the *IC* and *LT* models and computing the influence score is monotone and submodular which Chen et al. [3] prove to be #P-Hard under the *IC* model.

The propagation function f is sub-modular if it satisfies a natural diminishing returns property i.e., the marginal gain from adding an element v to a set S is at least as high as the marginal gain from adding the same element to a superset of S. Formally, a sub-modular function satisfies: $\forall S \subseteq T \subseteq \Omega$ and $x \in \Omega \backslash T, f(S \cup \{x\}) - f(S) \geq f(T \cup \{x\}) - f(T)$. This sub-modular property is essential as it guarantees that a *greedy* algorithm will have a $(1 - 1/e - \epsilon)$ approximation to the optimal value [22]. As it is presented previously, many IM algorithms rely on this theoretical guarantee to validate their strategy.

2.3 Computing Score

Influence Score: Computing the influence score requires solving Eq. (1), which is a generalization of the inclusion-exclusion principle from [28].

$$\sigma(\mathcal{S}) = \sum_{v_i \in V} a_S(v_i) = \sum_{v_i \in V} \mathsf{P}(\bigcup_{p_j \in P_{uv_i}} p_j),$$

$$P_{uv_i} = \{all\ paths\ event\ existence\ between\ u\ and\ v_i\}$$

(1)

In Eq. 1, the influence score of a seed set $\sigma(\mathcal{S})$ is the sum of activation probabilities $a_S(v)$ of any node $v \in V$ when users in $\mathcal{S}$ are targeted. The activation probability of a user is the probability that there exists a path between that user and any targeted user.

Computing $\sigma(\mathcal{S})$ is proven by [15] to be #P-Hard but [16] approximate the exact result by running $n = 10,000$ *Monte Carlo* simulations. In this article, the result of these simulations is written $\sigma_{MC}(\mathcal{S})$.

[16] prove that IM is an NP-Hard problem. In fact, it simply consists of two challenges. The first is finding the optimal seed set out of $2^{|V|}$ subsets of users or $\binom{N}{k}$ if we know k, and computing the influence score according to Eq. 1.

2.4 Algorithms

Clearly presented by Arora et al. [1] there are three main categories of IM algorithms: greedy, sampling and approximation.

GREEDY [16], CELF [17] and CELF++ [13] are all three lazy-forward algorithms which guarantee an approximation of $(1 - 1/e - \epsilon)$. To find $\mathcal{S}^*$ they start with $\mathcal{S} = \emptyset$ and incrementally add to $\mathcal{S}$ the node v which brings the largest marginal gain: $\sigma_{MC}(\mathcal{S} \cup v) > \sigma_{MC}(\mathcal{S})$, until $|\mathcal{S}| = k$. CELF, CELF++ improve computation time with the submodular property by storing temporary results.

Reverse Influence Sampling method [2] from Borg et al. like in IMM [25], TIM or TIM+ [26], use topological sampling. In the transpose graph, they generate a set $\mathcal{R}$ of size θ of random paths of greatest influence by picking users uniformly at random (Reverse Reachable sets). Using a greedy method, they build $\mathcal{S}^*$ by continuously adding to $\mathcal{S}^*$ the user who covers the greatest number of Reverse Reachable sets and removing them from $\mathcal{R}$.

Approximation algorithms such as EaSyIM [10], IRIE [15], SIMPATH [14], LDAG [5] or IMRANK [6], offer heuristics to compute $\sigma(\mathcal{S})$ Eq. (1), such as using the most probable path or the independence of paths.

Conclusions: Greedy solutions require hours or days of processing due to the repeated computation of $\sigma_{MC}(\mathcal{S})$. They perform poorly for seed sets larger than 50 and do not scale to large datasets [1]. Heuristics don't offer theoretical guarantee and often lack in precision. Sampling algorithms are significantly faster than greedy, are more precise than heuristics and offer a theoretical guarantee. In addition, none of these solutions are meant to dynamically generate a seed set with RTB constraints. There exists a great number of specific IM contributions which have been listed in [21]. It shows clearly, that no contributions have been made regarding the analysis of the IM challenge in a stream of online users.

To this end, we propose the RTIM algorithm: *Real-Time Influence Maximization*. It targets influential users, henceforth generating a seed set of influencers under RTB constraints. To ensure this, RTIM takes places in two stages: a pre-processing stage and a live stage which we present in the following.

3 RTIM Approach

RTIM is meant to perform in a RTB environment. The latter, consists of users who connect to a website which sells ad slots. The IM algorithm has to determine wether it is useful for targeting. As we know, these users, all belong to a very large social network through which they may influence neighbors.

The originality of our approach lies in its ability to target users who appear in this dynamic stream by estimating whether they will have a significant gain based on previously targeted users in the same stream and belonging to the same social network. While traditional approaches determine the best seed set of targeted users by processing a graph in which any user is considered online. Contrary to these solutions, RTIM allows us to adapt our IM strategy to take place in a RTB streaming environment.

Static algorithms, such as IMM [25], correspond to an optimistic approach which assumes that pre-computed users from their seed set will necessarily be online in the stream. However, many users of the pre-defined seed set won't be available to target during the advertisement campaign. In contrast, our approach, which can be considered as a pessimistic approach, allows us to dynamically fill our seed set with online users of interest for the advertisement campaign.

3.1 Step I: Pre-processing - Building the Influence Graph

First, an *influence graph* $\mathcal{G}_I(V, E, w_I)$ is built with weights on each edge to estimate the influence based on the number of incoming edges of a vertex. This influence estimation is commonly adopted in influence propagation [23]. It is defined formally as follows:

Definition 1 (Influence graph). *Consider $\mathcal{G}(V, E)$ the social graph where V is the set of vertices and $E \subseteq V^2$ is the set of oriented edges. The influence graph for $\mathcal{G}$ is the graph $\mathcal{G}_I(V, E, w_I)$ with the same sets of vertices and edges and a weighted function $w_I : E \to \mathbb{R}$ such that for an edge e_{ij} from vertice v_i to v_j: $w_I(e_{ij}) = \frac{1}{indegree(v_j)}$.*

Figure 1 depicts the influence graph for a social network between 5 users. For instance, user u_2 who follows or is influenced by users u_1, u_3 and u_5 has each of his incoming edge $e \in E$ weighted by: $w_I(e_{12}) = w_I(e_{32}) = w_I(e_{52}) = 1/3 = 0.33$.

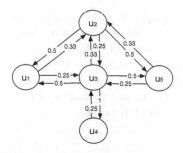

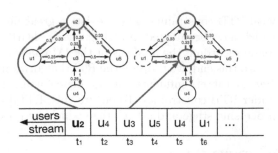

Fig. 1. Influence graph $\mathcal{G}_I$

Fig. 2. Ex. of the live stream ($\mathcal{T}$) of available users (Color figure online)

To estimate the influence score, we use the *Monte Carlo* approach by running n simulations, where n is a large number ([16] set $n = 10,000$). The influence score of each user u is the average number of users activated for all simulations.

Each simulation randomly test each outgoing edge of a user against the edge weight $w_I(e_{ij})$. When a neighbor is activated we can then recursively test neighbors until no more nodes are activated.

Since the simulations are all independent and the graph data structure is only read during the process, we can run the n simulations in parallel. However running 10,000 *Monte Carlo* simulations for each user $u \in \mathcal{G}$ remains extremely costly. Consequently, this computation must be performed offline and all influence scores are stored in a vector I: $\forall u_i \in \mathcal{G}, I_i = \sigma_{MC}(u_i)$.

3.2 Step II: User Targeting at Runtime

With the influence score computed in the pre-processing step, RTIM is able to select, during the stream, users to target. Consider the temporal stream of users $\mathcal{T}$ in which appears every online connection events of users from $\mathcal{G}$. Since a user can only be targeted when he appears in the stream, we need to decide in real-time whether he is worth targeting or not. To make this decision, our RTIM algorithm takes into consideration the influence score of users and the probability that they have already been targeted by neighbors.

To verify these two criteria, we set two thresholds, θ_I and θ_A, respectively the minimum influence score and the activation probability. Whenever a user is online, we check whether his influence score is important enough (above θ_I) to be a potential target for the ad campaign or not. We also check the probability for him to be activated by users he follows (above θ_A). If θ_I is validated and not θ_A, the user is targeted and added to the seed set. His activation probability is set to 1. This activation is propagated in the neighbourhood. This will enable us to make better targeting decisions for future users who appear in the stream.

Figure 2 illustrates the stream of users that are online and their correlation with the graph. $\mathcal{T}$ is a basic example of a RTB stream where users appear one at a time in discrete steps (in red/bold) and can only be targeted when available.

When the first user u_2 appears (time t_1), we verify his influence score I_2. If $I_2 > \theta_I$ then we consider that u_2 tries to activate his followers u_1, u_3 and u_5, and propagate probabilities to their own neighbors. Assume that u_1 is activated ($A_1 > \theta_A$) while u_3 and u_5 are not. When user u_4 is online (t_2), his influence score is insufficient to be targeted and is avoided. Then, when user u_3 appears in the stream (t_3), he is considered to be an influencer ($I_3 > \theta_I$) and not activated ($A_3 < \theta_A$). Thus the activation probability is propagated to u_1, u_2, u_5 and u_4. When u_5 appears in $\mathcal{T}$, even if his influence score is higher than θ_I, $A_5 > \theta_A$ since he has been influenced by u_2 and u_3. Thus it is not worth targeting him.

By applying the whole stream of users $\mathcal{T}$, our approach generates the seed set $\mathcal{S}^*$ where every user $u \in \mathcal{G}$ verifies θ_I and θ_A. The key point resides in the fact that RTIM maximizes the influence of connected users while removing those who are too close to users already targeted.

4   RTIM Model

Traditional influence maximization algorithms have an optimistic approach since they determine statically the users to target based on the final global influence score of the set of targeted users. If the advertisement campaign is not time-limited (infinite stream), these solutions potentially maximize the total score.

RTIM's strategy is quite different since it decides to target a user in real-time when he is available. So RTIM can be considered as a pessimistic algorithm since we decide to add him to the final seed set instantaneously, even if a "better" user to add to the seed set appears later in the stream.

Activation Probability Graph. At time t_0, when $\mathcal{T}$ starts, we create the activation probability graph as the influence graph $\mathcal{G}_I$ described in Sect. 3.1. We can adopt the matrix representation for the graph in the following equation: $M_{\mathcal{G}_I}(V, E) = A_{\mathcal{G}} \times InDeg_V$, where $A_{\mathcal{G}}$ is the adjacency matrix, i.e., $A_{\mathcal{G}}[i, j] = 1$ if there exists an edge from user u_i to user u_j, 0 otherwise, and $InDeg_V$ is the indegree vector, with $InDeg_V[i] = \frac{1}{indegree(u_i)}$. The activation probability vector AV is initialized as the vector with only 0 values.

Activation Probability Updates. Consider we have at time $t_{k-1} > t_0$, an activation probability vector $AV(t_{k-1})$. Then assume that at time t_k, a user u_i connects and we decide to target him. So his activation probability $AV(t_{k-1})[i]$ is now set to 1. This probability update impacts other probabilities in the graph. Indeed, users who follow u_i are now more likely to see this ad and we avoid targeting them in the future. We must update other activation probabilities through influence propagation to obtain the $AV(t_k)$ probability vector.

Definition 2 (Activation probability propagations). *Consider $\mathcal{G}(V, E)$ the social graph and its influence graph $\mathcal{G}_I(V, E, w_I)$ as defined in Sect. 3. The activation probability vector $AV(t_k)$ for $\mathcal{G}$ at instant t_k is recursively defined as:*

$$\begin{cases} AV^{(0)}(t_k) = M_{\mathcal{G}_I}(V, E) \times AV(t_{k-1}) \\ AV^{(i+1)}(t_k) = M_{\mathcal{G}_I}(V, E) \times AV^{(i)}(t_k) \end{cases}$$

Algorithm 1. Updating activation probabilities

Require: graph $\mathcal{G}$, nodes u and v, v's activation probability $A[v]$, u's neighbors $\mathcal{N}_u$, current path weight p, depth d
1: **procedure** ACTIVATIONSCORES($\mathcal{G}, u, p, d$)
2:    **for** $v \in \mathcal{N}_u$ **do**
3:        $A[v] \leftarrow 1 - (1 - A[v]) * (1 - p * w_{uv})$
4:        **if** d > 1 **then**
5:            ACTIVATIONSCORES($\mathcal{G}, v, p * w_{uv}, d - 1$)

So, after targeting a user u_i ($AV(t_{k-1})[i] = 1$) the vector is recursively combined with the activation probability graph $M_{\mathcal{G}_I}$ in order to propagate the activation while obtaining a convergence after i iterations:

$$AV(t_k) = AV^{(\infty)}(t_k) = M_{\mathcal{G}_I}(V, E) \times AV^{(\infty)}(t_k)$$

Since this model corresponds to a *Matrix population model* [18][1], we can guarantee its convergence due to the fact that the Eigenvalues of $M_{\mathcal{G}_I}(V, E)$ are real strictly positive (the matrix is real, asymmetric and non-diagonal). Moreover the propagation is an increasing and monotone function bounded to $\overrightarrow{1}$.

The aim of RTIM is to determine in real-time if a user is a good influencer while not already having been influenced by other users. To target influencers, we need to determine users worth targeting but also when users are considered activated by influencers. For this we define the threshold θ_I as the minimum influence score to reach, set to the influence score of the kth influencer. We also define the activation probability threshold θ_A, set by default to 0.5. Any user whose activation probability is greater than θ_A is considered to be activated and therefore will have attempted himself to propagate the information provided by an influencer and is therefore not worth targeting.

During the live stage we need to update users' activation probability. To achieve this, we propagate probabilities at depth less than d. For a user, if his influence score is above θ_I and his activation probability is below θ_A, the user is targeted. Otherwise we ignore him.

Equation 1 gave the activation probability of a user v. Since we consider paths of length 2 all paths between u and v are independent:

$$A[u] = \mathsf{P}(\bigcup_{p_j \in P_{uv}} p_j) = 1 - \prod_{w_i \in \mathcal{P}_{uv}^d} (1 - w_i), \tag{2}$$

Algorithm 1 illustrates activation probabilities updates. For each neighbor v of user u we propagate his activation probability (line 3). While the depth of propagation is sufficient we follow the propagation recursively (line 4&5). In the worst case it runs in $O(|V|^d)$ when all users are interconnected. However, updates can take place in a separate thread during the live stage.

For the live stage of RTIM, we consider that if any neighbor (of depth d) of a user is targeted then we update his activation probability. First, Algorithm 2

---

[1] Our model is not a Markov chain since the sum of a column can exceed 1.

Algorithm 2. RTIM Live

Require: graph $\mathcal{G}$, user u, vector of influence scores I, influence threshold θ_I, u's
    activation probability $A[u]$, depth d, stream of users $\mathcal{T}$, seed set $\mathcal{S}$ and max size k
1: Initialize $A \leftarrow \vec{0}$
2: **while** $|\mathcal{S}| < k$ **do**
3:     $u \leftarrow next(\mathcal{T})$
4:     **if** $I[u] \geq \theta_I$ and $A[u] \leq \theta_A$ **then**
5:        ACTIVATIONSCORES($\mathcal{G}$, u, 1, d)
6:        $\mathcal{S} \leftarrow \mathcal{S} \cup u$

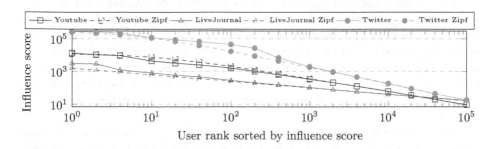

Fig. 3. Datasets' influence score distributions

initializes the activation probabilities to the 0 vector (line 1). Then, while the
seed set is not filled (line 2) we check each new incoming user u if he validates
both θ_I and θ_A (line 3 & 4). Deciding to target a user (line 4) is done in $O(1)$
and is thus instantaneous. If he does we add u to the seed set and propagate the
activation by applying Algorithm 1 (line 5 & 6).

5 Influence Analysis

Our model is experimented with empirical datasets of different sizes: `Youtube`,
`LiveJournal` and `Twitter`. We need their characteristics to understand the
impact of our approach. We can see in Table 1 the global statistics: # nodes, #
edges, and node degrees (mean, variance and standard deviation).

    We can see that `Youtube` is the "smallest" graph with less connections (mean
degree of 10) but with a high variation compared to its size. `LiveJournal` is
highly connected with a high number of edges and a mean degree of 34. However,
users are more homogeneously connected (low variance and STD). `Twitter`, on
the other hand, is the biggest graph in which users can have varying numbers of
connections with a mean degree of 70 but a variance of 6.4M.

    Figure 3 shows the distribution of influence scores for our graph datasets.
These distributions can be characterized by a standard *Zipf-Mandelbrot* distribution [20], traditionally used for distribution of ranked data. It is defined by:
$ZM(r) = \frac{B}{(r_0+r)^\alpha}$ where r is, here, the rank of the influencer. r_0 is a constant
representing the number of top influencers. B corresponds to the starting score
modifier and α is the decreasing speed of scores.

Table 1. Datasets characteristics

	# of nodes	# of edges	Deg. Mean	Degree Var.	STD
Youtube	1.13M	5.97M	10.53	10,304	101
LiveJournal	3.99M	69.3M	34.70	7,381	85
Twitter	41M	1.46B	70.50	6,426,184	2,534

Table 2. Zipf-Mandelbrot parameters for graph datasets

B	r_0	α	$\mathcal{X}^2$-Pearson value
8×10^4	11	0.78	0.976
1.55×10^3	0	0.395	0.971
1.7×10^6	6	0.99	0.969

Table 2 gives the corresponding values for those Zipf-Mandelbrot distributions and the $\mathcal{X}^2$-Pearson values (observation probabilities) found for each distribution. Youtube and Twitter behave similarly with a high r_0 (resp. 11 and 6) leading to 100 to 750 top-influencers. We can see that Twitter has more top influencers and then drops faster than Youtube. LiveJournal behaves differently with very few top influencers compared to Youtube or Twitter. The low value r_0 shows that top-influencers' score decreases faster at the beginning of the curve.

However, the decreasing speed of the influence score α witnesses really high values (resp. 0.78 and 0.99) which means that it is harder to become a top influencer on Twitter than Youtube. Likewise, the absolute number of influencers is really high with a B value between 10^4 and 10^6 leading to a long tail which only starts after more than 10^5 for Youtube and 2×10^5 users for Twitter. On the other hand, LiveJournal's scores curve decreases more slowly than the others with an α of only 0.395 giving the idea that the number of connections between users are closer to the average than Twitter or Youtube. Consequently, the long tail is reached more slowly than the others (4×10^5 users).

This conclusion is interesting in order to understand the impact of these social networks on influence maximization. Indeed, targeting top influencers in real-time requires choosing influencers according to their estimated score. For instance, users from the long tail are pretty identical and cannot be differentiated from each other, thus the decision to target or not an influencer depends on α which tells us how much an influencers score evolves.

6   Experiments

We wish to show in this section the impact of choosing influencers in a real-time stream of users. In order to do this, we need to compute the influence scores of users, generate multiple streams of users with varying distributions and compare the final solution of each algorithm for different graph datasets.

6.1   Experimental Process

Since RTIM is an IM algorithm which runs under RTB constraints, we want to compare it to an existing IM algorithm. We choose IMM [25] since it is the best compromise between computation speed, scalability and accuracy, especially on large datasets. The code for IMM is provided by [1] in C++.

Table 3. Update activation probabilities time

	Average	Median	Max
Youtube	70.3 ms	6.30 ms	193.4 ms
LiveJournal	61.1 ms	6.03 ms	192.0 ms
Twitter	85.9 ms	44.5 ms	411.1 ms

Stage I: Pre-processing. First, IMM is run in its entirety and adds k users to its seed set $\mathcal{S}_{IMM}$ ($k = 10,000$ and optimal seed $\epsilon = 0.1$). Recall that IMM relies on the fact that every user in the graph has the same probability to appear.

RTIM uses the *Monte Carlo* approach to compute the influence score of each user in the graph. We run n parallel simulations per node ($n = 10,000$) with a limit depth of 3 for scalability purposes. Thanks to the graph topology with a high connectivity (see Sect. 5), those *Monte Carlo* simulations converge faster. The influence scores of each user are stored in a vector I for future use.

Stage II: Live Stream Generation. It's during this stage that we read our stream and both algorithms have the opportunity to target influencers. Since no real streams of connected users are available online, we simulate users' behavior in the social networks with different distributions. Streams contain 10% of the total number of users and can appear several times with two different distributions: *Uniform* and *Log*.

The *Uniform* distribution supposes that all users have an equal probability of appearing in the stream. This distribution can be considered to be the worst case where top influencers can appear as frequently as low influencers.

The *Log* distribution supposes that users who have more in/out edges in the graph are more likely to be connected. The probability of user u_i being in the stream is therefore $P(u_i \in \mathcal{S}) = \log(deg_{u_i})/\sum_{u_i \in V} \log(deg_{u_i})$, where deg_{u_i} is the degree of u_i. We apply a logarithm on deg_{u_i} in order to give users with a low influence score a reasonable probability of showing up in the stream. This *Log* stream can be considered to be the best case where highly linked users are more likely to be present in the stream, and potentially top-influencers.

To ensure the stability of our model, we generated multiple random versions of each stream. Our results are averages of all versions.

Stage III: Live Stream Process. For IMM, during the live stage, if a user in the stream belongs to $\mathcal{S}_{IMM}$, he is targeted and immediately added to the final seed set $\mathcal{S}^*_{IMM}$. Regardless of the number of times he appears in the stream, the user is only targeted once. At the end of the stream: $\mathcal{S}^*_{IMM} = \mathcal{S}_{IMM} \cap \mathcal{T}$.

For each user u_i in the stream, if $ap(u_i) < \theta_A$ and $\sigma_{MC}(u_i) > \theta_I$, RTIM targets u_i instantaneously. u_i is added to the final seed set $\mathcal{S}^*_{RTIM}$ and we update the activation probability of his neighbors in a separate thread.

Table 3 gives the time spent to update propagation probabilities in the network. It shows that the average update time is less than 100 ms which satisfies our real-time requirement, and in most of the case far less (median). However,

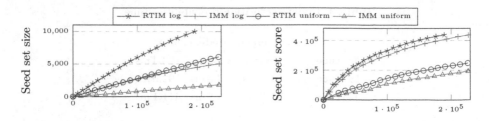

Fig. 4. Seed set score & size evolution with `Youtube`

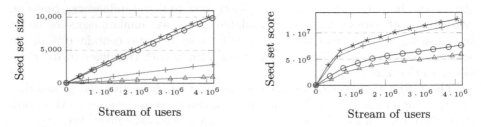

Fig. 5. Seed set score & size evolution with `LiveJournal`

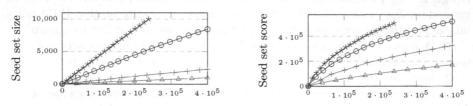

Fig. 6. Seed set score & size evolution with `Twitter`

some updates, especially on large dense graphs can take up to 411 ms. This is still negligible since a user cannot influence another in less than half a second.

6.2   Experimental Results

In the following experiments, we see the real-time evolution of the seed set influence score and size on the `Youtube`, `LiveJournal` and `Twitter` datasets.

Youtube. Figure 4 gives the results produced for a stream of 2.27×10^5 connected users over the `Youtube` dataset. The left-hand side shows the evolution of the seed set size where IMM hardly finds pre-defined influencers, especially for the *Uniform* distribution. RTIM evolves almost linearly with twice as many seeds for the *Uniform* distribution and 3.3 times more for the *Log* one. According to the *Log* distribution, RTIM finds more influencers and reaches k more quickly. The sudden stop of the RTIM seed set at 1.89×10^5 users is due to the fact that the marketing campaign is over with a full seed set of $k = 10,000$ users.

The other side shows that RTIM produces seed sets with higher scores than IMM. We can see different evolutions from the *Uniform* and *Log* streams. In fact, IMM targets few users in the *Uniform* distribution, since highly connected users are less often available online. On the other hand, RTIM targets more users according to their local influence on the graph. According to the *Log* distribution, IMM is closer to RTIM since top-influencers are more present in the stream. Consequently, it takes time for IMM to reach this goal by the end of the stream with a similar score (1,105 less), while RTIM stopped earlier when the seed set size reached k. This confirms the fact that IMM is better at maximizing k than RTIM in an infinite stream, however in a finite ad campaign this is not the case.

LiveJournal. Figure 5 focuses on `LiveJournal`. The stream contains almost 4×10^5 online users. On the left part, we can see the evolution of seed set sizes for which IMM finds very few expected influencers and produces 8.5 times less seeds for the *Uniform* stream (respectively 7 for the *Log* stream) than RTIM. In fact, RTIM targets influencers more easily than IMM which can be explained by the distribution of scores (see Sect. 5) with a very slow decreasing of the scores ($\alpha = 0.395$). This can be confirmed by the fact that `LiveJournal` has a low degree standard deviation and variance. Thus RTIM adapts locally to the users connection with similar scores while IMM only focuses on pre-chosen seeds.

We can see the evolution of seed set scores on the right part where scores are really different from the `Youtube` dataset. The impact of pre-determined seeds have a huge impact on the final seed set score since very few influencers appears in the stream while RTIM can choose a "similar" score in the neighbourhood.

Moreover, RTIM obtains a lower seed set score for the *Uniform* distribution than the *Log* one. It is due to the fact that RTIM Log fills the seed set more quickly after only 2.38×10^5 users in the stream. The impact of the specific distribution of scores of `LiveJournal` and the fact that users have high mean degrees (with a low variance) give more chances for common users (lower scores).

Twitter. The seed sets produced for `Twitter` are presented in Fig. 6. The stream is composed of 4.17×10^6 connected users. We observe first that RTIM seed sets evolves very quickly for both *Uniform* and *Log* streams. This is due to the huge amount of high score users of the `Twitter` distribution (see Sect. 5 with $B = 1.7 \times 10^6$), consequently RTIM targets any user in the stream that reaches the threshold θ_I. On the other hand, IMM evolves more slowly, 10 times less for *Uniform* and 3.5 times less for the *Log* stream. RTIM fulfills the marketing campaign $k = 10,000$ after 4×10^6 users in the stream.

For seed set scores, they evolve similarly to the `Youtube` dataset with close scores for the *Log* stream, even if the gap is higher due to huge seed set scores (600,000 less). The effect of the high decrease of the influence score ($\alpha = 0.99$ in Table 1) is observable here where IMM targets high influencers that have sufficient impact to grow rapidly while RTIM targets good influencers to guarantee a global impact in a minimum amount of time.

Conclusions. Our experiments showed that RTIM provides better seed sets score while maximizing the score in a minimum of time while IMM succeeds in maximizing on the whole dataset. The impact of the live stream distribution between *Uniform* and *Log* is such that both methods behave clearly better on users with very high degrees however IMM is more sensitive to this setting.

The seed set score curve is logarithmic since IM is sub-modular. Indeed, the more users we target the smaller the marginal gain to the overall seed set. First, the decrease of those scores is in favor of RTIM when α is low (LiveJournal) where IMM makes a choice on similar influencers while RTIM targets only available ones. Second, graphs with very high influence scores (induced by B) give RTIM more choices of influencers and so it fills up the seed set quickly.

7 Conclusion

In this article we have shown, that it is possible to answer the influence maximization problem in a real-time bidding environment, that up to now have not been applied to IM algorithms. We have shown, that static IM algorithms, such as IMM, that pre-compute the best seed set of size k can solve this problem so long as they are capable of generating in reasonable time a large seed set with $k \geq 10,000$. We have shown, in addition, that it is possible, in this setting, to compete with these powerful static IM algorithms by using a dynamic IM algorithm, such as RTIM, based on the local influence of each user. In fact, we have proven, that dynamic IM algorithms such as RTIM can outperform static algorithms when the stream of users is finite in size (or a fixed period of time).

It is important to note, that the RTB environment is more complex than the constraints which we used. It is for instance, not guaranteed that a user having been displayed an advertisement will see it, click on it, or even convert. For future works, we propose to extend RTIM to fully answer RTB constraints. Contrary to IM algorithms, RTIM could choose to target another user if a previous targeted one was not considered as activated. We can therefore, make RTIM much more interactive with dynamic user behavior while static solutions like IMM cannot.

In addition, should the graph be updated we can recompute local influence scores, if necessary, or keep targeting users in the live stream. Whereas, static IM algorithms need to recompute the best possible seed set for each new graph. We can also improve RTIM by adapting dynamically the θ_I threshold when processing the live stream. In fact, online user behavior, such as periodicity of connection during the day or week, has an impact on the final seed set score.

Acknowledgments. This work was supported by Kwanko.

References

1. Arora, A., Galhotra, S., Ranu, S.: Debunking the myths of influence maximization: an in-depth benchmarking study. In: SIGMOD, pp. 651–666. ACM (2017)
2. Borgs, C., Brautbar, M., Chayes, J., Lucier, B.: Maximizing social influence in nearly optimal time. In: SODA, pp. 946–957. Society for Industrial and Applied Mathematics, Philadelphia (2014)

3. Chen, W., Wang, C., Wang, Y.: Scalable influence maximization for prevalent viral marketing in large-scale social networks. In: SGKDD, pp. 1029–1038. ACM, Washington, DC (2010)

4. Chen, W., Wang, Y., Yang, S.: Efficient influence maximization in social networks. In: SIGKDD, Paris, France, pp. 199–208 (2009)

5. Chen, W., Yuan, Y., Zhang, L.: Scalable influence maximization in social networks under the linear threshold model. In: ICDM, Sydney, Australia, pp. 88–97 (2010)

6. Cheng, S., Shen, H., Huang, J., Chen, W., Cheng, X.: IMRank: influence maximization via finding self-consistent ranking. In: SIGIR, pp. 475–484 (2014)

7. Cheng, S., Shen, H., Huang, J., Zhang, G., Cheng, X.: Static greedy: solving the apparent scalability-accuracy dilemma in influence maximization. CoRR abs/1212.4779 (2012)

8. Cohen, E., Delling, D., Pajor, T., Werneck, R.F.: Sketch-based influence maximization and computation: scaling up with guarantees. In: CIKM, Shanghai, China, pp. 629–638 (2014)

9. Domingos, P., Richardson, M.: Mining the network value of customers. In: SIGKDD, pp. 57–66. ACM, New York (2001)

10. Galhotra, S., Arora, A., Roy, S.: Holistic influence maximization: combining scalability and efficiency with opinion-aware models. In: SIGMOD, pp. 743–758. ACM, New York (2016)

11. Galhotra, S., Arora, A., Virinchi, S., Roy, S.: ASIM: a scalable algorithm for influence maximization under the independent cascade model. In: WWW, Florence, Italy, pp. 35–36 (2015)

12. Goyal, A., Bonchi, F., Lakshmanan, L.V.: Learning influence probabilities in social networks. In: WSDM, pp. 241–250. ACM, New York (2010)

13. Goyal, A., Lu, W., Lakshmanan, L.V.: CELF++: optimizing the greedy algorithm for influence maximization in social networks. In: WWW, pp. 47–48 (2011)

14. Goyal, A., Lu, W., Lakshmanan, L.V.: SIMPATH: an efficient algorithm for influence maximization under the linear threshold model. In: ICDM, Vancouver, BC, Canada, pp. 211–220 (2011)

15. Jung, K., Heo, W., Chen, W.: IRIE: scalable and robust influence maximization in social networks. In: ICDM, Brussels, Belgium, pp. 918–923 (2012)

16. Kempe, D., Kleinberg, J., Tardos, E.: Maximizing the spread of influence through a social network. In: SIGKDD, pp. 137–146. ACM, New York (2003)

17. Leskovec, J., et al.: Cost-effective outbreak detection in networks. In: SIGKDD, pp. 420–429. ACM, NY (2007)

18. Leslie, P.H.: On the use of matrices in certain population mathematics. Biometrika **33**(3), 183–212 (1945)

19. Li, Y., Zhang, D., Tan, K.L.: Real-time targeted influence maximization for online advertisements. PVLDB **8**(10), 1070–1081 (2015)

20. Mandelbrot, B.B.: The Fractal Geometry of Nature, vol. 1. Freeman, New York (1982)

21. Sumith, N., Annappa, B., Bhattacharya, S.: Influence maximization in large social networks: Heuristics, models and parameters. Future Gene. Comp. Syst. **89**, 777–790 (2018)

22. Nemhauser, G.L., Wolsey, L.A., Fisher, M.L.: An analysis of approximations for maximizing submodular set functions. Math. Prog. **14**(1), 265–294 (1978)

23. Ohsaka, N., Akiba, T., Yoshida, Y., Kawarabayashi, K.: Fast and accurate influence maximization on large networks with pruned Monte-Carlo simulations. In: AAAI, Québec City, Québec, Canada, pp. 138–144 (2014)

24. Spencer, S., O'Connell, J., Greene, M.: The arrival of real-time bidding. Technical report, Google (2011)
25. Tang, Y., Shi, Y., Xiao, X.: Influence maximization in near-linear time: a martingale approach. In: SIGMOD, pp. 1539–1554. ACM, NY (2015)
26. Tang, Y., Xiao, X., Shi, Y.: Influence maximization: near-optimal time complexity meets practical efficiency. In: SIGMOD, pp. 75–86. ACM, NY (2014)
27. Yuan, S., Wang, J., Zhao, X.: Real-time Bidding for Online Advertising: Measurement and Analysis. CoRR abs/1306.6542 (2013)
28. Zhang, M., Dai, C., Ding, C., Chen, E.: Probabilistic solutions of influence propagation on social networks. In: CIKM, pp. 429–438. ACM, NY (2013)

LSCMiner: Efficient Low Support Closed Itemsets Mining

Yifeng Lu[(✉)], Florian Richter, and Thomas Seidl

Database Systems and Data Mining Group, LMU Munich, Munich, Germany
{lu,richter,seidl}@dbs.ifi.lmu.de

Abstract. Itemsets with relatively low support values are important since they usually suggest highly confident association rules, which are useful in applications such as recommendation systems and medical data analysis. However, most existing algorithms are mainly designed to mine frequent patterns and thus are time consuming in generating low support patterns. There are also a few algorithms focus on low support patterns but not efficient enough. Therefore, we propose here a low support closed pattern mining algorithm, utilizing top-down lattice traversing and novel closeness checking/pruning techniques. Extensive experiments show that our method is much more efficient to mine low support closed patterns than available alternatives.

1 Introduction

Itemset mining is an important topic in data mining for decades. Nowadays, it is still actively applied in many areas such as recommendation systems, financial data and medical data mining. Recent researches show that, compared to the advanced deep learning based recommendation systems, pattern based approaches are still competitive [8].

Most existing pattern mining algorithms are designed to mine frequent itemsets as they represent mainstream behavior. However, low support patterns or infrequent itemsets are also important since they usually imply highly confident association rules with solid support. In a large dataset, a low support pattern may actually occur hundred times. Their corresponding rules are useful in recommendation systems for better accuracy. For instance, fewer people in Europe will buy "rice" and "nori" together. Then the pattern "rice and nori" will not be included infrequent patterns, i.e., "sushi maker set" would be recommended only when low support patterns are considered. Low support patterns also play essential roles in other applications. In medicine area, they are crucial in identifying rare diseases. For domain expert, untypical responses to medications are more interesting than frequent and expected ones. In the analysis of traffic accidents, causes of accidents might hide in less frequent and abnormal behaviors. Low support patterns are also helpful in finding significant discriminative patterns [3]. Process mining approaches make use of them as well in identifying deviations in significant process [13].

R. Cheng et al. (Eds.): WISE 2019, LNCS 11881, pp. 293–309, 2019.
https://doi.org/10.1007/978-3-030-34223-4_19

Conventionally, low support patterns are achieved by executing frequent itemset mining algorithms with a small minimum support threshold. Thus, frequent patterns are accessed inevitably. If the huge number of frequent patterns are not of interest, such as the medication example mentioned above, this solution is time wasting. Even if in applications where both frequent and less frequent patterns are needed, the ability to mine low support patterns directly is still necessary. For instance, in dynamic environment, it is expensive for stream pattern mining approaches to track both frequent and less frequent patterns. When low support patterns are more stable over time, efficient low support pattern mining makes it possible to maintain them separately, so that we can build a more adaptive system which only tracks frequent patterns while less frequent ones are updated by user request. Therefore, a few approaches aimed at low support patterns are proposed [1,5,9,16,17]. However, they are either inefficient or with additional constraints.

In this work, we focus on mining low support itemsets. It is well known that redundancy is always a problem in itemset mining. Varies condensed representations and corresponding algorithms are proposed for frequent patterns. However, similar studies are still missing for low support patterns. Closed itemset is one of the most popular lossless condensed representations [14]. We propose a new top-down based algorithm which extracts low support closed patterns without traversing frequent ones. Our approach uses a very efficient tree-based structure. Novel closeness checking and pruning techniques are employed. We show that our approach can achieve the same level of complexity per itemset as other efficient frequent pattern mining algorithms.

2 Preliminaries

2.1 Problem Definition

Let $\mathcal{I}$ be the universe of items, a subset of $\mathcal{I}$ that contains l items is a l-itemset, denoted as $X = \{x_1, x_2, \ldots, x_l\}$. A transaction dataset $\mathcal{T}$ contains a set of transactions where each transaction $T \in \mathcal{T}$ is an itemset over $\mathcal{I}$. Let $\mathcal{T}(X) = \{T | T \in \mathcal{T}, X \subseteq T\}$ be the set of transactions in $\mathcal{T}$ that contains X, the (absolute) support of X on $\mathcal{T}$ is defined as $|\mathcal{T}(X)|$.

In this work, we tend to find less frequent or low support patterns, i.e., $|\mathcal{T}(X)| \ll |\mathcal{T}|$. Formally speaking, given two user-defined threshold: *minimum support* α and *maximum support* β, we are going to mine patterns X such that $\alpha \leq |\mathcal{T}(X)| < \beta$, where $\alpha \geq 1 \wedge \beta \ll |\mathcal{T}|$. In general, our mining task is the same as infrequent itemset mining since $\beta \ll |\mathcal{T}|$. The parameter α is introduced for more flexibility as users might consider patterns occurred less than α as noise. Conventional frequent itemset mining algorithms can also extract low support patterns by setting their minimum support threshold to α and then removing all frequent patterns with support larger than β.

An itemset X is a *closed* itemset in dataset $\mathcal{T}$ if and only if there is no other itemset Y in $\mathcal{T}$ such that $X \subset Y \wedge |\mathcal{T}(X)| = |\mathcal{T}(Y)|$. The closed itemset concept was first proposed in [14] to address the redundant problem in frequent itemset mining problem. It is a lossless condensed representation: user can determine the support of any frequent itemsets from closed frequent itemsets. The set of closed low support patterns is $\mathcal{LP}$.

A frequent border set $\mathcal{FB}$ is defined as the set of longest patterns such that $|\mathcal{T}(X)| \geq \beta$, which is also known as the maximal frequent itemset [2]. $\mathcal{FB}$ is necessary to make $\mathcal{LP}$ complete. For example, given a pattern $\{ab\}$, if $\exists X \in \mathcal{LP}$ such that $\{ab\} \subseteq X$ but $\not\exists X' \in \mathcal{LP}$ such that $X' \subseteq \{ab\}$, then the pattern $\{ab\}$ can be either frequent or not frequent. The border $\mathcal{FB}$ helps in this case to identify whether $\{ab\}$ is frequent or not.

2.2 Lattice Traversing and Related Works

Itemset mining is a process of itemset lattice traversing. Frequent itemset mining algorithms traverse the lattice bottom-up, i.e., starting from empty itemset. Thus, they must waste time on accessing frequent patterns before extracting less frequent itemsets. Similarly, frequent closed itemset mining algorithms also suffer the same problem when the user only want low support patterns.

Infrequent itemset mining algorithms [5,7,16] are proposed to mine patterns with support smaller than a given threshold. However, they still utilize the bottom-up traversing strategy. Rarity [17] algorithm uses the top-down traversing strategy which extracts low support patterns first. It is an apriori-like approach so that an expensive candidate generation step is necessary. These early-stage algorithms are indeed slower than well optimized frequent itemset mining approaches.

RP'Tree [18] suggests that there are three types of patterns: frequent patterns; infrequent patterns with infrequent items; infrequent patterns without infrequent items. It is designed only to return the second type of patterns. A negative item tree is proposed in [11] mine infrequent patterns top-down. We adopt this tree structure to mine closed patterns.

A bi-directional traversing framework is proposed in [12] to extract closed low support patterns by separating the dataset into a sparse and a dense part. Bottom-up and top-down traversing strategies are applied respectively. However, its top-down traversing part is slow due to duplication problems, which limits the performance of this framework. In this work, we make use of this framework to achieve better memory performance.

Some algorithms extracting descriptive patterns based on information theory [15] or top-k patterns of each item [10]. In theory, they could also return some low support patterns. However, the majority are left behind.

2.3 Support Counting on Negative Itemset Tree

The ni-tree [11] is initially proposed to mine all infrequent patterns. It stores support information of *negative represented (neg-rep) itemsets*. Neg-rep itemsets are itemsets represented by symbol of items that do not exist in the original itemsets. For example, given $\mathcal{I} = \{a, b, c, d, e, f\}$, an itemset $X = \{a, b, c\}$ can also be represented using the symbol of items not in X, denoted as $\overline{X} = \{d, e, f\}$.

Obviously, X and $\overline{X}$ represent the same information since $X \subseteq T \Leftrightarrow \overline{X} \supseteq \overline{T}$. Let $\overline{T}$ be the negative dataset formed by neg-rep itemsets, the support of $\overline{X}$ can be defined as the number of neg-rep transactions in $\overline{T}$ that covered by $\overline{X}$ such that:

$$|\overline{T}(\overline{X})| := |\{\overline{T}|\overline{T} \in \overline{T}, \overline{T} \subseteq \overline{X}\}| \Leftrightarrow |T(X)| = |\overline{T}(\overline{X})| \tag{1}$$

Tid	Itemset
1	b c
2	d e
3	a e
4	c d e
5	b d e
6	a d e

(a)

Tid	Negative Itemset
1	a d e
2	a b c
3	b c d
4	a b
5	a c
6	b c

(b)

Fig. 1. Transaction dataset and its negative dataset.

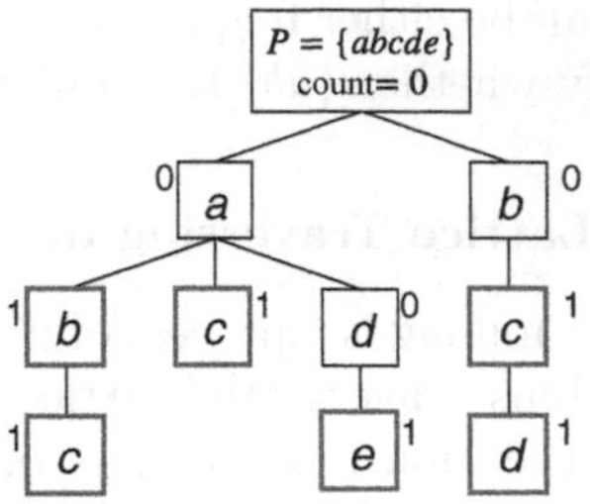

Fig. 2. The initial ni-tree of dataset in Fig. 1. Red nodes are *t-nodes*. (Color figure online)

The ni-tree is a prefix tree, as shown in Fig. 2. Items are sorted in ascending order concerning their frequency $\mathcal{T}$. Each node n is a triplet $\langle i, c, l \rangle$, where i and c are the item label and its count, l is the list of child nodes. c is initialized to 0. The root node $r = \langle P, c, l \rangle$ stores the current pattern P.

Each transaction $T \in \mathcal{T}$ is converted to $\overline{T}$ and inserted to the ni-tree. The last node, known as the *termination node* or *t-node* for short, will increase its count by 1. Thus, the count of a node n is the number of its corresponding transactions, i.e. $n.c = |\{\overline{T} \in \overline{T}, \overline{T} = n.L\}|$, where $n.L$ is the set of items on the path from root to n. According to Eq. 1, $|\mathcal{T}(X)|$ can be computed by aggregating all nodes whose path from the root is fully covered by $\overline{X}$:

$$|\mathcal{T}(X)| = |\overline{T}(\overline{X})| = \sum_{n.L \subseteq \overline{X}} n.c \tag{2}$$

For example, to identify the support of itemset $X = \{de\}$, the count of nodes on paths that covered by $\overline{X} = \mathcal{I} \setminus X = \{abc\}$ are aggregated, which equals to 4. Therefore, the ni-tree can be used to compute the support of a given pattern.

Moreover, given patterns X and X', $X \subset X' \Leftarrow \overline{X} \supset \overline{X'}$, the set of nodes for computing $|\mathcal{T}(X)|$ can be decomposed as: $\{n|n.L \subseteq \overline{X}\} = \{n|n.L \subseteq \overline{X'}\} \cup \{n|n.L \subseteq \overline{X}, n.L \nsubseteq \overline{X'}\}$. Thus, the aggregating process can be decomposed and computed recursively. For example, let $X = \{de\}, X' = \{bcde\}$, the support of X can be obtained by removing nodes on the path covered by $\overline{X'} = \{a\}$, which leads to a new ni-tree that represents the pattern $\{bcde\}$, as shown in Fig. 3. Then, removing nodes covered by $\overline{X}$ from the second ni-tree will generate the pattern $\{de\}$. Such process is in top-down style. In practice, we only need to create a new root node rather than a brand new ni-tree.

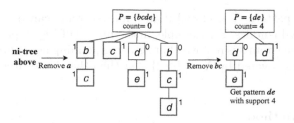

Fig. 3. Counting support on ni-tree.

3   Closed Itemset on Negative Itemset Tree

Though the ni-tree is not initially designed for closed pattern mining, we found
that the closeness can be determined readily by using t-nodes.

3.1   Closed Itemset Determination

According to the definition, an essential property of a closed pattern X is that
its support must be different from its supersets. In ni-tree, the count value of
any node, except t-nodes, is 0. Thus, if the count of all removed nodes is 0, the
generated pattern is not closed. A closed pattern can only be achieved if at least
one t-node is involved in the aggregating process. Formally speaking:

Theorem 1. *Given $\mathcal{I}$ and the initial ni-tree, let N_X be the set of nodes been
removed from the initial ni-tree to achieve the pattern X. Let $N_X^t \subseteq N_X$ be the
set of t-nodes been removed. Then pattern X is closed if and only if the set of
items been removed $(\overline{X})$ equals to the set of items on paths to t-nodes:*

$$\mathcal{I} \setminus X = \overline{X} = \bigcup_{n \in N_X^t} n.L \tag{3}$$

Proof. Obviously, $N_X = \{n | n.L \subseteq \overline{X}\}, N_X \supseteq N_X^t$. Thus,

$$\overline{X} = \bigcup_{n \in N_X} n.L \supseteq \bigcup_{n \in N_X^t} n.L \tag{4}$$

As non-terminated nodes are counted at 0 in the ni-tree, the support of pattern
X is the sum of all t-nodes:

$$|\mathcal{T}(X)| = |\overline{\mathcal{T}}(\overline{X})| = \sum_{n.L \subseteq \overline{X}} n.c = \sum_{n \in N_X^t} n.c \tag{5}$$

Let $M = \overline{X} \setminus (\bigcup_{n \in N_X^t} n.L)$. Thus, any node $n' \in N_X$ with item $n'.i \in M$ is not
on the path to a t-node in N_X^t. Removing such nodes or not won't affect the
support value, i.e. $|\mathcal{T}(X)| = |\mathcal{T}(X \cup M)|$. By closeness definition, $M = \emptyset \Leftrightarrow X$
is closed. $\square$

In short, a pattern X is closed if all removed items can be found on paths towards removed t-nodes. For example, in the ni-tree of Fig. 2, pattern $X = \{be\}$ is not closed since item d is removed but its corresponding nodes are not on a path towards t-nodes covered by $\overline{X} = \{acd\}$. On the other hand, itemset $X = \{de\}$ is closed.

3.2   Naïve Method

According to Theorem 1, top-down closed pattern mining can be realized by simply enumerating and removing all combinations of paths towards t-nodes. The ni-tree is slightly adapted. The root node and each t-node stores a list of pointers (l_t) linked to their child t-nodes, as shown in Fig. 4. In each step, nodes on the path from one t-node (excluding) to its child t-node (including) are removed together, which guarantees that only closed patterns are generated. Figure 4 illustrates an example. By removing all nodes on the path to t-node 1, a new ni-tree is generated, and the corresponding closed pattern $\{de\}$ is returned. Then removing t-node 2 in the new ni-tree lead to another closed pattern $\{e\}$. Enumerating all removing combinations generate all closed patterns.

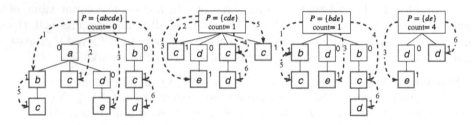

Fig. 4. The adapted initial ni-tree with t-node links (blue) and corresponding ni-tree by removing t-node 1, 2 or 1, 2 together. Each link is marked with the t-node id. In each step, only child t-nodes of root are considered (e.g. node 6 can only be removed after node 4). (Color figure online)

The main disadvantage of this naive approach is the duplicate accessing problem. For example, given $\mathcal{I} = \{abcde\}$, the pattern $X = \{ab\}$ can be achieved by either removing $\{cd\}$ and $\{ce\}$ or removing $\{cd\}$ and $\{de\}$. A pattern X might be accessed repeatedly up to $O(2^{|\mathcal{T}(X)|})$ times. Extra duplicate checking and pruning step are necessary. By examining patterns discovered so far, we can avoid the majority of duplicates. However, the overhead of the pruning step plus remaining duplicates are still time-consuming. Indeed, this naïve method is the top-down part used in the bi-directional traversing framework [12].

4   Algorithm: LSCMiner

4.1   Divide-and-Conquer Paradigm

The naive approach described above is a top-down based algorithm. However, it is not efficient due to the expensive duplicate accessing problem. To take advantage of top-down traversing, we propose our **L**ow **S**upport **C**losed Miner (LSCMiner), which employs the depth-first traversing strategy with novel closeness checking and pruning steps. The general mining process employed a divide-and-conquer paradigm, which is commonly used in bottom-up based algorithms. The main difference is that we remove items recursively, rather than grow patterns.

First of all, let operators $\prec$ and $\succ$ denote the concept of *"before (smaller)"* and *"after (larger)"* with respect to the ascending frequency order used by the ni-tree. Given $\mathcal{I} = \{a \prec b \prec c \prec \ldots\}$, the top-down mining process removes items recursively, which can be represented as a tree as shown in Fig. 5. We call the tree above as the *deletion tree*. Each node in the tree is the set of items to be removed, known as the *deletion set*. Given a node in the deletion tree, we say that deletion sets in its sub-tree and right to it are *under* or *after* the deletion set in the node, as shown in Fig. 5.

The first challenge is to combine the closeness checking process with the divide-and-conquer paradigm. According to Theorem 1, we need to check if every removed item can be found on paths to removed t-nodes. To solve the problem, we let each t-node n^t contains a list $n^t.L$, which stores items on the path from itself (including) to its proceeding t-node (excluding), as shown in Fig. 6. During the removing process, a set U is maintained to track items that are not covered by paths towards removed t-nodes yet. In one recursive step, we first add the current item to U. If a t-node n^t is removed, all items exist in $n^t.L$ are removed from U. When $U = \emptyset$, we knew that the current pattern is closed.

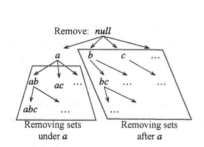

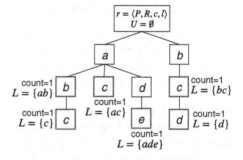

Fig. 5. We solve the mining task by removing items in a recursive way. Such process can be represented as a tree.

Fig. 6. The adapted ni-tree used in LSCMiner. R is the set of items been removed so far.

Figure 7 gives an example of the closed pattern mining process. We first remove item a, no t-node is removed right now. Thus, $U = \{a\}$ and the pattern $\{bcde\}$ is not closed. Then, we recursively remove b from the current ni-tree. There is a t-node of b is removed and $U = \{ab\} \setminus \{ab\} = \emptyset$. Pattern $\{cde\}$ is closed and should be added to the result set.

Algorithm 1 illustrates the pseudo code of the LSCMiner. Each iteration step removes one item (Line 5). If there are t-nodes in removed nodes list l_i, we remove

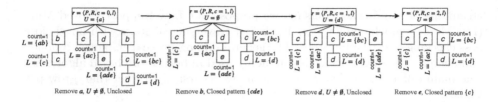

Fig. 7. Recursive steps of the removing process $a \rightarrow ab \rightarrow abd \rightarrow abde$.

items from U and aggregate counts (Line 10–13). If the current candidate pattern is not frequent, we attach nodes with larger item to the new ni-tree root for the next recursive call (Line 18). The recursive mining process is continued until the aggregated count is larger than the given maximum threshold β. Variables iM_1 and iM_2 are pruning thresholds as described later.

4.2 Pruning

An efficient algorithm should be able to prune unclosed itemsets as early as possible, known as the *"look ahead"* ability [14,19]. In our LSCMiner, we fully utilized the closeness property of the ni-tree. Two types of pruning methods are utilized.

Trial-and-Error Pruning. Our first pruning method (Line 27, Algorithm 1) is based on the following observation:

Theorem 2. *Given the current ni-tree root r and the current unclosed items set $U \neq \emptyset$. Let R be the set of items been removed so far. If $\exists i \in r.l$ such that no closed pattern in deletion sets under $\{R \cup i\}$, then there is also no closed pattern in deletion sets after $\{R \cup i\}$.*

Proof. The sub-ni-tree under item i must contain at least one t-node. Let l be the set of items from i (including) to a t-node n^t (including) in its sub-ni-tree. Obviously, we have $n^t.L \supseteq l$ and $i \in n^t.L$.

Let deletion sets after i be $R_{\succ i}$. Assuming the deletion set of i and deletion sets under i are not closed. If $\exists p \in R_{\succ i}$ which will lead to a closed pattern, then removing all items in $\{p \cup l\}$ will also lead to a closed pattern (by further removing the node n^t mentioned above). Obviously, $\{p \cup l\}$ is a deletion set of i or under i, which is contradict to our assumption (Fig. 8). □

In short, if removing i does not generate a closed pattern, the recursion call will be executed (Line 21, Algorithm 1). This recursive call will try all possible combinations of items with respect to i. If no closed pattern is generated, iterations on items (Line 5, Algorithm 1) after i can be canceled.

Input: Ni-tree root r, Minimum support α, Maximum support β
Output: Infrequent Itemset List $\mathcal{LP}$, Frequent Border List $\mathcal{FB}$

```
 1  LP ← ∅, FB ← ∅;
 2  LSCMiner(r,∅,+∞, −∞) ;
 3  return LP, FB ;
 4  Function LSCMiner(r, U, iM₁, iM₂)
 5  │   foreach Item i ∈ r.l ∧ i ⪯ iM₁ do
 6  │   │   lᵢ ←List of nodes in r.l with label i ;
 7  │   │   U' ← U ∪ {i}, P' ← r.P \ {i}, c' ← r.c;
    │   │   /* Closeness checking                              */
 8  │   │   foreach Termination node n ∈ lᵢ do
 9  │   │   │   U' ← U' \ n.is, c' ← c' + n.c ;
10  │   │   end
11  │   │   if c' < β then
12  │   │   │   if c' ≥ α ∧ U' = ∅ then
13  │   │   │   │   Add P' to LP
14  │   │   │   end
15  │   │   │   l' ← {n' ∈ r.l|n'.i ≻ i} ∪ {⋃_{n∈lᵢ} n.l}, r' ← {P',c',l'};
    │   │   │   /* Initial new end index                       */
16  │   │   │   if P' is closed then
17  │   │   │   │   iM₁', iM₂' ← +∞, −∞;
18  │   │   │   else
19  │   │   │   │   iM₁', iM₂' ←UpperBound(l, iM₂);
20  │   │   │   end
21  │   │   │   LCSMiner(r', U', iM₁', iM₂') ;
22  │   │   │   if No closed pattern generated in the recursive call above then
23  │   │   │   │   Break;                        // Trial-and-Error Pruning
24  │   │   │   end
25  │   │   else
26  │   │   │   if U' = ∅ then
27  │   │   │   │   Add P' to FB;
28  │   │   │   end
29  │   │   end
30  │   end
31  end
```

Algorithm 1: LSCMiner

Upper-Bound Pruning. The second pruning technique computes the largest possible item as an upper bound for the next recursion step. Given the current removed item i and the list of nodes to be removed l_i, assuming all nodes in l_i are not terminated, then the largest possible item iM_1' that can be removed in the next recursion step is the largest item among all children of nodes in l_i.

The reason is straightforward: the item i will be covered by a t-node if and only if at least one of its child is removed. If we remove an item $i' \succ iM_1'$ in the next recursion step, all items to be removed in the future are also larger than iM_1'. Thus, it is impossible to reach a t-node that covers i. For example, given the left ni-tree in Fig. 9, assuming now we are removing item a, which results

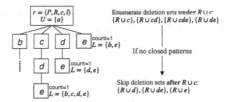

Fig. 8. Given the current unclosed ni-tree, Trial-and-Error pruning will try all deletion sets under $R \cup c$. If no closed patterns exists, then later deletion sets can be skipped.

in the right ni-tree in Fig. 9. However, t-nodes that cover a only exist in the sub-ni-tree of a. Thus, the upper bound for item removing on the second ni-tree is b. Further removing process on items after b is pruned since the item a will never be covered.

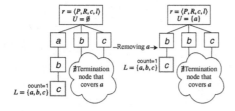

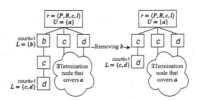

Fig. 9. The maximum item under nodes of a is $iM_1' = b$. Later removing process on the right ni-tree must removing b first since otherwise, t-nodes that cover a will not be removed.

Fig. 10. There is a t-node that covers b. Then the maximum upper bound up to now (iM_2'), which equals to the upper bound when removing a, is used as the upper bound for further removing process on the right ni-tree.

The above upper bound assumes that $\nexists n \in l_i$, i.e., item i can only be covered by children of nodes in l_i. However, if one node of item i is terminated, then i is covered by a node of itself. Removing items larger than the upper bound iM_1' can still lead to closed patterns. Another weaker upper bound iM_2' is introduced for this case, which is defined as the largest upper bound, except for infinity, among all previous recursion steps. For example, assuming the left ni-tree in Fig. 10 is achieved by removing item a, and the right ni-tree is achieved by further removing item b. Since node b is terminated, removing items larger than its children is valid. However, the previously removed item a needs to be covered so that the upper bound iM_2' is the upper bound when removing a. Algorithm 2 computes both upper bounds described here.

4.3 Complexity

Pattern mining is an NP-hard problem. The overall runtime is highly dependent on the number of desired patterns. For instance, one of the most efficient frequent

```
 1  Function UpperBound(l_i, iM_2)
 2  │   iM'_1 ← -∞
 3  │   foreach n ∈ l_i ∧ n is not t-node do
 4  │   │   x_last ←Last item in n.l
 5  │   │   if iM'_1 ≺ x_last then
 6  │   │   │   iM'_1 ← x_last
 7  │   │   end
 8  │   end
 9  │   if iM'_1 ≠ -∞ then
10  │   │   iM'_2 ← max(iM'_1, iM_2)
11  │   end
12  │   if ∃n ∈ l_i, n is terminated then
13  │   │   iM'_1 ← iM'_2
14  │   end
15  │   return iM'_1, iM'_2
16  end
```

Algorithm 2: Compute the new upper bound.

closed pattern mining algorithms, LCM [19], declares that it extracts each closed pattern in polynomial time: $O(P(|\mathcal{T}|))$. Let $\mathcal{U}$ and $\mathcal{UC}$ be the set of desired and undesired patterns, the time complexity per itemset of the LCM algorithm can be written as: $O(\frac{|\mathcal{U}|+|\mathcal{UC}|}{|\mathcal{U}|}P(|\mathcal{T}|))$, where $\mathcal{UC}$ contains frequent patterns in the low support closed pattern mining scenario.

Our approach can also achieve the same level of complexity. Given the current ni-tree root r and the current unclosed items set U, removing item i from the child list $r.l$ involves the following steps:

1. aggregate counts in removed nodes, which requires $O(|l_i|)$ time, where l_i is the list of nodes in $r.l$ labeled with i.
2. closeness checking if t-nodes exist, which requires $O(|U|\log(|n^t.L|))$ time, where $n^t.L$ is the set of items in a t-node and binary search is employed
3. add children of nodes in l_i to the new root node r', which takes $O(\sum_{n\in l_i}|n.l|)$ time.
4. add all nodes in $r.l$ with label larger than i to the new root node r', which requires $O(|l_{\succ i}|)$ time, where $l_{\succ i}$ is the list of nodes.

The total complexity is $O(|l_i|+|U|\log|n^t.L|+\sum_{n\in l_i}|n.l|+|l_{\succ i}|)$. $|U|$ and $|n^t.L|$ are limited to the size of a single transaction so that the second term can be seen as a constant. The length of $l_i, l_{\succ i}$ and $n.l$ are limited to the size of the dataset. Thus, the complexity of removing item i is polynomial. A closed itemset X is achieved by removing items in $\overline{X}$. The complexity to extract X is $O(\sum_{i\in\overline{X}}P(|\mathcal{T}|)) \in O(P(|\mathcal{T}|))$ since $|\overline{X}|$ is small compared to $|\mathcal{T}|$. Considering that our approach also accesses some unclosed patterns, its complexity per itemset is also $O(\frac{|\mathcal{U}|+|\mathcal{UC}|}{|\mathcal{U}|}P(|\mathcal{T}|))$, where $\mathcal{UC}$ are those unclosed patterns.

In terms of memory complexity, it is obvious that our approach is limited by the size of the dataset, similar to algorithms such as FPGrowth [6].

However, LSCMiner has to store the negative dataset such that a scale factor $s = \frac{|\mathcal{I}|}{\text{avg. transaction length}}$ exists, known as the sparsity of the dataset. s can be huge on sparse dataset. In this case, the bi-directional traversing framework proposed in [12] can be used so that our LSCMiner only need to handle the densest part of a dataset.

5   Experiments

We first conduct the runtime performance of our LSCMiner. In experiments, the naive approach described in Sect. 3.2 represents the performance of a simple top-down based algorithm. The LCM [19] algorithm represents the most efficient bottom-up based algorithm in solving low support pattern mining problem. We also conduct the bi-directional traversing framework [12] by combining the LCM algorithm with our LSCMiner. Other infrequent pattern mining algorithms are not included since they are either represented by LCM (bottom-up based) or too memory expensive to finish (apriori alike).

Database	Size (N)	Items $(\mathcal{I})$	Avg. length (L)
mushrooms	8k	119	23
chess	3k	75	37
connect	67k	129	43
accident	340k	468	33.8
kddcup99	1000k	135	16
BMS1	59k	497	2.5

Fig. 11. Real-life datasets in our experiments.

Algorithms are implemented using Java. The LCM implementation comes from [4]. 6 real-world datasets obtained from the fimi repository (http://fimi. ua.ac.be/data/) are used as our test datasets. Necessary information of those datasets are listed in Fig. 11. There are three small dense datasets, two large dense datasets, and one sparse dataset. First N transactions and first L items in each transaction are used in our experiments. We are interested in the time difference in accessing patterns on a certain level of support from different directions. Thus, we set $\alpha = \beta - 10$. The default value of N, L and β are provided in each experiment. The splitting threshold for the bi-directional framework is set to $\delta = 1\%$.

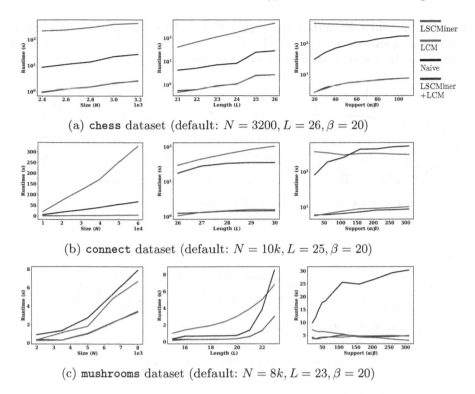

(a) `chess` dataset (default: $N = 3200, L = 26, \beta = 20$)

(b) `connect` dataset (default: $N = 10k, L = 25, \beta = 20$)

(c) `mushrooms` dataset (default: $N = 8k, L = 23, \beta = 20$)

Fig. 12. Runtime on small dense dataset.

Dense Data. Figure 12 illustrates the performance on small dense dataset. On these datasets, the bottom-up algorithm, LCM, is up to two order of magnitude slower than our top-down LSCMiner on the first two datasets. It is even slower than the naive approach under some settings. This is mainly because the bottom-up LCM algorithm has to traverse all frequent patterns. When β increased, i.e., we become more interested in frequent patterns, the runtime of LCM is reduced and may surpass top-down approaches since it needs to traverse less frequent patterns. On the `mushrooms` dataset, top-down approach is slower with $\beta > 150$. This is mainly because that the `mushrooms` dataset has less number of patterns. Our LSCMiner is very efficient. Its runtime grows similar to the LCM approach with increasing dataset size, which indicates that the time complexity of both approaches is on the same level. The combined approach is also efficient under most settings. Our LSCMiner under the bi-directional framework only need to handle the densest part of the dataset, which reduces the memory consumption, as discussed in Sect. 4.3. However, the slowness of the bottom-up part under some settings drag down its performance. The performance gap between top-down and bottom-up approaches is further enlarged on large dense datasets, as shown in Fig. 13. Both `accidents` and `kddcup99` datasets have larger size and longer transactions. The LCM algorithm is up to 3 order of magnitude slower than our LSCMiner. Even the naive approach is better under most cases. Though increasing β slows down our LSCMiner, it is still hard for LCM algorithm to overtake in the low support pattern mining scenario.

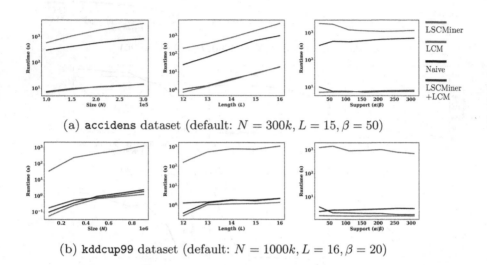

(a) `accidens` dataset (default: $N = 300k, L = 15, \beta = 50$)

(b) `kddcup99` dataset (default: $N = 1000k, L = 16, \beta = 20$)

Fig. 13. Runtime on large dense datasets.

Sparse Data. In theory, bottom-up algorithms should perform better than our top-down approach on a sparse dataset. According to our analysis above, both LCM and our LSCMiner have a complexity of $O(\frac{|\mathcal{C}|+|\mathcal{UC}|}{|\mathcal{C}|}P(|\mathcal{T}|))$. In the case of sparse datasets, $|\mathcal{UC}|$ in LCM is much smaller since the number of frequent patterns is tiny. On the other hand, a sparse dataset leads to a huge ni-tree, which is a substantial overhead for LSCMiner. Figure 14 illustrates the runtime performance on a sparse dataset BMS2. The results fulfill our expectations. The combined approach performs the best since we can take advantage of both bottom-up and top-down algorithms.

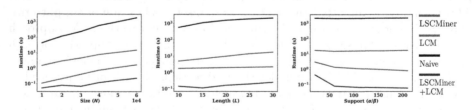

Fig. 14. Runtime on BMS1 dataset (default: $N = 60k, L = 30, \beta = 20$)

Memory-Performance Trade-Off. As discussed above, the performance of our LSCMiner is limited by the dataset size. An extra constant scale factor s exists since the ni-tree stores the negative dataset. Sparsity is a crucial factor that affects the memory consumption of LSCMiner. By applying the bidirectional traversing framework and adjusting the value of splitting threshold δ, the LSCMiner only need to traverse the densest part of the dataset, which costs much fewer memories. We can still benefit from the efficient top-down traversing since top-down traversing is powerful on dense dataset while bottom-up traversing is good at the sparse dataset, as shown in experiments above. In this section, we study the relation between memory consumption and runtime performance by investigating how the trade-off behaviors with respect to different (relative) dividing threshold δ.

Two dense datasets, **chess** and **connect**, are selected as representatives since they have both sparse and dense parts. We measure the memory consumption using the first 1k transactions in each dataset. The runtime value for **connect** dataset is measured with the first 10k transactions instead. When the value of δ is close to 0, all patterns are extracted by the LSCMiner. When the value of δ is close to the relative support of the most frequent item, the bottom-up approach extracts all patterns. Thus, by increasing δ, the bi-directional traversing is moving from purely LSCMiner to purely LCM algorithm (Fig. 15).

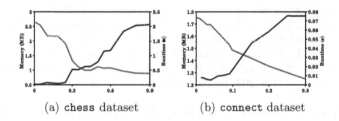

(a) **chess** dataset   (b) **connect** dataset

Fig. 15. Memory consumption and runtime under different δ values. δ is set up to 0.4 on **connect** dataset since almost all items occurred less than 40%.

On the **chess** dataset, the runtime of the bi-directional framework increased about 20 times while the memory consumption decreased about 2.5 times when moving from LSCMiner to LCM approach. On the **connect** dataset, the runtime increased about 7 times while the memory consumption decreased about 30%. LSCMiner is beneficial on both datasets: we spend some memory but get much better performance.

6   Conclusion

We present a very efficient low support closed pattern mining algorithm, LSCMiner, which avoids traversing undesired frequent patterns. It is particularly effective on datasets with huge amounts of frequent patterns. Though it is memory expensive to store the ni-tree, much better runtime performance is achieved in return. Furthermore, we can balance the memory consumption and runtime performance by using the bi-directional traversing framework. If only those dense datasets are considered, our LSCMiner guarantees to provide the best performance in time complexity.

References

1. Adda, M., Wu, L., Feng, Y.: Rare itemset mining. In: Sixth International Conference on Machine Learning and Applications, ICMLA 2007, pp. 73–80. IEEE (2007)
2. Burdick, D., Calimlim, M., Gehrke, J.: Mafia: a maximal frequent itemset algorithm for transactional databases. In: Proceedings of the 17th International Conference on Data Engineering, pp. 443–452. IEEE (2001)
3. Fang, G., Pandey, G., Wang, W., Gupta, M., Steinbach, M., Kumar, V.: Mining low-support discriminative patterns from dense and high-dimensional data. IEEE Trans. Knowl. Data Eng. **24**(2), 279–294 (2012)
4. Fournier-Viger, P., et al.: The SPMF open-source data mining library version 2. In: Berendt, B., et al. (eds.) ECML PKDD 2016. LNCS (LNAI), vol. 9853, pp. 36–40. Springer, Cham (2016). https://doi.org/10.1007/978-3-319-46131-1_8
5. Gupta, A., Mittal, A., Bhattacharya, A.: Minimally infrequent itemset mining using pattern-growth paradigm and residual trees. In: Proceedings of the 17th International Conference on Management of Data, p. 13 (2011)
6. Han, J., Pei, J., Yin, Y.: Mining frequent patterns without candidate generation. In: ACM Sigmod Record, vol. 29, pp. 1–12. ACM (2000)
7. Hoque, N., Nath, B., Bhattacharyya, D.: An efficient approach on rare association rule mining. In: Bansal, J., Singh, P., Deep, K., Pant, M., Nagar, A. (eds.) BIC-TA 2012, vol. 201, pp. 193–203. Springer, India (2013). https://doi.org/10.1007/978-81-322-1038-2_17
8. Kamehkhosh, I., Jannach, D., Ludewig, M.: A comparison of frequent pattern techniques and a deep learning method for session-based recommendation. In: RecTemp@ RecSys, pp. 50–56 (2017)
9. Koh, Y.S., Ravana, S.D.: Unsupervised rare pattern mining: a survey. ACM Trans. Knowl. Discov. Data (TKDD) **10**(4), 45 (2016)
10. Leroy, V., Kirchgessner, M., Termier, A., Amer-Yahia, S.: TopPi: an efficient algorithm for item-centric mining. Inf. Syst. **64**, 104–118 (2017)
11. Lu, Y., Richter, F., Seidl, T.: Efficient infrequent itemset mining using depth-first and top-down lattice traversal. In: Pei, J., Manolopoulos, Y., Sadiq, S., Li, J. (eds.) DASFAA 2018. LNCS, vol. 10827, pp. 908–915. Springer, Cham (2018). https://doi.org/10.1007/978-3-319-91452-7_58
12. Lu, Y., Seidl, T.: Towards efficient closed infrequent itemset mining using bi-directional traversing. In: DSAA 2018, pp. 140–149. IEEE (2018)
13. Mannhardt, F., De Leoni, M., Reijers, H.A., Van Der Aalst, W.M.: Balanced multi-perspective checking of process conformance. Computing **98**(4), 407–437 (2016)
14. Pasquier, N., Bastide, Y., Taouil, R., Lakhal, L.: Efficient mining of association rules using closed itemset lattices. Inf. Syst. **24**(1), 25–46 (1999)
15. Smets, K., Vreeken, J.: Slim: directly mining descriptive patterns. In: Proceedings of SIAM International Conference on Data Mining, pp. 236–247. SIAM (2012)

16. Szathmary, L., Napoli, A., Valtchev, P.: Towards rare itemset mining. In: 19th IEEE International Conference on Tools with Artificial Intelligence, ICTAI 2007, vol. 1, pp. 305–312. IEEE (2007)

17. Troiano, L., Scibelli, G.: A time-efficient breadth-first level-wise lattice-traversal algorithm to discover rare itemsets. Data Min. Knowl. Disc. **28**(3), 773–807 (2014)

18. Tsang, S., Koh, Y.S., Dobbie, G.: RP-tree: rare pattern tree mining. In: Cuzzocrea, A., Dayal, U. (eds.) DaWaK 2011. LNCS, vol. 6862, pp. 277–288. Springer, Heidelberg (2011). https://doi.org/10.1007/978-3-642-23544-3_21

19. Uno, T., Kiyomi, M., Arimura, H.: LCM ver. 2: Efficient mining algorithms for frequent/closed/maximal itemsets. In: Fimi, vol. 126 (2004)

Pattern Filtering Attention for Distant Supervised Relation Extraction via Online Clustering

Min Peng[1]([✉]), Qingwen Liao[1], Weilong Hu[1], Gang Tian[1], Hua Wang[2],
and YanChun Zhang[2]

[1] School of Computer Science, Wuhan University, Wuhan, China
{pengm,qingwen.liao,huweilong,tiang2008}@whu.edu.cn
[2] Centre for Applied Informatics, Victoria University, Melbourne, Australia
{hua.wang,yanchun.zhang}@vu.edu.au

Abstract. Distant supervised relation extraction has been widely used to extract relational facts in large-scale corpus but inevitably suffers from the wrong label problem. Many methods use attention mechanisms to address this issue. However, the attention weights in these models are not discriminative and precise enough to fully filter out noise. In this paper, we propose a novel Pattern Filtering Attention (PFA), which can filter noise effectively. Firstly, we adopt an online clustering algorithm on the instances labeled with the same relation to extract potential semantic centers (positive patterns) of each relation, and these patterns have less noise statistically. Then, we build a sentence-level attention based on the similarities of instances and positive patterns. Due to the large differences between these similarities, our model can assign more discriminative weights to instances to reduce the influence of noisy data. Experimental results on the New York Times (NYT) dataset show that our model can effectively improve the performance of relation extraction compared with state-of-the-art methods.

Keywords: Relation extraction · Distant supervision · Deep learning · Knowledge graph

1 Introduction

Relation extraction (RE) is a very important subtask in natural language processing (NLP). It aims to extract structured relational data that are generally in the form of triplet $\langle h, r, t \rangle$ from unstructured text. Previous RE models focus mostly on supervised classification and require a large number of labeled training data, which is labor-intensive for open domains. Thus, distant supervision [15]

This material is supported partially by National Key R&D Program of China under Grant No. 2018YFC1604000 and No. 2018YFC1604003, partially by National Science Foundation of China (NSFC) under Grant No. 61872272 and No. 61772382.

R. Cheng et al. (Eds.): WISE 2019, LNCS 11881, pp. 310–325, 2019.
https://doi.org/10.1007/978-3-030-34223-4_20

heuristically aligns text with knowledge bases (KBs) to automatically generate annotations. It assumes that a sentence related to an entity pair may express the same relation as that the entity pair expresses in KBs. Although distant supervision is effective in labeling training data, it suffers from the *wrong label problem*, which introduces noise into generated instances [19]. As shown in Table 1, S1 and S2 both mention entity pair *Donald Trump* and *New York*, so they are considered as positive instances for relation *born_in*. However, it is obvious that sentence S1 is a true positive example, while S2 is a false positive example.

Table 1. An example of instances for the triplet ⟨*Donald Trump, born_in, New York*⟩

KB fact	Relation instances
⟨*Donald Trump, born_in, New York*⟩	**S1**: Donald Trump was born in 1946 in New York
	S2: President Donald Trump has countless ties to New York
	...

In recent years, most methods concentrate on the combination of neural networks and multi-instances learning to overcome above weakness [9,11,22,23,26]. These methods treat a set of instances related to a specific entity pair as a bag and attempt to select valid instances in a bag to reduce noise in training data. Specifically, some methods only utilize one sentence that is most likely to be valid based on the *at-least-one* assumption [22], ignoring the rich information contained in neglected instances. Therefore, various sentence-level selective attention (ATT) mechanisms are proposed to automatically learn weights over multiple instances [2,6,11,26], and these strategies are regarded as soft selection strategies. However, the attention weights in these models are **not discriminative and not precise** enough to fully filter out noise. We prove this experimentally, and Fig. 1 shows two examples of attention weights learned by PCNN+ATT [11] models. Instead of the soft selection, some methods use reinforcement learning to redistribute false positive instances into the negative set [4], which deal with the noise data effectively. However, these models are difficult to train and reproduce, so their performances are not stable.

Different from these methods, we propose a novel Pattern Filtering Attention to overcome the above problems. Our model can learn more discriminative and more precise attention weights, thus greatly reducing the impact of noise, and the performance of our model is stable. As illustrated in Fig. 2, our method firstly extracts the features of sentences via CNN. Considering the constraints between entity types and relation, we introduce entity type to extract better instance features in this process. Then we explore positive patterns for each relation, and we believe that the positive patterns can be extracted from the instances through clustering because the true positive instances in the data set still account for a large proportion statistically [18]. Since the clustering process makes full

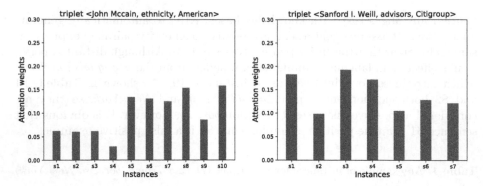

Fig. 1. Two examples of the attention weights learned by PCNN+ATT. We use si instead of each instance in a bag. Obviously, in these two figures, the attention weights are not discriminative enough between the valid and invalid instances.

use of the distribution information of instances, the positive patterns have less noise statistically. After that, we build a sentence-level attention over multiple instances based on the similarities between instances and positive patterns. In our model, the similarities of the valid instances and positive patterns are greatly different from those of the invalid instances and positive patterns, which makes the attention weight more discriminative. Finally, we generate better representations of the entity bags based on these attention weights. We evaluate our model on the New York Times dataset, and experimental results demonstrate that our model achieves competitive performance compared with baselines.

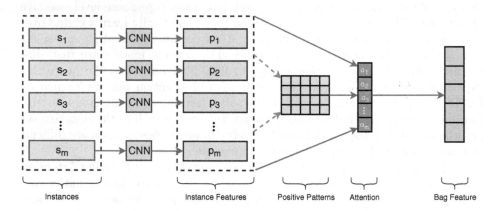

Fig. 2. The architecture of our model, where s_i indicates the original sentences. p_j indicates the sentence features encoded by CNN. α_i is the weight given by our Pattern Filtering Attention.

The contributions of this paper can be summarized as follows:

- We propose an online clustering algorithm to extract potential semantic centers (positive patterns) for each relation from labeled instances. These patterns contain the information of entity types and keep away from noise statistically.
- We combine sentence-level attention with pattern filtering, which filters noise more effectively via more discriminative and precise attention.
- Experimental results show that our model, combining sentence-level attention with pattern filtering, makes full use of valid instances and achieves competitive performance in relation extraction.

2   Related Work

Relation extraction is a fundamental task in NLP, which has been widely applied to many applications, such as question answering and information retrieval. Most previous models [5,16] require lots of labeled data to train a supervised classifier and achieve high performance. However, acquiring enough labeled data is time-consuming and labor-intensive. So Mintz et al. [15] align plain text with Freebase [1] to automatically generate training examples, while introducing massive wrong labels into training. To tackle this problem, Riedel et al. [19] adopt multi-instance learning based on the assumption that if two entities participate in a relation, at least one sentence that mentions these two entities might express that relation. Surdeanu et al. [20] expand multi-instance learning to multi-instance multi-label learning, which jointly models all the instances of a pair of entities in text and all their labels using a graphical model with latent variables. Yet these models using traditional NLP tools have weaknesses of error propagation in feature extraction, which limits the performance.

In recent years, deep neural network has become an excellent feature extraction tool and is widely used in relation extraction. Zeng et al. [22,23] use CNN and piecewise CNN to extract features of relation instances. Zhou et al. [26] combine bidirectional long short-term memory networks with two-level attention for relation classification. Moreover, Zhang et al. [25] use graph convolutional networks to pool information over dependency structures efficiently. Furthermore, Zhang et al. [24] explore the capsule network in multi-instance multi-label learning framework for relation extraction.

To alleviate the wrong label problem, many studies combine attention strategies with neural networks to select valid instances. Du et al. [2] explore a multi-level structured self-attention for better context representation and better sentence selection. Han et al. [6] propose a hierarchical attention scheme to better identify valid instances. There are other methods focus on reducing noisy labels, which is also a momentous topic in this area. Peng et al. [17] solve wrong label problem by achieving a more precise alignment through the similarity between the relation mentions. Liu et al. [12] use a soft-label method to correct wrong labels dynamically during training. Luo et al. [13] utilize dynamic transition matrix to characterize the noise and protects relation distribution from noise.

Rather than modeling the noise, other strategies try to filter out invalid instances from training data. Takamatsu et al. [21] propose a generative model that directly models the heuristic labeling process of distant supervision and predicts whether assigned labels are correct or wrong via its hidden variables. Qin et al. explore a sentence-level true-positive generator to choose true positive instances [18], and Feng et al. [4] select valid instances by reinforcement learning.

Although these methods achieve great success, they still have flaws in selecting valid instances due to lack of discriminative attention or complicated processes in reducing noise. Compared to them, we build sentence-level attention with pattern filtering to further reduce noise. Our model can assign more discriminative and precise attention based on the similarities between instances and positive patterns, thus better filtering the noise.

Table 2. Observations about the effect of entity types. The entity type of *Donald Trump* and *Obama* is *PERSON*, and the entity type of *New York* is *LOCATION*.

S_0	**Donald Trump** *was born in* **New York**
S_1	**Obama** *was born in* **New York**
S_2	**Donald Trump** *was born in* **Obama**
S_3	**Donald Trump** *is working in* **New York**

3   Methodology

In this section, we present the main parts of our model including instance feature extraction, positive patterns extraction and sentence-level attention combined with pattern filtering. In the training process, the positive patterns are adjusted by our online clustering algorithm and will be fine-tuned by the back propagation of the network. Using the semantic positive patterns extracted by clustering on instance features, our model builds more discriminative and precise attention over instances, which reduces the influence of noise effectively.

3.1   Instance Feature Extraction

It has been proved that neural networks show effectiveness in extracting features [23]. Hence, we adopt CNN as an instance feature encoder for relation instances to extract relational features. We firstly discuss the features of instances utilized for relation extraction and then describe how to extract these features.

The Effect of Entity Types. Most neural models only exploit the word-level information of entities, while ignoring the effect of entity types. For instance, as shown in Table 2, S_0 is the original sentence and S_1, S_2, S_3 are three variants. Due to the same entity types as S_0, sentence S_1 is possible to express the relation *born_in*. However, S_2 may express another relation because it has different entity types compared to that of S_0. Nevertheless, relations between entities can't be simply derived from entity types. S_3 doesn't express the *born_in* relation, although containing the same entity pair compared with S_0. In other words, there may exist relevance between entity types and entity relations. Therefore, our instance feature encoder is designed to extract instance features including the information of entity types.

Instance Feature Encoder. Given a sentence $s = \{w_1, w_2, \cdots, w_{|s|}\}$ where w_i is a word, the encoder generates the feature vector **p** of the sentence. The instance feature encoder mainly consists of vector representation, convolution layer and max-pooling layer.

Each word w_i is mapped to a low-dimensional vector $\mathbf{x}_i$ consisting of the pre-trained embeddings of the word, position embeddings and type embedding. We employ the method proposed by [14] to train a matrix $V \in \mathbb{R}^{d_w \times |V|}$, where V is the vocabulary. Similar to [23], we adopt two position embeddings to emphasize the importance of w for predicting relation. They are defined as the combination of the relative distances from w to head or tail entities. To utilize the information of entity types, we append a type embedding that indicates the entity type w belongs to. Figure 3 shows an example of relative distances and entity types for sentence '*Donald Trump was born in New York.*'. Here coarse-grain entity types including *LOCATION, PERSON, ORGANIZATION, OTHER* are considered. Assuming the dimensions of the three embeddings are d_w, d_p, d_t, then the dimension of the concatenated input word **x** is $d = d_w + 2d_p + d_t$. A sentence is transformed into a matrix $\mathbf{S} = [\mathbf{x}_1; \mathbf{x}_2; \cdots; \mathbf{x}_{|s|}]$.

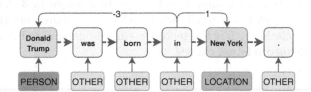

Fig. 3. An example of relative distances and entity types. The relative distances from word *in* to entities *Donald Trump* and *New York* are −3 and 1. The entity types of *Donald Trump* and *New York* are *PERSON* and *LOCATION*.

We employ n filters to extract local features of sentence s and the convolution matrix is $\mathbf{W} \in \mathbb{R}^{n \times (l \times d)}$, where l is the size of filter window. Let us define $\mathbf{x}_{i:j}$ as the concatenation of $\mathbf{x}_i$ to $\mathbf{x}_j$. The convolution operation of i-th filter is defined as follow:

$$c_{ij} = \mathbf{W}_i \mathbf{x}_{j-l+1:j} \tag{1}$$

where j ranges from 1 to $|s| + l - 1$. All out-of-range words $\mathbf{x}_i \notin \{\mathbf{x}_1, \mathbf{x}_2, \cdots, \mathbf{x}_{|s|}\}$ are set as zero vector for convolution.

The objective of max-pooling layer is to down-sample result vectors so as to reduce the computational cost and avoid over-fitting. Max-pooling operation of $\mathbf{c}_i$ is defined as:

$$p_i = \max(\mathbf{c}_i) \tag{2}$$

$\mathbf{c}_i \in \{\mathbf{c}_1, \ldots, \mathbf{c}_n\}$, where n is the number of filters. p_i is the i-th element of the vector $\mathbf{p} \in \mathbb{R}^n$. Hence the final representation of sentence s is $\mathbf{p}$.

Finally, a non-linear function such as *Rectified Linear Unit* (ReLU) is applied at the output to introduce nonlinearity in the final representation.

3.2   Positive Patterns Extraction

To select valid instances in a bag, we conduct an online clustering algorithm [3] on the feature vectors to extract semantic positive patterns for each relation statistically. Each cluster centroid is regarded as a positive pattern of one relation. We assume that instances similar to the patterns are more likely to be valid.

Given a subset $\mathbf{P}_r = \{\mathbf{p}_{r1}, \mathbf{p}_{r2}, \cdots, \mathbf{p}_{rN_r}\}$, in which all instance feature vectors in $\mathbf{P}_r$ are labeled with the same relation r and N_r indicates the number of instances labeled by relation r in training data, we aim to gather positive patterns from it via a clustering process. In detail, we cluster $\mathbf{P}_r$ into T groups as positive patterns, where T is a hyper-parameter. Considering computational efficiency, we conduct an online clustering algorithm on $\mathbf{P}_r$. The algorithm for sampling positive patterns is presented in Algorithm 1. Step 7 reflects the concept of online clustering. If $\mathbf{p}$ belongs to the t-th group of relation r, the cluster centroid $\mathbf{M}_{rt}$ is updated according to step 7, without recalculating the average of all feature vectors of the t-th group of relation r. After performing this algorithm on each relation, we obtain a positive pattern tensor (a 3-channel tensor) $\mathbf{M} = \{\mathbf{M}_{ij} | 1 \le i \le R, 1 \le j \le T\}$, where $\mathbf{M}_{ij} \in \mathbb{R}^L$ represents the j-th positive pattern for relation i, L is the sentence embedding size and R indicates the number of predefined relation classes.

Algorithm 1. Positive Patterns Extraction Algorithm

Input: The instances feature vector set $\mathbf{P} = \mathbf{P}_1, \mathbf{P}_2, \cdots, \mathbf{P}_R$
Output: The positive pattern tensor $\mathbf{M}$

1: Initialize pattern tensor $\mathbf{M}$
2: Set counting matrix $\mathbf{C} \leftarrow \mathbf{0}, \mathbf{C} \in \mathbb{R}^{R \times T}$
3: **for** $\mathbf{P}_r$ in $\mathbf{P}$ **do**
4:     **for** each instance feature $\mathbf{p}$ in $\mathbf{P}_r$ **do**
5:         **if** $\mathbf{p}$ is closest to $\mathbf{M}_{rt}$ **then**
6:             Increase $\mathbf{C}_{rt}$
7:                 $\mathbf{M}_{rt} \leftarrow \mathbf{M}_{rt} + \frac{1}{\mathbf{C}_{rt}} \times (\mathbf{p} - \mathbf{M}_{rt})$
8:         **end if**
9:     **end for**
10: **end for**
11: **return** $\mathbf{M}$

Because only a small amount of data is sent to the algorithm at a time, some patterns may be incorrect. To correct the deviation in the clustering process, we allow the pattern tensor to be adjusted in the end-to-end learning procedure.

3.3 Sentence-Level Attention Combined with Pattern Filtering

Based on the generated positive patterns and feature representations of instances, we enhance sentence-level attention mechanism using positive patterns.

Pattern Filtering. To select valid instances from the whole bag of relation instances related to the same entity pair, we employ the network presented in Fig. 2 for relation extraction. We assume that a valid instance should be similar to the positive patterns of its assigned relation. The similarity between instance s and the positive patterns can be defined as follows when the relation label of s is unknown:

$$sim(s) = \sum_{i=1}^{R} \sum_{j=1}^{T} \frac{1}{d(\mathbf{p}_s, \mathbf{M}_{ij})} \tag{3}$$

where $d(\cdot)$ is the Euclidean distance, $\mathbf{p}_s$ is the feature vector of s, T is the number of positive patterns of each relation.

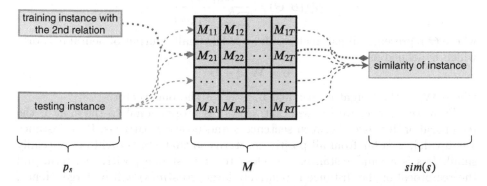

Fig. 4. An example of calculating the similarity of an instance. $\mathbf{p}_s$ represents instance features of relation instances or sentences, $\mathbf{M}$ represents our positive pattern tensor. $sim(s)$ represents the final output, i.e., the similarity of instances s.

Specifically, given the relation label r during the training process, instances are only filtered by positive patterns related to r. The similarity of instance s can be simplified as:

$$sim(s) = \sum_{j=1}^{T} \frac{1}{d(\mathbf{p}_s, \mathbf{M}_{rj})} \tag{4}$$

Figure 4 illustrates an example of acquiring similarities. The relation label of training instance is known, and it's the 2nd relation. Hence we only calculate the distances between the instances features and patterns of the 2nd relation to produce similarity. While the relation label of testing instance is unknown in our model, we then calculate the distances between the instance feature and all patterns to produce its similarity.

Sentence-Level Attention. We perform a sentence-level attention after calculating the similarity, and then we use the attention weights to de-emphasize the noisy sentences. Suppose there is a bag B containing $|B|$ sentences related to entity pair (head, tail), i.e., $B = \{s_1, s_2, \cdots, s_{|B|}\}$, the attention weight α_i for s_i is defined as:

$$\alpha_i = \frac{\exp(sim(s_i))}{\sum_k \exp(sim(s_k))} \tag{5}$$

Then the representation of B can be calculated as:

$$\mathbf{q} = \sum_{i=1}^{|B|} \alpha_i \mathbf{p}_i \tag{6}$$

We use a softmax layer to calculate the probabilities of all relation types that B may express. Thus the conditional probability of relation r is defined as follows:

$$p(r|B, \Theta) = \frac{\exp(o_r)}{\sum_{i=1}^{R} \exp(o_i)} \tag{7}$$

where Θ represents all parameters, and $\mathbf{o}$ is the finally output of neural network, which is defined as:

$$\mathbf{o} = \mathbf{W}_s \mathbf{q} + \mathbf{b}_s \tag{8}$$

where $\mathbf{W}_s$ is the weight matrix and $\mathbf{b}_s \in \mathbb{R}^R$ is the bias vector.

From the above, we can see that $sim(s)$ is proportional to the sum of the reciprocal of distances between sentence s and positive patterns. If an instance s is invalid, it is far from all positive patterns so that $sim(s)$ will be especially small. If s is a valid instance, it is close to at least one positive pattern, and the reciprocal of the distance is relatively large, so $sim(s)$ will be large. Hence, similarities of valid and invalid instances are greatly different. Attention weights based on the similarities are more discriminative, which can reduce noise in training data effectively.

3.4   Training Objective

Suppose there are N bags $\{B_1, B_2, \cdots, B_N\}$ in training data, and their labels are relations $\{r_1, r_2, \cdots, r_N\}$. we define the objective function of relation classification using cross-entropy as follows:

$$J = \sum_{i=1}^{N} \log p(r_i|B_i, \Theta) + \lambda ||\mathbf{M}||_2^2 \tag{9}$$

where Θ is all parameters of our model, $\mathbf{M}$ is the positive pattern tensor, $\lambda \in [0, \infty)$ is the weight of the L^2 norm of $\mathbf{M}$.

In experiments, we use Adam [10] optimizer to minimize the objective function and adopt dropout strategy [7] on the output layer to prevent overfitting.

4   Experiments

In this section, we firstly introduce the dataset and evaluation metrics, followed by the baseline models, experimental settings and results. Finally, we present analysis of the results.

4.1   Dataset and Evaluation Metrics

We evaluate our model on the NYT dataset developed by [19], which aligns Freebase[1] relations with the NYT corpus. The aligned sentences from years 2005–2006 of the NYT corpus are regarded as training data and those from year 2007 are for testing. Its entity mentions are found by Stanford named entity tagger and linked to Freebase. The dataset contains 53 relations including no relation "NA". There are 522,611 sentences linked to 281,270 entity pairs for training and 172,448 sentences linked to 96,678 entity pairs for testing.

Similar to previous work [11,15], we evaluate our model in the held-out evaluation. We compare the relation instances extracted from the test data against Freebase relations data. In experiments, we present both precision/recall curves and Precision@N (P@N).

To illustrate the effect of sentences number for our selective attention, we present Precision@N result in three test settings:

- **One:** For each testing entity pair, we randomly select one sentence and use this sentence to predict relations.
- **Two:** For each testing entity pair, we randomly choose two sentences and use them for relation prediction.
- **All:** For each testing entity pair, we use all sentences of it for relation extraction.

4.2   Baseline Models

Both traditional feature-based and neural methods are considered as our baselines. The brief introductions of these baselines are as follow:

- **Mintz:** This is a traditional distant supervised model proposed by [15], which extracts features from all instances.
- **MultiR:** A probabilistic graphical model of multi-instance learning by [8], which handles overlapping relations.

---

[1] freebase.com.

- **MIML:** A graphical model that jointly models both multiple instances and multiple labels [20].
- **CNN/PCNN+ATT:** Neural network models based on selective attention proposed by [11]. These methods use CNN or piecewise CNN (PCNN) for feature extraction instead of NLP tools, which avoids error propagation.
- **APCNN+soft-label:** A Soft-label method proposed by [12]. It corrects wrong labels at entity-pair level during training by exploiting semantic/syntactic information from correctly labeled instances.
- **CNN+RL:** Reinforcement learning is introduced to choose high-quality sentences at the sentence level, which deals with the noise of data [4].

4.3   Experimental Settings

We train word embeddings on the NYT corpus with word2vec tool.[2] Following previous work [11], we tune our model using three-fold validation on the training set. In our experiments, we select the dimension of type embedding among $\{3, 5, 10, 15\}$, the window size l among $\{3, 4, 5\}$, the number of feature maps or filters n among $\{128, 230, 256\}$, batch size among $\{32, 64, 128\}$, the learning rate among $\{0.001, 0.005, 0.01, 0.1\}$. We select the number of relation patterns for each relation T from 5 to 100 in steps of 5. We choose Adam [10] method as the optimizer to minimize the objective function. The best configurations are presented in Table 3.

Table 3. Main parameters setting in our experiments

Parameter name	Parameter value
Filter window size l	3
The number of filters n	230
Sentence size L	80
Word dimension d_w	50
Position dimension d_p	5
Type dimension d_t	5
The number of relation patterns T	15–50
Batch size	64
Learning rate	0.01
Dropout probability	0.5
Weight parameter λ	0.001

---

[2] https://code.google.com/p/word2vec/.

Table 4. P@N results for relation extraction with different settings

Test settings	One				Two				All			
P@N(%)	100	200	300	Avg	100	200	300	Avg	100	200	300	Avg
CNN+AVE	75.2	67.2	53.8	60.9	70.3	62.7	55.8	62.9	67.3	64.7	58.1	63.4
CNN+ATT	76.2	65.2	60.8	67.4	76.2	65.7	62.1	68.0	76.2	68.6	59.8	68.2
PCNN+AVE	71.3	63.7	57.8	64.3	73.3	65.2	62.1	66.9	73.3	66.7	62.8	67.6
PCNN+ATT	73.3	69.2	60.8	67.8	77.2	71.6	66.1	71.6	76.2	73.1	67.4	72.2
APCNN+soft-label	**84.0**	75.5	68.3	75.9	86.0	77.0	73.3	78.8	87.0	**84.5**	**77.0**	82.8
CNN+RL	–	–	–	–	–	–	–	–	80.0	73.5	70.6	74.7
CNN+PFA-type	82.0	75.0	66.3	74.4	80.0	77.5	70.7	75.8	82.0	79.5	73.3	78.2
CNN+PFA	80.0	**76.5**	**72.3**	**76.3**	**87.0**	**81.5**	**73.7**	**80.7**	**89.0**	83.3	76.4	**82.9**

4.4   Experimental Results

Table 4 shows the P@N results for three test settings the same as [11]. We can find that: (1) Compared to APCNN+soft-label, our CNN+PFA has better performance in One and Two test settings and comparable performance in All test setting. Besides, all the average performance of our model outperforms APCNN+soft-label, so we can conclude that our method can reduce the impact of noise more effectively. (2) When removing entity types, our CNN+PFA-type has weaker performance compared with APCNN+soft-label but it still outperforms CNN/PCNN+ATT models in three test settings. It indicates that our attention strategy based on positive patterns filtering is better in selecting more useful instances for relation extraction. However, its performance depends on whether the positive patterns we learned are accurate. Hence, we add entity type information to learn the constraints of entity types for relation patterns, which is useful for filtering noisy instances. (3) Compared with average strategies, attention-based models generally obtain higher performance with the same network. Therefore, we can conclude that the importance of each instance related to an entity pair is not consistent for relation extraction. (4) CNN+PFA performs better than CNN+RL. It is because RL strategy may wrongly remove some true positive instances, which leads that the rich information contained in those removed instances is discarded. But our model doesn't suffer from this problem and our model can make full use of valid instances based on the patterns filtering attention.

Figure 5(a) presents the results of precision/recall curves of multiple models. We follow the same experimental settings as [11]. CNN+PFA yields better performance compared with feature-based methods and neural models. It proves that our model combining sentence-level attention with pattern filtering is able to learn more precise weights, because they are discriminative. Additionally, without the consideration of entity types, the performance of CNN+PFA-type decreases by 9.5% compared with CNN+PFA, but is comparable to that of CNN+RL. It shows that the information of entity types can promote the performance of relation extraction.

As mentioned above, the performance of our model outperforms the existing methods in aggregate precision/recall metric and achieves comparable perfor-

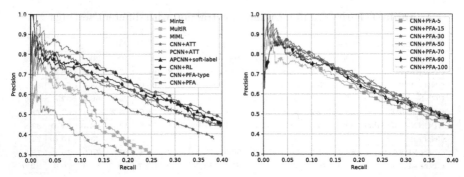

(a) Comparison of CNN+PFA and baseline.   (b) Comparison of different values(T).

Fig. 5. The performance of CNN+PFA

mance in P@N metric. It shows that discriminative attention is more beneficial to denoise the training data. Compared to many existing methods that use more complex networks such as PCNN or DCNN, our model combines a simpler CNN with efficient pattern filtering and produce good performances. We believe that the performance will be better when adopting complex networks in our model.

4.5   Detailed Analysis

In this section, we present the effect of the number of positive patterns and case study for our model.

The Effect of the Number of Positive Patterns. To explore how the number of relation patterns affects the extraction performance, we conduct a set of experiments using different numbers of patterns. Intuitively, the number of positive patterns for each relation is not particularly large, so we select it from 5 to 100. The experimental results are shown in Fig. 5(b). CNN+PFA-T is

Table 5. An example of attention weights assigned by CNN+ATT and CNN+PFA

Bag:⟨Mel Karmazin, company, Sirius Satellite Radio⟩	CNN+ATT	CNN+PFA
When Howard Stern was preparing to take his talk show to Sirius Satellite Radio, following his former boss, Mel Karmazin, Mr. Hollander argued that the show should ...	0.17	5.7E−07
... Mel Karmazin, the chief executive of Sirius, vowed last Wednesday that prices would not be raised and ...	0.54	0.99

our model which has T patterns for each relation. We can see that our model achieves a stable performance over different numbers of patterns.

The performance of our models improves with the increase of the number of relation patterns T at the beginning. Then models' performance tends to be stable when the value of T changes in a large range. For the reason, we believe that each relation may have a different number of relation patterns, but we simply set them to the same value T, which makes T have a wider range of values. However, when T is a larger value, such as 90, the performance drops gradually. We believe that this is because too many relation patterns increase the likelihood that each instance will be similar to the positive patterns, which leads to indiscriminative attention weights and poor guidance for relation extraction.

Case Study. Table 5 demonstrates an example for selective attention from the testing data. We present two sentences and their attention weights assigned by CNN+ATT and CNN+PFA.

We can find that both CNN+ATT and our model can correctly assign attention weights for the sentences according to their relevance to the relation *company*. Yet the former sentence obtains the weight of 0.17, which means CNN+ATT has still learned 17% noise from this sentence. Instead, CNN+PFA assigns more discriminative weights according to the similarities between instances and relation patterns, which weakens the influence of noisy instances effectively.

5   Conclusion and Future Works

In this paper, we introduce a novel model combining sentence-level attention with pattern filtering for relation extraction under distant supervision. The pattern filtering can select more valid instances in a bag by calculating their similarities with positive patterns that are gathered via an online clustering algorithm. Based on it, the sentence-level attention can make full use of valid instances and further reduce noise effectively, via assigning more discriminative and precise attention according to the similarities. We conduct experiments on NYT dataset and our model achieves competitive performance.

In the future, we will explore the appropriate number of patterns for each relation and how to extract positive patterns for relation from external text. Since entity types are conducive to relation extraction, fine-grained entity types are also worth investigating.

References

1. Bollacker, K., Evans, C., Paritosh, P., Sturge, T., Taylor, J.: Freebase: a collaboratively created graph database for structuring human knowledge. In: Proceedings of the 2008 ACM SIGMOD International Conference on Management of Data, pp. 1247–1250. AcM (2008)

2. Du, J., Han, J., Way, A., Wan, D.: Multi-level structured self-attentions for distantly supervised relation extraction. In: EMNLP (2018)
3. Duda, R.: Sequential k-means clustering. http://www.cs.princeton.edu/courses/archive/fall08/cos436/Duda/C/sk_means.htm
4. Feng, J., Huang, M., Zhao, L., Yang, Y., Zhu, X.: Reinforcement learning for relation classification from noisy data. In: Thirty-Second AAAI Conference on Artificial Intelligence (2018)
5. GuoDong, Z., Jian, S., Jie, Z., Min, Z.: Exploring various knowledge in relation extraction. In: Proceedings of the 43rd Annual Meeting on Association for Computational Linguistics, pp. 427–434. Association for Computational Linguistics (2005)
6. Han, X., Yu, P., Liu, Z., Sun, M., Li, P.: Hierarchical relation extraction with coarse-to-fine grained attention. In: Proceedings of the 2018 Conference on Empirical Methods in Natural Language Processing, pp. 2236–2245 (2018)
7. Hinton, G.E., Srivastava, N., Krizhevsky, A., Sutskever, I., Salakhutdinov, R.R.: Improving neural networks by preventing co-adaptation of feature detectors. arXiv preprint arXiv:1207.0580 (2012)
8. Hoffmann, R., Zhang, C., Ling, X., Zettlemoyer, L., Weld, D.S.: Knowledge-based weak supervision for information extraction of overlapping relations. In: Proceedings of the 49th Annual Meeting of the Association for Computational Linguistics: Human Language Technologies, vol. 1, pp. 541–550. Association for Computational Linguistics (2011)
9. Ji, G., Liu, K., He, S., Zhao, J., et al.: Distant supervision for relation extraction with sentence-level attention and entity descriptions. In: AAAI, pp. 3060–3066 (2017)
10. Kingma, D.P., Ba, J.: Adam: a method for stochastic optimization. arXiv preprint arXiv:1412.6980 (2014)
11. Lin, Y., Shen, S., Liu, Z., Luan, H., Sun, M.: Neural relation extraction with selective attention over instances. In: Proceedings of the 54th Annual Meeting of the Association for Computational Linguistics (Volume 1: Long Papers), pp. 2124–2133 (2016)
12. Liu, T., Wang, K., Chang, B., Sui, Z.: A soft-label method for noise-tolerant distantly supervised relation extraction. In: Proceedings of the 2017 Conference on Empirical Methods in Natural Language Processing, pp. 1790–1795 (2017)
13. Luo, B., et al.: Learning with noise: enhance distantly supervised relation extraction with dynamic transition matrix. In: Proceedings of the 55th Annual Meeting of the Association for Computational Linguistics (Volume 1: Long Papers), pp. 430–439 (2017)
14. Mikolov, T., Chen, K., Corrado, G., Dean, J.: Efficient estimation of word representations in vector space. arXiv preprint arXiv:1301.3781 (2013)
15. Mintz, M., Bills, S., Snow, R., Jurafsky, D.: Distant supervision for relation extraction without labeled data. In: Proceedings of the Joint Conference of the 47th Annual Meeting of the ACL and the 4th International Joint Conference on Natural Language Processing of the AFNLP, vol. 2, pp. 1003–1011. Association for Computational Linguistics (2009)
16. Mooney, R.J., Bunescu, R.C.: Subsequence kernels for relation extraction. In: Advances in Neural Information Processing Systems, pp. 171–178 (2006)
17. Peng, M., et al.: Improving distant supervision of relation extraction with unsupervised methods. In: Cellary, W., Mokbel, M.F., Wang, J., Wang, H., Zhou, R., Zhang, Y. (eds.) WISE 2016. LNCS, vol. 10041, pp. 561–568. Springer, Cham (2016). https://doi.org/10.1007/978-3-319-48740-3_42

18. Qin, P., Xu, W., Wang, W.Y.: DSGAN: generative adversarial training for distant supervision relation extraction. arXiv preprint arXiv:1805.09929 (2018)
19. Riedel, S., Yao, L., McCallum, A.: Modeling relations and their mentions without labeled text. In: Balcázar, J.L., Bonchi, F., Gionis, A., Sebag, M. (eds.) ECML PKDD 2010. LNCS (LNAI), vol. 6323, pp. 148–163. Springer, Heidelberg (2010). https://doi.org/10.1007/978-3-642-15939-8_10
20. Surdeanu, M., Tibshirani, J., Nallapati, R., Manning, C.D.: Multi-instance multi-label learning for relation extraction. In: Proceedings of the 2012 Joint Conference on Empirical Methods in Natural Language Processing and Computational Natural Language Learning, pp. 455–465. Association for Computational Linguistics (2012)
21. Takamatsu, S., Sato, I., Nakagawa, H.: Reducing wrong labels in distant supervision for relation extraction. In: Proceedings of the 50th Annual Meeting of the Association for Computational Linguistics: Long Papers-Volume 1, pp. 721–729. Association for Computational Linguistics (2012)
22. Zeng, D., Liu, K., Chen, Y., Zhao, J.: Distant supervision for relation extraction via piecewise convolutional neural networks. In: Proceedings of the 2015 Conference on Empirical Methods in Natural Language Processing, pp. 1753–1762 (2015)
23. Zeng, D., Liu, K., Lai, S., Zhou, G., Zhao, J.: Relation classification via convolutional deep neural network. In: Proceedings of COLING 2014, the 25th International Conference on Computational Linguistics: Technical Papers, pp. 2335–2344 (2014)
24. Zhang, N., Deng, S., Sun, Z., Chen, X., Zhang, W., Chen, H.: Attention-based capsule networks with dynamic routing for relation extraction. In: Proceedings of the 2018 Conference on Empirical Methods in Natural Language Processing, pp. 986–992 (2018)
25. Zhang, Y., Qi, P., Manning, C.D.: Graph convolution over pruned dependency trees improves relation extraction. In: Proceedings of the 2018 Conference on Empirical Methods in Natural Language Processing, pp. 2205–2215 (2018)
26. Zhou, P., et al.: Attention-based bidirectional long short-term memory networks for relation classification. In: Proceedings of the 54th Annual Meeting of the Association for Computational Linguistics (Volume 2: Short Papers), pp. 207–212 (2016)

Adaptive Rule Adaptation in Unstructured and Dynamic Environments

Alireza Tabebordbar[1]([⊠]), Amin Beheshti[2], Boualem Benatallah[1], and Moshe Chai Barukh[1]

[1] University of New South Wales, Sydney, Australia
{alirezat,boualem,mbarukh}@cse.unsw.edu.au
[2] Macquarie University, Sydney, Australia
amin.beheshti@mq.edu.au

Abstract. Rule-based systems have been used to augment machine learning based algorithms for annotating data in unstructured and dynamic environments. Rules can alleviate many of shortcomings inherent in pure algorithmic approaches. Rule adaptation is a challenging and error-prone task: in a rule-based system, there is a need for an analyst to adapt rules in order to keep them applicable and precise. In this paper, we present an approach for adapting data annotation rules in unstructured and constantly changing environments. Our approach offloads analysts from adapting rules and autonomically identifies the optimal modification for rules using a Bayesian multi-armed-bandit algorithm. We conduct experiments on different curation domains and compare the performance of our approach with systems relying on analysts. The experimental results show a comparative performance of our approach compared to analysts in adapting rules.

Keywords: Rule adaptation · Data annotation · Rule based systems · Data curation

1 Introduction

Data curation indicates processes and activities related to the integration, annotation, publication, and presentation of data throughout its lifecycle [5,8]. One category of data curation is data annotation, which aims to label the raw data to generate value and increase productivity. Data annotation has been studied widely in various computational machine learning algorithms for information extraction, item classification, and record-linkage [3,4,6,7,22,23]. However, in dynamic environments, e.g., Twitter, Facebook, data is constantly changing. Accordingly, relying on algorithmic approaches does not scale to the time constraints and business needs. Algorithms make predictions based on the historical data only, whereas in dynamic environments the distribution of data is constantly changing. Therefore, algorithms need to be re-trained to capture changes, which is expensive and time-consuming. Several solutions [2,11,16,19,26,27,30], have

© Springer Nature Switzerland AG 2019
R. Cheng et al. (Eds.): WISE 2019, LNCS 11881, pp. 326–340, 2019.
https://doi.org/10.1007/978-3-030-34223-4_21

been proposed to augment algorithms with rule-based techniques. Rules can alleviate many of the shortcomings inherent in pure algorithmic approaches. Rules can be written by non-technician analysts, which can be cost effective [11].

Typically, in rule-based systems [2,11,19,26,30] there is a need for an analyst to adapt rules, i.e., the process of modifying a rule to become better suited to its execution environment, to keep them applicable and precise. However, the successful adaptation of a rule requires an analyst to understand the context of data. Rule adaptation is particularly challenging and error-prone as an improper modification of a rule makes the rule inapplicable or ineffective. This problem exacerbated in dynamic environments as adaptation is not a one shot rule modification task. Accordingly, the analyst needs to modify the rule over its life cycle. In this paper, we propose an approach for adapting rules in dynamic and constantly changing environments, to offload analysts from adapting rules and to keep rules applicable and precise based on changes in the execution environment. We utilize a Bayesian multi-armed-bandit algorithm, an online learning algorithm that adapts a rule by estimating the performance of a set of candidate feature, e.g., keyword, extracted from the execution environment. Each time a rule annotates a set of items, the algorithm receives feedback over the number of items that the rule correctly/incorrectly annotated. Then, it updates the performance of candidate features based on the collected feedback. Over time, by collecting more feedback the algorithm better learns the performance of features and adapts the rule more precisely. This paper makes the following contributions:

- We propose a self-adaptive approach for adapting rules. We utilize a Bayesian multi-armed-bandit algorithm to adapt rules based on changes in the execution environment.
- To frame the problem as a Bayesian multi-armed-bandit algorithm, we propos a reward and demote schema. The schema assigns a reward if the algorithm identifies a rule that correctly tagged an item, and demotes a rule if it tags an irrelevant item. Over time, the algorithm (by observing rewards and demotes) learns a better adaptation for the rule.

We conduct experiments on different curation domains and compare the performance of our approach with systems relying on analysts. The experiment results show our approach could significantly improve the precision of rules in annotating data by as much as 29% precision compared to the initial results. The rest of this paper is organized as follows: We discuss the preliminaries and problem in Sect. 2. In Sects. 3 and 4, we describe our proposed approach. Section 5 presents the performance of our approach on three different curation domains: mental health, domestic violence and budget. Finally, we discuss related works in Sect. 6 before concluding the paper in Sect. 7.

2 Preliminaries and Problem Statement

Feature. We express a rule R in forms of features, where each feature $f \in R$ corresponds to a function in forms of "$\langle Dataset.Function.Operator \rangle \rightarrow value$", where *Dataset* is the data source suchas Twitter and Facebook, *Function* performs a curation task (e.g., feature extraction), *Operator* represents the condition for a feature to curate the data, and *value* is the output of a feature. Examples of a feature can be extraction functions, e.g., identifying named-entities, identifying similarity in keywords of text. Expressing a feature as a function allowing us to leverage the standard data-types as the feature's operator. For example, if a feature operates over textual data the operator for the feature will include string operators, such as *contains* and *exact*. Similarly, if a feature curates integer data the feature will include integer operators, such as *equals* and *less-than*. As an example, consider the feature "$f_1 = \langle Tweet.Keyword.Contains('Mental')\rangle$", feature f_1 curates Tweets that contains 'Mental' keyword. In this example, *Tweet* represents the dataset the feature operates for curating the data, *Keyword* represents the function of the feature, and *contains('Mental')* is the operator, and represents the condition for curating a Tweet.

Rule. We represent a rule R as a tree of features, where each feature $f \in R$ can have K children. We denote a path p in the tree as a sequence of features $f_1, ..., f_m$, where f_1 represents the root feature and f_m represents the last feature in the path. More precisely, a path p is a conjunction of a set of features is of the form $f_1 \wedge ... \wedge f_m$. To curate an item with a rule, the item should be annotated with all features within a path. Notice that, we do not require inventing our own rule language. Rather, the benefit of rules being expressed as features, we can adopt any suitable functional or rule-expression language for our purposes.

Tag. A Tag is the label, e.g., 'Mental Health', a rule assigns to a curated item, e.g., Tweet, to describe the item. In this paper, we use the terms *tag* and *annotate* interchangeably. As an example, consider the rule presented in Fig. 1. This rule is made up of three features $\{f_1, f_2, f_3\}$, and tags a Tweet with 'Mental Health', if the tweet curated with features in paths $p_1 = f_1 \wedge f_2$, or $p_2 = f_1 \wedge f_3$. More clearly, $Rule_1$ tags a Tweet with 'Mental Health', if the Tweet contains 'Mental' and 'Health', or 'Mental' and 'Medical' keywords.

Problem Statement. Given a rule, an analyst needs to adapt the rule to an adequate form to annotate data precisely. Typically, to adapt a rule, the analyst examines items that the rule annotated and identifies the potential modifications that make the rule more precise [11,16,18,19]. Over time, by examining the accuracy of annotated items, the analyst learns a better adaptation for the rule. Such problem is typically categorized under the category of *online learning* problem, where an analyst does not have access to the entire knowledge that is required to craft the most adequate type of rule. Instead, she incrementally learns to better adapt a rule through examining the rule performance over an extended period of time. To offload analysts from adapting rules, we formulated the problem through a Bayesian multi-armed-bandit algorithm. In our approach, each time a rule annotates a set of items, the algorithm receives feedback over items the rule correctly/incorrectly annotated. By collecting more feedback the algorithm learns to better adapt the rule.

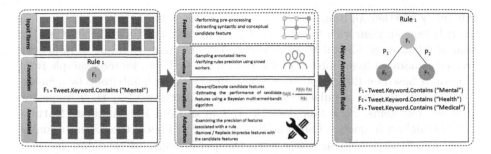

Fig. 1. The overview of our proposed approach for adapting rules.

3   Adaptive Rule Adaptation

The overview of our proposed solution shown in Fig. 1. The approach consists of four steps, feature extraction, observation, estimation, and adaptation. Each time a rule annotates a set of items, the algorithm extracts a set of candidate features from annotated items as the potential modifications for the rule (Sect. 3.1). Then, it takes a sample of annotated items and gathers feedback over the number of items correctly/incorrectly annotated (Sect. 3.2). Next, it utilizes a Bayesian multi-armed-bandit algorithm to estimate a probability distribution for candidate features based on the collected feedback, representing the performance of features in adapting the rule (Sect. 3.3). Finally, the algorithm determines the optimal modification for the rule based on the candidate features performance (Sect. 3.4).

3.1   Feature Extraction

The initial step in our workflow is feature extraction. Each time a rule annotates a set of items $I = \{i_1, i_2, i_3, ..., i_n\}$, the algorithm extracts a set of candidate features $T = \{t_1, t_2, ..., t_n\}$ to identify the potential modifications for the rule. For extracting candidate features we perform a preprocessing task on annotated items. The preprocessing task consists of three subtasks, tokenization, normalization, and noise removal. Tokenization splits each item into smaller tokens. Normalization removes stop words, and performs stemming. In noise removal, we skip certain characters, e.g., imoji, urls, that occur in items. We consider the remaining tokens as the candidate feature type of *keyword*.

3.2   Observation

The second step in the workflow is observation, which gathers feedback to update a Bayesian multi-armed-bandit algorithm about changes in the execution environment. For gathering feedback, we rely on crowd workers. Each time a rule annotates a set of items, the sampling algorithm takes a sample of annotated items $S = \{i'_1, i'_2, i'_3, ..., i'_n\}$, where $S \subset I$ and sends to the crowd. The crowd will verify whether an item $i' \in S$ correctly tagged with the rule or not, e.g., if the rule tags an item with 'Mental Health'. The task was to verify whether the item is relevant to 'Mental Health' or not. For selecting a sample of items, we divided annotated items into several subgroups [14]. Then, we took sample from subgroups based on their population. We represented each subgroup by a candidate feature, and the population of subgroups is determined by their frequency in annotated items. Each time, we take a sample from 100 candidate features with the highest frequency. Our sampling strategy will boost the Bayesian multi-armed-bandit algorithm to better learn the performance of candidate features in adapting a rule. For example, if we used more obvious techniques, such as random sampling, then our solution would consider all items equally likely.

3.3   Estimation

The third step in the workflow is estimation, which computes a probability distribution θ for candidate features to determine their performance in adapting the rule. This step consists of two components: (i) reward/demote schema, which calculates a reward/demote for candidate features using the workers feedback; and (ii) a Bayesian multi-armed-bandit algorithm, which estimates the performance of candidate features based on their rewards/demotes.

Reward/Demote Schema. As we have described previously, in dynamic environments, e.g., Twitter, Facebook, data is noisy and constantly changing, thus rules need adaptation to be applicable and precise. We rely on a Bayesian multi-armed-bandit algorithm to adapt rules. This algorithm is suitable for environments that need additional improvement to their decisions over time. The algorithm learns to improve its decision based on the feedback receives from the execution environment. To formulate the problem as a Bayesian multi-armed-bandit algorithm, we propose a reward and demote schema. The schema assigns a reward 'r' to a candidate feature $t \in T$ if it identifies the feature appeared in an item that verified as relevant. Conversely, it demotes 'd' a candidate feature if it identifies the feature appeared in an irrelevant item. Over time, as a rule annotates more items the schema updates candidate features reward/demote, allowing a Bayesian multi-armed-bandit algorithm to update its estimations about the performance of features in adapting the rule.

Bayesian Multi-armed-Bandit Algorithm. We utilized Thompson sampling [25], a Bayesian multi-armed-bandit algorithm that has shown the near optimal regret (given a period of time the regret is the difference between the probability distribution θ the algorithm estimated for the optimal action and the action selected by the algorithm) bound. Thompson sampling provides a dynamic policy for choosing which feature should be selected for adapting a rule, and an algorithm for incorporating new information to update this policy based on the candidate features reward/demote. Thompson sampling stores an estimated probability distribution θ for each candidate feature. This distribution indicates the performance of the feature in adapting the rule. The algorithm continuously observes the execution environment and gathers new feedback to

---

Algorithm 1: Estimating performance of candidate features

```
1 Function Est_Probability_Dist():
     Input: T
     Output: θ
2    r ← ∅, d ← ∅, θ ← {};
3    foreach t ∈ T do
4        foreach I' ∈ S do
5            if t ∈ I' AND I' is verified as Irrelevant then
6                │ d_t+ = 1;
7            else if t ∈ I' AND I' is verified as Relevant then
8                └ r_t+ = 1;
9            θ_t ← Beta(r_t, d_t)
10       return θ;
```

---

update the probability distribution estimated for candidate features, reflecting their performance in adapting the rule. Each time, the algorithm receives a set of candidate features $T = \{t_1, t_2, ..., t_n\}$ along with their reward/demote. Then, it updates probability distributions $\theta = \{\theta_1, \theta_2, ..., \theta_n\}$, where $0 < \theta < 1$ using the Bayesian formula:

$$P(\theta \mid t) = \frac{P(t \mid \theta) \times P(\theta)}{P(t)} \propto P(t \mid \theta) \times P(\theta)$$

where $P(t \mid \theta)$, represents the likelihood and $P(\theta)$, is the prior. The likelihood is a Bernulli distribution and the prior is a Beta distribution

$$P(t \mid \theta) = \theta^r (1 - \theta)^{n-r}, r = \sum_{r=0}^{n} t$$

$$P(\theta) = \frac{\theta_n^{\alpha_n - 1}(1 - \theta_n)^{\beta_n - 1}}{\beta(\alpha_n, \beta_n)}$$

α and β are the prior parameters. The initial value of α and β, indicates our initial belief about the performance of candidate features. We have chosen $\alpha = \beta = 1$, which means initially, we considered all features to have the same performance in adapting the rule. The prior is updated continuously based on the likelihood of feedback we gather from the execution environment. The posterior is proportional to the product of the prior and the likelihood, with the likelihood updated continuously after receiving the workers feedback. This update is easy to implement because the Beta and Bernoulli distributions are conjugate. Algorithm 1 shows pseudo code of estimating the value of θ for candidate features. As an example, consider the following rule:

$$Rule_1 = Tweet.keyword.contains(`Mental') : `MentalHealth'$$

which tags Tweets with 'Mental Health'. Assume, the following candidate features $T = \{t_1 : medical, t_2 : health, t_3 : wellbeing, t_4 : care, t_5 : qanda\}$

are extracted from annotated items as the potential modifications for the rule. First, to identify the performance of candidate features the algorithm computes their reward/demote using the workers feedback. Then, a Bayesian multi-armed-bandit algorithm estimates a probability distribution θ for candidate features. Each time the rule annotates a set of items, the algorithm updates the value of θ based on the feedback gathers from workers to better reflect features performance in adapting rules.

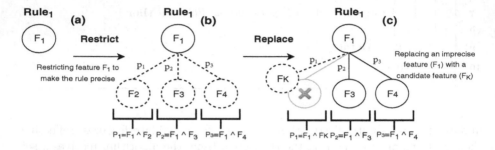

Fig. 2. Adapting a rule through replacing/ restricting its features

3.4 Adaptation

In this section, we explain how our proposed solution modifies a rule. Recall from Sect. 2, that we introduced a rule R as a tree of features, where each feature $f \in R$ can have K children. We also defined a path p in a rule as a conjunction of a set of features in forms of $p = f_1 \wedge f_2 \wedge \ldots \wedge f_n$. First, to adapt a rule we identify imprecise paths that annotate data with a precision below a threshold $\jmath$. The threshold represents the minimum precision a path should have to be considered as precise. We determine the precision of paths by calculating the number of relevant/irrelevant items their features annotated. After identifying imprecise paths, we determine whether to replace or further restrict their features. We replace a feature in an imprecise path if the number of annotated items was below the average number of items annotated with its siblings, indicating the feature is imprecise and incapable of adequately annotating data. Conversely, we restrict a feature if the number of annotated items was greater than or equal to average, indicating the feature is applicable, but should be restricted to be precise. For replacing or restricting features, we select candidate features that yielded the highest probability distribution θ estimated by a Bayesian multi-armed-bandit algorithm. Example 4 further explains how we replace or restrict a feature in the proposed solution.

Example. Consider $Rule_1$, which is made up of a feature (Fig. 2a). Suppose, after annotating a set of items at time τ_i the algorithm identifies that the rule is imprecise[1]. Thus, the algorithm examines the number of annotated items

---

[1] Annotates data with a precision below $\jmath$.

and adapts the rule by appending K candidate features[2] that yielded the highest probability distribution (restriction) (Fig. 2b). Thus, after adaptation $Rule_1$ annotates an item if the item curated with features in paths $p_1 = f_1 \wedge f_2$, or $p_2 = f_1 \wedge f_3$, or $p_3 = f_1 \wedge f_4$. Alternatively, the algorithm may replace a feature if it identifies the feature annotates data below the average number of items annotated with its siblings[3]. For example, suppose at time τ_{i+n} feature f_2 is identified as imprecise and incapable of annotating data adequately[4]. Thus, the algorithm removes feature f_2, and replaces the feature with a candidate feature that yielded the highest probability distribution value (Fig. 2c). To select a candidate feature, the algorithm performs a feature extraction task and estimates their probability distribution θ based on the reward/demote features accumulated from time τ_1 to τ_{i+n}. The proposed adaptation strategy allows to adapt a rule based on changes in the execution environment. For example, by replacing an imprecise feature with a candidate feature that obtained a high value of θ over an extended period of time, we keep the rule applicable as the rule would be adapted with a feature that better captures the salient aspect of data. Similarly, by restricting an imprecise feature that annotates a large number of items, we make the rule precise by filtering out the irrelevant items.

4   Gathering Workers Feedback

This section explains, how we contribute workers to verify items annotated with rules. For verifying items we created a task on Figure Eight microtasking market[5]. The workers task was to verify whether an item is relevant to the tag assigned by a rule or not. Workers could choose 'Yes' if they identify the item is relevant to the tag, and 'No' if they identify the item is irrelevant. In cases workers could not verify an item, they could choose 'I don't know'. For example, we present a Tweet to workers, which a rule tagged as relevant to 'Mental Health'. Then, workers task was to verify whether the Tweet is relevant to 'Mental Health' or not. In addition, we provide workers with a textual instruction to explain them how to verify items. We explained steps workers need to follow and provided them with three positive[6] and three negative[7] examples. For verifying each item we paid 1 cent, and each worker verified 10 items per page. At each round of the annotation task, we sent 3% of annotated items to workers.

---

[2] As feature f_1 is the root feature it annotates data above the average, thus satisfies the restriction condition.

[3] Features $\{f_3, f_4\}$ are siblings for feature f_2.

[4] Annotates data below the average number of items annotated with its siblings.

[5] https://www.figure-eight.com/.

[6] @lucianaberger: Mental health services facing serious shortages of mental health nurses decrease of 12% since 2010 psychiatrists.

[7] If i have to hire a car and drive home from belgium i am going to go mental stupid french air traffic control wanks on strike.

Stopping Condition. In the previous section, we explained how workers verify annotated items. However, continuously sending items to crowds increases the cost of adaptation task. Thus, there is a need to identify when a rule is stabilized to stop verifying more items. To address this problem, we developed a solution using the probabilistic policy defined in Thompson sampling algorithm to determine whether a path in a rule is stabilized or not. For each path, we estimate a probability distribution θ based on the number of relevant/irrelevant items annotated. Then, we define a smoothing window Q to record the value of θ. We set the size of smoothing window $Q = 3$ and average as the smoothing function. We consider a path as stabilized, if the value of Q increases or remains stable within 3ϵ, where $\epsilon = 0.01$[8]. More clearly, consider path $p_3 = f_1 \wedge f_4$ presented in Fig. 2. Each time the rule annotates a set of items the algorithm records the value of θ for the path. Then, we will compute the value of Q, where $Q_1 = AVG(\theta_1, \theta_2, \theta_3)$, and $Q_2 = AVG(\theta_2, \theta_3, \theta_4)$ and so forth. The algorithm stops sending items to workers, when the value of $Q_{i+1} + 3\epsilon \geq Q_i$, indicating the path is stabilized.

5   Experiments

First, in Sect. 5.1 we discuss the dataset was used for examining the performance of our approach. Then, in Sect. 5.2 we explain scenarios have been defined to show the applicability of our approach. Finally, we discuss results in Sect. 5.3.

5.1   Experiment Settings and Dataset

The core component of techniques described in the previous sections is implemented in Python. Three months of Twitter data (Australian region) were used as the input dataset (from May 2017 to August 2017) with $\approx$ 15 *millions* Tweets. MongoDB and ElasticSearch were used for storing and indexing the input dataset. We demonstrate the performance of our approach in three different curation domains (domestic violence, mental health, and budget). As the initial rules for annotating data we used rules that contains only one feature. For example, the initial rule for annotating Tweets in mental health domain was in form of $Tweet.keyword.contains('Mental')$: '$MentalHealth$', which tags Tweets that contains 'Mental' keyword. We demonstrate the performance of our approach within five rounds of annotation.

5.2   Experiment Scenarios

A set of experiments have been conducted to evaluate the performance of our approach. First, we show the precision of rules adapted based on the probability distribution θ, estimated for candidate features. We show that a Bayesian bandit algorithm by collecting more feedback better learns the performance of candidate

---

[8] We set the value of ϵ and Q, experimentally using simulated data.

features in adapting rules. In addition, we adapt rules with different number of features ($K = 10$ and $K = 20$) to show the applicability of our approach in different scenarios. Adapting a rule with a higher number of feature allows the rule to annotate a larger number of items, but with less precise ones.

The second scenario conducts a controlled experiment to compare the performance of our approach with an interactive system, which relies on an analyst for adapting rules. We reused the algorithm proposed by Suganthan et al. [11]. In this approach, the system sends a sample of items annotated with a rule to crowds, and receives feedback over the number of items correctly/incorrectly annotated. Then, the system tokenizes items, and weights every token using the TF/IDF weighting scheme. Each time an analyst selects a token for adapting the rule, the system incorporates the analyst feedback by adjusting the weight of other candidate tokens that co-occurred with the selected token using a relevance feedback algorithm [24]. The system continues showing tokens until the analyst is satisfied with the resulting rule.

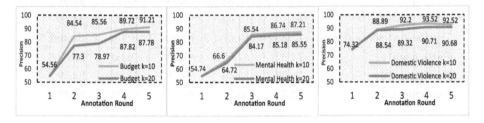

Fig. 3. The precision of our approach in adapting rules with different values of K ($K = 10$ and $K = 20$)

5.3   Results

1. Precision of adapted rules. This section demonstrates how a Bayesian multi-armed-bandit algorithm learns to formulate the workers (see Sect. 4) feedback to improve the precision of rules. We show the precision of rules adapted with 10 ($k = 10$) and 20 ($k = 20$) candidate features yielded the highest probability distribution θ. As presented in Fig. 3, by adapting rules with 10 candidate features our approach could significantly improve rules precision in all curation domains. For example, in budget domain the precision is improved by 36.65% from 54.56% to 91.21%. Similarly, in mental health and domestic violence domains the precision is improved by 32.47% and 18.20% respectively. Also, we repeated the experiment by adapting rules with a higher number of feature ($k = 20$). As presented in Fig. 3, in budget domain the precision is improved by 33.22, and in mental health and domestic violence domains the precision is improved by 30.81% and 16.36% respectively. In this experiment, we considered features that annotate data with a precision below 75% ($\jmath < 75\%$) as imprecise. Based on the obtained results, we concluded that a Bayesian multi-armed-bandit algorithm by collecting more feedback learns a better adaptation for rules over time. Although, by adapting rules with a higher number of feature the precision of rules drops, still the algorithm could improve rules precision and keep them applicable based on changes in the execution environment. In the next section, we compare the performance of our approach with interactive systems in adapting rules.

Table 1. Precision of rules adapted through participants in *Budget, Mental Health, and Domestic Violence Domains.*

Budget domain	Round 1	Round 2	Round 3	Round 4	Round 5
Participant 1	54.56	79.75	82.02	84.21	87.19
Participant 2	54.56	83.22	85.62	88.20	90.86
Participant 3	54.56	86.56	86.77	86.67	86.59
Mental health domain	Round 1	Round 2	Round 3	Round 4	Round 5
Participant 1	54.74	72.88	80.65	87.19	86.32
Participant 2	54.74	70.38	75.09	83.58	85.01
Participant 3	54.74	71.63	81.61	85.04	84.14
Domestic violence domain	Round 1	Round 2	Round 3	Round 4	Round 5
Participant 1	74.32	87.65	88.78	90.90	90.82
Participant 2	74.32	88.63	90.28	91.36	92.59
Participant 3	74.32	85.37	86.51	87.04	86.42

2. Comparing the Performance Against Analysts. To further understand the performance of our approach with systems relying on analysts for adapting rules, we performed a controlled experiment by implementing the interactive system proposed by Suganthan et al. [11]. We asked three PhD students in a lab that were familiar with the concept of learning algorithms, e.g., true positive rate, false positive rate, to participate in the experiment. We explained to them how the system works and how they can use the system to adapt rules. In addition, we allowed them to work with the system to gain the required understanding for adapting rules. To better compare the performance of our approach with the interactive system, we have asked participants to adapt rules in all domains. Then, in each curation domain, we selected the rule with highest obtained precision and compared it with rules adapted with our approach. In this experiments we asked participants to adapt rules with 20 features (k = 20). Table 1 shows the precision of rules obtained through participants. The results show our approach has a comparable performance to interactive systems. For example, in budget domain the highest obtained precision through participants is 90.86%, which is only 3.08% higher than our approach, while in domestic violence and mental health domains the highest obtained precision is 92.59% and 86.32% respectively, which shows a similar precision to our approach.

In addition, we compared the number of items annotated with rules adapted with our approach and participants. Figure 4 compares the number of annotated items. As presented, both approaches showed a similar result and could annotate a similar number of items. For example, in budget and domestic violence domains rules adapted by participants could slightly annotate more items than rules adapted with our approach. However, in mental health domain the rule adapted with our approach showed a better performance.

By comparing the precision and the number of items annotated with our approach, we believe that our adaptive approach outperforms current rule adaptation techniques. In particular, by considering the prohibitive cost of analysts for adapting rules, our proposed approach can boost companies that need to annotate data in unstructured and constantly changing environments with a limited budget.

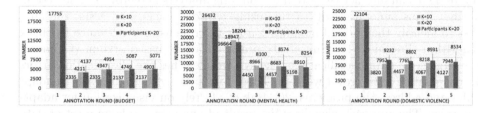

Fig. 4. Comparing the number of items annotated with participants and our approach

6 Related Works

In this section, we discuss prior works related to rule adaptation (Sect. 6.1), and online learning algorithms (Sect. 6.2).

6.1 Rule Adaptation

Rule adaptation is an evolutionary process focused on modifying a rule to better fit the rule to the execution environment. However, rule adaptation is a challenging and error-prone task, thus many solutions [11,13,16,19,28,30] have been proposed to assist analysts in adapting rules. Several solutions [13,16,18,19,28] focused on interactive approaches for adapting rules. In these solutions a system identifies the potential modifications for a rule and adapts the rule by interacting with the analyst. For example, Milo et al., proposed a cost-benefit approach for generalizing or specializing fraud detection rules. The approach developed a heuristic algorithm to interactively adapt rules with domain experts until a desired set of rules is obtained [19]. Volkovs et al., proposed a cost function to adapt integrity constraint (IC) rules. The approach relies on the analyst feedback to update the cost function and resolve inconsistencies in IC rules [28]. Liu et al., proposed an interactive approach for refining a rule using a set of positive and negative results. The approach uses a provenance graph to identify candidate changes that can eliminate the negative results [12,16]. However, these solutions have focused on adapting rules that operate in structured data, where a rule may adapted with only a limited number of features. In addition, many of these solutions hypothesis that the analyst has access to a ground truth, e.g., a dataset of items that are tagged with the correct label, to verify the effectiveness of an adaptation.

Alternatively, to adapt rules in both unstructured and dynamic environments some solutions [11,20,21,26,30] focused on augmenting interactive rule adaptation systems by coupling crowds with analysts. These solutions rely on crowd workers to provide feedback about the performance of rules and an analyst for identifying the optimal modification for rules. For example, Xie et al., proposed an approach for validating rules in information extraction applications. The approach proposed a voting technique to identify whether an adaptation of a rule produces positive impact in extracting information or not [30]. Suganthan et al., designed an interactive system by coupling analysts and crowds. The system verifies items annotated with a rule using crowd workers, and assist the analyst to identify the optimal modification for the rule using a relevance feedback algorithm (Rocchio) [11]. Sun et al., proposed a rule based system (Chimera) for large scale data classification. The system identifies the classification errors in cooperation with crowd workers. Then, forwards errors to analysts to adapt rules and address the problems [26]. Bak et al., rely on visualization for adapting rules. The approach visualizes the result of applying a rule on a set of data records. Then, asks the crowd workers to mark the effect of applying the rule on the data record, indicating the optimal adjustment for the rule [2].

Although coupling crowd workers with the interactive systems provides more flexibility for adapting rules in dynamic environments, these systems still relying on analysts for identifying the optimal modification of rules. In contrast, our approach observes changes in the execution environment and automatically adapts rules without relying on analysts.

6.2 Multi Armed Bandit Algorithm

In this section, we discuss how a Bayesian multi-armed-bandit algorithm has been used in dynamic and constantly changing environments. This algorithm increasingly used in large scale randomized A/B experimentation by technology companies [15]. One area of work that used a Bayesian multi-armed-bandit algorithm is educational learning to facilitate learners learning rate. For example, Williams et al., proposed a system (AXIS) to improve explanation generation for online learning materials by employing a combination of crowds and a Bayesian multi-armed-bandit algorithm [29]. Clement et al., used a multi armed bandit algorithm in intelligent tutoring systems to choose activities that provide better learning for students [10]. Other areas that relied on a Bayesian multi-armed-bandit algorithm are feature engineering [1], gaming [17], and online marketing [9]. In this context, we utilized a Bayesian multi-armed-bandit algorithm that adapts rules based on the feedback gathers from the execution environment. Over time, the algorithm by gathering more feedback learns a better adaptation for rules.

7   Conclusion

In this paper, we proposed an approach for adapting rules in unstructured and constantly changing environments. Our approach offloads analysts and automatically modifies rules based on changes in the execution environment. We utilized a Bayesian multi-armed-bandit algorithm, an online learning algorithm that learns the optimal modification for rules using the feedback gathers from the execution environment. We evaluated the performance of our approach on three months of Twitter data in three different curation domains: domestic violence, mental health, and budget. The evaluation results showed our approach has a comparable performance to systems relying on analysts for adapting rules.

Acknowledgements. We Acknowledge the AI-enabled Processes (AIP) Research Centre for funding part of this research.

We Acknowledge the Data to Decisions CRC (D2D CRC) and the Cooperative Research Centres Program for funding part of this research.

References

1. Anderson, M.R., Cafarella, M., Jiang, Y., Wang, G., Zhang, B.: An integrated development environment for faster feature engineering. Proc. VLDB Endowment **7**(13), 1657–1660 (2014)
2. Bak, P., Dolev, D., Yatzkar-Haham, T.: Rule adjustment by visualization of physical location data, 11 September 2014. US Patent App. 14/483,158
3. Beheshti, A., Benatallah, B., Nouri, R., Chhieng, V.M., Xiong, H., Zhao, X.: CoreDB: a data lake service. In: Proceedings of the 2017 ACM on Conference on Information and Knowledge Management, CIKM 2017, Singapore, 06–10 November 2017, pp. 2451–2454 (2017)
4. Beheshti, A., Benatallah, B., Nouri, R., Tabebordbar, A.: CoreKG: a knowledge lake service. PVLDB **11**(12), 1942–1945 (2018)
5. Beheshti, A., Benatallah, B., Tabebordbar, A., Motahari-Nezhad, H.R., Barukh, M.C., Nouri, R.: Datasynapse: a social data curation foundry. Distrib. Parallel Databases **37**(3), 351–384 (2019)
6. Beheshti, A., Vaghani, K., Benatallah, B., Tabebordbar, A.: Crowdcorrect: a curation pipeline for social data cleansing and curation. In: Proceedings of the Information Systems in the Big Data Era - CAiSE Forum 2018, Tallinn, Estonia, 11–15 June 2018, pp. 24–38 (2018)
7. Beheshti, S., Benatallah, B., Venugopal, S., Ryu, S.H., Motahari-Nezhad, H.R., Wang, W.: A systematic review and comparative analysis of cross-document coreference resolution methods and tools. Computing **99**(4), 313–349 (2017)
8. Beheshti, S., Tabebordbar, A., Benatallah, B., Nouri, B.: On automating basic data curation tasks. In: Proceedings of the 26th International Conference on World Wide Web Companion, Perth, Australia, 3–7 April 2017, pp. 165–169 (2017)
9. Burtini, G., Loeppky, J., Lawrence, R.: Improving online marketing experiments with drifting multi-armed bandits. In: ICEIS 1, pp. 630–636 (2015)
10. Clement, B., Roy, D., Oudeyer, P.-Y., Lopes, M.: Online optimization of teaching sequences with multi-armed bandits. In: 7th International Conference on Educational Data Mining (2014)

11. Paul Suganthan, G.C., et al.: Why big data industrial systems need rules and what we can do about it. In: Proceedings of the 2015 ACM SIGMOD International Conference on Management of Data, pp. 265–276. ACM (2015)
12. Hammoud, M., Rabbou, D.A., Nouri, R., Beheshti, S., Sakr, S.: DREAM: distributed RDF engine with adaptive query planner and minimal communication. PVLDB **8**(6), 654–665 (2015)
13. He, J., et al.: Interactive and deterministic data cleaning. In: Proceedings of the 2016 International Conference on Management of Data, pp. 893–907. ACM (2016)
14. Hunt, N., Tyrrell, S.: Stratified sampling. Retrieved November, 10:2012 (2001)
15. Kohavi, R., Longbotham, R., Sommerfield, D., Henne, R.M.: Controlled experiments on the web: survey and practical guide. Data Min. Knowl. Disc. **18**(1), 140–181 (2009)
16. Liu, B., Chiticariu, L., Chu, V., Jagadish, H., Reiss, F.: Refining information extraction rules using data provenance. IEEE Data Eng. Bull. **33**(3), 17–24 (2010)
17. Liu, Y.-E., Mandel, T., Brunskill, E., Popovic, Z.: Trading off scientific knowledge and user learning with multi-armed bandits. In: EDM, pp. 161–168 (2014)
18. Milo, T., Novgorodov, S., Tan, W.-C.: Rudolf: interactive rule refinement system for fraud detection. Proc. VLDB Endowment **9**(13), 1465–1468 (2016)
19. Milo, T., Novgorodov, S., Tan, W.-C.: Interactive rule refinement for fraud detection. In: EDBT (2018)
20. Ortona, S., Meduri, V.V., Papotti, P.: Robust discovery of positive and negative rules in knowledge bases. In: 2018 IEEE 34th International Conference on Data Engineering (ICDE), pp. 1168–1179. IEEE (2018)
21. Panahi, F., Wu, W., Doan, A., Naughton, J.F.: Towards interactive debugging of rule-based entity matching. In: EDBT, pp. 354–365 (2017)
22. Ratner, A., Bach, S.H., Ehrenberg, H., Fries, J., Wu, S., Ré, C.: Snorkel: rapid training data creation with weak supervision. arXiv preprint arXiv:1711.10160 (2017)
23. Ratner, A.J., Bach, S.H., Ehrenberg, H.R., Ré, C.: Snorkel: fast training set generation for information extraction. In: Proceedings of the 2017 ACM International Conference on Management of Data, pp. 1683–1686. ACM (2017)
24. Rocchio, J.J.: Relevance feedback in information retrieval. The SMART retrieval system: experiments in automatic document processing, pp. 313–323 (1971)
25. Russo, D., Van Roy, B., Kazerouni, A., Osband, I.: A tutorial on Thompson sampling. arXiv preprint arXiv:1707.02038 (2017)
26. Sun, C., Rampalli, N., Yang, F., Doan, A.: Chimera: large-scale classification using machine learning, rules, and crowdsourcing. VLDB Endowment **7**(13), 1529–1540 (2014)
27. Tabebordbar, A., Beheshti, A.: Adaptive rule monitoring system. In: Proceedings of the 1st International Workshop on Software Engineering for Cognitive Services, SE4COG@ICSE 2018, Gothenburg, Sweden, 28–2 May 2018, pp. 45–51 (2018)
28. Volkovs, M., Chiang, F., Szlichta, F., Miller, R.J.: Continuous data cleaning. In: 2014 IEEE 30th International Conference on Data Engineering (ICDE), pp. 244–255. IEEE (2014)
29. Williams, J.J., et al.: Axis: generating explanations at scale with learnersourcing and machine learning. In: ACM Conference on Learning@ Scale, pp. 379–388. ACM (2016)
30. Xie, J., Sun, C., Yang, F., Rampalli, N.: Automatic rule coaching, 2 September 2014. US Patent App. 14/475,470

Shadowed Authorization Policies - A Disaster Waiting to Happen?

Ehtesham Zahoor[1(✉)], Uzma Bibi[1(✉)], and Olivier Perrin[2(✉)]

[1] Secure Networks and Distributed Systems Lab (SENDS),
National University of Computer and Emerging Sciences, Islamabad, Pakistan
{ehtesham.zahoor,uzma.bibi}@nu.edu.pk
[2] LORIA, Université de Lorraine, BP 239,
54506 Vandoeuvre-lès-Nancy Cedex, France
olivier.perrin@loria.fr

Abstract. Information security has been in the mainstream of computing for the last few decades and our increasing reliance on the large scale distributed systems, such as the Cloud, has put greater emphasis on the security capabilities of these systems. The security concerns are amongst the important factors affecting adoption of Cloud. This paper identifies and addresses issues concerning management of hierarchical authorization policies in the Cloud. These policy models pose the risk of policy shadowing where the decision taken at higher levels mask the possibly erroneous or conflicting policies specification at the lower levels. We introduce the notion of shadowed policies and present a model which is based on formal Event-Calculus (EC); for the identification of shadowed policies. The results show that our proposed approach is scalable and practical.

1 Introduction

The need to protect valuable information has always been there. During the early ages, the information about food and shelter was of utmost importance. We live in a digital world now and our valuable digital information, such as business plans, images and confidential documents, needs to be protected from the malicious access. Information security has thus been in the mainstream of computing for the last few decades and would remain same for the foreseeable future. As our information security capabilities have matured and increased, so are the challenges we are being faced with. In this context, the widespread use of internet and adoption of large scale distributed systems, such as Cloud, we are faced with more challenging environments to enforce security principles. The past decade has seen significant increase in the use of Cloud Computing, as many organizations either have private Cloud deployments or they are using services from public Cloud providers. All major Cloud providers thus provide advanced security mechanisms to handle security challenges. One approach to implement security principles within an organization is by the use of a policy. The authorization or access control policy of an organization specifies which users can access

© Springer Nature Switzerland AG 2019
R. Cheng et al. (Eds.): WISE 2019, LNCS 11881, pp. 341–355, 2019.
https://doi.org/10.1007/978-3-030-34223-4_22

which resources, under what conditions, and what actions can they perform on the resources. All major Cloud providers thus provide the support of authorization policies and users can be organized into groups and policies can be assigned to them to specify their access permissions. Even with the advanced security capabilities provided by major Cloud providers, security breaches still happen resulting in loss of revenue and trust. In this context, authorization policies specification and management is a critical task and erroneous specification of even a single policy can lead to undesired consequences. The scale of the services from the Cloud providers makes the authorization management process challenging and complex and this can result in inducing more human errors.

One approach to improve the manageability of authorization process is the hierarchical access control model. In general, such policy models are evaluated from top to bottom, with each level providing more fine-grained access policies. This is the approach taken by major Cloud providers including AWS where policy specification and evaluation concerns different levels ranging from AWS organizations Service Control Policies (SCPs) to permission boundaries for a user or role. Hierarchical policy models pose the risk of policy shadowing where the decision taken at higher levels mask the erroneous or conflicting policies specification at the lower levels. Let us consider an example of hierarchical policy design having three levels, L0, L1 and L2, with the level L0 being the top most. If a permission is denied at the level L0, then any policy specification at the lower levels would be masked. The access control policy on the whole may produce the intended behavior but it poses serious challenges to policy management and any future change may result in erroneous behavior. As per the IBM sponsored 13th annual cost of a Data Breach study, more than one fourth of all breaches are triggered by human error. We believe that the shadowed policy's behavior is masked and not directly evident and thus a policy designer may tend to under constrain the rules assuming them to be handled at the higher level. The problem is amplified for the environments where the hierarchy structure itself can be dynamic, as we will highlight the case of AWS IAM policies where an Organization Unit can be moved within the AWS Organization tree.

We have identified the issues in our paper that are concerned to the management of authorization policies in the Cloud. We introduce the notion of shadowed policies and map their existence in the Amazon Web Services (AWS) Identity and Access Management (IAM) policies. We have highlighted the side-effects of having shadowed policies and how they induce more human errors. We introduce the notion of shadowed policies and present a model which is based on formal Event-Calculus (EC); for the identification of shadowed policies. The results show that our proposed approach is scalable and practical.

2   Background and Related Work

There are many facets of implementing information security within an organization and one way is through the use of security policies. After authentication, the decision of access control to the users is done by the authorization policies. The decision of authorization process is either to permit or deny the access and this decision may be evaluated in a certain context or conditions. One of the major studied area these days is access control and authorization policies management. One major subdomain has been the authorization models being used to implement the authorization process. In the Role Based Access Control (RBAC) model [1,2] authorization policies are based on roles of the users. RBAC has remained popular since its inception and all the major Cloud providers support RBAC. It does suffer from some scalability limitations including the role explosion. In order to ease management, the hierarchical RBAC introduces the concept of role hierarchy and inheritance and a parent role inherits all the permissions of inherited role [3]. For instance, there can be a role named *staff* and above in the role hierarchy can be a role named *manager* which implicitly inherits the permissions of the staff member. Hierarchical RBAC does provide ease of management but it makes difficult to enforce separation of duty (SoD) constraints. In Attribute-based access control (ABAC), the subjects, objects and the environment have attributes and the access decisions are made based on the boolean function on these attributes. ABAC can subsume RBAC and a detailed discussion on tradeoffs and characteristics of both models can be found in [4].

A number of approaches have addressed the need for formally modeling the authorization policies and verifying their consistency. In this context, authors have proposed a verification framework for conflicts detection in policies modeled through event-driven RBAC in [5]. In [6] authors have proposed a privacy preserving policy model and approach for handling policy conflicts. In [7] authors have proposed an approach to specify and verify authorization policies in the composition of Web services. A formal approach based on Fusion Logic for the specification and verification of properties such as consistency and SoD is discussed in [8]. In [9] authors have proposed an approach based on interval temporal logic for the specification and verification of temporal access control policies. In [10] authors have proposed an approach for the specification of policies in first order logic and then they use Prover9 theorem prover for proving proposed identity constraints. The verification of access control policies for SGAC is addressed in [11]. The authors have used *Alloy* and *ProB*, two first order logic model checkers. In [12] authors have introduced the concept of policy quality in terms of consistency, completeness, and minimality dimensions. There are many approaches that have addressed modeling and verifying the consistency of existing authorization languages. One such language is XACML (eXtensible Access Control Markup Language). Many approaches have been proposed to model and verify the consistency of XACML based policies [13–15]. In [16] authors have analyzed the specifications that handles the combination of authorization and management policies that detects inconsistencies and conflicts in policies. They have also modelled authorization in the behavior of system including policy specifications which is based on Event Calculus, but have not addressed shadowed authorization policies and their conflicts in specific. Authors have also addressed the problem of inconsistencies and conflicts in policies. Abnormal behavior is caused in a system due to these conflicting policies, hence, resolving these policy conflicts is highly significant. Chomicki and Lobo in [17] detect and resolve conflicts in ECA policies through a formal logic-based framework. Bandara in [18] analyzes and manages policies through a tool. The tool helps to query policies for validation and review. In [19] authors present a set of algorithms to check consistency among policies. None have addressed the policy conflicts and policy management in a hierarchal structure specifically targeting policy shadowing.

We believe that there is a research gap concerning both the consistency checking of authorization policies for the real world large scale distributed systems, such as Cloud, and the resolution of conflicts that concern policy management and thus stem from the implementation of policies at a larger scale. A formal ABAC based framework modeling policies and identifying inter and intra policy conflicts that exists between them is presented in [20] and authors have also proposed a model where policies from different cloud providers (AWS, GCP and Microsoft Azure) can be combined in a Multi-Cloud project [21]. This work addresses the conflicts related to policy management in the large scale distributed systems. One such conflict is policy redundancy and the redundant rules in an access control policy increase the size of the policy and would affect the performance and management of policies [22]. To best of our knowledge, there exists no approach that considers the case of shadowed authorization policies in large scale distributed systems. We have motivated the problem by presenting the case of AWS Identity and Access Management (IAM) and AWS Organizations, where policies exist at different levels and are evaluated based on a detailed and complex evaluation logic. We have justified the need for a formal approach and have both modified existing models for performance and correctness and presented new models needed for identifying shadowed policies. We have highlighted the side-effects of having shadowed policies and how they induce more human errors.

3   Case Study - AWS Policies Management

The authentication and authorization management service provided by AWS is called the Identity and Access Management (IAM) service. AWS IAM gives the opportunity of users management and their permissions and thus handles both authentication and authorization aspects. It provides a broad set of services ranging from managing users and their permissions to enabling multi-factor authentication (MFA) including auditing services. Policies are created and assigned to users, groups, roles, and resources, to achieve access management using IAM. When a request is made to access a resource, AWS evaluates different policies to reach a decision. The format of policies storage is JSON documents and follow specific syntax and structure. An example AWS policy is shown in Fig. 1. On a high level, AWS policies contain a set of statements and each statement represents a specific access control rule. Each statement contains the *service and resources* element which specify the AWS service, for instance *Amazon S3* and corresponding resource, for instance a S3 bucket, to which this statement applies. Further, using the *action* element of a statement one can specify what service-specific actions one is willing to perform on the resource specified earlier. The *conditions* element of the statement allows to further specify the conditions

```
{
      "Statement": [
            {
                  "Sid": "Stmt01",
                  "Effect": "Allow",
                  "Action": [
                        "aws-portal:ViewBilling"
                  ],
                  "Resource": [
                        "*"
                  ]
            }
      ]
}
```

Fig. 1. Single statement example from AWS IAM policy

under which the statement applies, for instance specifying the IP addresses from where the request arrives. Finally, the statement *effect* element specifies if the statement outcome is either *Allow* or *Deny* access. AWS supports different types of policies, these include the *Identity-based policies* which are attached to users, groups or roles and these policies grant permissions. The *resource-based policies* are attached to resources such as Amazon S3 buckets. *Permission boundaries* are assigned to users and roles and specify the maximum permissions can be granted to a user or role. In contrast to identity-based policies, permission boundaries on their own does not grant access but only limit the maximum access of identity-based policies. In order to discuss the *service control policies (SCPs)*, we first provide a brief introduction to AWS organizations.

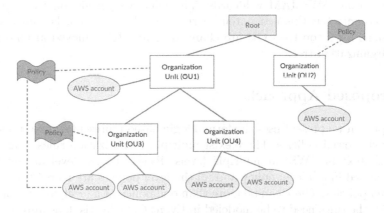

Fig. 2. An example AWS organization structure

AWS Organizations is an account management services by Amazon to provide services such as hierarchical grouping of accounts. Some key concepts include root, which is the top most container for all the accounts within the organization. Within a root, accounts are associated with the *Organization Units (OUs)*. Organization units can be organized in a tree-like structure (with fixed depth) and an *OU* can thus be part of another *OU*. Figure 2 shows an AWS organization with four organization units (*OUs*) each having associated accounts. The root is at the top. The type of access control policies applied at the account level granularity for AWS organizations are called Service Control Policies (SCPs). As similar to permission boundaries SCPs do not grant permissions but rather can be considered as a filter applied to the capabilities of an account. SCPs can be applied to different entities within an organization. If the SCP is applied to the root, it implicitly applies to all the *OUs* and associated accounts, as root is the top of the hierarchy. Similarly, if the SCP is applied to an Organization Unit (*OU*) it applies to that *OU* and any sub *OUs* associated with it.

AWS IAM is an example of hierarchical policy management as many such levels exists and there is a detailed and complicated policy evaluation process to check policies at each level in a top-down order. As the number of levels increase so does the risk of shadowed policies and in case of AWS Organizations, the *OUs* can themselves be nested and the maximum nesting supported by AWS is five *OUs* under the root. The SCPs associated with the higher level OUs are inherited at the sub OUs and thus any services blacklisted at the root results in shadowing SCPs at the lower levels. Before presenting the proposed approach, highlighting the risks of shadowed policies is highly significant. The shadowed policy's behavior is masked and not directly evident and thus a policy designer may tend to under specify the rules assuming them to be handled at the higher level. One such example can be of a permission boundary applied to a user that allows read only access to some resources. As the permission boundary is at the higher level, a policy designer may accidentally allow all actions on these resources using AWS IAM wildcards. The effective permission would remain to be read-only but this is a potential risk bound to happen. For an instance, permissions for a top level OU are changed or an *OU* is moved around in the AWS Organization tree.

4   Proposed Approach

The approach we present uses the formal logic based representation of policies to identify shadowed policies. There exist multiple authorization rules (statements in the context of AWS) at multiple levels. Rules at each level are evaluated and combined into a level policy and all the level policies are merged to identify shadowed policies. There can be more than one policy at a level, as supported by AWS. All the rules need to be modeled in Event-Calculus (is done automatically as the proposed models are generic) and we use a reasoner for Event-Calculus to evaluate rules and level policies. The algorithm below presents an abstract view of our proposed work. The algorithm is not optimized for performance but rather resembles the approach used in our EC models and the one taken by the EC reasoner. For simplicity, we assume that there exists a single policy at a level, however, the proposed approach handles multiple policies at same level, as shown later in Sect. 4. This can be the case when multiple sibling OUs have associate SCPs.

Algorithm 1. Identification of shadowed policies

Require: $rContext$ is the context to identify applicable policies.

```
 1: procedure DETECTSHADOWING(rSet, rContext)
 2:    for each level ∈ levelSet do                          ▷ levelSet is a set of all levels
 3:       for each rule ∈ rSet_level do            ▷ rSet_level is a set of rules at some level
 4:          rdecSet_level ← evaluateRule(rule, rContext)
 5:       end for
 6:       policyDecision ← NotApplicable
 7:       for each rdecision ∈ rdecSet_level do
 8:          if rdecision = deny then
 9:             policyDecision ← Deny
10:             break
11:          end if
12:          pdecSet_level ← policyDecision
13:       end for
14:    end for
15:    for each level ∈ levelSet do
16:       for each nextLevel ∈ levelSet do
17:          if pdecSet_level = Deny & pdecSet_nextLevel = Permit then
18:             pShadow = pShadow ∪ (pdecSet_nextLevel, pdecSet_level)
19:          end if
20:       end for
21:    end for
22: end procedure
```

We iterate through every set of levels (named $levelSet$) and then for each level we iterate through all the rules associated with that level ($rSet_{level}$), lines 2–3. Then, at line 4, we evaluate each $rule$ within the $rSet_{level}$ to identify if it permits or denies the access, or if it is not applicable. The decisions are based on the context and all the decisions are added to the decisions set for a level, $rdecSet_{level}$. Once we have a set of decisions for all rules within a level, we can combine them in a level policy and use a rule combining algorithm to conclude a policy decision. The rule combining algorithms include permit-overrides, deny-overrides and others. In practice, the approach taken by AWS is always deny-overrides, that is a single deny rule can cause the complete policy to be considered denied and we have thus used the same case in our algorithm, lines 7–12. We discuss other combining algorithms at the end of this section. We store the decisions of all level policies in $pdecSet$ and the decision of policy at a $level$ is represented by $pdecSet_{level}$, line 13. We can then identify the shadowed policies by iterating through level policy decisions, lines 15–20. We term a level policy to be shadowing if its level decision is 'deny' and there is a permit decision at the lower level, line 17, and we add the tuples of shadowing and shadowed policies to the set *pShadow*. Once we have identified tuples of shadowing and shadowed policies, it is further possible to identify the rules within the shadowing policy which are causing shadowing. This is needed as the rule combining algorithm being used is deny-overrides and it is possible that a single rule within the level policy is responsible. It can be accomplished by iterating through the rules within the shadowing policy and identifying the rules having deny decision. Let us conclude this section by highlighting the effect of rule combining algorithms and the shadowed policies. The rule combining algorithms specify how the combined individual decisions from multiple rules (and policies) reach a decision. The Deny-overrides combining algorithm considers a policy to be Denied even if it contains a single rule denying the access. Similarly, the Permit-overrides algorithm permits the access if a permit rule is there in the policy. Other algorithms include First-applicable and Only-one-applicable, but we limit our discussion to Permit and Deny-overrides. Let us consider the case of two level policies, pL0 and pL1 and each containing some rules permitting and denying the access. If we consider the deny-overrides to be the combining algorithm, both policies would evaluate to deny decision and even though they have the same decision, there are some permitting rules in pL1 that are being shadowed by pL0. The permit-overrides case is somewhat similar. We address this issue by considering a policy containing rules with multiple decisions to have intra-policy conflict [20] and resolving the conflicts before identifying the shadowed policies.

5   Event-Calculus Formalism

The proposed approach uses Event Calculus. It is a formal language used to represent events and their effects with reasoning. The choice of a formal approach is motivated by a number of factors. First, the authorization rules are evaluated not only based on syntactically matching subjects, objects, actions and decisions but rather on the environment or context as well, which may contain temporal (for example, the timing set for an access policy can be from 9am - 5pm) and other aspects. The subjects and other attributes may themselves have relations (for instance, Alice is member of group Users and some AWS resources is indeed part of S3 bucket). Then, the rules may be combined based on some rule combining algorithms into a policy. The policy shadowing itself can be based on a number of related aspects, such as the rule combining algorithms we discussed earlier or it may be the case that shadowing occurs only for specified time intervals or in some delegated scenario. The use of a formal expressive approach helps in collectively addressing these related aspects. In addition, EC has open-source tool support, *DECReasoner*[1]. The basic elements in EC are *events* (or actions), *fluents* (whose value can change on different time-points based on occurrence of events), and a set of predicates. Some predicates used in our models include the *Initiates(e, f, t)* predicate which specifies that if event e happens at time-point t then the fluent f holds after t. Similarly the *Terminates(e, f, t)* predicate

---

[1] http://decreasoner.sourceforge.net.

specifies that if event e happens at t then the fluent f does not hold after t. The *Happens(e, t)* predicate specifies that event e happens at timepoint t and the *HoldsAt(f, t)* is true iff fluent f holds at timepoint t. We use Event-Calculus [23] and we will only deal with the models that are simple and shows important aspects, without including the supporting axioms[2].

5.1 Rules Specification

The *rules* construct specifies one access rule. Each rule includes *Target*, an *Effect* and the *Conditions* associated with it. The EC meta-model for rules specification is shown below. The basic idea is to first decide if the rule is applicable in some context (achieved using RuleTargetHolds fluents and Match/Mismatch events) and then decide if the rule has any of the following conditions: permit, deny or not applicable, using fluents *RuleIsPermitted/Denied/NotApplicable* and Approve/DenyRule/RuleDsntApply events. The fluents are initialized in a way that when time=0, they do not hold and the reasoner should try to find a solution leading from initial state to reach the goal.

Model 1 (Meta-model for IAM Rules)

```
;Sorts for rules and their elements
sort rule, subject, object, action

;Fluents for Rules evaluation
fluent RuleTargetHolds(rule), RuleConditionHolds(rule)
fluent RuleEffectIsPermit(rule), RuleIsPermitted/Denied/NotApplicable(rule)
;Events for Rules evaluation
event (Mis)Match(rule), Approve/DenyRule(rule), RuleDsntApply(rule)

;These axioms link fluents with events
Initiates/Terminates (Match/Mismatch(rule), RuleTargetHolds(rule), time).
Initiates(Approve/DenyRule(rule), RuleIsPermitted/Denied(rule), time).
Initiates(RuleDsntApply(rule), RuleIsNotApplicable(rule), time).

;Conditions on events occurrence
Happens(ApproveRule(rule), time) → HoldsAt(RuleTargetHolds(rule), time) &
& HoldsAt(RuleEffectIsPermit(rule), time).
Happens(RuleDsntApply(rule), time) → !HoldsAt(RuleTargetHolds(rule), time).

;Initial state of the Fluents
!HoldsAt(RuleIsPermitted/Denied/NotApplicable(rule),0).
;The goal for the reasoner
HoldsAt(RuleTargetHolds(rule),1) / !HoldsAt(RuleTargetHolds(rule),1).
HoldsAt(RuleIsPermitted/Denied/NotApplicable(rule),2).
```

The model given above shows some EC sorts which can be considered as types for instantiating individual elements. For instance, *rule SomeRule* in an EC model would declare *SomeRule* to have type *rule*. We have used Initiates along with the definition of fluents, events and Terminates axioms to link them together. For instance, the Initiates axiom for RuleTargetHolds axiom specify that if the event Match happens at time point t, the fluent RuleTargetHolds

---

[2] Complete models along with setup and execution instructions are available at https://www.icloud.com/iclouddrive/0E4u-NuXGiGkpoql5BMamWhGQ#wise19.

would hold at t+1 and afterwards. The Terminates axiom for Mismatch event has opposite effect. The model above (and all other core models) are organized into files to be included in an EC model for any specific rule. This helps us in both manageability and have allowed us to develop automated tools to directly convert policies fetched from the Cloud (in JSON) to EC models. For instance, we modelled a specific rule, *L0Rule1*, to show the usage of generic model, given below:

```
Model 2 (Level0 rule specification)
load includes/rules/... ;generic model files
load includes/input.e ;Contextual attributes would be specified in input.e

rule L0Rule1
;Specifying when the rule target holds
Happens(Match(L0Rule1),time) →
{subject, object, action} subject = Alice & object = SomeRsrc & action = SomeActn.
Happens(Mismatch(L0Rule1),time)→subject!=Alice /object!=SomeRsrc /action!=SomeActn.
!HoldsAt(RuleEffectIsPermit(L0Rule1),0).
```

At first, the meta-model files are included, as shown earlier. The contents of file input.e will provide the context under which this rule needs to be evaluated such as the value of subject, object and action attributes. The rule is named as *L0Rule1* and we then a conditional axiom is stated that the event *Match* only holds true if there exists any match in the attribute name value pairs. We define the fluent *RuleEffectIsPermit* not to hold true at time=0, that is the rule denies the access and not permits. The rule model itself is very simple thanks to separating the core meta-model.

```
Solution 1 (Rule evaluation using DECReasoner)
0
RuleEffectIsPermit(L0Rule1).
Happens(Match(L0Rule1), 0).
1
+RuleTargetHolds(L0Rule1).
Happens(DenyRule(L0Rule1), 1).
2
+RuleIsDenied(L0Rule1).
```

If EC reasoner, *DECReasoner* are invoked, the solution shown above is returned. The reasoner first encodes the EC model in a SAT problem invoking an off the shelf SAT-solver (relsat in this case). The models found are then formatted to show events occurrence and fluents state at specific time-points. In this case, the event match happens at time-point 0 (as the value species in the input.e match the ones specify in the rule) and thus the RuleTargetHolds fluent holds at time-point 1 (shown with a + sign). Further, as the rule effect is not specified to be permit, the event *DenyRule* happens and the rule is considered denied. We can similarly model multiple rules in a level. We consider following additional rules, L0Rule2 concerns the user Bob and thus it does not apply. Similarly, the rule L0Rule3 concerns *SomeOtherActn* instead of *SomeActn* as specified in the input.e, so it does not apply as well. As per our shadowing identification logic, the rules at every level are grouped in a level policy. Instead of detailing policy meta-model and instantiation, we present the solution of level L0 policy that groups the three rules at level L0.

Solution 2 (Level policy evaluation using DECReasoner)

model 1:
0
Happens(Match(L0Rule1)/Mismatch(L0Rule2)/Mismatch(L0Rule3), 0).
1
+RuleTargetHolds(L0Rule1). Happens(DenyRule(L0Rule1), 1).
Happens(RuleDoesntApply(L0Rule2), 1). Happens(RuleDoesntApply(L0Rule3), 1).
2
+RuleIsDenied(L0Rule1). +RuleIsNotApplicable(L0Rule2).
+RuleIsNotApplicable(L0Rule3).Happens(DenyPolicy(L0Policy), 2).
3
+PolicyIsDenied(L0Policy).

The policy at level L1 is thus denied as it contains at least one rule as denying access. The policy evaluation algorithms in our case is chosen to be deny-overrides, permit-overrides and not applicable otherwise. As discussed earlier a policy cannot contain both permit and deny rules and is considered a conflict. In order to present policy shadowing identification models, let us consider another level L1, having two policies *L11Policy* and *L12Policy* (as per our naming convention, L11 means first policy of level 1). Both policies have three separate rules but intentionally tailored to have L11Policy resulting in permitting the access, while the L12Policy is not applicable. For the identification of shadowed policies, we can group multiple level policies and use EC axioms to identify shadowed policies.

Model 3 (Meta-model for the identification of shadowed policies)

predicate ParentOf(policy, policy) fluent PolicyShadowed(policy, policy)

event Shadowing(policy, policy) event NoShadowing(policy, policy)

Initiates(Shadowing(policy1,policy2), PolicyShadowed(policy1,policy2), time).
Terminates(NoShadowing(policy1,policy2), PolicyShadowed(policy1,policy2), time).

!HoldsAt(PolicyShadowed(policy1,policy2),0).
Happens(Shadowing(policy1,policy2),time) & ParentOf(policy1,policy2) →
HoldsAt(PolicyIsDenied(policy1), time) & HoldsAt(PolicyIsPermitted(policy2), time).

Happens(NoShadowing(policy1,policy2),time) & ParentOf(policy1,policy2) →
(!HoldsAt(PolicyIsDenied(policy1), time) /
(HoldsAt(PolicyIsDenied(policy1),time) & HoldsAt(PolicyIsDenied(policy2),time)) / (Holds
At(PolicyIsNotApplicable(policy1),time) / HoldsAt(PolicyIsNotApplicable(policy2),time))).

In the model above, we first define a predicate *ParentOf*, which defines the relationship amongst policies. Then, we define a fluent named *PolicyShadowed*, whose state would eventually represent if a policy is shadowing another policy. We then define some events and Initiates and Terminates axioms to link these events with the fluent. We further define some axioms to define that the event

Shadowing can only happen amongst policies if there exists a *ParentOf* predicate amongst them and if the parent policy is denying the access and the child policy is permitting the access. Similarly, we define that the event *NoShadowing* can only happen if there exists a *ParentOf* predicate amongst policies and there is no deny or permit relation amongst policies.

Model 4 (Model for aggregating level policies)

```
load includes/rules/... ;generic model files
load includes/policy/defined/L0Policy/L11Policy/L12Policy.e
load includes/input.e
;Contextual attributes would be specified in input.e

[policy1, policy2] (policy1 = L0Policy & policy2 = L11Policy) /
(policy1 = L0Policy & policy2 = L12Policy) <-> ParentOf(policy1, policy2).
```

The meta-model can be instantiated to specify a specific model for identifying shadowed policies and in the model above, we first include policy files for different policies and then define the *ParentOf* relations amongst them. More specifically, we define that *L0Policy* is parent of both *L11Policy* and *L12Policy*. The complete solution showing the evaluation results for policies and associated rules (if they are permitted, denied or not applicable) is returned, when the DECReasoner is invoked for the instantiated model above. The solution shows that the *L0Policy* does not shadow *L12Policy*, as one is denying access and the other is not applicable. The solution also shows that the *L0Policy* does shadow *L11Policy*, as one is denying the access and other is permitting the access. We have thoroughly tested our models on a number of complex configurations.

Solution 3 (Shadow policies identification using DECReasoner)

```
model 1:
0
RuleEffectIsPermit(L11Rule... L12Rule3).Happens(Match(L0Rule1/L11Rule1), 0).
Happens(Mismatch(L0Rule2/...L12Rule3), 0).
1
+RuleTargetHolds(L0Rule1/L11Rule1). Happens(ApproveRule(L11Rule1), 1).
Happens(DenyRule(L0Rule1), 1).
Happens(RuleDoesntApply(L0Rule2...L12Rule3), 1).
2
+RuleIsPermitted(L11Rule1). +RuleIsDenied(L0Rule1).
+RuleIsNotApplicable(L0Rule2...L12Rule3).
Happens(ApprovePolicy(L11Policy), 2). Happens(DenyPolicy(L0Policy), 2).
Happens(PolicyDoesntApply(L12Policy), 2).
3
+PolicyIsDenied(L0Policy). +PolicyIsNotApplicable(L12Policy).
+PolicyIsPermitted(L11Policy).
Happens(NoShadowing(L0Policy, L12Policy), 3).
Happens(Shadowing(L0Policy, L11Policy), 3).
4
+PolicyShadowed(L0Policy, L11Policy).
```

---

6   Performance Evaluation

For testing the correctness and scalability of our approach, we have created different test cases and evaluated them on Amazon EC2 c5.xlarge instance having

4 vCPUs and 8 GiB memory running Ubuntu Server 16.04 LTS. We have setup *DECreasoner* on the EC2 instance and have used the modified and improved version as proposed in [24]. Three test cases were evaluated; first we increase the number of levels with each level having maximum of two children each. As per our shadowed policies identification algorithm, multiple rules at a level are organized in a level policy. However, there can be multiple policies directly assigned to a level (as is the case with AWS Organizations) and these policies are at the same level. So, in the first test case, we assume the maximum policies at a level to be two. For the second test case we increase the number of policies at a level to a maximum of five and finally as an extreme case, we create a full binary tree having $2^h - 1$ policies, where h is the tree height (or implicitly the number of levels + 1). All the policies contain three rules and they are intentionally tailored to make the policies deny, permit and non applicable. Thus, if we have three policies, one is denied, the second one is permitted and the last one is not applicable.

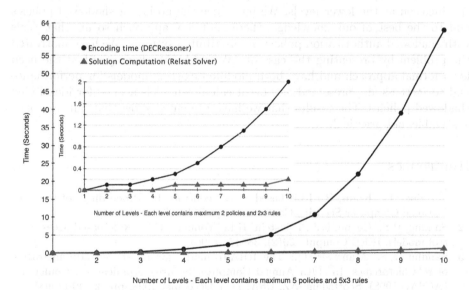

Fig. 3. Performance evaluation results

The performance evaluation results are shown in Fig. 3. In order to manage space limitations, we have merged the results of two different cases in a single figure. For both the cases, the Y-axis shows the time-taken in seconds while the X-axis shows the increase in the number of levels. The performance results are encouraging even though the EC to SAT encoding process does not scale well, a well known limitation of *DECReasoner*. The shadowing identification process does not require strict response time guarantees and is a occasional process used by policy designers to better manage the security policies of an organization. For simpler models, having two policies at a level and probably more common occurring use case, even at ten levels time taken by both encoding process and the SAT solver is around 2 s. For complex models, having five policies at a level and with 10 levels, the performance results are acceptable. We have evaluated further complex scenarios, which are rare to experience in practice, where we have a full binary tree and thus at a height of 6 the total number of policies is 63, each having three distinct rules. In this extreme scenario, not shown in the Fig. 3 due to space limitations, the time taken for EC to SAT encoding process is 3.5 min and the solution takes 2.9 s.

7   Conclusion

The proposed approach in the paper identifies and addresses issues concerning the management of hierarchical authorization policies in the Cloud based systems. These policy models pose the risk of policy shadowing where the decision taken at higher levels mask the possibly erroneous or conflicting policies specification at the lower levels. We introduce the notion of shadowed policies and to the best of our knowledge, there isn't any approach so far that deals with shadowed authorization policies in distributed systems. We have motivated the problem by presenting the case of AWS IAM. We have justified the need for a formal approach and have both modified existing models for performance and correctness and presented new event-calculus models needed for identifying shadowed policies. The results that we have presented show that our approach is scalable and practical.

References

1. Ferraiolo, D., Kuhn, R.: Role-based access controls. In: Proceedings of the 15th National Computer Security Conference, pp. 554–563 (1992)
2. Sandhu, R.S., Coyne, E.J., Feinstein, H.L., Youman, C.E.: Role-based access control models. IEEE Comput. **29**(2), 38–47 (1996)
3. Sandhu, R.S., Munawer, Q.: The RRA97 model for role-based administration of role hierarchies. In: 14th Annual Computer Security Applications Conference (ACSAC 1998), Scottsdale, AZ, USA, 7–11 December 1998, pp. 39–49 (1998)
4. Coyne, E., Weil, T.R.: ABAC and RBAC: scalable, flexible, and auditable access management. IT Prof. **15**(3), 14–16 (2013)
5. Shafiq, B., Vaidya, J., Ghafoor, A., Bertino, E.: A framework for verification and optimal reconfiguration of event-driven role based access control policies. In: 17th ACM Symposium on Access Control Models and Technologies, SACMAT 2012, Newark, NJ, USA, 20–22 June 2012, pp. 197–208 (2012)
6. Wang, H., Sun, L., Bertino, E.: Building access control policy model for privacy preserving and testing policy conflicting problems. J. Comput. Syst. Sci. **80**(8), 1493–1503 (2014)
7. Rouached, M., Godart, C.: Specification and verification of authorization policies for web services composition. In: CAiSE 2007 Forum, Proceedings of the CAiSE 2007 Forum at the 19th International Conference on Advanced Information Systems Engineering, Trondheim, Norway, 11–15 June 2007 (2007)

8. Cau, A., Janicke, H., Moszkowski, B.C.: Verification and enforcement of access control policies. Formal Methods Syst. Design **43**(3), 450–492 (2013)
9. Janicke, H., Cau, A., Siewe, F., Zedan, H.: Dynamic access control policies: specification and verification. Comput. J. **56**(4), 440–463 (2013)
10. Sabri, K.E.: Automated verification of role-based access control policies constraints using prover9. CoRR abs/1503.07645 (2015)
11. Huynh, N., Frappier, M., Mammar, A., Laleau, R.: Verification of SGAC access control policies using alloy and prob. In: 18th IEEE International Symposium on High Assurance Systems Engineering, HASE 2017, Singapore (2017)
12. Bertino, E., Jabal, A.A., Calo, S.B., Verma, D.C., Williams, C.: The challenge of access control policies quality. J. Data Inf. Qual. **10**(2), 6 (2018)
13. Turkmen, F., den Hartog, J., Ranise, S., Zannone, N.: Formal analysis of XACML policies using SMT. Comput. Secur. **66**, 185–203 (2017)
14. Nguyen, T.N., Thi, K.T.L., Dang, A.T., Van, H.D.S., Dang, T.K.: Towards a flexible framework to support a generalized extension of XACML for spatio-temporal RBAC model with reasoning ability. In: ICCSA, vol. 5 (2013)
15. Kolovski, V., Hendler, J.A., Parsia, B.: Analyzing web access control policies. In: WWW, pp. 677–686 (2007)
16. Bandara, A.K., Lupu, E., Russo, A.: Using event calculus to formalise policy specification and analysis. In: 4th IEEE International Workshop on Policies for Distributed Systems and Networks (POLICY 2003), Lake Como, Italy (2003)
17. Chomicki, J., Lobo, J., Naqvi, S.: A logic programming approach to conflict resolution in policy management. In: 7th International Conference on Principles of Knowledge Representation and Reasoning (KR 2000), Morgan Kaufman (2000)
18. Bandara, A.K.: Formal approach to analysis and refinement of policies. PhD thesis, University College London, University of London (2005)
19. Agrawal, D., Giles, J., Lee, K.W., Lobo, J.: Policy ratification. In: Sixth IEEE International Workshop on Policies for Distributed Systems and Networks (POLICY 2005), pp. 223–232 (2005)
20. Zahoor, E., Asma, Z., Perrin, O.: A formal approach for the verification of AWS IAM access control policies. In: De Paoli, F., Schulte, S., Broch Johnsen, E. (eds.) ESOCC 2017. LNCS, vol. 10465, pp. 59–74. Springer, Cham (2017). https://doi.org/10.1007/978-3-319-67262-5_5
21. Zahoor, E., Ikram, A., Akhtar, S., Perrin, O.: Authorization policies specification and consistency management within multi-cloud environments. In: Gruschka, N. (ed.) NordSec 2018. LNCS, vol. 11252, pp. 272–288. Springer, Cham (2018). https://doi.org/10.1007/978-3-030-03638-6_17
22. Guarnieri, M., Neri, M.A., Magri, E., Mutti, S.: On the notion of redundancy in access control policies. In: 18th ACM Symposium on Access Control Models and Technologies, SACMAT 2013, Amsterdam, The Netherlands, 12–14 June 2013, pp. 161–172 (2013)
23. Mueller, E.T.: Commonsense Reasoning. Morgan Kaufmann Publishers Inc., Burlington (2006)
24. Zahoor, E., Perrin, O., Godart, C.: An event-based reasoning approach to web services monitoring. In: ICWS (2011)

Web-Based Applications

A Dynamic Decision-Making Method Based on Ensemble Methods for Complex Unbalanced Data

Dong Chen[1], Xiao-Jun Wang[2(✉)], and Bin Wang[1(✉)]

[1] Key Laboratory of Advanced Design and Intelligent Computing,
Ministry of Education, Dalian University, Dalian 116622, China
wangbinpaper@gmail.com
[2] School of Management Science and Engineering,
Dongbei University of Finance and Economics (DUFE), Dalian 116025, China
wxjjessicaxj0903@126.com

Abstract. Class imbalance has been proven to seriously hinder the precision of many standard learning algorithms. To solve this problem, a number of methods have been proposed, for example, the distance-based balancing ensemble method that learns the unbalanced dataset by converting it into multiple balanced subsets on which sub-classifiers are built. However, the class-imbalance problem is usually accompanied by other data-complexity problems such as class overlap, small disjuncts, and noise instance. Current algorithms developed for primary unbalanced-data problems cannot address the complex-data problems at the same time. Some of these algorithms even exacerbate the class-overlap and small-disjuncts problems after trying to address the complex-data problem. On this account, this study proposes a dynamic ensemble selection decision-making (DESD) method. The DESD first repeats the random-splitting technique to divide the dataset into multiple balanced subsets that contain no or few class-overlap and small-disjunct problems. Then, the classifiers are built on these subsets to compose the candidate classifier pool. To select the most appropriate classifiers from the candidate classifier pool for the classification of each query instance, we use a weighting mechanism to highlight the competence of classifiers that are more powerful in classifying minority instances belonging to the local region in which the query instance is located. Tests with 15 standard datasets from public repositories are performed to demonstrate the effectiveness of the DESD method. The results show that the precision of the DESD method outperforms other ensemble methods.

Keywords: Dynamic ensemble selection · Unbalanced dataset · Classification

1 Introduction

Class imbalance refers to the scenario in which the number of training instances in different categories varies greatly, and it is a common phenomenon in classification tasks [1–3]. For instance, if some software were tested for detecting a specific fault, only a small portion of them would be faulty and the remaining software would be fault free. In real-world applications, the accuracy of detecting the minority class is vitally

© Springer Nature Switzerland AG 2019
R. Cheng et al. (Eds.): WISE 2019, LNCS 11881, pp. 359–372, 2019.
https://doi.org/10.1007/978-3-030-34223-4_23

important because the cost of misclassifying the minority class is more expensive compared to misclassifying the majority class. For example, in comparing the cost of misclassifying a cancer patient as a non-cancer patient to the cost of misclassifying a non-cancer patient as a cancer patient, in the former case the misclassification may result in one person dying, but in the latter case more diagnosis and treatment will be required. In summary, the accuracy of predicting the minority class is critical.

Generally speaking, learning from unbalanced data is a challenging task in data mining and machine learning [1–3]. In conventional methods of solving the classification problem, a classifier is generally trained on a balanced dataset. As such, this classifier will strongly favor the majority class and ignore the minority class when learning unbalanced datasets. Therefore, many algorithms have been proposed to address the class-imbalance problem [1, 3]. For example, the distance-based balancing ensemble (DBE) algorithm learns the unbalanced data by converting it into multiple balanced subsets on which sub-classifiers are built, and then integrating the outputs of the sub-classifiers using the distance-based combination rule (DCR) [4].

Most of these algorithms focus on solving the primary unbalanced-data problem. However, the class-imbalance problem is usually accompanied by other data-complexity problems, such as class overlap, small disjuncts, and noise instances [1, 5]. Current algorithms developed for the primary unbalanced-data problem cannot address the complex-data problem at the same time. Some of these algorithms even exacerbate the class-overlap and small-disjunct problems after trying to address the complex-data problem. These complex-data problems will deteriorate the performance of the existing algorithms developed for the primary unbalanced-data problem and finally result in the loss of performance of these algorithms [6–8].

To solve the problem of class imbalance with complex data, in this paper we propose a dynamic ensemble selection decision-making (DESD) method. The DESD consists of three components:

(1) Generation of the candidate classifiers pool. The DESD repeats the random-splitting technique to divide the dataset into multiple balanced subsets that contain no or few class-overlap and small-disjuncts problems. Then, the classifiers are built on these subsets to compose the candidate classifier pool.

(2) Dynamic selection of the most appropriate ensemble of classifiers. To select the most appropriate classifiers from the candidate classifier pool for the classification of each query instance, the competence of candidate classifiers is evaluated using the weighted instances in the neighborhood surrounding the query example; that is, we intend to select the classifiers that correctly classify more minority-class instances belong to the local region where the query instance is located.

(3) Integration of the selected candidate classifiers. Each selected classifier will give the individual classification result for predicting the query instance. Here, the outputs are integrated using the DCR.

To verify the effectiveness of the DESD method, we carried out an extensive experimental design. The proposed DESD method was evaluated on 15 publicly available standard datasets. G-mean and area under receiving operator characteristics graph (AUC) are used as the performance indicators.

The remainder of this study is organized as follows: In Sect. 2, the description of complex unbalanced data, dynamic ensemble selection, and combination rule are presented. In Sect. 3, the DESD method is proposed. In Sect. 4, experimental investigations of the DESD are carried out. In Sect. 5, conclusions are summarized.

2   Related Work

Classification of unbalanced data has attracted attention for some time and has been widely studied. However, most studies only focused on the primary unbalanced-data problem and ignored other complex-data problems. Meanwhile, DES has been a very active area of research in machine learning [9]. The combination rule in DES that we took here was the DCR in the DBE [4]. To clearly present our findings, we briefly review these works in the following subsections.

2.1   Difficulties in Learning Complex Unbalanced Data

Data complexity is a broad term that includes the lack of representative instances, class overlap, small disjuncts, and noise instance. To clarify the data-complexity problem, Fig. 1 shows artificially created two-dimensional unbalanced data in which the number of star points is significantly less than that of circular points; that is, the circular points represent the majority class and the star points the minority class.

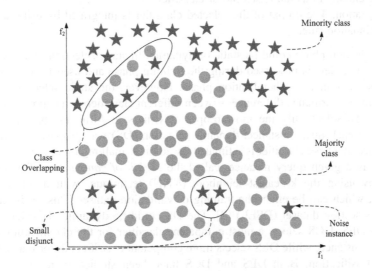

Fig. 1. Example of complex unbalanced data (figure is adapted from [5]).

The first problem of data complexity is the lack of representative instances because the minority class is very limited. In this case, the lack of representative data about the target concept makes the learning task difficult and affects the final classification result. The second problem of data complexity is class overlap. Some different class instances have similar values on the same attribute. As shown in Fig. 1, in the feature space, some minority (majority) instances are located in the region of the majority (minority) instances. Researchers have found that class overlap often occurs near decision boundaries. The third and fourth problems are small disjuncts and noise instance. It is worth noting the difference between them. Ideal classifiers should cover the small disjuncts, while the noise instances should be ignored because they have a negative impact on the performance of the classifier. However, it is difficult to learn from noise and unbalanced data, because noisy instances may induce a set of small disjuncts. For example, in Fig. 1, there are two small disjuncts and one noisy instance. Suppose a classifier generates disjunct for the noisy instance; then, this disjunct would be illegitimate with respect to the other disjuncts that are legitimately formed from severely under-represented sub-concepts.

2.2   Dynamic Ensemble Selection

Multiple classifier systems (MCSs) [10], which aim to classify the query instances by using a pool of base classifiers, have been proven to be superior to any base classifiers in various classification problems. A MCS consists of three parts.

(1) Generation: The candidate classifier pool is generated by using the learning dataset.
(2) Selection: This part finds the set of appropriate classifiers from the candidate classifier pool for the classification of the query instances. We refer the subset of base classifiers as the ensemble of classifiers.
(3) Integration: The output of the selected classifier is integrated by using a specific combination rule.

The selection phase could be static or dynamic. In static selection, the selection of ensemble classifiers is processed during the training phase and used for classifying all query instances in the generalization phase. However, the static selection does not consider the performance difference between different classifiers when predicting query instances. To select only the most proper classifiers to form an ensemble for the selection of each query instance, dynamic selection (DS) was proposed. Usually, the competence of a base classifier is estimated based on a small region in the feature space surrounding a given query instance, called the *region of competence*. This region is formed by using the k-nearest-neighbors (KNN) technique, with a set of labeled instances, which can be either the training or validation dataset. This set is called the dynamic selection dataset (DSEL) [9]. The DS could be dynamic classifier selection (DCS) or DES. DCS selects a most competent classifier for the classification of each new query instance, while DES selects more competent classifiers to form an ensemble for the classification. Both DES and DCS have been studied in recent years, and numerous papers are available examining them [11–17]. Among these, Ko *et al.* [12] introduced four DES methods inspired by Oracle, and further demonstrated the superiority of DES over DCS.

2.3   Distance-Based Combination Rule

Each of selected classifiers will give the individual classification result for predicting the query instance. It is then necessary to combine the results given by these sub classifiers. Assume that there are K binary class data and for each data the class labels are C_1 and C_2. Then, we could obtain K classifiers by applying a specific classification algorithm to each of these K binary class data. The ith classifier ($1 \leq i \leq k$) classifies query instance as C_j with the probability P_{ij} ($1 \leq j \leq 2$). The DCR [4] classifies the query instance as label C_j according to

$$C_j = argmax_{1 \leq j \leq 2} max_{1 \leq i \leq K} \frac{P_{ij}}{exp(D_{ij})}, \qquad (1)$$

where $D_{ij}(1 \leq i \leq k, 1 \leq j \leq 2)$ is the average Euclidean distance from the query instance to the data with class C_j in the ith subset. The DCR employs a distanced-based mechanism because the query instance is likely to be classified into the class with the shortest average Euclidean distance to the query instance.

3 Dynamic Ensemble Selection Decision-Making

In this section, we present our proposed DESD method for complex unbalanced data, expanding on the DBE and dynamic selection of classifiers, which consists of the following three steps.

Generation of Candidate Classifiers Pool. DESD repeats the random-splitting technique to divide the dataset into multiple balanced subsets. After data splitting, the complex dataset is decomposed into multiple subsets containing no or few class-overlap and small-disjuncts problems. Then, the classifiers are built on these subsets to compose the candidate classifier pool.

Dynamic Selection of Most Appropriate Ensemble of Classifiers. In classic DES methods, the distribution of instances within the region of competence is not considered, and it makes the final decision lean towards the majority class. To extend the DES approach to a complex unbalanced data scenario, we develop a novel weighting method to outstand the competence of a candidate classifier with more power in classifying the minority-class instances; that is, the instances belonging to the minority classes within the neighborhood of a query example have greater weights when evaluating the level of competence of a classifier in the candidate classifier pool.

Integration of Selected Candidate Classifiers. Each selected candidate classifier will give an individual classification result for predicting the query instance. Obviously, it is necessary to integrate these classification results. Here, these results are integrated by the DCR.

Since the third step is nothing but directly applying the DCR to integrate the selected classifiers, we will introduce the first two steps in the following sections.

3.1 Generation of Candidate Classifiers Pool

Data complexity is an obstacle in learning from unbalanced data. The instances in different categories should be balanced without complex data. At present, most of the data pre-processing techniques in DES use the bagging strategy [14, 16]; that is, sampling the original learning set with a replacement, the classifiers are built on these sampled datasets to generate the candidate classifiers pool. However, the datasets obtained by the bagging sampling strategy on the unbalanced dataset are still unbalanced, and the generalization performance of the ensemble model is still poor. The DBE [4] divides the majority instances into multiple majority subsets with sizes equal to twice the number of the minority-class instances. The algorithm combines each majority subset with the minority-class instances to form multiple unbalanced subsets, and then applies a modified adaptive semi-unsupervised weighted oversampling (A-SUWO) method in each subset to obtain balanced subsets for building classifiers.

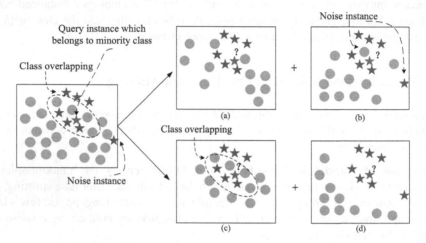

Fig. 2. Complex data after random splitting is decomposed into multiple subsets that contain no or few class-overlap and small-disjunct problems.

However, the subset generated by the DBE method may still have complex data in the feature space, as shown in Fig. 2. Figure 2(a) and (b)–(d) show the results of random splitting twice, respectively. We find that the subsets (b) and (c) still contain the noise instances and class-overlap problem, while subsets (a) and (d) contain no complex data. It is reasonable for us to repeat the random-splitting technology many times to obtain multiple subsets in which contain no or few class-overlap and small-disjuncts problems, and finally obtain multiple subsets with an imbalance ratio of 2 between the majority and minority instances. The modified A-SUWO method is also applied in these subsets to balance the subset. Afterwards, the selected base learner is trained using the generated balanced dataset to compose the candidate classifiers pool.

3.2   Dynamic Selection of Most Appropriate Ensemble

The subsets generated by random splitting may still have complex data. Involving the classifiers built on such subsets in the ensemble of classifiers will directly affect the final classification result, so it is necessary to select the competent classifiers from the candidate classifier pool for the classification of each query instance x_{query}. This considered, the most important step is how to measure the competence of candidate classifiers. Although many methods have been proposed to estimate the classifier's competence, all of them were presented based on the premise of a balanced scenario.

To select the proper ensemble of classifiers that are more powerful in classifying minority-class instances, we propose a dynamic selection algorithm, the detailed procedure of which is described in Algorithm 1. The candidate classifiers that correctly classify more minority-class instances belonging to the region of competence will have a higher competence.

Thus, our main goal is to describe the procedure for selecting a proper ensemble for each query instance in the unbalanced-data scenario. Here, the key step is to estimate the performance of candidate classifiers belonging to the region of competence for each query instance that needs to be classified. The region of competence is defined by using the k nearest neighbors of the query instance (lines 1–3 of Algorithm 1). Our purpose is to develop a DES classifier system to handle the issues present in complex imbalanced datasets. Hence, we intend to select the classifiers that are more powerful when classifying the minority-class instances in the region of competence.

This considered, we have developed a weighting mechanism that is described in lines 7–10 of Algorithm 1, in which the classification result of the instances in the region of competence are weighted. In this way, the ability of classifying the minority instances is emphasized while ensuring the accuracy of classification. Then, the competence of each candidate classifier is calculated in line 10, in which an adaptive weight adjustment procedure is embedded in the DES criteria; that is, classifiers with higher precision in classifying minority instances in the region of competence are associated with higher competence when the classification accuracy in the region of competence is the same. Finally, the ensemble of classifiers is combined by the DCR.

Algorithm 1 Dynamic Selection Algorithm

Input	C, candidate classifier pool; $D_{te,}$ test dataset; $D_{va,}$ validation dataset
	k, number of nearest neighbors
	$P\%$, percentage of classifiers to be selected
Output	ensemble of classifiers for each query instance $C^{'}$

1:   *For* each query instance x_{query} in D_{te}

2:      $C^{'} = \emptyset$

3:      find Ψ as the k nearest neighbors of the instance x_{query} in D_{va}

4:      x_{min} is the minority-class instances belonging to the region of Ψ

5:      y_{min} is the minority class label of x_{min}

6:      num_{min} is obtained by counting the number of x_{min}

7:      *For* each classifier h_i in C

8:
$$recall_{min} = \begin{cases} 1, \Psi \text{ consist of one class} \\ \frac{\sum_{j=1}^{num_{min}} I(h_i(x_{min(j)})=y_{min(j)})}{num_{min}}, otherwise \end{cases}$$

9:      $acc = \frac{\sum_{j=1}^{k} I(h_i(x_j)=y_j)}{k}$

10:     $W_i = recall_{min} * acc$

11:     *End*

12:     select $P\%$ most competent classifiers in C to compose the ensemble of classifiers $C^{'}$ for the query instance x_{query}

13: *End*

---

4   Experimental Study

To verify the effectiveness and usefulness of the proposed DESD in the scenario of complex unbalanced data, we designed an experimental study consisting of two investigations.

(1) Can the proposed dynamic selection algorithm be an effective method to improve the classification performance when dealing with the complex unbalanced-data problem?

(2) Are the classification results provided by the DESD method more precise than those provided by the state-of-the-art methods developed for unbalanced data?

Table 1. Statistical summary of the 15 complex unbalanced datasets.

ID	Dataset	ATT	Instances	Minority	Majority	Imbalanced ratio
1	ecoli3	8	336	35	301	8.60
2	ecoli034vs5	8	200	20	180	9.00
3	ecoli0234vs5	8	202	20	182	9.10
4	yeast0256vs3789	9	1004	99	905	9.14
5	ecoli046vs5	7	203	20	183	9.15
6	ecoli0347vs56	8	257	25	232	9.28
7	glass016vs2	10	192	17	175	10.29
8	glass0146vs2	10	205	17	188	11.06
9	glass2	10	214	17	197	11.59
10	ecoli0146vs5	7	280	20	260	13.00
11	ecoli4	8	336	20	316	15.80
12	pageblocks13vs4	11	472	28	444	15.86
13	glass5	10	214	9	205	22.78
14	yeast4	9	1484	51	1433	28.10
15	yeast1289vs7	9	947	30	917	30.57

We performed these experiments on 15 publicly standard datasets from the Keel dataset repository [18]. Table 1 presents the characteristics of these datasets, including the number of attributes (ATT), the number of datasets (Instances), the number of minority instances, the number of majority instances, and the imbalance ratio for each dataset. For more detailed information on the adopted datasets, interested researchers can refer to http://sci2s.ugr.es/keel/imbalanced.php.

In this study, G-mean and AUC are selected to use as the performance indicators. To introduce these performance indicators, we first review the confusion matrix (Table 2) to indicate the correct and incorrect classifications of the instance. Minority classes are positive and majority classes are negative in the confusion matrix.

G-mean is determined as follows:

$$G_{mean} = \sqrt{\frac{TP}{TP+FN} * \frac{TN}{TN+FP}}.$$ (2)

Table 2. Confusion matrix.

	Predicted positive class	Predicted negative class
Actual positive class	TP	FN
Actual negative class	FP	TN

AUC is the area under the ROC graph and is not sensitive to the distribution of the two classes, which makes it suitable as a performance measure for the unbalanced-data problem. The ROC graph is obtained by plotting the true positive rate (TPR) over the false positive rate (FPR), which are defined as follows:

$$TPR = \frac{TP}{N_p} , FPR = \frac{FP}{N_n},$$ (3)

where N_p is the number of positive (minority) instances and N_n the number of negative (majority) instances.

The parameter k in Algorithm 1 is set to 7, which was chosen by cross-validation. The random-splitting technique was repeated until the candidate ensemble consisted of 100 base learners. The percentage of classifiers to be selected in the candidate classifiers pool is set to 15 ($P = 15$ in Algorithm 1). With respect to the methods selected to carry out the comparison, the same parameter settings were used as in the literature.

To obtain objective results for the performance measure, we used a 5-fold cross-validation strategy in the experiment. To avoid misleading results due to occasional bad division of the training and testing datasets, we repeated the 5-fold cross-validation 10 times while shuffling the instances each time.

4.1 Analyzing Effectiveness of Dynamic Selection Procedure Proposed for Complex Unbalanced Data

In this subsection, we try to answer the question "Can the proposed dynamic selection algorithm be an effective method with which to improve the classification performance when dealing the complex unbalanced-data problem?" To this end, we carried out the experimental analysis to compare the proposed DESD method with the representative benchmark methods, that is, the initial base learner (C4.5 was selected as the base learner in this study), synthetic minority over-sampling (SMOTE) [19], dynamic selection of a single most competent classifier, and ensemble of all the candidate classifiers. Detailed descriptions of the representative benchmark methods follow:

- The base learner (denoted Base): This method builds the classifier by using C4.5 as the learning algorithm on the original learning set.
- Balancing the dataset using SMOTE (denoted SMOTE): In this method, the training dataset is balanced by using the SMOTE algorithm. Then, the classifier is built on the balanced dataset by using the learning algorithm.
- Dynamic classifier selection for complex unbalanced datasets (denoted DCSD): The generation of the candidate classifiers pool in this method is the same as that in our proposed method, but DCSD selects the most proper classifier rather than an ensemble for each query instance.
- Distance-based combination rule (denoted DCR): In this method, an ensemble is obtained by combining all the candidate classifiers with the DCR.

The detailed AUC and G-mean values of each dataset are presented in Table 3, in which the best results for each performance measure are emboldened. Observing the results, we can easily obtain that the generalization performance of the proposed DESD method outperforms the other methods in both performance measures. The average value in the last row in both performance measures is remarkable, as is the number of datasets in which the DESD method is the best method. However, we cannot extract meaningful conclusions without using the proper statistical analysis.

Table 3. Average AUC and G-mean results of DESD method and representative approaches.

ID	AUC					G-mean				
	Base	SMOTE	DCSD	DCR	DESD	Base	SMOTE	DCSD	DCR	DESD
1	0.8230	0.7527	**0.9330**	0.8806	0.9274	0.6538	0.6674	0.8555	0.8741	**0.8908**
2	0.8611	0.9063	**0.9660**	0.9222	0.9653	0.8083	0.8989	0.8967	0.9228	**0.9388**
3	0.8771	0.8710	0.9649	0.9266	**0.9782**	0.8219	0.8540	0.9020	0.9250	**0.9252**
4	0.7852	0.7417	0.8769	0.8115	**0.8938**	0.6909	0.7259	0.8034	0.7922	**0.8193**
5	0.7318	0.8136	0.9507	0.9192	**0.9699**	0.6936	0.8125	0.9025	0.9172	**0.9202**
6	0.9045	0.8115	0.9163	0.9111	**0.9587**	0.7570	0.8336	0.8660	0.9052	**0.9455**
7	0.5838	0.6545	0.8552	0.7229	**0.8800**	0.2225	0.6019	0.6995	0.664	**0.8656**
8	0.6051	0.6944	**0.8771**	0.7151	0.8676	0.4206	0.6180	0.5863	0.6469	**0.8105**
9	0.5897	0.7140	0.8608	0.7589	**0.9138**	0.2085	0.6375	0.6981	0.7339	**0.8563**
10	0.7654	0.8279	**0.9894**	0.9067	0.9606	0.7222	0.8410	**0.9845**	0.9051	0.9206
11	0.8652	0.8290	0.8945	0.9194	**0.9826**	0.8371	0.8485	0.8778	0.9154	**0.9824**
12	0.9757	0.9977	0.9966	0.9865	**0.9978**	0.9755	0.9977	0.9966	0.9864	**0.9977**
13	0.9707	0.8878	**0.9902**	0.9171	0.9805	0.7287	0.6769	**0.9901**	0.9129	0.9802
14	0.5924	0.6664	**0.8492**	0.7916	0.8362	0.4236	0.5806	0.7044	0.7806	**0.8152**
15	0.9210	0.8524	0.9545	0.8087	**0.9844**	0.7598	0.8684	0.9192	0.7761	**0.9818**
Avg	0.7901	0.8014	0.9250	0.8599	**0.9398**	0.6483	0.7642	0.8455	0.8439	**0.9100**

Table 4. Results of Wilcoxon tests for comparing DESD method with Base, SMOTE, DCSD, and DCR.

Performance measure	Comparison	R+	R−	Hypothesis	p-value
AUC	DESD vs Base	120	0	**Rejected for DESD at 5%**	0.000061
	DESD vs SMOTE	120	0	**Rejected for DESD at 5%**	0.000061
	DESD vs DCSD	86	34	Not rejected	0.1514
	DESD vs DCR	120	0	**Rejected for DESD at 5%**	0.000061
G-mean	DESD vs Base	120	0	**Rejected for DESD at 5%**	0.000061
	DESD vs SMOTE	105	15	**Rejected for DESD at 5%**	0.000123
	DESD vs DCSD	109	11	**Rejected for DESD at 5%**	0.0034
	DESD vs DCR	120	0	**Rejected for DESD at 5%**	0.000061

Therefore, the Wilcoxon signed-ranks test [20] is performed to compare the proposed DESD method with the selected methods on both performance measures. The Wilcoxon signed-ranks test is a non-parametric test that can be used to determine whether two dependent samples were selected from populations having the same distribution. There is a significant difference between the proposed algorithm and the others if this statistical value is greater than 0.05. The completed statistical results are shown in Table 4, where R+ is the sum of the positive signed ranks and R− is the sum of the absolute values of the negative signed ranks. It can be seen from Table 4 that the DESD method is significantly better than Base, SMOTE, and DCR, since the corresponding p-values obtained are lower than our α-value (0.05). Although there are no significant differences obtained in the comparison of DESD and DCSD for the AUC performance measure, we can still highlight the robust performance of the proposed DESD method, since in AUC the values of R+ are considerably greater than that of R−. Therefore, we can highlight the effectiveness of the proposed dynamic procedure in addressing complex unbalanced datasets.

4.2 Comparison of DESD Method with State-of-the-Art Methods

As we have mentioned above, this investigation aims to compare the proposed DESD methods with the state-of-the-art methods developed for unbalanced data. A set of methods have already been proposed in the literature for addressing unbalanced-data problems. We have selected two dynamic selection ensemble techniques from the literature as reference methods for our proposal, that is, DES-IM [5] and the combination of Ramo and KNU [21], two static ensemble methods, that is, DBE-DCR [4] and clustering-based undersampling in class-imbalanced data (named CUC by us) [22]. The results of the comparison of the proposed DESD method with these state-of-the-art methods are shown in Table 5.

From Table 5, it can be seen that the proposed DESD method performs considerably better than the selected state-of-the-art methods both in terms of the AUC and G-mean performance measures. The DESD method distinctly achieves the best results on nine out of 15 datasets when AUC is used as the performance measure, whereas with G-mean as the performance measure the DESD method performs best on all 15 datasets. With respect to the average performance, the DESD method also outperforms the selected state-of-the-art methods.

Table 5. Average AUC and G-mean results of DESD and state-of-the-art methods.

ID	AUC					G-mean				
	DES-IM	Ramo-KNU	DEB-DCR	CUC	DESD	DES-IM	Ramo-KNU	DEB-DCR	CUC	DESD
1	0.8859	0.9243	0.8983	0.9426	**0.9274**	0.7550	0.7794	0.8886	0.8580	**0.8908**
2	0.9736	0.9451	0.8646	**0.9660**	0.9653	0.8802	0.9054	0.8857	0.8824	**0.9388**
3	0.9356	0.9254	0.9057	0.9752	**0.9782**	0.8405	0.8855	0.9015	0.8599	**0.9252**
4	0.8610	0.8445	0.8400	0.8498	**0.8938**	0.7504	0.7240	0.8125	0.7778	**0.8193**
5	0.9579	0.9547	0.8889	**0.9714**	0.9699	0.9004	0.8819	0.8985	0.8590	**0.9202**
6	0.9376	0.9479	0.913	0.9514	**0.9587**	0.8438	0.8783	0.9051	0.8403	**0.9455**
7	0.7676	0.8140	0.6862	0.7910	**0.8800**	0.6166	0.6201	0.6445	0.6943	**0.8656**
8	0.7494	0.8337	0.733	0.7484	**0.8676**	0.3850	0.5285	0.6669	0.6373	**0.8105**
9	0.8945	0.7669	0.7241	0.7481	**0.9138**	0.6572	0.4586	0.6137	0.6332	**0.8563**
10	0.9240	**0.9760**	0.8702	0.9553	0.9606	0.8193	0.8566	0.8853	0.8277	**0.9206**
11	0.9409	0.9655	0.9282	0.9667	**0.9826**	0.8797	0.8862	0.9262	0.8606	**0.9824**
12	0.9989	**1.0000**	0.9831	0.9989	0.9978	**0.9977**	**0.9977**	0.9829	0.9898	**0.9977**
13	0.9427	**0.9878**	0.8598	0.9659	0.9805	0.8787	0.9390	0.8510	0.9317	**0.9802**
14	0.6972	0.7581	0.6184	0.7700	**0.8362**	0.4167	0.3574	0.5091	0.7314	**0.8152**
15	0.9558	0.9806	0.9549	**0.9913**	0.9844	0.8691	0.8859	0.9531	0.9484	**0.9818**
Avg	0.8948	0.9083	0.8446	0.9061	**0.9398**	0.7660	0.7723	0.8216	0.8221	**0.9100**

Table 6. Wilcoxon signed-ranks test results for comparing DESD method with Base, SMOTE, DCSD and DCR.

Performance measure	Comparison	R+	R−	Hypothesis	p-value
AUC	DESD vs DES-IM	120	0	**Rejected for DESD at 5%**	0.000061
	DESD vs Ramo-KNU	94	26	Not rejected	0.0536
	DESD vs DEB-DCR	112	8	**Rejected for DESD at 5%**	0.0015
	DESD vs CUC	108	12	**Rejected for DESD at 5%**	0.0043
G-mean	DESD vs DES-IM	120	0	**Rejected for DESD at 5%**	0.000061
	DESD vs Ramo-KNU	120	0	**Rejected for DESD at 5%**	0.000061
	DESD vs DEB-DCR	105	15	**Rejected for DESD at 5%**	0.000012
	DESD vs CUC	105	15	**Rejected for DESD at 5%**	0.000012

Again, one cannot obtain any meaningful conclusions without the appropriate statistical analysis. Therefore, we followed the same methodology used in the preceding section. We also employed the Wilcoxon signed-ranks test to compare the proposed DESD method with the state-of-the-art methods for both performance measures. The results of these tests are shown in Table 6. From Table 6, we can see that all the null hypotheses are rejected for the DESD method except the comparison between it and CUC. With respect to the pairwise comparison between the DESD method and CUC, one can observe the significant difference between them when G-mean is used as the performance measure, whereas considering AUC the null hypothesis of equivalence is not rejected with the p-value of 0.0536. However, the value of $R+$ (94.00) corresponding to the sum of the ranks for the DESD method is obviously greater than $R-$ (26.00) for CUC and the obtained p-value is low, which indicates the superiority of the DESD method with respect to CUC. Therefore, we can highlight the goodness of the proposed method using DES-based frameworks for complex unbalanced data.

5 Conclusions

The main contribution of this study is to propose a DESD method that improves the precision of classifying complex unbalanced data. Compared with the existing methods developed for unbalanced data, the dynamic decision framework can well solve complex unbalanced-data classification problems. Complex unbalanced data are decomposed into multiple balanced subsets that contain no or few class-overlap and small-disjuncts problems by repeating random splitting. Moreover, the dynamic selection algorithm is designed to select the proper ensemble of classifiers that are more powerful in classifying minority-class instances. The first experiment validated the effectiveness of the proposed dynamic selection algorithm in dealing with complex unbalanced data. In addition, the experimental results obtained by comparing the proposed DESD method with two dynamic ensemble selection methods and two static ensemble methods demonstrated that the precision of the proposed DESD method outperforms the previous methods in most of the datasets. As future work, the application of the DESD method to multi-class classification problems will be studied.

Acknowledgments. This work is supported by the National Natural Science Foundation of China (Nos. 61702070, 61751203, 61772100, 61672121, 61572093, 61802040), Program for Changjiang Scholars and Innovative Research Team in University (No. IRT_15R07), the Program for Liaoning Innovative Research Team in University (No. LT2015002), the Basic Research Program of the Key Lab in Liaoning Province Educational Department (No. LZ2015004).

References

1. He, H., Garcia, E.A.: Learning from imbalanced data. IEEE Trans. Knowl. Data Eng. **21**(9), 1263–1284 (2009)
2. Krawczyk, B.: Learning from imbalanced data: open challenges and future directions. Prog. Artif. Intell. **5**(4), 221–232 (2016)
3. Guo, H., et al.: Learning from class-imbalanced data: review of methods and applications. Expert Syst. Appl. **73**, 220–239 (2017)

4. Chen, D., Wang, X., Zhou, C., Wang, B.: The distance-based balancing ensemble method for data with a high imbalance ratio. IEEE Access **7**, 68940–68956 (2019)
5. García, S., et al.: Dynamic ensemble selection for multi-class imbalanced datasets. Inf. Sci. **445**, 22–37 (2018)
6. Weiss, G.M.: The impact of small disjuncts on classifier learning. In: Stahlbock, R., Crone, S., Lessmann, S. (eds.) Data Mining, pp. 193–226. Springer, Boston (2010)
7. García, V., Mollineda, R.A., Sánchez, J.S.: On in the k-NN performance in a challenging scenario of imbalance and overlapping. Pattern Anal. Appl. **11**(3–4), 269–280 (2008)
8. García, V., Sánchez, J.S., Ochoa Domínguez, H.J., Cleofas-Sánchez, L.: Dissimilarity-based learning from imbalanced data with small disjuncts and noise. In: Paredes, R., Cardoso, J.S., Pardo, X.M. (eds.) IbPRIA 2015. LNCS, vol. 9117, pp. 370–378. Springer, Cham (2015). https://doi.org/10.1007/978-3-319-19390-8_42
9. Cruz, R.M.O., Sabourin, R., Cavalcanti, G.D.C.: Dynamic classifier selection: recent advances and perspectives. Inf. Fusion **41**, 195–216 (2018)
10. Woźniak, M., Graña, M., Corchado, E.: A survey of multiple classifier systems as hybrid systems. Inf. Fusion **16**, 3–17 (2014)
11. Roy, A., et al.: Meta-learning recommendation of default size of classifier pool for META-DES. Neurocomputing **216**, 351–362 (2016)
12. Ko, A.H.R., Sabourin, R., Britto Jr., A.S.: From dynamic classifier selection to dynamic ensemble selection. Pattern Recogn. **41**(5), 1718–1731 (2008)
13. Cavalin, P.R., Sabourin, R., Suen, C.Y.: Dynamic selection approaches for multiple classifier systems. Neural Comput. Appl. **22**(3–4), 673–688 (2013)
14. Cruz, R.M.O., et al.: META-DES: a dynamic ensemble selection framework using meta-learning. Pattern Recogn. **48**(5), 1925–1935 (2015)
15. Cavalin, P.R., Sabourin, R., Suen, C.Y.: LoGID: An adaptive framework combining local and global incremental learning for dynamic selection of ensembles of HMMs. Pattern Recogn. **45**(9), 3544–3556 (2012)
16. Cruz, R.M.O., Sabourin, R., Cavalcanti, G.D.C.: META-DES. H: a dynamic ensemble selection technique using meta-learning and a dynamic weighting approach. In: International Joint Conference on Neural Networks (IJCNN), pp. 1–8. IEEE (2015)
17. Pérez-Gállego, P., et al.: Dynamic ensemble selection for quantification tasks. Inf. Fusion **45**, 1–15 (2019)
18. Alcalá-Fdez, J., et al.: Keel data-mining software tool: data set repository, integration of algorithms and experimental analysis framework. J. Multiple Valued Logic Soft Comput. **17**, 255–287 (2011)
19. Chawla, N.V., et al.: SMOTE: synthetic minority over-sampling technique. J. Artif. Intell. Res. **16**, 321–357 (2002)
20. Wilcoxon, F.: Individual comparisons by ranking methods. Biomed. Bull. **1**(6), 80–83 (1945)
21. Roy, A., et al.: A study on combining dynamic selection and data preprocessing for imbalance learning. Neurocomputing **286**, 179–192 (2018)
22. Lin, W.C., et al.: Clustering-based undersampling in class-imbalanced data. Inf. Sci. **409**, 17–26 (2017)

Collaborative Wireframing for Model-Driven Web Engineering

Peter de Lange[✉] [iD], Petru Nicolaescu, Mario Rosenstengel,
and Ralf Klamma [iD]

Chair for Information Systems and Databases, RWTH Aachen University,
Ahornstr. 55, 52074 Aachen, Germany
{lange,nicolaescu,rosenstengel,klamma}@dbis.rwth-aachen.de

Abstract. Today's Model-Driven Web Engineering (MDWE) approaches automatically generate Web applications from conceptual, domain-specific models. This enhances productivity by simplifying the design process through a higher degree of abstraction. Due to this raised level of abstraction, the collaboration based on conceptual models also opens up new use cases, such as the end user's tighter involvement into Web development. However, especially in the early design stages of Web applications, common practices for requirement elicitation mostly rely on paper prototypes or wireframes instead of MDWE, created usually in analog settings. The digitization of this process, combined with the benefits of model-driven development, bears a lot of potential for improving MDWE practices. In this contribution, we enhance an existing MDWE approach with wireframing capabilities, realized through real-time synchronization of models, wireframes and code. We present the conceptual considerations of our approach, the realization of the synchronous wireframing tool and the synchronization between wireframe and model. Our evaluation results for collaborative Web development tasks are promising and open the gate towards novel, collaborative and agile MDWE techniques.

Keywords: Model-Driven Web Engineering · Collaborative wireframing · Model-to-model synchronization

1 Introduction

Current Model-Driven Web Engineering (MDWE) approaches try to increase productivity by enabling the generation of Web applications based on information usually specified in the form of a conceptual model [7]. Corresponding to a certain domain-specific metamodel, the models reflect the structure of Web frontends and abstract the pagination and the navigation of applications. Based on certain templates and incorporated, framework-specific best practices, the resulting applications can be specified and instantiated accordingly. However, this abstraction still requires a rather good and specific development knowledge,

© Springer Nature Switzerland AG 2019
R. Cheng et al. (Eds.): WISE 2019, LNCS 11881, pp. 373–388, 2019.
https://doi.org/10.1007/978-3-030-34223-4_24

in order to be able to model and generate the software artifacts. Due to their complexity and adoption challenges, the initial design stages of Web applications cannot easily be included into the MDWE life-cycle, as they require intensive communication with end users, in order to elicitate requirements.

On the other hand, software prototyping is a popular software engineering method to quickly conceive the most important aspects of a software application in the very beginning stages of software development. In Web engineering, a wireframe is an agile prototyping technique to sketch the skeletal structure of a Website or Web application [2]. It is a collaborative and social process, that involves designers, end users, developers and other stakeholders. In contrast to a conceptual model, that consists of rather abstract nodes and edges, a wireframe provides a closer representation of the final Web application. Consequently, a wireframe is more intuitive and feels more familiar to end users. Therefore, especially users with a non-technical background can benefit from an integrated collaborative wireframing application into the MDWE process, that allows the design of Web frontends in an intuitive way. Such an application promises a lower learning curve, with less required knowledge about Web development. In order to achieve such a novel collaborative frontend development practice, live synchronization between models and wireframes have to be implemented.

In this contribution, we investigate the role of collaborative near real-time (NRT) wireframing for MDWE processes. Therefore, we adapt related conceptual mappings of MDWE and wireframes and develop both a conceptual mapping for the co-evolution of model and wireframe in NRT, as well as we integrate a Web-based collaborative wireframing editor into an existing MDWE framework to evaluate our conceptual mapping. Currently, there are no MDWE techniques we are aware of, that use collaborative, synchronous wireframing together with collaborative, synchronous modeling.

The paper is structured as follows. We start by presenting the background and related work needed for our approach in Sect. 2. Then in Sect. 3, we shortly introduce the existing MDWE approach and describe how the wireframing editor developed in this contribution integrates with it. We continue by explaining the formal concept of this integration in Sect. 4, before we describe the technical realization of both the wireframing editor and the synchronization of models with wireframes in Sect. 5. The results of our investigative end user evaluation are presented in Sect. 6. Finally, Sect. 7 concludes the paper and offers a perspective for future work.

2   Background and Related Work

Model-Driven (Web) Engineering. Most MDWE approaches follow the philosophy of separation of concerns [5]. Based on a comprehensive metamodel, certain viewpoints are defined to reflect specific aspects of a Web application. One of the first MDWE approaches that obeyed the separation of concerns paradigm in MDWE was OOHDM [18], with the goal of dealing with the increasing complexity of Web applications. It described a methodology for systematic guidance to design large scale, dynamic Web applications. The main activities of the OOHDM methodology comprised a conceptual-, navigational- and abstract user interface design and proposed how they are implemented in the final Web application. A slightly more recent, as well as ongoing, MDWE methodology is the UML-based Web Engineering (UWE) [6], which was conceived as a conservative extension of the UML. Thus, already existing concepts of the UML are not modified, the new extensions are just related to existing concepts. The first extensions are UML stereotypes, which are used to define new semantics for model elements, e.g. a navigation link. The Object Constraint Language (OCL) is used to define constraints and invariants for classes. UWE follows the separations of concerns principle to split up the modeling process into the conceptual-, navigational- and presentation modeling part. WebML is another MDWE approach developed in 2000. It does not propose another language for data modeling, but also extends the UML and is compatible with classical notations of ER-diagrams and others [4]. WebML as well emphasizes the concept of separation of concerns. Therefore, the development process is divided into four distinct modeling phases: The structural model represents the content of the site expressed as UML class- or ER diagram, the hypertext model consists of a composition- and navigation model. The former one describes which entities of the structural model are composed by a certain page and the latter one specifies the links between pages. The third one is the presentation model which expresses the layout and graphical appearance of pages. Finally, the personalization model defines user and/or user group specific content. In 2013, WebML emerged into the Interaction Flow Modeling Language (IFML) [3] and was adopted as standard by the Object Management Group (OMG).

ArchiMate is an enterprise architecture modeling language [11]. Although not a direct MDWE approach, it is relevant related work for our approach, because of its interpretation of the separation of concerns paradigm. It separates the content and visualization of the view. The main advantage of this is the usage of different visualizations on the same modeling approach and vice versa. The content of a view is derived from the base model and is expressed in the same modeling concept. The visualization on the other hand can be completely different from the actual representation of the model. ArchiMate allows to define a set of modeling actions, that alter the content of the model. These modeling actions are mapped to operations on a specific visualization of the view. This additional abstraction level allows to define any sort of visualization, like videos or dynamic charts. In this contribution, we use this concept of view separation to map certain operations on the wireframing editor to operations on the modeling canvas, which alter the current state of the wireframing-, respectively the modeling viewpoint.

Wireframing. There exist a plethora of wireframing and mockup tools on the Web. We here exemplary introduce Balsamiq[1] (as one of the most used ones) and Mockingbird[2] (as it features NRT collaboration and is Web-based). The idea

---

[1] https://balsamiq.com.
[2] https://www.gomockingbird.com.

behind Balsamiq is not to build large and fully interactive prototypes, which take hundreds of hours to develop and may lead to costly refinements if something can not be realized as intended. Instead, Balsamiq follows a more rapid development philosophy. This has the advantage that developers gain experience and evaluate components of the wireframe directly on a very early version of the Website, which can also involve end user feedback. The feedback is used to tweak the wireframes and the implementation process starts again. Therefore, Balsamiq offers only limited interactivity features on a wireframe. Mockingbird is a Web-based wireframing application that offers NRT collaborative editing. The graphical editor offers the most common UI elements of today's Web pages which can be rearranged and resized freely on a page. Similar to Balsamiq, it is possible to link pages and preview them to demonstrate the Website's interactivity flow.

Mockup Driven Development - Merging MDWE and Wireframing. *Mockup Driven Development* (MockupDD) is a hybrid, model-based and agile Web engineering approach [17]. The main goal of MockupDD is to extract and combine the best of MDWE methodologies and the rapid collaborative design process of wireframing, to add agility to existing MDWE approaches. MockupDD describes a transformation approach from a mockup to a comprehensive model that is further transformed to the specific models of an arbitrary MDWE approach. In most related approaches, wireframes are not considered as models and their impact declines in later development stages. MockupDD tackles this with a generic approach to integrate mockups directly into the whole MDWE development process. An additional computational instance builds the bridge from the output of an arbitrary wireframing tool to an arbitrary MDWE approach. The MockupDD methodology begins by creating UI mockups with an arbitrary tool, e.g. Balsamiq or Mockingbird. The resulting mockup file is then parsed, validated and analyzed with regards to a *Structural UI* (SUI) metamodel, which denotes each UI control element, their compositions and hierarchical structures. The goal is to obtain a "sufficient enough" structural model of the UI. Based on this SUI model, another transformation approach to the specific model of the used MDWE methodology is required. To further enrich the representational strength of a SUI model, MockupDD includes a tagging mechanism. A tag is simple specification that is applied over a concrete node of the SUI model and consists of a name and an arbitrary number of attributes. The main purpose of a tag is to define functional or behavioral aspects of a certain UI element. It allows the designer to construct more complex wireframe specifications. An UI element may have an arbitrary number of tags assigned to it. A SUI model enriched with tags is also called a *SUIT model*. The concept of MockupDD has been adapted to various modeling languages and domains (WebML [17], UWE [15], IFML [16], and specifically focusing on mockups of touch user interfaces [1]). We base this contribution partly on the conceptual findings of MockupDD, and take the approach one step further by co-evolving the wireframe and MDWE artifacts throughout the whole Web application development process.

3   Agile MDWE with Collaborative Wireframing Support

We integrated our wireframing approach with an existing Web-based collaborative MDWE framework, called the Community Application Editor (CAE) [9]. It follows the separation of concerns and defines three orthogonal viewpoints (or views) for the modeling of Web applications, based on a comprehensive metamodel. An application of the CAE can be authored using these three views, which we supplemented with an additional fourth wireframing view in the scope of this contribution.

- The *frontend component view*, represented by a model of a Web component
- The *backend component view*, represented by a model of a microservice
- The *application view*, represented by a model of a the communication and interactions between frontend and backend components
- The *wireframing view*, a visual representation of a frontend component

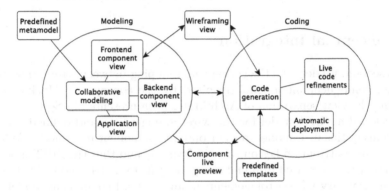

Fig. 1. Overview of the MDWE approach with integrated wireframing support.

Figure 1 gives an overview of the different views, as well as their connections with each other and the code refinements/deployment. It can be split up into two main phases, namely the *Modeling* and the *Coding* phase. Based on this modeling – coding cycle and on both a predefined Web application metamodel and templates, the CAE enables the collaborative, model-driven creation of Web applications. To illustrate this concept, Fig. 2 depicts four representations of the same sample frontend component. Figure 2a shows the conceptual model. Figure 2c depicts the wireframe visualization of the frontend component model. Both the wireframe model and the conceptual model are used as input for the code generation to generate the code artifacts depicted in the live code editor of Fig. 2b. This live code editor enables NRT code refinements directly in the browser, synchronized with the corresponding frontend/backend component model [10]. Finally, Fig. 2d shows a live preview of the resulting application, based on the code artifacts shown in the live code editor.

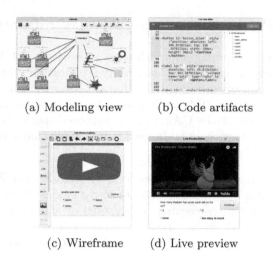

(a) Modeling view   (b) Code artifacts

(c) Wireframe   (d) Live preview

Fig. 2. Different representations of the same frontend component.

4   Conceptual Integration

In this section, we describe the conceptual mapping between wireframes and the frontend component metamodel, inspired by the concepts of MockupDD and following the view separation of ArchiMate, both presented in Sect. 2. For this, we developed a SUIT model for the wireframing editor and defined the transformations of this metamodel to the frontend component metamodel. While the focus of this contribution lies on the wireframing, and thus the SUIT model, two characteristics of the used frontend component metamodel need to be described. First, an arbitrary UI control element in our frontend component model can be defined as static or dynamic. In contrast to static HTML elements, a dynamic element is created through an event- or function call and thus is not necessarily part of the initial Web page rendering. Second, our MDWE approach relies on the generation of so-called *widgets* as base elements. Each frontend component consists of exactly one widget that contains all further elements. Thus, an application consisting of multiple frontend components exists of multiple (interacting) widgets. While this is certainly a specific of our used MDWE approach, the following conceptual integration is not generally limited to it and could easily be modified to provide a mapping between other wireframes and frontend component metamodels. The remainder of the frontend component metamodel, as well as the microservice and application metamodels of the used MDWE approach have no direct relation to the wireframing integration presented in this contribution and are thus not covered here.

Figure 3 illustrates the viewpoint concept of the CAE, extended with the wireframing viewpoint. The SUIT model of the wireframing editor comprises the most common HTML elements of the current HTML5 standard. It offers simple structural elements like buttons, text boxes and containers. Furthermore,

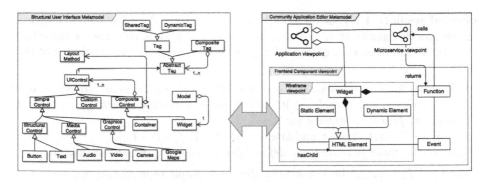

Fig. 3. Mapping of the SUIT model to the MDWE metamodel.

media elements like the HTML5 video and audio player and custom Polymer elements are supported. Also compositions of elements are defined, like a checkbox with label. Each UI control element of the SUIT model has its own set of attributes defined, according to the HTML5 standard. We also introduce a so called 'SharedTag', that can be assigned to any UI control element to add NRT collaborative behavior to it, using the Yjs shared editing framework [13]. In the following, we want to define the conceptual relationship between the SUIT model and the frontend component viewpoint metamodel. We use similar definitions for viewpoints and views as presented in [14].

Mapping Between SUIT - and Frontend Component Viewpoint Model. Conceptually, an instance of the SUIT model is a labeled tree. We formally define such a tree as a connected, acyclic and labeled graph. An arbitrary element $v \in V$ always has the signature $v = (l, t, A)$, where $l \in \Sigma$ is the label, with Σ being a finite alphabet of vertex and edge labels. $t \in T$ is the type of the node, where T is either a UI control element or a tag defined in the SUIT metamodel, as depicted in Fig. 3. For example T might consist of the following elements: $UI = \{Text, Button, Video, Canvas, ..\}$ and $Tag = \{SharedTag, DynamicTag, ..\}$ with $T = UI \cup Tag$. A is a finite set of properties related to an UI control element or tag and each $a \in A$ is a key-value-pair (k, v) with $k, v \in \Sigma$. The tree always consists of a distinguished vertex r, which is also called the root. The root is always of type $Widget$. The $parent(v)$ function is a helper function that yields the parent-vertex for a vertex of the SUIT tree. If the vertex v is the root, the root will be returned.

Definition 1. *A SUIT model is a labeled tree with $SUIT = (V, E)$. V is a finite, non-empty set of vertices. V is always initialized with the root r. E is a set of unordered pairs of distinct vertices $(v1, v2)$ with $v1 \neq v2$, which constitutes the edges of the tree.*

For the integration into our MDWE approach, a SUIT model is mapped to an instance of the frontend component viewpoint. Let $VP = (V, E)$ be an acyclic, directed graph that represents an arbitrary viewpoint. An edge $e \in E$ of such a graph has the signature $(l, t, v1, v2, A)$, where $l \in \Sigma$, t is the type of the edge, $v1, v2 \in V$ and A is a set of key-value-pairs that constitute the attributes of the edge.

Definition 2. *An instance M of a viewpoint of VP is an acyclic, directed graph with $M = (V', E')$. For each $v \in V'$ holds $type(v) \in label(V)$, with type and label being helper functions defined as:*
$type : V \mapsto \Sigma : (l, t, A) \mapsto t$ and $label : V \mapsto \Sigma : (l, t, A) \mapsto l$.
Analogously, these functions are defined for an edge $e \in E'$ of a viewpoint.

Now let $VP_{wireframe}$ be the acyclic directed graph representing an arbitrary instance of the wireframe viewpoint and W_{SUIT} a SUIT model representing a concrete wireframe. An instance of the SUIT model is mapped to an instance of the wireframe viewpoint with the following function φ:

$$\varphi : W_{SUIT} \mapsto VP_{wireframe} = (V, E) \mapsto (\phi(V), \gamma(E))$$

where ϕ is defined as follows:

$$\phi(V) = \{\phi(v)|v \in V\} \text{ with}$$

$$\phi : V \mapsto V' = (l, t, A) \mapsto (l', t', A') :$$

$$\begin{cases} (l, t, A) \mapsto (l, \text{HTML Element}, & \text{for } t \in UI \\ \quad A \cup \{(type, t), (static, true), (collaborative, false)\}), & \\ (l, \text{SharedTag}, \emptyset) \mapsto (l', \text{HTML Element}, shared(A')), & \text{for } l' = parent(l) \\ (l, \text{DynamicTag}, \emptyset) \mapsto (l', \text{HTML Element}, dynamic(A')), & \text{for } l' = parent(l) \\ (l, \text{Widget}, A) \mapsto (l, \text{Widget}, A), & \text{otherwise} \end{cases}$$

where *shared* and *dynamic* are functions that are applied to every attribute in A of the referenced 'HTML Element'-node. These helper functions change the value of the 'collaborative'- respectively 'static'-attribute for the referenced 'HTML Element'-node. All other attributes are left untouched.

$$shared(A) = \{shared'(a)|a \in A\}$$

$$shared' : A \mapsto A : \begin{cases} (k, false) \mapsto (k, true), & \text{for } k = collaborative \\ (k, v) \mapsto (k, v), & \text{otherwise} \end{cases}$$

and

$$dynamic(A) = \{dynamic'(a)|a \in A\}$$

$$dynamic' : A \mapsto A : \begin{cases} (k, true) \mapsto (k, false), & \text{for } k = static \\ (k, v) \mapsto (k, v), & \text{otherwise} \end{cases}$$

The relationships between the nodes in the wireframe viewpoint are generated with function γ:

$$\gamma(E) = \{\gamma(e)|e \in E\} \text{ with}$$

$$\gamma : E \mapsto E' = (v_1, v_2) \mapsto (l, t, v'_1, v'_2, A) :$$

$$\begin{cases} (v_1, v_2) \mapsto (l, \text{Wid. To El.}, v_1, v_2, A), & \text{for } v_1 = r, type(v_2) \in UI \\ & \text{for } type(v_1) \in UI, type(v_2) \in UI \\ (v_1, v_2) \mapsto \{(l, \text{hasChild}, v_1, v_2, A)\}, & \text{and } v_1 \neq v_2 \neq r \end{cases}$$

Summary. With φ, we only map the UI elements of the SUIT model to the wireframe viewpoint. An 'HTML element' node of the frontend component-, respectively wireframe viewpoint, consists of the four properties id, type, static and collaborative. The id of the HTML element is automatically generated by the mapping approach. The value of the type attribute is an element from the UI. The static and collaborative attributes are the only attributes represented as tags in the SUIT model. Furthermore, they are simple Boolean attributes and therefore have no own attributes defined. Additionally, the tags are unique and thus they only appear once for a certain UI element.

A node of the SUIT tree is mapped with ϕ to a certain 'HTML Element' or 'Widget' node. An arbitrary UI element of the SUIT model is always mapped to an instance of the 'HTML Element' node class, where the label of the UI element is the label of the node. The type of the UI element is mapped to the type-attribute of the node. By default, the 'static' attribute is true and the 'collaborative' attribute is false. To change the values of these attributes, a DynamicTag- respectively SharedTag element is mapped to the corresponding attribute in the 'HTML Element' node. For each tag a function is required, which alters a certain aspect of the signature of an 'HTML Element' node (e.g. type or attribute). For the definition of the current mapping approach, the two helper functions *shared* and *dynamic* are defined, which change the Boolean value of the associated attribute. The root-element of the SUIT tree is always mapped to the 'Widget'-node, where the label of the root is also the label of the 'Widget'-node. The same holds for the attributes. With function γ, the relationships between nodes are generated. The function comprises two cases. First the UI element is a direct child from the root. In such a case a single 'Widget To HTML Element' edge is created (abbreviated in the function with 'Wid. To El.', due to space restrictions). For the second case, we assume that v_1 is a parent of v_2 and v_1 and v_2 are not the root. In such a case, the 'hasChild' relationship is generated.

5   Realization

The wireframing editor is realized as a separate and independent part of the frontend component modeling space of the CAE[3]. We extended the persistence functionality of the MDWE framework, such that it also stores the wireframe models as XML files. Frontend component models and wireframe models are stored next to each other, because the SUIT model enriches the HTML elements of the frontend component model with additional metadata and type-specific attributes. Nevertheless, for code generation, a frontend component model is always required, while the SUIT model is optional. The microservice- and application modeling space are not affected by the extension and integration of the wireframing editor and are thus not further covered here.

---

[3] https://github.com/rwth-acis/CAE-WireframingEditor.

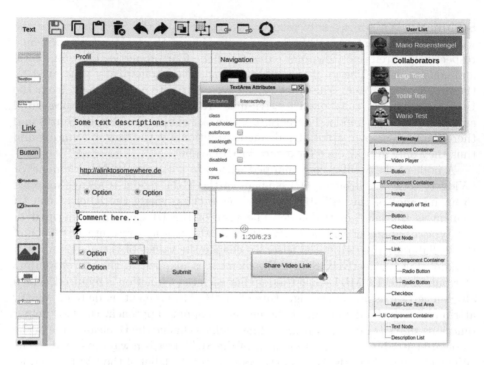

Fig. 4. Screenshot of the wireframing editor.

5.1 The Web-Based NRT Collaborative Wireframing Editor

Figure 4 depicts a screenshot of the wireframing editor. With it, an arbitrary number of users can collaboratively work on the graphical user interface of their frontend in NRT. The context menu of the drawing canvas and the menu bar at the top offer all common utility functions for the elements of the drawing canvas (copy&paste, delete an arbitrary number of selected elements and an undo&redo functionality). Additionally, designers are able to group and ungroup an arbitrary number of elements using the UI component container. All these features are available in NRT by using the Yjs shared editing framework. The editor uses an automatic save functionality. Each altering on the drawing canvas, attribute- or tag editor saves the current state of the wireframe model to the shared editing framework as XML string. Nevertheless, this shared data space is not a persistent storage, which is needed to persist the wireframe, once a collaborative wireframing session has ended. The editor offers an interface for using the persistence functionality of the MDWE framework, such that the wireframe model gets stored alongside other modeling artifacts of the MDWE framework in a relational database.

The editor provides certain awareness features to support the collaboration. The user list shows all collaborators currently working on the wireframe. If a remote user selects one or more elements on the drawing canvas, each element is highlighted with a surrounding frame and marked with the image of their OpenId

Connect profile they used to log into the modeling environment. If multiple users select the same element, their profile images appear on the bottom of the element.

The palette, depicted on the left, consists of all elements supported by the wireframing editor. Via drag&drop, an element from the palette can be added to the drawing canvas. The attributes of an element can be edited with the attribute editor. The screenshot depicts an attribute editor for a multi-line text field element. Currently, three types of attributes are supported by the editor. The first type is a string attribute which is edited with a simple text input field. The second type is a Boolean attribute that is represented by a checkbox in the attribute editor. The last type is the combo attribute, realized by a drop-down menu. The tag editor is integrated as a separate tab in the window where the attribute editor is located (not shown in the screenshot).

The hierarchy tree is a small helper tool, that visualizes the ordering of the elements of the drawing canvas. This ordering is important for the wireframe layout algorithm used in the code generation functionality of the MDWE framework (not further covered in this work, see [10]). The wireframe layout iterates through the wireframe model and calculates the position of the elements according to their position in the order of the corresponding container. Therefore the hierarchy tree also allows to change the order of the elements in a group via a simple context menu.

5.2   Transformation Algorithms

For the integration of the wireframing editor in the MDWE framework, two *model transformation algorithms* were developed to transform the SUIT wireframe model to a frontend component model and vice versa. The two transformations are only needed, if one of the two frontend component representations is not existing. After that, the two model states are kept synchronized by the *live mapper*. Both algorithms, as well as the live mapper, follow the conceptual mapping approach we described in Sect. 4.

Wireframe to Model Transformation. The wireframe to model transformation takes as input an instance of a wireframe model and the frontend component metamodel. The output of the transformation is a JSON object of the frontend component model. The implementation uses templates of a node-, edge- and attribute-representation in JSON of the frontend component model. We use the lodash template engine to compile the templates and generate nodes, edges and attributes of the resulting model. First, the transformation algorithm generates the 'Widget'-node, which represents the root element of the frontend component model. Then, it recursively runs over the wireframe model and creates a corresponding 'HTML Element'-node for each UI control element. The 'type'-attribute of the node is set to the value of the 'HTML Element'-node name of the corresponding UI control element. Furthermore the 'HTML Element'-node is marked as static and the 'id'-attribute of the node is automatically generated. The value of the id is composed of the 'type'-attribute value and unique. The identifier of the UI control element is reused for the resulting node, which allows to trace back an 'HTML Element'-node to the UI control element. This is necessary for the awareness features and the live mapper. For each node, a 'Widget to HTML Element'-edge is generated, because each node has a connection to the 'Widget'-root node. If the parent of the UI control element is not the root, additionally a 'hasChild'-edge is added to the set of edges. This edge type denotes the hierarchical structure of the wireframe. It connects the parent UI control element to one of it's child elements. If a UI control element has the 'shared'-tag assigned to it, the 'collaborative'-attribute of the corresponding 'HTML Element'-node is set to true as well. Since the frontend component metamodel allows every HTML Element to be collaborative, the wireframing editor allows this as well. The result of this transformation is a valid instance of the frontend component metamodel. However, the HTML attributes specified for a certain UI control element are lost, because the frontend component metamodel does not offer a way to represent them. Additionally the width, height and position of the UI control element in the wireframing editor are not related in any way to the position and dimension of corresponding 'HTML Element'-node. Therefore it is necessary to apply an auto-layout for directed graphs to the model, so that it is displayed correctly in the frontend component viewpoint modeling canvas.

Model to Wireframe Transformation. The input for this transformation is a JSON representation of the frontend component model and an instance of the wireframe editor. The latter one is required to map the 'type'-attribute of an 'HTML Element' node to the correct UI control element. Since the wireframe only represents the HTML elements of the frontend component model, we only have to consider the 'Widget' node (for the size of the whole frontend component) and those 'HTML Element' nodes that are connected to the 'Widget'-node and marked as static. All other node and edge types of the frontend component model can be ignored for this transformation. As already described in the previous transformation algorithm, certain UI layout information (for example the size and position of elements) is not present in the frontend component model. Thus, we initialize these attributes with default values defined in the wireframe model. Finally, the transformation algorithm assigns the 'shared'-tag to every 'HTML Element'-node which has the 'collaborative'-attribute set to true. The result of the transformation approach is an XML document that represents the wireframe model. The resulting model is then stored in the shared data space alongside with the frontend component model.

Live Mapper. The live mapper listens to events on the canvas of the frontend component modeling viewpoint and to the wireframing editor. In contrast to the two previously described transformations, the live mapper directly applies changes to the wireframe and frontend component model and visualizes the results in NRT. Additionally, the live mapper provides awareness features for the selection of entities in both the canvas of the frontend component modeling viewpoint and the wireframing editor. To give an example of the live mapping, the creation of a button element in the wireframe canvas leads to five to six operations on the frontend component viewpoint modeling canvas. First, the node is created on the canvas, the 'type'-, 'id'-, and 'static'-attributes are set and the new node is connected to the 'Widget'-node. If the button is placed in a container, an additional edge is created between the 'HTML Element'-node representing the container and the new node that represents the button. Furthermore, it is possible to edit the wireframe model through the frontend component model viewpoint. For example one can create any UI control element in the wireframe though the frontend component modeling canvas by creating an 'HTML Element'-node, connect it to the 'Widget'-node and set the 'static'-attribute to true. After each action on the wireframe, an auto layout algorithm for directed graphs is applied to the frontend component modeling viewpoint's canvas, only manipulating those elements that were updated.

6   Evaluation

The main intention of our exploratory evaluation was to gain user feedback about the integration of a wireframing editor into the MDWE process. In total we recruited eight participants, which were split up into groups of two, resulting in four evaluation sessions that each lasted about 60 min. All of the participants had a background in Web development or at least some programming experience. The participants were asked to develop a frontend for an already existing microservice backend. A specification for both the existing RESTful API, as well as the desired Web frontend was handed out to the participants at the beginning of the session. In the end, we asked the participants to answer a questionnaire, of which we depict the most interesting results in Fig. 5.

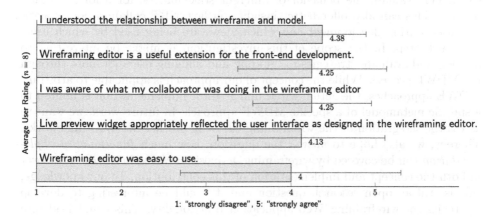

Fig. 5. Results of our exploratory user evaluation.

In general, we received positive feedback from the participants. With an average of 4, most participants found the wireframing editor was easy to use and with an average of 4.13 and 4.25, the participants found both their application reflected in the live preview widget as it was designed by them, as well as that they were aware of what their collaborator did in the wireframing editor. To conclude our user evaluation, with an average rating of 4.25, the majority of the participants thought the wireframing editor a useful extension for MDWE frontend development and the integration of the wireframe into the process was understood quite well (4.38).

7 Conclusions and Future Work

In this contribution, we presented a NRT collaborative wireframing editor, that offers an easy to use interface together with many HTML5 elements to design Web applications. The wireframing editor is successfully integrated into a view-based modeling concept of an existing MDWE approach. The integration is achieved through a formally defined and implemented live mapping approach that transforms the SUIT model of the wireframing editor to a frontend component model and vice versa in NRT. Following the WYSIWYG ("What You See Is What You Get") principle, the wireframes reflect the resulting Web application. Our first user evaluation shows the wireframe editor is perceived a useful and intuitive extension to prototype Web frontends. Most importantly, the wireframe editor correctly reflects the designed user interface in the final Web application. In general, the integration of a wireframing approach was acknowledged as a useful enhancement for the agile MDWE framework.

As future work, we need to expand the currently limited and investigative user evaluation. Even though the majority of the evaluation participants successfully solved the tasks and deployed the final application, it would be interesting to further evaluate the behavior of different stakeholders, for a longer period of time. This can also offer better insights on how NRT collaboration is actually used during development and which views are being used by which users for which step. In the scope of this investigation, it might be interesting to develop and evaluate strategies for guiding and nudging users [8] more through the MDWE process. While the concept of our approach is applicable to arbitrary MDWE approaches, the actual wireframing editor implementation is limited to a specific metamodel of a specific MDWE approach. In future work, we want to assess the integration of the wireframing editor into other MDWE frameworks. Thereby, we also hope to answer the question, how much *functionality implementation* can be covered by wireframing, a question we explicitly did not cover in both the concept and implementation of this contribution. To our knowledge, this is still an open research question and it would be interesting to develop strategies for wireframing Web application functionality. This could lead into the domain of visual programming languages like Scratch [12] and provide new opportunities to further integrate MDWE into agile development processes.

Acknowledgments. The authors would like to thank the German Federal Ministry of Education and Research (BMBF) for their kind support within the project "Personalisierte Kompetenzentwicklung durch skalierbare Mentoringprozesse" (tech4comp) under the project id 16DHB2110.

References

1. Angelaccio, M.: MetaPage-a data intensive MockupDD for agile web engineering. In: WEBIST, vol. 1, pp. 315–317 (2016)
2. Arnowitz, J., Arent, M., Berger, N.: Effective Prototyping for Software Makers. Elsevier Science, Interactive Technologies (2010)
3. Brambilla, M., Fraternali, P.: Interaction Flow Modeling Language: Model-Driven UI Engineering of Web and Mobile Apps with IFML. Morgan Kaufmann, Burlington (2014)
4. Ceri, S., Fraternali, P., Bongio, A.: Web modeling language (WebML): a modeling language for designing web sites. Comput. Networks **33**(1), 137–157 (2000)
5. Kent, S.: Model driven engineering. In: Butler, M., Petre, L., Sere, K. (eds.) IFM 2002. LNCS, vol. 2335, pp. 286–298. Springer, Heidelberg (2002). https://doi.org/10.1007/3-540-47884-1_16
6. Koch, N., Kraus, A.: the expressive power of UML-based web engineering. In: Second International Workshop on Web-Oriented Software Technology, vol. 16, pp. 105–119 (2002)
7. Koch, N., Meliá-Beigbeder, S., Moreno-Vergara, N., Pelechano-Ferragud, V., Sánchez-Figueroa, F., Vara-Mesa, J.: Model-driven web engineering. Upgrade Eur. J. Inform. Prof. **2**, 40–45 (2008)
8. Kosters, M., van der Heijden, J.: From mechanism to virtue: evaluating nudge theory. Evaluation **21**(3), 276–291 (2015)
9. de Lange, P., Nicolaescu, P., Klamma, R., Jarke, M.: Engineering web applications using real-time collaborative modeling. In: Gutwin, C., Ochoa, S.F., Vassileva, J., Inoue, T. (eds.) CRIWG 2017. LNCS, vol. 10391, pp. 213–228. Springer, Cham (2017). https://doi.org/10.1007/978-3-319-63874-4_16
10. de Lange, P., Nicolaescu, P., Winkler, T., Klamma, R.: Enhancing model-driven web engineering with collaborative live coding. In: Modellierung 2018, pp. 199–214. Gesellschaft für Informatik eV (2018)
11. Lankhorst, M.: Enterprise Architecture at Work: Modelling Communication and Analysis. Springer, Heidelberg (2009). https://doi.org/10.1007/978-3-662-53933-0
12. Maloney, J., Resnick, M., Rusk, N., Silverman, B., Eastmond, E.: The scratch programming language and environment. ACM Trans. Comput. Educ. **10**(4), 16:1–16:15 (2010)
13. Nicolaescu, P., Jahns, K., Derntl, M., Klamma, R.: Near real-time peer-to-peer shared editing on extensible data types. In: Proceedings of the 19th International Conference on Supporting Group Work, pp. 39–49. ACM (2016)
14. Nicolaescu, P., Rosenstengel, M., Derntl, M., Klamma, R., Jarke, M.: Near real-time collaborative modeling for view-based web information systems engineering. Inf. Syst. **74**, 23–39 (2018)
15. Rivero, J.M., Grigera, J., Rossi, G., Robles Luna, E., Koch, N.: Towards agile model-driven web engineering. In: Nurcan, S. (ed.) CAiSE Forum 2011. LNBIP, vol. 107, pp. 142–155. Springer, Heidelberg (2012). https://doi.org/10.1007/978-3-642-29749-6_10
16. Rivero, J.M., Rossi, G.: MockupDD: facilitating agile support for model-driven web engineering. In: Sheng, Q.Z., Kjeldskov, J. (eds.) ICWE 2013. LNCS, vol. 8295, pp. 325–329. Springer, Cham (2013). https://doi.org/10.1007/978-3-319-04244-2_31

17. Rivero, J.M., Rossi, G., Grigera, J., Robles Luna, E., Navarro, A.: From interface mockups to web application models. In: Bouguettaya, A., Hauswirth, M., Liu, L. (eds.) WISE 2011. LNCS, vol. 6997, pp. 257–264. Springer, Heidelberg (2011). https://doi.org/10.1007/978-3-642-24434-6_20
18. Schwabe, D., Rossi, G.: An object oriented approach to web-based applications design. Theory Pract. Object Syst. 4(4), 207–225 (1998). Special Issue Objects, Databases, and the WWW

Handling Disagreement in Ontologies-Based Reasoning via Argumentation

Said Jabbour[1], Yue Ma[2], and Badran Raddaoui[3]($\boxtimes$)

[1] CRIL-CNRS, Université d'Artois, Lens, France
[2] LRI, Univ. Paris-Sud, CNRS University Paris-Saclay, Saint-Aubin, France
[3] SAMOVAR, CNRS, Télécom SudParis, Institut Polytechnique de Paris, Évry, France
badran.raddaoui@telecom-sudparis.eu

Abstract. Ontologies are at the heart of the Semantic Web technologies. This paper introduces a framework for reasoning under uncertainty in the context of ontologies represented in description logics; these ontologies could be inconsistent or incoherent. Conflicts are addressed through a form of logic-based argumentation. We examine how the number of attacks and the weights of arguments can be used to define various labelling functions that identify the justification statuses of arguments. Then, different inference relations are distinguished to obtain meaningful answers to queries from imperfect ontologies without extra computational costs compared to classical DL reasoning. Lastly, we study the properties of these new entailment relations and their relationships with other well-known existing ones.

Keywords: Semantic Web · Ontologies · Argumentation · Uncertainty

1 Introduction

Ontologies play a central role in the development of the Semantic Web. They provide a precise definition of shared terms in web resources via concepts, entities, properties, and their inter-relations. Description logics (DLs) [3] are a family of knowledge representation languages and are the underlying logical foundation of the Ontology Web Language (OWL) [39]. Because of the dynamic nature of the Web, one can hardly expect to rely on ontologies without any errors.

There are two main strategies to deal with inconsistency in DL ontologies. A traditional approach is to diagnose and repair it when we encounter inconsistency [36]. This might not be practical or feasible in many contexts, e.g., when dealing with large ontologies, or integrating heterogeneous ontologies. The second approach is simply to accept the inconsistency and to apply a non-standard reasoning method to find meaningful answers from inconsistent ontologies. In this paper, we focus on the latter, which is more suitable for the setting in the web

© Springer Nature Switzerland AG 2019
R. Cheng et al. (Eds.): WISE 2019, LNCS 11881, pp. 389–406, 2019.
https://doi.org/10.1007/978-3-030-34223-4_25

area [26]. In particular, we consider the argumentative reasoning methods [13]. The basic idea of argumentation is that each plausible conclusion inferred from the knowledge base is justified by some reasons, called *arguments*, for believing in it. Due to inconsistency, those arguments may be attacked by other *counter-arguments*. The problem is thus to evaluate the arguments in order to select the most *acceptable* ones. Using argumentation-based ontology reasoning instead of classical approaches has interesting features: (1) the *maxcon* approach (selecting a maximal consistent subset) results in a loss of useful information, as it may not be certain which subset to choose, and therefore an arbitrary choice is made; (2) the *oracle* approach (constructing a consistent ontology by getting extra information to help resolve the conflicts) involves a lot of work that may not be necessary if for example a query can be answered from a small part of the agents' knowledge, and furthermore that this knowledge may not even be in conflict; and (3) argumentation-based approaches can be used for explanatory purposes. For example, if one wants to know why a conclusion is accepted, an argument having that conclusion can be presented. That argument can be attacked by other arguments and so on. Also, it might be possible to construct only a part of the graph related to the argument in question, thus having a better representation.

There are several proposals to deal with inconsistencies in DL ontologies through argumentation (see Sect. 6). Different from existing approaches, in this paper, we consider the scenario that our knowledge is both uncertain and inconsistent and/or incoherent, and we propose a logic-based argumentation framework to deal with incomplete and conflicting DL ontologies. We do so by adopting a distinct notion of attack [11,18] among arguments to encompass different forms of conflicts in DL ontologies. The paper presents the following major contributions: (1) a general framework for reasoning with uncertain, inconsistent and/or incoherent ontologies with the use of logic-based argumentation; (2) a general labelling method, sensitive to the numbers of attacks and the weights of arguments, with different interesting instantiations to identify the justification statuses of each argument; and (3) a number of inference relations derived from our framework in order to obtain meaningful answers without increasing the computational complexity of the reasoning process compared to classical DL reasoning. We also study the logical properties of these new entailment relations.

2   Preliminaries

In this paper, we focus on ontologies which are represented in $\mathcal{ALC}$ instead of the more expressive logics $\mathcal{SHOIN}(\mathcal{D})$ and $\mathcal{SHROIQ}(\mathcal{D})$ to keep the explanation simple. However, the results can be naturally extended to more expressive DLs belonging to the Ontology Web Language OWL2[1].

The *concept descriptions* of $\mathcal{DL}$ are built from a set of concept names $\mathcal{NC}$ and a set of role names $\mathcal{NR}$ using the constructors *top concept* $\top$, *empty concept* $\bot$, *negation* $\neg$, *conjunction* $\sqcap$, *disjunction* $\sqcup$, *value restriction* $\forall$, and *existential*

---

[1] http://www.w3.org/TR/owl2-profiles/.

restriction $\exists$. An interpretation is in the form of $I = (\Delta^I, \cdot^I)$ where Δ^I is the domain and $C^I \subseteq \Delta^I$ for a concept $C \in \mathcal{NC}$ and $r^I \subseteq \Delta^I \times \Delta^I$ for a role $r \in \mathcal{NR}$. The semantics can be extended to complex concepts by interpreting each constructor in the standard way [3]. An ontology is composed of a TBox and an ABox, where a TBox is a set of concept *inclusions* in the form of $C \sqsubseteq D$, and an ABox is a set of *assertions* in the form of $C(a)$ or $r(a, b)$. We call collectively $C \sqsubseteq D$, $C(a)$, or $r(a, b)$ an axiom. An ontology is therefore a finite set of axioms. An interpretation I satisfies a TBox axiom $C \sqsubseteq D$ (resp. an ABox assertion $C(a)$ or $r(a, b)$) if $C^I \subseteq D^I$ (resp. $a^I \in C^I$ or $(a^I, b^I) \in r^I$). For an ontology $\mathcal{O}$, we use $\mathsf{Mod}(\mathcal{O})$ to denote the set of models of $\mathcal{O}$, i.e., $\mathsf{Mod}(\mathcal{O}) = \{I \mid C^I \subseteq D^I \; \forall \; C \sqsubseteq D \in TBox, a^I \in C^I$ for all $C(a) \in ABox$, and $(a^I, b^I) \in r^I$ for all $r(a, b) \in ABox\}$.

We say that $\mathcal{O}$ is *satisfiable* if and only if $\mathsf{Mod}(\mathcal{O}) \neq \emptyset$. $\mathcal{O}$ is *inconsistent* if and only if $\mathsf{Mod}(\mathcal{O}) = \emptyset$. Moreover, $\mathcal{O}$ is *incoherent* if and only if there exists an unsatisfiable concept A, that is, $A^I = \emptyset$ for each $I \in \mathsf{Mod}(\mathcal{O})$. Notice that inconsistency and incoherence are recognized as two significant problems in managing ontological knowledge (e.g. [22, 26]). Given an ontology $\mathcal{O}$, an axiom α is said to be a consequence of $\mathcal{O}$ (denoted by $\mathcal{O} \vdash \alpha$) if for all models I of $\mathcal{O}$, I satisfies α. Two axioms α and β are called logically *equivalent*, denoted $\alpha \equiv \beta$, if I satisfies α iff I satisfies β for any interpretation I of $\mathcal{O}$.

Example 1. Let the ontology $\mathcal{O} = (TBox, ABox)$ with the $TBox = \{\alpha_1, \alpha_2, \alpha_3, \alpha_4\}$ and $ABox = \{\alpha_5\}$ given below:

$$\alpha_1 : \mathsf{PhdStudent} \sqsubseteq \mathsf{Student}, \qquad \text{(PhD students are students)}$$
$$\alpha_2 : \mathsf{PhdStudent} \sqsubseteq \exists \mathsf{hasSalary}.\top, \qquad \text{(PhD students are paid)}$$
$$\alpha_3 : \exists \mathsf{hasSalary}.\top \sqsubseteq \mathsf{Employee}, \qquad \text{(Those having salary}$$
$$\text{are employees)}$$
$$\alpha_4 : \mathsf{Student} \sqcap \mathsf{Employee} \sqsubseteq \bot \qquad \text{(Students are not employees)}$$
$$\alpha_5 : \mathsf{PhdStudent}(\mathsf{Peter}) \qquad \text{(Peter is a PhD student)}$$

$\mathcal{O}$ is incoherent because $\mathcal{O} \models \mathsf{PhdStudent} \sqsubseteq \bot$, that is, the concept PhdStudent is an unsatisfiable concept. Moreover, $\mathcal{O}$ is inconsistent because the ABox claims an instance of the unsatisfiable concept, which makes $\mathcal{O}$ without models.

Next, we recall the main notions of possibilistic $\mathcal{DL}$, denoted by $\mathcal{DL}_\pi$, as an adaptation of $\mathcal{DL}$ within a possibility theory setting [20]. More concretely, $\mathcal{DL}_\pi$ extends the classical DL $\mathcal{DL}$ in particular by possibilistic knowledge about concepts and roles as well as possibilistic knowledge about the instances of concepts and roles. $\mathcal{DL}_\pi$ provides a powerful way to represent and reason with inconsistent and uncertain ontologies. An $\mathcal{DL}_\pi$ ontology consists of a finite set of possibilistic TBox and ABox axioms. A possibilistic axiom is a pair (α, w) where α is an $\mathcal{DL}$ axiom and $w \in [0, 1]$ is a weight representing the confidence degree of α. Notice that the confidence degree of an axiom in an $\mathcal{DL}_\pi$ ontology is unique. For an $\mathcal{DL}_\pi$ ontology $\mathcal{O}$ and $\theta \in [0, 1]$, the θ-cut of $\mathcal{O}$ is defined as $\mathcal{O}_{\geq \theta} = \{\alpha \mid (\alpha, w) \in \mathcal{O}, w \geq \theta\}$. The semantics of $\mathcal{DL}_\pi$ logic [34] is defined by a possibility distribution π over the set Ω of all classical $\mathcal{DL}$ interpretations. An axiom α is called a *possibilistic consequence* of $\mathcal{O}$ to degree w, denoted $\mathcal{O} \vdash_p (\alpha, w)$, iff the following conditions are satisfied: (1) $\mathcal{O}_{\geq w}$ is consistent, (2) $\mathcal{O}_{\geq w} \vdash \alpha$, and (3) $\forall v > w, \mathcal{O}_{\geq v} \nvdash \alpha$.

Example 2 (Example 1 contd.). Let us consider an uncertain extension of the ontology given in Example 1, $\mathcal{O} = \{(\alpha_1, 0.9), (\alpha_2, 0.8), (\alpha_3, 0.7), (\alpha_4, 0.7), (\alpha_5, 1)\}$. Then, we have $\mathcal{O} \vdash_p (\text{Student}(\text{Peter}), 0.9)$, but $\mathcal{O} \nvdash_p (\text{PhdStudent} \sqsubseteq \text{Employee}, 0.7)$.

3   Prudent Argumentation Framework

To present our *prudent argumentation framework*, let us first review a standard notion called *justifications* for ontologies. The generation of justifications has been recognised as a highly desirable functionality of an ontology reasoner that is useful for both ontology development and ontology reuse, and many practical tools have been developed [2,6,31].

Definition 1. *[28] Let $\mathcal{O}$ be an $\mathcal{DL}$ ontology. A justification for an axiom α w.r.t. $\mathcal{O}$ is a subontology $\mathcal{O}' \subseteq \mathcal{O}$ s.t. (1) $\mathcal{O}'$ is consistent, (2) $\mathcal{O}' \vdash \alpha$, and (3) $\mathcal{O}'' \nvdash \alpha$ for any $\mathcal{O}'' \subset \mathcal{O}'$.*

Informally, a justification is a minimal consistent set of axioms in an ontology responsible for a particular entailment. Note that the first condition of Definition 1 is required since $\mathcal{O}$ is not necessarily consistent in this paper. Inspired by classical argumentation theory [1,12,38], the notion of justification can be naturally extended to define arguments in the light of both uncertainty and incoherence.

Definition 2. *Let $\mathcal{O}$ be an $\mathcal{DL}_\pi$ ontology. A prudent argument for an axiom α w.r.t. $\mathcal{O}$ is a triple $\langle \Phi, \alpha, w \rangle$ s.t. the following conditions hold: (1) Φ is coherent, (2) Φ is a justification for α w.r.t. $\mathcal{O}_{\geq 0}$, and (3) $w = \min\{w_i \mid (\phi_i, w_i) \in \Phi\}$.*

For the prudent argument $\langle \Phi, \alpha, w \rangle$, obviously, $\Phi \subseteq \mathcal{O}_{\geq 0}$. We call Φ the *support*, α the *conclusion*, and w the *weight* of the argument. w corresponds to the certainty degree with which α follows Φ. It is obtained by the *min* aggregation function [21].

Example 3. Some prudent arguments of $\mathcal{O}$ from Example 2 are: $\mathbb{A}_1 = \langle \{\alpha_1, \alpha_5\}, \text{Student}(\text{Peter}), 0.9 \rangle$, and $\mathbb{A}_2 = \langle \{\alpha_2, \alpha_3\}, \text{PhdStudent} \sqsubseteq \text{Employee}, 0.7 \rangle$. However, $\mathbb{A}_3 = \langle \{\alpha_1, \alpha_2, \alpha_3, \alpha_4\}, \text{PhdStudent} \sqsubseteq \bot, 0.7 \rangle$ is not a prudent argument because $\{\alpha_1, \alpha_2, \alpha_3, \alpha_4\}$ is incoherent.

Now, the relation between prudent arguments and the possibilistic consequence $\vdash_p$ is given below.

Proposition 1. *Let $\mathcal{O}$ be an $\mathcal{DL}_\pi$ ontology. If $\langle \Phi, \alpha, w \rangle$ is a prudent argument, then $\mathcal{O}^\Phi \vdash_p (\alpha, w)$, where $\mathcal{O}^\Phi = \{(\alpha, w) \in \mathcal{O} \mid \alpha \in \Phi\}$.*

Accordingly, $\mathcal{O}^{\Phi} \vdash_p (\alpha, \omega)$ does not imply $\mathcal{O} \vdash_p (\alpha, \omega)$. For example, $\{\alpha_2, \alpha_3\} \vdash_p$ (PhdStudent $\sqsubseteq$ Employee, 0.7). However, $\mathcal{O} \nvdash_p$ (PhdStudent $\sqsubseteq$ Employee, 0.7). Hence, prudent arguments do not suffer from the drowning problem of possibilistic reasoning (i.e., axioms whose degrees are less than a threshold are completely dropped).

The following result shows that the relation between the weights of two prudent arguments provides a relation between their supports, regardless the conclusions, which follows directly from Definition 2.

Proposition 2. *If* $\langle \Phi, \alpha, \omega_1 \rangle$ *and* $\langle \Phi', \beta, \omega_2 \rangle$ *are two prudent arguments and* $\Phi \subseteq \Phi'$, *then* $\omega_1 \geq \omega_2$. *In particular, if* $\Phi = \Phi'$, *then* $\omega_1 = \omega_2$.

Prudent arguments are not necessarily independent. The definition of more conservative prudent arguments captures a notion of subsumption between prudent arguments.

Definition 3. *A prudent argument* $\mathbb{A} = \langle \Phi, \alpha, \omega_1 \rangle$ *is more conservative than a prudent argument* $\mathbb{A}' = \langle \Psi, \beta, \omega_2 \rangle$, *denoted* $\mathbb{A} \succeq \mathbb{A}'$, *iff* $\Phi \subseteq \Psi$, *and* $\{\beta\} \vdash \alpha$. *We say* $\langle \Phi, \alpha, \omega_1 \rangle$ *and* $\langle \Psi, \beta, \omega_2 \rangle$ *are quasi-equivalent if* $\mathbb{A} \succeq \mathbb{A}'$ *and* $\mathbb{A}' \succeq \mathbb{A}$.

That is, a more conservative prudent argument can be seen as more general in the sense that it is less demanding on the support and less specific w.r.t. the conclusion. By Definition 3 and Proposition 2, the following conclusion holds. That is, a more conservative prudent argument has a higher confidence level.

Corollary 1. *Let* $\mathbb{A} = \langle \Phi, \alpha, \omega_1 \rangle$ *and* $\mathbb{A}' = \langle \Psi, \beta, \omega_2 \rangle$ *be two prudent arguments. If* $\mathbb{A} \succeq \mathbb{A}'$, *then* $\omega_1 \geq \omega_2$.

Corollary 2. *If* $\mathbb{A} = \langle \Phi, \alpha, \omega_1 \rangle$ *is quasi-equivalent with* $\mathbb{A}' = \langle \Psi, \beta, \omega_2 \rangle$, *then* $\Phi = \Psi$ *and* $\alpha \equiv \beta$.

The notion of quasi-equivalence captures situations where two prudent arguments can be said to make the same point on the same grounds. It will be used together with the notion of being more conservative to avoid some redundancy and infiniteness when counter-arguments need to be generated.

Interactions Between Prudent Arguments

Inconsistency and incoherence are two particular problems encountered in $\mathcal{DL}$ ontologies. These kinds of conflicts are brought together in a unique relation of conflict that we define as follows:

Definition 4. *Let* $\mathcal{O}$ *be an* $\mathcal{DL}$ *ontology and* α *be an axiom of the* $\mathcal{DL}$ *language.* α *is said to be conflictive with* $\Phi \subseteq \mathcal{O}$, *denoted as* $\alpha \bowtie \Phi$, *iff there exists* β *in* $\mathcal{DL}$ *language s.t.* $\Phi \vdash \beta$ *where (1)* $\mathsf{Mod}(\{\alpha, \beta\}) = \emptyset$, *or (2) there exists* $A \in \mathcal{N}_C$ *s.t.* $A^I = \emptyset$ *for all* $I \in \mathsf{Mod}(\{\alpha, \beta\})$.

It is important to stress that Definition 4 covers usual inconsistency-based conflicts and the conflicts through incoherence. Obviously, the relation $\bowtie$ is not reflexive, not irreflexive and symmetric.

Based on the conflict relation $\bowtie$, let us now explain how prudent arguments can be challenged in $\mathcal{DL}_\pi$ ontologies. The conclusions of some prudent arguments can be conflicting with the support of others, which leads to the notion of *defeater* defined in the following manner[2]:

Definition 5. *A defeater for a prudent argument* $\langle \Phi, \alpha, \omega_1 \rangle$ *is a prudent argument* $\langle \Psi, \beta, \omega_2 \rangle$ *where* $\beta \bowtie \Phi$ *and* $\omega_2 \geq \omega_1$.

Notice that in classical logic-based argumentation [12], if $\langle \Phi, \alpha \rangle$ is a defeater for $\langle \Psi, \beta \rangle$, $\langle \Psi, \beta \rangle$ is also a defeater for $\langle \Phi, \alpha \rangle$. Consequently, the notion of defeater is symmetric. However, this property does not hold w.r.t. the conflictive paradigm of $\mathcal{DL}_\pi$ logic. The notion of defeater defined for $\mathcal{DL}_\pi$ ontologies is thus asymmetric and anti-reflexive. Indeed, this relation depends on the weight associated to prudent arguments.

Since defeaters are prudent arguments, they can be ordered from more conservative to less conservative as follows.

Definition 6. *A prudent argument* $\langle \Psi, \beta, \omega_2 \rangle$ *is a maximally conservative defeater for* $\langle \Phi, \alpha, \omega_1 \rangle$ *iff* $\langle \Psi, \beta, \omega_2 \rangle$ *is a defeater for* $\langle \Phi, \alpha, \omega_1 \rangle$ *such that no defeaters for* $\langle \Phi, \alpha, \omega_1 \rangle$ *are strictly more conservative than* $\langle \Psi, \beta, \omega_2 \rangle$, *that is for all defeaters* $\langle \Psi', \beta', \omega_3 \rangle$ *for* $\langle \Phi, \alpha, \omega_1 \rangle$, *if* $\Psi' \subseteq \Psi$ *and* $\{\beta\} \vdash \beta'$ *(and hence,* $\omega_3 \geq \omega_2$), *then* $\Psi \subseteq \Psi'$ *and* $\{\beta'\} \vdash \beta$ *(and hence,* $\omega_2 \geq \omega_3$).

We assume that there exists an enumeration which we call *canonical enumeration* of all maximally conservative defeaters for $\langle \Phi, \alpha \rangle$.

As prudent arguments, maximally conservative defeaters are in an infinite number, as shown by the following results.

Proposition 3. *Let* $\langle \Psi, \beta, \omega_2 \rangle$ *be a maximally conservative defeater for* $\langle \Phi, \alpha, \omega_1 \rangle$. *There is an infinite set of maximally conservative defeaters for* $\langle \Phi, \alpha, \omega_1 \rangle$ *that are quasi-equivalent to* $\langle \Psi, \beta, \omega_2 \rangle$.

Now, it is possible to avoid some amount of redundancy among counter-arguments by ignoring the unnecessary variants of maximally conservative defeaters. To this end, we define a concept of *rational defeaters* as follows.

Definition 7. *Let* $\langle \Psi_1, \beta_1, \omega_1 \rangle, \ldots, \langle \Psi_n, \beta_n, \omega_n \rangle, \ldots$ *be the canonical enumeration of all maximally conservative defeaters for* $\langle \Phi, \alpha, \omega \rangle$. *Then,* $\langle \Psi_i, \beta_i, \omega_i \rangle$ *is a rational defeater for* $\langle \Phi, \alpha, \omega \rangle$ *iff* $\forall\, j < i$, $\langle \Psi_i, \beta_i, \omega_i \rangle$ *and* $\langle \Psi_j, \beta_j, \omega_j \rangle$ *are not quasi-equivalent.*

Henceforth, the rational defeaters gather all the possible attacks of a given prudent argument in the same ones: the ones which are representative of all defeaters for that prudent argument.

---

[2] In the argumentation literature, several attack relations were proposed (see [24]). Some of them, like the well-known rebutting, are encompassed by the defeater attack relation. We therefore focus on defeater relation.

From an $\mathcal{DL}_\pi$ ontology $\mathcal{O}$, several possibly interconnected prudent arguments can co-exist. They should be assembled to get a full understanding about the pros and cons, conducting a conclusion to be "accepted", "rejected" or "undecided". Argumentation trees are intended to gather those prudent arguments in the same structure.

Definition 8. *An argumentation tree for a conclusion α is a tree T whose nodes are prudent arguments s.t.:*

1. *The root of T is a prudent argument for α, i.e., $\langle \Phi, \alpha, \omega \rangle$,*
2. *The children of a node in T consist of all its rational defeaters $\langle \Psi_1, \beta_1, \omega_1 \rangle, \ldots,$ $\langle \Psi_n, \beta_n, \omega_n \rangle$ s.t., for $1 \leq i \leq n$, the following condition holds: there is $\gamma \in \Psi_i$ such that $\gamma \notin \Psi$, for every ancestor $< \Psi, \beta, \omega >$ of $\langle \Psi_i, \beta_i, \omega_i \rangle$.*

That is, an argumentation tree aims to exhaustively capture the way counter-arguments can take place as a dispute develops. Condition 2 requires that each counter-argument involves extra information thereby precluding cycles. The following result shows the finiteness of argumentation trees due to the fact that $\mathcal{O}$ is finite.

Proposition 4. *Given an $\mathcal{DL}_\pi$ ontology $\mathcal{O}$ and an axiom α, there is only a finite number of argumentation trees for α, and each argumentation tree is finite.*

Example 4 (Example 2 contd.). An argumentation tree for $\alpha = \text{Employee(Peter)}$ is given in Fig. 1.

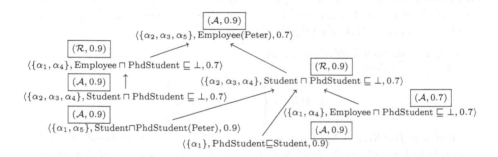

Fig. 1. An argumentation tree for the assertion Employee(Peter) w.r.t. the ontology $\mathcal{O}$. The boxed values are labels under the 3rd labelling instantiation (see Sect. 4), with which the intuitive assertion is acceptable unless the presence of conflict and uncertainty in the ontology.

Finally, given an $\mathcal{DL}_\pi$ ontology $\mathcal{O}$, it can have more than one argumentation tree for a given axiom. An *argument structure* aims to represent them in a global manner by considering all rational defeaters and all possible attacks for a given axiom and its contrary. More formally, we have:

Definition 9. *The argument structure for an axiom α is a pair of sets $\langle \mathcal{P}, \mathcal{S} \rangle$ with $\mathcal{P}$ the set of argumentation trees for α and $\mathcal{S}$ the set of argumentation trees for $\neg \alpha$.*

4   Prudent Argument Labelling

Given an argumentation tree, we want to determine whether the root argument wins (i.e., it is undefeated) or whether it loses (i.e., it is defeated), or undecided. Various developments in literature rely on exogenously given qualitative or quantitative information to decide, or give more refined accounts of, the justification status of arguments. Other works intuitively refine the notion of acceptability based on the number of (counter)-attacks on arguments (see [25] for an overview). In this section, we aim to define a more prudent notion of labelling sensitive to the number of attacks and the weights of arguments. More concretely, if an initiating prudent argument wins, then we may regard it as an acceptable inference. For this, each node is labelled as either $\mathcal{A}$ for accepted, $\mathcal{R}$ for rejected or $\mathcal{U}$ for undecided. To do so, we refine the judge function proposed in [12]. Let $N(T)$ be the set of nodes in an argumentation tree T. Consider a node $\mathbb{A} \in N(T)$, let $C(\mathbb{A})$, $C_{\mathcal{A}}(\mathbb{A})$, and $C_{\mathcal{R}}(\mathbb{A})$ denote the children, the accepted children, and the rejected children of $\mathbb{A}$, respectively.

Now, we introduce our general labelling function that allows to associate to each prudent argument the information concerning its possible justification statuses, depending both on the number of attacks and the certainty degree of their children in the argumentation tree. In other words, based on the propagation of numerical values assigned to prudent arguments, we determine the degree to which prudent arguments are justified. More formally, we have:

Definition 10. *Let T be an argumentation tree for an axiom α and $\mathbb{A} = \langle \Phi, \alpha, \omega \rangle$ be a prudent argument in T. A prudent argument labelling is a total function $Lab : \mathbb{A} \rightarrow \{(\mathcal{A}, \omega'), (\mathcal{R}, \omega'), \mathcal{U}\}$ defined as follows:*

1. *if $C(\mathbb{A}) = \emptyset$, then $Lab(\mathbb{A}) = (\mathcal{A}, \omega)$*
2. *if $C(\mathbb{A}) \neq \emptyset$, then*

$$Lab(\mathbb{A}) = \begin{cases} (\mathcal{R}, a) & \text{if } a > 0 \\ (\mathcal{A}, -a) & \text{if } a < 0 \\ \mathcal{U} & \text{if } a = 0 \end{cases}$$

where $a = f(g(N_1, \ldots, N_m))$ for $N_{i(1 \leq i \leq m)} \in C(\mathbb{A})$ is computed based on $f : \mathbb{R} \mapsto [-1, 1]$ and $g : \mathbb{A}^n \mapsto \mathbb{R}$. Note that we omit $\mathbb{A}$ in the signature of f, g when the referred prudent argument is clear.

Intuitively, Definition 10 states that all leaves in T are always acceptable as they have no attackers. The justified status of every other prudent argument depends on the status of its attackers (i.e. children), which in turn depends on the status of their attackers, and so on. This definition is a general bottom-up framework for labelling argumentation trees, which is based on the combination of the number of attacks and the weights of prudent arguments. Next, we give some interesting instantiations of this prudent labelling framework.

Instantiations of Prudent Argument Labelling

To instantiate the prudent argument labelling, we only need to give the definitions of the functions f and g used in Definition 10. Next, we consider the following cases:

1. $g(N_1, \cdots, N_m) = |C_{\mathcal{A}}(\mathbb{A})| - |C_{\mathcal{R}}(\mathbb{A})|$, and $f(x) = \frac{1-e^{-x}}{1+e^{-x}}$.
2. $g(N_1, \cdots, N_m) = \max_{i,j}\{\omega_i - \omega_j \mid \langle \Psi, \beta, \omega_i \rangle \in C_{\mathcal{A}}(\mathbb{A}), \langle \Theta, \gamma, \omega_j \rangle \in C_{\mathcal{R}}(\mathbb{A})\}$, and $f(x) = -x$.
3. $g(N_1, \cdots, N_m) = \max\{\omega_i \mid \langle \Psi, \beta, \omega_i \rangle \in C_{\mathcal{A}}(\mathbb{A})\} - \max\{\omega_j \mid \langle \Theta, \gamma, \omega_j \rangle \in C_{\mathcal{R}}(\mathbb{A})\}$, and $f(x) = x$.
4. $g(N_1, \cdots, N_m) = \sum_i\{\omega_i \mid \langle \Psi, \beta, \omega_i \rangle \in C_{\mathcal{A}}(\mathbb{A})\} - \sum_j\{\omega_j \mid \langle \Theta, \gamma, \omega_j \rangle \in C_{\mathcal{R}}(\mathbb{A})\}$, and $f(x) = \frac{1-e^{-x}}{1+e^{-x}}$.
5. $g(N_1, \cdots, N_m) = \frac{|C_{\mathcal{R}}(\mathbb{A})|}{|C(\mathbb{A})|}$, and $f(x) = -1$ if $x = 1$; $f(x) = 1$ if $0 \le x < 1$.
6. $f(g(N_1, \cdots, N_m)) = -1$.

The intuitive idea about the first instantiation is that the judgment value of choosing $\mathcal{R}, \mathcal{A}$ or $\mathcal{U}$ depends on the difference between the number of rejected children and that of accepted children. The intuition is that if a prudent argument is attacked by more defeaters that have been labelled as accepted than those labelled as rejected, this argument should be rejected because its defeaters are more often accepted. Otherwise, it can be accepted, unless if it has the same number of accepted and rejected defeaters, its labelling is $\mathcal{U}$. The second instantiation focuses on the values of children nodes of a prudent argument instead of their numbers. The intuition is that if the defeater of a prudent argument with the highest confidence value has been accepted (resp. rejected), then the prudent argument should be rejected (resp. accepted); Otherwise, it is undecided if the confidence values of accepted and rejected defeaters are the same. In the third instantiation, the decision of being accepted, rejected, or undecided is the same as the second instantiation, but the labelling value is different: for example, if a prudent argument has a child having the largest confidence 0.7 among its accepted children and the largest confidence 0.3 among its rejected children nodes, then the final labelling for this argument is $(\mathcal{R}, 0.4)$. In this way, we lower the confidence of labelling during the presence of accepted or rejected defeaters. Unlike the second or the third instantiation, the fourth variant aggregates all confidence values (by the sum function g) for the final decision. Then, the function f maps the summed number into $[-1, 1]$. The fifth instantiation says that the current node should be accepted iff all its defeaters are rejected, and rejected whenever. Notice that this corresponds to the complete labelling used for characterising the semantics of abstract argumentation theory [7]. Finally, the sixth function is indeed a simple case where a prudent argument is always accepted regardless their rational defeaters. We can see in Sect. 5 that such a simple labeling function allows us to encompass some existing inference relations.

Proposition 5. *The fifth labelling function gives a complete labelling. That is, $Lab(\mathbb{A}) = (\mathcal{A}, 1)$ iff $|C_{\mathcal{R}}(\mathbb{A})| = |C(\mathbb{A})|$, and $Lab(\mathbb{A}) = (\mathcal{R}, 1)$ iff $|C_{\mathcal{A}}(\mathbb{A})| \ne 0$.*

Notice that there are infinite possible instantiation manners, among which are the six above. We believe that the choice of an instantiation depends on specific application scenarios, which is under our future plan.

Now, with the notion of prudent labelling we can determine whether the conclusion of the root node is taken to hold.

Definition 11. *Let T be an argumentation tree for an axiom α where $\mathbb{A}_r$ is the root node of T. The judge function, denoted by $Judge()$, from T to $\{(Warranted, \omega), (Rejected, \omega), Undecided\}$ is defined as:*

$$Judge(T) = \begin{cases} (Warranted, \omega) & \text{if } Lab(\mathbb{A}_r) = (\mathcal{A}, \omega) \\ (Unwarranted, \omega) & \text{if } Lab(\mathbb{A}_r) = (\mathcal{R}, \omega) \\ Undecided & \text{if } Lab(\mathbb{A}_r) = \mathcal{U} \end{cases}$$

Example 5. Figure 1 gives the prudent argument labelling for the argumentation tree from Example 4 by using the third instantiation functions. Accordingly, $Judge(T) = (Warranted, 0.9)$.

5   Argumentative Inference Relations and Properties

In this section, we present several inference relations based on the argument structure and the labelling function. We also give the relationships of these new entailments with two existing inference relations defined by the notions of maximal consistent subset and argumentation (see Fig. 3).

Interestingly, our weighted inference relations can tell to what extent a given conclusion is acceptable, which make them often more suitable than classical reasoning in many situations like decision making, negotiation, persuasion, etc. To begin with, we need some additional definitions.

Definition 12. *An $\mathcal{DL}_\pi$ ontology $\mathcal{O}$ is conflict-free if $\mathcal{O}$ is consistent and coherent. The maximal conflict-free subontologies of $\mathcal{O}$ are defined as $MC(\mathcal{O}) = \{\mathcal{O}_1 \subseteq \mathcal{O} \mid \mathcal{O}_1 \text{ is conflict-free and } \forall\, \mathcal{O} \supseteq \mathcal{O}_2 \supset \mathcal{O}_1, \mathcal{O}_2 \text{ is not conflict-free}\}$.*

Intuitively, this definition states that no axiom from $\mathcal{O}$ can be added to $\mathcal{O}_1$ without losing consistency or/and coherence.

Now, we extend the inference relations proposed by [8, 10] to $\mathcal{DL}_\pi$ ontologies as follows.

Definition 13 ($\vdash_{MC}^\forall, \vdash_{MC}^\exists, \vdash_{MC}^{no}$)**.** *Given an $\mathcal{DL}_\pi$ ontology $\mathcal{O}$ and an axiom α:*

- $\mathcal{O} \vdash_{MC}^\forall \alpha$ *if $MC(\mathcal{O}) \neq \emptyset$ and for every $\mathcal{O}' \in MC(\mathcal{O})$, $\mathcal{O}' \vdash \alpha$;*
- $\mathcal{O} \vdash_{MC}^\exists \alpha$ *if there exists $\mathcal{O}' \in MC(\mathcal{O})$ s.t. $\mathcal{O}' \vdash \alpha$;*
- $\mathcal{O} \vdash_{MC}^{no} \alpha$ *if $\mathcal{O} \vdash_{MC}^\exists \alpha$ and $\forall\, \mathcal{O}' \in MC(\mathcal{O})$, $\alpha \not\bowtie \mathcal{O}'$.*

That is, $\vdash_{MC}$ is an inference relation totally determined by the set of maximal conflict-free subontologies. Notice that for the non-objection entailment $\vdash_{MC}^{no}$, an axiom is deduced from an $\mathcal{DL}_\pi$ ontology $\mathcal{O}$ if it follows from at least one maximal conflict-free subontoloy of $\mathcal{O}$ and the rest of the maximal conflict-free subontologies are not against (w.r.t. $\bowtie$).

An argumentative inference in the context of propositional logic has been proposed by [10]. We extend it naturally to $\mathcal{DL}_\pi$ by using prudent arguments as follows.

Definition 14. *Let $\mathcal{O}$ be an $\mathcal{DL}_\pi$ ontology and α be an axiom. α is said to be an argumentative consequence of $\mathcal{O}$, denoted by $\mathcal{O} \vdash_A \alpha$, iff:*

1. *there exists a prudent argument for α in $\mathcal{O}$, and*
2. *there is no prudent argument for $\neg\alpha$ in $\mathcal{O}$.*

Example 6 illustrates that $\vdash_p$ does not imply $\vdash_A$.

Example 6 (Example 1 contd.). A prudent argument for the claim ¬Employee(Peter) is $\langle\{\alpha_1, \alpha_4, \alpha_5\}, \neg\text{Employee}(\text{Peter}), 0.7\rangle$. However, we have $\mathcal{O} \nvdash_A$ ¬Employee(Peter) because there is also a prudent argument for $\langle\{\alpha_2, \alpha_3, \alpha_5\}, \text{Employee}(\text{Peter}), 0.7\rangle$. In contrast, it holds that $\mathcal{O}' \vdash_p$ (¬Employee(Peter), 0.7).

Observe that if there exist two prudent arguments $\langle\Phi, \alpha, \omega\rangle$ and $\langle\Psi, \beta, \omega'\rangle$ in $\mathcal{O}$ s.t. $\langle\Psi, \beta, \omega'\rangle$ is a rational defeater for $\langle\Phi, \alpha, \omega\rangle$, the conclusion $\mathcal{O} \vdash_A \alpha$ holds by the definition of $\vdash_A$, which is counterintuitive because the support of α is attacked. For example, consider the following ontology.

Example 7. Consider the $\mathcal{DL}_\pi$ ontology $\mathcal{O} = \{(\text{Student}(\text{Jim}), 0.9), (\neg\text{Student}(\text{Jim}), 0.8), (\text{Student} \sqcap \text{Employee} \sqsubseteq \bot, 0.7)\}$. So, $\mathcal{O} \vdash_A$ ¬Employee(Jim) since there is a prudent argument $\langle\{\text{Student}(\text{Jim}), \text{Student} \sqcap \text{Employee} \sqsubseteq \bot\}, \neg\text{Employee}(\text{Jim}), 0.7\rangle$ and no prudent argument for the conclusion Employee(Jim). However, it is intuitive to claim that this deduction is dangerous because this prudent argument is attacked by another prudent argument $\langle\{\neg\text{Student}(\text{Jim})\}, \neg\text{Student}(\text{Jim}), 0.8\rangle$, as shown in Fig. 2. Due to this fact, $\vdash_A$ is insufficient to reject this kind of careless entailment.

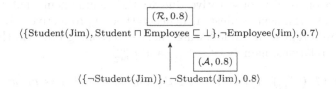

Fig. 2. A labelled argumentation tree w.r.t. $\mathcal{O}$

Now, to avoid the lack of generality and to elude the drawback of ignoring attacked arguments, in the following, we introduce several notions of consequence relations in the light of the argument structure and the labelling functions.

First, for a given argument structure $\langle\mathcal{P}, \mathcal{S}\rangle$ for α w.r.t. an $\mathcal{DL}_\pi$ ontology $\mathcal{O}$, let us consider the following conditions:

C1. $\mathcal{P} \neq \emptyset$ and $\mathcal{S} = \emptyset$.
C2. $\exists\, T \in \mathcal{P}, \, Judge(T) = (Warranted, \omega)$.
C3. $\forall\, T \in \mathcal{P}, \, Judge(T) = (Warranted, \omega)$, and $\mathcal{P} \neq \emptyset$.

C4. $\max_{T \in \mathcal{P}} \{\omega \mid Judge(T) = (Warranted, \omega)\} \geq d$, $d \in [0,1]$.

C5. $\forall\, T' \in \mathcal{S}$, $Judge(T') = (Unwarranted, \omega')$.

C6. $\forall\, T \in \mathcal{P}$, $Judge(T) \neq (Unwarranted, \omega)$.

For the first type of reasoning, we suggest that a conclusion follows from an ontology if the latter has an argument structure that supports this conclusion but no argument structure against that conclusion.

Definition 15 ($\vdash_c$). *Let $\mathcal{O}$ be an $\mathcal{DL}_\pi$ ontology and α is an axiom. We say α is credulously inferred from $\mathcal{O}$ with degree d, denoted $\mathcal{O} \vdash_c (\alpha, d)$, iff the argument structure $\langle \mathcal{P}, \mathcal{S} \rangle$ for α satisfies C1, C2, and C4.*

Example 8. Consider again the $\mathcal{DL}_\pi$ ontology $\mathcal{O}$ of Example 7. Then, we have $\mathcal{O} \not\vdash_c \neg\text{Employee(Jim)}$, although $\mathcal{O} \vdash_A \neg\text{Employee(Jim)}$.

Now, a more cautious inference relation can be defined as follows:

Definition 16 ($\vdash_s$). *Let $\mathcal{O}$ be an $\mathcal{DL}_\pi$ ontology and α is an axiom. Then, α is skeptically inferred from $\mathcal{O}$ with degree d, denoted $\mathcal{O} \vdash_s (\alpha, d)$, iff the argument structure $\langle \mathcal{P}, \mathcal{S} \rangle$ for α satisfies C1, C3, and C4.*

Example 9 (Example 7 contd.). It is not difficult to see that $\mathcal{O} \not\vdash_s \neg\text{Employee(Jim)}$.

We remark that the existence of a counter-argument for a conclusion does not imply that there must exist a prudent argument for its negation. For instance, in Example 7, there is no argument for Employee(Jim) even though the prudent argument for ¬Employee(Jim) is attacked.

It is important to stress that the above inference relations $\vdash_A, \vdash_c$, and $\vdash_s$ for a conclusion α are conservative due to the requirement that there must be no prudent argument against α, hence rather unproductive. To relax such constraint, we propose in the following another reasoning type via three logical consequence relations, namely $\vdash_{arg}^{\forall}$, $\vdash_{arg}^{\exists}$ and $\vdash_{arg}^{no}$.

Definition 17 ($\vdash_{arg}^{\forall}$). *Let $\mathcal{O}$ be an $\mathcal{DL}_\pi$ ontology and α is an axiom. Then, $\mathcal{O} \vdash_{arg}^{\forall} (\alpha, d)$ iff the argument structure $\langle \mathcal{P}, \mathcal{S} \rangle$ for α satisfies C3, C4, and C5.*

That is, Definition 17 allows us to take *all* attacks into account in order to judge if a given statement follows from $\mathcal{O}$.

Definition 18 ($\vdash_{arg}^{\exists}$). *Let $\mathcal{O}$ be an $\mathcal{DL}_\pi$ ontology and α is an axiom. Then, $\mathcal{O} \vdash_{arg}^{\exists} (\alpha, d)$ iff the argument structure $\langle \mathcal{P}, \mathcal{S} \rangle$ for α satisfies C2, C4, and C5.*

Clearly, the relation $\vdash_{arg}^{\exists}$ requires that there exists *at least* one warranted argumentation tree for α. For both $\vdash_{arg}^{\exists}$ and $\vdash_{arg}^{\forall}$, we require that the prudent arguments against the conclusion should be labelled as rejected.

We now investigate a new argumentative inference relation based on the notion of non-objection. The intuition behind is that no argumentation tree in the argument structure has an objection to the acceptance of the conclusion.

Definition 19 ($\vdash^{no}_{arg}$). *Let $\mathcal{O}$ be an $\mathcal{DL}_\pi$ ontology and α is an axiom. Then, $\mathcal{O} \vdash^{no}_{arg} (\alpha, d)$ iff the argument structure $\langle \mathcal{P}, \mathcal{S} \rangle$ for α satisfies C2, C4, C5 and C6.*

That is, α yields from $\mathcal{O}$ under the relation $\vdash^{no}_{arg}$ if it follows from at least one warranted argumentation tree and all the other argumentation trees for α are labelled as rejected or undecided.

Properties of Argumentative Inference Relations

This section summarizes the properties of our inference relations. It is easy to verify that they are all non-monotonic. Moreover, all the relations coincide with the classical definition when the ontology is consistent and coherent, which are two desired properties for dealing with conflicting ontologies. Next, we consider the following desired properties [26] of an inference relation $\vdash_x$:

- **Soundness:** If $\mathcal{O} \vdash_x (\alpha, d)$, then $\exists\, \mathcal{O}' \subseteq \mathcal{O}$ s.t. $\mathcal{O}' \nvdash_x \perp$, $\mathcal{O}' \vdash_x (\alpha, d)$, and $\mathcal{O}' \nvdash_x (\neg\alpha, d)$.
- **Consistency:** If $\mathcal{O} \vdash_x (\alpha, d)$, then $\mathcal{O} \nvdash_x (\neg\alpha, d)$.
- **Monotonicity w.r.t. degree:** If $\mathcal{O} \vdash_x (\alpha, d)$, then $\mathcal{O} \vdash_x (\alpha, d')$, where $0 \leq d' \leq d$.

By definition, all inference relations satisfy our first desiderata.

Proposition 6. $\vdash_c, \vdash_s,\; \vdash_A,\; \vdash^{\forall}_{MC},\; \vdash^{\exists}_{MC},\; \vdash^{no}_{MC},\; \vdash^{\forall}_{arg}, \vdash^{\exists}_{arg},$ *and* $\vdash^{no}_{arg}$ *satisfy Soundness.*

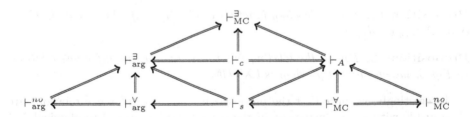

Fig. 3. Productivity comparison of inference relations, where $A \Rightarrow B$ means that the entailement relation A is more productive than B.

Interestingly, we can see that a difference between $\vdash^{\exists}_{MC}$ and the other eight relations is that it can hold that $O \vdash^{\exists}_{MC} \alpha$ and $O \vdash^{\exists}_{MC} \neg\alpha$ simultaneously, which is not the case for the other relations. This leads to the next result.

Proposition 7. $\vdash_c, \vdash_s, \vdash_A, \vdash^{\forall}_{MC}, \vdash^{no}_{MC}, \vdash^{\forall}_{arg}, \vdash^{\exists}_{arg},$ *and* $\vdash^{no}_{arg}$ *satisfy Consistency.*

The next result shows that the inference relations based on argumentation structures are monotonic w.r.t. degree.

Proposition 8. $\vdash_c, \vdash_s, \vdash_{arg}^{\forall}, \vdash_{arg}^{\exists}$, and $\vdash_{arg}^{no}$ satisfy Monotonicity w.r.t. degree.

Proposition 9 shows the productivity comparison of the different inference relations, as depicted by Fig. 3. We say that an inference relation A is *more productive* than B if each conclusion of A is also a conclusion of B. We observe that $\vdash_{MC}^{\forall}$ is the least productive one, and $\vdash_{MC}^{\exists}$ is the most productive one, with other 7 new inference relations among them.

Proposition 9. For an $\mathcal{DL}_\pi$ ontology $\mathcal{O}$, an axiom α, and a real number $d \in (0, 1]$, it holds:

$$\mathcal{O} \vdash_s (\alpha, d) \;\Rightarrow\; \mathcal{O} \vdash_c (\alpha, d) \;\Rightarrow\; \mathcal{O} \vdash_A \alpha.$$
$$\mathcal{O} \vdash_s (\alpha, d) \;\Rightarrow\; \mathcal{O} \vdash_{arg}^{\forall} (\alpha, d) \Rightarrow \mathcal{O} \vdash_{arg}^{\exists} (\alpha, d).$$
$$\mathcal{O} \vdash_c (\alpha, d) \;\Rightarrow\; \mathcal{O} \vdash_{arg}^{\exists} (\alpha, d) \Rightarrow \mathcal{O} \vdash_{MC}^{\exists} \alpha.$$
$$\mathcal{O} \vdash_{MC}^{\forall} \alpha \;\Rightarrow\; \mathcal{O} \vdash_{MC}^{no} \alpha \;\Rightarrow\; \mathcal{O} \vdash_A \alpha.$$
$$\mathcal{O} \vdash_{arg}^{\forall} (\alpha, d) \Rightarrow \mathcal{O} \vdash_{arg}^{no} (\alpha, d) \Rightarrow \mathcal{O} \vdash_{arg}^{\exists} (\alpha, d).$$
$$\mathcal{O} \vdash_{MC}^{\forall} \alpha \;\Rightarrow\; \mathcal{O} \vdash_s (\alpha, d).$$
$$\mathcal{O} \vdash_A \alpha \;\Rightarrow\; \mathcal{O} \vdash_{MC}^{\exists} \alpha.$$

The converses are false.

Proposition 10. $\mathcal{O} \vdash_{MC}^{\forall} \alpha \not\Rightarrow \mathcal{O} \vdash_s (\alpha, d)$ for the labeling functions 1–5. However, for the labeling function 6, we have $\mathcal{O} \vdash_{MC}^{\forall} \alpha$ iff $\mathcal{O} \vdash_s (\alpha, d)$.

Moreover, as given in Fig. 3, we have that $\mathcal{O} \vdash_c (\alpha, d)$ implies that $\mathcal{O} \vdash_{MC}^{\exists} \alpha$ but the converse is false in general. However, for a special labelling function, we obtain the equivalence between these two and other relations.

Proposition 11. For the labeling function 6, $\mathcal{O} \vdash_{MC}^{\exists} \alpha$ iff $\mathcal{O} \vdash_c (\alpha, d)$ iff $\mathcal{O} \vdash_A \alpha$ iff $\mathcal{O} \vdash_{arg}^{\exists} (\alpha, d)$.

Proposition 12. For all the labeling functions, and all the inference relations in Fig. 3, the entailment problem is EXPTIME.

Interestingly, we note that the proposed fine-grained inference relations are all solvable without an increment in the complexity compared to classical $\mathcal{DL}$ reasoning [4].

6   Related Work and Discussion

Reasoning with uncertainty and dealing with inconsistencies in ontologies is a long time studied topic. Some approaches have been proposed to extend DLs with uncertainty and are inconsistency-tolerant [32,34]. However, these work are insufficient to handle different types of inconsistencies, such as inconsistencies caused by assertional knowledge, or the drowning effect. The enhanced probabilistic DLs [35] can avoid the drowning effect, but still suffer from the inconsistencies caused by a set of conflicting knowledge sharing a same uncertain degree, which is unavoidable for many applications, which is avoided in our framework by prudent arguments that allow any conflict-free sub-ontologies with labeling functions over argument structures.

More recently, Inconsistency-Tolerant Query Answering has received a lot of attention [5,9,15,33,40]. The AR-semantics [29] together with its approximations, e.g. IAR, ICAR and ICR [14], has been widely studied [16,17] and extended to (c,l)no-semantics for a better productivity [8]. These work consider that inconsistency comes merely from the data, i.e., it occurs when some assertional ABox facts contradict some TBox axioms. In our framework, we handle conflicts that come from ABox, TBox, and from the interaction between them. Benefiting from deductive argumentation techniques, we emphasize on studying new semantics that can tolerate such general conflicts with different productivity between the credulous entailment ($\vdash_{MC}^{\exists}$) and the universal entailment ($\vdash_{MC}^{\forall}$, similar to AR-semantics) based on MC.

In addition, there are several proposals for argumentation with ontologies. In [41], the authors use argumentation to reason with possibly inconsistent rules on top of certain DL ontologies. In [30], the authors propose a reasoning method for inconsistent and uncertain ontologies. However, this method is insufficient to handle different types of faults in ontologies, i.e., incoherence. In [23], Gómez et al. present a decision support framework for ontology integration based on Defeasible Logic Programming (DeLP). More recently, in [19], the authors propose the concept of incoherency-tolerant semantics for certain Datalog$^\pm$, and show how incoherence affects classic inconsistency-tolerant semantics. Later, [37] introduce a probabilistic structured argumentation framework for handling inconsistency in uncertain data. These last three works are indeed relevant to but still different from ours from the following aspects: First, instead of Datalog$^\pm$ ontologies or probabilistic Presumptive Defeasible Logic Programming (PreDeLP) programs, we deal with DLs based conflicting (both inconsistent and incoherent) ontologies with an uncertainty degree associated to each axiom by means of a possibilistic logic. In particular, our approach spreads uncertainty degrees through argument structures, leading to novel entailment relations that account for weights attached to axioms. Second, [23,37] focus on logical consequences consisting of a single atom. In contrast, our approach is more general as it supports arbitrary DL axioms. Third, instead of using argumentation as the reasoning machinery (by transferring Datalog$^\pm$ ontologies to their correspondent defeasible ones), our approach allows to perform query-specific reasoning method for conflicting and uncertain ontologies without changing the original ontologies. Finally, instead of using an arbitrary argumentation tree to judge the acceptability of a query, our system allows to consider all argumentation trees for and against the query. By tacking the number of attacks and the weights of arguments, it can give a more prudent judgment for a given query, such as the distinction among warranted, unwarranted, and undecided conclusions.

To the best of our knowledge, it is the first framework that is capable of reasoning under uncertain and conflicting ontologies with the use of logic-based argumentation. Several labelling functions are provided and can be chosen depending on applications, which encompass an existing complete labelling as a special case. Lastly, our framework is also featured by several inference relations, independent of labelling functions, whose computations are without extra cost compared to the classical $\mathcal{DL}$ reasoning.

In future research, we plan to prune argumentation trees, by analysing the resonance of prudent arguments with the intended audience, in order to raise their impact as suggested by [27]. We will also build algorithms to compute different inference relations based on cutting edge DL justification algorithms [28] and approximations of argumentation trees with heuristics by adversarial search techniques.

References

1. Amgoud, L., Prade, H.: Reaching agreement through argumentation: a possibilistic approach. In: KR, pp. 175–182 (2004)
2. Arif, M.F., Mencía, C., Ignatiev, A., Manthey, N., Peñaloza, R., Marques-Silva, J.: BEACON: an efficient sat-based tool for debugging $E\hat{L}+$ ontologies. In: SAT, pp. 521–530 (2016)
3. Baader, F., Calvanese, D., McGuinness, D.L., Nardi, D., Patel-Schneider, P.F.: The Description Logic Handbook: Theory, Implementation and Applications (2010)
4. Baader, F., Calvanese, D., McGuinness, D.L., Nardi, D., Patel-Schneider, P.F. (eds.): The Description Logic Handbook: Theory, Implementation, and Applications, 2nd edn. Cambridge University Press, Cambridge (2010)
5. Baget, J., et al.: A general modifier-based framework for inconsistency-tolerant query answering. In: Principles of Knowledge Representation and Reasoning: Proceedings of the Fifteenth International Conference, KR 2016, pp. 513–516 (2016)
6. Bail, S., Glimm, B., Jiménez-Ruiz, E., Matentzoglu, N., Parsia, B., Steigmiller, A. (eds.): Informal Proceedings of the 3rd International Workshop on OWL Reasoner Evaluation, CEUR Workshop, vol. 1207 (2014)
7. Baumann, R.: Characterizing equivalence notions for labelling-based semantics. In: KR, pp. 22–32 (2016)
8. Benferhat, S., Bouraoui, Z., Croitoru, M., Papini, O., Tabia, K.: Non-objection inference for inconsistency-tolerant query answering. In: IJCAI, pp. 3684–3690 (2016)
9. Benferhat, S., Bouraoui, Z., Tabia, K.: How to select one preferred assertional-based repair from inconsistent and prioritized DL-Lite knowledge bases? In: IJCAI, pp. 1450–1456 (2015)
10. Benferhat, S., Dubois, D., Prade, H.: Argumentative inference in uncertain and inconsistent knowledge bases. In: UAI, pp. 411–419 (1993)
11. Besnard, P., Grégoire, É., Raddaoui, B.: A conditional logic-based argumentation framework. In: SUM, pp. 44–56 (2013)
12. Besnard, P., Hunter, A.: A logic-based theory of deductive arguments. Artif. Intell. **128**(1–2), 203–235 (2001)
13. Besnard, P., Hunter, A.: Elements of Argumentation. MIT Press, Cambridge (2008)
14. Bienvenu, M.: On the complexity of consistent query answering in the presence of simple ontologies. In AAAI, AAAI Press (2012)
15. Bienvenu, M.: Inconsistency-tolerant ontology-based data access revisited: taking mappings into account. In: IJCAI, pp. 1721–1729 (2018). ijcai.org

16. Bienvenu, M., Bourgaux, C., Goasdoué, F.: Computing and explaining query answers over inconsistent DL-Lite knowledge bases. J. Artif. Intell. Res. **64**, 563–644 (2019)
17. Bourgaux, C.: Inconsistency Handling in Ontology-Mediated Query Answering. PhD thesis, Université Paris Saclay (2016)
18. Bouzeghoub, A., Jabbour, S., Ma, Y., Raddaoui, B.: Handling conflicts in uncertain ontologies using deductive argumentation. In: Proceedings of IEEE/WIC 2017, pp. 65–72 (2017
19. Deagustini, C.A.D., Martinez, M.V., Falappa, M.A., Simari, G.R.: How does incoherence affect inconsistency-tolerant semantics for datalog±? Ann. Math. Artif. Intell. **82**(1–3), 43–68 (2018)
20. Dubois, D., Lang, J., Prade, H.: Possibilistic logic. In: Handbook of Logic in Artificial Intelligence and Logic Programming, pp. 439–513 (1994)
21. Dubois, D., Prade, H.: A possibilistic analysis of inconsistency. In: International Conference on Scalable Uncertainty Management, pp. 347–353 (2015)
22. Flouris, G., Huang, Z., Pan, J.Z., Plexousakis, D., Wache, H.: Inconsistencies, negations and changes in ontologies. In: AAAI, pp. 1295–1300 (2006)
23. Gómez, S.A., Chesñevar, C.I., Simari, G.R.: ONTOarg: a decision support framework for ontology integration based on argumentation. Expert Syst. Appl. **40**(5), 1858–1870 (2013)
24. Gorogiannis, N., Hunter, A.: Instantiating abstract argumentation with classical logic arguments: postulates and properties. Artif. Intell. **175**(9–10), 1479–1497 (2011)
25. Grossi, D., Modgil, S.: On the graded acceptability of arguments. In: IJCAI, pp. 868–874 (2015)
26. Huang, Z., van Harmelen, F., Ten Teije, A.: Reasoning with inconsistent ontologies. In: IJCAI, pp. 454–459 (2005)
27. Hunter, A.: Towards higher impact argumentation. In: AAAI, pp. 275–280 (2004)
28. Kalyanpur, A., Parsia, B., Horridge, M., Sirin, E.: Finding all justifications of OWL DL entailments. In: ISWC, pp. 267–280 (2007)
29. Lembo, D., Lenzerini, M., Rosati, R., Ruzzi, M., Savo, D.F.: Inconsistency-tolerant query answering in ontology-based data access. J. Web Semant. **33**, 3–29 (2015)
30. Liu, B., Li, J., Zhao, Y.: Repairing and reasoning with inconsistent and uncertain ontologies. Adv. Eng. Softw. **45**(1), 380–390 (2012)
31. Ludwig, M.: Just: a tool for computing justifications w.r.t. ELH ontologies. In: OWL/VSL, pp. 1–7 (2014)
32. Lukasiewicz, T.: Expressive probabilistic description logics. Artif. Intell. **172**(6–7), 852–883 (2008)
33. Lukasiewicz, T., Martinez, M.V., Pieris, A., Simari, G.I.: From classical to consistent query answering under existential rules. In: AAAI, pp. 1546–1552. AAAI Press (2015)
34. Qi, G., Ji, Q., Pan, J.Z., Du, J.: Extending description logics with uncertainty reasoning in possibilistic logic. Int. J. Intell. Syst. **26**(4), 353–381 (2011)
35. Riguzzi, F., Bellodi, E., Lamma, E., Zese, R.: Reasoning with probabilistic ontologies. In: IJCAI, pp. 4310–4316 (2015)
36. Schlobach, S., Cornet, R.: Non-standard reasoning services for the debugging of description logic terminologies. In: IJCAI, pp. 355–362 (2003)
37. Shakarian, P., et al.: Belief revision in structured probabilistic argumentation - model and application to cyber security. Ann. Math. Artif. Intell. **78**(3–4), 259–301 (2016)

38. Simari, G.R., Loui, R.P.: A mathematical treatment of defeasible reasoning and its implementation. Artif. Intell. **53**(2–3), 125–157 (1992)
39. W3C. OWL 2 Web Ontology Language. http://www.w3.org/TR/owl-overview/
40. Wan, H., Zhang, H., Xiao, P., Huang, H., Zhang, Y.: Query answering with inconsistent existential rules under stable model semantics. In: AAAI, pp. 1095–1101 (2016
41. Williams, M., Hunter, A.: Harnessing ontologies for argument-based decision-making in breast cancer. In: ICTAI, pp. 254–261 (2007)

A Cost-Efficient Multi-cloud Orchestrator for Benchmarking Containerized Web-Applications

Devki Nandan Jha[✉], Zhenyu Wen, Yinhao Li, Michael Nee, Maciej Koutny, and Rajiv Ranjan

Newcastle University, Newcastle upon Tyne, UK
{d.n.jha2,zhenyu.wen,y.li119,maciej.koutny,raj.ranjan}@ncl.ac.uk,
info@michael-nee.co.uk

Abstract. Benchmarking the containerized web-applications across multiple cloud gives web-application owners more chance to deploy their applications on cheaper host while meeting their performance requirements. However, benchmarking a large number of cloud hosts (about 267 cloud providers in the world) to find a flexible deployment option becomes a grand challenge. Users need to evaluate as many hosts as possible to find an option which offers expected performance at the lowest price. It is also necessary to benchmark the hosts for longer duration so that it can capture the uncertainty of cloud environment.

In this paper, we present **S**mart **D**ocker **B**enchmarking **O**rchestrator (**SDBO**), a general orchestrator that automatically benchmarks containerized web-applications in multi-cloud environment. At the same time, SDBO is able to maximize the numbers of evaluated cloud providers and type of hosts without exceeding users' budgets. Moreover, we propose a *flexible execution* module which enhances SDBO's ability to capture the performance variation of benchmark web-application for longer period of time in the defined users' budgets.

Keywords: Cloud computing · Benchmarking · Orchestrator · Web-application

1 Introduction

Evolution of microservice architecture that modularizes the application into smaller independent components gives the flexibility for developers to implement each component as a standalone service. Every microservice component in the web-application chain can communicate either via synchronous (HTTP/HTTPS) or asynchronous (AMQP) network communication protocols depending on the level of desired *component autonomy*. Note that many cloud providers such as Amazon and Microsoft offer containers virtualized at the operating system level which facilitates the deployment of microservices i.e. each component of the web-application can be encapsulated into a container. Since containers have

© Springer Nature Switzerland AG 2019
R. Cheng et al. (Eds.): WISE 2019, LNCS 11881, pp. 407–423, 2019.
https://doi.org/10.1007/978-3-030-34223-4_26

many advantages including light-weight, fast start up/shut down, packaged; as a result, users can move their web-applications fast and deploy them efficiently.

However, the multi-cloud environment provide diverse options for users to deploy their web-applications, which means users have more chance to find a cheaper host which still meets their deployment requirements such as cost, throughput, latency. To this end, the users need to test the performance in these hosts before actually deploying and publishing their web-applications. The common practice is to use the standard benchmarking applications to test the hosts instead of using users' own application. This is because these benchmarking applications have the standard procedures to evaluate the performance of the host, thereby obtaining more comprehensive results. Moreover, benchmarking all the hosts from different cloud providers is very challenging as each provider has their own architecture and programming interface [9]. Existing research [10,14] focuses mainly on evaluating the benchmark web-application on different host configurations alone. However, [3,6,13] discuss some frameworks that provide the automatic systems to perform the benchmark across multiple clouds.

Web-application is a long running system and its performance must be guaranteed all the time. On the other hand, the underlying cloud environment is very dynamic and resource preemption happens frequently in the virtualized environment [8]. The performance is also affected by the interference caused by other applications deployed on the same server [5]. Observing the performance variation for a longer duration is an important task for benchmarking web-application. Unfortunately, running the benchmark applications in various hosts over different clouds for a longer duration (say at least 24 h) is very costly. To the best of our knowledge, we could not find any study that considers *cost efficiency* for benchmarking i.e. maximize the number of evaluated hosts and benchmarking time within a defined budget.

In this paper, we aim to build a smart orchestrator for benchmarking containerized web-applications in multi-cloud. SDBO is designed to solve the complexity of deploying benchmark applications in multi-cloud environment that have different programming APIs and numerous ways to interact. To achieve the *cost efficiency*, first, we develop an algorithm that maximizes the number of evaluation hosts based on users' budgets and pre-defined benchmarking time. Then, the *flexible execution* module is designed to capture the performance variation of cloud environment by partitioning the pre-defined benchmarking time into a set of slots. In summary, this paper makes the following contributions:

- We developed a novel orchestrator, SDBO that automates the definition and execution of benchmarks for containerized web-applications. In particular, the orchestrator allows the user to choose the benchmark applications and hosts across different cloud providers.
- SDBO has a native feature of optimization that maximizes the utility of user's budget by maximizing the number of cloud providers and the hosts for benchmarking.

– Based on the optimized execution plans, we interact the plans with the *flexible execution* module to run the benchmarks in a set of time interval thereby capturing the performance variation for longer duration.

2   Related Work

The web-application benchmarks need to be deployed on various host configurations in the multi-cloud environment. Orchestrating the systematic deployment consists of following steps [16]: (i) defining the benchmark with their attributes and relationships, (ii) defining the host machine configuration (e.g. CPU cores, location), (iii) instantiating the cloud host complying the application requirements, (iv) monitoring the resources to ensure the QoS and SLA parameters, and (v) controlling the overall processes. Performing all these steps manually is tedious, error-prone and requires a lot of time and diverse knowledge of architecture and accessing mechanism of all these environments. There are different frameworks available that automate/semi-automate the orchestration steps. [7,12] evaluated the performance of containers for scientific applications where a few of them [17,18] evaluate for big data applications. However, most of these works are intended for single cloud environment, without considering the complexity of interacting with various APIs/SDKs provided by different cloud providers.

There are few existing frameworks that handle the orchestration of benchmarks in multi-cloud environment. CloudBench [13] and Smart CloudBench [3] automates the benchmark execution in multi-cloud environment. However, it is not easy to define the benchmarks using these frameworks. Also they are not specific for containerized environment. Additionally, Varghese et al. proposed a framework called DocLite [15] to evaluate the performance of VMs using containerized microbenchmarks. Microbenchmarks are executed on different VMs and the ranking is evaluated by using the set of weights provided by the user for different system parameters. This framework is specific for scientific application and may not be applicable for web-application. Our proposed SDBO orchestrates the benchmark for web-application while allowing users to define and deploy the benchmark in a very interactive and user-friendly way.

Additionally, there are some commercial tools, e.g. CloudHarmony[1] available that perform the benchmark for users but are not specific for particular application. Also, they do not provide all the required metrics specific to that particular application for making proper decisions before final resource selection and provisioning. The limitations of the existing work are briefly summarized as follows.

Limitations. The cloud providers offers shared computing resources to their customers, which makes the cloud environments dynamic and the SLA very hard to guarantee [11]. Moreover, the web-application is very sensitive to the

---

[1]   https://cloudharmony.com/.

dynamically changing environment that directly affect user's satisfaction. Capturing or monitoring the changing behavior of the cloud environments requires the users to run their benchmark applications over a considerable time, which is very costly. Existing benchmark frameworks are not able to solve the trade-off between the limited budgets and the long-time benchmarking experiments.

Additionally, the variety of cloud providers offer a massive configuration choices of host. For instance, Amazon EC2 provide 43 types of host for their customers excluding self customized hosts. It is not possible to run the benchmark applications over all available resources. The state-of-the-art systems do not consider this case that provides an optimized recommendation to help users in selecting the hosts from the massive number of available hosts spanned across multiple cloud providers.

3   System Overview

This section discusses the architecture and system design details of SDBO.

3.1   SDBO Architecture

Figure 1 illustrates the architecture of SDBO and the dependencies of each component. SDBO is implemented as a web-application that provides a *User Interface* for users to interact, explore and manage their benchmarking experiments. The *User Interface* allows the user to choose an existing benchmark application or customize a new application. Moreover, users can easily select the available hosts from different cloud providers, define the benchmarking time for each selected host, and specify the total budget for running the experiments. Next, this configuration information is stored in a relational Database.

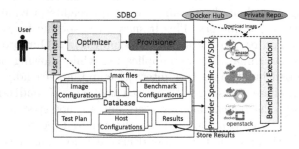

Fig. 1. System architecture of SDBO

The *Optimizer* is designed to create an optimized host list based on the information provided by the user. It retrieves the necessary information (host configurations, benchmark duration and budget) from the Database and applies a heuristic algorithm to generate an optimized host list for running the benchmarking experiments. More details about the *Optimizer* are given in Sect. 3.2.

The generated host list is automatically stored in the Database. Next, users can choose the *flexible execution* option for benchmark execution. If the user chooses to execute the benchmark experiments, the *Provisioner* will be triggered to provision the resources, deploy the benchmark applications and execute the applications based on the user entered information and optimized host list. The benchmark is executed for the specified interval of time and the completion is notified to the *Provisioner*. The results are stored in the Database in real-time for further evaluation and analysis. Finally, the user is notified after completion of the benchmarking experiment and following that cloud resources are released. The main steps of the execution workflow of SDBO is shown in Fig. 2.

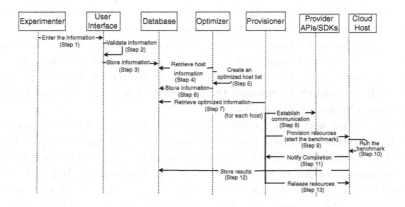

Fig. 2. SDBO execution workflow

3.2 SDBO Design

Optimizer and Its Formal Model. SDBO benchmarks containerized web-application in a multi-cloud environment. Let N represent the number of cloud providers $C_i | i \in \{1, N\}$ where each provider C_i has T type of hosts $v_{i,t} | t \in \{1, T\}$. In our model, we assume a one-to-one mapping between host and container. Consider $\mathcal{C}(v_{i,t})$ to be the unit cost of using $v_{i,t}$, $\tau_{i,t}$ is the time units for which $v_{i,t}$ is chosen to run and $\mathcal{B}$ is the user budget for the benchmark, finding an optimal set of hosts for the benchmark is modelled as a Binary Integer Linear Programming problem (BILP). The defined objective function is given in Eq. 1 subject to constraints as given in Eq. 1a–1c.

$$\text{maximize:} \sum_{i=1}^{N} \sum_{t=1}^{T} \mathbf{x_{i,t}} + \lambda \sum_{i=1}^{N} \sum_{t=1}^{T} (\sum \mathbf{x_{i,t}} - \mathbf{T}) \qquad (1)$$

$$\sum_{i=1}^{N}\sum_{t=1}^{T}(\mathcal{C}(v_{i,t}) \times \tau_{i,t}) \leq \mathcal{B} \tag{1a}$$

$$\forall i \: \forall t \: \tau_{i,t} \geq 0 \tag{1b}$$

$$\forall i \sum_{t=1}^{T} x_{i,t} \geq 1, \: \forall t \sum_{i=1}^{N} x_{i,t} \geq 1 \tag{1c}$$

Where, $x_{i,t}|x_{i,t} \in \{0,1\}$ is a binary variable which represents whether $v_{i,t}$ is selected or not. The first factor of the optimization problem is to comprehend maximum selection of hosts and the second considers a penalizing factor to boost the spanning of maximum number of cloud providers. λ is a tunable parameter which is incorporated to maintain a balance.

Constraint 1a states that the total cost of benchmarking different containers running inside the host must be less than the defined budget. Also, the cost is calculated only if $x_{i,t}$ is 1 with a positive execution time for host $v_{i,t}$ (constraint 1b). Finally constraint 1c enforces the selection of at least one cloud provider and at least one host configuration.

We developed and implemented a heuristic algorithm for the *Optimizer* to solve the problem formalized above. The algorithm generates an optimized list of hosts while satisfying all the defined constraints. The details about how to create an optimized list of hosts is discussed in Algorithm 1. It first calculates the total cost, $\mathcal{C}^{T}(v_{i1,t1})$ for each selected host, $v_{i1,t1}$ (line 4). It then performs a local sorting (using merge sort) for each selected cloud provider, $i1$ according to the increasing host cost and stores it in a temporary list, $List_{i1}$ (line 6). Following that it selects a host with minimum cost globally and adds to the final host list, $V2$ (line 14, line 24) until the final_cost is less than budget, $\mathcal{B}$ (line 8). To maintain the fairness and diversity among different cloud providers, there is a provision to add a penalty if the cloud has been selected (line 17). A host is selected only if the penalty imposed to that cloud is less than a defined value (100 for our case) or if there is no other providers left for selection (line 10).

Provisioner. Once the *Optimizer* generates a benchmark plan, the users can decide whether they want to submit the plan for execution via the user friendly web interface. If the user agrees to perform the experiment, the functions implemented in *Provisioner* will be triggered. First, the *Provisioner* will check the connection and the requirement of the resources on different clouds. Next, it uses a background process application, Hangfire to create and launch the hosts on the selected cloud providers.

Flexible Execution. SDBO offers two types of execution strategy (a) *solitary execution* and (b) *manifold execution*. *Solitary execution* is the basic strategy where users can set a particular time interval for evaluating the benchmark on the desired host configuration. The performance evaluation in this case is limited as it executes only for the particular time interval. We know that the host's QoS performance is highly dependent on the system parameters, e.g. current

---

Algorithm 1: *optimizer*

Input: $V1$ - list of hosts $v_{i1,t1}$ selected by the user, $\tau_{i1,t1}$ - time for executing the benchmark on host $v_{i1,t1}$, $\mathcal{C}(v_{i1,t1})$ - unit cost of using host $v_{i1,t1}$, $\mathcal{B}$ - budget

Output: $V2$ - optimized list of hosts

1 $\forall i1 \; fine_{i1} = 0, \; V2 = [], \; final\_cost = 0$
2 **for** *each selected provider* $i1$ **do**
3   **for** *each selected host type* $t1$ **do**
4        $\mathcal{C}^T(v_{i1,t1}) = \mathcal{C}(v_{i1,t1}) \times \tau_{i1,t1}$
5   **end**
6   Sort the host $v_{i1,t1}$ in ascending order of total cost $\mathcal{C}^T(v_{i1,t1})$ using Merge sort and store in a list, $List_{i1}$
7 **end**
8 **while** *(final_cost $\leq \mathcal{B}$)* **do**
9   Search the first element of all list and find the host $v_{i1',t1'}$ with smallest cost
10   **if** *(fine$_{i1'}$ > 100 & $\forall i1$ (!empty(List$_{i1}$)))* **then**
11   Skip $List_{i1}$ from current calculation
12   continue
13   **else if** *(fine$_{i1'}$ $\leq$ 100 & $\forall i1$ (!empty(List$_{i1}$)))* **then**
14   Add $v_{i1',t1'}$ to $V2$
15   Delete $v_{i1',t1'}$ from the list $List_{i1'}$
16       $final\_cost = final\_cost + \mathcal{C}^T(v_{i1',t1'})$
17       $fine_{i1'} = fine_{i1'} \times 10$
18   **for** *($\forall i1 <> i1'$)* **do**
19   **if** *(fine$_{i1}$ > 10)* **then**
20             $fine_{i1} = fine_{i1}/10$
21   **end**
22   **end**
23   **else**
24   Add $v_{i1',t1'}$ to $V2$
25   Delete $v_{i1',t1'}$ from the list $List_{i1'}$
26       $final\_cost = final\_cost + \mathcal{C}^T(v_{i1',t1'})$
27   **end**
28 **end**

---

workload, network state, etc. which may vary with time [4]. This variation is especially significant for the web-applications due to the continuous execution and the resource preemption in the virtualized environment.

To capture this variation, we propose *manifold execution* strategy that executes the benchmark application in the same host but in multiple time intervals. The user is asked to define the number of iterations along with other parameters for the optimizer. The optimizer then generates an optimized list of hosts which is associated with the execution timestamps. As a result, the *Provisioner* can schedule the deployment and execution based on the host configurations and its associated execution timestamps.

4   Metrics Profiling

SDBO can support benchmarking for different type of web-applications including e-commerce, social media and banking system. It does not only capture the basic web-application features, e.g. response time, throughput illustrated in Sect. 4.1, but also supports more complex and advanced metrics (see Sect. 4.2).

4.1   Basic Metrics

Response Time (ΔT). Response time is the total time taken by the web-application to process a request and generate its response. It is a basic metrics to evaluate the performance of any web-application. Normally, response time depends on many factors varying from the host infrastructure, scheduling policy and the current load on the system to the host capability and network capacity to handle a user's request. Average response time $\mu(T)$ and standard deviation of the response time $\sigma(T)$ are used frequently to measure the performance of the web-applications. Lower response time represents better performance.

Throughput (TP). Throughput represents the host performance in terms of number of requests that can be handled per unit time. Consider that there are total N number of sample requests which are successfully executed in Δt time interval where $\Delta t = (Start\ time - Finish\ time)$, throughput is calculated as $TP = N/\Delta t$.

CPU Usage (CPU). It gives the percentage of CPU used by the container while executing the process. We obtain this information from docker stats APIs [1] embedded with our orchestrator.

Memory Usage ($Memory$). Docker stats APIs also allow us to obtain percentage of memory used by the monitored container.

Network Throughput (Net). This metric indicates how much data can be transferred from a client to the target container in a unit time interval and is represented in Mega bits per seconds (Mbps).

Block I/O (I/O). Block input/output refers to the amount of data written to or read from the block storage devices in a unit time interval and is also represented in Mbps. We collect Net and I/O also from Docker stats APIs.

4.2   Advanced Metrics

Based on the collected basic metrics which are stored in our database, users can perform more complex queries to profile the complex systems.

Apdex Score. Apdex (Application Performance Index)[2] is considered as an open standard developed to standardize the methods for benchmarking, tracking

---
[2]   http://www.apdex.org/index.html.

and reporting the application performance. It utilizes the Response Time (ΔT) to check the user satisfaction level for an application's performance. Based on a defined threshold for the response time **T**, Apdex defines three acceptable zones namely Satisfied, Tolerated or Frustated.

An Apdex score is calculated using the number of requests satisfied and tolerated out of the total requests received. The contribution of satisfied and tolerated requests for the user satisfaction level is 100% and 50% respectively. Let NR, SR and TR be the total number, satisfied number and tolerated number of requests respectively, an Apdex score is calculated as given in Eq. 2. The value of an Apdex score lies between 0 and 1 with higher values representing better satisfaction levels.

$$Apdex\ Score = (SR + TR/2)/NR \tag{2}$$

Host Stability. Stability of host machine is the metric to measure the consistency of the system performance. It is defined as the inverse of variability experienced by different basic metrics. Given the average μ_i and standard deviation σ_i for ith basic system metric ($i \in M$) executed for time T, variability is calculated as given in Eq. 3.

$$Variability = 1/T \sum_{t=0}^{T} \sum_{i=0}^{M} (\sigma_{i,t}/\mu_{i,t}) \tag{3}$$

Thereby, host stability is calculated as $Host\ Stability = 1/Variability$. Hosts with smaller stability values show that the performance is inconsistent and is not suggested for execution.

Host Suitability. Host suitability metric represents the worthiness of a host in terms of performance and cost. It is computed using Eq. 4.

$$Host\ Suitability = TP/Cost \tag{4}$$

where, TP is the throughput, $Cost$ is the per unit execution cost for that particular host. The higher the value of host suitability the better is the host.

5   Evaluation

To illustrate the effectiveness of SDBO, we performed a case study using a simple web-application benchmark. The details are presented in the section below.

5.1   Experiment Setup

SDBO is tested both in simulation and on a real testbed. The simulation is to test the scalability of our proposed optimization algorithm, and the real testbed is to evaluate the system performance. The experiment setup is detailed as follows.

Scalability Evaluation. Our algorithm is tested on a Lenovo PC with Intel(R) Core(TM) i5-6200U CPU @2.3 GHz - 2.4 GHz with 16 GB memory and 512 GB

Table 1. Experiment host configuration

CSP	Host Type	CPU cores	Memory	Disk	Price/hr($)
AWS	t2.nano	1	0.5	EBS	0.0066
	t2.micro	1	1	EBS	0.0132
	t2.small	1	2	EBS	0.026
	t2.medium	2	4	EBS	0.052
	t2.large	2	8	EBS	0.1056
	m4.large	2	8	EBS	0.116
	t2.xlarge	4	16	EBS	0.2112
	c4.xlarge	4	7.5	EBS	0.237
	m4.2xlarge	8	32	EBS	0.464
	c4.2xlarge	8	15	EBS	0.476
Azure	Standard_B1s	1	1	2	0.0118
	Standard_B1ms	1	2	2	0.0236
	Standard_B2s	2	4	4	0.0472
	Standard_F2	2	4	8	0.119
	Standard_B2ms	2	8	4	0.0944
	Standard_D2_v3	2	8	4	0.116
	Standard_B4ms	4	16	8	0.189
	Standard_A4_v2	4	8	8	0.222
	Standard_B8ms	8	32	16	0.378
	Standard_D8_v3	8	32	16	0.464

SSD. We collected 20 host configurations from AWS and Azure as the input dataset as shown in Table 1.

Benchmark Application and Its Deployment. SDBO is published on Google Cloud App Engine (*B2 instance class*) London (europe-west2). Therefore, the users can access to the system from any place and run their benchmarking applications via the user interface. PostgreSQL Database is associated with the SDBO, and stores different configuration of hosts and benchmark images for running the experiments. The database is also deployed on a Google cloud *n1-standard-2* instance with 2 vCPUs, 7.50 GB memory, 128 GB disk. All these components are running independent following the microservice architecture.

We utilized a popular benchmark application, TPC-W[3] with SimplCommerce that emulate the activities of a sample e-commerce web-application. This application is containerized and used to benchmark various type of hosts on AWS and Azure as shown in Table 1. The load on the web-application is created by Apache JMeter[4] according to the test plans defined by the user. To emulate real traffic, JMeter is not configured on the same cloud where the benchmark applications are running. The containerized load generator is deployed on Digital Ocean cloud and the host is a *Standard droplet* machine with 6 vCPUs, 16 GB memory and 320 GB SSD disk.

---

[3] https://cs.nyu.edu/~totok/professional/software/tpcw/tpcw.html.

[4] https://jmeter.apache.org/.

5.2   Cost Optimization

In this section, we evaluate the performance of our optimizer which aims to maximize the number of hosts within the constraint of users' budgets and pre-defined benchmarking time.

To highlight the advantages of the optimizer, we considered 20 host configurations from AWS and Azure (see Table 1). Moreover, we assume that the user would like to run their benchmarking experiment for 3.5 h, with four different budgets \$ 0.5, \$ 1.0, \$ 1.5 and \$ 2.0. We compared the performance of our optimized selection method (*Opt*) with the random selection method (*Rand*).

Figure 3 demonstrates that the optimized option selects the higher number of host in all the cases, compared to the random selection method. The reason is because the optimizer always selects the host with lower price first and then it moves to higher cost host. This is based on the logic that a user wants to deploy their web-application on the cheapest hosts that can meet their QoS requirements. Our algorithm design fits to this logic very much. However, the random selection method selects any host which may not be cost optimized. In addition to this, our method can provide a more stable numbers of hosts as shown in Fig. 3, where the *Opt* has much smaller variance than *Rand*.

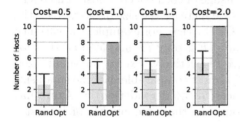

Fig. 3. Comparing the optimized result with random selected result

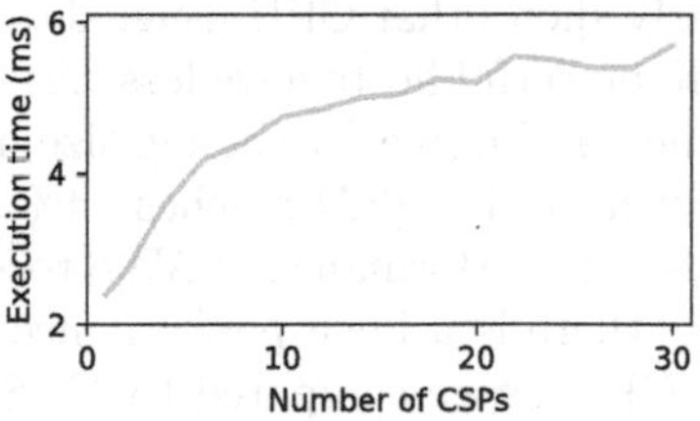

Fig. 4. Schematic diagram showing the execution time complexity of the Optimizer

We also evaluate the scalability of our algorithm by simulating a scenario with varying number of cloud providers with each provider having 50 different host configurations available. Figure 4 shows the execution time of different cases with increasing number of cloud providers varying from 1 to 30. The result shows that the execution time only increases linearly as the number of cloud providers increases. Moreover, the maximal execution time is 5.7 ms for 30 cloud providers, which is comparatively very small as compared to the deployment time.

5.3   Basic Metrics Profiling

In this subsection, we present the benchmark results of an optimized test case. We select a subset of the hosts from Table 1 as the input to our optimizer that then generates 6 hosts (highlighted with gray color in Table 1) for benchmarking experiments.

To obtain the throughput of each deployed benchmark application, we emulated the *bursty request*, i.e. we send the maximal number of requests to the web-applications simultaneously without causing any response error. In other words, the web-applications are fully saturated. Table 2 shows the maximal number of requests for each selected host. Figure 5 illustrates the value of basic metrics (as specified in Sect. 4.1) of the selected hosts, collected from the experiments.

Table 2. Number of requests to saturate the host

CSP	Seq	Host	No. of req (to saturate)	CSP	Seq	Host	No. of reg (to saturate)
AWS	A	t2.small	300	Azure	D	Standard_B1ms	300
	B	t2.medium	600		E	Standard_B2s	600
	C	t2.xlarge	1500		F	Standard_B4ms	1500

CPU Usage. Figure 5(a) shows the CPU usage of each selected host. The result clearly shows that CPU usage decreases as we increase the size of host. Also, the more powerful hosts have less variation of CPU usage. For example, the variance of the CPU usage for the big size hosts C and F is only 7% and 5% and that for small host A and D reaches 24% and 12% respectively. Except for small sized hosts, the performance of AWS to Azure is almost comparable. For the small size host, there is a huge performance difference (52.5% degradation) as Azure has less CPU usage compared to AWS for processing the same number of request.

Memory Usage. The memory usage (Fig. 5(b)) also shows the similar trend except the variation which is much less (highest is 0.76 for host E) as compared to CPU usage. Highest memory usage is noticed for host D followed by host A with 16.1% and 15.3% respectively. Note that the memory usage is varying only in the initial phase, after that the usage is almost constant. It is caused by the property of the Docker container, the memory once allocated is not released back until the container is terminated or restarted.

Network Throughput. The result in Fig. 5(c) shows that the hosts from Azure have about twice the network throughput, compared to the hosts from AWS. The hosts from the same cloud provider have the same network throughput except for host B from AWS where the throughput is much less (487.2 Mbps) than others A (889.2 Mbps) and C (869.2 Mbps).

I/O Throughput. As compared to the above three basic metrics, block I/O shows different trends (see Fig. 5(d)). The I/O throughput of AWS hosts is very random which is because of the selected hosts that offers EBS support. Thus, the throughput is also affected by the time elapsed between the I/O requests and EBS server [2]. Moreover, our benchmark application is block I/O intensive, so

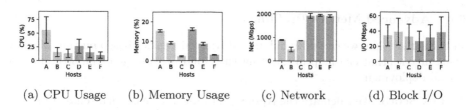

(a) CPU Usage   (b) Memory Usage   (c) Network   (d) Block I/O

Fig. 5. Basic container system metrics while specifying the workload to 300 requests per second with ramp up period as 0 s. CPU and memory usage are given in percentage while network and block I/O throughput are in Megabits per second (Mbps). Black bar on top represents the standard deviation.

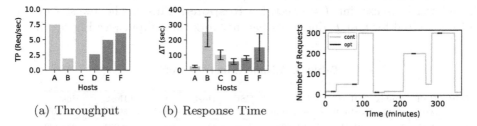

(a) Throughput   (b) Response Time

Fig. 6. System throughput and response time.

Fig. 7. Workload pattern for continuous and optimized execution

the collected statistics are not the maximal I/O throughput of each host. This is demonstrated very well in Azure hosts (see D, E, F in Fig. 5(d)) that the I/O throughput increases with the increase in the number of requests.

System Throughput. Figure 6(a) illustrates the throughput of different hosts. It is clearly depicted from the figure that the throughput increases linearly with the capacity of the host and AWS hosts have comparatively better throughput than Azure for similar sized machine except for host B. The reason of lower throughput for host B is the degraded network throughput as depicted in Fig. 5(c). Therefore, the network becomes the bottleneck in this case. The highest throughput achieved by AWS host is 8.93 requests per second (host C).

Response Time. Figure 6(b) shows the results of the response time. The results show that the response time is significantly affected by the network throughput and the number of requests. Figure 5(c) show that host B has the worst network throughput which causes the significantly higher response time as shown in Fig. 6(b), i.e. 252.2 s. If the network throughput is constant, the response time increases with the increase of number of requests. Note that we try to saturate the web-application until it reaches the maximal number of requests that it can handle without causing errors. Thus, a large number of requests are queuing and waiting for being processed, and this is the main reason that causes the high response time for many requests.

5.4   Advanced Metrics Profiling

In this subsection, we compute different advanced metrics based on the collected basic metrics that can help in selecting the cloud provider and the hosts for the actual deployment.

Apdex Score. We calculate the Apdex score for same case as discussed in Sect. 5.3. Since we are considering the case of a saturated system where the response time is high, we set the threshold for the response time to 50 s. The higher the Apdex score, the better users' satisfaction. Table 3 shows the Apdex score for all selected hosts. The result clearly shows that smaller hosts have better Apdex scores as compared to larger hosts. We have explained why the smaller hosts have lower response time (see Sect. 5.3 **Response Time**). The lowest score of 0.15 is noticed for host B due to its bad network throughput (see Fig. 5(c)).

Table 3. Advanced metrics profile

CSP	Host seq	Apdex score	Host stability	Host suitability	CSP	Host seq	Apdex score	Host stability	Host suitability
AWS	A	1	1.115	286.399	Azure	D	0.8	0.921	109.979
	B	0.15	0.790	36.313		E	0.5	0.772	104.470
	C	0.5	0.834	42.311		F	0.4	0.846	32.042

Host Stability. A host with a higher stability value is considered best as it signifies less performance variation with the elapsed time. The result in Table 3 shows that the stability of small and large host instances are higher. The highest value is for host A with stability index of 1.115 followed by host D with the index of 0.921. The worst stability index is for host E with the value of only 0.772.

Host Suitability. Host suitability is computed as discussed in Eq. 4. A host with higher suitability value is considered to be better as it provides better throughput to cost ratio. The suitability index for the selected hosts show a downward trend with increasing size. For both AWS and Azure, smaller machines have better suitability values as shown in Table 3.

5.5   Flexible Execution

Continuous execution of a benchmark for longer duration is the best way to capture the performance variation. However, the benchmarking cost in this case is very high. To capture the performance variation of changing environment in a defined budget, SDBO offers the *flexible execution* module.

   We do not have access to the cloud hypervisor, therefore, we are not able to emulate the resource changing or preemption of the hosts. As an alternative, we emulate the performance variation of our web-application by changing the number of requests in a period of time. If our tool can observe the performance variation with the changing number of requests, it can also capture the variations that may be caused by other reasons.

To this end, we define 9 test plans, each plan is defined with a timestamp and the number of requests need to be sent as shown in Fig. 7 depicted by *Cont.* For example, the first plan is to send 15 requests starting at 00:00 min timestamp. Following that, the second plan sends 50 requests starting at 30:00 min timestamp. We keep the web-application (benchmark application) running for 360 min to cover all timestamps from the test plans. For the *Opt* case, we randomly selected 5 test plans and sort them based on the timestamp as shown in Fig. 7 and Table 4. The web-application (benchmark application) is executed for 10 min, if and only if the timestamp is reached. The above described two experiments were executed simultaneously with the same host configuration (AWS *t2.medium*).

Table 4 shows response time and throughput collected from both scenarios. The result clearly shows that SDBO can capture the same performance with a maximal variation of 15% in Case I for response time and 5.9% in Case III for throughput. The cost for the optimized method is much less than continuous way of deployment as the total time of deployment for *Opt* is only 50 min as compared to 360 for *Cont*.

Table 4. Comparison of Optimized; *Opt* and Continuous; *Cont* method for Response time and Throughput. Values in [] represent standard deviation.

Case	No. of req	Response time (sec)		Throughput (req/sec)	
		Cont.	Opt.	Cont.	Opt.
Case	10	0.684 [±0.29]	0.789 [±0.36]	6.71	6.54
Case II	15	2.114 [±0.10]	2.122 [±0.15]	6.23	6.17
Case III	50	4.105 [±0.13]	4.441 [±0.18]	10.49	9.87
Case IV	200	21.381 [±1.34]	22.482 [±2.04]	6.90	6.59
Case V	300	23.463 [±6.13]	24.649 [±4.18]	7.56	7.25

6   Conclusion

To facilitate web-application benchmarking in multiple cloud with cost efficiency and flexibility, we proposed SDBO which is the first cost-efficient web-application benchmarking orchestrator. SDBO provides the smart user interface that expedites the handling of benchmark even for a non-expert user. Also, the cost optimization offered by the orchestrator helps the user to select a variety of hosts while *flexible execution* captures the long time performance variation in a limited budget.

Future work. The *flexible execution* module allow users to execute their benchmark applications at any pre-defined timestamp. We can leverage this feature and develop an advanced sampling method to collect the system metrics that can feed to some machine learning methods to have a better observation of the uncertainty in cloud environments. Moreover, we will extend our orchestrator to benchmark other applications such as stream processing and big data applications.

References

1. Docker stats. https://docs.docker.com/engine/reference/commandline/stats/. Accessed 17 Apr 2019
2. I/o characteristics and monitoring. https://docs.aws.amazon.com/AWSEC2/latest/UserGuide/ebs-io-characteristics.html/. Accessed 17 Apr 2019
3. Chhetri, M.B., et al.: Smart cloudbench—a framework for evaluating cloud infrastructure performance. Inf. Syst. Front. **18**(3), 413–428 (2016)
4. Duan, Q.: Cloud service performance evaluation: status, challenges, and opportunities - a survey from the system modeling perspective. Digit. Commun. Networks **3**(2), 101–111 (2017)
5. Jha, D.N., Garg, S., Jayaraman, P.P., Buyya, R., Li, Z., Morgan, G., Ranjan, R.: A study on the evaluation of HPC microservices in containerized environment. Concurrency Comput. Pract. Experience e5323 (2019)
6. Jha, D.N., Nee, M., Wen, Z., Zomaya, A., Ranjan, R.: SmartDBO: smart docker benchmarking orchestrator for web-application. In: The World Wide Web Conference, pp. 3555–3559. ACM (2019)
7. Kozhirbayev, Z., Sinnott, R.O.: A performance comparison of container-based technologies for the cloud. Future Gener. Comput. Syst. **68**, 175–182 (2017)
8. Leitner, P., Cito, J.: Patterns in the chaos - a study of performance variation and predictability in public iaas clouds. ACM Trans. Internet Technol. **16**(3), 15:1–15:23 (2016)
9. Ranjan, R., Benatallah, B., Dustdar, S., Papazoglou, M.P.: Cloud resource orchestration programming: overview, issues, and directions. IEEE Internet Comput. **19**(5), 46–56 (2015)
10. Scheuner, J., Cito, J., Leitner, P., Gall, H.: Cloud workbench: benchmarking IaaS providers based on infrastructure-as-code. In: Proceedings of the 24th International Conference on World Wide Web, pp. 239–242. ACM (2015)
11. Serrano, D., et al.: SLA guarantees for cloud services. Future Gener. Comput. Syst. **54**, 233–246 (2016)
12. Sharma, P., Chaufournier, L., Shenoy, P., Tay, Y.C.: Containers and virtual machines at scale: a comparative study. In: Proceedings of the 17th International Middleware Conference, Middleware 2016, pp. 1:1–1:13 (2016)
13. Silva, M., Hines, M.R., Gallo, D., Liu, Q., Ryu, K.D., Da Silva, D.: CloudBench: experiment automation for cloud environments. In: 2013 IEEE International Conference on Cloud Engineering (IC2E), pp. 302–311. IEEE (2013)
14. Sobel, W., et al.: Cloudstone: multi-platform, multi-language benchmark and measurement tools for web 2.0. In: Proceedings of CCA (2008)
15. Varghese, B., Subba, L.T., Thai, L., Barker, A.: Doclite: a docker-based lightweight cloud benchmarking tool. In: 2016 16th IEEE/ACM International Symposium on Cluster, Cloud and Grid Computing (CCGrid), pp. 213–222. IEEE (2016)
16. Weerasiri, D., Barukh, M.C., Benatallah, B., Sheng, Q.Z., Ranjan, R.: A taxonomy and survey of cloud resource orchestration techniques. ACM Comput. Surv. (CSUR) **50**(2), 26 (2017)

17. Xavier, M.G., Neves, M.V., De Rose, C.A.F.: A performance comparison of container-based virtualization systems for mapreduce clusters. In: 2014 22nd Euromicro International Conference on Parallel, Distributed and Network-Based Processing (PDP), pp. 299–306. IEEE (2014)
18. Zhang, Q., Liu, L., Pu, C., Dou, Q., Wu, L., Zhou, W.: A comparative study of containers and virtual machines in big data environment. In: 2018 IEEE 11th International Conference on Cloud Computing (CLOUD), pp. 178–185. IEEE (2018)

Highlighting Weasel Sentences for Promoting Critical Information Seeking on the Web

Fumiaki Saito[1(✉)], Yoshiyuki Shoji[2], and Yusuke Yamamoto[1(✉)]

[1] Shizuoka University, Johoku 3-5-1, Naka-ku, Hamamatsu, Shizuoka, Japan
{saito,yamamoto}@design.inf.shizuoka.ac.jp
[2] Aoyama Gakuin University, Fuchinobe 5-10-1, Chuo-ku, Sagamihara-shi,
Kanagawa, Japan
shoji@it.aoyama.ac.jp

Abstract. This paper proposes a system that highlights weasel sentences while browsing webpages. The term weasel sentence is defined in the context of this paper as a quotation with an unknown or unidentifiable source. Following this definition, the system automatically detects weasel sentences in browsed webpages. Then, we investigate how highlighting weasel sentences affects the search behaviors and decision making of the users searching for information on the web. An online user study yielded the following results: (1) Highlighting the weasel sentences encouraged participants to invest more time in web browsing and to view a larger number of webpages. (2) The effect of (1) was more significant when participants were familiar with the search topics. (3) Web browsing elicited less change in the confidence of the search answers when participants were familiar with the given topics. The findings provide insights into how users can avoid gathering misleading on the web.

Keywords: Web browsing · Information credibility · Critical information seeking · Human factor · User interface

1 Introduction

The credibility of online digital content is becoming a social problem. With the evolution of digital library technologies, people can create digital contents and casually search for them on the web. Currently, many people often rely on digital contents such as webpages for decision making on their future actions. Therefore, as stated by the ACRL Information Literacy Competency Standards [1], the credibility of online digital contents must be carefully checked. However, people frequently accept online digital contents as credible and also trust the technologies that search for them. For example, Nakamura and their colleagues found that people often believe that web search engines rank webpages by their credibility scores [18]. Pan et al. conducted a similar survey and reported that

© Springer Nature Switzerland AG 2019
R. Cheng et al. (Eds.): WISE 2019, LNCS 11881, pp. 424–440, 2019.
https://doi.org/10.1007/978-3-030-34223-4_27

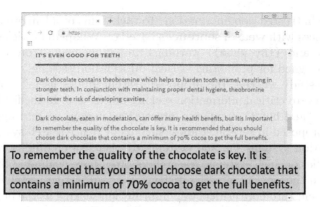

Fig. 1. Overview of our prototype system.

a lot of people choose a higher ranked webpage based on a greater trust in web search engine algorithms [19].

In general, people often think of the credibility of information as an objective quality like authenticity or accuracy. However, as researchers in social psychology indicate, information credibility is a subjective quality, and its interpretation depends on information receivers [6]. To prevent gathering of inaccurate information, users must be made aware of the existence of misinformation or misleading information among web contents and appreciate the value of critical information seeking.

In this paper, we propose a system that highlights weasel sentences in browsed webpages as a low-credibility signal that prompts critical information seeking (see Fig. 1). Information credibility can be assessed by various measures such as authority, date of publication, coverage, and documentation[1]. In this paper, We focus on *the anonymous authority of claims*, highlighting the sentences which seem specific or meaningful but have unknown or unidentifiable sources. For example, in Wikipedia, weasel phrases such as "some people say" and "researchers believe" should be rewritten because they appear to be authorized without substantial evidence, and may therefore bias the readers[2]. In the Japanese Wikipedia, 6430 articles are manually tagged with the {{Weasel}} and {{By whom}} warnings and with rewrite recommendations[3]. In reality, a larger number of documents with weasel phrases reside on the web. Furthermore, web information is often viewed without sufficiently considering the information source [17], although researchers in information literacy call attention to information source for obtaining credible information [1]. This situation carries a severe risk of wrong decision making based on low-quality

---

[1] Berkeley Library's Evaluating resources: http://guides.lib.berkeley.edu/evaluating-resources.

[2] Unsupported attributions: https://en.wikipedia.org/wiki/Wikipedia:Manual_of_Style/Words_to_watch.

[3] The number is at the time of April 19th, 2018.

information in trusted web information. To tackle this problem, we build a detector of sentences with **weasel sentences** based on Wikipedia articles tagged with {{Weasel}} and {{By whom}} warnings and implement it on a web browser extension designed for promoting critical information seeking.

We also study whether the proposed system guides users' browsing/search behaviors toward critical information seeking on the web. In the fields of information retrieval and human-computer interaction, various search support systems have been proposed in several studies, such as evidence search systems [14], dispute suggestion systems [5], and systems to visualize credibility-related scores [20]. In studies on such systems, researchers often have evaluated their proposed systems from the aspects of user satisfaction or subjective usefulness. However, how the systems promote critical web information seeking at the behavioral level and the attitude level has been rarely explored. We discuss how the proposed system can affect the dwell time, the number of viewed pages, and confidence in decision making through an online user study.

The main contributions of this paper are as follows:

- We propose a method that detects weasel sentences in browsed webpages based on weasel annotation data in Wikipedia.
- Through an online user study, we show that highlighting weasel sentences increases both the time expended in reading webpages, and the number of webpages viewed.
- The highlighting effect depends on the search-topic familiarity of the users.

2   Related Works

2.1   Evaluation of Quality and Credibility of Web Information

Various methods that evaluate the quality and credibility of web information have been proposed. Dong et al. developed a method to evaluate web page information credibility, assuming that credible webpages have few false facts or claims [4]. Hasan et al. proposed a machine learning approach that evaluates the quality of Wikipedia articles from their readabilities and editing histories [7]. The model of Wang et al. classifies web information as true or false, using fact-checking websites on political topics [21]. Differently from these related works, we target the sentences in webpages, not the webpages themselves. Moreover, we adopt the anonymous authority of claims as a measure of sentence credibility.

2.2   Attitude for Critical Information Seeking on the Web

Even if people are aware of suspicious information, they often fail to judge its credibility due to the use of incorrect heuristics, i.e., cognitive bias [10]. Ieong et al. revealed the existence of domain bias whereby search users believe that relevant webpages are authorized by specific domains [9]. White et al. studied the relationship between beliefs about search topics and search behaviors [22]. They suggested that if a searcher's belief in a search topic is strong, it is difficult to shift the belief after search and browsing.

2.3   Promoting Critical Information Seeking

Several methods that promote critical information seeking have also been proposed in the fields of human-computer interaction and information retrieval. SEARCH DASHBOARD, proposed by Bateman et al. provides a UI that reflects search behaviors and summarizes search histories [2]. Liao et al. revealed that suggesting the opinion stance and expertise of the information sender can mitigate the so-called echo chamber effect, in which recommender systems present users only with contents that reflect their own views [15]. Yamamoto et al. proposed a "query priming" system that implicitly activates critical information seeking in web searches [24]. By highlighting the insufficiently accurate sentences, we aim to encourage users to consider the credibility of the information found in their document browsing.

3   Highlighting Weasel Sentences

This section describes our approach for automatically judging the anonymous authority of claims in documents. It then proposes a prototype system that highlights the weasel sentences during web browsing.

3.1   Classifier

Weasel sentence detection is treated as a binary classification problem on textual documents. Our proposed system vectorizes a given sentence and detects its anonymous authority status using a trained model.

Below we describe the (1) training data, (2) classification features, and (3) classification performance.

Training Data. Our weasel sentence detection focuses on the {By whom} tag in Wikipedia. Wikipedia volunteers tag the sentences in Wikipedia articles with various warnings that encourage other volunteers to improve the article quality. The {By whom} tag is one of such warning tags. The training dataset was derived from Japanese Wikipedia articles. After dividing the entire set of Wikipedia articles into sentences, we extracted the sentences labeled with the {By whom} tag as *weasel sentences (positive examples)*. Sentences not annotated with the {By whom} tag were then randomly extracted from the articles containing sentences with *weasel* tags. These sentences were accumulated as *non-weasel sentences (negative examples)*. We finally collected 2,236 positive examples and 2,236 negative examples for building an weasel sentence classifier.

Table 1. Examples of weasel expressions used for weasel sentence detection.

Weasel expression
People are saying..., It has been claimed that..., Some people believe..., It is known that..., It has been mentioned that..., Researchers claim..., Critics claim..., I heard that..., There is criticism to...

Features. For weasel sentence classification, our proposed system vectorizes the Wikipedia sentences as two feature types:

Bag-of-Words (BOW): Binary vectors representing the presence (1) or absence (0) of specific nouns, adjectives, and verbs in the sentences. In this study, the BOW vectors were created from terms appearing in more than two sentences. The target terms were extracted from the sentences by a Japanese morphological analyzer, MeCab[4].

Existence of weasel expressions: Binary vectors representing the presence (1) or absence (0) of specific weasel expressions in the sentences. Examples of weasel expressions are "some people think" and "researchers believe." To extract this feature, we prepared 27 weasel expressions listed in a Japanese Wikipedia article on weasel expressions[5]. Some of these expressions are listed in Table 1.

Performance Evaluation. The usefulness of the above features in weasel sentence detection was experimentally evaluated on the dataset described in Subsect. 3.1. For this evaluation, we trained a support vector machine (SVM) classifier with a radial basis function kernel.

The best parameters C and γ of the SVM classifier, and the classification performance were estimated by 5-fold cross-validation. The precision, recall, F1-value, and accuracy were evaluated as 0.772, 0.748, 0.760, and 0.764, respectively.

3.2   Prototype System

Our prototype system was developed as a web browser extension for the user study. Once the extension is installed, any weasel sentences in the browsed webpages are highlighted by the system. Figure 1 is a screenshot of the system overview. The system flow is as follows:

1. When a user visits a webpage, the client system (browser) sends the webpage URL to our application programming interface (API) server.
2. The API server extracts the texts from the webpage and divides them into sentences. The end of each sentence is detected by the ".", "?", and "!" symbols.

---

[4] MeCab: [http://taku910.github.io/mecab/].

[5] Weasel expression on Wikipedia (Japanese version): https://bit.ly/33lJ1hC.

3. The server classifies the sentences into weasel and non-weasel sentences using the trained model.
4. The server sends the weasel sentences to the client.
5. The client system highlights the weasel sentences in the browsed webpage. The highlights are presented as bold font against a yellow background.

3.3 Hypotheses

If the proposed system promotes critical information seeking during web browsing, then users should change their search/browsing behaviors to improve the integrity of their final decisions. This study poses the following hypotheses:

H1. The proposed system extends the time expended in web information seeking.
H2. The proposed system increases the number of visited webpages.
H3. The proposed system improves the users' confidence in their decisions made through web information seeking.
H4. The above-mentioned effects vary with users' familiarity with the search topics.

4   User Study

This section describes the experimental design and the evaluation procedure of the proposed system.

Table 2. Search-task questions and topic familiarity. Numbers in parentheses represent the standard deviations among 188 participants.

Question	Mean of familiarity (SD)
Does cinnamon help improve diabetes?	−2.53 (0.93)
Does vitamin C help prevent pneumonia?	−2.28 (1.00)
Can cocoa decrease blood pressure?	−2.05 (1.24)
Does garlic help improve and prevent a common cold?	−1.51 (1.33)

4.1 Participants

We recruited 250 participants through a Japanese crowdsourcing service, Crowd-Works[6]. After excluding the data of 67 participants who did not complete all tasks or spent an exceptionally long time on the tasks, the data from 188 participants were eligible for the analysis. Each participant received approximately $2 for their time.

---

[6] CrowdWorks: https://crowdworks.jp/.

4.2 Tasks

Each participant performed search tasks on four contentious medical questions (see Table 2). Medical issues were chosen because they are crucial to our life and require careful decision making. We re-used the four questions used in the user study of [24]. The participants were asked to search for the answers to each question using our experimental system, and provide a *yes* or *no* response to each question.

Before starting each task, we asked the participants how familiar they were with the task question. The answers were provided on a six-point Likert scale (from -3: completely unfamiliar to $+3$: completely familiar). Table 2 lists the mean topic familiarities of the participants with the four tasks. On average, the mean familiarity was under -1.50. This result confirms that most participants lacked sufficient knowledge to answer the task questions before participating in the user study.

4.3 Design and Procedure

The effects of two factors (*UI condition* and *topic familiarity*) were examined by a between-subjects design. The UI condition (see Subsect. 4.4 for details) consisted of a *proposed* level, which highlighted the weasel sentences while viewing the webpages, and a *controlled* level, which did not highlight the weasel sentences. The topic familiarity factor consisted of six levels as described in Subsect. 4.2. The participants were randomly allocated to one of the UI conditions. Consequently, 105 participants were allocated into the *proposed* group and 83 participants were categorized into the *controlled* group.

After signing a consent form on the crowdsourcing website, the participants progressed to our experimental website for the user study. As discussed above, the user study comprised four search tasks (see Table 2). The search task order was randomized for each participant to control the task ordering effect. Each search task consisted of three phases.

In the first phase, we asked the participants to indicate their familiarity with the target topics. The participants provided their responses on a six-point Likert scale as described in Subsect. 4.2. The participants then guessed their answers to the task questions and assessed their confidence in their answers (again on a six-point Likert scale from -3: completely unconfident to $+3$: completely confident). We defined this confidence assessment as the *pre-condidence* level.

The second phase of the user study was the *search phase*. In this phase, the participants researched their answers to each task using the experimental search system. Each search task was introduced by a brief instruction; for example,

> *Does cinnamon help improve diabetes? Click the "Start search" button and search for an answer using our search system. When you find a satisfactory answer, come back to this webpage and report the answer.*

After reading the description, the participants began searching by clicking the "Start search" button. Upon clicking the button, the participants were directed to our pre-prepared search engine result pages (SERPs) (see next subsection for details). The participants were asked to visit the webpages on each SERP, and to report their answers when satisfied with the result. The search process was not time-limited.

After the search phase, the participants reported their confidence levels in their answers (on a six-point Likert scale from −3: completely unconfident to +3: completely confident). We defined this confidence assessment as the *post-condidence* level.

4.4 Experimental System

When preparing the SERPs for the task topics, we imitated commonly used search engines such as Google and Bing. Each SERP displayed 30 fixed search results on the target topic, each comprising a title, a snippet, and a URL. Note that our SERPs did not provide a search box for modifying the initial query. The search results of the SERPs were displayed through a Bing web search API[7]. The search results on the tasks were pre-selected by inserting queries of the form "{medical symptom name} AND {possible treatment}" into the Bing API. Finally, we inserted the top-30 results for each task (e.g. "blood pressure AND cocoa") into the associated SERP.

When a participant clicked on a SERP title, the system opened a webpage in a different browser tab. All webpages listed on the SERPs were cached before disseminating the user study. The participant behaviors in the user study were monitored by a Javascript code embedded in each webpage. Furthermore, under the *proposed* UI condition, the weasel sentences in the webpages were highlighted in yellow.

The weasel sentences to be highlighted by the system were manually selected in advance. The manual selection was necessary because our classifier for weasel sentence detection was imperfect. By avoiding the incorrect classifications, we could purely examine the effects of highlighting the weasel sentences on the participants' behavior and decision making.

5 Analysis

The user study provided the behavioral data and pre/post-confidence levels in 752 sessions returned by 188 participants. This section describes the statistical analysis of the data.

5.1 Statistical Approach

The collected data were analyzed by the Bayesian approach. Although the Bayesian approach is less familiar than the frequentist approach, it was selected because it better handles the data uncertainty, yielding the probability distributions of the target parameters [11]. The Bayesian models were constructed

---

[7] https://azure.microsoft.com/services/cognitive-services/bing-web-search-api/.

using generalized linear mixed models (GLMMs) [13], extensions of linear models that model the target responses even when they follow a nonlinear distribution. GLMMs also distinguish the *fixed effects* caused by the experimental conditions from the *random effects* caused by variations among the random samples, such as participants and tasks. The Bayesian GLMMs have become popular tools for modeling various user behaviors in information-retrieval research and human-computer interactions, where they are replacing traditional ANOVA analysis [8, 11, 13].

In this study, we conducted the Bayesian GLMMs to study the effects of weasel sentence highlight, tasks, and participants, with probability interval. We developed Bayesian GLMMs of five response variables: *session time, dwell time in the SERPs, average dwell time per webpage, number of viewed pages*, and *confidence change* (described in Sect. 5.2)[8]. The fixed effects were *UI condition* and *topic familiarity*. The interactions between *UI condition* and *topic familiarity* also considered. The random effects were introduced by the subjects and tasks.

The *session time, dwell time in SERPs* and *average dwell time per webpage*, were assumed to follow Weibull distributions. Previously, Liu et al. reported that the dwell time in webpages follows this distribution [16]. Meanwhile, the *number of viewed pages* was assumed as Poisson-distributed, as appropriate for count data. As the link function for modeling the above-fixed variables on the GLMMs, we selected the log function. After observing the histogram, the *confidence change* was assumed to follow a normal distribution. As the link function for modeling the *confidence change*, we selected the identity function. The fixed and random effects were described by non-informative prior distributions.

We validated our hypotheses through two approaches. The first approach adopted the *high density interval (HDI)*, which summarizes the posterior distribution of a parameter such that every point inside the interval has higher credibility than any point outside the interval [12]. When zero lies outside the 95% (sometimes 90%) HDI of the coefficients of the target variable, Bayesian statistics interprets that the variable exerts a significant effect on the outcome. Meanwhile, frequentists conduct a significance test of the null hypothesis at the specified significance level (typically $\alpha = 0.05$, sometimes $\alpha = 0.1$) [12]. The other approach directly examines the posterior probabilities that our hypotheses are correct, given the data and their non-informative prior distributions.

Table 3. Coefficients for the generalized linear mixed model for session time, with mean, standard error, 90% HDI, and 95% HDI.

Variable	Mean	SE	90% HDI	95% HDI
Intercept	4.38	0.19	[4.02, 4.75]	[3.87, 4.87]
UI	0.29	0.16	[0.02, 0.56]	[−0.04, 0.61]
Familiarity	−0.14	0.04	[−0.19, −0.08]	[−0.20, −0.06]
UI * Familiarity	0.08	0.05	[0.003, 0.16]	[−0.01, 0.17]

---

[8] For Bayesian GLMMs, we used the R package brms [3].

5.2 Response Variables

To investigate hypotheses **H1**, **H2**, **H3**, and **H4** in Subsect. 3.3, we constructed GLMMs of the following five response variables.

Session Time. This variable measures the time expended by a participant during searching and browsing the webpages assigned to the task. The session time may lengthen when the searching and browsing tasks are more carefully performed. The session time of a participant completing a task was obtained by summing his/her dwell time in the associated SERP and his/her total dwell time in the webpages during the task. This measure was investigated in hypothesis **H1** and **H4**.

Dwell Time in SERPs. This variable measures the time expended by a participant in the SERP of a given task. This time may lengthen if the useful webpages are more carefully selected from the list of search results. This measure was evaluated in hypothesis **H1** and **H4**.

Average Dwell Time per Webpage. This variable measures the time expended (on average) by a participant on each webpage of the assigned task. This time may lengthen when any webpage is carefully browsed for quality-judgment information or evidence collection. This measure was investigated on hypothesis **H1** and **H4**.

Number of Viewed Webpages. This variable measures the number of webpages viewed by a participant during a task. The webpage count will increase when the participant visits more webpages to collect and compare multiple pieces of evidence. This measure was evaluated in hypothesis **H2** and **H4**.

Confidence Change. This variable measures the extent to which the participants' confidence in their task answers (beliefs) changes after the web search. The post-confidence will increase if participants consider that the evidence strengthened or weakened their prior belief. This measure was evaluated in hypothesis **H3** and **H4**.

6 Results

6.1 Session Time

Table 3 shows the results of the Bayesian GLMM analysis on session time. The 90% HDI of the *UI condition* variable excluded zero (equivalent to $p < 0.10$ in the frequentist approach). Furthermore, under the *proposed* condition, the session time was extended with a very high posterior probability (a 96.0% chance that the coefficient exceeded zero over the posterior distribution).

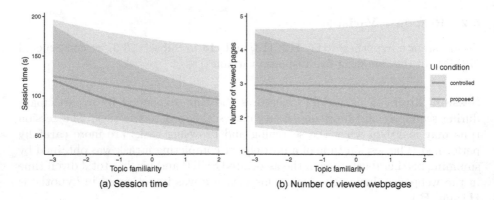

(a) Session time   (b) Number of viewed webpages

Fig. 2. Marginal effects of UI conditions at specific levels of topic familiarity for (a) session time and (b) number of viewed webpages. Green and orange lines indicate the predicted response values under *proposed* and *controlled* conditions, respectively. Colored backgrounds delineate the 95% credible intervals (CIs) of the predicted values. (Color figure online)

Table 4. Coefficients of the generalized linear mixed model for dwell time on the SERPs, with mean, standard error, 90% HDI, and 95% HDI.

Variable	Mean	SE	90% HDI	95% HDI
Intercept	3.10	0.17	[2.78, 3.42]	[2.67, 3.53]
UI	0.19	0.16	[−0.09, 0.45]	[−0.12, 0.51]
Familiarity	−0.11	0.04	[−0.17, −0.04]	[−0.18, −0.03]
UI*Familiarity	0.06	0.05	[−0.02, 0.15]	[−0.04, 0.17]

The 90% HDI of the interaction between *UI condition* and *topic familiarity* was also greater than zero. We confirmed a 96.1% chance that the coefficient exceeded zero over the posterior distribution. Figure 2a illustrates the marginal effects of the *UI condition* and topic familiarity. This figure indicates that (1) the participant session times were longer in the *proposed* condition than in the *controlled* condition, and (2) when participants were more familiar with a search topic, the *UI condition* exerted a larger effect than when participants were unfamiliar with the topic.

6.2   Dwell Time on SERP and Webpage

As indicated in Table 4, the 90% HDIs of the *UI condition* and the interaction between *UI condition* and *topic familiarity* contained zero (equivalently, the null hypothesis cannot be rejected at the $\alpha = 0.1$ significance level). In addition, there was insufficient evidence (insufficiently many high chances) that the coefficients of *UI condition* and the interaction between *UI condition* and *topic familiarity* were positive (87.3% for the *UI condition*; 86.9% for the interaction).

Table 5. Coefficients of the generalized linear mixed model for average dwell time per webpage, with mean, standard error, 90% HDI, and 95% HDI.

Variable	Mean	SE	90% HDI	95% HDI
Intercept	3.65	0.16	[3.36, 3.95]	[3.26, 4.03]
UI	−0.09	0.16	[−0.36, 0.18]	−0.42, 0.23]
Familiarity	−0.04	0.04	[−0.11, 0.03]	[−0.12, 0.05]
UI * Familiarity	−0.02	0.05	[−0.11, 0.08]	[−0.13, 0.09]

Table 6. Coefficients of the generalized linear mixed model for the number of viewed webpages, with mean, standard error, 90% HDI, and 95% HDI.

Variable	Mean	SE	90% HDI	95% HDI
Intercept	0.84	0.18	[0.48, 1.19]	[0.36, 1.35]
UI	0.23	0.14	[−0.01, 0.46]	[−0.06, 0.51]
Familiarity	−0.07	0.04	[−0.14, −0.004]	[−0.15, 0.01]
UI * Familiarity	0.07	0.05	[−0.01, 0.15]	[−0.03, 0.17]

As indicated in Table 5, the 90% HDIs of the *UI condition* and the interaction between *UI condition* and *topic familiarity* contained zero (in the frequentist approach, the null hypothesis could not be rejected at the $\alpha = 0.1$ significance level). Besides, there was insufficient evidence that the coefficients of *UI condition* and the interaction were positive (28.7% for the *UI condition*; 38.7% for the interaction).

6.3   Number of Viewed Webpages

Table 6 shows the GLMM results on the number of visited webpages. The 90% HDIs of *UI condition* and the interaction between *UI condition* and *topic familiarity* contained zero (in the frequentist approach, the null hypothesis could not be rejected at the $\alpha = 0.1$ significance level). However, the *proposed* condition encouraged the participants to visit more webpages with high probability (a 94.5% chance that the coefficient exceeded zero over the posterior distribution). Furthermore, there was a marginally high chance that the coefficient of the interaction between *UI condition* and *topic familiarity* exceeded zero over the posterior distribution (90.4%).

Figure 2b illustrates the marginal effects of *UI condition* and *topic familiarity* for the number of viewed webpages per task. This figure indicates that (1) the participants under the *proposed* condition viewed more webpages than those under the *controlled* condition, and (2) for participants familiar with the search topic, the *UI condition* exerted a larger effect than for participants unfamiliar with the topic.

Table 7. Coefficients of the generalized linear mixed model for confidence change, with mean, standard error, 90% HDI, and 95% HDI.

Variable	Mean	SE	90% HDI	95% HDI
Intercept	1.27	0.23	[0.85, 1.66]	[0.78, 1.78]
UI	−0.12	0.27	[−0.56, 0.32]	[−0.64, 0.40]
Familiarity	−0.37	0.07	[−0.49, −0.25]	[−0.51, −0.23]
UI * Familiarity	−0.17	0.09	[−0.32, −0.02]	[−0.35, 0.01]

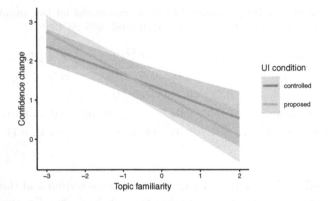

Fig. 3. Marginal effects of UI conditions at specific levels of topic familiarity for evaluating the confidence change. Green and orange lines indicate the predicted response values under *proposed* and *controlled* conditions, respectively. Colored backgrounds delineate the 95% CIs of the predicted values. (Color figure online)

6.4   Confidence Change

Table 7 shows the GLMM results of the confidence changes after the tasks. As indicated in the table, the 90% HDI of the *UI condition* variable included zero. Moreover, the probability of the coefficient exceeding zero was low over the posterior distribution (32.8%). However, as previously observed for the interaction between *UI condition* and *topic familiarity*, the 90% HDI of the interaction variable was less than zero (equivalent to $p < 0.1$ in the frequentist approach). Furthermore, we confirmed a 96.7% chance that the coefficient of the interaction was below zero over the posterior distribution.

Figure 3 illustrates the marginal effects of *UI condition* and *topic familiarity*. This figure indicates that when the participants were not familiar with the search topics, the confidence change was larger under the *proposed* condition than under the *controlled* condition. On the contrary, when the participants were familiar with the search topics, the confidence change was lower under the *proposed* condition than under the *controlled* condition.

7   Discussion

7.1   Detection of Weasel Sentences

As described in Subsect. 3.1, the precision, recall, F-value, and accuracy of our classifier all exceeded 0.7, indicating that our classifier outperformed a simple classifier that randomly judges whether or not sentences are weasel ones.

Rather than discussing the development of a high-performance classifier, we investigated the effect of the classifier on user behaviors during web browsing. Therefore, we simply employed the SVM classifier as a baseline method. However, the performance of weasel-sentence detection could be improved. For example, our classifier neglected the order and semantics of the terms in the sentences. In the future, we have a plan to employ a deep neural network-based method for higher performance. A further study of how the performance of weasel sentence detection affect user behavior should be conducted.

7.2   Effect of Weasel Sentences During Web Browsing

Here, we discuss hypotheses **H1**, **H2**, **H3**, and **H4** in terms of the study results.

For **H1**, we examined the session time, dwell time in SERPs, and average dwell time per webpage during each search task. According to the analytical results, the *UI condition* affected the session time (Subsect. 6.1). However, the behavioral analysis revealed no influence of *UI condition* on the dwell time in SERPs or on the average dwell time per webpage (Subsect. 6.2). From these results, we cannot conclude whether the participants spent longer in the SERPs or the webpages under the proposed condition. However, we can conclude that highlighting the weasel sentences promoted longer web browsing in a particular session level. Therefore, we consider that **H1** was supported.

For **H2**, we examined the number of viewed webpages during each search task. Highlighting the weasel sentences influenced the number of viewed webpages (Subsect. 6.3), meaning that participants using the proposed system visited more webpages than those using the controlled system. From this result, we consider that highlighting the weasel sentences promotes browsing for comparison with other webpages or additional verification, thus supporting **H2**.

For **H3**, we examined the confidence difference in the participants' answers before and after the search tasks. The effect of highlighting the weasel sentences depended on the participants' familiarity with the researched topic. According to Fig. 3, if the participants were unfamiliar with the search topic, their confidence change was more influenced by web browsing under the *proposed* condition than under the *controlled* condition. On the other hand, if the participants were familiar with the search topic, their confidence change was lower under the *proposed* condition than under the *controlled* condition. From these results, we consider that **H3** was partially supported; specifically, it was supported when the participants were unfamiliar with the search topics.

In the **H4** evaluation, participants who were more familiar with the search topics were more influenced by weasel-sentence highlighting than those with less knowledge of the topic (Fig. 2a and b). Furthermore, as discussed for Fig. 3, highlighting the weasel sentences exerted less effect on the confidence change when the participants were familiar with the topic. These results support hypothesis **H4**.

From the above discussion, we conclude that highlighting the weasel sentences encouraged the participants to view more webpages, and consequently extend their time in web browsing, when they were somewhat familiar with the search topics. Before the user study, we expected that when the participants saw the highlighted weasel sentences in a webpage, they would read the text more carefully over a longer time than when the weasel sentences was not highlighted. However, this expectation was not supported by our results: the average dwell time per webpage was no longer under the proposed condition than under the controlled condition. We surmised that the participants judged the highlighted text as poor-quality, and consequently abandoned it. The actual interpretations and usages of the highlighted sentences should be investigated through participant interview or eye-tracking analysis.

Interestingly, we found that when the participants reported more knowledge of the search topic, there was less difference between their pre- and post-confidence levels than those of participants reporting low knowledge of the topic, despite using our proposed system. One interpretation is that when the participants with high topic familiarity looked at highlighted weasel sentences, the sentences might strength their prior beliefs in their task answers. This interpretation indicates that if people use our system with incorrect prior beliefs, they may be prompted into wrong decision making. As discussed in [23,25], several studies have claimed that user characteristics such as topic familiarity and critical thinking capability can affect critical information seeking on the web. Therefore, we should conduct a deeper analysis of the relationship between user characteristics and the effect of the proposed method. Moreover, we should study a better method to support unbiased and critical information seeking on the web.

8   Conclusion

We proposed a system that highlights the weasel sentences on webpages during web browsing. Furthermore, we studied how highlighting such sentences affected the search behaviors and decision making of people searching for web nformation.

The findings of the online user study are summarized as follows: (1) Weasel-sentence highlighting encouraged the participants to extend their web browsing time and view more documents. (2) The effect of finding (1) was more significant when the participants were familiar with the search topics. (3) When participants with more topic familiarity used our system, their prior belief in their search-task answers was less influenced by web browsing than that of users unfamiliar with the topic. The proposed search-interaction design can enhance user engagement in critical information seeking on the web.

Acknowledgments. This work was supported in part by Grants-in-Aid for Scientific Research (18H03243, 18H03244, 18H03494, 18KT0097, 18K18161, 16H02906) from MEXT of Japan.

References

1. American Library Association, Association for College and Research Libraries: Information Literacy Competency Standards for Higher Education. Technical report (2000)
2. Bateman, S., Teevan, J., White, R.W.: The search dashboard: how reflection and comparison impact search behavior. In: Proceedings of CHI (2012)
3. Bürkner, P.C.: brms: an R package for Bayesian multilevel models using Stan. J. Stat. Softw. **80**(1), 1–8 (2017)
4. Dong, X.L., et al.: Knowledge-based trust: estimating the trustworthiness of Web sources. In: Proceedings of the VLDB Endowment, vol. 8, no. 9, pp. 938–949 (2015)
5. Ennals, R., Trushkowsky, B., Agosta, J.M.: Highlighting disputed claims on the Web. In: Proceedings of WWW (2010)
6. Fogg, B.J., Tseng, H.: The elements of computer credibility. In: Proceedings of CHI (1999)
7. Hasan Dalip, D., André Gonçalves, M., Cristo, M., Calado, P.: Automatic quality assessment of content created collaboratively by Web communities: a case study of wikipedia. In: Proceedings of JCDL (2009)
8. Hofmann, K., Mitra, B., Radlinski, F., Shokouhi, M.: An eye-tracking study of user interactions with query auto completion. In: Proceedings of CIKM (2014)
9. Ieong, S., Mishra, N., Sadikov, E., Zhang, L.: Domain bias in Web search. In: Proceedings of WSDM (2012)
10. Kahneman, D.: Thinking, Fast and Slow. Macmillan, London (2011)
11. Kay, M., Nelson, G.L., Hekler, E.B.: Researcher-centered design of statistics: why Bayesian statistics better fit the culture and incentives of HCI. In: Proceedings of CHI (2016
12. Kruschke, J.: Doing Bayesian Data Analysis: A Tutorial with R, JAGS, and Stan. Academic Press, Cambridge (2014)
13. Lee, J., Walker, E., Burleson, W., Kay, M., Buman, M., Hekler, E.B.: Self-experimentation for behavior change: design and formative evaluation of two approaches. In: Proceedings of CHI (2017)
14. Leong, C.W., Cucerzan, S.: Supporting factual statements with evidence from the Web. In: Proceedings of CIKM (2012)
15. Liao, Q.V., Fu, W.T.: Expert voices in echo chambers: effects of source expertise indicators on exposure to diverse opinions. In: Proceedings of CHI (2014
16. Liu, C., White, R.W., Dumais, S.: Understanding web browsing behaviors through Weibull analysis of dwell time. In: Proceedings of SIGIR (2010)
17. Metzger, M.J.: Making sense of credibility on the Web: models for evaluating online information and recommendations for future research. J. Am. Soc. Inf. Sci. Technol. **58**(13), 2078–2091 (2007)
18. Nakamura, S., et al.: Trustworthiness analysis of Web search results. In: Proceedings of ECDL (2007)
19. Pan, B., Hembrooke, H., Joachims, T., Lorigo, L., Gay, G., Granka, L.: In Google we trust: users' decisions on rank, position, and relevance. J. Comput. Mediated Commun. **12**(3), 801–823 (2007)

20. Schwarz, J., Morris, M.: Augmenting Web pages and search results to support credibility assessment. In: Proceedings of CHI (2011)
21. Wang, W.Y.: Liar, liar pants on fire: a new benchmark dataset for fake news detection. In: Proceedings of ACL (2017)
22. White, R.: Beliefs and biases in web search. In: Proceedings of SIGIR (2013)
23. Yamamoto, T., Yamamoto, Y., Fujita, S.: Exploring people's attitudes and behaviors toward careful information seeking in Web search. In: Proceedings of CIKM (2018)
24. Yamamoto, Y., Yamamoto, T.: Query priming for promoting critical thinking in Web search. In: Proceedings of CHIIR (2018)
25. Yamamoto, Y., Yamamoto, T., Ohshima, H., Kawakami, H.: Web access literacy scale to evaluate how critically users can browse and search for Web information. In: Proceedings of WebSci (2018)

Generating an Evolving Skills Network from Job Adverts for High-Demand Skillset Discovery

Elisa Margareth Sibarani[1,2]([⊠]) and Simon Scerri[3]([⊠])

[1] University of Bonn, Bonn, Germany
sibarani@cs.uni-bonn.de
[2] Institut Teknologi Del, Toba Samosir, Indonesia
[3] Fraunhofer IAIS, Sankt Augustin, Germany
simon.scerri@iais.fraunhofer.de

Abstract. Understanding the needs of highly-dynamic job market sectors is of crucial importance to job seekers, employers, and educational bodies alike. This paper describes efforts to identify skill demand composition and dynamics by constructing and interpreting a time series of skills networks that are routinely identified through an established agglomerative hierarchical clustering with breadth-first search order based on co-word occurrences. We focus on Data Science as an example of a highly dynamic sector. Data collected from job adverts between 2016–2017 is pre-processed to identify distinct evolving skills networks observed over at least 12 months. These result in 40 time-series that are used to track the evolving skills clusters and to define the skillsets in high-demand. To return a quantitative scientific result, we implement three traditional statistical models (Naive, Simple Exponential Smoothing (SES), and Holt's linear trend) to forecast future skills cluster composition. The analysis is done based on the centrality and density indices generated for each evolving cluster within the skills networks. Forecasts based on the previous quarter(s) are then checked against actual observations in terms of positioning within a density- and centrality-based strategic quadrant. The F-measures observed (75% and 73% for two top methods) demonstrate the suitability of our approach to identify core skillsets in the near future based on recent data with a high level of accuracy.

Keywords: Graph mining · Network clustering · Time-evolving network

1 Introduction

Many employment sectors have experienced rapid changes due to the increasing availability of data and a corresponding surge in data value extraction through data science, artificial intelligence, and automation. These changes have introduced new opportunities for employment and education, but also challenges due

© Springer Nature Switzerland AG 2019
R. Cheng et al. (Eds.): WISE 2019, LNCS 11881, pp. 441–457, 2019.
https://doi.org/10.1007/978-3-030-34223-4_28

to a discrepancy between the number of individuals possessing relevant skills (supply) and the amount of highly-specific job positions that have come into play (demand). For example, the widening gap between supply and demand in the European data economy is expected to reach 2 million by 2020[1]. This extreme skills gap translates into a loss running into (Euro) billions. Therefore, it is paramount to up-skill and re-skill the existing talent pool to meet the need of fast-evolving employment sectors.

Our investigation considers the hypothesis that skill demand can, to an extent, be discovered and predicted, by tracking the skills network evolution over a series of observances derived from web-posted job adverts. It offers a large-scale, dynamic, and freely available dataset representing job market skill demand; offering a possibility to identify trends in the evolution of requested skill-sets for a specific sector. Our literature survey (Sect. 2) indicated a lack of publicly available tools or methods addressing this possibility. Moreover, the type of analysis we consider differs from studies relying on a frequency value [1,2], since we are not primarily concerned with studying the change in the number of job postings per skill. Evolution-based skill-set forecasting is equally valuable for:

1. job candidates seeking to optimize their skill-set by acquiring complementary new skills to maximize their competitiveness in the job market.
2. academic institutions and training bodies, seeking to offer adaptive and forward-looking curricula based on the observed demands.
3. employment bodies, providing skills-based job matching, career consultation, etc.
4. industry seeking to make informed decisions on future investments.

Our method relies on a large number of job adverts that were published on several online channels between 2016–17. The method for collection, transformation, and curation of the dataset was adapted from the earlier study by Sibarani et al. [3] and is therefore not a contribution of our paper. However, we implement the whole process for data collection and curation since we cover a longer period (2016–2017) and more job adverts from more web job portal sources. Afterward, the dataset is represented as an undirected and weighted graph that has a topology of interconnected skills and weight indices representing the strength between skills based on their observed job advert co-occurrences. We then implement the Coulter method [4], which performs agglomerative hierarchical clustering with breadth-first search to decompose the networks into sub-graphs or clusters, which are sets of highly interconnected skills. Using that data, we create a series of skills clusters for each country under observation. To ensure consistency, we select those distinct but evolving clusters that are observed in at least four consecutive yearly quarters. Although we can summarize their trend and tendency in the near future according to the stability and consistency of the skills clusters structure, nevertheless we need to provide a quantitative result for this

---

[1] https://ec.europa.eu/digital-single-market/en/news/final-results-european-data-market-study-measuring-size-and-trends-eu-data-economy.

inherently empirical question. Therefore time series forecasting is then employed on the resulting 40 cluster series in an attempt to forecast their density and centrality within the entire skills graphs.

2 Related Work

There is little to no published results applying network clustering to monitor market-driven skill-set demand. Network clustering (or graph partitioning) can detect hidden network structures by decomposing them into clusters (or communities) such that there is a dense set of edges within each.

Newman and Girvan [5] introduced one key network clustering algorithm. It implements divisive hierarchical clustering based on breadth-first search. Xu et al. [6] proposed a structural clustering algorithm for networks (SCAN) that is based on the hierarchical agglomerative method. Despite the high effectiveness at discovering community structure in networks, both algorithms [5] and [6] were purposely designed for an unweighted graph. Another clustering algorithm was presented by Blondel et al. [7], best known as Louvain algorithm, which is a heuristic method based on modularity optimization that unfolds a complete hierarchical community structure for a network. The accuracy is excellent for the top-level hierarchy and extremely fast for networks of unprecedented sizes. Unlike [5] and [6], Louvain algorithm tends to deal with an undirected and weighted graph. Similarly, Coulter et al. [4] introduced an agglomerative hierarchical clustering algorithm with breadth-first search based on patterns of co-occurrence between pairs of items (i.e., words or noun phrases) in a corpus of text, in order to identify links and to give weight to each link. Although both [7] and [4] are purposely designed for weighted graphs, the approach that was undertaken by each algorithm is different, which, in return, generates distinct results as well.

The Louvain algorithm [7] places all vertices and links that are available in the graph into the new clusters; therefore, there is a chance that not every result detected by the Louvain method is meaningful. Despite its capability to make sure every cluster has relatively dense internal and sparse external connections, it cannot guarantee that every node in the cluster is important and has sufficiently high connectivity with other nodes in the same cluster. It is highly possible that a low-degree node belongs to a cluster only because it has zero connectivity with other clusters. This primary concern influences the outcome of the Louvain algorithm, which might lead to failure in identifying the core of clusters, for which extracting the highly demanded skills would be impractical. In contrast, the Coulter algorithm [4] is a weight-based method with a tendency to find the cores of clusters. The core nodes in a cluster often have substantial similarity, and hence are connected early in the agglomerative process. Thus, this approach is, in a way, comparable and appropriate for the dataset and the goal of our study.

Most studies related to the time-based analysis for identifying skill-set demand based on the job market only address the frequency of job adverts about a specific skill. In 2014, Smith and Ali [2] conducted a study to assess the employable skills in programming to guide curriculum decisions. Utilizing the search by keyword function in one job portal, they collected data for several years, comprising the date, keywords identifying a skill, and a total number of jobs found about it. They concluded that skill is still in demand based on the trend line that remains strong and continues to grow relative to the total number of jobs. Kobayashi et al. [10] also applied text mining to automatically classify job information from vacancies collected from various employment websites; to analyze the extracted worker attributes (i.e., skills), summarize them and use them to cluster specific jobs; thus this effort had jobs at the focus. Sodhi and Son [1] also analyzed online job ads from online sources, focusing on skills demand related to Operational Research (OR). They compiled a dictionary (hierarchical structure of categories, sub-categories, and keywords) of OR-related words and phrases, which is used with content analysis software to analyze the ads. The proportion of ads that have keywords associated with the skills categories reflects the skills demanded by the industry. On the contrary, our objective is to learn patterns of skills from job postings using the data collection approach that is routinely collecting data posted by job advertisers from multiple sources. Moreover, our results are skill cluster-based, rather than focused on a particular skill. Our approach uses a vocabulary that is generated semi-automatically (in this paper, we focus on Data Science) in place of a dictionary; thus, we eliminate the need for tedious dictionary compilation.

3   The Proposed Framework

This section includes an extensive overview of our main contribution: a Skills Network Evolution Tracking framework that is capable of observing skills network from online-posted job adverts and discovering the highly on-demand skills by tracking the changes in job markets. Therefore, the focus is on the reliance on changes in sub-graph structure to predict which labor market skills will be in-demand, creating a process which is tractable and can improve the visibility of changes in job markets for end-users, e.g., policymakers, employers and workers. Although it is not the primary goal of this study, we nevertheless implement several traditional statistical prediction models to offer a proof-of-concept of how the prediction results on highly-demanded skills can look like, so we can validate and return a quantitative scientific result, despite its nature as an inherently empirical question.

The proposed framework, as depicted in Fig. 1, comprises all the required steps, starting from data curation and transformation, graph construction, subgraph identification (or skills network clustering), sub-graph quadrant and class' type identification, skill series generation, and network evolution tracking for the prediction task. As for the data curation and transformation, our proposed framework re-used the pipeline from an earlier study by Sibarani et al. [3], thus we omit this process from the detail description below, as we assume that there is data in the knowledge base. However, to make it as clear as possible, we nevertheless start by summarizing the data collection and briefly introduce the information extraction and the transformation process.

3.1 Dataset Compilation

We collected and analyzed 620,760 job adverts related to 'Data Science' from 17 European countries. The current results are limited by this data, which covers the period 2016–17, as shown in Table 1. Adverts were obtained from Adzuna[2], Indeed[3], and Trovit[4]. The disparity between data for the two years is due to job adverts only having been available from four countries (Germany, France, United Kingdom, and the Netherlands) between Jan–Sep 2017.

Table 1. Distribution of job adverts each country by year

Country	2016	2017	Country	2016	2017
Denmark	7315	–	Switzerland	26653	1738
Portugal	28886	2187	Belgium	12358	623
Germany	107227	10856	Spain	16786	–
Hungary	5881	764	France	68450	7989
Italy	18396	1773	Czech Republic	9032	–
Austria	10879	1291	Sweden	16460	1932
The Netherlands	65307	4344	Romania	15239	1549
Ireland	38159	4946	Poland	9804	2437
United Kingdom	111377	10122			

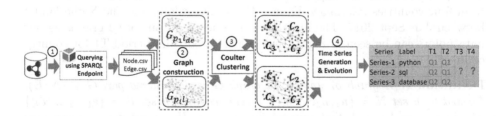

Fig. 1. Skills network evolution tracking framework architecture.

The adverts were indexed by a total of 1,287,994 keywords (a mean of 2.07 per job advert), identified using the Ontology-Based Information Extraction

---

[2] https://www.adzuna.com/.
[3] https://de.indeed.com/?r=us.
[4] https://de.trovit.com/.

pipeline [3] guided by the *Skills and Recruitment Ontology*[5] (SARO) [13]. Collected job adverts are transformed to comply with the Resource Description Framework (RDF) W3C standard for data representation, in which Fig. 3 depicts one job advert represented in RDF format. All job ad-related information that consists mainly of `JobPosting` and `Skill` is stored as RDF instances adhering to the concept in the SARO ontology.

3.2   Graph Construction and Clustering

The data used for the objectives in this paper, therefore, consists of RDF graphs connecting skills using an equivalence index to measure the strength of association between two keywords, utilizing its co-occurrence frequency and is calculated using (1)

$$s_{ij} = \frac{(C_{ij})^2}{(C_i \cdot C_j)}, 0 \leq s_{ij} \leq 1 \tag{1}$$

where, C_{ij} is the number of job adverts in which the skill pair appears; C_i is the number of times that the keyword i is used for indexing a job advert from the dataset; C_j is the number of times that the keyword j is used for indexing a job advert from the dataset.

A graph was initially available for each country, for the whole period covered. To support our goal, that is to generate a time series of skill clusters, we needed to partition these graphs into time period-based observations further. Since the country-specific data coverage during the two years for which data is available was not always consistent at a weekly- or even monthly-level, therefore in our experiment we decided to group the data per trimester to smoothen out the irregular volume of job adverts, which gives in total of 92 out of 136 (complete version: 17 countries for eight periods) data points to be observed. Notice the discrepancy amount that happened due to job adverts only having been available from four countries (Germany, France, United Kingdom, and the Netherlands) between Jan–Sept 2017. The illustration of one graph out of 92 graphs as our data points can be seen in Fig. 2, that is generated from period T1 for country code 'cz'.

Definition 1. *A graph or undirected graph G is an ordered pair $G = (N, E)$, formed by a set $N = \{n_1, n_2, ..., n_N\}$ of vertices and a set $E = \{e_1, e_2, ..., e_E\}$ of edges $e_k = \{n_i, n_j\}$ that connect the vertices. Hence we define a subgraph at period p and country l as a weighted graph $G_{p,l}(N_{p,l}, E_{p,l})$, where an edge $e(u, v) \in E_{p,l}$ connects nodes u, v in $N_{p,l}$ and s_{uv} is the similarity between them, represented by their strength of association.*

Then we applied the Coulter algorithm to generate skill clusters for each graph; reducing a vast network of related keywords into multiple related smaller clusters that are easier to observe. The algorithm uses two passes through the data. Pass-1 identifies clusters by choosing the link with the highest strength

---

[5] https://elisasibarani.github.io/SARO/.

value, and its co-occurrence frequency is higher than t, a predefined threshold, thus making the link as a starting point as well. For each graph, we set a different threshold t based on the median value of the co-occurrence frequency of a particular graph. Other links and their corresponding nodes are added to the cluster in the decreasing order of their strength value based on a breadth-first search until there are no more links that exceed the co-occurrence threshold. Therefore, the Pass-1 builds clusters that can identify areas of intense focus, by generating the primary associations among skills (*internal nodes*) and its corresponding skills (*internal links*). In Pass-2, each Pass-1 cluster is extended by adding links with the highest strength and exceed the co-occurrence frequency threshold, and both nodes of the link must be included in some Pass-1 generated-clusters. The steps will continue for each identified cluster in Pass-1 until no remaining link meets the co-occurrence threshold. Pass-2 can identify keywords that associate in more than one cluster, and thereby indicate the substantial issues, by generating links between Pass-1 nodes across clusters, which are called *external nodes* and *external links*.

The application of network clustering algorithm to the graph depicted in Fig. 2, resulting in total 11 clusters, represented by nodes with the same background colors are in the same cluster and vice versa, except for the nodes without any background colors, are not belong to any clusters. Therefore, from 174 nodes and 1651 edges that belong to graph T1 country code 'cz', the clustering algorithm chose 106 nodes and 197 edges as the focus of the skills network, which will be used further to construct a series of skills cluster for each country. Also, for each resulted cluster, we identify the *density* and *centrality* index. The density index measures the cluster's internal strength (local context) by calculating the average (mean) of the weights of the internal links, while the centrality value measures the strength of a cluster's interaction with other clusters (global context), that is calculated with the square root of the sum of the squares of all external link values.

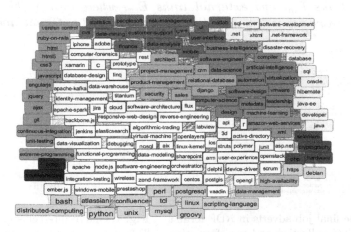

Fig. 2. One graph that is observed from period T1 in country code 'cz'.

Figure 4 zooms-in one resulted cluster (Cluster-3) from graph T1 of country code 'cz' that is located in the bottom part of Fig. 2, with skill *mysql* as the cluster's label. We assigned a label for each resulting cluster by choosing the internal or external node a of a cluster Cl with the highest weight W_{Cl}. This is calculated using (2), which is adapted from [14]

$$W_{Cl}(a) = \frac{K_{Cl}(a)}{n_{Clin} + n_{Clex}}, 0 < K_{Cl}(a) \leq n_{Clin} + n_{Clex}, 0 < W_{Cl}(a) \leq 1 \quad (2)$$

where, n_{Clin} is the sum of the weights of the internal links of cluster Cl; n_{Clex} is the sum of the weights of the external links of cluster Cl; and $K_{Cl}(a)$ is sum of the weights of the edges where an internal or external node a is incidental to the internal or external links of Cl. Hence the keyword with most weight serves to label cluster Cl.

In this study, we apply the clustering algorithm on the skills network for which the clusters are not known ahead of time, thus, to know how good the clusters found by the algorithm are, we calculate the *modularity*, as a measure of the quality of a particular division of a network, that is adapted from [5]. For the clustering of a graph with k clusters, the *modularity* is defined as (3):

$$Q = \sum_{s=1}^{k} \left[\frac{l_s}{L} - \left(\frac{d_s}{2L} \right)^2 \right] \quad (3)$$

where, L is the sum of the weights of all edges in the graph, l_s is the sum of the weights of edges between vertices within cluster s, d_s is the sum of the weights of all incidental edges to the vertices in cluster s. We define skills clusters as follows. We fix an input data of 92 graphs $G_{p,l}$, on which the clustering algorithm then produces a set of subgraphs (clusters) $C = \{C_1, C_2, ..., C_n\}$ for each of the original network $G_{p,l}$.

Definition 2. *A cluster C_1 of graph $G_{p,l}$ comprises* ***internal nodes*** *N_{int},* ***internal links*** *E_{int} and* ***external links*** *E_{ext} where each of the link's co-occurrence frequency is $\geq$ **t**, a predefined threshold, and* ***external nodes*** *N_{ext},*


```
<?xml version="1.0" encoding="UTF-8"?>
<RDF>
  <Description about="http://www.edsa-project.eu/adzuna/eyJhbGciOiJIUzI1NiJ9.eyJzIjoidDl
    <type resource="http://www.edsa-project.eu/edsa#JobPosting"/>
    <url resource="https://www.glassdoor.de/Job/prague-data-analytics-jobs-SRCH_IL.0,6_
    <jobLocation>Prague, Czech Republic</jobLocation>
    <datePosted datatype="http://www.w3.org/2001/XMLSchema#Date">2016-01-03T13:56:30+00
    <source>Adzuna</source>
    <hiringOrganization>Cofidis</hiringOrganization>
    <jobTitle>Data Scientist (Data Analytics) </jobTitle>
    <requiresSkill resource="http://www.edsa-project.eu/skill/data-science"/>
    <requiresSkill resource="http://www.edsa-project.eu/skill/statistics"/>
    <requiresSkill resource="http://www.edsa-project.eu/skill/linux"/>
    <requiresSkill resource="http://www.edsa-project.eu/skill/scripting-language"/>
    <requiresSkill resource="http://www.edsa-project.eu/skill/python"/>
    <requiresSkill resource="http://www.edsa-project.eu/skill/matlab"/>
    <requiresSkill resource="http://www.edsa-project.eu/skill/scala"/>
    <requiresSkill resource="http://www.edsa-project.eu/skill/apache-spark"/>
    <requiresSkill resource="http://www.edsa-project.eu/skill/hadoop"/>
    <requiresSkill resource="http://www.edsa-project.eu/skill/mysql"/>
    <description>In Cofidis, we are expanding our team dedicated to the back-end of our
    platform, designed for creating analytical data products for our customers. We are
    operations engineer to join our Applications Back-end DevOps team. The successful c
```


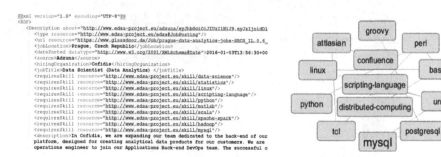

Fig. 3. The final job adverts in RDF format after the data collection and transformation has been done.

Fig. 4. One resulted cluster (Cluster-3) from period T1 in country code 'cz'

where each external node belongs to another clusters of the same graph. The cluster's attributes are: **density** *value Den,* **centrality** *value Cen, the* **label** *Lab, and* **modularity** *value mod. The number of clusters, k is not known a priori.*

Furthermore, a new round of classification is needed to identify clusters with a strong ability to structure the general network. We distinct each resulted clusters from the clustering task based on four categories [8,9]: *isolated, secondary, principal,* and *crossroads.* A cluster is either connected to another cluster(s), or it is *isolated,* which is characterized by an absence of links with other clusters.

Definition 3. *A cluster C_k is* **isolated** *iff its external links $E_{ext} = \emptyset$.*

The distinction between principal and secondary clusters must be retained because the former designates the core of a network. In fact, within the set of principal clusters, we highlight those clusters that have at least y secondary clusters, namely crossroads, because, by their power to connect, crossroads clusters play an essential role in the transformation of a network. As a result, principal clusters – and in particular crossroads clusters – identify the focal point of the object being studied.

Definition 4. *A cluster C_k is the* **secondary** *of cluster C_x, which therefore makes cluster C_x as a* **principal** *of cluster C_k iff there exist p number of external links $e_{1:p}$ in cluster C_x with strength value that is $\geq$ the minimum strength value from all internal links in cluster C_x, and each of this external link $e_1, e_2, ..., e_p$ connects nodes (u, v) where $u \in$ internal nodes N_{int} of cluster C_k and $v \in$ internal nodes N_{int} of cluster C_x, or vice versa, where p has a value greater than δ, a predefined link threshold.*

Definition 5. *Cluster C_x is categorized as a* **crossroads** *cluster iff it is a* **principal** *cluster of y* **secondary** *clusters, where y has a value greater than α, a predefined qualified secondary threshold.*

The parameter $\delta = 2$ is defined based on the average of the amount of external link per cluster, and $\alpha = 2$ is set according to the median value of total number of other clusters that are linked to a cluster x.

We illustrate cluster classification process for Cluster-3 in Fig. 5, in which Cluster-3 (in the middle, same as cluster in Fig. 4), is connected to two other clusters (Cluster-7 on the left-side and Cluster-2 on the right-side) from graph

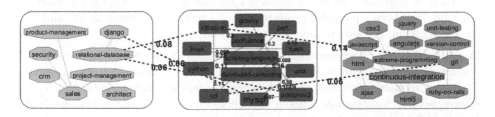

Fig. 5. Crossroads identification of Cluster-3 from T1 in country code 'cz'.

T1 in country code 'cz'. Cluster-3 is not a *isolated* since it has links with other clusters, and it is classified as a *principal* cluster, because it has at least two links to another cluster where its strength value is higher than the minimum strength value of Cluster-3 internal links, in this case, 0.049. Since Cluster-3 is connected to two other clusters with this specification, three links to Cluster-7 (each has strength 0.06 on two links and 0.08 on the other one) and two links to Cluster-2 (each on 0.14 and 0.06), therefore it is eligible to be a **crossroads** cluster.

3.3   Time Series Generation and Evolution

The final step that is crucial in our proposed framework is the time series generation. Our time series is different from the common literature, in which a time series is usually defined as a series of (floating) values, at regular timestamps. In our proposed framework, we identify a series of similar skills clusters in order to track the network's evolution. Thus we create a series of similar clusters for each location/country, for example, Cluster-10 from T1 corresponds to Cluster-8 from T2; Cluster-8 corresponds to Cluster-4 from T3, which corresponds to Cluster-2 from T4 and so on. One series might come to an end; another might only start after a specific time (for example, at T2). To quantify the similarity of clusters, we apply the *Similarity Index* (SI) [4], by measuring the intersection of the keywords in two clusters, as can be seen in (4)

$$SI(W_i, W_j, W_{ij}) = 2 \times \left(\frac{W_{ij}}{W_i + W_j}\right), 0 < SI \leq 1 \tag{4}$$

where, W_{ij} is the number of skills common to cluster-i C_i and cluster-j C_j; W_i is the number of skills in C_i; and W_j is the number of skills in C_j. In order to characterize the most significant changes and interactions in our time series, this is where the cluster's type class comes at handy, by limiting our comparisons to *crossroads* clusters, and to include at least one crossroads cluster at one or some point in a generated series. Hence our algorithm to create a series is, for each country, starting from the T1, we obtain the crossroads clusters and make it as the point of departure of each series. Then, for the next period (T2), using the *SI* index and a word threshold, we compare each cluster in T2 with the cluster in the series, that is of T1, and choose cluster with the highest *SI* as the next series. This process must be executed until the end of the whole period (T8). If, for example, a series x has a cluster at times T1, and no defined cluster in T2; then to find the cluster for times T3 for series x, the comparison analysis must be done between cluster candidates in T3 to the data point in T1, and the cluster with the highest *SI* index will be chosen.

Definition 6. *For each location l, a series of skills cluster S_i is a period-ordered sequence of clusters from location l, $(C_{1,l}, C_{2,l}, ..., C_{p,l})$, created based on comparison analysis using similarity index between clusters on each period instance. To ensure notable intersection between two clusters, we set a minimum similarity threshold θ to define the least number of similar words (i.e., skills) based on the average of the number of keywords in a job advert.*

To support the network evolution tracking task in order to define the highly in-demand skills, we focus on the cluster's attributes called *centrality* and *density* indices. Using both indices, we allocate a cluster and identify their quadrant location in the strategic diagram, which consists of the horizontal and vertical axes, which represents centrality and density. The origin of the diagram is at the median of the respective axis values and is drawn by plotting centrality and density of each skills network within a two-dimensional space divided into four quadrants [8, 9]:

- *Q1* contains clusters that are both central (strongly connected to other clusters) and have intense internal links reflecting a high degree of development; indicating the *core* of the network.
- *Q2* comprises central clusters, but the density of their internal links is relatively low; indicating their growing maturity and is already shown its importance, classified as *emerging* clusters.
- *Q3* contains less central clusters – or peripheral – while their density is high; indicating a well-developed but less interest with weak interactions with other clusters.
- *Q4* contains clusters that are both peripheral and little developed, indicating possible marginal clusters that might gradually *disappear* from the sector.

Based on four quadrants above, we then could track the quadrant location of each cluster in each series in order to follow its evolution and transformation over several periods, especially for the cluster which is located in Q1, since it represents the highly in-demand skills. More specifically, we fix our investigation on the series, which shows a stable transformation during the period being studied. By manually observing the resulted series and its evolving trends, we define four categories of transition curves, which can broadly be described as follows:

1. Category A represents clusters exhibiting a steady and stable curve.
2. Clusters in Category B exhibit a simple curve (gradual changes in quadrant location) with a trend that continues during the forecasting period.
3. Category C includes clusters exhibiting a stable curve, with a sudden quadrant location shift during the last observed period.
4. Category D exposes a complex curve, with frequent and repeated changes in the clusters' quadrant location.

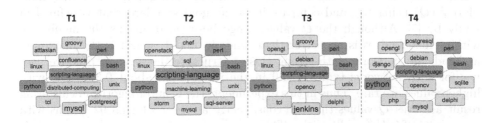

Fig. 6. One series of skills cluster (Series ID '2') that is successfully generated for country code 'cz'.

One resulted series of country code 'cz' is depicted in Fig. 6, where the starting data point of the series is a crossroads cluster, that is Cluster-3 in Fig. 5.

4   Evaluation and Discussion

This section presents the results of the skills in-demand discovery through the time series generation and skills clusters evolution tracking. The objectives of our study are two-fold:

1. to evaluate the adequacy of the Coulter algorithm in finding the relevant clusters and its performance based on the modularity gain.
2. to identify whether the proposed method has the potential to correctly track established high-demanded skills clusters identified by their 'quadrant' location.

Considering each of the above will enable us to determine whether we can confirm our hypothesis – that skill demand can be discovered and projected based on the evolution tracking of skills clusters with a sufficient amount of earlier observations. Although to show that clusters that are included in the series are having occurring changes and thus could be plausibly interpreted in terms of the evolution of the labor market skills demand, is inherently an empirical question, we nevertheless provide a proof-of-concept of our proposed framework and return a quantitative scientific result for the discovery and prediction of high demanded skills. To achieve a quantitative scientific answer, we implement three traditional statistical prediction models that are sufficiently diverse to cope with the ranges of variability and adjustment to trend, namely Naive, Simple Exponential Smoothing (SES), and Holt's linear trend method [15]. Then we employ walk-forward validation to compare forecasted values with the actual values that are observed later for the period in question. Finally, we consider the forecasted quadrant location of established clusters against their next known location and compute an F-measure to determine the forecast accuracy.

Clustering Result Evaluation. We implemented Coulter algorithm and determined the clusters for every 92 graphs constructed from our curated job adverts dataset, resulting in a total of 944 clusters. We then evaluated the clustering results of Coulter algorithm by calculating each graph's clustering modularity (Q) using (3), and compare it to the gold standard that is defined in study by [5]. Although the Q values range between 0, means the number of within-cluster edges is no better than random, and 1, which is the maximum and indicate networks with strong cluster structure, however in practice, values for such networks typically fall in the range from about 0.3 to 0.7, with higher values are rare. Thus in this experiment, the Coulter algorithm clustering result achieves Q values that are consistently above 0.3 for all 92 graphs with an average of 0.41 for all graphs. Thus we conclude that our clustering result is sufficiently good and eligible to discover a sensible division of our skills graphs dataset.

Table 2. Prediction F-measure for each quadrant location in all investigated series

Quadrant label	Precision			Recall			F-measure		
	Naive	SES	Holt	Naive	SES	Holt	Naive	SES	Holt
Q1	65%	63%	62%	75%	94%	88%	70%	*75%*	*73%*
Q2	23%	33%	32%	25%	35%	40%	24%	34%	36%
Q3	0%	0%	0%	0%	0%	0%	–	–	–
Q4	33%	40%	44%	24%	19%	19%	28%	26%	27%

Evaluation of the Skills Cluster-Based Demand Prediction. We establish the list of the series of skills clusters containing at least one crossroads cluster to determine what themes played a role in structuring the skills in demand over time. For the data and period (T1–T8) in consideration, this generates 195 series. However, only 40 series have consistently one cluster in four quarterly periods, and just four series have a cluster in each period T1–T8. Since four series was considered too low to derive significant interpretations, we opted to focus on the 40 skills cluster series and their evolution over four quarters. Forecast calculations were conducted based on the 40 series (each containing four versions of an evolving skills cluster), starting by using two data points (T1–T2) to train the model, then the previous two to forecast the third, and the previous three to forecast the fourth. Thus, a total of 80 cluster forecasts are generated. Since unlike Naive, the SES and Holt methods weight distant and recent observations differently, for their generation, we set the first estimate to the first observation value.

In the second experiment, the forecasted centrality and density values resulting from the three methods considered were projected onto the strategic diagram (refer to Sect. 3.3) to get the quadrant location of each cluster. We compare the quadrant location based on the forecast result to the actual quadrant in order to assess the performance of the forecasting method. The precision, recall, and F-measure were computed for 80 clusters between the two datasets and is summarized in Table 2.

Overall the F-measure ranges between 70%–75% for *Q1*, 24%–36% for *Q2*, and 26%–28% for *Q4*, with no prediction potential observed for *Q3*. Based on the objective of this study, the evaluation section focuses on the *Q1* label since all clusters in this quadrant are both central to the global network (strongly connected to other clusters) and display a high degree of development (have strong internal links). Thus, these strategical clusters are considered an indication of skill sets that are in high-demand. The F-measure result for the *Q1* label is quite reliable and seems to indicate that the proposed approach is suitable for predicting highly-demanded skills. The reasons for the F-measure value decrease for other quadrant labels are not clear and are worth further exploration, as outlined in conclusion, since they can give useful heuristics to enhance the overall prediction potential of the chosen method.

Category-Based Prediction Optimization. In addition to the above results which hold for generic established clusters as observed in our dataset, we focus our investigation of the evolving clusters on the series which shows a stable transformation during the period being studied, as explained in Sect. 3.3. Table 3 shows three examples for each category (A–D) in four consecutive periods (T1–T4). Category A examples have a stable quadrant location throughout, while others have shifted to a different quadrant in only one period or have started in a different quadrant compare to the rest. All cluster series in category B displays a gradual shift from being central to the labor demand for the sector under consideration ($Q1$) to a less developed one ($Q2$); eventually reaching the margin of the whole network ($Q4$), or the other way around. Category C embodies series that have clusters with a steady and stable quadrant location during period T1 until T3, but then suddenly change track to other quadrants in T4. Finally, category D shows cluster series with less consistent trends.

Following the identification of these four categories, we re-calculate the F-measure based on these groupings. The results, shown in Table 4, focus on the predictions for $Q1$, i.e., we only consider the category-based predictions to determine which of the skill clusters are expected to shift to Q1 (most stable and developed). The second column shows the distribution (ratio) of clusters within the series assigned to these four categories (in total 80 clusters from 40 series). The results indicate that forecasts are most

Table 3. Examples for transition curve categories from the dataset

Category	Country code	Series ID	Period			
			T1	T2	T3	T4
A	ch	3	Q2	Q1	Q1	Q1
	gb	1	Q1	Q4	Q1	Q1
	pt	3	Q1	Q4	Q4	Q4
B	pt	1	Q3	Q3	Q4	Q4
	se	1	Q1	Q1	Q2	Q4
	dk	1	Q3	Q2	Q2	Q1
C	at	3	Q4	Q4	Q4	Q2
	nl	1	Q3	Q3	Q3	Q1
	nl	4	Q2	Q2	Q2	Q4
D	ch	4	Q4	Q2	Q4	Q2
	nl	2	Q2	Q4	Q2	Q3
	se	5	Q2	Q4	Q3	Q1

accurate when the clusters are already within a stable and steady series (category A), followed by those in category C, B, and finally, D. In summary, demand trends ascribed to category A (which represent 47.5% of the total), the highest F-measure is achieved by the SES and Holt method at 90%, followed by Naive method at 84%. The second highest F-measure is obtained for clusters assigned to category C at 60% for all three forecasting methods. The accuracy for category C probably achieves this performance due to the stability of the series' during several periods (they only shift in the last period). In contrast, the F-measure for categories B and D reach a high of 50% (SES) and 36% (SES and Holt), respectively. Therefore the forecasting by time series analysis is generally not appropriate for these kinds of trend curves because of their instability and non-periodic movement.

To explain the relevance of these results, we handpick a specific example and explain how the method can be used to support some of the intended users.

Table 4. F-measure results for **Q1** prediction on the series organized by identified category (and percentage of series belonging to each category)

Category	Percentage	Precision-Q1			Recall-Q1			F-measure-Q1		
		Naive	SES	Holt	Naive	SES	Holt	Naive	SES	Holt
A	47.5%	90%	85%	88%	79%	96%	92%	*84%*	*90%*	*90%*
B	15%	25%	40%	25%	33%	67%	33%	29%	50%	29%
C	12.5%	43%	43%	43%	100%	100%	100%	60%	60%	60%
D	25%	20%	22%	22%	50%	100%	100%	29%	36%	36%

Fig. 7. Other series of skills cluster (Series ID '1') that is successfully generated for country code 'cz'.

The example consists of a series ID '2' of an evolving skills cluster observed in the Czech Republic as depicted in Fig. 6. The series starts in *Q1* from T1 until T4. Our SES-based method determines that the cluster will persist within *Q1* with a 75% accuracy, as shown in Table 2. However, given the series's past behavior, we can associate its evolution with *Category A*, which could raise the accuracy further to 90%. Based on this result, end-users can consider including (in their curricula or personal studies) four skills (*python, scripting-language, perl,* and *bash*) that have consistently appeared within the highly-established skill set in the last periods. In contrast, they could reconsider the relevance of other skills (*atlassian, confluence, distributed computing*) that have not recently featured. To discover a comprehensive summary of the skillsets that are highly-demanded for sector 'data science' in the country code 'cz', we should do investigation on all series which are available on four consecutive periods, and their evolutions are classified as Category A for a more reliable result. Thus by looking at another series (series ID '1') for country code 'cz' which is depicted in Fig. 7, the end-users can justify that two skills (*.net* and *.net-framework*) are also highly-demanded by considering its consistent appearance in the whole period (T1–T4).

5   Conclusion

In this paper, we presented a framework for analyzing the temporal evolution of the community structure of large linked networks to discover re-inforced skillsets that emerge from the observation of co-occurring skill references elicited from a large amount of job adverts. The objective was to investigate the hypothesis that with sufficient observations, it is possible to identify specific skillsets and predict how they are expected to behave in the near future. The dynamics of the skills network over time is examined through the strategic positioning of each cluster in a series, to identify trends based on its centrality and density indices. The results prove our hypothesis partly—our method can determine which clusters will position themselves in $Q1$, with an accuracy of 75%, i.e., it can accurately forecast which skill sets will be the most influential (central, developed) in the near future. Furthermore, if these clusters are observed to exhibit a specific kind of trend associated with Category A and Category C, the accuracy rises to 90% and 60%, respectively. Thus, the method presented in this paper can cautiously be used to guide the decision-making processes faced by the four different kinds of identified end-users: prospective candidates, curricula developers, employment bodies, and forward-looking industry. Despite the good performance of the proposed approach to identify the most influential skillsets, future studies will also investigate other graph clustering algorithm and different clustering evaluation criteria to measure the quality of the resulting clusters.

Acknowledgement. This work has been co-funded by the European Union's Horizon 2020 research and innovation programme under the QualiChain project, Grant Agreement No 822404.

References

1. Sodhi, M.S., Son, B.-G.: Content analysis of OR job advertisements to infer required skills. J. Oper. Res. Soc. **61**(9), 1315–1327 (2010)
2. Smith, D., Ali, A.: Analyzing computer programming job trend using web data mining. Issues Informing Sci. Inf. Technol. **11**, 203–214 (2014)
3. Sibarani, E.M., Scerri, S., Morales, C., Auer, S., Collarana, D.: Ontology-guided job market demand analysis: a cross-sectional study for the data science field. In: SEMANTiCS 2017, pp. 25–32. ACM, New York (2017). https://doi.org/10.1145/3132218.3132228
4. Coulter, N., Monarch, I., Konda, S.: Software engineering as seen through its research literature: a study in co-word analysis. J. Am. Soc. Inform. Sci. **49**(13), 1206–1223 (1998)
5. Newman, M.E.J., Girvan, M.: Finding and evaluating community structure in networks. Phys. Rev. E **69**, 026113 (2004)
6. Xu, X., Yuruk, N., Feng, Z., Schweiger, T.A.J.: SCAN: a structural clustering algorithm for networks. In: KDD 2007, pp. 824–833. ACM, New York (2007). https://doi.org/10.1145/1281192.1281280
7. Blondel, V.D., Guillaume, J.-L., Lambiotte, R., Lefebvre, E.: Fast unfolding of communities in large networks. J. Stat. Mech: Theory Exp. **10**, P10008 (2008)
8. He, Q.: Knowledge discovery through co-word analysis. Libr. Trends **48**(1), 133–159 (1999)
9. Callon, M., Courtial, J.-P., Laville, F.: Co-word analysis as a tool for describing the network of interactions between basic and technological research: the case of polymer chemistry. Scientometrics **22**(1), 155–205 (1991)
10. Kobayashi, V.B., Mol, S.T., Berkers, H.A., Kismihók, G., Den Hartog, D.N.: Text mining in organizational research. Organ. Res. Methods **21**(3), 733–765 (2018)

11. Kotu, V., Deshpande, B.: Time series forecasting. In: Kotu, V., Deshpande, B. (eds.) Predictive Analytics and Data Mining, pp. 305–327. Morgan Kaufmann, Boston (2015)

12. Fu, T.-C.: A review on time series data mining. Eng. Appl. Artif. Intell. **24**(1), 164–181 (2011)

13. Dadzie, A.-S., Sibarani, E.M., Novalija, I., Scerri, S.: Structuring visual exploratory analysis of skill demand. Web Seman. Sci. Serv. Agents World Wide Web **49**, 51–70 (2018). https://doi.org/10.1016/j.websem.2017.12.004

14. Polanco, X.: Co-word analysis revisited: modelling co-word clusters in terms of graph theory. In: Proceedings of the 10th International Conference on Scientometrics and Informetrics, vol. 2, pp. 662–663 (2005)

15. Hyndman, R.J., Athanasopoulos, G.: Forecasting: Principles and Practice. https://otexts.com/fpp2/. Accessed 30 May 2019

Picture News Collection: A Dataset for Automatic Picture News Thumbnail Selection

Yi-Kun Tang[1,2], Heyan Huang[1,2(✉)], Xuewen Shi[1,2], and Xian-Ling Mao[1,2]

[1] School of Computer Science and Technology, Beijing Institute of Technology,
Beijing 100081, China
{tangyk,hhy63,xwshi,maoxl}@bit.edu.cn
[2] Beijing Engineering Research Center of High Volume Language Information
Processing and Cloud Computing Applications, Beijing, China

Abstract. Picture news has become more and more popular among online news in recent years. As the first impression to viewers, thumbnail plays a very important role in picture news. However, it is time consuming to manually select thumbnails for a huge amount of picture news. In this paper, we introduce a new task of automatic picture news thumbnail selection. Given a piece of picture news containing a set of images, this task is to select several appropriate images from the picture news as candidate thumbnails. To this end, we present a large publicly available image dataset for this task, called Picture News Collection(The Picture News Collection 0.1 version can be publicly available online at https://github.com/anonymity01/Picture-News-Collection.). The Picture News Collection contains more than 4 million images of 347,731 picture news from two famous news websites, Sina News and NetEase News. Selecting good enough thumbnails is complicated and needs to consider many aspects, such as attraction, hot topics, content integrity, etc. In order to select appropriate candidate thumbnails, we propose an attention-based thumbnail selection model, and the experimental results comparing with three image classification based baselines show that our proposed methods outperform the baselines. We introduce the automatic picture news thumbnail selection task and the dataset to encourage further studies of this challenge.

Keywords: Automatic picture news thumbnail selection · Picture News Collection · Image selection

1 Introduction

Picture news is a kind of online news that uses a set of images as its main body (see Fig. 1). Picture news thumbnail is an image meticulously selected from all images in the picture news by news editors. Figure 2 shows some examples of picture news, and ea ch thumbnail is signed in green square frame. As the first

© Springer Nature Switzerland AG 2019
R. Cheng et al. (Eds.): WISE 2019, LNCS 11881, pp. 458–472, 2019.
https://doi.org/10.1007/978-3-030-34223-4_29

Fig. 1. An example of picture news and its thumbnail. (Color figure online)

impression to viewers, thumbnail plays a very important role in picture news. A good thumbnail can attract more viewers to click and read the picture news.

As massive amounts of picture news are posted online every day, it requires news editors' extensive efforts to manually select satisfactory thumbnails. To this end, we introduce a new task of automatic picture news thumbnail selection in this paper. Given a piece of picture news containing a set of images, the goal of this task is to select several applicable images from the picture news as candidate thumbnails, and provide some possible thumbnail options for news editors to choose from.

There are some image selection problems similar to this task, such as selecting the best photos in personal albums [1,11], photo summarization [4,17] and selecting thumbnails from videos [13,19]. However, none of these traditional tasks focus on selecting an image from a set of high-quality aesthetic images, where the images all share a common theme to express a main idea of news.

It is complex to select suitable picture news thumbnails, since we should consider many aspects of the images, and then choose the best ones. The selected candidate thumbnails are based more on relative considerations among images in a piece of picture news, rather than absolute judgments. This process may contain consideration in attraction (line 2 and 4 in Fig. 2), hot topics (line 1 and 2 in Fig. 2), content integrity (line 1 and 3 in Fig. 2), etc.

In this paper, we present a large publicly available image dataset, called Picture News Collection. The current version of Picture News Collection contains 347,731 picture news downloaded from two famous news websites, Sina News[1] and NetEase News[2]. Both of the two parts are consist of five hot news topics, including topic fashion, history, Internet, sports and star in Sina News part, and topic education, military, news, sports and star in NetEase News part. In this dataset, each picture news contains 2 to 50 images, one of which is labeled as the thumbnail. There are a total of more than 4 million images in the dataset.

---

[1] https://news.sina.com.cn.

[2] https://news.163.com.

Fig. 2. Examples of picture news. Each line represents all the images in a piece of picture news, and each thumbnail is signed in green square frame. (Color figure online)

To select appropriate thumbnail for picture news, we propose an attention-based thumbnail selection model. This model takes all the images in a piece of picture news as inputs, and maps them into a global feature encoding. Then, a score function is defined to measure the recommendation score. We use N-pair loss [18] to optimize the model parameters. In the experiments, we compare our proposed methods with three image classification based baselines. Experimental results shows that our proposed methods outperform the baselines in term of thumbnail recommendation accuracy.

This paper makes the following important contributions:

- We put forward a task of automatic picture news thumbnail selection, and release a publicly available dataset for this task, Picture News Collection.
- To select appropriate thumbnails, we propose an attention-based thumbnail selection method including a two-pass attention encoder (TAE) and a function measuring the recommendation scores.
- Experimental results show that our proposed two methods can tackle this task preliminarily, and outperform the baselines.

The rest of the paper is organized as follows. In Sect. 2, we review some related work. The proposed Picture News Collection is introduced and analyzed in Sect. 3. In Sect. 4, we introduce our proposed attention-based thumbnail selection model. In Sect. 5, we present the experiments of the thumbnail selection task among the baselines and our two methods on Picture News Collection. Finally, we show our conclusions and future work in Sect. 6.

2   Related Work

Image selection has drawn much attention in recent years. Typical applications include tasks such as selecting the best photos in personal albums [1,6,11,14], photo summarization [4, 15–17, 22] and selecting thumbnails from videos [8, 13, 19, 23]. Although our picture news thumbnail selection task has some similarities with these tasks, there are also many significant differences among them.

The task of selecting the best photos in personal albums is to select subsets of photos from large collections for easy storage and sharing. The purpose of this task is to pick out beautiful or enjoyable pictures, and reduce the number of pictures. For this task, the main focus is generally the quality and aesthetics of the images, such as focal aperture, exposure, sharpness, and other external photo properties [1, 2, 12, 14]. There are also some work considering individual interests or expectations of users [5–7, 21]. The main difference between this task and our picture news thumbnail selection is that the input images of thumbnail selection are all good aesthetical images of high quality. Because they have already been selected by news editors from a larger set of original photos taken by news photographers.

Similar to selecting the best photos in personal albums task, photo summarization [4, 15–17, 22] also selects a set of photos from many photos. Besides image quality, photo summarization also lays emphasis on diversity and coverage of images to constitute an informative summary. Differently, as the most representative image in a piece of picture news, each single recommended thumbnail is supposed to summarize the global picture news in thumbnail selection task.

Automatic video thumbnails selection is to select a video frame as the video thumbnail. Lots of previous work regards videos as frame sequences and aims at finding good-quality, meaningful and attractive frames as video thumbnails [8, 13, 19, 23]. However, due to the time and space span of videos, there are usually great differences between key frames of the same video in theme and content. In contrast, to express a main idea of news, images in the same picture news all share a common theme, and are quite related in meaning.

To sum up, picture news thumbnail selection is different from the existing image selection tasks in many ways. Therefore, in this paper, we introduce the picture news thumbnail selection task, and propose the Picture News Collection for this task.

3   Picture News Collection

In this section, we introduce our proposed dataset in detail, the Picture News Collection. We begin with the method of collecting and pre-processing the Picture News Collection (Sect. 3.1), and then we analyze the construction of this newly released dataset (Sect. 3.2).

3.1   Dataset Collecting and Pre-processing

Picture News Collection is a large image dataset crawled from several picture news sections on two famous news website, Sina News and NetEase News, respectively. So Picture News Collection contains two parts: Sina News and NetEase News, each containing picture news in five popular news topics.

Sina News is a news website that timely accesses to global news information, domestic and foreign news, wonderful sports events, financial trends, film and television entertainment events, etc. In Sina News part, we crawled some picture news in five hot news topics: fashion, history, Internet, sports and star[3] before 2018.

NetEase News is also a famous news website in China, covering news in entertainment, sports, finance, technology and many other information contents. Similarly, in NetEase News part, we crawled some picture news in five popular news topics: education, military, news, sports and star[4] before 2018.

Before building the Picture News Collection, we did several steps to pre-process the original picture news dataset crawled from the websites. For both the above two parts of the dataset, we discarded picture news with pixel damaged images, and we also removed a small part of the picture news whose thumbnail is different from any images in it.

In addition, most of the images in the original dataset have high pixels, which requires a lot of storage space. Besides, the size of the pixel matrix of raw images is usually different, making it difficult to directly use them as inputs of neural network methods. For these reasons, we reshaped each image in our Picture News Collection to a 224×224 pixels color image.

Table 1. The statistics of the Picture News Collection.

Number	Sina News				
	fashion	history	Internet	sports	star
Total news	20749	27646	41019	61077	68862
Average images / news	14.31	16.64	16.28	13.50	8.22
Number	NetEase News				
	education	military	news	sports	star
Total news	6195	13179	21232	22577	65195
Average images / news	11.76	14.71	6.79	15.03	8.92

In the Picture News Collection, each of the picture news contains several images, and one of them is tagged as the thumbnail of that piece of picture news. The statistics of the filtered dataset is shown in Table 1, including the total number of news in each topic, and the average number of images per picture news.

---

[3] http://slide.{fashion, history, tech, sports, ent}.sina.com.cn{ /, /, /Internet,/, /star}.

[4] http://{edu, war, news, sports, ent}.163.com.

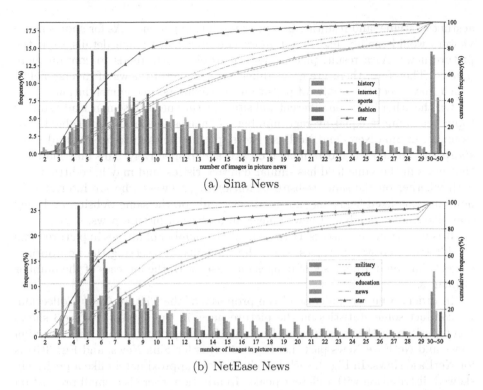

(a) Sina News

(b) NetEase News

Fig. 3. A comparison of the frequency and cumulative frequency of each topic in the dataset.

3.2   Dataset Analysis

The Picture News Collection version 0.1 contains 347,731 picture news in total. The content of the picture news covers fashion, history, Internet, sports, star, education, military and social news. In this dataset, each picture news contains 2 to 50 images, one of which is labeled as the thumbnail of the picture news. The total number of images exceeds 4 million in the Picture News Collection version 0.1. By analyzing Table 1, we can find out two interesting phenomena.

On the one hand, picture news on the same news website but in different topics may contain quite a bit different number of images. For instance, considering Sina News part, picture news in topic star contains only 8.22 images on average, while in other four topics of Sina News part, each picture news contains more than 13 images on average. Comparing with topic star, picture news in these topics contains nearly two times of images in number. This might be because of the natural attributes of news itself and the preference of news editors or viewers in different fields. Considering star news and sports news, the former mainly focuses on delivering entertainment information, such as promotion of new movies and attendance of stars in some activities, while the latter usually reports competition details. Viewers tend to relax themselves via reading star news. Thus, too many images in the same picture news may cause the viewers' aesthetic fatigue, and resulting in a poor reading experience. As for sports news, viewers would mainly like to get more comprehensive game details from the picture news. As a result, picture news in topic sports tends to contain more images. Therefore, there are great differences in picture news of diverse fields, which may affect the results of picture news thumbnail recommendations.

On the other hand, on different websites, picture news in semantically similar topics has almost equal average number of images, according to Table 1. For example, in Sina News and NetEase News, picture news in topic star and topic sports has around 8.5 and 14 images on average, respectively. This may manifest that news in the same field has similar characteristics, and may have little to do with whether on the same website. Additionally, viewers who are interested in news in some fixed fields may not read news only on the same website, and they may like similar number of news images in a piece of picture news.

It is important and necessary to analyze the number of images in picture news on different websites or topics, since the difficulty of selecting the appropriate correct thumbnail from a set of images increases with the increase of the number of images.

In order to better understand the property of the Picture News Collection, we conduct some statistics on the picture news in the dataset. Figure 3 shows a comparison of the frequency and cumulative frequency of each topic in Sina News and NetEase News part, i.e. Figure 3(a) is for Sina News, and Fig. 3(b) is for NetEase News. In Fig. 3, each bar chart looks approximately like a positively skewed distribution with different peaks. In addition, there is a small proportion of picture news with too many or too few images in each topic, especially when the picture news contains a lot of images. This inspires us to set a threshold for the maximum number of images in the picture news of training set during training, regardless of the picture news with too many images. For the picture news whose image number is less than the threshold, we set a mask to record the true number of images (see Sect. 5 for more details).

4   Attention-Based Thumbnail Selection Method

In this section, we describe our attention-based thumbnail selection approach to select appropriate thumbnails for picture news. Figure 4 gives an illustration of the proposed model architecture. Given a piece of picture news, $M = \{m_i\}_{i=1}^{N}$, including N images, the goal of this task is to select appropriate candidate thumbnails from the set of images, and provide a thumbnail reference for news editors to choose from.

Our proposed attention-based thumbnail selection model mainly contains: (i) a two-pass attention encoder (TAE) that maps a set of images into a global feature encoding c_f, and (ii) a score function to measure the recommendation score for an image with the global encoding c_f as the condition. In our approach, we leverage ResNet [10] to get the representations $R = \{r_i\}_{i=1}^{N}$ of images, where $r_i \in \mathbb{R}^{d_m}$. The initial encoding c_0 of the news pictures set is performed by 3D ResNets [9], which is a ResNet-based architecture with 3D convolutions.

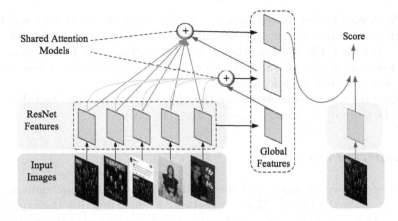

Fig. 4. Illustrations of the proposed the attention-based thumbnail selection model. (Color figure online)

4.1   Two-Pass Attention Encoder (TAE)

Intuitively, images in picture news make differentiated contributions. To explore those information explicitly, we introduce a two-pass attention encoder architecture. The encoder performs a two-pass procedure: (i) during the first pass, TAE takes c_0 and R as inputs, getting a primary encoding c_1; (ii) then, at the second pass, we replace c_0 with c_1 and re-use TAE to get the final encoding $c_f = c_2$. The second pass can be seen as a amplifier which amplifies weight differences.

Specifically, we apply Scaled Dot-Product Attention [20] to calculate attention weights, so the c_t is computed by:

$$c_t = \text{LayerNorm}(W \times (\text{softmax}(\frac{c_{t-1}R^T}{\sqrt{d_m}}) \times R)), t \in \{1,2\}, \qquad (1)$$

where $W \in \mathbb{R}^{d_m \times d_m}$ and $c_t \in \mathbb{R}^{d_m}$. LayerNorm$(\cdot)$ is the layer normalization [3] function, and we set $d_m = 512$ in this work.

4.2   Loss Function

As we all know, there are several non-thumbnails and only one thumbnail in each picture news. Contrastive loss and triplet loss are inappropriate in this task, due to the slow convergence. Therefore, we need to jointly compare more than one negative examples. We chose to use multi-class N-pair loss [18] as our loss function, which can be given by:

$$\mathcal{L} = \mathcal{L}_{n-pair}(\{c_f, r^+, \{r_i^-\}_{i=1}^{N-1}\}) = \log(1 + \sum_{i=1}^{N-1} \exp(c_f^\top r_i^- - c_f^\top r^+)), \qquad (2)$$

where r^+ represents the feature vector of the thumbnail in M, and r_i^- denotes the feature vector of the i_{th} non-thumbnail images. Unlike the multi-class logistic loss, the N-pair loss objective encourages options similar to the ground truth answers to score better than the dissimilar ones, which means options that are nearly correct but different from the golden standard may not be over penalized. In the following section (see Sect. 5), we will give a detailed comparison between N-pair loss and multi-class logistic loss.

During the test process, we choose the i_{th} image as the thumbnail that maximizes the score $c_f^\top r_i$, i.e. $\mathrm{argmax}_i c_f^\top r_i$, $\forall i \in \{1, .., N\}$.

5   Experiments

In this section, we explore the challenge of recommending thumbnails for picture news using the Picture News Collection. We first describe the experiment settings. Then, we introduce three baselines and our proposed two methods in the experiments. Finally, we analyze the experimental results.

5.1   Experiment Settings

We train our baselines and methods using stochastic gradient descent with a batch size of 128 examples, dropout of 0.5, and initial learning rate of 0.0005.

We divide the Picture News Collection into three parts in a scale of about 8:1:1, i.e. approximately 80% for training set, 10% for development set and 10% for testing set. Table 2 shows the statistics of training set, development set and testing set in the experiments, where Table 2(a) is the statistics of Sina News, and Table 2(b) is that of NetEase News.

In order to speed up the training procedure, picture news with too many images are discarded during training. Considering properties of the sub-datasets in different topics, we set each of them a threshold of maximum image number, respectively. Each remaining training data is ensured to cover at least 80% of the original data in the corresponding topic, according to Table 1 and Fig. 3. Besides, we do not have any limitation on the number of images in every sample of the development set and the testing set.

5.2   Baselines and Methods

In the experiments, we first extract feature for each image using Resnet [10]. As for feature extraction for the global picture news, we have two strategies: (1) we simply compute the mean value of all the image feature vectors in a piece of picture news, (2) we extract feature for the global picture news based on 3D ResNets [9], i.e. we use all the image matrices in a piece of picture news as inputs and then directly get a global feature vector for the picture news.

Accordingly, we compare the following three neural network based image classification baselines and our proposed two methods in our experiments, where the abbreviations of the baselines or the methods are in the brackets:

Single image based classification (S+C): In this baseline, we transform the picture news thumbnail selection task to a binary classification problem. We

Table 2. Statistics of training set, development set and testing set in the experiments.

(a) Sina News.

Topic	Training set			Development set			Testing set		
	total	average	max	total	average	max	total	average	max
fashion	13388	10.58	20	2000	14.29	50	2000	14.20	50
history	19083	12.24	25	2000	16.49	50	2000	16.26	50
Internet	26951	11.92	25	4000	16.53	50	4000	16.51	50
sports	39412	10.18	19	6000	13.50	50	6000	13.49	50
star	46223	6.14	10	6000	8.06	49	6000	8.06	50

(b) NetEase News.

Topic	Training set			Development set			Testing set		
	total	average	max	total	average	max	total	average	max
education	4060	8.50	17	600	11.40	49	600	11.49	49
military	9065	10.41	21	1000	15.35	50	1000	15.53	50
news	15413	5.77	10	2000	6.91	48	2000	6.84	48
sports	14978	10.59	22	2000	15.47	50	2000	15.51	50
star	42787	5.76	10	6000	9.26	50	6000	9.26	50

consider each thumbnail in the dataset as class A, and non-thumbnail images as class B during training. As for the development set and the test set, we recommend images with the highest k probability for class A as candidate thumbnails in each picture news. Obviously, there is a defect in this baseline that images in the same picture news are regarded to be independent. Besides, in **S+C**, what we only care about is whether a single images can be a thumbnail individually, rather than considering every image in a piece of picture news comprehensively and selecting the most suitable one as the thumbnail.

Mean value based classification (M+C): On the basis of the **S+C** baseline, we add the global picture news feature to each single image. Concretely, we concatenate the mean value based global picture news feature vector and each single image in the picture news, respectively. Similar to the **S+C** baseline, we then use the connected vectors as inputs of the classification algorithm.

3D ResNets based classification (3D+C): The only difference between this baseline and the **M+C** baseline is that in **3D+C**, we use 3D ResNets to get the global picture news feature vector. Apart from this, other steps of the experiments are consistent with the **M+C** baseline.

Mean value based N-pair (M+NP): Instead of simply connecting the features, we use the mean value based global picture news feature vector and each image feature vector as the inputs of our proposed attention-based thumbnail selection model. The detailed model is described in Sect. 4.

3D ResNets based N-pair (3D+NP): 3D ResNets is used to get the global picture news feature vector in this method, and as for other steps, follow the **3D+NP** method.

5.3   Results and Analyze

We run the three baselines and our proposed two methods in the experiments. Given a piece of picture news with several images, the algorithm is supposed to determine which image is suitable to be a thumbnail for the picture news. We measure the performance of the methods on picture news thumbnail recommendation with accuracy.

The top-1, top-2 and top-5 accuracy results on recommending appropriate thumbnails for picture news in the Picture News Collection can be seen in Table 3. Table 3(a)–(e) show the development step and testing step results on topic fashion, history, Internet, sports and star in Sina News, and results on topic education, military, news, sports and star in Netease News are shown in Table 3(f)–(j). The best performances on each test set compared among the five algorithms are in bold. From Table 3, we can see that our proposed methods **M+NP** and **3D+NP** significantly outperform the baselines, especially on top-1 accuracy. In some subsets with more images in average number, such as topic fashion and Internet in Sina News and topic star in Netease News, the accuracy results using **M+NP** and **3D+NP** are nearly twice that of the baselines. Comparing the results of **M+NP** and **3D+NP**, we can see that **3D+NP** performs better than **M+NP** in most cases, since 3D ResNets algorithm considers spatial information among images in the same picture news.

From the structure of the Picture News Collection (see Sect. 3 for more details) and the experimental results in Table 3, we can find that the picture news thumbnail selection task is very complex, and factors affecting its accuracy are also quite complicated. We think the accuracy of this task is mainly affected by three factors: the size of the dataset, image number in a piece of picture news in the dataset, and the characteristic difference and editors' or viewers' preference among picture news in various news topics or websites.

We then only take the results of top-1 accuracy as an example to analyze the impact of different factors on this task. Obviously, for the first factor, model parameters can be better optimized on large data, and this may improve the model performance. For example, the data size of the topic star in Netease News is much larger than education and military, and its results are much better. It is also obvious that the number of images in each picture news can influence the recommendation accuracy. For example, the average number of images in topic star in Netease News is about 9, while for topic military and sports in Netease News is about 15. The picture news thumbnail recommendation accuracy on the former is significantly higher than others. The third factor, has been discussed in Sect. 3.2. This makes it more difficult to correctly recommend thumbnail for picture news.

Figure 5 shows six recommendation result examples. The **3D+NP** method successfully recommends the correct thumbnails in the 2nd, 3rd and 5th examples. In addition, the top-1 recommendation score in each example is much larger than others. Although the top-1 results of the rest examples are not very good, the score of each true thumbnail is not much lower than each top-1 score.

Table 3. Thumbnail recommendation accuracy on the Picture News Collection.

(a) Topic fashion in Sina News.

Method		Top-1(%)	Top-2(%)	Top-5(%)
S+C	Dev	13.50	23.60	53.20
	Test	13.50	24.80	52.75
M+C	Dev	13.70	23.75	53.30
	Test	12.45	24.25	52.50
3D+C	Dev	13.70	23.90	53.25
	Test	12.30	23.90	52.20
M+NP	Dev	23.00	33.95	61.70
	Test	24.00	**36.45**	62.50
3D+NP	Dev	22.50	33.45	62.75
	Test	**24.75**	35.95	**63.05**

(b) Topic history in Sina News.

Method		Top-1(%)	Top-2(%)	Top-5(%)
S+C	Dev	12.10	20.85	47.85
	Test	12.95	23.80	49.25
M+C	Dev	12.00	20.85	47.60
	Test	13.25	23.85	49.25
3D+C	Dev	12.20	21.10	48.25
	Test	12.85	24.10	49.55
M+NP	Dev	20.15	30.90	56.20
	Test	19.10	29.95	57.15
3D+NP	Dev	19.75	30.00	55.20
	Test	19.10	**30.85**	**57.25**

(c) Topic Internet in Sina News.

Method		Top-1(%)	Top-2(%)	Top-5(%)
S+C	Dev	11.25	21.35	48.25
	Test	12.60	23.35	49.23
M+C	Dev	11.65	21.83	48.98
	Test	12.50	23.75	49.53
3D+C	Dev	11.48	21.80	49.05
	Test	12.48	23.45	49.70
M+NP	Dev	21.40	32.13	55.65
	Test	20.43	**31.93**	56.80
3D+NP	Dev	21.48	31.90	56.38
	Test	20.63	31.45	**57.00**

(d) Topic sports in Sina News.

Method		Top-1(%)	Top-2(%)	Top-5(%)
S+C	Dev	11.85	21.93	49.10
	Test	11.50	22.73	50.83
M+C	Dev	11.65	22.42	49.47
	Test	11.52	22.28	51.05
3D+C	Dev	11.67	22.40	49.45
	Test	11.60	22.43	51.02
M+NP	Dev	13.72	24.23	52.58
	Test	12.72	24.12	52.63
3D+NP	Dev	13.53	24.87	53.30
	Test	12.83	**24.78**	**53.40**

(e) Topic star in Sina News.

Method		Top-1(%)	Top-2(%)	Top-5(%)
S+C	Dev	17.42	33.23	74.22
	Test	16.65	32.58	74.08
M+C	Dev	17.70	33.50	74.67
	Test	16.65	33.42	74.27
3D+C	Dev	17.43	33.65	74.68
	Test	16.77	33.27	74.20
M+NP	Dev	18.47	34.65	75.47
	Test	18.80	35.63	75.85
3D+NP	Dev	18.17	34.23	75.53
	Test	**19.08**	**36.35**	**75.98**

(f) Topic education in Netease News.

Method		Top-1(%)	Top-2(%)	Top-5(%)
S+C	Dev	14.67	26.33	59.33
	Test	15.83	28.83	56.83
M+C	Dev	14.83	26.33	59.50
	Test	**16.00**	26.50	56.67
3D+C	Dev	14.67	26.00	59.67
	Test	15.50	27.00	56.83
M+NP	Dev	16.67	29.83	60.67
	Test	14.33	29.50	58.83
3D+NP	Dev	16.00	30.50	59.83
	Test	14.17	**29.67**	**61.83**

(g) Topic military in Netease News.

Method		Top-1(%)	Top-2(%)	Top-5(%)
S+C	Dev	10.10	20.50	46.10
	Test	10.60	21.40	46.20
M+C	Dev	10.20	20.30	44.90
	Test	10.10	20.40	46.70
3D+C	Dev	11.00	20.80	45.20
	Test	10.10	20.50	47.40
M+NP	Dev	13.70	25.40	51.70
	Test	**10.70**	**21.90**	**48.70**
3D+NP	Dev	13.60	23.80	53.60
	Test	9.80	20.00	47.90

(h) Topic news in Netease News.

Method		Top-1(%)	Top-2(%)	Top-5(%)
S+C	Dev	19.70	37.80	78.85
	Test	18.90	38.50	81.00
M+C	Dev	19.55	38.25	79.50
	Test	19.45	37.75	80.75
3D+C	Dev	19.50	37.80	79.95
	Test	18.85	38.35	81.30
M+NP	Dev	22.50	41.85	83.25
	Test	22.05	41.65	81.80
3D+NP	Dev	22.25	40.85	81.65
	Test	**23.20**	**42.25**	**82.45**

(i) Topic sports in Netease News.

Method		Top-1(%)	Top-2(%)	Top-5(%)
S+C	Dev	10.30	20.70	47.65
	Test	9.35	20.25	47.75
M+C	Dev	10.60	20.50	48.75
	Test	9.75	20.90	48.40
3D+C	Dev	10.80	20.35	48.40
	Test	9.70	19.90	48.95
M+NP	Dev	14.85	28.15	57.40
	Test	**13.85**	**25.90**	57.15
3D+NP	Dev	15.60	27.30	56.55
	Test	13.25	24.65	**57.20**

(j) Topic star in Netease News.

Method		Top-1(%)	Top-2(%)	Top-5(%)
S+C	Dev	17.12	33.92	74.82
	Test	17.45	33.47	74.32
M+C	Dev	17.62	33.87	74.73
	Test	17.40	33.07	74.35
3D+C	Dev	17.35	33.87	74.63
	Test	17.25	33.07	74.47
M+NP	Dev	29.88	49.85	81.70
	Test	29.23	**49.70**	82.35
3D+NP	Dev	29.88	48.67	81.82
	Test	**29.30**	49.23	**82.58**

Fig. 5. Recommendation result examples in different news topics using **3D+NP**. Each line represents images in a piece of picture news. In each picture news, the bottom of the actual thumbnail is marked with a red heart graphic. The number below the images indicates the top five highest recommendation scores, where higher scores represent greater confidence for thumbnail selection. The thickness of the green square frames outside the images has positive correlation to the recommendation score. (Color figure online)

In summary,our proposed two methods perform significantly better than the baselines. In some cases, the top-1 accuracy of our proposed methods is nearly 30%, and the top-5 accuracy rate is over 80%. Affected by many factors, the recommendation of thumbnail for picture news is a challenging task.

6   Conclusion and Future Work

In this paper, we introduce a new task of automatic thumbnail selection for picture news. With regard to this task, we present a large publicly available image collection, Picture News Collection. We provide details of collecting and pre-processing method and analysis for the Picture News Collection. In this task, we propose an attention-based thumbnail selection model to select candidate thumbnails for picture news. Although the experimental results show that our proposed methods outperform the three baselines, picture news thumbnail recommendation still remains a challenging task.

In the future, we will first extend the range and scale of the dataset. Moreover, we will add manual annotations henceforth. Next, we will add title and caption information of each image for the picture news in the dataset. The extended dataset will be used for better solving the picture news thumbnail recommendation challenge and other interesting tasks, like attractive news title generation.

Acknowledgments. We thank all anonymous reviewers for their valuable comments. This work is supported by National Key R&D Plan (No. 2016QY03D0602), BIGKE (No. 20160754021), NSFC (No. 61772076), NSFB (No. Z181100008918002), Major Project of Zhijiang Lab (No. 2019DH0ZX01), and CETC (No. w-2018018).

References

1. Aiello, L.M., Schifanella, R., Redi, M., Svetlichnaya, S., Liu, F., Osindero, S.: Beautiful and damned. combined effect of content quality and social ties on user engagement. IEEE Trans. Knowl. Data Eng. **29**(12), 2682–2695 (2017)
2. Aydın, T.O., Smolic, A., Gross, M.: Automated aesthetic analysis of photographic images. IEEE Trans. Visual Comput. Graphics **21**(1), 31–42 (2015)
3. Ba, J.L., Kiros, J.R., Hinton, G.E.: Layer normalization. arXiv preprint arXiv:1607.06450 (2016)
4. Ceroni, A.: Personal photo management and preservation. In: Mezaris, V., Niederée, C., Logie, R.H. (eds.) Personal Multimedia Preservation. SSCC, pp. 279–314. Springer, Cham (2018). https://doi.org/10.1007/978-3-319-73465-1_8
5. Ceroni, A., Solachidis, V., Niederée, C., Papadopoulou, O., Kanhabua, N., Mezaris, V.: To keep or not to keep: an expectation-oriented photo selection method for personal photo collections. In: Proceedings of the 5th ACM on International Conference on Multimedia Retrieval, pp. 187–194. ACM (2015)
6. Ceroni, A., Solachidis, V., Niederée, C., Papadopoulou, O., Mezaris, V.: Expo: an expectation-oriented system for selecting important photos from personal collections. In: Proceedings of the 2017 ACM on International Conference on Multimedia Retrieval, pp. 452–456. ACM (2017)
7. Fu, M., et al.: Learning personalized expectation-oriented photo selection models for personal photo collections. In: 2015 IEEE International Conference on Multimedia & Expo Workshops, pp. 1–6. IEEE (2015)
8. Gao, Y., Zhang, T., Xiao, J.: Thematic video thumbnail selection. In: 2009 16th IEEE International Conference on Image Processing, pp. 4333–4336. IEEE (2009)
9. Hara, K., Kataoka, H., Satoh, Y.: Can spatiotemporal 3D CNNS retrace the history of 2D CNNS and imagenet? In: Proceedings of the IEEE Conference on Computer Vision and Pattern Recognition, pp. 6546–6555 (2018)
10. He, K., Zhang, X., Ren, S., Sun, J.: Deep residual learning for image recognition. In: Proceedings of the IEEE Conference on Computer Vision and Pattern Recognition, pp. 770–778 (2016)
11. Kuzovkin, D., Pouli, T., Cozot, R., Le Meur, O., Kervec, J., Bouatouch, K.: Image selection in photo albums. In: Proceedings of the 2018 ACM on International Conference on Multimedia Retrieval, pp. 397–404. ACM (2018)
12. Li, C., Loui, A.C., Chen, T.: Towards aesthetics: a photo quality assessment and photo selection system. In: Proceedings of the 18th ACM International Conference on Multimedia, pp. 827–830. ACM (2010)
13. Liu, W., Mei, T., Zhang, Y., Che, C., Luo, J.: Multi-task deep visual-semantic embedding for video thumbnail selection. In: Proceedings of the IEEE Conference on Computer Vision and Pattern Recognition, pp. 3707–3715 (2015)

14. Marchesotti, L., Murray, N., Perronnin, F.: Discovering beautiful attributes for aesthetic image analysis. Int. J. Comput. Vision **113**(3), 246–266 (2015)
15. Rabbath, M., Sandhaus, P., Boll, S.: Automatic creation of photo books from stories in social media. ACM Trans. Multimedia Comput. Commun. Appl. **7**(1), 27 (2011)
16. Seah, B.S., Bhowmick, S.S., Sun, A.: Prism: concept-preserving social image search results summarization. In: Proceedings of the 37th International ACM SIGIR Conference on Research & Development in Information Retrieval, pp. 737–746. ACM (2014)
17. Sinha, P., Mehrotra, S., Jain, R.: Summarization of personal photologs using multidimensional content and context. In: Proceedings of the 1st ACM International Conference on Multimedia Retrieval, p. 4. ACM (2011)
18. Sohn, K.: Improved deep metric learning with multi-class n-pair loss objective. In: Advances in Neural Information Processing Systems, pp. 1857–1865 (2016)
19. Song, Y., Redi, M., Vallmitjana, J., Jaimes, A.: To click or not to click: automatic selection of beautiful thumbnails from videos. In: Proceedings of the 25th ACM International on Conference on Information and Knowledge Management, pp. 659–668. ACM (2016)
20. Vaswani, A., et al.: Attention is all you need. In: Advances in Neural Information Processing Systems, pp. 5998–6008 (2017)
21. Walber, T.C., Scherp, A., Staab, S.: Smart photo selection: interpret gaze as personal interest. In: Proceedings of the SIGCHI Conference on Human Factors in Computing Systems, pp. 2065–2074. ACM (2014)
22. Wang, Y., Han, B., Li, D., Thambiratnam, K.: Compact web video summarization via supervised learning. In: 2018 IEEE International Conference on Multimedia & Expo Workshops, pp. 1–4. IEEE (2018)
23. Zhang, W., Liu, C., Wang, Z., Li, G., Huang, Q., Gao, W.: Web video thumbnail recommendation with content-aware analysis and query-sensitive matching. Multimedia Tools Appl. **73**(1), 547–571 (2014)

A Graph-Based Approach to Explore Relationship Between Hashtags and Images

Zhiqiang Zhong[1], Yang Zhang[2], and Jun Pang[1,3(✉)]

[1] Faculty of Science, Technology and Communication, University of Luxembourg,
Esch-sur-Alzette, Luxembourg
[2] CISPA Helmholtz Center for Information Security, Saarland Informatics Campus,
Saarbrücken, Germany
[3] Interdisciplinary Centre for Security, Reliability and Trust,
University of Luxembourg, Esch-sur-Alzette, Luxembourg
jun.pang@uni.iu

Abstract. Online social networks are playing a great role in our daily life by providing a platform for users to present themselves, articulate their social circles, and interact with each other. Posting image is one of the most popular online activities, through which people could share experiences and express their emotions. Intuitively, there must exist a connection between images and their associated hashtags. In this paper, we focus on systematically describing this relationship and using it to improve downstream tasks. First, we use a two-sample *Kolmogorov-Smirnov* test on an Instagram dataset to show the existence of the relationship at a significance level of $\alpha = 0.001$. Second, in order to comprehensively explore the relationship and quantitatively analyse it, we adopt a graph-based approach, utilising the semantic information of hashtags and graph structure among images, to mine meaningful features for both hashtags and images. At last, we apply the extracted features about the relationship to improve an image multi-label classification task. Compared to a state-of-the-art method, we achieve a 12.0% overall precision gain.

1 Introduction

The last decade has witnessed the rapid development of online social networks (OSNs). To certain extent, OSNs have mirrored our society: people perform various activities in OSNs as they do in the offline world, such as establishing social relations, interacting with their friends, sharing life moments, and expressing opinions about various topics.

Image is one of the most popular information being shared in OSNs. For instance, 300 million photos are uploaded to Facebook on a daily base.[1] Moreover, there exist several popular OSNs dedicated to image sharing, including

---

[1] https://zephoria.com/top-15-valuable-facebook-statistics/.

© Springer Nature Switzerland AG 2019
R. Cheng et al. (Eds.): WISE 2019, LNCS 11881, pp. 473–488, 2019.
https://doi.org/10.1007/978-3-030-34223-4_30

Table 1. Two example images from Instagram. Hashtags are generated by users, and labels are given by Google's Cloud Vision API.

Hashtags	#wintersport **#skiing** #piste #travelgram #snowboarding #mountain #blue_sky #travel **#snow** #tyrol #winter #outdoors #austria #snowboard #scenery #lechtal	**#skiing** #utah **#snow** #equipment # feedtheyouth #findyourgreatest
Labels	helicopter, piste, mode of transport, mountain, snow, geological phenomenon, winter, cable car, mountainous landforms, mountain range	fir, snow, winter, geological phenomenon, mountain range, winter sport, ski equipment, ski, fun, tree

Instagram and Flickr. Images themselves are a rich source of information. Previously, researchers have studied images in OSNs from various perspectives [6,23, 26]. These works mainly concentrate on the contents of the images, thus adopting computer vision techniques as the main instrument. Different from images hosted on other platforms, images in OSNs are often affiliated with other types of user-shared information, such as image captions and hashtags. Such information can contribute to understanding OSN images as well. However, the relationship between images and user-shared information has been left mostly unexplored. We aim to fill this gap by analysing the relationship between images and hashtags.

A hashtag is a single word or short phrase prefixed by the "#" symbol [6]; it is initially created to serve as a metadata tag for people to efficiently search for information in OSNs. Interestingly, hashtags themselves have evolved to convey far richer information than expected and provide an incredibly varied and nuanced method for describing images. Some hashtags describe precise objects in the images, e.g., #glass, #window, #building, and #sky; some are related to the feelings and intent of the users, such as #lovelyday, #whyme, and #celebrating; others refer to some event or geographic position, e.g., #paris, #rio, and #newyork [25]. Besides, users also create many hashtags to convey meanings which previously did not exist in natural languages, e.g., #tbt (an abbreviation for "Throw Back Thursday"), or hashtags without specific meanings, e.g., #igphoto. Therefore, how to accurately describe and understand the relationship between hashtags and image contents is a significant issue.

Contributions. In this paper, we perform an empirical study on the relationship between hashtags and image contents. Our experiments are conducted on a real-world dataset collected from Instagram. It is worth noting that as it is time-consuming to tag the image contents for all the images in our dataset manually

(148,106 images), we use the image *labels* obtained from an automatic image detection tool, i.e., Google's Cloud Vision API, to represent the image contents.

Relationship Verification and Quantification. We first verify the relationship between hashtags and image contents (represented by their labels) using the two-sample *Kolmogorov-Smirnov* (KS) test. Experiments demonstrate that hashtags are indeed related to image contents with a significance level $\alpha = 0.001$.

Furthermore, we model the relationship between hashtags and images (i.e., their labels) as bi-directional prediction tasks, i.e., using an image's associated hashtags to predict the image's labels (H2L) and using an image's labels to predict its hashtags (L2H). The prediction performance is then used to describe the strength of the relationship between images and hashtags. For the H2L task, a straightforward approach is to use word embedding methods [10] to transform hashtags into continuous vectors, representing hashtag semantics, which are later used as features to train a machine learning classifier. A similar approach can be applied to the L2H task as well, namely, to use the obtained label vectors from word embedding methods to predict image's hashtags. However, this approach only considers the semantic meaning of hashtags (and labels), while neglecting connections among the images. As demonstrated by the example in Table 1, if two images share a few hashtags (i.e., #skiing, #snow), then their contents may have certain similarity as well (i.e., both are about outdoor winter sports in the mountain). To this end, we propose a graph-based approach, which can explore both semantic information of hashtags (and labels) and the graph structure among the images, to measure the relationship between hashtags and images.

Through extensive experiments, we show that our approach has better prediction performance – 34.85% overall precision (O-P) for the H2L task and 23.88% O-P for the L2H task on our Instagram dataset. Compared with the approach based on word embedding, it achieves 70.1% and 17.9% O-P gain for the two tasks, respectively.

Application. After verifying and quantifying the relationship between hashtags and images, we further explore this relationship to improve one downstream task – image multi-label classification. Experiments on the *NUS-WIDE* dataset [4] show that we can achieve a 12.0% O-P gain over a state-of-the-art method. This result further shows that there is indeed a significant relationship between hashtags and image contents.

Overall our current paper makes the following two contributions:

1. We statistically demonstrate and quantify the relationship between hashtags and images. In particular, we propose a new graph-based approach which can extract comprehensive information from both hashtags and images.
2. We further apply the above-identified relationship with our new approach to improve the performance of image multi-label classification.

2   Image-Hashtag Relationship Verification

Instagram is one of the most popular OSNs and a major platform for hashtag- and image-sharing. Therefore, we resort to Instagram to collect our dataset relying on its public API.[2] Our data collection follows a similar strategy as the one proposed by Zhang et al. [25]. Concretely, we sample users from New York by their geo-tagged posts. Then, for each user, we collect all her/his images. In total, we obtain 10,605,399 images from 25,658 users. Then, we perform some pre-processing filtering out those images with less than 3 hashtags. Table 2 gives the top 20 most frequent hashtags together with their frequencies.

Table 2. The set of most frequent hashtags in our Instagram dataset.

nyc	21553	manner	3341
new_york	7918	nofilter	3265
love	6695	summer	3107
brooklyn	4454	food	2889
instagood	4224	photooftheday	2857
travel	3754	foodporn	2717
newyorkcity	3617	latergram	2689
manhattan	3610	sunset	2427
tbt	3608	picoftheday	2369
art	3505	friend	2336

As mentioned before, we represent image contents as labels. Manual labelling can be an option but not scalable. Instead, we adopt Google's Cloud Vision API[3] to label images. The Cloud Vision API is supported by pre-trained machine learning models; it describes an image's content as a list of labels. The detected labels cover various aspects of an image ranging from the contained objects to personal feelings as well, e.g., happiness. It is worth noticing that this API has been already used in social media image analysis before [15]. Table 1 depicts two images labelled by Google's Cloud Vision API.

In total, we have spent 227$ on labelling 148,106 images. There are 255,298 different hashtags associated with these images. On average, each image has 6.46 hashtags, and each hashtag can appear in 4.19 images. Figure 1(a) presents the distribution of the number of hashtags associated with each image. We can see that the images with 3 hashtags have the largest count, and most images have less than 10 hashtags.

For all our images, Google's Cloud Vision API provides 6,327 different labels. On average, each image contains 8.27 labels. Figure 1(b) presents the distribution of the number of labels for each image. Google's Cloud Vision API gives at most

---

[2] The dataset was collected in January 2016 when Instagram's API was still publicly available.

[3] https://cloud.google.com/vision/.

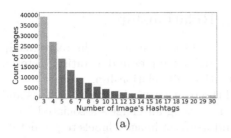 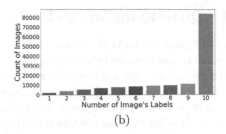

(a)   (b)

Fig. 1. (a): Distribution of the number of hashtags associated with each image in our Instagram dataset. (b): Distribution of the number of labels for each image in our Instagram dataset.

10 labels for one image, thus the amount of images with 10 labels is much more than the amount of images with other numbers of labels (<10).

From the example in Table 1, we can confirm that the labels given by Google's Cloud Vision API can sufficiently describe the image contents. It can find the objects (e.g., cable car, piste, ski equipment) in the images, and detect the subject (e.g., winter sport) and feeling (e.g., fun) of images. Besides, we also find out that some hashtags have a close relationship with image contents, e.g., #snowboard, #piste, #skiing, and some of them describe additional information, e.g., #utah, #austria, #travelgram. However, there are also some other hashtags which do not have too much relation with the image's contents, e.g., #findyourgreatest.

To verify the existence of the relationship between hashtags and image labels, we perform a two-sample KS test. We construct two vectors hc_c and hc_d with equal number of elements, where each element in hc_c is obtained by calculating the appear ratios of labels in images that have one specific hashtag and similarly each element in h_d is the appear ratio score of labels in images that don't have this hashtag. We perform a two-sample KS test on vectors hc_c and hc_d. The null hypothesis here is that the appear ratio of labels in images with one specific hashtag does not differ from images without this hashtag, i.e., these two vectors are the same, $H_0 : hc_c = hc_d$. Another hypothesis is that the appear ratio of labels in images with one specific hashtag differs from images without this hashtag. Therefore, we have the following two-sample KS test:

$$H_0 : hc_c = hc_d, H_1 : hc_c \neq hc_d$$

The two-sample KS test result suggests a strong evidence with a significance level $\alpha = 0.001$ (p-value $= 1e - 91$) to reject the null hypothesis. As a result, we confirm that there exists a relationship between hashtags and image contents.

3   Quantifying Image-Hashtag Relationship

In the previous section, we have demonstrated the existence of the relationship between hashtags and image contents (through examples and a statistical test). In this section, we will systematically quantify this relationship.

Our idea for quantification is to model the relationship between hashtags and images as bi-directional prediction tasks, i.e., using an image's associated hashtags to predict the image's labels (H2L) and using an image's labels to predict its hashtags (L2H). The prediction results can be used to quantify the relationship strength – higher prediction performance indicates a stronger relationship.

In the rest of the section, we first discuss how to use word embedding methods to extract semantic meaning for hashtags and labels for our prediction tasks (Sect. 3.1). Then, we present a graph embedding based approach in Sect. 3.2. The experimental results are presented in the end (Sect. 3.3). For presentation purposes, we use H2L as an example task, similar approaches can be described for the L2H task as well.

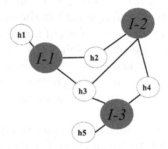

Fig. 2. The example graph of $\mathcal{G}$. Graph consists of two types of nodes: image (I) and hashtag (h), where each image node connects with the hashtag nodes that appear with the image, and each hashtag node connects with those image nodes with which the hashtag is tagged.

3.1   Word Embedding Based Approach

We use $\mathcal{I}$ to represent the set of images. Each image i is associated with a list of hashtags $H_i = \{h_1, h_2, \ldots, h_{m_i}\}$ and a list of labels $L_i = \{\ell_1, \ell_2, \ldots, \ell_{n_i}\}$. We use m_i and n_i to denote the number of hashtags and labels in an image i, respectively. Furthermore, we use $\mathcal{H}$ to represent the set of all the hashtags and $\mathcal{L}$ to represent the set of all the labels.

For our H2L task, one intuitive approach is to use hashtags' semantic meaning as the features to train a machine learning classifier to predict image labels. We apply word embedding to transform each hashtag into a continuous vector, and average the vectors of all hashtags of an image as its feature. To train hashtag embedding, we adopt the Word2vec model [10], meaning that we treat each image's associated hashtags as a "phrase", and all these phrases form a "corpus". The learning process follows the same objective function as Skip-Gram, by applying stochastic gradient descent.

3.2   Graph Embedding Based Approach

The above word embedding based approach only considers the semantic meaning of hashtags (and labels) while neglecting connections among the images. In the example depicted in Table 1, if two images share some hashtags, then their contents share certain similarities as well. We hypothesise that connections among images also possess a strong signal for our prediction task, thus we aim for a method to summarise this relationship as new features.

Our idea of feature extraction is to organise images in a graph according to the connections among them and extract images' connection information represented in the graph. The graph we construct is $\mathcal{G} = (\mathcal{H}, \mathcal{I}, \mathcal{E}_{HI})$. $\mathcal{G}$ contains two types of nodes: hashtag ($\mathcal{H}$) and image ($\mathcal{I}$), each image node connects with its hashtags and each hashtag node connects with images that it appears with (edges in $\mathcal{E}_{HI}$). The graph in Fig. 2 depicts an example of $\mathcal{G}$.

The state-of-the-art method to extract information from a graph is graph embedding, which aims to learn a mapping that embeds nodes as points in a low-dimensional vector space [8]. Through optimising this mapping, geometric relation in this learned space reflects the attributes of the original graph.

The graph embedding method we adopt is DeepWalk [13], it is inspired by the idea of word embedding. We treat a graph as a "document" and sample sequence of nodes by random walk on the graph as a "phrase". Then, word embedding methods can be applied to these phrases as a traditional document task to return us the feature vectors of image nodes. The main reason for adopting this method is that it is relatively efficient and suitable for a large dataset, and its idea has been successfully used in other hashtag-related work [2,24].

3.3   Experiments

We evaluate the two approaches proposed in Sects. 3.1 and 3.2 on the bi-directional prediction tasks (H2L and L2H) on our Instagram dataset to quantify the relationship between hashtags and images.

Evaluation Metrics. We adopt those overall evaluation metrics that are widely used in multi-label image classification fields [20], including overall precision (O-P), overall recall (O-R) and overall F1 score (O-F1).

The precision is the number of correctly predicted labels (or hashtags) divided by the number of predicted labels (or hashtags); the recall is the number of correctly predicted labels (or hashtags) divided by the number of ground-truth labels (or hashtags); the F1 score is the geometrical average of the precision and recall scores. Overall means the average is taken over all testing examples. Moreover, we only consider the top 3 predictions for both tasks in our evaluation.

Preprocessing. We adopt the following steps to prepare our dataset. We first convert hashtags into lowercase and delete punctuation. Second, as multiple hashtags may refer to the same underlying concept, we apply a simple process that utilises WordNet [11] synsets to merge some hashtags into a single canonical form, such as "coffeehouse" and "coffeeshop" to "cafe". Third, for the H2L task:

we select the most frequent 100 labels from the dataset and keep images with these labels. For the L2H task: we similarly select 100 most frequent hashtags from the dataset and keep images with these hashtags. To study the influence of the number of hashtags (or labels) of images for these two tasks respectively, we set the minimum number of hashtags (or labels) of the image as a hyperparameter n, then we can filter images with different n. After the preprocessing, we randomly select 20,000 images with different settings.

Implementation Details. For fairness, the default embedding dimension d in this paper is set to 256. For the approach based on word embedding, we adopt the Skip-Gram implementation provided by *gensim* [14], and keep the default parameters provided by the software. For the approach based on graph embedding, i.e., DeepWalk, we set the length of each walk to 80 and the number of walks per node to 10.

In the end, we need to feed these extracted features into a logistic classifier to make predictions. In this way, we evaluate the following two approaches to our prediction tasks: Word2vec+logistic and DeepWalk+logistic.

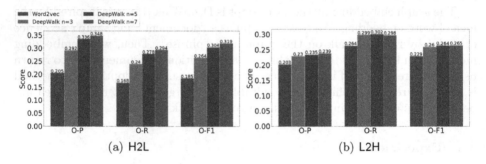

(a) H2L   (b) L2H

Fig. 3. (a): Experimental results of the task H2L with Word2vec embedding, and DeepWalk embeddings with the different minimum number of hashtags per image ($n = 3, 5, 7$). (b): Experimental results of the task L2H with Word2vec embedding, and DeepWalk embeddings with the different minimum number of labels per image ($n = 3, 5, 7$).

Results. The results for the task H2L are listed in Fig. 3(a). We can see that all the O-P scores are no less than 20% for all four settings. Moreover, the results of different DeepWalk embeddings are better than the results of Word2vec embedding (for example, a 70.1% O-P gain and a 72.4% O-F1 gain for DeepWalk with $n = 7$). This indicates that DeepWalk could explore a more comprehensive relationship between hashtags and image contents than only considering the hashtag semantics. Moreover, the results of DeepWalk get better when increasing n. When compare the results of DeepWalk embedding with $n = 7$ and the DeepWalk embedding with $n = 3$, there is a 19.3% O-P gain and a 20.9% O-F1 gain. This indicates an image's content has a more significant relationship with hashtags, when the image are tagged with more hashtags.

The results of the task L2H are listed in Fig. 3(b). We can find that all the O-P scores are more than 20% and the O-F1 scores are more than 22% for these four settings. Similarly, DeepWalk embeddings achieve an improvement when compared with Word2vec embedding. For the O-P scores, the performance gain of DeepWalk embedding with $n = 7$ is 17.9% compared with the Word2vec embedding. But the improvement of DeepWalk embeddings in the L2H task is less significant than in the H2L task. This indicates that for the L2H task, the information provided by the graph relationship among images has a similar strength as only exploring the label's semantic meaning. Besides, comparing the results of DeepWalk embedding with $n = 7$ with the DeepWalk embedding with $n = 3$, there is a 4.0% O-P gain and a 2.0% O-F1 gain. It indicates that more knowledge of image contents could not significantly help us to predict hashtags for the image.

4   Application

After verifying and quantifying the relationship between hashtags and images, we focus on whether this relationship can be used to improve a downstream task. In particular, we aim to use hashtags' information summarised by the approach based on DeepWalk to improve the performance of a baseline model on the multi-label classification task.

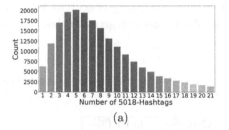
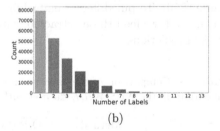

(a)   (b)

Fig. 4. (a): Distribution of the number of 5018-Hashtags associated with each image in the *NUS-WIDE* dataset. (b): Distribution of the number of labels for each image in the *NUS-WIDE* dataset.

In order to make sure the reliability of images' labels and to prove the universality of our method, we use the *NUS-WIDE* dataset, which contains human-generated labels and hashtags shared by real users, for this task. *NUS-WIDE* is a web image dataset [4], and it contains 269,648 images from Flickr. It has two types of hashtags: (i) 5018 unique hashtags (5018-hashtags); (ii) 1000 cleaner hashtags without noisy and rarely-appearing hashtags. Figure 4(a) presents the distribution of the number of 5018-hashtags associated with each image. We could see that the most frequent numbers of hashtags with images are 4, 5, and 6, and this dataset has quite some images with less than three hashtags.

The images in the dataset are also manually annotated using 81 labels by human annotators, which cover different aspects including object classes, scenes, and attributes. The labels on each image are considered as ground truth to represent the image contents.[4] On average, each image contains 1.87 such labels. The Fig. 4(b) presents the distribution of the numbers of labels. We can find that images with only one label have the largest count, and there are only a few images with more than 8 labels.

Preprocessing. To demonstrate the application of the relationship between hashtags and images to improve the performance of image multi-label classification, we use a pre-trained convolution neural network (CNN) as the baseline approach to extract the image features (or image embedding). This technique has been successfully used for many image-related tasks, i.e., image classification [1,20], image recognition [16], etc. Then, we use the returned image embeddings to train a classifier to make predictions. Second, we use the 5018-hashtags, in this way we keep all the information provided by users. Third, while building the graph structure, we use the same settings as in Sect. 3.3, and we use 81 labels and set $n = 1$, i.e., we keep all available images from the *NUS-WIDE* dataset.

Implementation Details. For the baseline CNN, we use 16 layers VGG network [18] pre-trained on ImageNet 2012 classification challenge dataset [5] using Pytorch deep learning framework. For our DeepWalk-based approach, we use the graph structure $\mathcal{G}$, the same as discussed in Sect. 3.3. The dimensions of the CNN embeddings, Word embeddings and DeepWalk embeddings are set as the same (256). To put different embeddings together, we simply concatenate them. In the end, we feed these extracted features into a logistic regression classifier to make predictions.

Table 3. Comparison of the experimental results of the top 3 image multi-label classification on the *NUS-WIDE* dataset with 5018-hashtags.

Methods	O-P (%)	O-R (%)	O-F1 (%)
CNN	48.3	59.5	53.8
Word2vec	40.3	49.4	44.3
DeepWalk	51.3	62.9	56.5
CNN+DeepWalk	**55.9**	**68.5**	**61.6**
CNN+RNN	49.9	61.7	55.2

Results. We use the same evaluation metrics as discussed in Sect. 3.3. Table 3 presents the classification results of approaches using the CNN embeddings, the Word2vec embeddings, the DeepWalk embeddings and the CNN+DeepWalk embeddings, respectively. From the results in Table 3, we could first find that the

---

[4] This explains why we cannot directly use our Instagram dataset as we don't have such ground truth.

CNN+DeepWalk embeddings can improve the classification performance when only using the CNN embeddings for multi-label classification (with 17.0% O-P gain and 16.2% O-F1 gain). Second, the performance of the DeepWalk embeddings is still better than using the CNN embeddings (with 6.2% O-P gain and 5.0% O-F1 gain). This indicates the relationship between hashtags and image contents is significantly useful for image multilabel classification.

Moreover, we list the results of one state-of-the-art approach CNN+RNN [20], which combines image features and the corresponding hashtags for image multilabel classification. We can find that the results of DeepWalk and CNN+DeepWalk embeddings are better than the CNN+RNN embeddings (a 2.8% O-P gain and a 2.4% O-F1 gain for DeepWalk, and a 12.0% O-P gain and a 11.6% O-F1 gain for CNN+DeepWalk). It further indicates that our approach for relationship exploration is more comprehensive.

Observations. In this section, we present detailed examples to understand the different predictions given by the CNN embeddings and the CNN+DeepWalk embeddings.

In Table 4, there are two images with their associated labels and hashtags from the *NUS-WIDE* dataset, as well as the predictions made by the two approaches based on the CNN embeddings and the CNN+DeepWalk embeddings, respectively.For the image on the left,the CNN embeddings give one correct prediction ("person") and two incorrect predictions ("sky", "water").We notice that this image is somehow unclear and over light. Since the CNN embeddings come from the image itself, it somehow mistakes this strong light in the background as "sky" or "water". Besides, the correctly predicted label "person" is one of the most popular labels in the dataset (24.6% images contain this label), so this label could not provide precise information to identify the image contents. On the other hand, the CNN+DeepWalk embeddings correctly predict the two labels "military" and "person". This indicates that this approach can capture more comprehensive information about this image itself.

Table 4. Two example predictions by the CNN approach and the CNN+DeepWalk approach on the *NUS-WIDE* dataset.

5018-Hashtags	#film, #army #war, #historic	#fish, #photography #underwater
Labels (ground truth)	military, person	animal, coral, fish
Prediction (CNN)	person, sky, water	animal, coral, water
Prediction (CNN+DeepWalk)	military, person, sky	animal, coral, fish

For the image on the right, the CNN embeddings give two correct labels ("animal", "coral") and one incorrect label ("water"). However, this incorrect label is different from those two incorrect labels for the left image, as it is still relevant to the contents of the image. We could recognise that the image presents an underwater environment, so "water" is not wrong even it does not appear as one of the truth labels. The fish in the right image disguises itself in the environment. In this case, the visual CNN embeddings are not sufficient in capturing small objects (i.e., "fish") in the image. On the other hand, CNN+DeepWalk embeddings succeed in predicting all three labels.

From these two example predictions on the *NUS-WIDE* dataset, we can confirm that our hashtag features through the DeepWalk embeddings can provide useful information to improve the image multi-label classification even when the image quality is not good enough, or the objectives in the image are not easy to be found by visual features.

We are also interested in knowing whether the DeepWalk embeddings could embed images into the correct position in the embedding space.More specifically, whether they can keep images with similar contents to be close in the space. For this aim, we select 9 groups of labels and each group to have 2 different labels and collect sample images only containing one of the groups of labels. We then transform these image embeddings obtained with DeepWalk into a 2D space using the dimensionality reduction algorithm t-SNE [9].

We visualise the result in Fig. 5, and observe the existence of clustering structure in images' embeddings. These images with different groups of labels are separated into different clusters, and related clusters are close in the space. For instance, in the figure, we can first find that images with the labels related to the animal ("cat", "birds", "fish") are in the left side while the images with labels related to plants ("flowers", "plants") are in the right side and the images about

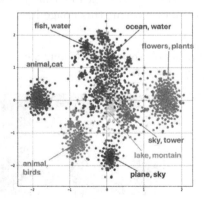

Fig. 5. Visualisation of our DeepWalk embeddings. The images information are mapped to the 2D space using the t-SNE package with learned DeepWalk image embeddings as input. We select some labels: ["animal", "cat"], ["ocean", "water"], ["flowers", "plants"], ["fish", "water"], ["airport", "clouds"], ["lake", "mountain"], ["plane", "sky"], ["animal", "birds"] and ["sky", "tower"] and collect images have these labels.

the natural scene ("water", "ocean", "mountain") are in the middle.Second, we can also find that images labelled by ["lake", "mountain"] and images labelled by ["fish", "water"] are mixed with the images labelled by ["ocean", "water"]. This is due to the contents of the two images have similar semantics.

5   Related Work

There has been a diverse array of academic works on exploring the information contained in hashtags. Tsur et al. try to explore what information are contained in hashtags based on a massive dataset from Twitter [19], and they view hashtags as ideas that could express users. As a result, they present the richness of information in hashtags. Furthermore, some work use hashtags to detect the topic of tweets on Twitter [21] and predict hashtags based on tweet contents [7,17]. These work indicate there is a strong relationship between hashtags and text contents, and it is possible to make two-way predictions between them.

Focusing on the relationship between hashtags and images, Niu et al. propose a semi-supervised Relational Topic Model (ss-RTM) to use hashtags information to recognise social media images [12]. They first organise images into a network if they share some hashtags. Then, they treat this network as a document and use a statistical model RTM, which is widely used in natural language processing tasks to extract the topic relationship among documents, to extract images' relationships into representative vectors. Compared with our work, they only use hashtags' information to build up the network but ignore their semantic meaning in the final features. Besides, due to the computational cost of RTM, they cannot involve a large number of images in one network, and there might be a strong influence from noisy hashtags. Wang et al. propose a framework (CNN-RNN) which combines hashtags and image features to perform classification [20]. CNN-RNN mainly contains two parts – a CNN model for extracting semantic representations from the images, and an RNN (recurrent neural network) to model image/labels relationship and hashtags dependency. Due to the advantages of RNN, this framework can utilise the order information among hashtags, and it can predict a long sequence of labels.It achieves better performance compared with ss-RTM, but it neglects the connections among images. Recently, Wang et al. utilise a hashtag-related knowledge graph to improve image multi-label classification [22]. They first build a large knowledge graph, which contains millions of hashtags and their semantic relationships. Then they apply the deep graph embedding methods to extract hashtags' relationship to representative vectors and use the representative vectors to assistant the classification task. But it is a high-cost work to build up a knowledge graph with millions of hashtags, and they only consider the hashtags semantic information but neglect the graph structure among the images.

6   Conclusion and Future Work

In this paper, we have performed an empirical study on verifying and quantifying the relationship between hashtags and images based on real-world datasets collected from Instagram and Flickr, and we successfully applied the verified relationship to improve a downstream task.

We have implemented a statistical test to verify the existence of the relationship between hashtags and images on the Instagram dataset. Then, we designed bi-directional prediction tasks (H2L and L2H) and used the prediction performance to quantify the relationship. In particular, we proposed a new graph-based approach to integrate both the semantic meaning of hashtags (and labels) and the graph structure of the images, which indeed help to extract more comprehensive information for hashtags (and labels). In the end, we adopted a widely used dataset *NUS-WIDE* which has tags given by users and manual labels, and successfully applied the extracted features of hashtags from the H2L task to improve the performance of image multi-label classification and achieved a 12.0% overall precision gain compared to a state-of-the-art method.

Hashtags can be naturally organised into different categories, according to their semantics. In the future, we will first focus on the influence of hashtag categories, i.e., investigating the different relationship strength between each category of hashtags and images. Second, on OSNs different users have different habits of using hashtags, and we hypothesise that the richness of the semantic meaning contained in their hashtags could be different. How to explore this, e.g., to perform link prediction as in [2,3,24], is part of our future work. Third, so far we have only applied the extracted hashtag features for an image multi-label classification task in this work. In the future, we want to utilise the extracted label features (from the L2H task) to perform hashtag recommendation in OSNs.

Acknowledgements. This work is partially supported by the Luxembourg National Research Fund through grant PRIDE15/10621687/SPsquared.

References

1. Akata, Z., Reed, S., Walter, D., Lee, H., Schiele, B.: Evaluation of output embeddings for fine-grained image classification. In: Proceedings of the 2015 Conference on Computer Vision and Pattern Recognition (CVPR), pp. 2927–2936. IEEE (2015)
2. Backes, M., Humbert, M., Pang, J., Zhang, Y.: walk2friends: Inferring social links from mobility profiles. In: Proceedings of the 2017 ACM SIGSAC Conference on Computer and Communications Security (CCS), pp. 1943–1957. ACM (2017)
3. Cheng, R., Pang, J., Zhang, Y.: Inferring friendship from check-in data of location-based social networks. In: Proceedings of the 2015 Workshop on Social Network Analysis in Applications (SNAA), pp. 1284–1291. ACM (2015)
4. Chua, T.S., Tang, J., Hong, R., Li, H., Luo, Z., Zheng, Y.: NUS-WIDE: a real-world web image database from National University of Singapore. In: Proceedings of the 2009 International Conference on Image and Video Retrieval (CIVR). ACM (2009)
5. Deng, J., Dong, W., Socher, R., Li, L.J., Li, K., Fei-Fei, L.: Imagenet: A large-scale hierarchical image database. In: Proceedings of the 2009 Conference on Computer Vision and Pattern Recognition (CVPR). pp. 248–255. IEEE (2009)

6. Denton, E., Weston, J., Paluri, M., Bourdev, L.D., Fergus, R.: User conditional hashtag prediction for images. In: Proceedings of the 2015 ACM Conference on Knowledge Discovery and Data Mining (KDD), pp. 1731–1740. ACM (2015)
7. Godin, F., Slavkovikj, V., Neve, W.D., Schrauwen, B., de Walle, R.V.: Using topic models for Twitter hashtag recommendation. In: Proceedings of the 2013 International Conference on World Wide Web (WWW), pp. 593–596. ACM (2013)
8. Hamilton, W.L., Ying, R., Leskovec, J.: Representation learning on graphs: methods and applications. IEEE Data Eng. Bull. **40**, 52–74 (2017)
9. van der Maaten, L., Hinton, G.: Visualizing data using t-SNE. In: Eurographics Conference on Visualization (EuroVis), pp. 2579–2605. Eurographics Association (2008)
10. Mikolov, T., Chen, K., Corrado, G., Dean, J.: Efficient estimation of word representations in vector space. In: Proceedings of the 2013 International Conference on Learning Representations (ICLR) (2013)
11. Miller, G.A.: WordNet: a lexical database for english. Commun. ACM **38**(11), 39–41 (1995)
12. Niu, Z., Hua, G., Gao, X., Tian, Q.: Semi-supervised relational topic model for weakly annotated image recognition in social media. In: Proceedings of the 2014 Conference on Computer Vision and Pattern Recognition (CVPR), pp. 4233–4240. IEEE (2014)
13. Perozzi, B., Al-Rfou, R., Skiena, S.: DeepWalk: Online learning of social representations. In: Proceedings of the 2014 ACM Conference on Knowledge Discovery and Data Mining (KDD), pp. 701–710. ACM (2014)
14. Řehůřek, R., Sojka, P.: Software framework for topic modelling with large corpora. In: Proceedings of the LREC 2010 Workshop on New Challenges for NLP Frameworks, pp. 45–50. ELRA (2010)
15. Richards, D.R., Tunçer, B.: Using image recognition to automate assessment of cultural ecosystem services from social media photographs. Ecosyst. Serv. **31**, 318–325 (2018)
16. Schroff, F., Kalenichenko, D., Philbin, J.: FaceNet: A unified embedding for face recognition and clustering. In: Proceedings of the 2015 Conference on Computer Vision and Pattern Recognition (CVPR), pp. 815–823. IEEE (2015)
17. She, J., Chen, L.: TOMOHA: topic model-based hashtag recommendation on twitter. In: Proceedings of the 2014 International Conference on World Wide Web (WWW), pp. 371–372. ACM (2014)
18. Simonyan, K., Zisserman, A.: Very deep convolutional networks for large-scale image recognition. In: Proceedings of the 2018 International Conference on Learning Representations (ICLR) (2015)
19. Tsur, O., Rappoport, A.: What's in a hashtag? content based prediction of the spread of ideas in microblogging communities. In: Proceedings of the 2012 ACM International Conference on Web Search and Data Mining (WSDM), pp. 643–652. ACM (2012)
20. Wang, J., Yang, Y., Mao, J., Huang, Z., Huang, C., Xu, W.: CNN-RNN: a unified framework for multi-label image classification. In: Proceedings of the 2016 Conference on Computer Vision and Pattern Recognition (CVPR), pp. 2285–2294. IEEE (2016)
21. Wang, X., Wei, F., Liu, X., Zhou, M., Zhang, M.: Topic sentiment analysis in twitter: a graph-based hashtag sentiment classification approach. In: Proceedings of the 2011 ACM International Conference on Information and Knowledge Management (CIKM), pp. 1031–1040. ACM (2011)

22. Wang, X., Ye, Y., Gupta, A.: Zero-shot recognition via semantic embeddings and knowledge graphs. In: Proceedings of the 2018 Conference on Computer Vision and Pattern Recognition (CVPR), pp. 6857–6866. IEEE (2018)
23. Wu, J., Yu, Y., Huang, C., Yu, K.: Deep multiple instance learning for image classification and auto-annotation. In: Proceedings of the 2015 Conference on Computer Vision and Pattern Recognition (CVPR), pp. 3460–3469. IEEE (2015)
24. Zhang, Y.: Language in our time: an empirical analysis of hashtags. In: Proceedings of the 2019 International Conference on World Wide Web (WWW), pp. 2378–2389. ACM (2019)
25. Zhang, Y., Humbert, M., Rahman, T., Li, C.T., Pang, J., Backes, M.: Tagvisor: a privacy advisor for sharing hashtags. In: Proceedings of the 2018 Web Conference (WWW), pp. 287–296. ACM (2018)
26. Zhou, B., Lapedriza, À., Khosla, A., Oliva, A., Torralba, A.: Places: a 10 million image database for scene recognition. IEEE Trans. Pattern Anal. Mach. Intell. **40**(6), 1452–1464 (2018)

Entity Linkage and Disambiguation

RDF Graph Anonymization Robust to Data Linkage

Remy Delanaux[1]([✉]), Angela Bonifati[1], Marie-Christine Rousset[2,3], and Romuald Thion[1]

[1] Université Lyon 1, LIRIS CNRS, Villeurbanne, France
{remy.delanaux,angela.bonifati,romuald.thion}@univ-lyon1.fr
[2] Université Grenoble Alpes, CNRS, INRIA, Grenoble INP, Grenoble, France
marie-christine.rousset@imag.fr
[3] Institut Universitaire de France, Paris, France

Abstract. Privacy is a major concern when publishing new datasets in the context of Linked Open Data (LOD). A new dataset published in the LOD is indeed exposed to privacy breaches due to the linkage to objects already present in the other datasets of the LOD. In this paper, we focus on the problem of building *safe* anonymizations of an RDF graph to guarantee that linking the anonymized graph with any external RDF graph will not cause privacy breaches. Given a set of privacy queries as input, we study the data-independent safety problem and the sequence of anonymization operations necessary to enforce it. We provide sufficient conditions under which an anonymization instance is safe given a set of privacy queries. Additionally, we show that our algorithms for RDF data anonymization are robust in the presence of sameAs links that can be explicit or inferred by additional knowledge.

Keywords: Linked Open Data · Data privacy · RDF anonymization

1 Introduction

Since its inception, the Linked Open Data (LOD) paradigm has allowed to publish data on the Web and interconnect uniquely identified objects by realizing widely open information exchange and data sharing. The LOD cloud is rapidly growing and contains 1,231 RDF graphs connected by 16,132 links (as of June 2018). Since 2007, the number of RDF graphs published in the LOD has seen an increase of about two orders of magnitude. Nevertheless, the participation of many organizations and institutions to the LOD movement is hindered by individual privacy concerns. Personal data are ubiquitous in many of these data sources and recent regulations about personal data, such as the EU GDPR General Data Protection Regulation (GDPR) make these organizations reluctant to publish their data in the LOD.

While there has been some effort [15, 22] to bring data anonymization techniques from the relational database world to the LOD, such as variations of

© Springer Nature Switzerland AG 2019
R. Cheng et al. (Eds.): WISE 2019, LNCS 11881, pp. 491–506, 2019.
https://doi.org/10.1007/978-3-030-34223-4_31

k-anonymity [16,18,23], most of the state of the art is mainly based on differential privacy techniques for relational data [8,17]. However, differential privacy is not a perfect match for Linked Data, focusing more on statistical integrity rather than accurate, qualitative query results which represents the main usage of Linked Data through SPARQL endpoints [2,19]. Differential privacy is indeed useful whenever the aggregate results of data analysis (such as statistics about groups of individuals) can be seamlessly published. Whereas this is highly desirable in many applications, it becomes not sufficient in privacy-preserving data publishing (PPDP) [9] scenarios where the privacy of individuals need to be protected while at the same time ensuring that the published data can be utilized in practice. Whereas the underpinnings of PPDP under its most prominent form, such as anonymization, have been widely studied for relational data (see [9] for a comprehensive survey), the theoretical foundations for PPDP in the context of Linked Data have only been recently laid out in [12] by focusing on the theoretical study of its computational complexity.

In this paper, we build upon the foundations of [12] by focusing on the linkage safety requirement and present practical algorithms to compute the anonymization operations needed to achieve such a requirement when a graph G is linked to external graphs in the LOD. By relying on the computational complexity of the linkage safety problem, which is AC_0 in data complexity under the *open-world* assumption, we address the problem of actually computing a safety-compliant sequence of anonymization operations setting up their guarantees against linkage attacks. In doing this, we also devote special care to :sameAs links (i.e. links expressed in RDF syntax) that can be either explicit in the original graph G linking to entities in external graphs or derived by inference mechanisms on G itself. In particular, this approach exhibits two distinguishing features. First, it is *query-based* since the privacy policies as well as the anonymization operations are specified by means of conjunctive queries and updates in SPARQL, respectively. Second, our approach is *data-independent* since, given a privacy policy (specified as a set of privacy queries), our algorithms produce anonymization operations (under the form of delete and update queries) with the guarantee that their application to *any RDF graph* will satisfy the safety requirement. Our contributions can be summarized as follows: we first ground the linkage safety problem to the sequence of anonymization operations necessary to enforce it by providing a novel *data-independent* definition of safety; such a definition considers a set of privacy queries as input and does not look at the actual graph instances (Sect. 4); as such, it departs from the basic definition of linkage safety of [12]. We then provide sufficient conditions under which an anonymization instance is safe given a set of privacy queries and design an anonymization algorithm that solves the above query-based safety requirement and study its runtime complexity (Sect. 5). Next, we introduce :sameAs links and show that slight modification of our algorithm is robust to them (Sect. 6) and finally, we provide a quick discussion about the evaluation of our algorithms and the remaining utility of the anonymized graphs (Sect. 7) that confirms the good behavior of our framework in practice. Related work is discussed in Sect. 2, and we provide the necessary background in Sect. 3. We conclude in Sect. 8. All proofs and implementations are available online in a companion appendix.[1]

2   Related Work

A query-based approach to privacy-preserving RDF data publishing has been presented in [6], in which the focus was to check the compatibility between a privacy policy and an utility policy (both specified as queries) and to build anonymizations preserving the answers to a set of utility queries (when compatibility is ensured). However, the above approach suffers from the lack of resilience against privacy breaches caused by linking external datasets, which is clearly a recurrent situation in the LOD.

In line with existing works [7,11,20] on safety models defined in terms of secret or privacy queries for relational data. A query-based safety model for RDF data has been introduced in [12] where linking RDF graphs is reduced to their union, several results are provided on the computational complexity of the decision problems. In our paper, we slightly extend the considered safety model and we address the data-independent construction problem underpinning safety, i.e. how to produce a sequence of update operations that are safe *for any RDF graph*, given a privacy policy expressed as queries.

Graph specific, but non RDF, declinations of privacy criteria and related attacks have been proposed such as l-opacity [21] or k-isomorphism [5], the typical use cases being social networks graphs. In this paper, we follow a complementary direction where the privacy criteria is declared by the data protection officer and not fixed, with a concrete and efficient procedure that uses standard and efficient SPARQL engines to enforce them. Compared to existing approaches based on k-anonymity in RDF graphs [15,22], we focus on generalizations that replace constants by blank nodes. We have also shown that in some cases, triple suppressions are required in addition to generalizations for guaranteeing *safe* anonymizations.

Privacy-preserving record linkage has been recently considered in [24] as the problem of identifying and linking records that correspond to the same real-world entity without revealing any sensitive information about these entities. For preserving privacy while allowing the linkage, masking functions are proposed to transform original data in such a way that there exists a specific functional relationship between the original data and the masked data. The problem of privacy-preserving record linkage is a difficult problem that is significantly different from the privacy-preserving data publishing problem considered in this paper, in which sameAs links are input of the anonymization process.

3   Formal Background

We recall the usual concepts for RDF graphs and SPARQL queries as formalized in [13]. Let $\mathbf{I}$, $\mathbf{L}$ and $\mathbf{B}$ be countably infinite pairwise disjoint sets representing

---

[1] See https://perso.liris.cnrs.fr/remy.delanaux/papers/WISE2019appx.pdf.

respectively *IRIs*, *literals* and *blank nodes*. IRIs (Internationalized Resource Identifiers) are standard identifiers used for denoting any Web resource described in RDF within the LOD. We denote by $\mathbf{T} = \mathbf{I} \cup \mathbf{L} \cup \mathbf{B}$ the set of *terms*, in which we distinguish *constants* (IRIs and literals) from *blank nodes*, which are used to model unknown IRIs or literals like in [3,10] and correspond to *labeled nulls* [1].

We also assume an infinite set $\mathbf{V}$ of variables disjoint from the above sets. Throughout this paper, we adhere to the SPARQL conventions: variables in $\mathbf{V}$ are prefixed with a question mark (?), IRIs in $\mathbf{I}$ are prefixed with a colon (:), blank nodes in $\mathbf{B}$ are prefixed with an underscore and a colon ($\_:$).

Definition 1 (RDF graph and graph pattern). *An* RDF graph *is a finite set of* RDF triples (s, p, o), *where* $(s, p, o) \in (\mathbf{I} \cup \mathbf{B}) \times \mathbf{I} \times (\mathbf{I} \cup \mathbf{L} \cup \mathbf{B})$. *A* triple pattern *is a triple* $(s, p, o) \in (\mathbf{I} \cup \mathbf{B} \cup \mathbf{V}) \times (\mathbf{I} \cup \mathbf{V}) \times (\mathbf{I} \cup \mathbf{L} \cup \mathbf{B} \cup \mathbf{V})$. *A* graph pattern *is a finite set of triple patterns.*

We can now define the three types of queries that we consider. Definition 2 corresponds to *conjunctive queries* and will be the basis for formalizing the sensitive information that must not be disclosed. Definition 6 corresponds to *counting queries* which will model a form of utility that it may be useful to preserve for analytical tasks. Finally, Definition 7 describes *update queries*, modeling the anonymization operations handled in our framework.

Definition 2 (Conjunctive query). *A* conjunctive query Q *is defined by an expression* SELECT $\bar{x}$ WHERE $GP(\bar{x}, \bar{y})$ *where* $GP(\bar{x}, \bar{y})$, *also denoted* body(Q), *is a graph pattern without blank nodes and* $\bar{x} \cup \bar{y}$ *is the set of its variables, among which* $\bar{x}$ *are the* result *variables, and the subset of variables in predicate position is disjoint from the subset of variables in subject or object position. A conjunctive query Q is alternatively written as a pair* $Q = \langle \bar{x}, GP \rangle$. *A* boolean query *is a query of the form* $Q = \langle \emptyset, GP \rangle$.

Conjunctive queries with variables in predicate position are allowed, if such variables do not appear in a subject or object position. This ensures that within a conjunctive query, all occurrences of a given variable are in the same connected component (see Definition 3).

Example 1. The conjunctive query SELECT ?p WHERE {?s ?p ?o. ?s a :VIP.} conforms to Definition 2. Intuitively, this query selects all properties of subjects who are "VIP".

Definition 3 (Connected components of a query). *Given a conjunctive query* $Q = \langle \bar{x}, GP \rangle$, *let* $G_Q = \langle N_Q, E_Q \rangle$ *be the undirected graph defined as follows: its nodes* N_Q *are the distinct variables and constants appearing in subject or object position in* GP, *and its edges* E_Q *are the pairs of nodes* (n_i, n_j) *such that there exists a triple* (n_i, p, n_j) *or* (n_j, p, n_i) *in* GP.

Each subgraph SG_Q *of* G_Q *corresponds to the subgraph of* body(Q) *made of the set of triples* (s, p, o) *such that either* (s, o) *or* (o, s) *is an edge of* SG_Q. *By slight abuse of notation, we will call the connected components of the query Q the (disjoint) subsets of* $GP =$ body(Q) *corresponding to the connected components* *of G_Q. A connected component GP_C of the query Q is called* boolean *when it contains no result variable.*

Example 2. Let Q be the following query in the SPARQL syntax where a is a shorthand for rdf:type, Q has two connected components GP_1 and GP_2:

```
SELECT ?x ?y WHERE { ?x  :seenBy ?z.      ?z :specialistOf   ?y.
                     ?v  a        :VIP.    ?v :isHospitalized true. }
GP1 = { ?x :seenBy ?z.       ?z :specialistOf   ?y. }
GP2 = { ?v a         :VIP.   ?v :isHospitalized true. }
```

Definition 4 (Critical terms). *A variable (resp. constant) in subject or object position having several occurrences within the body of a query is called a* join variable *(resp.* join constant*). We name join variables, join constants and result variables of a query as its* critical terms.

Example 3. The query SELECT ?p WHERE { ?s ?p ?o.   ?s a :VIP.} has two critical terms: ?s, which has two occurrences, and ?p which is a result variable. Critical terms are computed in Algorithm 1 (Lines 5 to 10).

The evaluation of a query $Q = \langle \bar{x}, GP \rangle$ over an RDF graph G consists in finding mappings μ assigning the variables in GP to terms such that the set of triples, denoted $\mu(GP)$, obtained by replacing with $\mu(z)$ each variable z appearing in GP, is included in G. The corresponding answer is defined as the tuple of terms $\mu(\bar{x})$ assigned by μ to the result variables.

Definition 5 (Evaluation of a conjunctive query). *Let $Q = \langle \bar{x}, GP \rangle$ be a conjunctive query and let G be an RDF graph. The* answer set *of Q over G is defined by :* $\text{Ans}(Q, G) = \{\mu(\bar{x}) \mid \mu(GP) \subseteq G\}$.

Definition 6 (Counting query). *Let Q be a conjunctive query. The query* $\text{Count}(Q)$ *is a* counting query, *whose answer over a graph G is defined by:* $\text{Ans}(\text{Count}(Q), G) = |\text{Ans}(Q, G)|$.

We now define an additional ingredient: *update queries*. Intuitively, an update query DELETE $D(\bar{x})$ INSERT $I(\bar{y})$ WHERE $W(\bar{z})$ isNotBlank($\bar{b}$) executed on a graph G searches for the instances of the graph pattern $W(\bar{z})$ in G, then deletes the instances of $D(\bar{x})$ and finally inserts the $I(\bar{y})$ part. The isNotBlank operator will be used in Algorithm 1 to avoid replacing the images of critical terms that are already blank nodes.

Definition 7 (Update query). *An* update query *(or update operation) Q_u is defined by DELETE $D(\bar{x})$ INSERT $I(\bar{y})$ WHERE $W(\bar{z})$ isNotBlank($\bar{b}$) where D (resp. W) is a graph pattern whose set of variables is $\bar{x}$ (resp. $\bar{z}$) such that $\bar{x} \subseteq \bar{z}$; and I is a graph pattern where blank nodes are allowed, whose set of variables is $\bar{y}$*

such that $\bar{y} \subseteq \bar{z}$. isNotBlank$(\bar{b})$ is a parameter where $\bar{b}$ is a set of variables such that $\bar{b} \subseteq \bar{z}$. The evaluation of Q_u over an RDF graph G is defined by:

$$\mathsf{Result}(Q_u, G) = G \setminus \{\mu(D(\bar{x})) | \mu(W(\bar{z})) \subseteq G \wedge \forall x \in \bar{b}, \mu(x) \notin \mathbf{B}\}$$
$$\cup \{\mu'(I(\bar{y})) | \mu(W(\bar{z})) \subseteq G \wedge \forall x \in \bar{b}, \mu(x) \notin \mathbf{B}\}$$

where μ' is an extension of μ renaming blank nodes from $I(\bar{y})$ to fresh blank nodes, i.e. a mapping such that $\mu'(x) = \mu(x)$ when $x \in \bar{z}$ and $\mu'(x) = b_{new} \in \mathbf{B}$ otherwise. The application of an update query Q_u on a graph G is written $Q_u(G) = \mathsf{Result}(Q_u, G)$. This notation is extended to a sequence of operations $O = \langle Q_u^1, \ldots Q_u^n \rangle$ by $O(G) = Q_u^n(\ldots (Q_u^1(G)) \ldots)$.

4   Safety Model

We generalize the definition of a *safe anonymization* introduced in [12] as follows: an RDF graph is safely anonymized if it does not disclose any *new* answer to a set of privacy queries when it is joined with any external RDF graph. Additionally, compared to [12], we define a notion of *data-independent* safety for a sequence of anonymization operations independently of *any* RDF graph. Given an RDF graph G, a sequence O of update queries called *anonymization operations* and a set $\mathcal{P}$ of conjunctive *privacy queries*, the safety of the *anonymization instance* $(G, O, \mathcal{P})$ is formally defined as follows.

Definition 8 (Safe anonymization instance). *An anonymization instance $(G, O, \mathcal{P})$ is safe iff for every RDF graph G', for every $P \in \mathcal{P}$ and for every tuple of constants $\bar{c}$, if $\bar{c} \in \mathsf{Ans}(P, O(G) \cup G')$ then $\bar{c} \in \mathsf{Ans}(P, G')$.*

Notice that the *safety* property is stronger than the *privacy* property defined in [6,12] which requires that for every privacy query P, $\mathsf{Ans}(P, O(G))$ does not contain any tuple made only of constants. In contrast with [12], the safety problem that we consider is data-independent and is a construction problem. Given a set of privacy queries, the goal is to build anonymization operations guaranteed to produce a safe anonymization when applied to any RDF graph, as follows.

Definition 9 (Safe sequence of anonymization operations). *Let O be a sequence of anonymization operations, let $\mathcal{P}$ be a set of privacy queries, O is safe for $\mathcal{P}$ iff $(G, O, \mathcal{P})$ is safe for every RDF graph G.*

Problem 1. The data-independent SAFETY problem.
    Input : $\mathcal{P}$ a set of privacy queries.
    Output: A sequence O of update operations such that O is safe for $\mathcal{P}$.

Our approach to solve Problem 1 is to favour *whenever possible* update operations that replace IRIs and literals by blank nodes over update operations that delete triples. We exploit the standard semantics of blank nodes that interprets them as existential variables in the scope of local graphs. As a consequence, two blank nodes appearing in two distinct RDF graphs cannot be equated.

The privacy-preserving approach described in [6] is also data-independent but is based on deleting operations that may lead to unsafe anonymizations, as shown in Example 4.

Example 4. Let consider the following privacy query P stating that IRIs of people seen by a specialist of a disease should not be disclosed.

```
SELECT ?x WHERE { ?x :seenBy ?y.   ?y :specialistOf ?z. }
```

Let the RDF graph to anonymize be G that is made of the following triples:

```
:bob :seenBy :mary.   :mary :specialistOf :cancer.   :mary :worksAt :hospital1.
:ann :seenBy :mary.   :jim :worksAt :hospital1.
```

Let O_1 be the update query deleting all the :seenBy triples, written as `DELETE { ?x :seenBy ?y. } WHERE { ?x :seenBy ?y. }` in SPARQL. The resulting anonymized RDF graph $O_1(G)$ is as follows:

```
:mary :specialistOf :cancer.   :mary :worksAt :hospital1.
:jim   :worksAt :hospital1.
```

O_1 preserves privacy (the evaluation of P against $O_1(G)$ returns no answer). However, O_1 is not safe since the union of $O_1(G)$ with an external RDF graph G' containing the triple (:bob, :seenBy, :mary) will provide :bob as an answer. This example shows that the problem for safety comes from a possible join between an internal and an external constant (:mary here). This can be avoided by replacing critical constants by blank nodes. Example 5 illustrates the strategy considered in Algorithm 1 to enforce safety (see Sect. 5).

Example 5. Consider the following update query O_2:

```
DELETE {?x   :seenBy ?y.   ?y   :specialistOf ?z.}
INSERT {_:b1 :seenBy _:b2. _:b2 :specialistOf ?z.}
WHERE  {?x   :seenBy ?y.   ?y   :specialistOf ?z.}
```

The result RDF graph $O_2(G)$ is made of the following triples and is *safe*:

```
_:b1 :seenBy _:b2.   _:b2 :specialistOf :cancer. :mary :worksAt :hospital1.
_:b3 :seenBy _:b4.   _:b4 :specialistOf :cancer. :jim   :worksAt :hospital1.
```

It is worth noticing that the result of the counting query $\mathsf{Count}(P)$ is preserved i.e. it returns the same value as when evaluated on the original RDF graph G. Many other utility queries are preserved, such as for instance the one asking for who works at which hospital.

5   Safe Anonymization of an RDF Graph

In this section, we provide an algorithm that computes a solution to the SAFETY problem. We first prove a *sufficient condition* (Theorem 1) guaranteeing that an anonymization instance is safe, then we define an algorithm based on this condition. We extend the definition of a mapping, which is now allowed to map constants to blank nodes: an *anonymization mapping* μ is a function $\mathbf{V} \cup \mathbf{I} \cup \mathbf{L} \to \mathbf{T}$. For a triple $\tau = (s, p, o)$ we write $\mu(\tau)$ for $(\mu(s), \mu(p), \mu(o))$. Theorem 1 is progressively built on two conditions that must be satisfied by all the connected components of the privacy queries.

Theorem 1. *An anonymization instance* $(G, O, \mathcal{P})$ *is safe if the following conditions hold for every connected component* GP_c *of all privacy queries* $P \in \mathcal{P}$:

(i) for every critical term x of GP_c, for every triple $\tau \in GP_c$ where x appears, for each anonymization mapping μ s.t. $\mu(\tau) \in O(G)$, $\mu(x) \in \mathbf{B}$ holds;

(ii) if GP_c does not contain any result variable, then there exists a triple pattern of GP_c without any image in $O(G)$ by an anonymization mapping.

The intuition of condition (i) is that if all the images of *critical terms* are blank nodes, it is impossible to graft external pieces of information to the anonymized graph as they cannot have common blank nodes. Condition (ii) deals with boolean connected components with no result variable. We are now able to design an anonymization algorithm that solves the SAFETY problem. Algorithm 1 computes a sequence[2] of operations O for a privacy policy $\mathcal{P}$ such that O is safe for $\mathcal{P}$. Operations are computed for each connected component of each privacy query from $\mathcal{P}$, the crux being to turn conditions (i) and (ii) into update queries.

Algorithm 1. Find update operations to ensure safety

Input   : a set $\mathcal{P}$ of privacy conjunctive queries $P_i = \langle \bar{x}_i, GP_i \rangle$
Output : a sequence O of operations which is safe for $\mathcal{P}$

```
1  function find-safe-ops(P):
2      Let O = ⟨ ⟩;
3      for Pᵢ ∈ P do
4          forall the connected components GPc ⊆ GPi do
5              Let I := [ ];
6              forall the (s, p, o) ∈ GPc do
7                  if s ∈ V ∨ s ∈ I then I[s] = I[s] + 1;
8                  if o ∈ V ∨ o ∈ I ∨ o ∈ L then I[o] = I[o] + 1;
9              Let x̄c := {v | v ∈ x̄i ∧ ∃τ ∈ GPc s.t. v ∈ τ};
10             Let Tcrit := {t | I[t] > 1} ∪ x̄c;
11             Let SGPc = {X | X ⊆ GPc ∧ X ≠ ∅ ∧ X is connected} ordered by
                  decreasing size;
12             forall the X ∈ SGPc do
13                 Let X' := X and x̄' = {t | t ∈ Tcrit ∧ ∃τ ∈ X s.t. t ∈ τ};
14                 forall the x ∈ x̄' do
15                     Let b ∈ B be a fresh blank node;
16                     X' := X'[x ← b];
17                 O := O + ⟨DELETE X INSERT X' WHERE X isNotBlank(x̄')⟩
18             if x̄c = ∅ then
19                 Let τ ∈ GPc // non-deterministic choice
20                 O := O + ⟨DELETE τ WHERE GPc⟩
21      return O;
```

---

[2] The $+$ operator denotes the concatenation of sequences.

The starting point of Algorithm 1 is to compute joins variable, constants and then critical terms (Lines 5 to 10). The update queries that replace the images of critical terms by blank nodes are built from Line 14 to Line 17. The subtle point is that as many update queries as the connected subsets of the component GP_c need to be constructed. Considering these subsets in decreasing order of cardinality (Line 11) and using the isNotBlank($\bar{x}'$) construct guarantees that all the images of a critical term in a given RDF graph will be replaced *only once*. Non-connected subsets of GP_c are skipped because their own connected components are handled afterwards. Finally, if the connected component under scrutiny is boolean, one of its triple is deleted (Line 20).

When applied to the privacy query considered in Example 4, operation O_2 reported in Example 5 is the first one generated at Line 17. When applied to Example 2, Algorithm 1 will sequentially generate anonymization operations starting from those replacing the images of all the variables by blank nodes (since all its variables are critical), followed by those deleting all the triples corresponding to one of the triple patterns (?v a :VIP or ?v :isHospitalized true) in the boolean connected component. Note that anonymizations may create an RDF graph where some properties have been replaced by blank nodes. In this case, the output is a *generalized* RDF graph. Theorem 2 states the soundness and computational complexity of Algorithm 1. Since Algorithm 1 is data-independent, its exponential worst-case complexity (due to the powerset SGP_c computed on Line 11) is not necessarily an important limitation in practice, as it will be demonstrated in Sect. 7.

Theorem 2. *Let $O = $ find-safe-ops$(\mathcal{P})$ be the sequence of anonymization operations returned by Algorithm 1 applied to the set $\mathcal{P}$ of privacy queries: O is safe for $\mathcal{P}$. The worst-case computational complexity of Algorithm 1 is exponential in in the size of $\mathcal{P}$.*

Algorithm 2 (reported in the companion appendix) is a polynomial approximation of Algorithm 1 obtained as follows: instead of considering all possible subsets of triple patterns of SG_c (Line 12), we simply construct update queries that replace, in each triple pattern $\tau \in GP_c$, every critical term with a fresh blank node. As a result, there does not exist anymore any equality between images of join variables, literals or IRIs (while in Algorithm 1 all occurrences of each critical term were replaced by the same blank node). For instance, Algorithm 2 generates a sequence of three update queries, one for each triple, more general than O_2 from Example 5 of Sect. 4. Theorem 3 states that Algorithm 2 is sound but leads to anonymizations that are more general than those produced by Algorithm 1.

Theorem 3. *The worst-case computational complexity of Algorithm 2 is polynomial in the size of $\mathcal{P}$. Let O and O' be the result of applying respectively Algorithm 1 and Algorithm 2 (with the same non deterministic choices) to a set $\mathcal{P}$ of privacy queries: for any RDF graph G, $(G, O, \mathcal{P})$ is safe and $G \models O(G)$ and $O(G) \models O'(G)$.*

Theorem 4 establishes that the anonymization operations computed by Algorithm 1 preserve some information on $\mathsf{Count}(P)$ for privacy queries P with no boolean connected component. It is not necessarily the case for Algorithm 2.

Theorem 4. *Let $O = \mathtt{find\text{-}safe\text{-}ops}(\{P\})$ be the output of Algorithm 1 applied to a privacy query P with no boolean connected component. For every RDF graph G, $O(G)$ satisfies the the condition $\mathsf{Ans}(\mathsf{Count}(P), O(G)) \geq \mathsf{Ans}(\mathsf{Count}(P), G)$.*

6   Safe Anonymization Robust to :sameAs links

One of the fundamental assets of the LOD is the possibility to assert that two resources are the same by stating `owl:sameAs` triples (shortened to `:sameAs` later), also known as *entity linking*. We do not consider `:sameAs` between properties and we interpret `:sameAs` triples (called `:sameAs` links) as equality between constants (including blank nodes) that are in subject or object position. With this interpretation, `:sameAs` links can also be inferred by a logical reasoning on additional knowledge known on some properties (e.g. that a property is functional). In this section, we study the impact of both explicit and inferred `:sameAs` links on safety.

We extend Definition 5 to the semantics of query answering in presence of a set sameAs of `:sameAs` links. Let closure(sameAs) be the transitive, reflexive and symmetric closure of sameAs. This set can be computed in polynomial time. We write $G[b_0 \leftarrow b_0', \ldots, b_k \leftarrow b_k']$ for denoting the graph obtained from G by replacing each occurrence of b_i by b_i' for every $i \in [1..k]$.

Definition 10 (Answer of a query modulo sameAs). *Let Q be a conjunctive query, G an RDF graph and sameAs a set of `:sameAs` links. A tuple $\bar{a}$ is an answer to Q over G modulo sameAs iff there exists $(b_0, \mathtt{:sameAs}, b_0')$, ..., $(b_k, \mathtt{:sameAs}, b_k')$ in closure(sameAs) s.t. $\bar{a} \in \mathsf{Ans}(Q, G[b_0 \leftarrow b_0', \ldots, b_k \leftarrow b_k'])$. We note $\mathsf{Ans}_{\mathsf{sameAs}}(Q, G)$ the answer set of Q over G modulo sameAs.*

Hence, we extend Definition 8 to handle a set sameAs of `:sameAs` links.

Definition 11 (Safety modulo sameAs). *An anonymization instance $(G, O, \mathcal{P})$ is safe modulo sameAs iff for every RDF graph G', for every $P \in \mathcal{P}$ and for any tuple of constants $\bar{c}$, if $\bar{c} \in \mathsf{Ans}_{\mathsf{sameAs}}(P, O(G) \cup G')$ then $\bar{c} \in \mathsf{Ans}_{\mathsf{sameAs}}(P, G')$.*

O is safe modulo sameAs for $\mathcal{P}$ if $(G, O, \mathcal{P})$ is safe modulo sameAs for every RDF graph G and for every set sameAs of `:sameAs` links.

We first study how to build anonymization operations that are robust to explicit `:sameAs` links. Then, we focus on handling the case of inferred `:sameAs` links through knowledge.

Theorem 5 establishes that Algorithm 1 (and thus Algorithm 2) computes safe anonymizations even in presence of a set sameAs of explicit `:sameAs` links.

Theorem 5. *Let O be the result of applying Algorithm 1 to a set $\mathcal{P}$ of privacy queries: for any set* sameAs *of explicit* :sameAs *links, O is safe modulo* sameAs *for $\mathcal{P}$.*

We address two cases in which knowledge on properties may infer equalities. The first case occurs in the ontology axiomatization of the OWL language [3] when some of the properties are functional or inverse functional, as in Definition 12, where we model equalities by :sameAs links.

Definition 12. *A property p is* functional *iff for every $?x$, $?y_1$, $?y_2$:*
$(?x, p, ?y_1) \wedge (?x, p, ?y_2) \Rightarrow (?y_1, :sameAs, ?y_2)$.
A property p is inverse functional *iff for every $?x$, $?y_1$, $?y_2$:*
$(?y_1, p, ?x) \wedge (?y_2, p, ?x) \Rightarrow (?y_1, :sameAs, ?y_2)$.

For example, declaring that property :bossOf as inverse functional expresses the constraint that every person has only one boss. As shown in Example 6, exploiting this knowledge may lead to re-identifying blank nodes that have been produced by the previous anonymization algorithms.

Example 6. Let P be the following privacy query written in SPARQL syntax:

```
SELECT ?x  WHERE { ?x :seenBy ?y. ?x :bossOf ?z. }
```

Let G, $O(G)$ and G' be the following RDF graphs where O is an update operation returned by Algorithm 1:

```
G     = {:bob :seenBy :mary. :bob :bossOf _:b1. _:b1 :bossOf :ann.}
O(G)  = {_:b :seenBy :mary. _:b :bossOf _:b1. _:b1 :bossOf :ann.}
G'    = {:bob :bossOf :jim. :jim :bossOf :ann.}
```

From $O(G) \cup G'$ and the inverse functionality of :bossOf, it can be inferred first (:jim, :sameAs, _:b1) and second (:bob, :sameAs, _:b). Consequently, _:b is re-identified as :bob, which is returned as answer of P over $O(G) \cup G'$ modulo sameAs, and the anonymization operation O is not safe.

One solution is to add a privacy query for each functional property p and for each inverse functional property q, respectively SELECT ?x WHERE {?x p ?y.} and SELECT ?x WHERE {?y q ?x.}. By doing so, the update queries returned by our algorithms will replace each constant in subject position of a functional property by a fresh blank node, and each constant in an object position of an inverse functional property by a fresh blank node. In the previous example, the constant :ann in (_:b1, :bossOf, :ann) would be replaced by a fresh blank node.

The second case that we consider may lead to infer equalities (modeled as :sameAs links) when a property is completely known, i.e., when its closure is available in an external RDF graph. For instance, suppose that the closure of the property :seenBy is stored in an extension of the external RDF graph G' containing the following triples:

---

[3] See OWL 2 RDF-Based Semantics, notably section 5.13. https://www.w3.org/TR/2012/REC-owl2-rdf-based-semantics-20121211#Semantic_Conditions.

```
:bob   :seenBy :mary.   :alice :seenBy :ann.
:john  :seenBy :ann.    :tim   :seenBy :ann.
```

Knowing that G' is the complete extension of the seenBy predicate allows to infer (_:b, :sameAs, :bob) and thus to re-identify the blank node _:b.

One solution is to add a privacy query SELECT ?x ?y WHERE { ?x p ?y } for each property p for which we suspect that a closure could occur in the LOD. Then, the update queries returned by our algorithms will replace each constant in the subject or object position of such a property by a fresh blank node. For instance, in the Example 6, the constant :mary in (_:b, :seenBy, :mary) would be replaced by a fresh blank node.

7   Experimental Evaluation

We have evaluated the runtime performance of the anonymization process produced by Algorithm 1 and the resulting loss of precision on three real RDF graphs for which we have designed a *reference privacy query* as a union of privacy conjunctive queries. Table 1 provides the indicators characterizing each RDF graph used in the experiments: *#Triples* (respectively *#IRIs* and *#Blanks*) denotes the number of triples (respectively unique IRIs and unique blank nodes) in the graph, and *#PrivQuery* denotes the size of the reference privacy query (i.e., the sum of the triple patterns in each conjunctive privacy query). The reference privacy queries are reported in the companion appendix. The source code of our prototype is openly available on GitHub[4].

Table 1. RDF graphs and privacy queries used in our experiments.

RDF graph	#Triples	#IRIs	#Blanks	#PrivQuery
TCL	6,443,256	1,020,580	705,030	19
	Synthetic transportation data			
Drugbank[a]	517,023	109,494	0	6
	Real-world data about approved drugs			
(Swedish) Heritage[b]	4,970,464	1,687,452	0	6
	Real world Europeana Swedish heritage data			

[a]http://wifo5-03.informatik.uni-mannheim.de/drugbank/
[b]General Europeana portal: https://pro.europeana.eu/page/linked-open-data

7.1   Runtime Performances

The average runtime of Algorithm 1 has been measured over 10 executions for each graph, using the reference privacy queries as input. Table 2 reports the results: *T-Algo*1 denotes the time in seconds for computing the sequence of

---

[4] https://github.com/RdNetwork/safe-lod-anonymizer.

anonymization operations by Algorithm 1 (which depends only of the reference privacy query) whereas *T-Anonym* denotes the time in seconds required for applying them on the different RDF graphs to compute their anonymized version, and *#UpdateQueries* is the number of update queries returned by Algorithm 1.

The anonymization time is reasonable in all cases. The cost for anonymizing the Swedish Heritage graph is due to a few update queries with graph patterns having many occurrences in the graph.

Table 2. Running time (in seconds) of anonymization process.

RDF graph	T-Algo 1	#UpdateQueries	T-Anonym
TCL	0.207	16	3.5
Drugbank	0.012	6	1.7
Swedish Heritage	0.013	14	53.6

7.2 Evaluation of the Precision Loss

We have evaluated the *precision loss* and how it depends on the *privacy query specificity* defined relatively to the reference privacy query. We distinguish the *absolute precision loss* which is the *the number of blank nodes* introduced by the anonymization process, from the *relative precision loss* which the ratio of it with the total number of IRIs in the input graph.

From each reference privacy query P, we create more specific privacy queries by applying a set of random *mutations* that replace a variable by a constant in one of the privacy conjunctive queries. The specificity of a privacy query P' obtained by this mutation process is defined by $\mathsf{specif}(P') = |\mathsf{Ans}(P',G)| / |\mathsf{Ans}(P,G)|$.

By construction, $\mathsf{specif}(P')$ is a normalized value between 0 and 1, as any mutated privacy query P' is more specific than the reference privacy query P.

Results displayed on Figs. 1a to c show that the precision loss grows linearly with the policy specificity: the less precise the privacy policy is in its selection of data, the more blank nodes will be inserted in the graph.

We can also observe that precision is very dependent on the input: if the privacy policy only cover a specific part of the whole data (e.g. only the subscriptions in the data of whole transportation network) then its impact is quite small: for the TCL graph (Fig. 1a), this precision value only drops marginally (99.9% to 99.4%). The trend is similar for other graphs: precision drops when the privacy policy gets more general. It drops to 85% in the case of the Swedish Heritage graph, and 96% for the Drugbank graph. This confirms that in general, using plausible privacy policy semantics, the number of IRIs lost in the anonymization process is not huge.

However, Fig. 1c for the Swedish Heritage graph have a quite large spread on the $x = 0$ line. Indeed, the privacy policy forbids the disclosure of very general

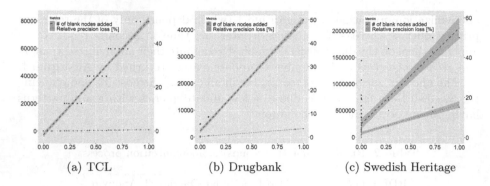

(a) TCL   (b) Drugbank   (c) Swedish Heritage

Fig. 1. Loss of precision depending on privacy query specificity for each graph.

pieces of information such as the description of objects in the graph. Thus, this leads to many replacements by blank nodes in such a situation.

8   Conclusion

We have tackled the safety problem for Linked Open Data by providing a data-independent version and grounding it in a set of privacy queries (expressed in SPARQL). Our algorithms let seamlessly construct sequences of anonymization operations in order to prevent privacy leakages due to external RDF graphs along with explicit or inferred knowledge (under the form of sameAs links). We have proved the soundness of our anonymization algorithms and shown their runtime complexity. We have conducted experiments showing the quality of our anonymization and the performance of its operations.

Our approach can be seamlessly combined with existing privacy-preserving approaches. Once the RDF graph is transformed according to the operations generated by our algorithms, one could apply any other method to the obtained RDF graph. In particular, it could be verified whether the resulting anonymized RDF graph verifies some desired k-anonymity property. The adaptation of k-anonymity approaches for a more fine-tuned generalization of literals, see for instance [22], is planned as future work. Our approach can be also combined with ontology-based query rewriting for first-order rewritable ontological languages such as RDFS [3], DL-Lite [4] or EL fragments [14], by providing as input to Algorithm 1 the rewritings of the privacy queries.

We envision several other directions of future work. The first is to study the potential risk for re-identification of delegating the generation of fresh blank nodes to a standard SPARQL engine. Next, since the conditions provided in Theorem 1 for guaranteeing that a anonymization instance is safe are sufficient but not necessary, it would be beneficial to explore both sufficient and necessary conditions. We also plan to extend our safety model to handle additional knowledge, like for instance that some properties are equivalent. Finally, we plan to study whether considering the data-dependent version of the safety problem could lead to more specific anonymization operations while guaranteeing safety.

Acknowledgements. This work has been supported by the Auvergne-Rhône-Alpes region through the ARC6 research program funding Remy Delanaux's PhD; by the LabEx PERSYVAL-Lab (ANR-11-LABX-0025-01); by the SIDES 3.0 project (ANR-16-DUNE-0002) funded by the French Programme Investissement d'Avenir (PIA); and by the Palse Impulsion 2016/31 program (ANR-11-IDEX-0007-02) at UDL.

References

1. Abiteboul, S., Hull, R., Vianu, V.: Foundations of Databases. Addison-Wesley, Boston (1995)
2. Bonifati, A., Martens, W., Timm, T.: An analytical study of large SPARQL query logs. PVLDB **11**(2), 149–161 (2017)
3. Buron, M., Goasdoué, F., Manolescu, I., Mugnier, M.L.: Reformulation-based query answering for RDF graphs with RDF ontologies. In: ESWC (2019, to appear)
4. Calvanese, D., De Giacomo, G., Lembo, D., Lenzerini, M., Rosati, R.: Tractable reasoning and efficient query answering in description logics: the DL-Lite family. J. Autom. Reasoning **39**(3), 385–429 (2007)
5. Cheng, J., Fu, A.W., Liu, J.: K-isomorphism: privacy preserving network publication against structural attacks. In: SIGMOD, pp. 459–470. ACM (2010)
6. Delanaux, R., Bonifati, A., Rousset, M.C., Thion, R.: Query-based linked data anonymization. In: Vrandečić, D., et al. (eds.) ISWC 2018. LNCS, vol. 11136, pp. 530–546. Springer, Cham (2018). https://doi.org/10.1007/978-3-030-00671-6_31
7. Deutsch, A., Papakonstantinou, Y.: Privacy in database publishing. In: ICDT, pp. 230–245 (2005)
8. Dwork, C.: Differential privacy. In: Bugliesi, M., Preneel, B., Sassone, V., Wegener, I. (eds.) ICALP 2006. LNCS, vol. 4052, pp. 1–12. Springer, Heidelberg (2006). https://doi.org/10.1007/11787006_1
9. Fung, B.C.M., Wang, K., Chen, R., Yu, P.S.: Privacy-preserving data publishing: a survey of recent developments. ACM Comput. Surv. **42**(4), 14:1–14:53 (2010)
10. Goasdoué, F., Manolescu, I., Roatis, A.: Efficient query answering against dynamic RDF databases. In: EDBT, pp. 299–310 (2013)
11. Grau, B.C., Horrocks, I.: Privacy-preserving query answering in logic-based information systems. In: ECAI, pp. 40–44 (2008)
12. Grau, B.C., Kostylev, E.V.: Logical foundations of privacy-preserving publishing of linked data. In: AAAI, pp. 943–949. AAAI Press (2016)
13. Gutiérrez, C., Hurtado, C.A., Mendelzon, A.O.: Foundations of semantic Web databases. In: PODS, pp. 95–106. ACM (2004)
14. Hansen, P., Lutz, C., Seylan, I., Wolter, F.: Efficient query rewriting in the description logic EL and beyond. In: IJCAI, pp. 3034–3040. AAAI Press (2015)
15. Heitmann, B., Hermsen, F., Decker, S.: k-RDF-neighbourhood anonymity: combining structural and attribute-based anonymisation for linked data. In: PrivOn@ISWC, vol. 1951 (2017). CEUR-WS.org
16. Li, N., Li, T., Venkatasubramanian, S.: t-Closeness: privacy beyond k-anonymity and l-diversity. In: ICDE, pp. 106–115. IEEE Computer Society (2007)
17. Machanavajjhala, A., He, X., Hay, M.: Differential privacy in the wild: a tutorial on current practices & open challenges. PVLDB **9**(13), 1611–1614 (2016)

506   R. Delanaux et al.

18. Machanavajjhala, A., Kifer, D., Gehrke, J., Venkitasubramaniam, M.: L-diversity: privacy beyond k-anonymity. TKDD **1**(1), 3 (2007)
19. Malyshev, S., Krötzsch, M., González, L., Gonsior, J., Bielefeldt, A.: Getting the most out of Wikidata: semantic technology usage in Wikipedia's knowledge graph. In: ISWC, pp. 376–394 (2018)
20. Miklau, G., Suciu, D.: A formal analysis of information disclosure in data exchange. J. Comput. Syst. Sci. **73**(3), 507–534 (2007)
21. Nobari, S., Karras, P., Pang, H., Bressan, S.: L-opacity: linkage-aware graph anonymization. In: EDBT, pp. 583–594 (2014). OpenProceedings.org
22. Radulovic, F., García-Castro, R., Gómez-Pérez, A.: Towards the anonymisation of RDF data. In: SEKE, pp. 646–651. KSI Research Inc. (2015)
23. Sweeney, L.: k-anonymity: a model for protecting privacy. Int. J. Uncertainty Fuzziness Knowl. Based Syst. **10**(5), 557–570 (2002)
24. Vatsalan, D., Sehili, Z., Christen, P., Rahm, E.: Privacy-preserving record linkage for big data: current approaches and research challenges. In: Zomaya, A.Y., Sakr, S. (eds.) Handbook of Big Data Technologies, pp. 851–895. Springer, Cham (2017). https://doi.org/10.1007/978-3-319-49340-4_25

WebEL: Improving Entity Linking with Extra Web Contexts

Yiting Wang[1], Zhixu Li[1,2(✉)], Qiang Yang[4], Zhigang Chen[3], An Liu[1], Guanfeng Liu[5], and Lei Zhao[1]

[1] School of Computer Science and Technology, Soochow University, Suzhou, China
ytwang123@stu.suda.edu.cn, {zhixuli,anliu,zhaol}@suda.edu.cn
[2] iFLYTEK Research, Suzhou, China
[3] State Key Laboratory of Cognitive Intelligence, iFLYTEK,
Hefei, People's Republic of China
zgchen@iflytek.com
[4] King Abdullah University of Science and Technology, Jeddah, Saudi Arabia
qiang.yang@kaust.edu.sa
[5] Macquarie University, Sydney, Australia
guanfeng.liu@mq.edu.au

Abstract. Entity Linking is the task of determining the identity of textual entity mentions given a predefined Knowledge Graph (KG). Plenty of existing efforts have been made on this task using either "local" information (contextual information of the mention in the text), or "global" information (relations among candidate entities). However, either local or global information might be insufficient especially when the given text is short. To get richer local and global information for entity linking, we propose to enrich the context information for mentions by getting extra contexts from the web through Web Search Engines. Based on the intuition above, two novel attempts are made. The first one adds web-searched results into an embedding-based method to expand the mention's local information, where an attention mechanism is applied to help generate high-quality web contexts, while the second one uses the web contexts to extend the global information, i.e., finding and utilizing more extra relevant mentions from the web contexts with a graph-based model. Finally, we could combine the two models we proposed to use both extended local and global information from the extra web contexts. Our empirical study based on six real-world datasets shows that using extra web contexts to extend the local and global information could effectively improve the performance of entity linking.

Keywords: Entity Linking · WSE · Attention mechanism

1 Introduction

Entity Linking (EL), also known as Named Entity Disambiguation, aims at assigning ambiguous textual mentions of entities to their referent entities in a

© Springer Nature Switzerland AG 2019
R. Cheng et al. (Eds.): WISE 2019, LNCS 11881, pp. 507–522, 2019.
https://doi.org/10.1007/978-3-030-34223-4_32

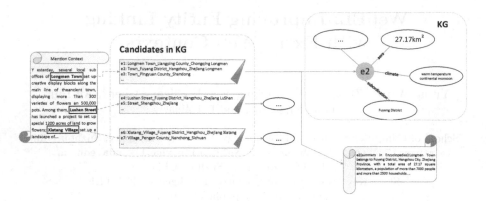

Fig. 1. An Entity Linking example

predefined Knowledge Graph (KG) [35]. EL has been studied extensively [2,3, 5,30], due to its importance as a fundamental component in various tasks such as question answering [36], knowledge base expansion [19,27], and information extraction [17]. An example of entity linking is illustrated in Fig. 1, where the mention *Longmen Town* in the mention context should be linked to the entity e_2 in KG.

In order to link the ambiguous mentions to correct entities, early approaches work on modeling *textual* context by measuring the "local" similarities between the contextual words of a mention and the encyclopedic descriptions of every of its candidate entities [4,28]. Developed from the *textual*-based approaches, the *embedding*-based methods find ways to map both mentions and candidate entities into the same continuous vector space, and then calculate the similarity between every (mention, candidate entity) pair to decide the linking entity for each mention [9,18,26]. Initially, only word embedding based on the input text is performed for mentions, while entity embedding based on the KG is performed for candidate entities [9]. More recent works [18,26] tend to use more sophisticated embedding techniques to learn the characteristics of the mentions and the candidate entities better. For example, Huu et al. propose a joint learning approach to simultaneously model the local and global features for entity linking [18]. In addition, Luo et al. apply word embedding and Bi-LSTM (Bi-directional Long Short-Term Memory) model to capture distributed representations [26]. However, the context information of each mention in a text may not be rich enough to fully denote the semantic meaning of the mention, which will greatly decrease the quality of the linking results.

Another line of works prefer to utilize the "global" *coherence* among all the mentions in the input text, where all the mentions in the text could be simultaneously disambiguated [8,16,31,34]. Typically, a global mention-entity graph could be constructed, based on which various probabilistic models [1] or graph models such as PageRank [37] could be adopted to find the most probable linking entities for each entity mention. However, as the number of mentions and candidate entities increases, the calculation quantity increases exponentially, and the matrix generated from the graph becomes more and more sparse. Also, if the text is short, the number of mentions will be small, which results in the difficulty in building effective graphs.

To get richer local and global information for entity linking, especially for short texts, this paper proposes a so-called **WebEL** approach to improve entity linking with extra web contexts. More specifically, we enrich the context information for mentions by getting extra contexts from the web through Web Search Engines. To achieve this, we search the context or the mentions in the same context together in the Web Search Engines. After getting the WSE results, we preprocess them to filter those noisy words, and then use a self-attention mechanism to help select the most relevant words in the context as the web-based extra contexts for the given mentions.

Based on the intuitions above, two novel attempts are made. The first one adds web-searched results into an embedding-based method to expand the mention's local information, where an attention mechanism is applied to help generate high-quality web contexts. More specifically, we use the context selected from Web Search Engine and mention's context, with attention mechanisms to do a selection to generate a specified number of candidate entities. After that, we use the context from WSE to supplement the target mention's context in training our embedding-based EL model if the context is short. The second one considers to use the web-searched results to extend the global information with a graph-based model. Firstly, we find more relevant mentions in the selected context from WSE. Then we obtain candidate entities from the knowledge graph. After that, we use the triple information in the knowledge graph to build our graph-based EL model. Finally, we combine the two models we propose above to use both extended local and global information from the extra web contexts for EL.

We summarize our contributions as follows:

- We novelly propose to enrich the context information for mentions in doing EL by getting extra contexts from the web through Web Search Engines.
- We propose to add web-searched results into an embedding-based method to expand the mention's local information, where an attention mechanism is applied to help generate high-quality web-based contexts.
- We also propose to use the web-searched results to extend the global information with a graph-based model, such that we could combine the two models we put forward to use both extended local and global information from the extra web contexts for EL.

Our empirical results on six real-world datasets show that WebEL could effectively improve the performance of state-of-the-art EL methods.

Roadmap. The rest of the paper is organized as follows: We cover the related work in Sect. 2, and then we define the problem in Sect. 3. After presenting our model in Sect. 4, we report our experiments in Sect. 5. We finally conclude in Sect. 6.

2   Related Work

Entity Linking (EL) is an important task which is a necessary step in many scenarios, such as Information Extraction, Information Retrieval, Content Analysis, Question Answering and so on. So far, plenty of work has been done on EL which can be roughly divided into two categories: (1) using local information, i.e., contextual information of the mention in the text, with embedding-based models, or (2) using global information, i.e., relations among candidate entities, with graph-based models.

2.1   Graph-Based Models

As soon as graph model is put forward, it is applied to various tasks, such as path selection in social networks [24,25]. Of course, this model is also widely used in EL, and has achieved good results. Han et al. first put forward the referent graph model to infer entity linking relationship [15]. Specifically, they represent all mentions and candidate entities as nodes and compute weighed edges between each other. A transfer matrix is obtained by using Referent Graph and the random walk algorithm [15], which is used to get the linking results. Afterwards, many researchers use similar graph models to solve entity linking problems [1,37]. Alhelbawy et al. give each node an initial confidence score and use Page-Rank to rank nodes. Zhang et al. use greedy search and an adjusted Monte Carlo random walk to improve both accuracy and efficiency of the graph model. Besides, Ganea et al. propose a probabilistic graph model to resolve EL problems which does not require extensive feature engineering, nor an expensive training procedure [11].

2.2   Embedding-Based Models

In Natural Language Processing (NLP), word embedding is one kind of widely used technology including One-Hot representation of word bag (BOW) and n-gram [23,33]. Han et al. apply BOW to construct the vectors of the contexts containing mentions and entities respectively and then use these vectors to compute the weighted edges among them [15]. However, the relationship between the relative positions of words is ignored and the embedding tends to be very long and sparse. In addition, n-gram and co-occurrence matrix are considered to structure embedding [13,22,29]. Unfortunately, these methods are confined to the time costs due to the high dimension. So as to overcome above difficulties, Word2Vec model is put forward which involves two algorithms, one is CBOW and the other is Skip-Gram. The advantage of this model is that it can measure the semantic similarity of words and can be analogized according to the semantics of words [20]. Yamada et al. use Word2Vec to construct embedding and calculate the semantic similarity between texts containing mention and the candidate entity through embedding [35].

　More recent works tend to use more sophisticated embedding techniques to learn the characteristics of the mentions and the candidate entities better.

Deep learning technologies are widely used in these methods. Francis-Landau et al. use convolutional neural networks to capture topic information between the mention's context and the target entity [10]. They use rich parameterizations n-gram to capture different topics as many as possible which can be computational complexity and hard to adjust parameters. In addition, Sun et al. present a neural network approach which takes consideration of the semantic representations of mention, context and entity. Specifically, the model encodes them into continuous vector space and effectively leverages them for entity disambiguation. The problem of their method is not suitable for long context [32]. Luo et al. apply Bi-LSTM (Bi-directional Long Short-Term Memory) model to mine semantic information between mentions and targeted entities [26]. Their method does not consider the implication relations between mentions in the same context. In view of the weakness of above methods, multiple depth learning models attract more and more attention to resolve EL problem. Huu et al. not only use CNN to get the distribution of mentions and target candidates, but also use RNN to provide global constraints of mentions [18]. Le et al. mainly uses different latent relations between different mentions in the same context to link entities, which also exists the problem of insufficient context [21]. In this paper, we mainly compare with the method that Le et al. propose.

3   Problem Definition

Definition 1. *Given a knowledge graph kg with a set of entities $E = \{e_1, e_2, ..., e_m\}$ having corresponding entity contexts in $S = \{s_1, s_2, ..., s_m\}$ and a set of relations among entities in $R = \{r_1, r_2, ..., r_n\}$, and a mention context collection $C = \{c_1, c_2, ..., c_k\}$ in which a set of named mentions $M = \{m_1, m_2, ..., m_p\}$ are identified in advance, the goal of entity linking is to map each named mention $m_i \in M$ in text collection to its corresponding entity $e_j \in E$ in the knowledge graph kg using C and (or) S and (or) R.*

4   Our Approaches

We propose to enrich both the local and global information for mentions by getting extra contexts from the web through Web Search Engines (WSE). In the following of this section, we first introduce how we generate extra contexts from the web-searched results, and add them into an embedding-based method to expand the mention's local information in Sect. 4.1. After that, we present how to use the web-searched results to extend the global information with a graph-based model in Sect. 4.2.

4.1   WebEL: Embedding-Based Approach

In the following of this subsection, we first present how we generate high-quality web contexts for mentions with the help of WSE, and then introduce how we do EL with web-extended local information.

4.1.1   Generating High-Quality Web Contexts

The basic workflow of getting selected WSE context is depicted in Fig. 2. After fetching the raw web-searched results with WSE, we consider to use self-attention mechanism to reduce noise for the contexts getting from WSE. We give details as follows.

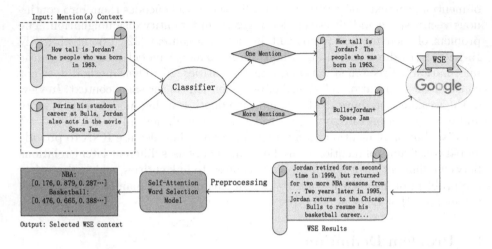

Fig. 2. Generating high-quality web contexts with WSE

(1) Searching Mentions with WSE. We use Web Search Engine to get extra contexts. We divide the data into two situations. One case is that there is only one mention in a text and the other is that there are more than one mention in a text. For the case one, we search for whole text in search engines to get extra contexts and for the other one, we join mentions in the one text together and put those mentions in search engine. For each data, after getting the extra contexts, we divide the contexts into some words. After removing the stop words, we use self-attention to select WSE context containing 50 words called WSE_{words} preparing for the next steps. Here 50 is decided on window length.

(2) Self-Attention Mechanism. Due to WSE results coming from different documents, and after word segmentation, it is difficult to capture the semantic relevance between words from WSE. So we use Self-Attention mechanism to mine interactive information hidden between the words. Here we introduce the *Self-Attention Word Selection Model* in detail. For each mention m, first, we get embedding of the words from WSE results $V_{words} = \{v_1, v_2, v_3, \cdots, v_n\}$. Next, we feed them into the self-attention layer to better capture the semantic relevance between words from WSE:

$$V_{self} = softmax(\frac{(V_{words}W^Q(V_{words}W^K)^{\mathrm{T}})}{\sqrt{d}})(V_{words}W^V) \qquad (1)$$

where $W^Q, W^K, W^V \in \mathbb{R}^{d \times d}$ are the projection matrices, which can make the model more flexible. V_{self} contains the new vectors of the words from WSE after adjustment. At last, in order to get the similarity between the mention and words, we use the inner product to get the final score.

$$score_{i,j} = \tilde{v}_i \cdot m_j \qquad (2)$$

where $\tilde{v}_i \in V_{self}$, and m_j is the embedding of the mention m_j.

We select top 50 words and their new embeddings to make up selected WSE context through the final score and put them to the next step. Here 50 is decided on the length of window.

4.1.2 EL with Extra Web Contexts

After getting high-quality extended contexts from the web for mentions, we would like to use two attention mechanisms and prior probability to presort the candidate entities and get the designated quantity of candidates with the highest scores. Finally, we put the candidates to the training model to get the final result. We can see the whole process from Fig. 3.

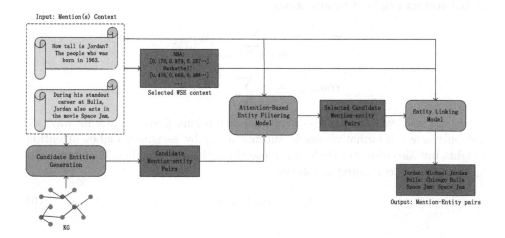

Fig. 3. WebEL: Embedding-based model

(1) Candidate Selection. For each mention m_i, we use a prior probability to select the *top-r* candidate entities first and then retain *top-k* candidates with the highest prior probability. For the remaining $r - k$ candidate entities, we will do further screening. We get embeddings of the words from the selected WSE context obtained in the previous step and the mention context. Also, we get the embeddings of candidate entities. We add an attention mechanism on the words from WSE and words from mention context respectively. After that, we can get two new vectors to represent the mention. For each mention-candidate pair, we use inner product to get two scores. And we sum up the two scores through weighted method. Finally, we select another *top-m* candidates with the highest scores.

Here, we focus on the method of obtaining *top-m* candidates. For the mention m_i, we get the embeddings of the selected WSE context $V_{new} = \{v_{w_1}, v_{w_2}, v_{w_3}, \cdots, v_{w_k}, \cdots, v_{w_{50}}\}$ obtained from last step. For each candidate e_j, we get its embedding v_{e_j}, so, the embeddings of all the candidates of the mention m_i can be represented as $V_e = \{v_{e_1}, v_{e_2}, \cdots, v_{e_j}, \cdots, v_{e_n}\}$. Here $n = r - k$. In addition, we set a *window* = 25 to get fixed number of words on the left and right of the mention m_i in the mention context, which is called con_{words}. We get the embeddings of these words $V_{con} = \{v_{c_1}, v_{c_2}, v_{c_3}, \cdots, v_{c_k}, \cdots, v_{c_{50}}\}$ as the same. If the mention context is too short, we use "NIL" to make up for it.

We measure the compatibility of e_j with m_i by computing their similarities based on $sim(v_{e_j}, w_{m_i,e_j})$ and $sim(v_{e_j}, con_{m_i,e_j})$.

$$Score_{fin} = \beta sim(v_{e_j}, w_{m_i,e_j}) + (1 - \beta)sim(v_{e_j}, con_{m_i,e_j}) \qquad (3)$$

where β is the parameter to balance the weights of $sim(v_{e_j}, w_{m_i,e_j})$ and $sim(v_{e_j}, con_{m_i,e_j})$. w_{m_i,e_j} is the WSE embedding while con_{m_i,e_j} is the context embedding of m_i conditioned on candidate e_j and are defined as the average sum of word global vectors weighted by attentions.

$$w_{m_i,e_j} = \sum_{w_k \in WSE_{words}} \gamma_{kj} v_{w_k} \qquad (4)$$

$$con_{m_i,e_j} = \sum_{c_k \in con_{words}} \theta_{kj} v_{c_k} \qquad (5)$$

where γ_{kj} and θ_{kj} are the $k - th$ word's attentions from e_j. In this way, we not only select informative words automatically by assigning higher attention weights but also filter out irrelevant noise through small weights. The attentions γ_{kj} and θ_{kj} are computed as follows.

$$\gamma_{kj} \propto sim(v_{w_k}, v_{e_j}) \qquad (6)$$

$$\theta_{kj} \propto sim(v_{c_k}, v_{e_j}) \qquad (7)$$

where $sim(.,.)$ is the similarity measurement and we use inner product here.

Note that for datasets in Chinese, we should first make use of the particularity of the geographic data set to select candidate entities by matching. And then use the same method with attention mechanisms as English datasets to get the *top k+m* candidates.

(2) Model Training. After the above three steps, for each mention m_i, we obtained $k + m$ candidates and an extra context from WSE results, except for words in mention's own context. Next, we train our Entity Linking model.

We assume that there are K latent relations [21]. Each relation k is assigned to a mention pair (m_i, m_j) with a non-negative weight μ_{ijk}. The pairwise score (m_i, m_j) is computed as a weighted sum of relation-specific pairwise scores.

$$\Phi(e_i, e_j, D) = \sum_{k}^{K} \mu_{ijk} \Phi_k(e_i, e_j, D) \tag{8}$$

$\Phi_k(e_i, e_j, D)$ is a pairwise score function.

$$\Phi_k(e_i, e_j, D) = v_{e_i}^{\mathrm{T}} R_k v_{e_j} \tag{9}$$

$R_k \in \mathbb{R}^{d \times d}$ is a diagonal matrix. The weights μ_{ijk} are normalized scores:

$$\mu_{ijk} = \frac{1}{W_{ijk}} exp \left\{ \frac{f^{\mathrm{T}}(m_i, con_i) D_k f(m_j, con_j)}{\sqrt{d}} \right\} \tag{10}$$

where W_{ijk} is a normalization factor, $f(m_i, con_i)$ is a function mapping (m_i, con_i) onto $\mathbb{R}^d$ and $D_k \in \mathbb{R}^{d \times d}$ is also a diagonal matrix. Here, we use a single-layer neural network as f where con_i is a concatenation of the average embedding of words in the left context with the average embedding of words in the right context of the mention m_i. We set a *window* $= 25$ here and if the left or right context is not enough, we add the extra words from selected WSE context in order to make complete.

After that, we use the local score function identical to Ganea et al. (2017) [12] and the pairwise scores defined as explained. In addition, we use max-product loopy belief propagation (LBP) to estimate the max-marginal probability. Finally, for each mention m_i, the final score function is given by:

$$\rho_i(e) = g(\hat{q}_i(e|D), \hat{p}(e|m_i)) \tag{11}$$

where g is a two-layer neural network, $\hat{q}_i(e|D)$ is the max-marginal probability and $\hat{p}(e|m_i)$ is the probability of selecting e conditioned only on m_i. Finally, following Le et al. (2018), we use the same ranking loss as them [21].

$$L(\varphi) = \sum_{D \in \mathcal{D}} \sum_{m_i \in D} \sum_{e \in C_{m_i}} h(m_i, e) \tag{12}$$

$$h(m_i, e) = max(0, \delta - \rho_i(e_i^*) + \rho_i(e)) \tag{13}$$

where φ and δ are the model parameters. $\mathcal{D}$ is a training dataset, D is the data in the dataset, C_{m_i} is a set of the candidates of m_i and e_i^* is the ground-truth entity.

4.2   WebEL: Graph-Based Approach

Although the *WebEL: embedding-based* approach could achieve good results for mentions with insufficient text information, it does not considers to use the relationship between candidate entities of mentions in the same contexts. In

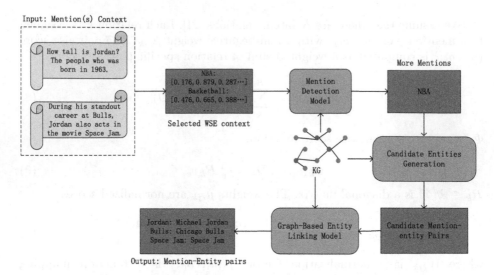

Fig. 4. WebEL: graph-based model

this subsection, we propose a *WebEL: graph-based* approach with a graph-based model. Finally we could combine the EL results of embedding-based model and global-based model to generate the final EL results.

Specifically, we select the relevant named entities as the new mentions from the selected WSE context to build graph. These named entities should satisfy a condition that their selected candidate entities must have a triple relationship with at least one of the original mentions' candidate entities. The model is shown in Fig. 4.

We refer to the graph-based model [15] which makes use of the graph structure to mine the semantic information of the mentions in the context and the relationship between the candidate entities. This method models the local compatibility as a compatible relation between the name mention and the candidate entities, and the strength of the compatible relation is calculated based on the Bag of Words model. This is equivalent to give a weight to each edge between mention and candidate entities in the graph model called $P(m_i \rightarrow e_m)$. In addition, this method also calculates the scores between the candidate entities. The calculation formula is as follows.

$$ER(m,n) = 1 - \frac{log(max(|M|,|N|)) - log(|M \cap N|)}{log(|R|) - log(min(|M|,|N|))} \tag{14}$$

where m and n are the two entities from the same piece of data. And M and N are the sets of all entities that link to m and n in knowledge graph respectively. R is the total number of links in the entire knowledge graph. The weight $Q(e_m \rightarrow e_n)$ is computed as the proportion of $ER(m,n)$ in the sum of the scores between the candidates in the same piece of data. In addition, each mention in the same data has an initial value which is computed through TF-IDF recorded as $R(m_i)$. Through the computation of $P(m_i \rightarrow e_m)$, $Q(e_m \rightarrow e_n)$ and $R(m_i)$, we can give each node in the graph a score. So we can represent the graph as a matrix. Finally, through matrix calculation and weight transferring, each candidate entity gets a graph model score S_{gra}. We combine the score S_{em} obtained by *WebEL: embedding-based* mentioned above with S_{gra} in the way of weighted summation to get a final score S_{fin}.

$$S_{fin} = \lambda S_{em} + (1 - \lambda) S_{gra} \tag{15}$$

where λ is the parameter to balance the weights of *WebEL: embedding-based* approach score and *WebEL: graph-based* approach score. We rank the candidate entities according to S_{fin}, and the highest score is considered to be the linking entity.

5   Experiments

We now report our experimental study in this section on the six real-world datasets.

5.1   Dataset and Parameter Setting

The parameters mentioned in Sect. 4 except for β and λ follow the work of Le et al. (2018) [21]. Here $r = 30$, $k = 3$, $m = 4$ and $\delta = 0.01$. The parameter β varies with different datasets and we set λ as 0.6.

– **Chinese dataset.** We establish a Chinese data set called Geographic dataset. This database contains a lot of geographic information which is crawled from Baike[1] and several major geographic websites, like xzqh[2]. There are total 8985 mentions most of which are some places with the different types of synonyms from 2378 mention contexts and we altogether collect 38436 geographically related entities. Noticeably, the dataset annotates the data to its own knowledge base. The local knowledge base utilized in this dataset is part of Baidu Baike and merely takes advantage of the information in InfoBoxes.
– **English datasets.** we use AIDA-CoNLL dataset. This dataset contains aida-train, aida-A and aida-B, having 946, 216, and 231 documents respectively. We use aida-train for training and test the models on aida-B and four popular test sets: MSNBC, AQUAINT, ACE2004 and WNEDCWEB(CWEB). The first three are small sets, only having 20, 50, and 36 documents whereas CWEB is much larger with 320 documents. Specially, we consider only mentions that have entities in the Wikipedia.

We basically use $F1$ score to evaluate the effectiveness of the methods. $F1$ **Score:** a combination of precision and recall, which is calculated by $F1 = \frac{2 * precision * recall}{precision + recall}$. In this formula, Precision: the percentage of correctly linked instance pairs among all linked instance pairs, Recall: the percentage of correctly linked instance pairs among all instance pairs that should be linked.

---

[1] https://baike.baidu.com/.
[2] http://www.xzqh.org/html/.

5.2   Comparison with Previous Methods

In this section, we compare the effectiveness of our two Entity Linking algorithm, i.e., *WebEL: embedding-based* and *WebEL: combined* methods, with several previous methods.

- The *Hoffart et al. (2011)* method makes use of context from knowledge bases to build a coherence graph for collective disambiguation [16].
- The *Ratinov et al. (2011)* method utilizes the information from Wikipedia link structure to arrive at coherent sets of disambiguations for a given document [31].
- The *Cheng and Roth (2013)* method provides an Integer Linear Programming formulation of Wikification to rich relational analysis of the text which incorporates the entity-relation inference problem [6].
- The *Chisholm and Hachey (2015)* method gets entity information from pages which have links to Wikipedia [7].
- The *Globerson et al. (2016)* method explores an attention-based approach to collective entity resolution, motivated by the evidence for each candidate is based on a small set of strong relations, rather than relations to all other entities in the document [14].
- The *Yamada et al. (2016)* method proposes an embedding method jointly maps words and entities into the same continuous vector space [35].
- The *Ganea and Hofmann (2017)* method puts forward a neural attention mechanism over local context windows, and a differentiable joint inference stage for disambiguation [12].
- The *Le et al. (2018)* method makes use of different latent relationships between mentions in the context to solve the problem of Entity Linking [21].
- The *WebEL: embedding-based* method makes use of web search engines to get extra contexts for mention and then uses attention mechanisms to process all the contexts mention have to extend and improve the embedding-based model.
- The *WebEL: combined* method gets more named entities from selected WSE context to extend and improve the graph model and combines with *WebEL: embedding-based* method.

5.3   Experimental Results

Here we talk about results of different methods on different datasets. We test *WebEL: embedding-based* method on all the English datasets and the Chinese geographic dataset. However, to build a graph model, we need to consider the triple relationship of candidate entities, and we do not have the relevant data information of the English datasets. So for *WebEL: combined* method, we only do the research on the Chinese data.

As is shown in Table 1, we can see F1 scores on aida-B of the previous methods and ours, which all use Wikipedia mention-entity index. This table shows that our method outperforms any other previous methods. After adding Web

Table 1. F1 scores on aida-B set.

Methods	aida-B
Chisholm and Hachey (2015)	88.7
Globerson et al. (2016)	91.0
Yamada et al. (2016)	91.5
Ganea and Hofmann (2017)	92.2
Le et al. (2018)	93.0
WebEL: embedding-based	93.2

Table 2. F1 scores on four out-domain test sets.

Methods	MSNBC	AQUAINT	ACE2014	CWEB	Avg
Hoffart et al. (2011)	79	56	80	58.6	68.4
Ratinov et al. (2011)	75	83	82	56.2	74.1
Cheng and Roth (2013)	90	**90**	86	67.5	83.4
Ganea and Hofmann (2017)	93.7	88.5	88.5	77.9	87.2
Le et al. (2018)	93.9	88.4	89.3	77.6	87.3
WebEL: embedding-based	**94.4**	89.2	**90.1**	**78.1**	**88.0**

Search Engine results, our method achieves 0.2 % higher than that of Le et al. (2018) [21], which proves the validity of introducing context through WSE. In addition, as you can see in Table 2, F1 scores that perform best on different datasets are represented bold. Our approach outperforms on MSNBC, ACE2014, CWEB. On all the four datasets, our *WebEL: embedding-based* method achieves higher F1 score than the Le et al. (2018) [21]. Also, from the average F1 value, our method is at least 0.7 % points higher than the previous methods.

For Chinese dataset, on the basis of the *WebEL: embedding-based* method, we propose a *WebEL: combined* method. You can see in Table 3, after adding more named entities and their candidate entities to the graph model and combining with the *WebEL: embedding-based* method, the *WebEL: combined* method achieves higher F1 score, increased by 0.9% points.

Experiments show that our proposed methods in this paper can achieve better results on different real datasets than some state-of-art methods. Furthermore,

Table 3. F1 scores on geographic dataset.

Methods	Geographic dataset
Le et al. (2018)	83.7
WebEL: embedding-based	84.2
WebEL: combined	85.1

compared with the model of Le et al. (2018) [21] and our model, we can see expanding mention context through web search engines can make the result of entity linking better. In addition, combined entity linking method can combine the advantages of different methods more effectively.

6   Conclusions

In this paper, we present two models WebEL: embedding-based model and WebEL: graph-based model, which enrich either local or global information with extra web contexts for EL. Experiments on six real-world datasets show that our model can get higher F1 score than the state-of-the-art methods. In our future work, we will continue to research on how to get rid of noise data and get more effective contexts from WSE so that we can further improve EL results. Furthermore, the relationship between candidate entities is also an aspect that can be better used in the future.

Acknowledgments. This research is partially supported by National Natural Science Foundation of China (Grant No. 61632016, 61402313, 61472263), and Natural Science Research Project of Jiangsu Higher Education Institution (No. 17KJA520003), and this is a project funded by the Priority Academic Program Development of Jiangsu Higher Education Institutions.

References

1. Alhelbawy, A., Gaizauskas, R.: Graph ranking for collective named entity disambiguation. In: Meeting of the Association for Computational Linguistics, pp. 75–80 (2014)
2. Basile, P., Caputo, A.: Entity linking for tweets. In: Meeting of the Association for Computational Linguistics, pp. 1304–1311 (2017)
3. Blanco, R., Ottaviano, G., Meij, E.: Fast and space-efficient entity linking in queries. In: Eighth ACM International Conference on Web Search and Data Mining, pp. 179–188 (2015)
4. Bunescu, R.C., Pasca, M.: Using encyclopedic knowledge for named entity disambiguation. In: Conference of the European Chapter of the Association for Computational Linguistics, pp. 9–16 (2006)
5. Cai, R., Wang, H., Zhang, J.: Learning entity representation for named entity disambiguation. In: Meeting of the Association for Computational Linguistics, pp. 30–34 (2013)
6. Cheng, X., Roth, D.: Relational inference for wikification. In: Proceedings of the 2013 Conference on Empirical Methods in Natural Language Processing, pp. 1787–1796 (2013)
7. Chisholm, A., Hachey, B.: Entity disambiguation with web links. Trans. Assoc. Comput. Linguist. **3**, 145–156 (2015)
8. Cucerzan, S.: Large-scale named entity disambiguation based on Wikipedia data. In: EMNLP-CoNLL 2007, Proceedings of the 2007 Joint Conference on Empirical Methods in Natural Language Processing and Computational Natural Language Learning, Prague, Czech Republic, 28–30 June 2007, pp. 708–716 (2007)

9. Fang, W., Zhang, J., Wang, D., Chen, Z., Li, M.: Entity disambiguation by knowledge and text jointly embedding. In: SIGNLL Conference on Computational Natural Language Learning., pp. 260–269 (2016)

10. Francislandau, M., Durrett, G., Klein, D.: Capturing semantic similarity for entity linking with convolutional neural networks. In: North American Chapter of the Association for Computational Linguistics, pp. 1256–1261 (2016)

11. Ganea, O.E., Ganea, M., Lucchi, A., Eickhoff, C., Hofmann, T.: Probabilistic bag-of-hyperlinks model for entity linking. In: International World Wide Web Conferences, pp. 927–938 (2015)

12. Ganea, O.E., Hofmann, T.: Deep joint entity disambiguation with local neural attention. arXiv preprint arXiv:1704.04920 (2017)

13. Gang, Z., Zong-Min, M.A., Kan, H.M., Niu, L.Q.: Texture feature extraction approach using co-occurrence matrix. J. Shenyang Univ. Technol. $32(2)$, 192–195+211 (2010)

14. Globerson, A., Lazic, N., Chakrabarti, S., Subramanya, A., Ringaard, M., Pereira, F.: Collective entity resolution with multi-focal attention. In: Proceedings of the 54th Annual Meeting of the Association for Computational Linguistics (Volume 1: Long Papers), vol. 1, pp. 621–631 (2016)

15. Han, X., Sun, L., Zhao, J.: Collective entity linking in web text: a graph-based method. In: International ACM SIGIR Conference on Research and Development in Information Retrieval, pp. 765–774 (2011)

16. Hoffart, J., et al.: Robust disambiguation of named entities in text. In: Proceedings of the Conference on Empirical Methods in Natural Language Processing, pp. 782–792. Association for Computational Linguistics (2011)

17. Hoffmann, R., Zhang, C., Ling, X., Zettlemoyer, L.S., Weld, D.S.: Knowledge-based weak supervision for information extraction of overlapping relations. In: Meeting of the Association for Computational Linguistics: Human Language Technologies, pp. 541–550 (2011)

18. Huu, T., Fauceglia, N., Muro, M.R., Hassanzadeh, O., Gliozzo, A.M., Sadoghi, M.: Joint learning of local and global features for entity linking via neural networks. In: The International Conference on Computational Linguistics (2016)

19. Ji, H., Grishman, R., Dang, H.T., Griffitt, K., Ellis, J.: Overview of the TAC 2010 knowledge base population track. In: Text Analysis Conference (2009)

20. Landgraf, A.J., Bellay, J.: word2vec skip-gram with negative sampling is a weighted logistic PCA. Computation and Language (2017)

21. Le, P., Titov, I.: Improving entity linking by modeling latent relations between mentions. arXiv preprint arXiv:1804.10637 (2018)

22. Lei, K., Deng, Y., Zhang, B., Shen, Y.: Open domain question answering with character-level deep learning models. In: International Symposium on Computational Intelligence and Design, pp. 30–33 (2018)

23. Liu, F.H., Gu, L., Gao, Y., Picheny, M.: Use of statistical n-gram models in natural language generation for machine translation. In: IEEE International Conference on Acoustics, Speech, and Signal Processing, 2003, Proceedings, vol. 1, pp. I-636-I-639 (2003)

24. Liu, G., Wang, Y., Orgun, M.A.: Finding k optimal social trust paths for the selection of trustworthy service providers in complex social networks. IEEE Trans. Serv. Comput. $6(2)$, 152–167 (2013)

25. Liu, G., Yan, W., Orgun, M.A.: Optimal social trust path selection in complex social networks. In: Twenty-Fourth AAAI Conference on Artificial Intelligence (2010)

26. Luo, A., Gao, S., Xu, Y.: Deep semantic match model for entity linking using knowledge graph and text. Procedia Comput. Sci. **129**, 110–114 (2018)
27. Mcnamee, P., Dang, H.T.: Overview of the TAC 2009 knowledge base population track. In: Text Analysis Conference (2009)
28. Mihalcea, R., Csomai, A.: Wikify! linking documents to encyclopedic knowledge. In: Sixteenth ACM Conference on Conference on Information and Knowledge Management, pp. 233–242 (2007)
29. Pauls, A., Dan, K.: Faster and smaller n-gram language models. In: The Meeting of the Association for Computational Linguistics: Human Language Technologies, Proceedings of the Conference, Portland, Oregon, USA, 19–24 June 2011, pp. 258–267 (2012)
30. Phan, M.C., Sun, A., Yi, T., Han, J., Li, C.: NeuPL: attention-based semantic matching and pair-linking for entity disambiguation. In: CIKM (2017)
31. Ratinov, L.A., Dan, R., Downey, D., Anderson, M.: Local and global algorithms for disambiguation to wikipedia. In: The Meeting of the Association for Computational Linguistics: Human Language Technologies, Proceedings of the Conference, Portland, Oregon, USA, 19–24 June 2011, pp. 1375–1384 (2011)
32. Sun, Y., Lin, L., Tang, D., Yang, N., Ji, Z., Wang, X.: Modeling mention, context and entity with neural networks for entity disambiguation. In: International Conference on Artificial Intelligence, pp. 1333–1339 (2015)
33. Wallach, H.M.: Topic modeling: beyond bag-of-words. In: International Conference on Machine Learning, pp. 977–984 (2006)
34. Witten, I.H., Milne, D.N.: An effective, low-cost measure of semantic relatedness obtained from Wikipedia links. In Proceedings of the First AAAI Workshop on Wikipedia and Artificial Intelligence (2008)
35. Yamada, I., Shindo, H., Takeda, H., Takefuji, Y.: Joint learning of the embedding of words and entities for named entity disambiguation. In: Conference on Computational Natural Language Learning, pp. 250–259 (2016)
36. Yih, W.T., Chang, M.W., He, X., Gao, J.: Semantic parsing via staged query graph generation: question answering with knowledge base. In: Meeting of the Association for Computational Linguistics and the International Joint Conference on Natural Language Processing, pp. 1321–1331 (2015)
37. Liu, M., Chen, L., Liu, B., Zheng, G., Zhang, X.: DBpedia-based entity linking via greedy search and adjusted Monte Carlo random walk. ACM Trans. Inf. Syst. **36**(2), 16 (2017)

Entity Disambiguation Based on Parse Tree Neighbours on Graph Attention Network

Kexuan Xin, Wen Hua$^{(\boxtimes)}$, Yu Liu, and Xiaofang Zhou

School of Information Technology and Electrical Engineering,
The University of Queensland, Brisbane, Australia
ke.xin@uqconnect.edu.au, {w.hua,yu.liu}@uq.edu.au, zxf@itee.uq.edu.au

Abstract. Entity disambiguation (ED) aims to link textual mentions in a document to the correct named entities in a knowledge base (KB). Although global ED model usually outperforms local model by collectively linking mentions based on the topical coherence assumption, it may still incur incorrect entity assignment when a document contains multiple topics. Therefore, we propose to extract global features locally, i.e., among a limited number of neighbouring mentions, to combine the respective superiority of both models. In particular, we derive mention neighbours according to the syntactic distance on a dependency parse tree, and propose a tree connection method CoSimTC to measure the cross-tree distance between mentions. Besides, we extend the Graph Attention Network (GAT) to integrate both local and global features to produce a discriminative representation for each candidate entity. Our experimental results on five widely-adopted public datasets demonstrate better performance compared with state-of-the-art approaches.

Keywords: Entity linking · Dependency parse tree · Cross-sentence distance · Graph Attention Network

1 Introduction

Entity disambiguation (ED), which is also known as entity linking (EL), is one of the fundamental preprocessing tasks in Natural Language Processing (NLP), which can benefit various applications such as information retrieval, question answering, machine translation, etc. ED aims to resolve the semantic ambiguity of a mention and link it to the correct entry in a given knowledge base (KB), as illustrated in Fig. 1.

Generally, ED approaches can be classified into two categories: local model and global model. Local model [4,22,32] resolves mention ambiguity by utilising features such as surface form and local context, while global model (also called collective entity linking) [8,10,13,16] achieves better performance by finding the best alignment of all the mentions in a document to maximize topical coherence. However, collective linking may end up with incorrect entity assignment

© Springer Nature Switzerland AG 2019
R. Cheng et al. (Eds.): WISE 2019, LNCS 11881, pp. 523–537, 2019.
https://doi.org/10.1007/978-3-030-34223-4_33

Fig. 1. An illustration of entity linking. The correct candidate entity for each mention is labelled by red dash arrow. (Color figure online)

due to the extremely strong assumption of topical coherence at document level. Consider the example in Fig. 1. Global model prefers to link both "England" to a rugby union team which is obviously incorrect for the former one. Moreover, it usually incurs high computational complexity when the document contains a large number of mentions. Therefore, the main trend of current ED approaches [2,9,19,29] is to combine both local and global features, or in other words, consider coherence locally. In fact, [2] claimed that topical coherence only need to hold among neighbouring mentions.

1.1   Challenges and Contributions

Determining mention adjacency is a non-trivial task. Traditional sequence-based approach [2,26] measures the distance between mentions by simply counting the number of words in between, which results in semantically inconsistent neighbours sometimes. In this work, we resort to *dependency parse tree* to incorporate the syntactic structure of a sentence for more accurate distance estimation. It is commonly believed that the closer two mentions are on a parse tree, the higher linguistic relatedness they have. However, there are still two issues we need to address carefully. First, it is highly possible for multiple mentions to have the same tree-distance to a target mention, since each word can connect to several other words on a parse tree. This calls for a neighbour selection mechanism to deal with such situation. Considering that sequence-distance evaluates word closeness from a different perspective with tree-distance and it causes less distance conflict among words, we propose to combine both distance measures when determining mention neighbours. Furthermore, parse tree is constructed for a single sentence, which makes the cross-sentence tree-distance unmeasurable directly. Therefore, we introduce an algorithm *CoSimTC* to connect the parse trees of adjacent sentences, which derives a whole tree structure for each document, and then extract mention neighbours based on the document-level parse tree.

Another problem that needs to be considered is how to combine local features and global coherence together. Graph-based methods have been successfully applied in this area, such as the Graph Convolutional network (GCN) [18] utilised in [2]. However, GCN is highly dependent on the graph structure, which usually leads to low generation ability. We believe that Graph Attention Network (GAT) [31] is an ideal alternative in our case to integrate neighbour coherence into local features. Specifically, based on the mention neighbours extracted from the document parse tree, we utilise *distance decay attention* to encode the mention distance information into deep neural network. In this way, the graph attention layer can successfully extract the discriminative features and explore relatedness between entities to infer the best entity linking result.

The main contributions of our ED method can be summarised as below:

- We derive meaningful local neighbours for each mention in a more linguistic way by utilising dependency parse tree. We propose a tree connection method CoSimTC to generate a whole tree structure for each document, which can measure cross-sentence distance between mentions. Our distance metric further combines both sequence-distance and tree-distance to reflect mention closeness from different perspectives. The proposed tree-based neighbours can help each mention to make more accurate linking decision.
- Our neural ED approach combines basic deep neural network model with Graph Attention Network (GAT), which integrates the discriminative features for each candidate entity on both local and global aspects. The distance decay attention produces better representation for each candidate entity.
- We evaluate our method on five widely-adopted public datasets and compare with existing state-of-the-art ED models. The experimental results verify the superiority of our proposal.

2   Related Work

Feature Extraction. is a key step in ED systems. Early local ED models mainly applied string features (e.g., edit distance [2,21], Dice coefficient score [20], Hamming distance [5], etc.) to perform lexical match between candidate entity and context words around the mention. These features are effective, fundamental and still commonly used today. Based on the intuition that all the mentions in the same document should be related to the main topic of the document [8], many recent ED systems began to measure topical coherence within documents, which is known as global approach or collective linking. For example, [13] used dense sub-graph estimation algorithms to identify the only one mention-entity edge, and other models disambiguate mentions collectively via PageRank or Random walk [1,10,13,14]. However, these global methods usually have a high computational complexity because they consider relatedness among all mentions within a document. [2,17,26] claimed that topical coherence only need to be maintained within the neighbor mentions to alleviate computational cost. Similarly, [28] proposed that each linking only needs to be coherent with another linking within the document and introduced a Pair-Linking algorithm.

Neural network approach is becoming increasingly popular these years owing to the development and advancement of deep learning. [15] utilized Stacked Denoising Auto-encoders to learn document representation, and [6] adopted CNN to model context information at multiple granularities. [12] applied

CNN and LSTM-encoders from multiple sources of information, without any domain-specific data and hand-engineered features in their ED systems. However, the above models do not consider the difference of importance among context words and candidate entities. Therefore, attention mechanism was introduced and successfully applied in some ED systems. For example, [28] used entity profile as the attention vector to produce local context embedding. [25] proposed co-attention to capture the most discriminative components from mention contexts and entity descriptions. Moreover, [24] performed effective usage of context via attention-based GRU encoding along with some sparse features of context words.

Past effort on entity disambiguation provides a variety of valuable methods to perform ED. Based on these approaches, we propose a new neural network model, which makes use of local context and builds a mention-entity graph based on dependency parse tree to measure the syntactic structure distance and capture more accurate semantic information. Besides, we apply Graph Attention Network to capture entity-entity coherence and thus improve ED performance.

3   Entity Disambiguation Model

A typical ED framework consists of three main modules: candidate generation, feature extraction, and neural network model. For each mention m_i, a set of candidate entities denoted as $\phi(m_i)$ are selected based on prior probability $p(e_j|m_i)$. Discriminative features for each $e_j \in \phi(m_i)$ are extracted and fed into the neural network model to learn the final probability of linking m_i to e_j. We elaborate on our approaches to feature extraction and neural network model in the following.

3.1   Feature Extraction

Local Features
Local features reflect the compatibility between candidate entity and local information (e.g., the mention, surrounding context words). Inspired by [2,24,30], we include below local features for each $e_j \in \phi(m_i)$:

- String features based on surface form and entity title, including edit distance and Boolean features which show if they are identical, if one is the prefix, suffix of the other and vice versa.
- Entity-context features reflecting the lexical matching between entity title and context, i.e., if the title overlaps the context words in specific locations.
- Attention-based context information measuring the compatibility between e_j and context, which is computed as $cos(e_j, c_{m_i,e_j})$. cos is cosine similarity and $c_{m_i,e_j} = \sum_{\omega_k \in C(m_i)} cos(\omega_k, e_j)$ where ω_k reflects the long-reliability of e_j from mention context $(C(m_i))$.

Global Features
Unlike traditional collective ED models, our idea is based on the intuition that only mentions close to current mention, which we call **mention neighbours**

hereafter, need to be considered for topical coherence. Our global features are extracted based on mention neighbours. We define mention neighbors of m_i as mentions that have top-k minimum distance to m_i, denoted as $\mathcal{N}(m_i) = \{m_{i1}, m_{i1}, ..., m_{ik}\}$.

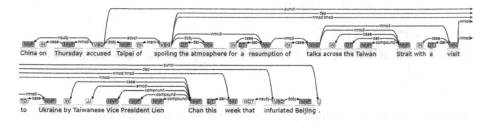

Fig. 2. The dependency parse tree of a sample sentence

Normally, we calculate the number of words between mentions as their distance, which is called *sequence distance*. However, the sequence distance cannot accurately reflect the closeness between mentions sometimes. For example, considering mention *Lien Chan* in Fig. 2, *Beijing* has a smaller sequence distance than *Taiwan Strait*, but is obviously less related to *Lien Chan*. Hence, we resort to dependency parse tree for a better distance measure that can reflect the syntactic closeness between words.

Tree Distance and Tree Connection. Dependency parse tree is a tree-based structure that reflects the syntactic of a sentence. We define the **tree distance** between two nodes in a sentence as their shortest path distance, namely the minimum number of hops to traverse from one node to the other. For example, the tree distance between "like" and "Jack" is 2. Note that parse tree is constructed on top of the collection of words rather than mentions. Hence, to measure the distance between mentions consisting of multiple words (e.g., *University of Sydney*), we choose the LCA (Lowest Common Ancestor) as the representative node of the mention (e.g., *University*). Figure 2 is the parse tree structure produced by Stanford CoreNLP Tool. Here, the tree distance between *Lien Chan* and *Taiwan Strait* is 3, closer than the tree distance between *Lien Chan* and *Beijing*, which is 6. Obviously, tree distance is usually more meaningful in measuring the semantic relatedness among mentions than sequence distance.

However, since parse tree is constructed for each single sentence, we cannot directly measure the tree distance for mentions in different sentences. An effective **tree connection** method is indispensable in order to address such issue. In this work, we introduce five possible solutions to connect parse trees of consecutive sentences, as explained below.

1. *Root-based*: Each parse tree has a single root node to represent the syntactic centre of the corresponding sentence. Therefore, a straightforward strategy

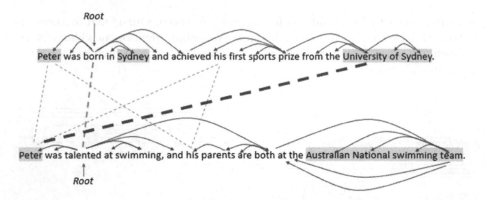

Fig. 3. An illustration of multiple parse trees for two sentences where highlighted words are mentions and blue arcs are dependency edges. Green lines represent cross-sentence co-reference connection, orange dash lines represent root connection, and purple dash lines represent the adjacent cross-sentence mention connection. (Color figure online)

for tree connection is to add an edge between the root nodes of adjacent dependency parse trees, as illustrated in Fig. 3 with orange lines.

2. *Mention-based*: Inspired by the traditional sequence distance used to measure mention adjacency, another heuristic solution for tree connection is based on the sequentially adjacent mentions of consecutive sentences. In particular, as shown in Fig. 3, the last mention of the former sentence is linked to the first mention of the latter sentence.

3. *Coreference-based*: Co-reference is one of the most popular measures of semantic correlation between textual expressions. That is, expressions linked through co-reference relationship are claimed to represent the same real-world entity in a given context. For example, in Fig. 3, "Peter" and "his" in both sentences refer to the same person. Therefore, we can also connect trees by adding edges between co-referenced expressions, as depicted in Fig. 3 with green lines. Since our purpose is to connect adjacent trees, we only add edges to co-reference in adjacent sentences, while ignoring the intra-sentence co-reference. In this way, semantically related mentions can be pushed closer even if they lie in different sentences.

4. *Similarity-based*: Co-reference relation does not always exist in adjacent sentences, and hence another cross-sentence relatedness measure is needed sometimes. Given all the possible mention pairs between two sentences, we estimate their semantic similarities and add an edge between mention pair which has the highest similarity. In this work, we infer mention similarity based on both surface form and context words. Hence, we define the similarity score function as:

$$Sim(m_i, m_j) = \alpha * cos(m_i, m_j) + \frac{1-\alpha}{2} * ctx\_sim(m_i, m_j) \qquad (1)$$

where $ctx\_sim(m_i, m_j)$ is related to mentions' left context embeddings $(l_c tx(m))$ and right context embeddings $(l_c tx(m))$. The embeddings of mention and context are the average of the word vectors contained in surface form and context words respectively. The context similarity is:

$$ctx\_sim(m_i, m_j) = cos(l\_ctx(m_i), l\_ctx(m_j)) + cos(r\_ctx(m_i), r\_ctx(m_j)) \quad (2)$$

Algorithm 1. CoSimTC

Input: a set of parse trees $T = (t_1, t_2, \ldots, t_n)$, and their corresponding
       sentences $S = (s_1, s_2, \ldots, s_n)$
Output: the whole Document Tree DT with all trees connected
for *each adjacent sentence pair* (s_i, s_{i+1}) **do**
    | $coref_{s_i, s_{i+1}} \leftarrow$ cross_coref(s_i, s_{i+1}) ;
    | **if** $coref_{s_i, s_{i+1}} \neq \emptyset$ **then**
    | | DT.add_edge$(coref_{s_i, s_{i+1}})$
    | **else**
    | | $sim_{s_i, s_{i+1}} \leftarrow cross\_sim(s_i, s_{i+1})$;
    | | **if** $sim_{s_i, s_{i+1}} \neq \emptyset$ **then**
    | | | DT.add_edge$(sim_{s_i, s_{i+1}})$
    | | **else**
    | | | $adj_{s_i, s_{i+1}} =$ cross_adj(s_i, s_{i+1}) ;
    | | | DT.add_edge$(adj_{s_i, s_{i+1}})$
    | | **end**
    | **end**
end

---

5. *CoSimTC*: Our final algorithm to connect adjacent parse trees, which is CoSimTC (Coreference-Similarity Tree Connection), is based on a combination of the above coreference-based and similarity-based approaches. We first link together all the co-references between adjacent sentences. For sentences that do not have any co-reference, we connect the cross-sentence mentions that have pair-wise highest similarity score computed as Eq. 1. Finally, for sentences with no mentions, we connect their adjacent words to link these sentences. The details of CoSimTC algorithm is shown in Algorithm 1. A whole document tree for a sequence of continuous sentences can be produced at the end of the algorithm.

After connecting all sentences of a document into a whole tree DT, we calculate the tree distance between each pair of mentions based on DT, and select mentions with the top-k minimum distances as neighbors of the current mention.

Neighbour-Based Global Features. Inspired by [2], given mention neighbors $\mathcal{N}(m_i)$ for each mention m_i, we extract two types of global features for each candidate $e^i_j \in \phi(m_i)$ based on the assumption that any neighbouring mention $m \in \mathcal{N}(m_i)$ is topically coherent with m_i.

Intuitively, the correct candidate entity of a mention should be compatible with all the neighbor mentions, and the higher ranking a neighbor has, the more important it is for disambiguating the target mention. Therefore, we introduce a **distance-decay neighbour mention similarity** between the candidate entity and the surface form of each mention neighbour:

$$\{cos(e_j^i, m_j) * p^r | m_j \in \mathcal{N}(m_i)\} \tag{3}$$

where p is the distance decay factor and $r \in [0, k-1]$ is the ranking of the corresponding neighbor. For each candidate entity, we concatenate its local features described in Sect. 3.1, entity embedding and the distance-decay neighbor mention similarity to form the feature vector $F \in \mathbb{R}^{d_o}$, which is the initial representation of the candidate entity node.

Finally, we integrate global information for an entity by constructing its subgraph based on mention neighbors. We define the subgraph as $G = (E, R)$, where $E = \{e_1, e_2, ..., e_n\}$ is the set of entities which serves as the vertices, and $R = \{r_j^i | r_j^i = Rel(e_i, e_j)\}$ which serves as the collection of edges in the graph reflecting the relatedness between connected entities. Since we assume that m_i is coherent to its mention neighbors, their corresponding entities should also be coherent with each other. Therefore, the nodes connected to $e_j^i \in \phi(m_i)$ are the candidates of each $m \in \mathcal{N}(m_i)$, which can be defined as:

$$\mathcal{N}(e_j^i) = \{e_k^l | e_k^l \in \phi(m_k), m_k \in \mathcal{N}(m_i)\} \tag{4}$$

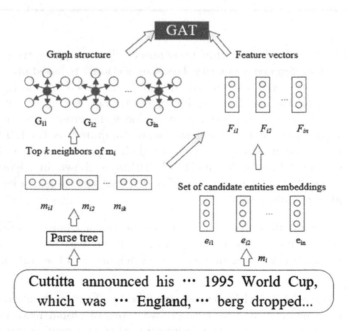

Fig. 4. Framework overview.

3.2 Neural Network Model

The overall ED framework is illustrated in Fig. 4. We extract mention neighbors for each mention using the parse tree, based on which the entity subgraph is constructed as global features. We also extract a series of local discriminative features. We then implement and extend the Graph Attention Network (GAT) [31] to explore the coherence between entities, which integrates both local information and global structure information for entity disambiguation. The key difference with the original GAT structure in [31] is that our input graph is the subgraph structure built in Sect. 3.1, instead of the whole graph containing all mentions within the document.

The objective of the neural network model is to find the best linking candidate for each mention, which satisfies that:

$$\Gamma(m) = \arg\max_{e_i \in \phi(m)} p(e_i|m) \tag{5}$$

where $p(e_i|m)$ is the probability that m refers to candidate e_i, and $p(e_i|m) \propto exp(score(m, e_i))$. $score(m, e_i)$ is the score function which is learned by our neural network model. We introduce the detailed computation process below.

Encoding features: In our model, each entity serves as a node and the relations between two entity serve as edges. For any $e_i \in \phi(m)$, we have feature vector $F_i \in \mathbb{R}^{d_o}$, which is the initial representation of the entity. Firstly, we will encode the feature vector using a perception (MLP): $h_i^1 = g(F_i)$, where h_i^1 is the hidden state of entity e_i, g is a two-layer MLP.

Entity relation scores: Then, we use graph attention layer to compute the relation score between e_i and each of $e_j \in \mathcal{N}(e_i)$. Different from conventional GAT, we use multiplicative attention to compute the relation between two entities:

$$Rel(e_i, e_j) = p^r\left(\frac{\mathbf{e}_i^T \mathbf{R} \mathbf{e}_j}{|\mathbf{e}_i||\mathbf{e}_j|}\right) \tag{6}$$

where $\mathbf{e}_i, \mathbf{e}_j \in \mathbb{R}^{d_e}$ is the embedding of the two entities, $R \in R^{d_e \times d_e}$ are diagonal matrix that represent the relation between the two entity node. Here, we also add weight decay p^r which is the same value as in Sect. 3.1. This relation score contains the relatedness between e_i and e_j in KB. It also reflects the closeness in text, because it weaken the weight for candidates that has long distance between their corresponding mentions. After calculating $Rel(e_i, e_j)$ for each $e_j \in \mathcal{N}(e_i)$, we normalize them to guarantee that all the $Rel(e_i, e_j)$ of neighbor nodes sum to one. Then we can derive the new representation of e_i as:

$$h_i' = \sigma\left(\sum_{e_j \in N(e_i)} Rel(e_i, e_j) W h_j\right) \tag{7}$$

Decoding: After above computations, the hidden state of e_i now contains the features of itself, integrating with the features of all $e_j \in \mathcal{N}(e_i)$. Finally, we map the hidden state h_i^t to the confidence score of choosing e_i:

$$score = W^t h_i^t + b^t \tag{8}$$

Training Objective: The training objective of our neural network model is to minimize the cross-entropy loss. The objective can be described as:

$$\mathcal{L}_m = -\sum_{j=1}^{n} P(e_g|m)log(P(e_j|m)) \tag{9}$$

where e_g is the ground truth candidate entity. Suppose that we have D mentions per document and there are totally $\mathcal{D}$ documents in our training dataset. The overall objective function is the loss of all mentions within the training corpus:

$$\mathcal{L} = \sum_{D \in \mathcal{D}} \sum_{m \in D} \mathcal{L}_m \tag{10}$$

4 Experiments

In this experiment, we evaluated 4 types of distance measures used to derive mention neighbors: seq-dist, root-based, mention-based, which are introduced in Sect. 3.1 and our proposed distance measure CoSimTC.

4.1 Experimental Setting

Baselines and Datasets. We compare our experimental results with seven SOTA entity linking systems: *Milne and Witten* [23] is an early classical approach; *Hoffart et al.* [16], *Guo and Barbosa* [11] and *Ratinov et al.* [29] are the effective graph-based approaches; *Ganea and Hofmann* [9] is the novel collective linking approach using Loopy Belief Propogation (LBP) [9]; *Cheng and Roth*[3] and *Phong and Ivan* [19] explore relatedness between entities.

We verify our ED model on five popular publicly-available datasets: MSNBC, AUIAINT, and ACE2004 [29] are cleaned versions released by [11]; WNED-CWEB [7] and WNED-WIKI [11] are two larger but less reliable datasets that are automatically extracted from ClueWeb and Wikipedia, respectively. The statistics of these datasets are shown in Table 1.

Table 1. Statistics of datasets in this experiment.

Dataset	Total mentions	Total docs	Mentions/doc	Gold recall
MSNBC	656	20	32.8	98.5%
AQUIAINT	727	50	14.5	94.2%
ACE2004	257	36	7.1	90.6%
WNED-CWEB	11154	320	34.8	91.1%
WNED-WIKI	6821	320	21.3	92%

Candidate Generation and Selection. We generate the candidates by keeping the top-30 entities with the largest $\hat{p}(e|m)$. After that, for each mention, we select eight candidates from its top-30 entities: (1) four entities with the highest $\hat{p}(e|m)$; (2) four entities with the highest context-entity similarity scores, which is measured as $\mathbf{e}^T(\sum_{\omega \in W} \mathbf{w})$, where $\mathbf{e}, \mathbf{w} \in \mathbb{R}^d$ are entity and word embeddings, and W is the set of context words within the 50-word window around m_i.

Training Details and Hyper-Parameter Settings. We use CoNLL-YAGO [16] as our training and validation corpus. We set $d = 300$ and use GloVe word embeddings [27] and entity embeddings from [19]. The window size of local context is 50 for $c_{e,m}$ and 3 for Eq. 1. The mention neighbor size is 6, the distance decay is 0.95, and the α used in Eq. 1 is 0.7. To generate the dependency parse tree for each sentence, we implement the dependency parsing tool of Stanford CoreNLP toolkit in our experiment. As for the neural network hyper-parameters, the two layers of MLP encoder have 2000 and 1 hidden units. We use two layers of graph attention layer, and the $diag(\mathbf{R})$ is sampled from $\mathcal{N}(1, 0.1)$. We also implement early-stop, with each document as a mini-batch.

4.2   Experimental Results

Table 2 shows micro F1 scores on the 5 public datasets. In our experiment, we only consider in-KB accuracy. The reported performance of baselines and our methods all use Wikipedia and YAGO mention-entity index. It can be shown that our model achieves comparable results with the listed SOTA methods, and it achieves the highest F1 score on MSNBC, AQUIANT, and ACE2004 datasets. On average of all datasets, our model produces the highest average F1 score. Guo and Barbosa [11] performs extremely well on WIKI dataset, but is slightly worse than ours on the first three datasets. Besides our model, Cheng and Roth [3], Phong and Ivan [19] also explore relatedness between mentions and the latter achieves excellent results on many datasets. Compared to [19], our model has worse performance on ClueWEB and WIKI datasets, but higher F1 score on the other three datasets and the average F1 score. Overall, our approach achieves better performance on the first three datasets, but the performance is relatively worse on the last two datasets. This is because parse trees are not robust to noise, and the annotation error in ClueWEB and WIKI datasets may lead to wrong parsing text structure. Whereas, the sentences in the first three datasets are shorter and cleaner, which is suitable in our semantic parsing scenario.

4.3   Component Analysis

Effect of Global Information from Neighbor Nodes. Fig. 5 shows the effect of integrating information from neighbor nodes using GAT. The local result means solely using the MLP described in Sect. 3.2 to encode the feature vector, while global model is our proposed ED neural network model. It can be shown that integrating information from neighbors can improve the performance in all 5 datasets by more than 4% on micro-F1. The improvement of the global model

Table 2. F1 scores on five public datasets, where the bold and underlined scores are the best and second best performance achieved in corresponding dataset, respectively.

Methods	MSNBC	AQUIAINT	ACE2004	ClueWEB	WIKI	Avg
Milne and Witten	78	85	81	64.1	81.7	77.96
Hoffart et al.	79	56	80	58.6	63	67.32
Ratinov et al.	75	83	82	56.2	67.2	72.68
Cheng and Roth	90	90	86	67.5	73.4	81.38
Guo and Barbosa	92	87	88	77	**84.5**	85.7
Ganea and Hofmann	93.7	88.5	88.5	**77.9**	77.5	85.22
Phong and Ivan	93.9	88.3	89.9	77.5	78.0	85.51
CoSimTC	**94.16**	**90.90**	**92.92**	76.96	75.02	**85.99**

in the first three datasets is smaller because our extracted features can already solve more than 85% entity linking cases. ClueWEB and WIKI are challenging datasets since they contain much noise and hence difficult entity linking cases. Therefore, global information becomes more important for these two datasets especially when the encoding features cannot handle those hard cases.

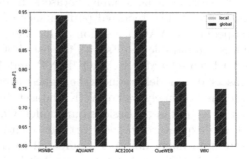

Fig. 5. Impact of using GAT to integrate neighbor information.

Comparison of Distance Measures. Fig. 6 illustrates the performance achieved by four types of distance measures in our neural network model. Except seq-dist measure, the other three measures all use parse tree distance for intra-sentence distance. It can be shown that CoSimTC has the overall best performance among these four types of distance measures because it captures both coreference relations and mention-context similarity across sentences. The performance of mention-based measure is also good, though slightly lower than CoSimTC, because it also considers the sequence distance for inter-sentence distance. The performance of root-based distance is not quite satisfactory and sometimes even worse than seq-dist, because it can only solve the intra-sentence distance correctly while naively connecting the root of each sentence brings mistakes in some cases. For example, words outside the current sentence may have smaller distance to current mention than words within the current sentence, if the outside words are closer to the root node of sentence's parse tree.

We can also see that the difference between these four types of measures for ACE2004 dataset is the most evident, because this dataset is smaller and cleaner compared to other datasets. Whereas, the performance of these four measures have a negligible difference in MSNBC dataset because it does not have many mentions within a single sentence, and in this case, using parse tree or not does not incur so much influence. In addition, the benefit of using parse tree is not quite obvious for ClueWEB and WIKI datasets. This is because these two datasets contain much noise, which may bring mistakes to parse trees. For example, some documents may have long sentences broken into several short sentences, or do not have linguistic structure. This type of sentences, such as website address, news titles, messy symbols, cannot be parsed into meaningful parse trees. Besides, the Stanford coreference parser may also involve some incorrect coreferences across sentence, which may weaken the superiority of using coreference relation to connect sentences.

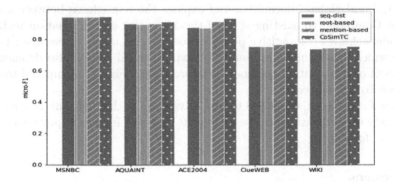

Fig. 6. Comparison of seq-dist, root-based, mention-based and CoSimTC measures.

Qualitative Illustration of CoSimTC. Figure 7 demonstrates the effect of using CoSimTC distance measure to form mention neighbors compared with using seq-dist as an alternative. The word *Kent* in red is our target mention. The other colored words are extracted mention neighbors, where the mentions in green are the same mention neighbors that these two methods produce. It can be shown that there is one different mention neighbor between these two methods, where *Somerset* is still closely related to the cricket discussed in the context, while *Grace Road* is not quite related to the topic of the context. The output probabilities of linking to the corresponding candidates are listed in the last three rows. We can see that both of these two methods can predict the correct result because the local context is evident enough to demonstrate the topic information. However, CoSimTC method produces higher probability of linking to the gold entity, because it can gather more accurate global information, with the help of generated mention neighbors.

1.	'West Indian all-rounder Phil Simmons took four for 38 on Friday as Leicestershire beat Somerset by an innings and 39 runs in two days to take over at the head of the county championship.'
2.	'Their stay on top, though, may be short-lived as title rivals Essex, Derbyshire and Surrey all closed in on victory while Kent made up for lost time in their rain-affected match against Nottinghamshire.'
3.	'After bowling Somerset out for 83 on the opening morning at Grace Road, Leicestershire extended their first innings by 94 runs before being bowled out for 296 with England discard Andy Caddick taking three for 83.'

Mention neighbors(CoSimTC): Essex, Nottinghamshire, Surrey, Derbyshire Somerset, Somerset		Mention neighbors(seq-dist): Surrey, Derbyshire, Essex Nottinghamshire, Somerset, Grace Road	
'Kent'	0.09%	'Kent'	0.09%
'Kent_County_Cricket_Club'	99.46%	'Kent_County_Cricket_Club'	97.68%
'Kent_(band)'	0.03%	'Kent_(band)'	0.03%

Fig. 7. A demonstration of the effect of using parse tree to produce mention neighbors.

5   Conclusion

In this paper, we propose a novel distance measure CoSimTC based on parse tree to produce mention neighbors. It combines the benefit of parse tree distance and sequence distance and solves the cross-tree problem at document level. The mention neighbors derived by CoSimTC are helpful for the target mention to integrate useful global information within the document. We also propose a novel neural network model based on graph attention network (GAT) to integrate both local and global information and explore the relatedness between entities flexibly. Compared to existing state-of-the-art entity disambiguation methods, our model achieves competitive performance and has the best average F1 score on five widely-used public datasets. The experimental results also demonstrate the benefit of the mention neighbors produced by CoSimTC, compared to using sequence distance directly.

In the future work, we intend to take advantage of the relations in parse trees to generate the semantic patterns within the text and further improve the entity linking performance.

References

1. Alhelbawy, A., Gaizauskas, R.: Graph ranking for collective named entity disambiguation. ACL **2**, 75–80 (2014)
2. Cao, Y., Hou, L., Li, J., Liu, Z.: Neural collective entity linking. In: COLING, pp. 675–686 (2018)
3. Cheng, X., Roth, D.: Relational inference for wikification. In: EMNLP, pp. 1787–1796 (2013)
4. Cucerzan, S.: Large-scale named entity disambiguation based on wikipedia data. In: EMNLP-CoNLL (2007)
5. Dredze, M., McNamee, P., Rao, D., Gerber, A., Finin, T.: Entity disambiguation for knowledge base population. In: COLING, pp. 277–285 (2010)
6. Francis-Landau, M., Durrett, G., Klein, D.: Capturing semantic similarity for entity linking with convolutional neural networks. In: NAACL, pp. 1256–1261 (2016)
7. Gabrilovich, E., Ringgaard, M., Subramanya, A.: FACC1: freebase annotation of clueweb corpora. http://lemurproject.org/clueweb09/FACC1/, Cited by 5 (2013)
8. Ganea, O.E., Ganea, M., Lucchi, A., Eickhoff, C., Hofmann, T.: Probabilistic bag-of-hyperlinks model for entity linking. In: WWW, pp. 927–938 (2016)
9. Ganea, O.E., Hofmann, T.: Deep joint entity disambiguation with local neural attention. In: EMNLP, pp. 2619–2629 (2017)

10. Guo, Z., Barbosa, D.: Robust entity linking via random walks. In: CIKM, pp. 499–508. ACM (2014)
11. Guo, Z., Barbosa, D.: Robust named entity disambiguation with random walks. Semant. Web **9**(4), 459–479 (2018)
12. Gupta, N., Singh, S., Roth, D.: Entity linking via joint encoding of types, descriptions, and context. In: EMNLP, pp. 2681–2690 (2017)
13. Hachey, B., Radford, W., Curran, J.R.: Graph-based named entity linking with wikipedia. In: Bouguettaya, A., Hauswirth, M., Liu, L. (eds.) WISE 2011. LNCS, vol. 6997, pp. 213–226. Springer, Heidelberg (2011). https://doi.org/10.1007/978-3-642-24434-6_16
14. Han, X., Sun, L., Zhao, J.: Collective entity linking in web text: a graph-based method. In: SIGIR, pp. 765–774. ACM (2011)
15. He, Z., Liu, S., Li, M., Zhou, M., Zhang, L., Wang, H.: Learning entity representation for entity disambiguation. In: ACL(Short Papers), vol. 2, pp. 30–34 (2013)
16. Hoffart, J., et al.: Robust disambiguation of named entities in text. In: EMNLP, pp. 782–792 (2011)
17. Hua, W., Wang, Z., Wang, H., Zheng, K., Zhou, X.: Short text understanding through lexical-semantic analysis. In: 2015 IEEE 31st International Conference on Data Engineering, pp. 495–506, April 2015
18. Kipf, T.N., Welling, M.: Semi-supervised classification with graph convolutional networks. In: ICLR (2016)
19. Le, P., Titov, I.: Improving entity linking by modeling latent relations between mentions. In: ACL, pp. 1595–1604 (2018)
20. Lehmann, J., Monahan, S., Nezda, L., Jung, A., Shi, Y.: LCC approaches to knowledge base population at TAC 2010. In: TAC (2010)
21. Liu, X., Li, Y., Wu, H., Zhou, M., Wei, F., Lu, Y.: Entity linking for tweets. In: ACL, vol. 1, pp. 1304–1311 (2013)
22. Mihalcea, R., Csomai, A.: Wikify!: linking documents to encyclopedic knowledge. In: CIKM, pp. 233–242. ACM (2007)
23. Milne, D., Witten, I.H.: Learning to link with wikipedia. In: CIKM, pp. 509–518. ACM (2008)
24. Mueller, D., Durrett, G.: Effective use of context in noisy entity linking. In: EMNLP, pp. 1024–1029 (2018)
25. Nie, F., Cao, Y., Wang, J., Lin, C.Y., Pan, R.: Mention and entity description co-attention for entity disambiguation. In: AAAI (2018)
26. Pappu, A., Blanco, R., Mehdad, Y., Stent, A., Thadani, K.: Lightweight multilingual entity extraction and linking. In: WSDM, pp. 365–374. ACM (2017)
27. Pennington, J., Socher, R., Manning, C.: Glove: Global vectors for word representation. In: EMNLP, pp. 1532–1543 (2014)
28. Phan, M.C., Sun, A., Tay, Y., Han, J., Li, C.: NeuPL: Attention-based semantic matching and pair-linking for entity disambiguation. In: CIKM, pp. 1667–1676. ACM (2017)
29. Ratinov, L., Roth, D., Downey, D., Anderson, M.: Local and global algorithms for disambiguation to wikipedia. In: ACL, pp. 1375–1384 (2011)
30. Vaswani, A., et al.: Attention is all you need. In: NIPS, pp. 6000–6010 (2017)
31. Veličković, P., et al.: Graph attention networks. In: ICLR (2018)
32. Yamada, I., Shindo, H., Takeda, H., Takefuji, Y.: Joint learning of the embedding of words and entities for named entity disambiguation. In: CoNLL, p. 250 (2016)

Bibliographic Name Disambiguation with Graph Convolutional Network

Hao Yan[1,2(✉)], Hao Peng[1,2], Chen Li[1,2], Jianxin Li[1,2], and Lihong Wang[3]

[1] Beijing Advanced Innovation Center for Big Data and Brain Computing,
Beihang University, Beijing, China
{yanhao,penghao,lichen,lijx}@act.buaa.edu.cn
[2] State Key Laboratory of Software Development Environment, Beihang University,
Beijing, China
[3] National Computer Network Emergency Response Technical Team/Coordination
Center of China, Beijing, China
wlh@isc.org.cn

Abstract. Name disambiguation, which aims to distinguish real-life person from documents associated with a same reference by partition the documents, has received extensive concern in many intelligent tasks, e.g., information retrieval, bibliographic data analysis and mining system. Existing methods implement name disambiguation utilizing linkage information or biographical feature, however, only a few work try to combine them effectively. In this paper, we propose a novel model that incorporates structural information and attribute features based on the Graph Convolutional Network to learn discriminating embedding, and achieves individual distinction by equipping a hierarchical clustering algorithm. We evaluate the proposed model on real-world academic networks Aminer, and experimental results show that the proposed method is competitive with the state-of-the-art methods.

Keywords: Name disambiguation · Graph Convolutional Network · Clustering

1 Introduction

While you are searching for academic publications by an author name, the response may disappoint you. For instance, sometimes you want to peruse masterpieces of "Tom Mitchell", a professor of Carnegie Mellon University, well-known in machine learning fields. After typing the name in search box, the query result is a long list of papers having a author named "Tom Mitchell". Unexpectedly, the topics ranging from computer science, biology to economics, and only several papers is relevant to the scholar you concerned. It would be better if you search author in digital libraries, for example, DBLP[1], Cite Seer[2] and Aminer[3]. These search engine will list candidates named "Tom Mitchell"

---

[1] http://dblp.uni-trier.de.
[2] http://citeseerx.ist.psu.edu.
[3] http://www.aminer.cn.

R. Cheng et al. (Eds.): WISE 2019, LNCS 11881, pp. 538–551, 2019.
https://doi.org/10.1007/978-3-030-34223-4_34

with corresponding institute. It is much easier to find the scholar you searching, clicking the target candidate, and you will acquire all his publications. Technology behind the convenience is a lot of machine learning algorithms including name disambiguation [6,26].

Name disambiguation is an important problem, which has numerous applications in information retrieval, bibliographic data analysis and other fields [4,20]. In information retrieval, name disambiguation is crucial for understanding query purpose. As mentioned before, while querying "Tom Mitchell", name disambiguation is necessary to split query result into different groups according to entities behind the name. In addition to the literature search facility, digital libraries also provide useful analysis that is being used for better decision making by funding agencies and academic institutions for grants and individual's promotion decisions [13]. If publications of different persons with same name can not be attributed accurately, the analysis would be misleading.

Due to its importance, the name disambiguation problem has attracted substantial attention from information retrieval and data mining communities. Many existing methods [2,11] used biographical features to distinguish people with same name, e.g, name, address, institutional affiliation, email address, etc. But biographical features are hard to obtain and liable to change. Usually, publications can reveal author research fields and interests, such as the similarity between two papers is a clue to find whether they have same authorship. Recently work [18] solved name disambiguation problem based on paper attributes similarity, e.g, keywords and title. Other methods uses relational data in the form of a collaboration graph, and solved name disambiguation by using graph topological features. For instance, Hermansson [10] used a classification model based on graphlet kernels, and Zhang [28] used a network embedding based method on anonymized graphs. Through previous studies, we find that both attribute features and graph structural information have contribution to solve name disambiguation problem. It's well known that Graph Convolutional Network (GCN) [15] is an efficient model to integrate both attribute features and structural information. Zhang [29] proposed a graph auto-encoders [14] based method involving graph topology and attribute features, but this method neglects the linkage between papers and authors and co-authorship.

To utilize information as much as possible and achieve better performance, we propose a novel graph structure and attribute features involved representation learning model. Specifically, we make use of two personalized GCNs embeddings of papers and authors into a low-dimensional space, and then maintain close linked entities proximate to each other in embedding space with minimizing the careful designed objective function. Then, a Hierarchical Agglomerative Clustering (HAC) algorithm [8][4] could be integrated to solve name disambiguation problem. The proposed method is evaluated on real-world large-scale academic networks Aminer dataset. The experimental results show that our proposed method is competitive with several state-of-the-art methods.

---

[4] https://github.com/mstrosaker/hclust.

The remainder of this paper is divided into five sections. In Sect. 2, we briefly reviewing previous representative works directly related to ours. Then we detailed formulate name disambiguation problem, define three types of graph in Sect. 3 and introduced our method in Sect. 4. In Sect. 5, we show experiment results on real-world large data and compare our model with several state-of-the-art methods. In the end, we draw our conclusions.

2 Related Work

2.1 Name Disambiguation

Recently, name disambiguation has been defined as clustering problem. Previous studies have focused on how to strike the balance between documents similarity quantization, determination of cluster size, and achieving better disambiguation. According to the selection of clustering basis, the existing literature can be roughly classified into three categories: attribute features based, linkage based, and hybrid methods.

Attribute features based methods generally focus on how to measure the similarity between documents. The work of Huang [12] introduces Support Vector Machine (SVM) to distinguish candidate documents which are initially grouped according to name similarity, and then utilizes DBSCAN to cluster documents. Yoshida [27] achieves better cluster results by using efficient feature from a two-stage clustering process. Different from the unsupervised methods described above, Han [9] employs SVM and Naive Bayes to implement name disambiguation in a classified manner. Similarly, Louppe [18] proposes a semi-supervised hierarchical clustering method based on a classifier to achieve more efficient document similarity metrics.

Linkage based methods are more focused on the graph structure (composed of articles and authors) information than the attribute features based methods. GHOST [5] achieves node clustering on a co-author-based graph through mining the relationships between documents more granularly. By considering the linkage between documents as a transfer process, Tang [25] uses Hidden Markov Random Fields (HMRFs) to model the document chain uniformly and solve the name disambiguation using probability model. The work of Zhang [28] embeds documents into low-dimensional space without involving private data, and implements name disambiguation through HAC.

Besides, there are some methods try to combine the advantages of the two methods above. Zhang [29] proposes a novel representation learning method that can contain both global and local information, achieves a good performance, and was applied in Aminer.

2.2 Graph Convolutional Networks

As a method for efficiently integrating attribute features of graph structural information, GCN has been widely studied and applied. Firstly, Bruna [1] define the convolution operation in an irregular graph structure by using convolution Fourier formula, which has achieved competitive results. Defferrard [3] advance GCN through multi-order information diffusion, and implements an approximate calculation using the Chebyshev formula. The work of Kipf [15] proposes a first-order approximation GCN, which defines a new information diffusion matrix to achieve efficient node feature learning, and achieved good results in semi-supervised classification tasks.

Recently, GCN has been applied to a large number of tasks including graph mining, text classification, traffic prediction and event mining [17, 21–23], and it has been verified that it can effectively combine structural information with node feature. Kipf [24] employ GCN to relation learning by equipping update module with an information passing component. Then apply GCN to the event detection with a novel pooling method and achieved better results [19]. To analyze the compositional principles of protein molecular networks, Alex [7] utilize GCN to obtain molecular embedding and model the composition of proteins. Besides, GCN is widely used in tasks, e.g., named entity recognition and relation extraction. To our best, we are the first to introduce GCN and triplet loss to solve name disambiguation problem.

3 Preliminaries

In this section, we first present the formulation of the problem, and then introduce the three types of graph used in our model.

3.1 Problem Formulation

Let a be a given name reference, and $\mathcal{D}^a = \{D_1^a, D_2^a, ..., D_N^a\}$ be a set of N documents associated with the author name reference a. $\{A_1, A_2, ..., A_M\}$ is the collaborator set of author named a in $\mathcal{D}^a$, denoted as $\mathcal{A}^a$ where $a \notin \mathcal{A}^a$. We assume there is no disambiguation in $\mathcal{A}^a$ that means each name reference could identify a collaborator. In real-life it is common that several person have same name. The goal of name disambiguation is to divide $\mathcal{D}^a$ into K disjoint sets $\mathcal{C}^a = \{C_1^a, C_2^a, ..., C_K^a\}$, in each set C_k^a, all documents belong to the same person p_k and documents associated with author p_k must in same set C_k^a. The problem could be formalized as follow.

Definition 1 Name Disambiguation. *Denote $\Theta(d_i^a)$ as a function to get the person p_k who named a and associated with d_i^a, the task of name disambiguation is to find a partition function Φ to divide $\mathcal{D}^a$ into K disjoint clusters, i.e.,*

$$\Phi(\mathcal{D}^a) \rightarrow C^a \tag{1}$$

every cluster in C^a meets $\forall d_i^a \in C_k^a, \Theta(d_i^a) = p_k$ and $\forall d_i^a \in \{d_i^a | \Theta(d_i^a) = p_k\}, d_i^a \in C_k^a$, that is equivalent to

$$C_k^a = \{d_i^a | d_i^a \in \mathcal{D}^a, \Theta(d_i^a) = p_k\} \tag{2}$$

3.2   Graphs in Bibliographic Domain

In bibliographic domain, such as dblp and Aminer, there are linked and attributed information we can utilize to solve this problem. For example, given two papers, the authors of the two papers collaborated closely with each other, there is a high probability that the two papers belong to the same author. This is appropriate that we assume the topic of papers associate to same author would be similar, because different scholar has his own interests and specific research fields. The paper's attribute information such as title, keywords, abstract and venue would reveal that. Besides, the author's attribute information is also useful to solve name disambiguation problem. We denote feature of i_{th} document associated with the author name a as f_i^d, similarly, feature of j_{th} person in collaborator set is represented as f_j^p. Document and person feature matrix is F_d and F_p respectively.

Definition 2 Person-Person Graph. *Given a name reference a, the person-person graph denoted as $G_{pp} = (\mathcal{A}^a, E_{pp})$, nodes in this graph is the collaborator set $\mathcal{A}^a$, the weight of e_{ij} is defined as the number of distinct documents in which A_i and A_j have collaborated.*

Definition 3 Document-Person Graph. *Given a name reference a, the person-document graph is represented as $G_{dp} = (\mathcal{D}^a \cup \mathcal{A}^a, E_{dp})$, a bipartite graph. $\mathcal{D}^a$ is documents associated with name reference a, $\mathcal{A}^a$ is the collaborator set. If a person node A_j is the author of a document node D_i, then the edge weight w_{ij} is 1, otherwise is 0.*

Definition 4 Document-Document Graph. *Given a name reference a, the graph is represented as $G_{dd} = (\mathcal{D}^a, E_{dd})$, if two documents are similar enough, build a edge between them. We measure the similarity of two documents based on the common features shared by the two documents, firstly, calculate IDF (Inverted Document Frequency) of each feature, and then sum up the IDF of common features shared by two documents, if the result above a threshold, set the weight w_{ij} between document i and document j to 1.*

In this study, we use two personalized GCNs embedding both structural and attribute features within three types of graph into a same d-dimensional space, and then we use document embedding matrix as input and applies HAC to partition $\mathcal{D}^a$ into K disjoint clusters. At this stage, K is a user-defined parameter which we match with the ground truth during the evaluation phase.

4   Methodology

In this section, we discuss the design and implementation of our method to solve author disambiguation problem in detail. Our work concentrate on the representation learning of documents and persons. As embedding acquired, HAC is applied as other clustering based methods did. The overview of our embedding model is shown as Fig. 1.

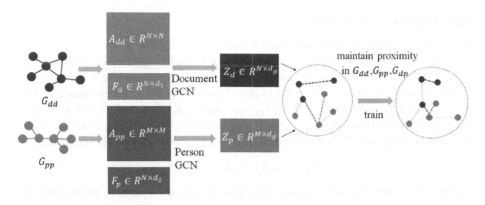

Fig. 1. The architecture of our proposed embedding model, G_{pp}, G_{dd}, G_{dp} is Person-Person Graph, Document-Document Graph and Document-Person Graph based on authorship and documents' similarity. A_{dd}, A_{pp} is adjacent matrix corresponding with G_{pp} and G_{dd}, and F_d, F_p is feature matrix. The documents and persons embedding (denoted as Z_d, Z_p) is acquired by Document-GCN and Person-GCN respectively, then keep close linked (bold dotted line) entities in G_{pp}, G_{dd} and G_{dp} proximate to each other in embedding space with minimizing triplet loss.

4.1 Graph Embedding

As mentioned in Sect. 3, there are two types of node in three types of graph. We use two personalized GCNs acquire embedding matrix for nodes in Person-Person Graph and Document-Document Graph respectively, due to its effectiveness for modeling networked data.

The goal of GCN is to learn a function of a graph $G = \{N, E\}$ which takes nodes feature matrix F and graph adjacent matrix A as input and produces a node-level output embedding matrix Z which incorporates both structural information and nodes feature. The function can be describe as follow:

$$H^{(l)} = g(H^{(l-1)}, A) = \sigma(\hat{A}H^{(l-1)}W^{(l)}) \tag{3}$$

where H^l is a feature matrix which is output of lth graph convolutional layer, when it is first layer, $H^0 = F$. The output of final layer is Z, $W^{(l)}$ is a weight matrix for l^{th} neural network layer, $\sigma(\cdot)$ is an activation function and $\hat{A}$ is a symmetrically normalized adjacency matrix, $\hat{A} = D^{\frac{1}{2}}AD^{-\frac{1}{2}}$, D is the degree matrix of G.

Denote document embedding matrix as Z_d and person embedding matrix as Z_p, we formulize the two personalized GCNs as follow:

$$Z_d = \sigma(\hat{A}_{dd}\sigma(\hat{A}_{dd}F_dW_d^{(1)})W_d^{(2)}) \tag{4}$$

$$Z_p = \sigma(\hat{A}_{pp}\sigma(\hat{A}_{pp}F_pW_p^{(1)})W_p^{(2)}) \tag{5}$$

where $\hat{A}_{dd}$ and $\hat{A}_{pp}$ is symmetrically normalized adjacency matrix of G_{dd} and G_{pp} respectively, F_d and F_p is document and person nodes feature matrix. We call the two personalized GCN as Document-GCN and Person-GCN.

4.2 Objective Function

In intuition, scholars who collaborate more often are more likely to have similar research interests than those who do not work together or seldom coauthor a paper. This relation should be maintained in embedding space. Given a triplet consists of three person node A_i, A_j and A_k, the corresponding embedding learned from Person-GCN is z_{pi}, z_{pj} and z_{pk}, if A_i collaborates with A_j more frequently than $A_k(w_{ij} > w_{ik})$, the distance between z_{pi} and z_{pj} should be smaller than the distance between z_{pi} and z_{pk}. All triplets should satisfy

$$\|z_{pi} - z_{pj}\|_2 \, < \, \|z_{pi} - z_{pk}\|_2 \tag{6}$$

where $\|\cdot\|_2$ is the Euclidean norm. The loss function to make triplets meet this condition is:

$$\mathcal{L}_{pp}(A_i, A_j, A_k) = max\{\|z_{pi} - z_{pj}\|_2 - \|z_{pi} - z_{pk}\|_2, 0\} \tag{7}$$

Similar to the relation between person, the more common features shared two documents, the closer their embeddings are to each other. Specifically, the distance between documents D_i and D_j in embedding space is smaller than the distance between documents D_i and D_k if $w_{ij} = 1$ and $w_{ik} = 0$. For Document-Document graph, the loss function is:

$$\mathcal{L}_{dd}(D_i, D_j, D_k) = max\{\|z_{di} - z_{dj}\|_2 - \|z_{di} - z_{dk}\|_2, 0\} \tag{8}$$

where z_{di} is embedding of document D_i learned by Document-GCN.

As documents embedding and persons embedding acquired from the two personalized GCNs, the linkage between document and person in Document-Person could restrict personalized GCNs to map the two types of nodes into a same space that distance between different types of nodes could be measured. In intuition, the distance between a document i and its author j is smaller than the distance between document i and another person k. With $w_{ij} = 1$ and $w_{ik} = 0$, the loss function is

$$\mathcal{L}_{dp}(D_i, A_j, A_k) = max\{\|z_{di} - z_{pj}\|_2 - \|z_{di} - z_{pk}\|_2, 0\} \tag{9}$$

The three loss functions has same structure, they all have an anchor node i, a positive node j and a negative node k, the objective is to maximize the distance between anchor and negative node and minimize the distance between anchor and negative node. For each document and person as anchor node, we sample positive node according to the weight w to anchor node, the bigger the weight, the more likely it is to be selected. The negative node is chosen by their distance to anchor node, the smaller the distance, the more likely it is to be selected.

To preserve all constrains simultaneously, we propose a model which combines all loss functions derived from three different graphs and joint minimizes the following objective function:

$$\mathcal{L} = \sum_{A_i \in \mathcal{A}} \mathcal{L}_{pp}(A_i, A_j, A_k) + \sum_{D_i \in \mathcal{D}} \mathcal{L}_{dd}(D_i, D_j, D_k) + \sum_{D_i \in \mathcal{D}} \mathcal{L}_{dp}(D_i, A_j, A_k) \tag{10}$$

$$s.t. \ w_{ij} > w_{ik} \quad \forall G \in \{G_{dd}, G_{pp}, G_{dp}\}$$

In training stage, we minimize $\mathcal{L}_{dd}$, $\mathcal{L}_{pp}$, $\mathcal{L}_{dp}$ and update their corresponding gradients successively. The complete learning algorithm is summarized in Algorithm 1.

Algorithm 1. Graph structure and attribute features based name disambiguation method

---

Require: name reference a associated $\mathcal{D}^a$, $\mathcal{A}^a$, cluster size K.
Ensure: K disjoint clusters.
 1: Construct G_{pp}, G_{dd}, G_{dp}, acquire F_d, F_p.
 2: **for** each epoch **do**:
 3:     get documents and persons embedding by Document-GCN and Person-GCN.
 4:     sample triplets from G_{pp}.
 5:     minimize $\mathcal{L}_{pp}(A_i, A_j, A_k)$ and update parameters in Person-GCN with Adagrad.
 6:     sample triplets from G_{dd}.
 7:     minimize $\mathcal{L}_{dd}(D_i, D_j, D_k)$ and update parameters in Document-GCN with Adagrad.
 8:     sample triplets from G_{dp}.
 9:     minimize $\mathcal{L}_{dd}(D_i, A_j, A_k)$ and update parameters in both Document-GCN and Person-GCN with Adagrad.
10: **end for**
11: Given K, perform HAC to partition $\mathcal{D}^a$ into K disjoint clusters $\mathcal{C}^a$ with documents embedding as input.
12: **return** $\mathcal{C}^a$

---

5 Experiments

In this section, we analyze the proposed model empirically on a challenging benchmark proposed by Aminer [29]. The benchmark consists of 70,258 documents from 12,798 authors. The document contains rich information such as title, author, keywords, published year and venue. We random select 20 name references from the dataset. There are 264 documents and 8 distinct persons for a name reference in average. For author's name "L.Song", there are 700 associated documents and 33 distinct real-life authors. This is a difficult disambiguation task.

5.1 Baseline Methods

To validate the effectiveness of our model, we compare it against three state of the art methods.

Aminer [29]: This is a two steps method, firstly train a model with a little mount data to map document feature into global embedding space in which documents associated with same author would be close to each other. And then a GCN based graph auto-encoder with global embedding as node feature is used to learn document representations. The objective is to minimize the reconstruction error between dot product of embedding and origin documents feature similarity based graph. Finally, the clustering result is generated by HAC.

Zhang [28]: This method constructs three types of graph based on coauthors and document similarity. A graph embedding is learned by minimizing the triplet loss which aims to make the distance between linked nodes is smaller than others, and then perform cluster algorithm. This method is similar to ours but it neglects nodes' attribute feature.

Louppe [18]: This method trains a function for measure distance between a document pair based on document feature, and then used a semi-supervised HAC to determine clusters.

5.2   Experimental Settings

In all experiments, we use Aminer proposed global embedding [29] as document feature, specifically, we sample 500 name references from Aminer dataset (as training data for Louppe's method [18] too), and then train a supervised model to learn document embedding with metric learning. Taking the document feature as input, the model's output is global embedding. We use one-hot embedding of author name as person feature due to author information is scarce in dataset. The IDF threshold to construct document-document graph is set as 32. For both document-GCN and person-GCN, the first GCN layer size is 64 and the second layer size is 128. Our model is trained with 0.01 learning rate and 1000 epochs. The parameters of baseline methods is set according to the origin paper or open source code. We run all the experiments on a 32 cores machine with 128G memory.

5.3   Effectiveness Evaluation

Table 1 shows the performance comparison of name disambiguation between our proposed model and other competing methods for all 20 name references. As commonly performed in name disambiguation research, we compare our model with baseline methods in pairwise precision, recall and F1 [16]. Each row is a name reference evaluated in our experiments, the columns (3, 4, 5) is various baseline methods, the last is the average of evaluate metrics of all 20 name references. For the accuracy of the experiment, we execute every method 5 times on each name reference.

As we observed, due to mainly modeling document similarity, AMiner's method and Louppe's method could distinguish real-world authors more precisely, for 17 names they are the best in precise. It's also worth noting that, although Zhang's method learns documents embedding with structural information only, it achieves the best for 5 name references' recall. Specifically, for "L.Song" and "J.Shao", it exceeds Zhang's and Louppe's methods for 15.9% to 30.9%, for average recall, the superiority is 10.4% and 9.9%. The significant improvement shows that the relation within authorship is helpful to gather together documents with same author as much as possible.

With combining structural information and documents attribute features, our method makes a better trade-off between precise and recall, performs the best for 9 name references in terms of recall and 10 names in terms of F1. Shown as Fig. 2, for average, the recall and F1 of our method is the best.

Table 1. Comparison of precision, recall and F1 between our proposed method and other baseline methods for name disambiguation task on 20 name references.

Name	Our method			AMiner			Zhang			Louppe		
	Prec	Rec	F1	Prec	Rec	F1	Prec	Rec	F1	Prec	Rec	F1
M. Chen	98.8	99.2	99.0	**98.9**	**99.9**	**99.4**	88.6	98.8	93.4	94.1	97.1	95.5
W. Zhang	45.3	61.6	52.1	**54.2**	50.2	52.1	44.5	**84.3**	**58.3**	47.2	67.9	55.7
J. Du	70.5	72.8	71.6	68.4	67.2	67.7	15.3	73.8	25.3	**81.4**	**75.4**	**78.3**
H.B. Li	56.0	**86.1**	67.7	63.4	75.2	68.3	13.7	61.6	22.4	**75.3**	66.9	**70.8**
Y.Y. Li	43.9	**93.3**	59.7	74.1	65.5	69.5	25.3	51.3	33.9	**72.7**	66.8	**69.6**
X. Zhang	84.0	81.9	**83.0**	**88.3**	64.0	74.2	60.0	58.0	59.0	62.9	**82.3**	71.3
J.M. Fu	**97.3**	99.4	**98.3**	**97.3**	50.6	66.6	**97.3**	98.9	98.1	94.2	**100**	97.0
J.G. He	76.3	**90.1**	82.6	**92.2**	82.7	**87.2**	36.8	89.5	52.1	82.4	88.8	85.4
B. Hong	79.5	82.9	**81.1**	76.2	72.9	74.5	17.2	**85.0**	28.6	**83.4**	71.6	77.1
W. Yang	81.5	97.5	87.5	**96.5**	**98.2**	**97.4**	48.5	95.7	64.3	91.2	76.5	83.2
R. Lu	69.7	**83.5**	75.8	77.7	83.0	**80.2**	11.6	80.7	20.3	**86.4**	65.5	74.5
J. Feng	91.2	**95.8**	**93.4**	92.0	90.8	91.2	13.9	88.0	23.9	76.2	82.9	79.4
X. Qin	91.9	**95.2**	**93.5**	92.1	94.6	93.3	51.9	93.8	66.8	81.4	94.5	87.5
S. Wang	**57.7**	92.8	**71.1**	56.8	64.4	60.3	20.2	84.9	32.7	56.0	85.4	67.6
L. Song	61.0	86.6	**71.6**	62.0	75.0	67.8	24.1	**93.1**	38.2	**69.2**	71.4	70.3
F. Teng	94.0	99.1	**96.5**	**99.5**	87.6	93.2	94.0	98.2	96.0	87.9	**100**	93.6
S. Song	81.4	91.3	**86.1**	**92.4**	78.4	84.8	28.5	**93.0**	43.6	88.0	74.1	80.5
K. Xu	**91.4**	**98.6**	**94.9**	91.3	75.7	81.1	71.3	94.4	81.2	82.9	97.1	89.4
J. Shao	65.9	90.4	74.8	**90.7**	63.2	74.5	51.5	**94.1**	66.6	88.7	78.2	**83.1**
J. Lu	77.4	**98.8**	86.7	**95.7**	66.5	78.5	70.8	96.3	81.6	83.4	89.7	**87.5**
Avg	75.7	**89.8**	**81.3**	**83.0**	75.3	78.1	44.3	85.7	54.3	76.6	75.8	74.7

5.4 Component Contribution Analysis

Our proposed model consists of three types of graphs. For each graph we design a triplet loss function for maintaining graph proximity in embedding space. In this section, we analysis the contribution of each of the three components for the name disambiguation task by incrementally adding the components in the embedding model. We first add Document-Person graph, followed by Document-Document graph, and Person-Person graph. Specifically, we evaluate $\mathcal{L}_{dd}$, $\mathcal{L}_{dd} + \mathcal{L}_{dp}$, $\mathcal{L}_{dd} + \mathcal{L}_{pp} + \mathcal{L}_{dp}$ three types of loss function combinations.

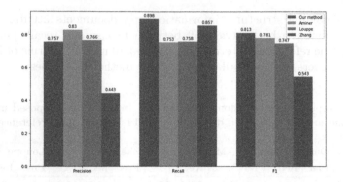

Fig. 2. Comparison of average pairwise precision, recall and F1

Table 2 shows the name disambiguation performance in terms of pairwise precision, recall and F1 using our proposed embedding model with different component combinations. As we see, after adding each component, we observe improvement for recall and decline for precise while F1 is rising, that means our model could make a better trade-off with more structural information.

Table 2. Component contribution analysis

Object function	Precision	Recall	F1
$\mathcal{L}_{dd}$	**77.52**	83.20	80.25
$\mathcal{L}_{dd} + \mathcal{L}_{dp}$	76.44	86.21	81.04
$\mathcal{L}_{dd} + \mathcal{L}_{pp} + \mathcal{L}_{dp}$	75.73	**89.84**	**81.34**

6   Conclusion

In this paper, we have proposed a novel representation learning based solution to address the name disambiguation problem. Our proposed representation learning model embed both document and person entities into a same space with two personalized GCNs and maintain proximity of close linked entities in embedding space by minimizing the careful designed objective function. Benefited from structural information and attribute features, the learned embedding could be effectively utilized for name disambiguation. Experimental results shows our proposed method makes a better trade-off between precise and recall, it is competitive with many of the existing state-of-the-arts for name disambiguation.

Learning embedding with same epochs for different graphs (different name reference) is likely to overfit, how to avoid and achieve a better performance could be future work.

Acknowledgements. This work is supported in part by National Key R&D Program of China 2017YFB0803305, NSFC 61772151&61872022, Beijing Advanced Innovation Center for Big Data and Brain Computing.

References

1. Bruna, J., Zaremba, W., Szlam, A., LeCun, Y.: Spectral networks and locally connected networks on graphs. In: 2nd International Conference on Learning Representations, ICLR 2014, Conference Track Proceedings, Banff, AB, Canada, 14–16 April 2014 (2014)
2. Cen, L., Dragut, E.C., Si, L., Ouzzani, M.: Author disambiguation by hierarchical agglomerative clustering with adaptive stopping criterion. In: The 36th International ACM SIGIR Conference on Research and Development in Information Retrieval, SIGIR 2013, Dublin, Ireland, July 28–August 01 2013, pp. 741–744 (2013)
3. Defferrard, M., Bresson, X., Vandergheynst, P.: Convolutional neural networks on graphs with fast localized spectral filtering. In: Advances in Neural Information Processing Systems 29: Annual Conference on Neural Information Processing Systems 2016, Barcelona, Spain, 5–10 December 2016, pp. 3837–3845 (2016)
4. Elmacioglu, E., Tan, Y.F., Yan, S., Kan, M., Lee, D.: PSNUS: web people name disambiguation by simple clustering with rich features. In: Proceedings of the 4th International Workshop on Semantic Evaluations, SemEval@ACL 2007, Prague, Czech Republic, 23–24 June 2007, pp. 268–271 (2007)
5. Fan, X., Wang, J., Pu, X., Zhou, L., Lv, B.: On graph-based name disambiguation. J. Data Inf. Qual. $2(2)$, 10:1–10:23 (2011)
6. Ferreira, A.A., Gonçalves, M.A., Laender, A.H.F.: A brief survey of automatic methods for author name disambiguation. ACM SIGMOD Rec. $41(2)$, 15–26 (2012)
7. Fout, A., Byrd, J., Shariat, B., Ben-Hur, A.: Protein interface prediction using graph convolutional networks. In: Advances in Neural Information Processing Systems 30: Annual Conference on Neural Information Processing Systems 2017, Long Beach, CA, USA, 4–9 December 2017, pp. 6533–6542 (2017)
8. Froud, H., Lachkar, A.: Agglomerative hierarchical clustering techniques for Arabic documents. In: Nagamalai, D., Kumar, A., Annamalai, A. (eds.) Advances in Computational Science, Engineering and Information Technology, vol. 225, pp. 225–267. Springer, Heidelberg (2013). https://doi.org/10.1007/978-3-319-00951-3_25
9. Han, H., Giles, C.L., Zha, H., Li, C., Tsioutsiouliklis, K.: Two supervised learning approaches for name disambiguation in author citations. In: ACM/IEEE Joint Conference on Digital Libraries, JCDL 2004, Tucson, AZ, USA, 7–11 June 2004, Proceedings, pp. 296–305 (2004)
10. Hermansson, L., Kerola, T., Johansson, F., Jethava, V., Dubhashi, D.P.: Entity disambiguation in anonymized graphs using graph kernels. In: 22nd ACM International Conference on Information and Knowledge Management, CIKM 2013, San Francisco, CA, USA, October 27–November 1 2013, pp. 1037–1046 (2013)

11. Hoffart, J., et al.: Robust disambiguation of named entities in text. In: Proceedings of the 2011 Conference on Empirical Methods in Natural Language Processing, EMNLP 2011, 27–31 July 2011, John McIntyre Conference Centre, Edinburgh, UK, A Meeting of SIGDAT, a Special Interest Group of the ACL, pp. 782–792 (2011)

12. Huang, J., Ertekin, S., Giles, C.L.: Efficient name disambiguation for large-scale databases. In: Fürnkranz, J., Scheffer, T., Spiliopoulou, M. (eds.) PKDD 2006. LNCS (LNAI), vol. 4213, pp. 536–544. Springer, Heidelberg (2006). https://doi.org/10.1007/11871637_53

13. Hussain, I., Asghar, S.: A survey of author name disambiguation techniques: 2010–2016. Knowl. Eng. Rev. **32**, e22 (2017)

14. Kipf, T.N., Welling, M.: Variational graph auto-encoders. CoRR (2016)

15. Kipf, T.N., Welling, M.: Semi-supervised classification with graph convolutional networks. In: 5th International Conference on Learning Representations, ICLR 2017, Toulon, France, 24–26 April 2017, Conference Track Proceedings (2017)

16. Levin, M., Krawczyk, S., Bethard, S., Dan, J.: Citation-based bootstrapping for large-scale author disambiguation. J. Am. Soc. Inf. Sci. Technol. **63**(5), 1030–1047 (2012)

17. Li, J., et al.: Graph CNNs for urban traffic passenger flows prediction. In: Smart-World/SCALCOM/UIC/ATC/CBDCom/IOP/SCI 2018, Guangzhou, China, 8–12 October 2018 (2018)

18. Louppe, G., Al-Natsheh, H.T., Susik, M., Maguire, E.J.: Ethnicity sensitive author disambiguation using semi-supervised learning. In: Ngonga Ngomo, A.-C., Křemen, P. (eds.) KESW 2016. CCIS, vol. 649, pp. 272–287. Springer, Cham (2016). https://doi.org/10.1007/978-3-319-45880-9_21

19. Nguyen, T.H., Grishman, R.: Graph convolutional networks with argument-aware pooling for event detection. In: Proceedings of the Thirty-Second AAAI Conference on Artificial Intelligence, (AAAI-2018), the 30th Innovative Applications of Artificial Intelligence (IAAI-2018), and the 8th AAAI Symposium on Educational Advances in Artificial Intelligence (EAAI-2018), New Orleans, Louisiana, USA, 2–7 February 2018, pp. 5900–5907 (2018)

20. On, B.W.: Social network analysis on name disambiguation and more. In: International Conference on Convergence & Hybrid Information Technology (2008)

21. Peng, H., et al.: Fine-grained event categorization with heterogeneous graph convolutional networks. In: Proceedings of the Twenty-Eighth International Joint Conference on Artificial Intelligence, IJCAI 2019, Macao, China, 10–16 August 2019 (2019)

22. Peng, H., et al.: Hierarchical taxonomy-aware and attentional graph capsule RCNNs for large-scale multi-label text classification. CoRR (2019)

23. Peng, H., et al.: Large-scale hierarchical text classification with recursively regularized deep graph-CNN. In: Proceedings of the 2018 World Wide Web Conference on World Wide Web, WWW 2018, Lyon, France, 23–27 April 2018 (2018)

24. Schlichtkrull, M., Kipf, T.N., Bloem, P., van den Berg, R., Titov, I., Welling, M.: Modeling relational data with graph convolutional networks. ESWC 2018. LNCS, vol. 10843, pp. 593–607. Springer, Cham (2018). https://doi.org/10.1007/978-3-319-93417-4_38

25. Tang, J., Fong, A.C.M., Wang, B., Zhang, J.: A unified probabilistic framework for name disambiguation in digital library. IEEE Trans. Knowl. Data Eng. **24**(6), 975–987 (2012)

26. Tran, H.N., Huynh, T., Do, T.: Author name disambiguation by using deep neural network. In: Nguyen, N.T., Attachoo, B., Trawiński, B., Somboonviwat, K. (eds.) ACIIDS 2014. LNCS (LNAI), vol. 8397, pp. 123–132. Springer, Cham (2014). https://doi.org/10.1007/978-3-319-05476-6_13

27. Yoshida, M., Ikeda, M., Ono, S., Sato, I., Nakagawa, H.: Person name disambiguation by bootstrapping. In: Proceeding of the 33rd International ACM SIGIR Conference on Research and Development in Information Retrieval, SIGIR 2010, Geneva, Switzerland, 19–23 July 2010, pp. 10–17 (2010)

28. Zhang, B., Hasan, M.A.: Name disambiguation in anonymized graphs using network embedding. In: Proceedings of the 2017 ACM on Conference on Information and Knowledge Management, CIKM 2017, Singapore, 06–10 November 2017, pp. 1239–1248 (2017)

29. Zhang, Y., Zhang, F., Yao, P., Tang, J.: Name disambiguation in AMiner: clustering, maintenance, and human in the loop. In: Proceedings of the 24th ACM SIGKDD International Conference on Knowledge Discovery & Data Mining, KDD 2018, London, UK, 19–23 August 2018, pp. 1002–1011 (2018)

Graph Learning

Semi-supervised Graph Embedding
for Multi-label Graph Node Classification

Kaisheng Gao, Jing Zhang$^{(\boxtimes)}$ (iD), and Cangqi Zhou

School of Computer Science and Engineering,
Nanjing University of Science and Technology, Nanjing 210094, China
{kaisheng_gao, jzhang, cqzhou}@njust.edu.cn

Abstract. The graph convolution network (GCN) is a widely-used facility to realize graph-based semi-supervised learning, which usually integrates node, features, and graph topologic information to build learning models. However, as for multi-label learning tasks, the supervision part of GCN simply minimizes the cross-entropy loss between the last layer outputs and the ground-truth label distribution, which tends to lose some useful information such as label correlations, so that prevents from obtaining high performance. In this paper, we propose a novel GCN-based semi-supervised learning approach for multi-label classification, namely ML-GCN. ML-GCN first uses a GCN to embed the node features and graph topologic information. Then, it randomly generates a label matrix, where each row (i.e., label vector) represents a kind of labels. The dimension of the label vector is the same as that of the node vector before the last convolution operation of GCN. That is, all labels and nodes are embedded in a uniform vector space. Finally, during the model training of ML-GCN, label vectors and node vectors are concatenated to serve as the inputs of the relaxed skip-gram model to detect the node-label correlation as well as the label-label correlation. Experimental results on several graph classification datasets show that the proposed ML-GCN outperforms four state-of-the-art methods.

Keywords: Graph convolution network · Graph embedding · Graph node classification · Multi-label classification

1 Introduction

There exist many graph-structured datasets in the real world, such as social networks, academic citation networks, and knowledge graph. Graph Representation Learning (GRL) methods that aim to learn the vector representations for graphs has attracted much attention in recent years. Because the dimension of every node vector could be very large it may suffer from the high computational complexity and huge memory space usage, if we merely use the one-hot encoding methods or a discrete adjacency matrix to present the nodes. Therefore, we usually embed a graph into a low-dimensional space, which not only preserves the structural information but also significantly reduces the computational costs. Within this low-dimensional space, the distance between two nodes with a close relation in the original graph will also be close in a measure derived from the embedding presentation. Here, the close relation of two

© Springer Nature Switzerland AG 2019
R. Cheng et al. (Eds.): WISE 2019, LNCS 11881, pp. 555–567, 2019.
https://doi.org/10.1007/978-3-030-34223-4_35

nodes means that they are directly connected with each other or share a set of common neighbors, which is often used to define the similarity of two users in a social network.

There are several graph embedding methods proposed in recent years. For example, GF [1] factorizes the adjacency matrix and minimizes the L2-norm of the embedding matrix. LINE [2] defines two joint probability distributions for each pair of nodes, one using the adjacency matrix and the other using the embedding vector. Then, LINE minimizes the KL divergence of these two distributions. DeepWalk [3] uses a random walk to generate a node sequence. Then, for each node sequence, it applies the Word2Vec model [4] to get the node embedding by treating each sequence as a word sentence. All the above methods can be classified as the shallow model, compared with the methods using deep learning technology. Recently, a kind of deep learning models, namely graph neural networks (GNN), has attracted much attention, including some typical methods such as GraphSage [5], graph attention networks (GAT) [6], and graph convolutional networks (GCN) [7], which use neural networks to train classification models on graph-structured datasets.

GCN is a deep neural network model to catch structural information in a graph, which has been widely used in several machine learning paradigms, such as semantic role labeling [8], event extraction [9], and recommendation tasks [10]. In addition, the GCN model also obtains good performance in graph-based semi-supervised learning because its structure is robust to the missing information in training sets [7]. In a semi-supervised learning task, GCN uses a graph convolution operation to integrate each node and its one-hop neighbor information in each layer. After conducting several layers of convolution, each node in the network can gather its k-hop neighbor information in the final layer, which is the embedded feature presentation of such a node. Eventually, we can use some supervised information to train a classifier based on these embedded features.

Usually, multi-label classification models are trained in a semi-supervised manner, because not all labels on every instance are obtained values. In multi-label graph datasets, one node may have several labels. i.e. the correlation between this node and these labels are high, we called it node-label correlation. if two labels are highly correlated, the nodes with these labels should be close in the embedding space. For example, in movie genres dataset, the genres (labels) "Western" and "Adventure" always appear in the same movie. Thus, two movies with labels "Western" and "Adventure" respectively should also be close to each other. We called this *label-label correlation*. Because this correlation is not reflected in the graph structure it cannot be captured in the original GCN models. Accordingly, for a multi-label graph dataset, some nodes may have several specific labels. That is, one node and some labels may be highly correlated, which is called the *node-label correlation* in this study.

To address this issue, we propose a novel GCN-based model for semi-supervised multi-label graph node classification, namely ML-GCN. To capture the high non-linear correlations among nodes, we use a two-layer neural network model, on each of which we conduct a series of graph convolution operations. To preserve the label-label correlation, we treat each label as a vector so that we can measure the relationship between two labels. After labels are embedded, we can shorten the distance between two nodes whose labels are highly correlated in the embedding space. After obtaining the representation of all nodes in a graph, we can train a multi-label classifier to make predictions to the unlabeled nodes. In the proposed ML-GCN method, we use a sigmoid layer as the downstream learning method. The contributions of this paper are three-fold:

- We first investigate the applicability of graph convolutional network applying to the multi-label learning and point out that the label-label correlation should be considered to improve the learning performance.
- We propose a novel learning method, namely ML-GCN, where labels on nodes and the nodes themselves are uniformly embedded into the same low-dimensional space. ML-GCN can capture both node-label and label-label correlations. To the best of our knowledge, it is the first that the labels of each node are embedded and fed into GCN.
- We conduct a comprehensive empirical study on three real-world multi-label graph node classification datasets, whose results demonstrate that ML-GCN outperforms four state-of-the-art methods.

The remainder of the paper is organized as follows: In Sect. 2, we briefly review the related work. Section 3 presents the novel ML-GCN method. In Sect. 4, we compare our ML-GCN with four state-of-the-art methods on three real-world datasets. Section 5 concludes the paper.

2 Related Work

A large number of application problems can be abstracted as a classification problem in a graph structure, where some attributes of nodes in the graph are being predicted. In recent years, various kinds of graph neural network models have been proposed [11], such as graph convolution networks (GCN) [7], graph attention networks (GAT) [6], graph autoencoder [12], graph generative networks [13], graph spatial-temporal networks [14], and so on. The principle of most of these approaches is the *neural message passing* proposed by Gilmer et al. [15]. In the message-passing framework, a GNN can be viewed as a message passing algorithm, where the representation of a node is iteratively computed from the features of its neighbor nodes using a differentiable aggregation function. For the identity of the principle, the GCN model can be considered as a fundamental structure of most GNN models [11], which aggregates each node with its neighbors and let the node receive messages from its neighbors. Therefore, in this paper, we mainly focus on GCN models. GCN models can be divided into two categories: spectral-based and spatial-based approaches. The spectral-based methods define convolution operations by introducing filters from the perspective of graph signal processing [16], where the convolution on the graph is interpreted as removing noise from graph signals and passing messages in the spectral domain. The spatial-based approaches formulate convolution operations on a node as aggregating feature derived from its neighbors and the information passing through it. In general, all the GNN-based methods attempt to embed the graph structural information into vectors and follow the same hypothesis that nodes with similar structure tend to be close in the embedding space.

Multi-label learning is usually semi-supervised because, in many situations, instances in the training set do not necessarily have all the potential labels been assigned values. The training process usually learns from fully-labeled, partly-labeled, and even unlabeled samples to form predictive models. For the multi-label learning in a graph structure, a straightforward method is to train multiple independent binary classifiers for each label. However, this simple method has several defects: It does not consider the correlations among labels; The number of labels to predict will grow exponentially as the number of label categories increases; It is essentially limited by ignoring the topological structure among nodes. In some recent studies, researchers attempted to capture label-label correlations in some classical deep learning models for multi-label classification. Gong et al. [17] used a ranking-based learning strategy to train deep convolutional neural networks for multi-label image recognition and found that the weighted approximated-ranking loss performs best. Wang et al. [18] utilized recurrent neural networks (RNNs) to transform labels into embedded label vectors so that the correlation between labels can be employed. Wang et al. [19] introduced a spatial transformer layer and long short-term memory (LSTM) units to capture label correlation.

In this study, our novel learning method is still based on the GCN model but first introduces the label matrix embedding to capture the label-label correlation among the graph nodes.

3   The Proposed Method

The key idea behind the proposed ML-GCN is that it embeds multiple labels and nodes in the same space, where label-label correlations and label-node correlations can be simultaneously considered. In this section, we first introduce the problem statement and some preliminaries. Then, we present the label embedding scheme of ML-GCN. Finally, we present the optimization algorithm of the ML-GCN model.

3.1   Problem Statement

We define graph $G = (V, E, X, Y)$ as an undirected graph, where $V = (V_l \cup V_u)$ is a finite node set that includes n_l labeled nodes (V_l) and n_u unlabeled nodes (V_u). There are totally $n = n_l + n_u$ nodes on the graph. E is an edge set and $X \in \mathbb{R}^{n \times d}$ is a feature matrix of all the graph nodes. $Y \in \mathbb{R}^{n_l \times c}$ is a 0–1 matrix that presents the labels of n_l labeled nodes, where c is the number of different label types in the dataset. The adjacency matrix of the graph is denoted by $A = [a_{ij}] \in \mathbb{R}^{n \times n}$, where a_{ij} is the weight assigned on the edge between nodes i and j. The degree matrix of A is denoted by a diagonal matrix $D = diag(d_1, \ldots, d_n)$, where $d_i = \sum_j a_{ij}$ is the degree of node i. The symmetric normalized Laplacian matrix is denoted by $L_{sym} = I - D^{-\frac{1}{2}} A D^{-\frac{1}{2}}$. Our goal is to build a multi-label classification model that can predict the labels of unlabeled graph nodes.

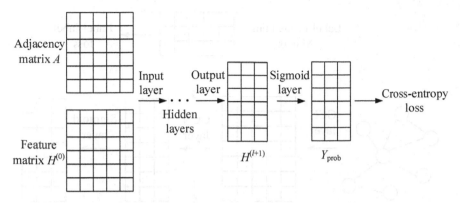

Fig. 1. The structure of graph convolutional networks (GCN).

3.2 Preliminaries: Graph Convolutional Network (GCN)

To embed features of nodes and their structural information, we first introduce the graph convolutional networks (GCN) [7]. Figure 1 shows a basic structure of graph convolutional networks. In particular, the core of GCN is the operation in each layer, which can be defined as follows:

$$H^{(l+1)} = \sigma\left(\tilde{D}^{-\frac{1}{2}}\tilde{A}\tilde{D}^{-\frac{1}{2}}H^{(l)}W^{(l)}\right). \tag{1}$$

Here, $\tilde{A} = A + I_{n_l + n_u}$ is an adjacency matrix with self-connections added. Matrix $I_{n_l + n_u}$ is an identity matrix. Diagonal matrix. Diagonal matrix $\tilde{D} = diag(\tilde{d}_1, \ldots \tilde{d}_n)$ is a degree matrix of $\tilde{A}$, where $\tilde{d}_i = \sum_j \tilde{A}_{ij} \cdot W^{(l)} \cdot W^{(l)}$ trainable parameters of the l-th layer. Function σ is an activation function. In this paper, the activation function of each layer is defined as $\sigma(x) = \max(0, x)$ as it used in other studies [7]. In the first layer, we have $H^{(0)} = X$. That is, we take the graph feature matrix as the input of GCN. In the last layer, we have

$$Y_{prob} = \text{sigmoid}\left(H^{(l+1)}\right) = 1/\left(1 + \exp\left(-H^{(l+1)}\right)\right), \tag{2}$$

where $H^{(l+1)} \in \mathbb{R}^{n \times c}$ and Y_{prob} is the probability distribution of labels for each node. Then, we minimize the cross-entropy loss between Y_{prob} and labeled nodes:

$$\min \sum_{i=1}^{n_l} y_i log\left(y_{prob}^{(i)}\right), \tag{3}$$

where y_i and y_{prob} denote the row vectors of Y and Y_{prob}, respectively. That is, we embed all nodes into a c-dimension space and use a sigmoid function to determine the predicted values of the labels. However, this simply model may be confronted with some drawbacks:

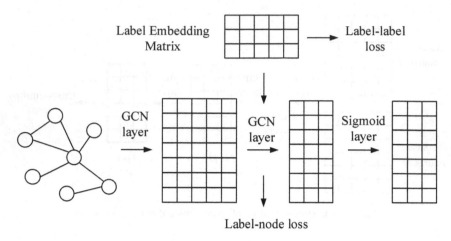

Fig. 2. The framework of the proposed ML-GCN.

- If we utilize fewer layers to construct GCN, the difference between the dimension of the last layer and the second to the last layer may be quite large. It may cause hidden feature loss and make the model difficult to optimize. For example, on the dataset in [20], whose input feature dimension is 3703 and the number of labels is six, if we use a two-layer GCN, we cannot let the dimension decrease smoothly regardless of the settings of the hidden layer dimension.
- As the article [21] pointed out, if we simply stack more layers, the model will mix the features of nodes from different labels and make them indistinguishable. This is because each layer of GCN applies Laplacian smoothing [22] to features, and every two nodes with a connected path tend to be close with Laplacian smooth.
- A multi-label classification model with a sigmoid layer cannot capture the label-label relationship because it treats each label individually. Thus, it may lose some information on the multi-label graph dataset.

To address these drawbacks, we propose a novel model ML-GCN for multi-label classification in the next subsection.

3.3 ML-GCN: Label Embedding Matrix

The ML-GCN introduces a label embedding matrix as well as the label-node co-embedding to GCN. Let $Z_Y \in \mathbb{R}^{c \times l}$ denote the label embedding matrix, where c is the number of different label classes and l is the dimension of label vectors. We generate the label embedding matrix randomly at the beginning of training. The dimension of the matrix is the same as the dimension of node features before the last graph convolution operation. Here, we set $H^{(l+1)}$ as the last output before the sigmoid layer. That is, the dimension l of the label embedding matrix is the same as the dimension of $H^{(l)}$. Then, we can calculate the label-label correlation and the label-node correlation using the Z_Y and $H^{(l)}$, respectively. Figure 2 shows the framework of the proposed ML-GCN. Here, each grid represents a matrix. We feed a graph into the first GCN layer and obtain

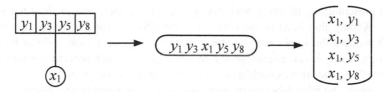

Fig. 3. Convert a node with several labels to a sentence.

the first embedding matrix as the output of this layer. Then, we use the randomly generated label embedding matrix to calculate the label-label loss, and together with the first embedding matrix to calculate the label-node loss. Then, we feed the first embedding matrix into the second GCN layer. Finally, we use the output of the sigmoid layer to calculate the cross-entropy loss against the ground truth.

Consider a node with several labels. Our goal is to maximize the occurrence probability of these labels given the node. The inputs are the node vectors and the corresponding label vectors. If we treat a node and its labels as a sentence, our goal also can be expressed as "given a center word (node), to predict the neighbor words (labels)," which is the essential idea of Skip-Gram [23]. For example, in Fig. 3, we have a node with four labels, and we can treat each element as a word and generate a sentence. Then, we utilize Skip-Gram for the next calculation.

In the Skip-Gram model, for a word w_i and window size c, we can extract w_i and its $c - 1$ neighbors with w_i at the center. Word w_i and each of its neighbor can form a pair as (w_i, w_j). The co-occurrence probability of w_j given w_i is defined as:

$$P(w_j|w_i) = \frac{\exp\left(w_j^T w_i\right)}{\sum_{t=1}^{M} \exp(w_t w_i)}, \tag{4}$$

where M is all the words in the corpus. Thus, we can obtain the word embedding by maximize such co-occurrence probability for all the word pairs.

Consider the node-label sentences. Given a node x_i and its labels $Y_{x_i} = \{y_1, y_2, \ldots, y_c\}$, the vector representation of x_i is the i-th row of $H^{(l)}$, denoted by h_i. The label vector of y_j is z_{y_j}. We only consider the node as the center word and remove the window size. We use each label to form a pair with the node because there is no predefined order of its labels. Therefore, we have a set of node-label pairs, denoted by $\{(x_iy_1), (x_iy_2), \ldots, (x_iy_c)\}$. For any node x_i, we can optimize the node and its label embedding by maximize the object function as follows:

$$\max \frac{1}{c} \sum_{y_j \in Y_{x_i}} \log P(z_{y_j}|h_i) \tag{5}$$

Since this function is operated in the second to the last layer of GCN and uses the features of layer $H^{(l)}$, we can better capture the node-label correlation in a high dimensional space before the feature dimension is reduced to the label-class wise. As we know GCN conducts the Laplacian smoothing on each node, whose consequence is that the presentations of many nodes may tend to be the same at the final stage of training. Adding this function prevents the side-effort of the Laplacian smoothing in GCN. It hinders the Laplacian smoothing which aggregates each node to be hard to distinguish. Thus, it can accelerate the training process and prevent the model from over smoothing that makes each node converge to the same point.

To capture the label-label correlation, we utilize the same model but get rid of the node vectors. That is, we only use the labels of a node to construct the sentence. For example, given a node x_i with labels $\{y_1, y_2, \ldots, y_c\}$, we only use labels to construct a sentence, which forms a set of label-label pairs with the combinations of all different labels, denoted by $\{(y_1 y_2), (y_1 y_3), \ldots, (y_c y_{c-1})\}$. Note that the pairs of $(y_i y_j)$ and $(y_j y_i)$ are different. similar to Eq. (5), we have the objective function as follows:

$$\max \frac{1}{c} \sum\nolimits_{y_i, y_j \in Y_{x_i}, i \neq j} \log P(z_{y_j} | z_{y_i}). \tag{6}$$

If the node only has one label, we omit the label-label relation and only calculate Eq. (5) on this node. To maximize the Eqs. (5) and (6), we can reserve the node-label correlation as well as label-label correlation in the embedding space.

3.4   Co-optimization and Negative Sampling

To calculate Eqs. (5) and (6), we need to calculate $P(z_{y_j} | h_i)$ and $P(z_{y_j} | z_{y_i})$, which requires the summation over all the labels. The calculation may cost too much running time because some multi-label graph datasets may have abundant label types. To accelerate the calculation of these two co-occurrence probabilities, we use a trick of negative sampling in the Skip-Gram model. First, we rewrite Eq. (1) as follow:

$$\min - \log \sigma(z_{y_j} h_i) - \sum\nolimits_{t=1}^{K} \mathbb{E}_{y_t \sim P(y)} \log \sigma(-z_{y_t} h_i), \tag{7}$$

where K is a hyperparameter that represents the number of sampled labels for one node-label pair. Therefore, the task becomes to distinguish the target label y_j from the K labels drawn from the noise distribution $P(y)$. The idea behind the negative sampling is: We will maximize the co-occurrence probability of z_{y_j} given h_i and minimize the probability of a randomly sampled labels z_{y_t} given the same node h_i. In practice, we define a noise distribution as chosen to be $U(y)^{3/4} / \sum_y U(y)^{3/4}$, where $U(y)$ is the unigram distribution of the labels. Here, we only consider the co-occurrence times of each label type on labeled data as the unigram distribution. If the sample process obtains the positive label $y_t = y_j$, we just resample y_t until the condition $y_t \neq y_j$ is satisfied.

Similar to Eq. (5), we sample K labels as the negative labels and rewrite Eq. (6) as follows:

$$\min - \log \sigma(z_{y_j} z_{y_i}) - \sum\nolimits_{t=1}^{K} \mathbb{E}_{y_t \sim P(y)} \log \sigma(-z_{y_t} z_{y_i}). \tag{8}$$

The goal is to distinguish the label y_j from K sampled negative labels on the condition of given y_i. To calculate Eqs. (7) and (8) in each labeled node, we can obtain the loss function L_{n-l} denoting the node-label loss calculated by Eq. (7) and l_{l-l} denoting the label-label loss calculated by Eq. (8). With the sigmoid loss of the last layer, we can have the final objective for optimization:

Algorithm 1: ML-GCN (Training and Predicting)

Input: Graph G, feature X, label Y_L, number of GCN layers $l+1$
Output: labels of unlabeled nodes Y_U
1: randomly generate the label matrix Z_Y
2: $H^{(0)} = X$
3: **for** epoch = $1, ..., n$ **do**:
3:     **For** i = $0, 1, ..., l$ **do**:
4:         calculate the output of $i+1$ GCN layer $H^{(i+1)}$ using $H^{(i)}$
5:     $L_{sigmoid} = \text{crossentropy}\left(Y_L, \text{sigmoid}(H^{(l+1)})\right)$
6:     calculate Eq. (7) using Z_Y and $H^{(l)}$, and obtain L_{n-l}
7:     calculate Eq. (8) using Z_Y, and obtain L_{l-l}
8:     optimize to minimize $L_{sum} = \lambda_1 L_{l-l} + \lambda_2 l_{n-l} + l_{sigmoid}$ using Adam [24]
9: **return** 1 **if** $Y_{prob} = \text{sigmoid}(H^{(l+1)}) > 0.5$, **else** 0, given Y_U

$$L_{sum} = \lambda_1 L_{l-l} + \lambda_2 l_{n-l} + l_{sigmoid}, \tag{9}$$

where $\lambda_1, \lambda_2 \in \mathbb{R}$ are the hyper parameters to weight three terms in the objective functions. We optimize the function with Adam optimizer [24]. We summarize all above contents with a pseudocode and list in Algorithm 1 (ML-GCN).

4   Experiments

In this section, we first present the datasets used in our experiments, methods in comparisons, and the experimental settings. Then, we focus on discussing the experimental results.

4.1   Datasets

Compared with the plenty of single-label classification datasets, there are only a few real-world multi-label graph node classification datasets that can be used in our experiments. We evaluate our ML-GCN model on three datasets collected from different domains of biology, movie, and social media. These datasets are chosen not only because they belong to different domains but also, they have different network topologic structures. The details of the datasets are listed in Table 1.

Table 1. The details of the datasets used in our experiments

Dataset	Domain	Nodes	Edges	Classes	Features
Facebook	Social	347	5038	24	224
Yeast	Biology	1240	1674	13	831
Movie	Movie	7155	404241	20	5297

The *Facebook* dataset [25] is a social network. The nodes represent users of Facebook and the edges represent the fan following relation. The feature of each node is the personal information of the corresponding user. The task is to determine the 'circles' tags of each user (node). One user can belong to multiple circles.

The *Yeast* dataset is part of the KDD Cup 2001 challenge[1]. The graph is constructed based on the interactions between proteins. Each node represents a gene. The gene code information is set as the feature of nodes. The task is to predict the function of these genes.

We constructed a movie dataset from Movielens-2k dataset[2]. The Movielens-2k dataset contains movies information such as actors, genres, and tags information. We set the tags information as the feature of movies and set a common director as an edge. For example, if two movies share the same director, we added an edge between these two movies, and set the weight of this edge to 1. The task is to predict the genres of the movies.

4.2 Methods in Comparison and Experimental Settings

Method in Comparison: We compared our ML-GCN with the following state-of-the-art methods:

- Multilayer perception (MLP) is a classical label classifier takes only node feature as input and ignores the graph structure.
- Deepwalk [3] learns node features by treating random walks in a graph as the equivalent of sentences.
- GCN [7] takes both node feature and graph structure as the input.
- Partly ML-GCN is a simpler ML-GCN without the calculation of the label-label loss. This method is added to evaluate the impact of the loss function on the performance of the learning models.

Experimental Settings: For fair comparisons, all the methods (MLP, GCN, Partly ML-GCN and ML-GCN) use two-layer models. For dataset *Facebook*, we set the middle layer dimension to 64 and use 100 nodes for training and 150 nodes for testing. For dataset *Yeast*, we set the middle layer dimension to 256 and use 200 nodes as training nodes and 500 nodes as testing nodes. For dataset Movie, we set the middle

---

[1] http://pages.cs.wisc.edu/~dpage/kddcup2001/.

[2] http://ir.ii.uam.es/hetrec2011/datasets.html.

layer dimension to 512 and use 500 nodes as training data and 2000 nodes as testing nodes. For all datasets, we set the number of negative sample to 5, set walk length to 40 for DeepWalk and set the window size to 10. All models are trained using Adam with a learning rate of 0.01. The parameters λ_1 and λ_2 are both equal to 0.25. We use the micro-F1 score (in percentage) as the evaluation metric in the paper.

4.3 Experimental Results

Experiment 1 (Overall Performance): The classification results of five methods on three datasets in terms of the micro-F1 score are summarized in Table 2. The best performance is in bold in the table. We have the following observations. Overall, our proposed ML-GCN method consistently outperforms the other methods on all datasets. Compared with the original GCN, on dataset *Facebook*, our ML-GCN achieves the improvement of 1.72 points, and on the datasets with stronger label-label correlations (i.e., datasets *Yeast* and *Movie*), the improvement of ML-GCN archived as high as 3 points. Thus, ML-GCN successfully captures the label-label correlations. Furthermore, our ML-GCN also outperforms the Partly ML-GCN on all dataset, which shows that the calculation of the label-label loss in the model training indeed improves the performance of the learning models.

Table 2. Experimental results in terms of micro-F1 score (in percentage)

Method	Facebook	Yeast	Movie
MLP	58.13	63.79	33.62
DeepWalk [3]	58.89	53.40	33.94
GCN [7]	58.13	63.16	35.72
Partly ML-GCN	59.51	65.27	37.75
ML-GCN	**59.85**	**66.06**	**37.96**

Experiment 2 (Performance Under Different Training Set Sizes): To investigate whether our ML-CGN is consistently superior to GCN under different training set sizes, we randomly selected different proportions of the instances from the original datasets to form the training sets. The experimental results are summarized in Table 3. We have the following observations. Overall, the proposed ML-GCN outperforms GCN under all different training set sizes on all datasets. On the *Movie* dataset, the advantage of ML-GCN over GCN will increase as the proportion of the training sets increases. That means, when the training instances increase, our ML-GCN is easier to capture the label-label correlations. On the *Yeast* dataset, the exceeding of ML-GCN to GCN is around 2 pinots, which is similar to that on the Movie dataset but better than that on the Facebook dataset. Again, it shows that on the datasets with stronger label-label correlations (i.e., datasets *Yeast* and *Movie*), our ML-GCN performs much better.

Table 3. Experimental results under different training set sizes in terms of micro-F1 score

Dataset	Method	10%	20%	30%	40%
Facebook	GCN [7]	57.25	58.45	59.95	60.05
	ML-GCN	58.13	59.63	60.14	60.98
Yeast	GCN [7]	61.23	62.45	62.73	63.68
	ML-GCN	63.03	64.54	64.04	65.77
Movie	GCN [7]	36.82	37.64	38.06	38.23
	ML-GCN	38.04	39.92	40.64	40.76

5   Conclusion

In this paper, we present a novel ML-GCN method for semi-supervised multi-label graph node classification. By embedding the label and node information into the same low-dimensional space, ML-GCN can jointly capture both node-label and la-bel-label correlations, which improves the performance of the learning models, com-pared with the state-of-the-art methods. In the future, we will consider embedding the contents of nodes to the learning models.

Acknowledgments. This work has been supported by the National Natural Science Foundation of China under grants 91846104 and 61603186, the Natural Science Foundation of Jiangsu Province, China, under grants BK20160843 and BK20180463, and the China Postdoctoral Science Foundation under grants 2017T100370.

References

1. Ahmed, A., Shervashidze, N., Narayanamurthy, S., Josifovski, V., Smola, A.J.: Distributed large-scale natural graph factorization. In: Proceedings of the 22nd International Conference on World Wide Web, pp. 37–48. ACM (2013)
2. Tang, J., Qu, M., Wang, M., Zhang, M., Yan, J., Mei, Q.: LINE: large-scale information network embedding. In: Proceedings of the 24th International Conference on World Wide Web, pp. 1067–1077. International World Wide Web Conferences Steering Committee (2015)
3. Perozzi, B., Al-Rfou, R., Skiena, S.: Deepwalk: online learning of social representations. In: Proceedings of the 20th ACM SIGKDD International Conference on Knowledge Discovery and Data Mining, pp. 701–710. ACM (2014)
4. Mikolov, T., Chen, K., Corrado, G., Dean, J.: Efficient estimation of word representations in vector space. arXiv preprint arXiv:1301.3781 (2013). https://arxiv.org/abs/1301.3781
5. Hamilton, W., Ying, Z., Leskovec, J.: Inductive representation learning on large graphs. In: Advances in Neural Information Processing Systems, pp. 1024–1034 (2017)
6. Veličković, P., Cucurull, G., Casanova, A., Romero, A., Lio, P., Bengio, Y.: Graph attention networks. arXiv preprint arXiv:1710.10903 (2017). https://arxiv.org/abs/1710.10903
7. Kipf, T.N., Welling, M.: Semi-supervised classification with graph convolutional networks. arXiv preprint arXiv:1609.02907 (2016). https://arxiv.org/abs/1609.02907

8. Marcheggiani, D., Titov, I.: Encoding sentences with graph convolutional networks for semantic role labeling. In: Proceedings of the 2017 Conference on Empirical Methods in Natural Language Processing, pp. 1506–1515 (2017)

9. Nguyen, T.H., Grishman, R.: Graph convolutional networks with argument-aware pooling for event detection. In: Thirty-Second AAAI Conference on Artificial Intelligence, pp. 5900–5907. AAAI (2018)

10. Ying, R., He, R., Chen, K., Eksombatchai, P., Hamilton, W.L., Leskovec, J.: Graph convolutional neural networks for web-scale recommender systems. In: Proceedings of the 24th ACM SIGKDD International Conference on Knowledge Discovery & Data Mining, pp. 974–983. ACM (2018)

11. Wu, Z., Pan, S., Chen, F., Long, G., Zhang, C., Yu, P.S.: A comprehensive survey on graph neural networks. arXiv preprint arXiv:1901.00596 (2019). https://arxiv.org/abs/1901.00596

12. Kipf, T.N., Welling, M.: Variational graph auto-encoders. arXiv preprint arXiv:1611.07308 (2016). https://arxiv.org/abs/1611.07308

13. You, J., Ying, R., Ren, X., Hamilton, W., Leskovec, J.: GraphRNN: generating realistic graphs with deep auto-regressive models. In: International Conference on Machine Learning, pp. 5694–5703 (2018)

14. Yan, S., Xiong, Y., Lin, D.: Spatial temporal graph convolutional networks for skeleton-based action recognition. In: Thirty-Second AAAI Conference on Artificial Intelligence, pp. 7444–7452. AAAI (2018)

15. Gilmer, J., Schoenholz, S.S., Riley, P.F., Vinyals, O., Dahl, G.E.: Neural message passing for quantum chemistry. In: Proceedings of the 34th International Conference on Machine Learning, pp. 1263–1272 (2017). http://www.JMLR.Org

16. Shuman, D., Narang, S., Frossard, P., Ortega, A., Vandergheynst, P.: The emerging field of signal processing on graphs: extending high-dimensional data analysis to networks and other irregular domains. IEEE Signal Process. Mag. 3(30), 83–98 (2013)

17. Gong, Y., Jia, Y., Leung, T., Toshev, A., Ioffe, S.: Deep convolutional ranking for multilabel image annotation. In: International Conference on Learning Representations (2014). https://arxiv.org/abs/1312.4894

18. Wang, J., Yang, Y., Mao, J., Huang, Z., Huang, C., Xu, W.: CNN-RNN: a unified framework for multi-label image classification. In Proceedings of the IEEE Conference on Computer Vision and Pattern Recognition, pp. 2285–2294 (2016)

19. Wang, Z., Chen, T., Li, G., Xu, R., Lin, L.: Multi-label image recognition by recurrently discovering attentional regions. In: Proceedings of the IEEE International Conference on Computer Vision, pp. 464–472 (2017)

20. Sen, P., Namata, G., Bilgic, M., Getoor, L., Galligher, B., Eliassi-Rad, T.: Collective classification in network data. AI Mag. 29(3), 93 (2018)

21. Li, Q., Han, Z., Wu, X.M.: Deeper insights into graph convolutional networks for semi-supervised learning. In: Thirty-Second AAAI Conference on Artificial Intelligence, pp. 3538–3545. AAAI (2018)

22. Taubin, G.: A signal processing approach to fair surface design. In: Proceedings of the 22nd Annual Conference on Computer Graphics and Interactive Techniques, pp. 351–358. ACM (1995)

23. Mikolov, T., Sutskever, I., Chen, K., Corrado, G.S., Dean, J.: Distributed representations of words and phrases and their compositionality. In: Advances in Neural Information Processing Systems, pp. 3111–3119 (2013)

24. Kingma, D.P., Ba, J.: Adam: a method for stochastic optimization. arXiv preprint arXiv:1412.6980 (2014). https://arxiv.org/abs/1412.6980

25. Leskovec, J., Mcauley, J.J.: Learning to discover social circles in ego networks. In: Advances in Neural Information Processing Systems, pp. 539–547 (2012)

Structural Role Enhanced Attributed Network Embedding

Zhao Li[1], Xin Wang[1,2]($\boxtimes$), Jianxin Li[3], and Qingpeng Zhang[4]

[1] College of Intelligence and Computing, Tianjin University, Tianjin, China
[2] Tianjin Key Laboratory of Cognitive Computing and Application, Tianjin, China
{lizh,wangx}@tju.edu.cn
[3] School of Information Technology, Deakin University, Melbourne, Australia
jianxin.li@deakin.edu.au
[4] School of Data Science, City University of Hong Kong, Hong Kong, China
qingpeng.zhang@cityu.edu.hk

Abstract. In recent years, network embedding methods based on deep learning to process network structure data have attracted widespread attention. It aims to represent nodes in the network as low-dimensional dense real-value vectors and effectively preserve network structure and other valuable information. Most network embedding methods now only preserve the network topology and do not take advantage of the rich attribute information in networks. In this paper, we propose a novel deep attributed network embedding framework (RolEANE), which can preserve network topological structure and attribute information well at the same time. The framework consists of two parts, one of which is the network structural role proximity enhanced deep autoencoder, which is used to capture highly nonlinear network topological structure and attribute information. The other part is that we proposed a neighbor optimization strategy to modify the Skip-Gram model so that it can integrate the network topological structure and attribute information to improve the final embedded performance. The experiments on four real datasets show that our method outperforms other state-of-the-art network embedding methods.

Keywords: Network embedding · Attributed network · Autoencoder · Structural role proximity

1 Introduction

In the real world, a natural network structure is formed between entities, such as social relationships between individuals, communication relationships between terminal devices, and capital transactions between merchants. The network is an important form and information carrier that expresses the connection between these entities. With the development of big data and Internet technologies, the scale and complexity of network structure data in the real world has gradually increased. This causes the traditional network representation algorithms based

© Springer Nature Switzerland AG 2019
R. Cheng et al. (Eds.): WISE 2019, LNCS 11881, pp. 568–582, 2019.
https://doi.org/10.1007/978-3-030-34223-4_36

on spectrum embedding, optimization, and other frameworks to face many problems when solving the corresponding problems. In recent years, the network representation learning method based on deep learning to process network structure has attracted the attention of both academia and industry. This new network representation learning is also called network embedding. It aims to represent nodes in the network as low-dimensional dense real-value vectors and effectively preserve network structure and other valuable information. These real-value vectors can easily and efficiently support downstream tasks such as node classification, link prediction, node clustering, network visualization, and more.

The earlier network embedding algorithm is mainly based on the method of matrix eigenvector calculation. This method generally converts the network into a matrix representation, and then performs dimensionality reduction to obtain a low-dimensional representation of the network by solving the form of the matrix feature vector. This type of algorithm obtains a k-dimensional node representation by computing the first k eigenvectors or singular vectors of the relation matrix. The relation matrix is generally the adjacency matrix or Laplace matrix of the network. Since the feature vector needs to store the relation matrix as a whole in the main memory during the calculation process, the eigenvector calculation of the large-scale matrix is time- and space-consuming. So this type of method is difficult to be applied on large-scale data. Therefore, focusing on the research of new network embedding algorithms is an important task.

In addition to the topological structure information, the network also has a lot of attribute information, which often contains very valuable information. For example, the financial network has very rich financial transaction information, customer basic information, financial product information, equipment fingerprint information and so on. This information plays a particularly important role in financial business applications such as transaction anti-fraud, personal credit risk assessment, personalized product recommendation, and precision marketing. These attributed networks is very common in real life, so it is meaningful to study the attributed network. In addition, the topological structure and attribute information of the network are complex and highly nonlinear, so capturing this highly nonlinear information during network embedding is a very difficult and challenging task.

2   Related Work

2.1   Plain Network Structure Embedding

DeepWalk [1] is inspired by the famous word representation learning algorithm word2vec. Perozzi et al. experimentally verified that the nodes in the random walk sequence and the words in the document all follow the power-law. Therefore, the famous word2vec algorithm is further applied to the random walk sequence to learn the node representation. DeepWalk first samples a large number of random walk sequences on the network, and then uses the Skip-Gram and Hierarchical Softmax models to model the probability of each pair of nodes in the random walk sequence. The other two representative algorithms based on the expansion of the Random Walk framework are node2vec [2] and LINE [3]. Node2vec introduces the Breadth-First Sampling (BFS) and Depth-First Sampling (DFS) into the generation process of random walk sequences by setting two parameters p and q. LINE is a network representation learning algorithm that can be applied to large-scale directed weighted graphs. The LINE algorithm proposes the concept of first-order proximity and second-order proximity, and probabilistic modeling of all first-order proximity and second-order proximity nodes. SDNE [4] designs the unsupervised learning part by reconstructing the neighborhood structure of each node to maintain the second-order proximity, and uses the first-order proximity as the supervised information to improve the representation in the potential space. There are many other classic network embedding methods, such as GraRep [5] for capturing high-order proximity, HOPE [6] for capturing asymmetric high-order proximity in directed networks, and so on.

2.2 Attributed Network Embedding

In addition to the topological structure information, the network also contains a wealth of attribute information. These information content usually has a lot of value, which not only has a huge impact on the formation of the network, but also directly affects the performance of network embedding. In recent years, the attributed network embedding attracted wide attention. TADW [7] proves that DeepWalk is equivalent to matrix decomposition, which combines DeepWalk and contextual text features into a matrix decomposition framework to implement attributed network embedding. AANE [8] enables the joint learning process to be done in a distributed manner, thereby accelerating the embedding of attributed networks. HSCA [9] integrates homophily, structural context, and vertex content to learn an effective network representation. LANE [10] learns vertex representations by embedding network structure proximity, attribute associations, and label proximity into a unified potential representation. pRBM [11] uses a Restricted Boltzmann Machine (RBM) to design a new model called paired RBM, which learns vertex representation by combining vertex attributes and link information. ASNE [12] proposes a general framework for embedding social networks by capturing structural proximity and attribute proximity. However, most of the existing attributed network embedding methods only retain the local network structure and attribute information, and it is difficult to obtain global information and highly nonlinear information.

2.3 Structural Role Proximity Network Embedding

At present, most of the existing network embedding methods rarely consider the information of the network node structural role. In addition to the first-order and second-order relationships, some nodes in the network usually have special structural role relationships. Structural role proximity is intended to embed nodes that are distant from each other but have similar structural roles to each other. Struct2vec [13] first encodes the vertex structural role proximity into a multilayer graph, where the weight of each layer's edges is determined by the corresponding proportional structural role difference. Then it performs Deep-Walk on the multilayer graph to learn the vertex representation. GraphWave [14] uses the spectral graph wavelet diffusion model to embed the vertex neighborhood structure into the low-dimensional space and maintain the proximity of the structural role. These network embedding methods that consider the network structural role proximity, only preserve the network topological structure information but do not preserve the network attribute information.

2.4 Our Contribution

It is a difficult problem to seamlessly integrate the topological structure information and attribute information in the network, so that the performance of network embedding is improved. In addition, it is also a challenge to effectively capture highly nonlinear information in the network. In order to solve the above key challenges, we propose a novel deep attributed network embedding framework called RolEANE. Our main contributions are as follows:

- We propose a network structural role proximity enhanced deep autoencoder, which can well capture highly nonlinear network topological structure and attribute information. It also captures the network high-order proximity and the structural role proximity, which can well preserve the global information of the network.
- We propose a neighbor optimization strategy to modify the Skip-Gram model, which can seamlessly integrate the network topological structure and attribute information to improve the final embedded performance.
- We have verified the effectiveness of this network embedding framework through a large number of experiments on four real-world datasets.

3   The Proposed Model

In this section, we will give a detailed description of each part of the RolEANE model framework.

3.1   Problem Definition

Definition 1. (Attributed Network) *Let $G = (V, E, A)$ be an attributed network, where V denotes the set of n nodes and E represents the set of edges. $A \in \mathbb{R}^{n \times m}$ is a matrix that encodes all node attribute information. $A_i. \in \mathbb{R}^m$ is the i-th row of A, which denotes the attributes of the i-th node.*

Definition 2. (Topological Structure Proximity) *denotes the proximity of two nodes in topological structure of network, including first-order proximity, second-order proximity, and high-order proximity. The first-order proximity captures network homogeneity, which means that the directly connected nodes are similar. The second order proximity captures the information of common neighbors, which means that nodes with more common neighbors are more similar to each other. The higher order proximity is used to capture more neighbor information.*

Definition 3. (Attribute Proximity) *denotes the similarity of network attribute information between nodes. Specifically, the attribute proximity of two nodes v_i and v_j is determined by similarity of A_i. and A_j.. By constraining the attribute proximity, we can model the homogeneity of attributes so that nodes with similar attributes will be close to each other in the space after embedding.*

Definition 4. (Structural Role Proximity) *denotes the structural role similarity of nodes in the network with each other. In addition to the direct connection and local neighbor information, the network structure usually has similar structural role nodes at the global level such as the edge of the chain, the center of the star, the group members, and the bridge between the two communities. The structural role proximity is intended to embed nodes that are far apart but have similar structural roles between each other, which can preserve more global information about the network.*

3.2   Structural Role Proximity Enhanced Autoencoder

In this paper, we use a node topological potential calculation method to measure the structural role proximity of network nodes. Analogous to the field concept in physics, we regard a network as a data field with a field effect, where the nodes of the network influence other nodes along the edges. The position of the node in the network structure is equivalent to the potential of the node in the data field, which is called topological potential. We believe that nodes in the network can affect other nodes and are also affected by other nodes. The importance of each node is different, and the degree of influence between nodes is also related to the distance between nodes. The interaction between nodes generally has local characteristics, and the influencing ability of each node will rapidly decline as the network distance increases. Therefore, this paper uses the node topological potential to quantify the structural role proximity of network nodes, and uses Gaussian potential function with good mathematical properties to describe the interaction between nodes. The structural role proximity of nodes in the network is defined as:

$$\varphi(v_i) = \frac{1}{n} \sum_{j=1}^{n} m_j \times e^{-(\frac{wd_{ij}}{\sigma})^2} \tag{1}$$

where m_j represents the quality of the node v_j, which is simply measured by the degree of the node $\deg(v_j)$. σ indicates the influence factor, which is an adjustable parameter that can affect the range of the node. wd_{ij} represents the network weighted shortest distance between the node v_i and v_j.

In the actual scenario, the interaction between network nodes is positively correlated with the weight of the edge and negatively correlated with the distance between the two nodes. So this paper defines wd_{ij} as:

$$wd_{ij} = \sum_{k=1}^{l} \frac{d_k}{w_k} \tag{2}$$

In particular, d_k represents the length of each edge in the shortest path between the node v_i and v_j, where the length of the edge is set to 1. When the network is unweighted, setting $w_k = 1$.

We use autoencoder to capture nonlinear information in the network. Autoencoder is a powerful and widely used deep learning model. The basic autoencoder includes input layer, hidden layer, and output layer, as follows:

$$h_i = \sigma \left(W^{(1)} x_i + b^{(1)} \right), \hat{x}_i = \sigma \left(W^{(2)} h_i + b^{(2)} \right) \tag{3}$$

where x_i is the input data, and h_i is the implicit layer representation of the encoder. $\hat{x}_i$ is the reconstructed result of the decoder for x_i. $\sigma(\cdot)$ represents a nonlinear activation function.

Let $\theta = \left\{ W^{(1)}, W^{(2)}, b^{(1)}, b^{(2)} \right\}$ be the model parameter set. We update the parameter θ by learning the following reconstruction error function:

$$\min_{\theta} \sum_{i=1}^{n} \| \hat{x}_i - x_i \|_2^2 \tag{4}$$

In order to capture the highly nonlinear topological structure and attribute information in the network, we propose a network structural role proximity enhanced deep autoencoder model. More specifically, the representation of each hidden layer is formulated as follows:

$$
\begin{aligned}
h_i^{(1)} &= \sigma \left(W^{(1)} x_i + b^{(1)} \right), \dots \\
h_i^{(k)} &= \sigma \left(W^{(k)} h_i^{(k-1)} + b^{(k)} \right)
\end{aligned}
\tag{5}
$$

where k denotes the number of layers for the encoder and decoder. Correspondingly, $h_i^{(k)}$ refers to the representation of the i-th node via the k layer encoder. $\sigma(\cdot)$ represents the activation functions. $W(k)$ and $b(k)$ are the transformation matrix and bias vector in the k-th layer, respectively.

We use two network structural role proximity enhanced deep autoencoder to embed network topological structure and network attributes, respectively.

In order to capture highly nonlinear information of network attributes, the attribute representation H^A is obtained by minimizing the following attribute reconstruction error function:

$$L_{ae} = \sum_{i=1}^{n} \left\| \hat{A}_{i\cdot} - \varphi(v_i) A_{i\cdot} \right\|_2^2 \tag{6}$$

where $A = [A_{ij}] \in \mathbb{R}^{n \times m}$ is the attribute matrix. $A_{i\cdot} \in \mathbb{R}^m$ is the i-th row of A, which denotes the attribute of the i-th node.

In order to capture the highly nonlinear information of the network topological structure, the topological structure representation H^M is obtained by minimizing the following topological structure reconstruction error function:

$$L_{se} = \sum_{i=1}^{n} \left\| \hat{M}_{i\cdot} - \varphi(v_i) M_{i\cdot} \right\|_2^2 \tag{7}$$

where M is the high-order proximity matrix [5]. This paper uses the random walk to obtain the high-order context of each node and then construct its corresponding adjacency matrix M.

3.3   Neighbor-Modified Skip-Gram Model

Up to now, we have successfully embedded network topological structure and attribute information through the proposed structural role proximity enhanced deep autoencoder. However, it is a challenge to seamlessly integrate the final representations H^A and H^M without generating noise. In order to solve this problem, we use a neighbor optimization strategy to modify the Skip-Gram model to seamlessly integrate network topological structure and attribute information.

The core idea of the Skip-Gram model is that two words with similar contexts have similar representations. Inspired by this idea, we expand it as follows:

- If two nodes have similar neighbor nodes, then the two nodes are similar.
- If two nodes have neighbor nodes with similar attribute information, the two nodes are similar.

Sampling of neighbor nodes is a key part of the above expansion. Drawing on the idea of DeepWalk, we first use random walk to generate a large number of paths on network $G = (V, E)$, and then get the neighbors of each node $v_i \in V$ from these paths. Specifically, given a window size b, a random walk sequence is generated. If the node v_j appears in the neighbor window of the node v_i, the node v_j is the neighbor of the node v_i. It is worth noting further that the node v_j does not have to be a directly connected node of the node v_i, as long as the node v_i can reach the node v_j in b steps.

Based on the above analysis, our goal is to maximize the following function:

$$\prod_{v_i \in V} \prod_{v_j \in N(i)} p(v_j | v_i) \tag{8}$$

The objective function above can be equivalent to minimize its negative logarithm as follows:

$$L_{as} = - \sum_{v_i \in V} \sum_{v_j \in N(i)} \log \left(p(v_j | v_i) \right) \tag{9}$$

where $N(i)$ is the set of neighbor nodes of node v_i, $p(v_j | v_i)$ is the joint probability of the topological structure representation of the node v_i and the attribute representation of the node v_j, which is defined as:

$$p(v_j | v_i) = \frac{\exp \left(\left(H_{j \cdot}^A \right)^T H_{i \cdot}^M \right)}{\sum_{k=1}^{n} \exp \left(\left(H_{k \cdot}^A \right)^T H_{i \cdot}^M \right)} \tag{10}$$

For large-scale networks, it is difficult to directly calculate the joint probability, so we uses negative sampling technology [15] to optimize:

$$\log \sigma \left(\left(H_{j \cdot}^A \right)^T H_{i \cdot}^M \right) + \sum_{i=1}^{k} E_{v_n \sim P_{n(v)}} \log \sigma \left(- \left(H_{n \cdot}^A \right)^T H_{i \cdot}^M \right) \tag{11}$$

where $P_{n(v)}$ is the negative sample set of node v_i, and empirically set $P_{n(v)} \propto \deg(v)^{3/4}$ [15]. k is the number of negative edges, and $\sigma(x) = 1/(1 + \exp(-x))$ is the sigmoid function.

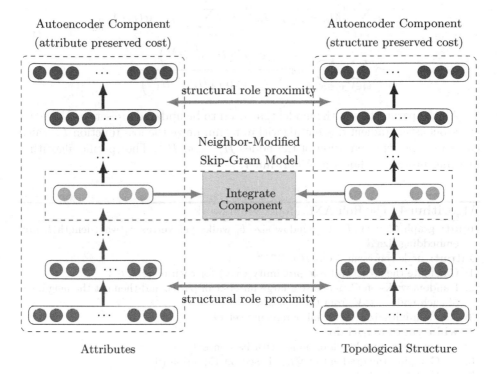

Fig. 1. The RolEANE model framework.

3.4 The RolEANE Model Framework

In this subsection, we will introduce how the RolEANE model framework uses network topological structure and attribute information to perform network embedding. As shown in Fig. 1, the RolEANE model framework uses two network structural role proximity enhanced deep autoencoder to embed network topological structure information and network attribute information. After the k-layer representation of the deep autoencoder, the network structure representation H^M and the network attribute representation H^A are obtained. Then we substitute H^M and H^A into the neighbor-modified Skip-Gram model, which can seamlessly integrate the two.

As a result, the objective function of the RolEANE model framework is defined as a linear combination of L_{se}, L_{ae}, and L_{as}:

$$
\begin{aligned}
L =& \alpha L_{se} + \beta L_{ae} + \gamma L_{as} \\
=& \alpha \sum_{i=1}^{n} \left\| \hat{M}_{i\cdot} - \varphi(v_i) M_{i\cdot} \right\|_2^2 + \beta \sum_{i=1}^{n} \left\| \hat{A}_{i\cdot} - \varphi(v_i) A_{i\cdot} \right\|_2^2 \\
& -\gamma \sum_{v_i \in V} \sum_{v_j \in N(i)} \log \frac{\exp\left(\left(H_{j\cdot}^A \right)^T H_{i\cdot}^M \right)}{\sum_{k=1}^{n} \exp\left(\left(H_{k\cdot}^A \right)^T H_{i\cdot}^M \right)}
\end{aligned}
\tag{12}
$$

All the parameters in the model that need to be updated are set to Θ. We use the stochastic gradient descent algorithm to minimize the loss function L, which can solve the representation of the nodes $H_{i\cdot}^M$ and $H_{i\cdot}^A$. The specific algorithm learning process is shown in Algorithm 1.

Algorithm 1. The RolEANE Model Framework

Input: graph $G = (V, E, A)$, window size b, walks per vertex γ, walk length t, and embedding size d

Output: node representations $H \in \mathbb{R}^{n \times d}$

1: Calculate the structural role proximity $\varphi(v_i)$ for each node $v_i \in V$
2: Random walks on G, generate a large number of paths, and then get the neighbors of each node $v_i \in V$ from these paths.
3: Random initialization for all paraments set Θ
4: **while** not converged **do**
5:     Sample a mini-batch of nodes with its context
6:     Compute the gradient of ∇L_{ae} based on Equation (6)
7:     Update attribute autoencoder module parameters
8:     Compute the gradient of ∇L_{se} based on Equation (7)
9:     Update network topological structure autoencoder module parameters
10:     Compute the gradient of ∇L_{as} based on Equation (11)
11:     Update neighbor-modified Skip-Gram module parameters
12: **end while**
13: Obtain node representations $H \in \mathbb{R}^d$

It is worth noting that our model framework is superior to the state-of-the-art network embedding methods. First, our model framework uses a deep autoencoder that captures highly nonlinear information from the network. Second, when we embed the network structure and attribute information, we capture the high-order proximity and structural role proximity of the network, which preserves the global information of the network. Furthermore, when we use the neighbor-modified Skip-Gram model to seamlessly integrate the network structure and attribute information, we can also preserve the local information of the network.

4 Experiments

To verify the validity of our proposed algorithm, we compare several state-of-the-art network embedding methods on four public benchmark datasets.

4.1 Datasets

Our experiments were conducted on the following public benchmark datasets.

- **Cora:** It contains 2,708 papers in the field of machine learning and is divided into 7 categories. The citation network consists of 5,429 links. Each of these papers is described by a 0/1-valued word vector of length 1,433, which represents the absence/presence of the corresponding word in the dictionary.
- **Citeseer:** It consists of 3,312 scientific publications and is divided into 6 categories. The citation network consists of 4,732 links, each of which is described by a word vector of 0/1 value of length 3,703.
- **PubMed:** It consists of 19,716 papers related to diabetes in the biomedical field and is divided into three categories. The citation network consists of 44,338 links. Each of these papers is described by a TF/IDF weighted word vector from a dictionary of 500 unique words.
- **Wiki:** Consists of Web pages in the real world and hyperlinks between them. There are 2,405 Web pages, divided into 17 categories. It has 12,761 hyperlinks between the Web page. The Web page text content is collected as feature information of the node.

The published public benchmark datasets information used in the experiments is summarized in Table 1.

Table 1. Description of benchmark datasets.

Dataset	Nodes	Edges	Attributes	Labels
Cora	2,708	5,429	1,433	7
Citeseer	3,312	4,732	3,703	6
PubMed	19,717	44,338	500	3
Wiki	2,405	12,761	4,973	17

4.2 Experiment Settings

Baseline Methods. We compared it with several state-of-the-art network embedding methods. We can divide these network embedding methods into two categories, one is the network embedding method that only considers the network topological structure information, and the other is the method that adds the node attribute information to the network embedding process. The former category of methods mainly includes DeepWalk [1], LINE [3], node2vec [2] based on the Random Walk framework, and GraRep [5] and HOPE [6] based on the matrix decomposition framework. The latter category of methods mainly includes TADW [7], LANE [10], and ASNE [12].

Experimental Parameter Settings. For all baseline methods, we adopted the parameters from the original paper. Specifically, we set the embedding size as 128, window size as 10, the walk length as 80, the number of walks as 10, and negative samples as 5. For GraRep, the maximum step is set to 4. For LINE, we concatenate the first-order and second-order result together as the final embedding result. For node2vec, two parameters $p = 0.5$ and $q = 0.5$. For RolEANE, we concatenate the structure embedding result and attribute embedding result together as the final embedding result. The node influence range parameter $\sigma = 1.3405$ [15]. Furthermore, the number of layers and dimensions of our proposed model on different data sets are shown in Table 2.

Table 2. Specific structure setting information.

Dataset	Number of neurons in each layer
Cora	2708-1000-500-128-500-1000-2708
	1433-500-128-500-1433
Citeseer	3312-1000-500-128-500-1000-3312
	3703-1000-500-128-500-1000-3703
PubMed	19717-1000-500-128-500-1000-19717
	500-200-128-200-500
Wiki	2405-1000-500-128-500-1000-2405
	4973-1000-500-128-500-1000-4973

4.3   Results and Analysis

In this paper, the node classification task of the learned node representation is used to measure the performance of the network embedding method. For the node classification task, we first randomly divide the network nodes into two parts, which are the training set and the testing set. For the comprehensive evaluation, we select a portion of nodes varying from 10% to 60% as the training set, and the remaining nodes as the testing set. We train a one-vs-rest logistic regression classier on the training set and evaluate it on the testing set. The Micro-F1 and Macro-F1 are used as metrics to measure the classification results. We repeat this process 10 times and report the average performance of Micro-F1 and Macro-F1.

The classification experimental results are shown in Tables 3, 4, 5, and 6, respectively. As can be seen from the results of these four tables, our proposed RolEANE model has achieved significant improvements over the state-of-the-art network embedding methods. The detailed experimental results are as follows:

– **Cora:** As can be seen from the results of the Cora dataset, our method has improved over the baseline methods. When the training rate is small, our method has obvious advantages over other methods. When the training rate is high, our method is less effective than other methods. It is worth noting that we can see that other attributed network embedding methods have no obvious advantage over plain network structure embedding methods.

Table 3. Node classification results of Cora.

Training sample ratio		10%	20%	30%	40%	50%	60%
Micro-F1	DeepWalk	0.7576	0.7850	0.8038	0.8135	0.8205	0.8223
	LINE (1st+2nd)	0.7358	0.7821	0.8056	0.8113	0.8219	0.8211
	node2vec	0.7650	0.7850	0.7917	0.8037	0.7991	0.8100
	GraRep	0.7670	0.7831	0.7874	0.7832	0.7932	0.7988
	HOPE	0.5643	0.5786	0.6281	0.6299	0.6344	0.6309
	TADW	0.7598	0.7962	0.8070	0.8215	0.8134	0.8213
	LANE	0.6910	0.7546	0.7856	0.8047	0.8089	0.8179
	ASNE	0.5863	0.5990	0.6310	0.6532	0.6797	0.6813
	RolEANE	**0.7920**	**0.8066**	**0.8207**	**0.8289**	**0.8360**	**0.8376**
Macro-F1	DeepWalk	0.7417	0.7750	0.7942	0.8053	0.8074	0.8045
	LINE (1st+2nd)	0.7149	0.7689	0.7899	0.7903	0.7936	0.8011
	node2vec	0.7413	0.7690	0.7828	0.7945	0.7838	0.7979
	GraRep	0.7564	0.7664	0.7758	0.7753	0.7838	0.7813
	HOPE	0.5301	0.5378	0.6129	0.6133	0.6165	0.6078
	TADW	0.7462	0.7853	0.7952	**0.8099**	0.8087	0.8065
	LANE	0.6877	0.7416	0.7701	0.7886	0.7910	0.7955
	ASNE	0.5532	0.5677	0.5804	0.5899	0.6166	0.6189
	RolEANE	**0.7759**	**0.7889**	**0.8015**	0.8089	**0.8194**	**0.8233**

Table 4. Node classification results of Citeseer.

Training sample ratio		10%	20%	30%	40%	50%	60%
Micro-F1	DeepWalk	0.5246	0.5327	0.5554	0.5575	0.5604	0.5605
	LINE (1st+2nd)	0.5106	0.5287	0.5302	0.5685	0.5685	0.5589
	node2vec	0.5421	0.5489	0.5579	0.5689	0.5754	0.5801
	GraRep	0.5310	0.5363	0.5373	0.5440	0.5447	0.5457
	HOPE	0.4139	0.4187	0.4217	0.4353	0.4438	0.4392
	TADW	0.5998	0.6210	0.6429	0.6474	0.6518	0.6464
	LANE	0.4990	0.5456	0.6171	0.6210	0.6379	0.6405
	ASNE	0.4232	0.4645	0.5237	0.5319	0.5410	0.5576
	RolEANE	**0.6016**	**0.6566**	**0.6876**	**0.7054**	**0.7004**	**0.7083**
Macro-F1	DeepWalk	0.4827	0.4905	0.5045	0.5069	0.5161	0.5159
	LINE (1st+2nd)	0.4821	0.4876	0.4899	0.4902	0.4932	0.4889
	node2vec	0.4964	0.4988	0.5121	0.5132	0.5155	0.5173
	GraRep	0.4689	0.4691	0.4665	0.4710	0.4722	0.4723
	HOPE	0.3383	0.3412	0.3487	0.3636	0.3797	0.3754
	TADW	0.5538	0.5633	0.5893	0.5957	0.6059	0.5999
	LANE	0.4781	0.5019	0.5949	0.5956	0.6005	0.6110
	ASNE	0.4019	0.4277	0.4975	0.5012	0.5123	0.5236
	RolEANE	**0.5650**	**0.6130**	**0.6438**	**0.6499**	**0.6554**	**0.6574**

Table 5. Node classification results of PubMed.

Training sample ratio		10%	20%	30%	40%	50%	60%
Micro-F1	DeepWalk	0.7523	0.7621	0.7654	0.7671	0.7681	0.7668
	LINE (1st+2nd)	0.7553	0.7669	0.7706	0.7732	0.7766	0.7800
	node2vec	0.7620	0.7689	0.7775	0.7805	0.7828	0.7832
	GraRep	0.7433	0.7589	0.7565	0.7599	0.7666	0.7676
	HOPE	0.7033	0.7054	0.7235	0.7330	0.7345	0.7506
	TADW	0.7789	0.8062	0.8192	0.8263	0.8321	0.8302
	LANE	0.7722	0.8056	0.8132	0.8179	0.8244	0.8278
	ASNE	0.7877	0.7942	0.8035	0.8240	0.8310	0.8379
	RolEANE	**0.8312**	**0.8412**	**0.8445**	**0.8494**	**0.8480**	**0.8542**
Macro-F1	DeepWalk	0.7156	0.7235	0.7234	0.7264	0.7289	0.7227
	LINE (1st+2nd)	0.7235	0.7266	0.7289	0.7293	0.7321	0.7365
	node2vec	0.7464	0.7467	0.7533	0.7633	0.7686	0.7689
	GraRep	0.7267	0.7279	0.7334	0.7345	0.7458	0.7465
	HOPE	0.7256	0.7282	0.7279	0.7333	0.7403	0.7453
	TADW	0.7721	0.8051	0.8181	0.8251	0.8293	0.8219
	LANE	0.7625	0.7863	0.8091	0.8100	0.8109	0.8132
	ASNE	0.7610	0.7787	0.7885	0.8019	0.8135	0.8178
	RolEANE	**0.8287**	**0.8383**	**0.8417**	**0.8473**	**0.8462**	**0.8525**

Table 6. Node classification results of Wiki.

Training sample ratio		10%	20%	30%	40%	50%	60%
Micro-F1	DeepWalk	0.5967	0.6210	0.6567	0.6687	0.6849	0.6902
	LINE (1st+2nd)	0.5866	0.6320	0.6569	0.6636	0.6801	0.6878
	node2vec	0.5796	0.6133	0.6306	0.6583	0.6733	0.6871
	GraRep	0.5939	0.6190	0.6347	0.6472	0.6492	0.6559
	HOPE	0.5284	0.5545	0.5700	0.5883	0.6084	0.6091
	TADW	0.5995	0.6312	0.6659	0.6712	0.7073	0.7112
	LANE	0.5678	0.6308	0.6639	0.6812	0.6951	0.7061
	ASNE	0.5632	0.6278	0.6679	0.6822	0.7061	0.7244
	RolEANE	**0.6966**	**0.7304**	**0.7354**	**0.7413**	**0.7638**	**0.7683**
Macro-F1	DeepWalk	0.4616	0.4863	0.5577	0.5632	0.5783	0.5871
	LINE (1st+2nd)	0.4579	0.4930	0.5589	0.5611	0.5678	0.5711
	node2vec	0.4145	0.4662	0.5235	0.5348	0.5449	0.5509
	GraRep	0.4053	0.4559	0.4769	0.4984	0.5061	0.5062
	HOPE	0.3731	0.3892	0.4108	0.4262	0.4381	0.4391
	TADW	0.4151	0.5310	0.5629	0.5667	0.5732	0.5757
	LANE	0.4601	0.5298	0.5647	0.5846	0.5913	0.5971
	ASNE	0.5377	0.6042	0.6411	0.6582	0.6734	0.6771
	RolEANE	**0.5528**	**0.6052**	**0.6426**	**0.6730**	**0.6921**	**0.7153**

- **Citeseer:** It can be seen from the results of the Citeseer dataset that the embedding method considering attribute information has advantages over the plain network structure embedding method, especially when the training rate is high. Our method has a significant improvement over other benchmark methods at any training rate.
- **PubMed:** It can be seen from the results of the PubMed dataset that the embedding method considering attribute information has obvious advantages at any training rate compared with the plain network structure embedding method. Our method has a significant improvement over other methods at any training rate.
- **Wiki:** It can be seen from the results of the Wiki dataset that the embedding method considering attribute information has advantages over the plain network structure embedding method. Our method has a significant improvement over other methods at any training rate.

In summary, the Micro-F1 and Macro-F1 values of our method are significantly improved at any training rate compared to other methods. In addition, we also found that, for smaller scale networks, most attributed network embedding methods have no obvious advantages compared to plain network structure embedding methods. When the network attribute information is sparse, the advantage of the attributed network embedding methods compared with the plain network structure embedding methods is not obvious. When the training rate is gradually increased, the performance of the attributed network embedding methods is significantly improved compared to the plain network structure embedding methods. We believe that the proposed method uses two deep autoencoders to preserve network topological structure information and network attribute information, respectively, so it has advantages over other network attribute embedding methods and plain network structure embedding methods.

5   Conclusion

In this paper, we propose a novel network embedding method framework called RolEANE in order to capture the network topological structure and attribute information. The method framework consists of two network structural role proximity enhanced deep autoencoders and a neighbor-modified Skip-Gram model. We have confirmed on four real-world datasets that the RolEANE model framework has achieved significant performance improvements over other state-of-the-art network embedding methods.

Acknowledgments. This work is supported by the National Natural Science Foundation of China (61572353), the Natural Science Foundation of Tianjin (17JCY-BJC15400), and the Australian Research Council Linkage Project (LP180100750).

References

1. Perozzi, B., Al-Rfou, R., Skiena, S.: Deepwalk: online learning of social representations. In: Proceedings of the 20th ACM SIGKDD International Conference on Knowledge Discovery and Data Mining, pp. 701–710. ACM (2014)
2. Grover, A., Leskovec, J.: node2vec: scalable feature learning for networks. In: Proceedings of the 22nd ACM SIGKDD International Conference on Knowledge Discovery and Data Mining, pp. 855–864. ACM (2016)
3. Tang, J., Qu, M., Wang, M., Zhang, M., Yan, J., Mei, Q.: Line: large-scale information network embedding. In: Proceedings of the 24th International Conference on World Wide Web, pp. 1067–1077. International World Wide Web Conferences Steering Committee (2015)
4. Wang, D., Cui, P., Zhu, W.: Structural deep network embedding. In: Proceedings of the 22nd ACM SIGKDD International Conference on Knowledge Discovery and Data Mining, pp. 1225–1234. ACM (2016)
5. Cao, S., Lu, W., Xu, Q.: GraRep: learning graph representations with global structural information. In: Proceedings of the 24th ACM International on Conference on Information and Knowledge Management, pp. 891–900. ACM (2015)
6. Ou, M., Cui, P., Pei, J., Zhang, Z., Zhu, W.: Asymmetric transitivity preserving graph embedding. In: Proceedings of the 22nd ACM SIGKDD International Conference on Knowledge Discovery and Data Mining, pp. 1105–1114. ACM (2016)
7. Yang, C., Liu, Z., Zhao, D., Sun, M., Chang, E.: Network representation learning with rich text information. In: Twenty-Fourth International Joint Conference on Artificial Intelligence (2015)
8. Huang, X., Li, J., Hu, X.: Accelerated attributed network embedding. In: Proceedings of the 2017 SIAM International Conference on Data Mining, pp. 633–641. SIAM (2017)
9. Zhang, D., Yin, J., Zhu, X., Zhang, C.: Homophily, structure, and content augmented network representation learning. In: 2016 IEEE 16th International Conference on Data Mining (ICDM), pp. 609–618. IEEE (2016)
10. Huang, X., Li, J., Hu, X.: Label informed attributed network embedding. In: Proceedings of the Tenth ACM International Conference on Web Search and Data Mining, pp. 731–739. ACM (2017)
11. Wang, S., Tang, J., Morstatter, F., Liu, H.: Paired restricted Boltzmann machine for linked data. In: Proceedings of the 25th ACM International on Conference on Information and Knowledge Management, pp. 1753–1762. ACM (2016)
12. Liao, L., He, X., Zhang, H., Chua, T.S.: Attributed social network embedding. IEEE Trans. Knowl. Data Eng. **30**(12), 2257–2270 (2018)
13. Ribeiro, L.F., Saverese, P.H., Figueiredo, D.R.: struc2vec: learning node representations from structural identity. In: Proceedings of the 23rd ACM SIGKDD International Conference on Knowledge Discovery and Data Mining, pp. 385–394. ACM (2017)
14. Donnat, C., Zitnik, M., Hallac, D., Leskovec, J.: Learning structural node embeddings via diffusion wavelets. In: Proceedings of the 24th ACM SIGKDD International Conference on Knowledge Discovery & Data Mining, pp. 1320–1329. ACM (2018)
15. Mikolov, T., Sutskever, I., Chen, K., Corrado, G.S., Dean, J.: Distributed representations of words and phrases and their compositionality. In: Advances in Neural Information Processing Systems, pp. 3111–3119 (2013)

Context-Aware Temporal Knowledge Graph Embedding

Yu Liu, Wen Hua$^{(\boxtimes)}$, Kexuan Xin, and Xiaofang Zhou

School of Information Technology and Electrical Engineering,
The University of Queensland, Brisbane, Australia
{yu.liu,w.hua}@uq.edu.au, ke.xin@uqconnect.edu.au, zxf@itee.uq.edu.au

Abstract. Knowledge graph embedding (KGE) is an important technique used for knowledge graph completion (KGC). However, knowledge in practice is time-variant and many relations are only valid for a certain period of time. This phenomenon highlights the importance of temporal knowledge graph embeddings. Currently, existing temporal KGE methods only focus on one aspect of facts, i.e., the factual plausibility, while ignoring the other aspect, i.e., the temporal consistency. Temporal consistency models the interactions between a fact and its contexts, and thus is able to capture fine-granularity temporal relationships, such as temporal orders, temporal distances and overlapping. In order to determine the useful contexts for the fact to be predicted, we propose a two-way strategy for context selection. In particular, we decompose the target fact into two parts, relation and entities, and measure the usefulness of a context for each part respectively. Furthermore, we adopt deep neural networks to encode contexts and score the temporal consistency. This consistency is used with factual plausibility to model a fact. Due to the incorporation of temporal information and the interactions between facts and contexts, our model learns a more representative embeddings for temporal KG. We conduct extensive experiments on real world datasets and the experimental results verify the effectiveness of our proposals.

Keywords: Knowledge graph embedding · Temporal consistency · Factual plausibility · Context-aware embedding

1 Introduction

Large-scale knowledge graphs (KGs) have been used in many real-world applications including entity linking [6,17], relation extraction [18,30] and question answering [15,29]. However, the coverage of these KGs are still far from complete [21,23], which limits their effectiveness and benefit. To tackle this issue, a new technique known as knowledge graph completion (KGC) has been proposed. KGC aims to derive new facts from an existing KG via knowledge graph embedding (KGE), or in other words, by transforming the entities and relations of the KG into a low-dimensional continuous space and predicting new facts based on

© Springer Nature Switzerland AG 2019
R. Cheng et al. (Eds.): WISE 2019, LNCS 11881, pp. 583–598, 2019.
https://doi.org/10.1007/978-3-030-34223-4_37

the embeddings [26]. Generally speaking, KGE aims to learn a score function to measure the plausibility of a fact. Classical embedding models, e.g., TransE [1], DistMult [28], and HolE [19], learn the score function using only facts observed in the KG. Additional information is also incorporated in many recent methods [12,20] to further improve the embedding performance.

However, existing methods regard relational facts as time-invariant and ignore the corresponding valid temporal period. Actually, many relations are changing and involving over time, i.e., they are only valid for a certain time period. For example, the relation instance *SpouseOf* ("Brad Pitt", "Angelina Jolie") holds true only over the temporal interval "[2014, 2016]". Therefore, it is important to incorporate the temporal information for knowledge graph embedding. To the best of our knowledge, research on temporal KGE is quite recent and only a few work [2–4,7,11,24] has attempted to address this issue. These methods incorporate temporal information in various ways, such as modelling entity evolutions [3,24] or temporal-aware relations [4,7,11], or projecting entities and relations into time-aware hyperplanes. Although the above models have successfully improved the embedding performance to some extent, they are still limited to simply measuring one aspect of facts, i.e., how well the entities and relation are composed together as the traditional static KGE methods. It is necessary to capture more interactions in the temporal dimension for learning a representative temporal-aware KGE model. One promising direction is to learn the temporal interactions between a fact and its contexts, namely other facts that share certain component with the target fact.

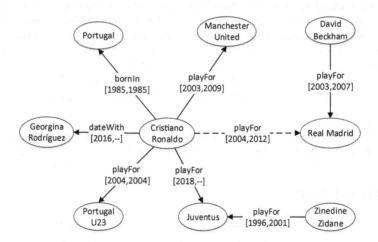

Fig. 1. An example of temporal interactions for predicting the fact ("Cristiano Ronaldo", "playFor", "Real Madrid", [2004, 2012])

In this work, we propose a temporal-aware KGE model that measures two aspects of a fact: factual plausibility and temporal consistency. Our main idea is summarised as follows: a fact is valid if (1) it is composed of plausible head entity, relation and tail entity; (2) the valid interval of the fact is temporally consistent with its contexts. We use an example to show how the temporal consistency works. Figure 1 is a small fragment of a temporal knowledge graph. There are five facts describing "Cristiano Ronaldo" which are regarded as the *contexts* for "Cristiano Ronaldo". Assume that we need to predict whether the fact ("Cristiano Ronaldo","playFor","Real Madrid", "[2004, 2012]") holds or not. It is possible for us, as human beings, to examine related contexts and conclude that the fact is false based on the temporal interactions between the given time period "[2004, 2012]" and each context. In particular, one interaction confirms our claim, namely the given period ("[2004, 2012]") has a very large overlapping with the period that Ronaldo plays for Manchester United ("[2003, 2009]"). However, it is still challenging for machines to directly adopt such strategy. First, not all contexts are useful for prediction and some can even be misleading. For example, the fact on "dateWith" is useless for determining the correctness of ("Cristiano Ronaldo","playFor","Real Madrid", "[2004, 2012]") and the fact "playFor Portugal U23" could be misleading. Second, the temporal interactions exist in various forms and in different categories, including temporal orders, intersections or temporal distances. For instance, temporal interval of "bornIn" relation is always prior to the interval of "playFor" relation. Meanwhile, facts of "playFor" relation have overlapping or some temporal distances with each other. These variants bring challenges to effectively modelling temporal interactions and measuring the temporal consistency.

To address these issues, we propose two novel modules to conduct temporal knowledge graph embedding, with one for context selection and the other for modelling the temporal interactions. In particular, we introduce a relation-entity-aware mechanism to determine useful contexts for the target fact. Besides, we use convolutional neural networks (CNNs) to extract high-level features of temporal interactions, based on which a fully connected layer is adopted to learn the temporal consistency score. This consistency score will be used together with factual plausibility score to predict the correctness of a fact. Existing temporal KGE models only focus on the first aspect, i.e., factual plausibility, while ignoring the second aspect, i.e., temporal consistency. By combining both features, our model can learn more representative embeddings for knowledge graph completion.

Our main contributions in this work can be summarised as below:

- We propose a context-aware KGE model which explicitly measures the temporal consistency of facts and contexts, and integrates with factual plausibility for learning a more representative embedding.
- We design a context selection strategy that considers the usefulness of a context from two perspectives, i.e., relations and entities.
- We introduce a novel mechanism to model the temporal interactions between the target fact and its contexts for scoring the temporal consistency.
- We conduct extensive experiments on real world datasets, and the experimental results verify the superiority of our proposals over existing state-of-the-art methods.

The remaining of this paper is organised as follows: We summarise the current literature of knowledge graph embedding in Sect. 2. The problem is formally defined in Sect. 3, and our solutions are described in detail in Sect. 4. Section 5 reports the experimental results, followed by a brief conclusion in Sect. 6.

2 Related Work

We classify existing knowledge graph embedding models into two categories depending on whether the temporal information is considered or not. We briefly summarise related work of each category in the following.

2.1 Traditional KG Embedding

Traditional knowledge graph embedding is to learn the representations of the components of a (static) knowledge graph, including entities and relations. TransE [1] is the first and a powerful transition-based model, which interprets a triplet (h, r, t) as a translation from head entity h to tail entity t via relation r in the continues vector space, i.e. $h + r \approx t$, where $h, r, t \in \mathbb{R}^d$ is the embeddings of h, r and t, respectively. TransE uses a distance-based score function $s(h, r, t) = \|h + r - t\|_{l_1/l_2}$ to measure the fact's plausibility. However, as the symmetry of TransE's score function, TransE is only able to model the *one-to-one* relations and fails to model *many-to-one*, *one-to-many*, *many-to-many* relations. Many other works have been proposed to solve this problem, such as transH [27], transR [13]. Recently, more powerful KGE models have been proposed, including HolE [19], ComplEx [25] and RotateE [22]. HolE uses the circular operation to score a fact, *i.e.*, $s(h, r, t) = r^\mathsf{T}(h \star t)$ where $\star$ is the circular operation. As circular operation is not commutative, HolE is able to model asymmetric relations. ComplEx extends embeddings into complex field so as to better model asymmetric relations, *i.e.*, h, r and t are complex-valued embeddings. RotatE is also defined on complex field and regards each relation as a rotation from the source entity to the target entity, *i.e.*, $s(h, r, t) = \|h \circ r - t\|_{l_1}$, where $\circ$ is the Hadmard product. Though these methods can model various relations, they still ignore the importance of incorporating temporal information.

2.2 Temporal KG Embedding

To our best knowledge, there are very few works that incorporate temporal information in knowledge graph embedding [2–4,7,11,24]. According to where they incorporate temporal information, we divide them into three categories. Previous works [3,24] assume entities are involving with time and model the entity representations over time. Recently, some works [4,7,11] shift to model the interactions between relations and time. [7] discovers there are certain temporal orders for different relations. For example, facts involving a person may follow the timeline: "bornIn" → "gradudateFrom" → "worksAt". After incorporating temporal orders via Integer Linear Program, [7] uses TransE model to measure factual plausibility. [4,11] target at directly embedding time into relations. [11] studies various ways to combine the time embedding vector with relation embedding vector, such as concatenate, sum or dot product operations. However, as [11] shows, the performance of these simple strategies are still limited. [4] regards timestamps as a sequence of digits, *i.e.*, from 0 to 9, then uses LSTMs to encode the relation vectors and the time digits. Besides, HyTE [2] directly projects entities and relations into time-specific hyperplanes and then models the factual plausibility via TransE. Generally, these methods transfer the time-aware facts into triplets (head entity, relation, tail entity), then use traditional KGE methods to measure the factual plausibility. As a result, these methods fail to capture more interactions in the temporal dimension, *e.g.*, the temporal consistency between facts and contexts.

3   Problem Definition

In this paper, we consider the task of learning the representations of a temporal knowledge graph via enforcing the consistency between contexts and valid temporal intervals. In the following, boldface upper-case and lower-case letters indicate matrices and vectors, respectively.

Let E, R and T denotes an entity set, a relation set and a timestamp set, perspectively. Our framework for temporal knowledge graph embedding can be defined as follows:

Definition 1 (Temporal Knowledge Graph). *A temporal KG is defined as a directed graph $G = (E, R, T)$ where (1) E is the set of entities (nodes); (2) R is the set of relations (edges); (3) T is the set of valid temporal intervals (labels).*

Definition 2 (Fact). *A fact is defined as a 4-tuple $f = (h, r, t, \tau)$, where $h, t \in E$ is the head entity and the tail entity, respectively, $r \in R$ is a relation between h and t, and $\tau = [\tau_s, \tau_e] \subseteq T \times T$ is the temporal interval when (h, r, t) holds in the real world. Facts observed in the KG are stored as a collection $D^+ = \{(h, r, t, \tau)\}$.*

Example: Fact ("Cristiano Ronaldo","playFor", "Manchester United", "[2003, 2009]") tells the truth that Ronaldo plays for Manchester from 2003 to 2009.

Definition 3 (Context). *The context of a target entity e is defined as the aggregate set of facts $C_e = \{f_1, \cdots, f_n\}$ such that each fact f_i contains e.*

Example: (1) Fact ("Cristiano Ronaldo","playFor" "Real Madrid", "[2009, 2018]") is a context for entity "Cristiano Ronaldo"; (2) Fact ("Wayne Rooney","playFor" "Manchester United", "[2004, 2017]") is a context for entity "Manchester United", given the fact in the previous example.

Definition 4 (Temporal Knowledge Graph Embedding). *Temporal KGE is the task to learn the representations of a temporal Knowledge Graph G. In other words, the temporal KGE task aims to embed the entity set E, relation set R and time set T into a low dimensional continuous vector space, say $\mathbb{R}^d$. Generally, KGE enforces the embeddings are compatible with observed facts [26].*

To evaluate the KGE performance, link prediction [1] is widely used. Link prediction is to predict the missing part of a given incomplete fact. As this missing part can be the head entity, tail entity or relation, the link prediction task is consist of three subtasks: head prediction, tail prediction and relation prediction. In particular, considering an observed fact $f = (h, r, t, \tau)$ from the test dataset, we replace f as $(?, r, t, \tau)$, $(h, ?, r, \tau)$ and $(h, r, ?, \tau)$ for the task of head prediction, relation prediction and tail prediction, respectively. Here, ? refers to the missing part. These prediction tasks are often reduced to a ranking problem through a learned score function which measures the plausibility of a candidate fact. Ideally, after scoring all possible candidate facts, the valid candidate facts should be placed at top of the ranking list. The higher rank indicates a better performance of the temporal KGE.

4   Methodology

In this section, we will introduce our proposed context-aware model for temporal KGE. In the following, we will present (1) an overview of our model, (2) the characterizing of the context for temporal consistency, (3) the characterizing of the factual plausibility, (4) the objective function and the training process.

4.1   Model Overview

Different from existing temporal KGE models which only measure the plausibility of facts, we explicitly incorporate the temporal consistency between facts and contexts. Instead of only calculating the factual plausibility, we consider two aspects of a fact, i.e., a fact is valid if (1) the fact is composed of plausible head entity, relation and tail entity; (2) The valid interval of the fact is temporally consistent with its contexts. Based on this intuition, we propose a context-aware model which measures these two aspects of a fact to predict its validation.

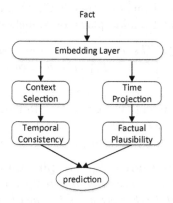

Fig. 2. An overview of the temporal-aware KGE model

From Fig. 2 we can see there are five modules in our proposed temporal KGE framework, i.e., one embedding layer, two modules for estimating temporal consistency, and two modules for measuring factual plausibility. The embedding layer transforms the one-hot representations of the components of a fact into low-dimensional dense vectors, i.e., embeddings. In particular, for a fact $(h, r, t, [\tau_s, \tau_e])$, we obtain its head entity embedding $h \in \mathbb{R}^d$, tail entity embedding $t \in \mathbb{R}^d$, relation embedding $r \in \mathbb{R}^d$ and time embeddings $\tau_s \in \mathbb{R}^d$ and $\tau_e \in \mathbb{R}^d$ through the embedding layer. Temporal consistency measures how well the fact and its contexts are compatible with each other. Since not all contexts are useful and some can even be misleading for temporal consistency, we propose a fact-aware context selection mechanism to aggregate useful contexts. We then introduce a two-layer architecture to model the temporal interactions between the selected contexts and the input fact. These interactions capture fine-granularity temporal relationships which are useful for temporal consistency. On the other hand, factual plausibility measures how well the head entity, relation and tail entity can be composed together. Many existing methods can be used for the traditional KGE task, such as TransE [1]. However, for temporal-aware KGs, we believe a necessary step is to encode temporal information into relation embedding since a fact is only valid during the given temporal interval. Therefore, we project relation embeddings on the time dimension to derive the time-aware embeddings. After that, we calculate the factual plausibility based on the new embeddings. Finally, we combine these two aspects together and predict whether a fact is true or false. Due to the incorporating of temporal consistency, we believe our model is able to learn more representative embeddings.

4.2 Characterizing of the Context for Temporal Consistency

A major novelty of our work is explicitly characterizing the contexts of head entity and tail entity for calculating the temporal consistency. In this part, we first study how to select useful contexts for a target fact, then investigate how to use the selected contexts to measure the temporal consistency.

Context Selection. Obviously, not all information for an entity is related to the fact to be predicted and some can even be noisy and misleading. Thus, an essential step is to select useful contexts for the input fact. Heuristically, contexts about different relations have different influences for a certain fact. For example, when encoding the fact ("Cristiano Ronaldo", "playFor", "Manchester United", [2003, 2009]), contexts about "Cristiano" in the soccer field are more useful than the information about his personal life, such as facts on "spouseOf" or "hasChildren". Meanwhile, some information about "Manchester United", such as facts on "locationOf", "hasCapacity", and "foundedOf", are noisy for the current fact. This observation indicates the importance of considering relation for context selection. As deep convolutional neural networks (CNNs) have achieved promising performance on many NLP tasks, such as relation extraction [14] and sentence classification [9], we also extend a CNN architecture to calculate the relation-aware usefulness of contexts. Formally, given a fact $f = (h, r, t, [\tau_s, \tau_e])$ and all contexts $C_e = \{f_1, \ldots, f_n\}$ for a core entity e ($e = h$ or $e = t$), we calculate the relation-aware usefulness of each context $f_i \in C_e$ for the fact f as follows:

$$\boldsymbol{u}_{i,r}^{(1)} = CNNs(\boldsymbol{r} \oplus \boldsymbol{r}_i; \Theta_1) \tag{1}$$

$$u_{i,r} = \sigma(\boldsymbol{W}_r^{(2)} \boldsymbol{u}_{i,r}^{(1)} + b_r^{(2)}) \tag{2}$$

where CNNs is a two-depth convolutional network with max-pooling layers, Θ_1 denotes all related parameters to the CNNs, $\boldsymbol{r}$ and $\boldsymbol{r}_i$ is the relation embedding of r (in fact f) and r_i (in context f_i), respectively. $\boldsymbol{W}_r^{(2)}$ is the weight matrix and $b_r^{(2)}$ is the bias in the second layer. $\oplus$ is the concatenation operation and $\sigma(\cdot)$ is the activation function. Here, we use TanH. As e can be either head entity or tail entity, we can calculate the relation-aware usefulness of each context in C_h and C_t for the input fact.

Besides, useful contexts are representative for the entity. For example, contexts of "Cristiano Ronaldo" are more useful if the contexts are able to distinguish "Cristiano" with some other football players, e.g. "Lionel Messi" and "Zinedine Zidane". This suggests useful contexts should be able to distinguish with similar entities. As entity embedding essentially encodes semantics, the similar entities are the nearest neighbours in the dense space. In particular, given an entity e, we obtain its top-k nearest neighbours $\{n_{e,i}\}_{i=1}^{k}$ via calculating the l_1 distance for all other candidates $e' \in E$, i.e., $||e - e'||_{l_1}$ where e and e' is the embeddings of e and e', respectively. After we obtain the top-k nearest neighbours of entity e, we use another similar CNN architecture to calculate the entity usefulness. Consider an observed fact $f = (h, r, t, [\tau_s, \tau_e])$ and the contexts $C_e = \{f_1, \ldots, f_n\}$ for core entity e. Formally, given the top-k nearest neighbours $\{n_{e,j}\}_{j=1}^{k}$ of e, the entity-aware usefulness of context $f_i = (h_i, r_i, t_i, [\tau_{i,s}, \tau_{i,e}])$ can be calculated as:

$$\boldsymbol{u}_{i,j,e}^{(1)} = CNNs((\boldsymbol{e} - \boldsymbol{n}_{e,j}) \oplus \boldsymbol{h}_i \oplus \boldsymbol{t}_i; \Theta_2) \tag{3}$$

$$u_{i,j,e} = \sigma(\boldsymbol{W}_e^{(2)} \boldsymbol{u}_{i,j,e}^{(1)} + b_e^{(2)}) \tag{4}$$

$$u_{i,e} = \frac{1}{k} \sum_{j=1}^{k} u_{i,j,e} \tag{5}$$

where $u_{i,e}$ is the entity usefulness score of ith context in C_e for distinguishing e and its neighbours, $\boldsymbol{e}$ is the entity embedding of core entity e (e can be head entity or tail entity), and $\boldsymbol{n}_{e,j} \in \mathbb{R}^d$ is the entity embedding of j-th neighbour. Because the difference between entity and its neighbours is critical for link predictions, we use the difference $(\boldsymbol{e} - \boldsymbol{n}_{e,j})$ as one of the input vectors. Based on the above two aspects, we obtain the usefulness score as follows:

$$u_i = u_{i,r} + u_{i,e} \tag{6}$$

Finally, we obtain top-w context facts $\tilde{C}_e = \{f_1, \cdots, f_w\}$ from C_e with highest scores. As e can be head entity or tail entity, given a fact $f = (h, r, t, [\tau_s, \tau_e])$, we select useful head contexts $\tilde{C}_h$ and tail contexts $\tilde{C}_t$ respectively for calculating temporal consistency.

Temporal Consistency. Now we introduce how to use the selected contexts, i.e., $\tilde{C}_h$ and $\tilde{C}_t$, to calculate the temporal consistency. We notice there are many kinds of temporal interactions between a fact and the contexts of the fact, such as temporal orders, temporal distances and temporal intersections. For example, the fact ("Cristiano","playFor", "Manchester", [2003, 2009]) temporally contains a context fact (he plays for Portugal U23 during 2004), and has some temporal distance to the context fact (he plays for Juventus since 2018), and must happen after his birth date. Obviously, these interactions are in various categorises and in different relations with numerous values. But, heuristically, these interactions should be temporally compatible with each other if the fact (and the contexts) is true. Based on this intuition, we first model the interactions between the fact and the contexts, and then measure its temporal consistency which indicates how the fact and contexts are compatible with each other.

Since we are modelling the temporal interactions, we only focus on the valid time periods of the input fact and of each context. Formally, given a fact $f = (h, r, t, [\tau_s, \tau_e])$ and one context fact $(h_i, r_i, t_i, [\tau_{i,s}, \tau_{i,e}])$ in the contexts $\tilde{C}_e$, the interaction vector used for calculating temporal consistency is defined as follows:

$$\tilde{t}_{i,e} = (\boldsymbol{\tau}_s \oplus \boldsymbol{\tau}_e \oplus \boldsymbol{\tau}_{i,s} \oplus \boldsymbol{\tau}_{i,e}) \tag{7}$$

where $\oplus$ is the concatenate operation, $\boldsymbol{\tau}_{(i),s} \in \mathbb{R}^d$ and $\boldsymbol{\tau}_{(i),e} \in \mathbb{R}^d$ is the time embedding vector of start time $\tau_{(i),s}$ and end time $\tau_{(i),e}$, respectively. By doing this concatenate operation for all contexts, we obtain the interaction matrices $\tilde{T}_h \in \mathbb{R}^{w \times 4 \times d}$ and $\tilde{T}_t \in \mathbb{R}^{w \times 4 \times d}$ for a fact $(h, r, t, [\tau_s, \tau_e])$. The next step is to measure the temporal consistency based on these interaction matrices.

For each interaction, i.e., $\tilde{t}_{i,e}$, we use a two-layer architecture to calculate the consistency score. First, we extract its high-level features via convolutional neural networks (CNNs), and then we use a fully connected layer to obtain the score according to these high-level features. This architecture can be shown as follows:

$$s_{i,e}^{(1)} = CNNs(\tilde{t}_{i,e}; \Theta_3) \tag{8}$$

$$s_{i,e} = \sigma(W_\tau^{(2)} s_{i,e}^{(1)} + b_\tau^{(2)}) \tag{9}$$

where CNNs is a two-depth convolutional layer with max-pooling, Θ_3 are all related parameters to the CNNs, $W_\tau^{(2)}$ is the weight matrix and $b_\tau^{(2)}$ is the bias in the second layer. $\sigma(\cdot)$ is the activation function. As before, we use TanH activation function. Since each interaction reflects the temporal consistency independently, we score each interaction separately and then select the maximum one as the temporal consistency score as follows:

$$s_{\tau,e} = max(\{s_{i,e}\}_{i=1}^w) \tag{10}$$

We use the max operation because false facts can be compatible with some interactions (i.e., has some small scores), but not all of them are consistent (i.e., also has some large scores). Using above Eq. 10, we obtain the temporal consistency scores $s_{\tau,h}$ and $s_{\tau,t}$ for the selected head context $\tilde{C}_h$ and tail context $\tilde{C}_t$ of a fact $(h, r, t, \tau_s, \tau_e)$, respectively. Finally, we use the average as the temporal consistency score as follows:

$$s_\tau = \frac{1}{2}(s_{\tau,h} + s_{\tau,t}) \tag{11}$$

4.3   Characterizing of the Factual Plausibility

Time Projection. As relational facts are only valid during the given temporal interval, we need to encode the temporal information as well. Since the temporal constraint modifies the relation (i.e., shows when the relation is valid), we only encode the temporal information into relation embeddings via the projection operation. Formally, given a fact (h, r, t, τ), we obtain the time-aware relation embedding $r_\tau \in \mathbb{R}^d$ as follows:

$$r_\tau = r - (\tau_s^\top \cdot r) \cdot \tau_e \tag{12}$$

where $\cdot$ is the element-wise dot product and $r \in \mathbb{R}^d$, $\tau_s \in \mathbb{R}^d$, $\tau_t \in \mathbb{R}^d$ is the embedding of r, τ_s, τ_e respectively.

Factual Plausibility. Inspired by translational distance model transE [1], we define the time-aware score function $s_f = s(h, r_\tau, t)$ as:

$$s_f = \gamma - \|h + r_\tau - t\|_{l_1} \tag{13}$$

where $\gamma > 0$ is a hyper-parameter, $h, r_\tau, t, \in \mathbb{R}^d$, respectively, and l_1 is the l_1 norm.

4.4   Objective Function and Training Process

Our model is trained in the prediction setting. In other words, given a fact (h, r, t, τ) with its labels (true or false), our model learns to predict whether this fact holds true. The prediction is calculated as follows:

$$\hat{y} = f(x) = sigmoid(s_f * s_\tau) \tag{14}$$

where x is a fact, s_f and s_τ is the factual plausibility score and temporal consistency score, respectively. We use the binary cross entropy as our objective function. The loss function can be calculated as follows:

$$L = - \sum_{x \in D^+ \cup D^-} y log(\hat{y}) - (1-y) log(1-\hat{y}) \tag{15}$$

where D^+ is the positive training dataset and D^- is the negative sampling dataset.

5 Experiments

In this section, we evaluate our proposed model for temporal knowledge graph embedding. In the following, we will first introduce the experimental settings, including datasets, parameters, evaluations and baselines. After that, we will introduce the experimental results which show the effectiveness of our proposed model. At last, we will also focus on the effectiveness of context selection.

5.1 Experimental Setting

Datasets. In our experiments, we follow [2] and choose two temporal knowledge graph datasets, i.e., YAGO11k and Wikidata12k. YAGO11k and Wikidata12k are the subsets of facts extracted from YAGO3 [16] and Wikidata [11], respectively. All facts in both datasets are temporal-aware facts, i.e., coupled with their valid time interval. Both datasets are well-organised into training set, validation set and test set. The statistics of these two datasets are shown in Table 1.

Table 1. Details of the two datasets

Dataset	#Entity	#Relation	Train/valid/test
YAGO11k	10,623	10	16.4k/2k/2k
Wikidata12k	12,554	24	32.5k/4k/4k

Link Prediction. To evaluate the embedding performance, we adopt widely-used link prediction [1]. We follow [2] to construct a triple dictionary $D_t = \{(h, r, t) | (h, r, t, *) \in D\}$ and apply the filtered protocol [1] to sample negative data. We rank all candidates in the increasing order of their scores via our score function and find the rank of the actual fact (h, r, t, τ) in the ranking list. We use this ranking list to evaluate the model.

Evaluation Metrics. We use two popular metrics Hit@10 (Hit@1 for relation prediction) and mean rank (MR). Hit@10 (or Hit@1) measures the proportion of the test entities (or relations) ranked in top10 (or top1) of the ranking list. Mean rank measures the average ranks of the test entities (and relations). The higher Hit@10, the better performance of the prediction. Meanwhile, the lower mean rank, the better performance.

Hyperparameters. We select the hyperparameters of our model via grid search on validation datasets. The hyperparameters are listed as follows: the embedding dimension $d \in \{64, 128\}$, the learning rate $lr \in \{0.1, 0.01, 0.001, 0.0001\}$, the context window size $w \in \{2, 4, 8, 16\}$, the size k for top-k nearest neighbours $\in \{3, 5, 10\}$ and γ in the score function $\in \{1, 5, 10\}$. The finally adopted settings for Yago dataset are: $\{d = 128, lr = 0.001, w = 8, k = 10, \gamma = 1\}$, and for Wikipedia dataset are $\{d = 128, lr = 0.0001, w = 16, k = 10, \gamma = 1\}$. We set batch size $b = 512$ and use Adam [10] as the optimizer, l_1 norm in the score function. We use uniform xavier [5] to initialize the embedding layer and adopt negative sampling as [1].

Baselines. We compare our proposed model with six baseline models, i.e., TransE [1], HolE [19], ComplEx [25], pRotatE [22], t-TransE [8] and HyTE [2]. These baseline models are in two categories. The first four baseline models (TransE, HolE, ComplEx and RotatE) are designed for (static) knowledge graph embedding, while the last two (t-TransE, HyTE) are designed for temporal knowledge graph embedding. RotatE and HyTE are the state-of-the-art models and achieved best performance for (static) KGE and temporal KGE, respectively.

5.2   Experimental Results

In this part, we show the effectiveness of our proposed model by comparing with six baselines. As Table 2 shows, for the entity prediction task, our model is consistently better than all baselines on two datasets. For entity predictions, our model gains a large boost compared with state-of-the-art temporal KGE model (HyTE) on both Mean Rank metric and Hit@10 metric. In particular, our model improves Hit@10 metric on Yago dataset from 16.0% to 26.9%, and improves Hit@10 from 25.0% to 50.1% on Wikipedia dataset. This significant improvement validates our claim that the temporal consistency helps to learn more representative embeddings for knowledge graph completion. Besides, we observe that for the Mean Rank metric, temporal-aware KGE models outperform the traditional static KGE models in a large margin, which proves our claim that temporal information can lead to better embeddings. We also notice that for the Hit@10 metric, some traditional models (CompleEx, RotatE) achieve much higher score than HyTE though all of them only measure the factual plausibility. This is because CompleEx and RotatE are built on the complex field which has a richer representative ability than the embeddings in real field. However, as our model is built on real field and achieves a larger margin than CompleEx and RotatE, this again shows the effectiveness of the context for temporal consistency. For relation prediction task we evaluate on the Hit@1 metric since the number of predicted relations is small (10 for Yago and 24 for Wiki). We can see that our model achieves a large improvement on all datasets compared with traditional KGE models, and is also quite competitive with HyTE. This shows that although we mainly focus on modelling the contexts for temporal consistency which rarely involves relations, the time projection for factual plausibility still works well.

Table 2. Prediction accuracy on two datasets. † denotes the results are take from [2].

Dataset	YAGO11K						Wikidata12K					
Metric	Mean Rank			Hit@10 (Hit@1)			Mean Rank			Hit@10 (Hit@1)		
	Tail	Head	Rel	Tail	Head	Rel	Tail	Head	Rel	Tail	Head	Rel
TransE†	504	2020	1.7	4.4	1.2	1.7	520	740	1.35	11.0	6.0	88.4
HolE†	1828	1953	2.57	29.4	13.7	69.3	734	808	2.23	25.0	12.3	84.0
ComplEx	1825	2556	3.59	29.3	17.8	61.3	411	409	2.24	53.8	50.3	86.7
RotatE	1535	2244	4.17	16.8	11.9	32.9	1283	1264	4.98	42.1	39.1	52.8
t-TransE†	292	1692	1.66	6.2	1.3	75.5	283	413	1.97	24.5	14.5	74.2
HyTE†	107	1069	1.23	38.4	16.0	81.2	179	237	1.13	41.6	25.0	92.6
Our model	90	670	1.21	42.4	26.9	79.9	109	193	1.29	59.4	50.1	93.1

5.3 The Effectiveness of Context Selection

In this part, we evaluate the effectiveness of context selection module for temporal consistency from (1) the context selection method and (2) the influence of context size. In order to have a deeper understanding of the influence of contexts, we collect and count the context length of each entity in the two datasets. Table 3 shows the average length of contexts, max length of contexts, and the number of entities that have no context. We can see that Wikipedia dataset has a richer but more complex context environment than Yago dataset (avg = 5.17 v.s. avg = 3.05). We also notice the number of relations in the Wikipedia dataset is much larger than that in the Yago dataset (24 v.s. 10), which also makes the context environment more complex in Wikipedia.

Table 3. Context statistics of the two datasets

Dataset	Avg length	Max length	# Empty
YAGO11k	3.05	254	3812
Wikidata12k	5.17	285	155

We first evaluate our proposed context selection method against its four variants, i.e., (1) randomly sample contexts; (2) use all contexts; (3) only use entity-usefulness score function for context selection; (4) only use relation-usefulness score function for context selection. The comparison results are listed in Table 4. We notice that all methods achieve comparative results compared with state-of-the-art temporal KGE model on two datasets. This empirically validates the significance of temporal consistency. Beside, on Yago dataset, our proposed model achieves the best performance on three metrics and is still very competitive on the rest three metrics. On Wikipedia dataset, though our complete model is not consistently better than relation-aware (or entity-aware) model, it is always better than all previous baseline models (e.g. HyTE) from Table 2. We also notice the last two models (i.e., "entity-aware" model and "relation-aware" model) generally have better performance than the first two models (i.e., "random" model and "all selection" model) on Yago dataset. This further empirically validates our claim that not all contexts are useful, and also verifies the effectiveness of our proposed relation (and entity) usefulness function.

Table 4. Prediction accuracy of context selection on two datasets.

Dataset	YAGO11K						Wikidata12K					
Metric	Mean Rank			Hit@10 (Hit@1)			Mean Rank			Hit@10 (Hit@1)		
	Tail	Head	Rel	Tail	Head	Rel	Tail	Head	Rel	Tail	Head	Rel
Random context	94	693	1.22	39.7	25.7	79.3	149	183	1.07	57.3	50.6	96.2
All context	95	701	1.21	40.7	25.8	80.5	126	143	1.10	58.6	51.9	96.7
Entity-aware model	91	696	1.21	41.1	25.5	79.6	122	134	1.13	59.2	50.8	92.8
Rel-aware model	85	673	1.22	41.4	26.0	79.0	111	203	1.34	58.3	49.4	91.3
Complete model	90	670	1.21	42.4	26.9	79.9	109	193	1.29	59.4	50.1	93.1

Now, we study the influence of context size on two datasets. From Table 5 we can see that context size plays different roles on different datasets. For Yago dataset, the variants of different size have similar performance. This is because Yago dataset has relatively simple context environment (e.g. avg = 3.05). When $w = 4$ or $w = 8$, our model generally has better performance than $w = 2$ or $w = 16$ on Yago dataset. This is consistent with our expectation because when context size is too small or too large, the representations of selected contexts are limited. However, for Wikipedia dataset which is in a richer and more complex environment of contexts, the results are very different from those in Yago dataset. The performance is very bad when the context size is very small ($w = 2$), but the results are getting better when the context size is relatively large (e.g. $w = 4$). Whereas our model achieves great improvement from 13.9% to 55.9% (and to 59.4%) for tail prediction on Hit@10 metric when $w = 8$ (and $w = 16$). This phenomenon highlights the importance of contexts for temporal consistency and empirically validates the claim that a proper context size is matched with the environment of a dataset.

Table 5. Prediction results on different context sizes.

Dataset	YAGO11K						Wikidata12K					
Metric	Mean Rank			Hit@10 (Hit@1)			Mean Rank			Hit@10 (Hit@1)		
	Tail	Head	Rel	Tail	Head	Rel	Tail	Head	Rel	Tail	Head	Rel
Context-2	92	685	1.22	41.4	25.8	79.7	1283	1284	9.72	11.1	9.4	46.7
Context-4	88	674	1.21	41.5	26.6	79.9	766	819	3.94	13.9	12.4	69.4
Context-8	90	670	1.21	42.4	26.0	79.9	150	294	1.58	55.9	47.2	91.6
Context-16	86	677	1.21	40.6	25.8	80.3	109	193	1.29	59.4	50.1	93.1

6 Conclusion

In this paper, we propose a context-aware model for the temporal KEG task. Our model explicitly measures the temporal consistency between a fact and its contexts and integrates with factual plausibility for learning a more representative embedding. Besides, we design a context selection strategy which considers the usefulness of a context from two perspectives. We also introduce a novel mechanism to model the temporal interactions between the target fact and its contexts for scoring the temporal consistency. Experiments on real world datasets verify the effectiveness of our proposed model. In the future, we will consider to leverage richer contexts such as paths, text and graph patterns, to improve the temporal KGE task.

References

1. Bordes, A., Usunier, N., Garcia-Duran, A., Weston, J., Yakhnenko, O.: Translating embeddings for modeling multi-relational data. In: Advances in Neural Information Processing Systems, pp. 2787–2795 (2013)
2. Dasgupta, S.S., Ray, S.N., Talukdar, P.: HyTE: hyperplane-based temporally aware knowledge graph embedding. In: Proceedings of the 2018 Conference on Empirical Methods in Natural Language Processing, pp. 2001–2011 (2018)
3. Esteban, C., Tresp, V., Yang, Y., Baier, S., Krompaß, D.: Predicting the co-evolution of event and knowledge graphs. In: 2016 19th International Conference on Information Fusion (FUSION), pp. 98–105. IEEE (2016)
4. García-Durán, A., Dumančić, S., Niepert, M.: Learning sequence encoders for temporal knowledge graph completion. arXiv preprint arXiv:1809.03202 (2018)
5. Glorot, X., Bengio, Y.: Understanding the difficulty of training deep feedforward neural networks. In: Proceedings of the Thirteenth International Conference on Artificial Intelligence and Statistics, pp. 249–256 (2010)
6. Hua, W., Zheng, K., Zhou, X.: Microblog entity linking with social temporal context. In: Proceedings of the 2015 ACM SIGMOD International Conference on Management of Data, pp. 1761–1775. ACM (2015)
7. Jiang, T., et al.: Towards time-aware knowledge graph completion. In: Proceedings of COLING 2016, the 26th International Conference on Computational Linguistics: Technical Papers, pp. 1715–1724 (2016)
8. Jiang, T., et al.: Encoding temporal information for time-aware link prediction. In: Proceedings of the 2016 Conference on Empirical Methods in Natural Language Processing, pp. 2350–2354 (2016)
9. Kalchbrenner, N., Grefenstette, E., Blunsom, P.: A convolutional neural network for modelling sentences. arXiv preprint arXiv:1404.2188 (2014)
10. Kingma, D.P., Ba, J.: Adam: a method for stochastic optimization. arXiv preprint arXiv:1412.6980 (2014)
11. Leblay, J., Chekol, M.W.: Deriving validity time in knowledge graph. In: Companion of the The Web Conference 2018, pp. 1771–1776. International World Wide Web Conferences Steering Committee (2018)
12. Lin, Y., Liu, Z., Luan, H., Sun, M., Rao, S., Liu, S.: Modeling relation paths for representation learning of knowledge bases. arXiv preprint arXiv:1506.00379 (2015)

13. Lin, Y., Liu, Z., Sun, M., Liu, Y., Zhu, X.: Learning entity and relation embeddings for knowledge graph completion. In: Twenty-Ninth AAAI Conference on Artificial Intelligence (2015)
14. Lin, Y., Shen, S., Liu, Z., Luan, H., Sun, M.: Neural relation extraction with selective attention over instances. In: Proceedings of the 54th Annual Meeting of the Association for Computational Linguistics (Vol. 1: Long Papers), pp. 2124–2133 (2016)
15. Luo, K., Lin, F., Luo, X., Zhu, K.: Knowledge base question answering via encoding of complex query graphs. In: Proceedings of the 2018 Conference on Empirical Methods in Natural Language Processing, pp. 2185–2194 (2018)
16. Mahdisoltani, F., Biega, J., Suchanek, F.M.: YAGO3: a knowledge base from multilingual wikipedias. In: CIDR (2013)
17. Mendes, P.N., Jakob, M., García-Silva, A., Bizer, C.: DBpedia spotlight: shedding light on the web of documents. In: Proceedings of the 7th International Conference on Semantic Systems, pp. 1–8. ACM (2011)
18. Min, B., Grishman, R., Wan, L., Wang, C., Gondek, D.: Distant supervision for relation extraction with an incomplete knowledge base. In: Proceedings of the 2013 Conference of the North American Chapter of the Association for Computational Linguistics: Human Language Technologies, pp. 777–782 (2013)
19. Nickel, M., Rosasco, L., Poggio, T.: Holographic embeddings of knowledge graphs. In: Thirtieth AAAI Conference on Artificial Intelligence (2016)
20. Oh, B., Seo, S., Lee, K.H.: Knowledge graph completion by context-aware convolutional learning with multi-hop neighborhoods. In: Proceedings of the 27th ACM International Conference on Information and Knowledge Management, pp. 257–266. ACM (2018)
21. Shi, B., Weninger, T.: Open-world knowledge graph completion. In: Thirty-Second AAAI Conference on Artificial Intelligence (2018)
22. Sun, Z., Deng, Z.H., Nie, J.Y., Tang, J.: RotatE: knowledge graph embedding by relational rotation in complex space. arXiv preprint arXiv:1902.10197 (2019)
23. Toutanova, K., Chen, D.: Observed versus latent features for knowledge base and text inference. In: Proceedings of the 3rd Workshop on Continuous Vector Space Models and their Compositionality, pp. 57–66 (2015)
24. Trivedi, R., Dai, H., Wang, Y., Song, L.: Know-Evolve: deep temporal reasoning for dynamic knowledge graphs. In: Proceedings of the 34th International Conference on Machine Learning-Volume 70, pp. 3462–3471 (2017). http://www.JMLR.org
25. Trouillon, T., Welbl, J., Riedel, S., Gaussier, É., Bouchard, G.: Complex embeddings for simple link prediction. In: International Conference on Machine Learning, pp. 2071–2080 (2016)
26. Wang, Q., Mao, Z., Wang, B., Guo, L.: Knowledge graph embedding: a survey of approaches and applications. IEEE Trans. Knowl. Data Eng. **29**(12), 2724–2743 (2017)
27. Wang, Z., Zhang, J., Feng, J., Chen, Z.: Knowledge graph embedding by translating on hyperplanes. In: Twenty-Eighth AAAI Conference on Artificial Intelligence (2014)
28. Yang, B., Yih, W.t., He, X., Gao, J., Deng, L.: Embedding entities and relations for learning and inference in knowledge bases. arXiv preprint arXiv:1412.6575 (2014)
29. Yih, S.W.t., Chang, M.W., He, X., Gao, J.: Semantic parsing via staged query graph generation: question answering with knowledge base (2015)
30. Zeng, D., Liu, K., Chen, Y., Zhao, J.: Distant supervision for relation extraction via piecewise convolutional neural networks. In: Proceedings of the 2015 Conference on Empirical Methods in Natural Language Processing, pp. 1753–1762 (2015)

Interaction Graph Neural Network for News Recommendation

Yongye Qian[1], Pengpeng Zhao[1,2]([✉]), Zhixu Li[1], Junhua Fang[1], Lei Zhao[1],
Victor S. Sheng[3], and Zhiming Cui[4]

[1] Institute of AI, School of Computer Science and Technology,
Soochow University, Suzhou, China
20175227011@stu.suda.edu.cn, {ppzhao,zhixuli,jhfang,zhaol}@suda.edu.cn
[2] Key Lab of IIP of CAS, Institute of Computing Technology, Beijing, China
[3] The University of Central Arkansas, Conway, USA
ssheng@uca.edu.cn
[4] School of Electronic and Information Engineering,
Suzhou University of Science and Technology, Suzhou, China
zmcui@mail.usts.edu.cn

Abstract. Personalized news recommendation has become a highly challenging problem in recent years. Traditional ID-based methods such as collaborative filtering are not suitable for news recommendation due to the extremely rapid update of candidate news. Various content-based methods have been proposed for news recommendation and achieved the state-of-the-art performance. Recently, knowledge-aware news recommendation further improves the performance through discover latent knowledge level connections among the news. However, we argue that the above content-based methods do not fully utilize the collaborative information latent in user-item interactions into user and news representation learning process. In this paper, we propose a new news recommendation model, Interaction Graph Neural Network (IGNN), which integrates a user-item interactions graph and a knowledge graph into the news recommendation model. Specifically, IGNN obtains the representation of users and items with two graphs. One is the knowledge graph, and another is the user-item interaction graph. It learns the content-based feature from knowledge-level and semantic-level with convolutional neural networks and fuses the high-order collaborative signals extracted from the user-item interaction graph into user and news representation learning process with a graph neural network. Extensive experiments are conducted on the two real-world news data sets, and experimental results show that IGNN significantly outperforms the state-of-the-art approaches for news recommendation.

Keywords: News recommendation · Graph Neural Network · Knowledge graph

© Springer Nature Switzerland AG 2019
R. Cheng et al. (Eds.): WISE 2019, LNCS 11881, pp. 599–614, 2019.
https://doi.org/10.1007/978-3-030-34223-4_38

1 Introduction

With the rapid development of the Internet, people's news reading habits have gradually shifted from traditional media such as newspapers and TV to the Internet. However, news applications provide massive news every day, which makes readers overwhelmed due to the information explosion. Therefore, to improve the users' satisfaction and stickiness, the key problem is how to pick out the news which the users are interested in and want to read.

Many researchers have proposed many approaches for news recommendation. Traditional ID-based methods such as collaborative filtering (CF) [1] are widely used in recommender systems. However, many facts indicate that ID-based methods are less effective for news recommendation owing to the highly time-sensitivity of news, and their relevance expires quickly within a short period. Relatively speaking, traditional semantic models [2] and topic models [3] are more effective, due to highly content-based of news. For example, ELSA [22] is an explicit localized sentiment analysis method for location-based news recommendation, and DSSM [23] is a content-based deep neural network to rank a set of documents for a given query, where the ranking is related to the relevance of query to each document calculated by cosine similarity between the low dimensional vectors of query and document. But these models only capture their relations based on the co-occurrence or the clustering structure of words, ignoring the latent knowledge-level connection.

With the success of knowledge graph, Deep Knowledge-aware Network (DKN) [4] is proposed to extract the knowledge-level connection by employing a knowledge graph, which obtains a knowledge-aware representation vector for each piece of news. To get dynamic user representation regarding current candidate news, DKN applies an attention module to aggregate the user's historical records with different weights according to candidate news. The latest news recommendations based on knowledge graph [24] is a graph traversal algorithm as well as a novel weighting scheme for cold-start content-based news recommendation utilizing the named entities. Although DKN pre-trains the embeddings by a words embedding model and a knowledge graph embedding model, we argue that the above knowledge-aware content-based methods do not incorporate the latent collaborative signals in user-item interactions into the learning process of user and news representation. The DKN model and most of other existing methods build the embedding function only base on the descriptive features and lack an explicit encoding of the collaborative signal. This collaborative signal information is significant for the recommendation and can be extracted from the high-order connectivity of user and news interaction graph, which can reveal the behavioral similarity between users and news.

In this paper, we propose an Interaction Graph Neural Network (IGNN) for news recommendation, which extracts the graph information of news-news and user-news, respectively. For the sub-graph of news, IGNN first stacks three embedding matrices of words, entities and the contextual information of entities embedded by transE [13], and then obtains the interaction information of content and knowledge with a convolutional neural network (CNN).

Furthermore, we encode the latent collaborative signal in the interaction graph into the learning process of user and news representation by exploiting the high-order connectivity with an embedding propagation layer to improve the embedding process. By stacking multiple embedding propagation layers, we can enforce the embeddings to capture the collaborative signal in high-order connectivities. Since each layer produces embeddings of all users and news, we combine the embeddings of all layers to aggregate the collaborative signal learned from different orders of connectivities to form the embedding function.

To summarize, this work makes the following main contributions.

(1) We conduct research on the latest progress of news recommendation and highlight the importance of integrating the collaborative signal in user-item interaction graph into the recommendation models.

(2) We propose a new model IGNN for news recommendation based on a knowledge graph and an interaction graph, from which our model IGNN can extract the content-based knowledge-aware connection and the collaborative signal by employing knowledge-aware neural network and embedding propagation layers respectively.

(3) We conduct empirical studies on two real-world datasets. Extensive experimental results demonstrate the state-of-the-art performance of IGNN and its effectiveness in improving the quality of embedding with embedding propagation.

2   Related Work

In this section, we will introduce the related work of this paper. First, we introduce the recommendations based on the knowledge graph, and then the graph neural network will be introduced briefly.

2.1   Knowledge Graph Recommendation

In literature works, the knowledge graph increases attention and has been applied to the recommender system. For example, PER [5] uses a heterogeneous information network to represent the attribute relationship of item-item in the knowledge graph and adopts bayesian collaborative filtering to solve the entity recommendation problem. A model [6] employs the diffusion activation technology to integrate the network structure characteristics of the knowledge graph into the recommender system. CKE [7] proposed a collaborative knowledge-based embedding framework, which integrates the semantic information of the knowledge graph into collaborative filtering as implicit feedback to enhance the performance of the collaborative filtering algorithm. CFRL [8] proposed to compute the semantic similarity between items with employing the representation learning of knowledge graph. Based on the research mentioned above, our model IGNN applies a knowledge graph to news representation learning by extracting keywords and entities in the news, which can get the best-fit representation with mining the potential knowledge-level information contained in the news.

2.2 Graph Neural Network

In recent years, many researchers have used graph convolutional network for accurate recommendations. For instance, GC-MC [9] applies the graph convolutional network (GCN) [10] on the user-item graph. However, it only employs one convolutional layer to exploit the direct connections between users and items. Hence, it fails to reveal the collaborative signal in high-order connectivity. Pin-Sage [11] is an industrial solution that employs multiple graph convolution layers on item-item graph for Pinterest image recommendation. As such, the CF effect is captured on the level of item relations, rather than the collective user behaviors. SpectralCF [12] proposed a spectral convolution operation to discover all possible connectivity between users and items in the spectral domain. Though the eigendecomposition of the graph adjacency matrix, it can discover the connections between a user-item pair. However, the eigendecomposition causes a high computational complexity, which is very time-consuming. GC-SAN [25] dynamically construct a graph structure for session sequences and capture rich local dependencies via a graph neural network, which improve the representation of session sequences. For the model proposed in this paper, we make IGNN effective in exploiting the collaborative signal in high-order connectivity, by designing a specialized graph convolution operation on the user-item interaction graph.

3   Interaction Graph Neural Network

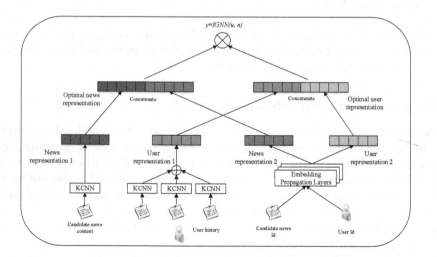

Fig. 1. Framework of Interaction Graph Neural Network. The KCNN layer and embedding propagation layers are illustrated in Figs. 3 and 4 respectively.

In this section, we will introduce our proposed model IGNN in details. The framework of IGNN is illustrated in Fig. 1. There are three components in the framework: (1) a knowledge-based convolutional neural network (KCNN), which can get the word structure information and potential knowledge-level connection of news content; (2) embedding propagation layers that inject high-order connectivity modeling into the embeddings; (3) a prediction layer, computing the score of target news after the model IGNN obtains the optimal representation of users and news. Figure 1 illustrates that the input is the target news content and the id of news. For a specific user, the input is the news content that the user has read before and the user's id. Please note that we get the knowledge-level information from an open knowledge graph: Wikipedia. After getting the keywords of news, we can get the related entity by searching the knowledge graph and downloading the triple relation meanwhile. Then IGNN extracts the semantic-level and knowledge-level information with the knowledge-aware convolutional neural network (introduced in Sect. 3.1), and obtains the high-order collaborative signal in the high-order connectivity by employing the embedding propagation layers (introduced in Sect. 3.2). At last, we concatenate the two representation vector of target news and users, and then compute the user preference on the target news in the prediction model (introduced in Sect. 3.3).

3.1   Knowledge-Aware Convolutional Neural Network

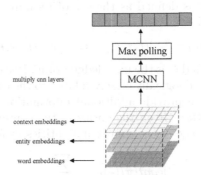

multiply cnn layers

context embeddings

entity embeddings

word embeddings

Fig. 2. Illustration of the knowledge-aware convolutional neural network

In this part, we will introduce the design of the knowledge-aware convolutional neural network (KCNN). Firstly we introduce the methods of knowledge graph embedding.

A typical knowledge graph consists of millions of entity-relational entity triples (h, r, t), where h, r and t represent the head, the relationship, and the tail of the triplet, respectively. Given all triples in a knowledge graph, the goal of knowledge graph embedding is to learn a low-dimensional representation vector for each entity and relationship, to maintain the structural information of the original knowledge graph. In recent years, knowledge graph embedding based on translation has attracted a wide attention due to its simple model and superior performance. In this paper, we apply the TransE model for knowledge graph embedding.

TransE [13] requires $\mathbf{h} + \mathbf{r} \approx \mathbf{t}$, when (h, r, t) holds, where $\mathbf{h}$, $\mathbf{r}$ and $\mathbf{t}$ are the corresponding vectors of h, r and t. Therefore, TransE assumes a score function defined as follows:

$$f_r(h, t) = \|\mathbf{h} + \mathbf{r} - \mathbf{t}\|_2^2, \tag{1}$$

is low if (h, r, t) holds, and is high otherwise.

To encourage the discrimination between correct triples and incorrect triples, for the TransE model, the following margin-based ranking loss is used for training:

$$\pounds = \sum_{(h,r,t)\in\Delta} \sum_{(h',r',t')\in\Delta'} max(0, f_r(h, t) + \gamma - f_r(h', t')), \tag{2}$$

where γ is the margin, Δ and Δ' are the set of correct triples and incorrect triples.

Then we introduce the KCNN model, which is illustrated in Fig. 2. There are three matrices at the bottom of Fig. 2. The bottom matrix is the word embedding matrix. We use $t = w_{1:n} = [w_1, w_2, ..., w_n]$ to denote the raw input sequence of a news title t of length n, and $w_{1:n} = [w_1, w_2, ..., w_n] \in R^{d \times n}$ to denote the word embedding matrix of news, and each word w_i may be associated with an entity embedding $e_i \in R^{k \times 1}$ and the corresponding context embedding $\bar{e}_i \in R^{k \times 1}$, which is pre-trained by transE, where k is the dimension of entity embedding. The context of entity e is defined as the set of its immediate neighbors in the knowledge graph as follows.

$$context(e) = \{e_i | (e, r, e_i) \in G \quad or \quad (e_i, r, e) \in G\} \tag{3}$$

where r is the relation, and G is the knowledge graph. Because contextual entities are often semantically and logically closely related to the current entity, the usage of contextual entities can provide additional information and contribute to the identification of the entity. Given the entity e, the context entity embedding $\bar{e}$ is calculated as the average of the contextual entities as follows.

$$\bar{e} = \frac{1}{|context(e)|} \sum_{e_i \in context(e)} \mathbf{e}_i \tag{4}$$

where $\mathbf{e}_i$ is the embedding vector of e_i learned by knowledge graph embedding (the TransE model in our work).

Now we stack the three embedding matrices together to be an image with three channels:

$$\mathbf{map} = [[w_1, w_2, ..., w_n], [e_1, e_2, ..., e_n], [\bar{e}_1, \bar{e}_2, ..., \bar{e}_n]] \in R^{d \times n \times 3}, \tag{5}$$

After getting the input of the convolutional neural network, we apply several multiple filters $\mathbf{m}$ to exploit the content-based information in the news as follows.

$$c^{m_i} = f(m_i * \mathbf{map} + b), \tag{6}$$

we employ a max pooling layer to get the largest feature from the output feature map:

$$c_{max} = max(c^{m_1}, c^{m_2}, ..., c^{m_l}), \tag{7}$$

where l is the number of filters.

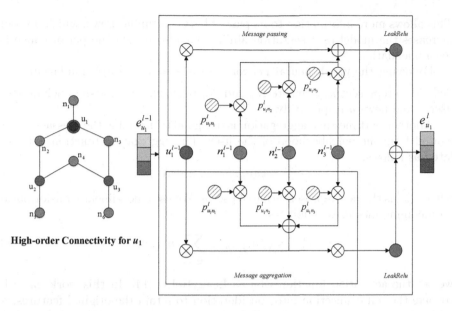

Fig. 3. The left part is the user-news interaction graph, and the right part exhibits how the embedding propagation layers act on the target user u_1

3.2 Embedding Propagation Layer

In propagation layers, we extract the collaborative signals along the graph structure and enrich the representation of users and items with the message-passing architecture of the graph neural network (GNN).

In fact, the historical items of users can reflect the preference of users to some extent. Similarly, user consuming the same news can be treated as the feature of the news, which can reflect the collaborative similarity of two news, based on which we perform the information propagation between the connected user and news.

The embedding propagation layers consists of two parts: message passing and message aggregation. In this work, the message from n to u for a connected user-news pair (u, n) is formulated as follows.

$$m_{u \leftarrow n} = w(e_n, e_u, p_{u,n}), \tag{8}$$

where $m_{u \leftarrow n}$ is the information being propagated, and $w(\cdot)$ is the information propagation function. The input of the $w(\cdot)$ is e_u, e_n and $p_{u.n}$, which controls the decay factor on the propagation of edge (u, n). $w(\cdot)$ is defined as follows.

$$m_{u \leftarrow n} = \frac{1}{\sqrt{|N_u| |N_n|}} (W_1 e_n + W_2(e_n \odot e_u)), \tag{9}$$

where $W_1 \in R^{d' \times d}$ and $W_2 \in R^{d' \times d}$ are the trainable weights to learn the useful information for propagation, and $e_n \odot e_u$ denotes the element-wise product, which is employed to encode the interaction between e_n and e_u into the message pass.

This allows more information to be passed between similar news, which not only increases the model representation ability but also boosts the performance for recommendation.

Following the graph neural network, $\frac{1}{\sqrt{|N_u||N_n|}}$ is a graph Laplacian norm defined to represent $p_{u,n}$, where N_u and N_n represent the first-hop neighbors of user u and news n respectively.

Another part message aggregation is designed to assemble the messages propagated from the neighborhood of the user. The aggregation function is formulated as follows.

$$e_u^{(1)} = \varphi(e_u, m_{u \leftarrow n}|i \in N_u), \tag{10}$$

where $e_u^{(1)}$ is the new representation of user u. We use a non-linear transformation on incoming messages as follows.

$$e_u^{(1)} = \psi(m_{u \leftarrow u} + \sum_{i \in N_u} m_{u \leftarrow i}), \tag{11}$$

we set the activation function $\psi(\cdot)$ as LeakyReLU [14]. In this work, in order to take the self-connection into consideration to retain the original features, we add $m_{u \leftarrow u}$ into the formulation. It is defined as $m_{u \leftarrow u} = W_1 e_u$. Analogously, the news n embedding representation can be identified as $e_n^{(1)}$ by propagating information from connected users.

For the high-order connectivity, we exploit it by stacking more propagation layers. Therefore, the user and news can be single represented by assembling the messages from high-hop neighbors. It is crucial for interaction learning to encode such a collaborative signal, which is helpful to estimate the relevance score between the user and news.

A user or news can obtain the message which is propagated from its k-hop neighborhood by stacking k embedding propagation layers. Therefore, the user u can be recursively represented as follows.

$$e_u^{(k)} = \psi(m_{u \leftarrow u}^{(k)} + \sum_{i \in N_u} m_{u \leftarrow i}^{(k)}), \tag{12}$$

and the message being propagated is formulated as:

$$m_{u \leftarrow n}^{(k)} = \frac{1}{\sqrt{|N_u||N_n|}}(W_1^{(k)} e_n + W_2^{(k)}(e_n^{(k-1)} \odot e_u^{(k-1)})), \tag{13}$$

where $e^{(k-1)}$ is the representation generated from the previous embedding propagation, memorizing the messages from the $(l-1)$-hop neighbors.

As the right part of Fig. 3, the collaborative signal like $u_1 \leftarrow i_2 \leftarrow u_2 \leftarrow i_4$ can be exploited in the embedding propagation process.

3.3   Prediction Model

After propagating with k layers and knowledge-aware convolutional neural network, for news n, we can obtain two parts of news, namely $(e_n^1, e_n^2, ..., e_n^k)$ and e_n^{cb}.

After that, we employ a concatenate function to constitute the final embeddings of the news as follows.

$$e_n^* = e_n^{(0)}||e_n^{(1)}||...||e_n^{(k)}||e_n^{cb} \tag{14}$$

where $||$ is the concatenation operation. Analogously, for user u, we use the historical news of the user to represent the user content-based information. The user content-based representation can be formulated as follows.

$$e_u^{cb} = mean(e_{n_1}^{(cb)}, e_{n_2}^{(cb)}, ..., e_{n_q}^{(cb)}) \tag{15}$$

where $mean(\cdot)$ denotes a mean pooling function, and $e_{n_q}^{(cb)}$ is the historical news that the user read before. At last, the final representation of the user can be defined as follows.

$$e_u^* = e_u^{(0)}||e_u^{(1)}||...||e_u^{(k)}||e_u^{cb} \tag{16}$$

where $e_u^{(k)}$ is generated by embedding propagation layers.

Finally, we conduct the inner product to estimate the user's preference towards the target news:

$$\hat{y}_{IGNN}(u,n) = e_u^{*\top} e_n^* \tag{17}$$

3.4 Optimization

Similar to other works [15], we opt for the BPR loss, which has been widely used to optimize recommendation models. For each pair (u, n_{pos}), we randomly choose one negative pair (u, n_{neg}) from the news that user hasn't read, and then compute the loss of the pairwise. Therefore, the objective function for optimizing our model is as follows.

$$loss_{IGNN} = \sum_{(u,n_{pos},n_{neg}) \in \Gamma} -ln\mu(\hat{y}_{IGNN}(u,n_{pos}) - \hat{y}_{IGNN}(u,n_{neg})) + \lambda \|\Theta\|_2^2 \tag{18}$$

where Γ denotes a training set; $\mu(\cdot)$ is the sigmoid function; Θ is the model parameter set. Additionally, we conduct L_2 regularization parameterized by λ on Θ to prevent overfitting.

Training. We adopt mini-batch Adam to optimize the model and update the model parameters. Although deep learning models have a strong representation ability, they usually suffer from overfitting. Following prior work, Dropout is an effective solution to prevent overfitting. In this work, we adopt two dropout techniques in embedding propagation layers: message dropout and node dropout. Message dropout drops out the messages information randomly. In particular, we employ message dropout on the messages propagated in Eq. (9) with a p_1 probability. Therefore, only part of the message information participates in the new representation. We also employ node dropout to randomly block a particular node and discard all its outgoing messages. In particular, we randomly drop p_2 percent on the nodes of the Laplacian matrix in the l-th propagation layer, where p_2 is the dropout ratio.

4   Experiment

In this section, we present our experiments and corresponding experimental results, including dataset description, parameter settings, and comparisons of models.

4.1   Dataset Description

To evaluate the effectiveness of our model, we conduct experiments on two real-world datasets. One is the DC competition news data from the Data Castle. The other is the Adressa news dataset [16]. All the statistics of the two datasets are shown in Table 1.

Table 1. Statistics of datasets.

Dataset	Users	Items	Interactions	Entities	Triples
DC	10,000	6385	116,225	15,284	39,167
Adressa	640,503	20,428	3,101,991	45,604	74,219

DC. The DC dataset consists of 10,000 user historical clicked news, which comes from a financial website. Each line contains a user id, a news id, timestamp, a news title, news content. In this work, we just select the user id, the news id, the title of news as our selection.

Adressa. It is published with the collaboration of Norwegian University of Science and Technology (NTNU) and Adressavisen (a local newspaper in Trondheim, Norway) as a part of the RecTech project on recommendation technology. It is an event-based news dataset that includes anonymized users with their clicked news articles. Each reading event corresponds to a user reading a particular news article that contains 18 attributes. These attributions are not all very useful in our recommender system. Thus, we just select the user id, the news id, the title of news as our selection, for generating our datasets.

Sub Knowledge Graph. We search all occurred entities in the dataset as well as the ones within their one-hop neighborhood entity in the Wikipedia knowledge graph and extract all edges (triples) from the Wikipedia knowledge graph too.

4.2   Experiment Settings

Evaluation Metrics. For each user in the test set, we treat all the news that the user has not interacted with as negative news. Then each method outputs the user's preference scores over all the news, except the positive ones used in the training set. To evaluate the effectiveness of Top-K recommendation, we employ two widely-used evaluation protocols [17,18]: Recall@K and NDCG@K. In this work, we set K = 20. The results we report are the average metrics for all users in the test set.

Baselines. We use the following state-of-the-art methods as baselines in our experiments:

- MF [15]: This is a matrix factorization model optimized by a Bayesian personalized ranking (BPR) loss, which exploits the user-item direct interactions only as the target value of interaction function.
- LibFM [19]: This model is a state-of-the-art feature-based factorization model and widely used in CTR scenarios.
- DeepWide [20]: The model is a general deep model for recommendation, combining a linear channel with a non-linear channel.
- NeuMF [17]: This model is a state-of-the-art neural CF model, which uses multiple hidden layers above the element-wise and concatenation of user and item embeddings to capture their non-linear feature interaction.
- GC-MC [9]: This model adopts a GCN encoder to generate the representations for users and items, where only the first-order neighbors are considered. Hence one graph convolution layer, where the hidden dimension is set as the embedding size, is used as suggested in the original paper.
- DKN [4]: This is a deep recommendation framework, which devises a multichannel CNN to fuse semantic-level and knowledge-level representations of news, and introduces an attention mechanism for click through rate prediction.

Parameter Settings. We implement our model in Tensorflow. The embedding size of entity embedding, words embedding, context embedding and embedding propagation layers are all 64, same to all baselines. The learning rate is set as 0.01, the batch-size is set 1024, and the coefficient of L_2 normalization is 10^{-2}. The message dropout and the node dropout are all set 0.1. We use the default Xavier initializer to initialize the model and pretrain the entity embedding and context embedding by the TransE model.

4.3   Performance Comparisons

To demonstrate the recommendation performance of our model IGNN, we compare it with other state-of-the-art models. Our experimental results of all methods on two datasets are shown in Table 2. From Table 2, we can observe that:

First, the MF model performs the worst. This indicates that models based on the deep neural network have a better performance on recommendation, due to the stronger representation ability of the deep neural network. Besides MF can't exploit the high-order connectivity in the user-item interaction graph.

Second, we can find that the GC-MC model performs better than MF, Deepwide, and LibFM. This indicates that incorporating the first-order neighbors can improve the representation learning, which verifies the importance of extracting the high-order connectivity in the user-items interaction graph.

Table 2. Overall performance comparisons.

Method	DC dataset		Adressa	
	Recall@20	NDCG@20	Recall@20	NDCG@20
MF	0.58742	0.39962	0.10384	0.09046
LibFM	0.59834	0.44839	0.14758	0.11248
Deep-wide	0.60287	0.44894	0.13857	0.10295
NeuMF	0.65892	0.52183	0.16842	0.12485
GC-MC	0.63572	0.48834	0.17864	0.12047
DKN	0.67927	0.55856	0.19210	0.13385
IGNN	**0.73180**	**0.59315**	**0.20728**	**0.15729**

Third, DKN performs the best on the two datasets, comparing with other baselines. This is because DKN extracts the knowledge-level information and fuses the information into the semantic-level information. Meanwhile, DKN represents the user preference with an attention network.

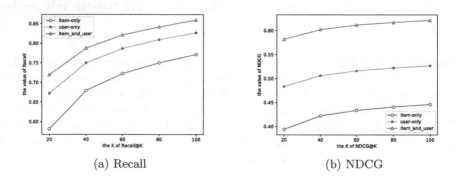

(a) Recall   (b) NDCG

Fig. 4. The effectiveness of different combination ways.

Finally, our model IGNN performs the best for the Top-K recommendation task, compared with all baseline models. Specifically, IGNN has a maximum improvement of 16.8% in terms of Recall@20, and a maximum 30.6% improvement in terms of NDCG@20, compared with the MF model. Analogously, IGNN has a minimum improvement of 7.9% in terms of Recall@20, and a minimum 17.5% improvement in terms of NDCG@20, compared with the DKN model. The reason why IGNN has improved significantly on both datasets is that we not only take the semantic-level information and knowledge-level connection of news into account, but also employ the high-order connectivity between the user-item interaction graph.

4.4   Model Analysis and Discussion

Variation of IGNN. In this part, We will further investigate the performance of the variation of IGNN to verify the contribution of its various components. We investigate first the performance of under different combinations of the knowledge-aware CNN and embedding propagation layers. Specifically, we conduct experiments on three different combinations. First, the embedding propagation layers only enhance the representation of the user. Second, the embedding propagation layers only enhance the representation of news. Third, the embedding propagation layers enhance the user and news at the same time. Our experimental results are shown in Fig. 4. From Fig. 4 we can see that the way of combining user and item at the same time is an out-of-the-art way. The reason is that user-only or item-only may lose the collaborative signal with the computation of the prediction model. Overall, our final model IGNN performs the best.

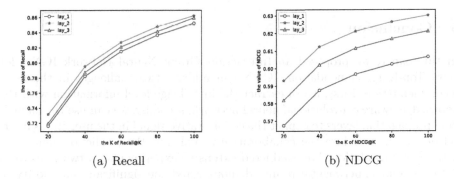

(a) Recall   (b) NDCG

Fig. 5. The effectiveness of the number of the embedding propagation layers.

Impact of the Number of Embedding Propagation Layers. In addition, we also investigate whether IGNN can benefit from multiple embedding propagation layers. We search the number of embedding propagation layers in the range

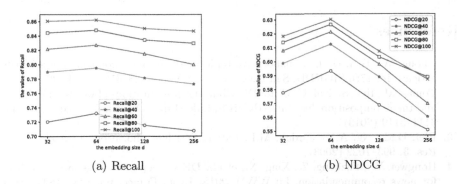

(a) Recall   (b) NDCG

Fig. 6. The effectiveness of the number of embedding propagation layers.

of $(1, 2, 3)$. Figure 5 shows our experiment results. We can find that IGNN with two embedding propagation layers has the best performance on both Recall and NDCG metrics, and IGNN with only one embedding propagation layer performs worst. This is because the one embedding propagation layer only considers the first-hop neighbors. It can't capture the collaborative signal well. The reason why IGNN with 3 embedding propagation layers performs worse than IGNN with 2 embedding propagation layers may be overfitting.

Impact of the Embedding Dimension D. In Fig. 6, we investigate the effectiveness of the embedding size d in $\{32, 64, 128, 256\}$ on DC news dataset. From the results showed in Fig. 6, we can observe that our model IGNN performs best with the embedding dimension set as 64, and the performance of IGNN degrades as the increment of the embedding dimension after 64. This demonstrates that overfitting may occur when the implicit factor dimension of IGNN is too high.

5   Conclusion

In this paper, we proposed an Interaction Graph Neural Network IGNN for news Top-K recommendation. IGNN can address two challenges in the news recommendation. First, it can extract the knowledge-level information with the knowledge-aware convolutional neural network; secondly, we can use the embedding propagation layers to exploit the collaborative signal in the user-items interaction graph, and fuse the collaborative signal into the prediction model for news recommendation. We conducted extensive experiments on two real-world datasets. Our experimental results demonstrated the significant superiority of IGNN, compared with the baselines. In the future, we will further study how to employ the social network [21] and extracting other useful information to improve the performance of IGNN on news recommendation.

Acknowledgements. This research was partially supported by NSFC (No. 61876117, 61876217, 61872258, 61728205), Open Program of Key Lab of IIP of CAS (No. IIP2019-1) and PAPD.

References

1. Chong, W., Blei, D.M.: Collaborative topic modeling for recommending scientific articles. In: 17th SIGKDD, San Diego, CA, USA, pp. 448–456. ACM (2011)
2. Tomas, M., Ilya, S., Kai, C., et al.: Distributed representations of words and phrases and their compositionality. In: 27th NIPS, Lake Tahoe, Nevada, United States, pp. 3111–3119 (2013)
3. Blei, D.M., Ng, A.Y., Jordan, M.I.: Latent Dirichlet allocation. J. Mach. Learn. Res. **3**, 993–1022 (2003)
4. Hongwei, W., Fuzheng, Z., Xing, X., et al.: DKN: deep knowledge-aware network for news recommendation. In: WWW 2018, Lyon, France, pp. 1835–1844. ACM (2018)

5. Xiao, Y., Xiang, R., Yizhou, S., et al.: Personalized entity recommendation: a heterogeneous information network approach. In: 7th WSDM, pp. 283–292. ACM, New York, NY, USA (2014)
6. Grad-Gyenge, L., Filzmoser, P., Werthner, H.: Recommendations on a knowledge graph. In: 1st International Workshop on Machine Learning Methods for Recommender Systems, pp. 13–20 (2015)
7. Fuzheng, Z., Nicholas, J., Defu, L., et al.: Collaborative knowledge base embedding for recommender systems. In: 22nd SIGKDD, San Francisco, California, USA, pp. 353–362. ACM (2016)
8. Xiyu, W., Qimai, C., Hai, L.: Collaborative filtering recommendation algorithm based on representation learning of knowledge graph. Comput. Eng. **2**(44), 226–232 (2018)
9. van den Berg, R., Kipf, T.N., Welling, M.: Graph convolutional matrix completion. CoRR (2017)
10. Kipf, T.N., Welling, M.: Semi-supervised classification with graph convolutional networks. In: 5th ICLR, Toulon, France. OpenReview.net (2017)
11. Rex, Y., Ruining, H., Kaifeng, C., et al.: Graph convolutional neural networks for web-scale recommender systems. In: 24th SIGKDD, London, UK, pp. 974–983. ACM (2018)
12. Lei, Z., Chun-Ta, L., Fei, J., et al.: Spectral collaborative filtering. In: 12th Conference on Recommender Systems, Vancouver, BC, Canada, pp. 311–319. ACM (2018)
13. Antoine, B., Nicolas, U., Alberto, G., et al.: Translating embeddings for modeling multi-relational data. In: 27th NIPS, Lake Tahoe, Nevada, United States, pp. 2787–2795 (2013)
14. Petar, V., Guillem, C., Arantxa, C., et al.: Graph attention networks. In: 6th ICLR, Vancouver, BC, Canada (2018)
15. Steffen, R., Christoph, F., Zeno, G., et al.: BPR: Bayesian personalized ranking from implicit feedback. In: 25th UAI, Montreal, QC, Canada, pp. 452–461. AUAI Press (2018)
16. Jon Atle, G., Lemei, Z., Peng, L., et al.: The Adressa dataset for news recommendation. In: WI 2017, Leipzig, Germany, pp. 1042–1048. ACM (2017)
17. Xiangnan, H., Lizi, L., Hanwang, Z., et al.: Neural collaborative filtering. In: WWW 2017, Perth, Australia, pp. 173–182. ACM (2017)
18. Jheng-Hong, Y., Chih-Ming, C., Chuan-Ju, W., et al.: HOP-rec: high-order proximity for implicit recommendation. In: 12th Conference on Recommender Systems, Vancouver, BC, Canada, pp. 140–144. ACM (2018)
19. Steffen, R.: Factorization machines with libFM. ACM TIST **3**(3), 57:1–57:22 (2012)
20. Heng-Tze, C., Levent, K., Jeremiah, H., et al.: Wide & deep learning for recommender systems. In: 1st Workshop on Deep Learning for Recommender Systems, Boston, MA, USA, pp. 7–10. ACM (2016)
21. Xiang, W., Xiangnan, H., Liqiang, N., et al.: Item silk road: recommending items from information domains to social users. In: 40th SIGIR, Shinjuku, Tokyo, Japan, pp. 185–194. ACM (2017)
22. Jeong-Woo, S., A-Yeong, K., Seong-Bae, P.: A location-based news article recommendation with explicit localized semantic analysis. In: 36th SIGIR, Dublin, Ireland, pp. 293–302. ACM (2013)
23. Po-Sen, H., Xiaodong, H., Jianfeng, G.: Learning deep structured semantic models for web search using clickthrough data. In: 22nd CIKM, San Francisco, California, USA, pp. 2333–2338. ACM (2013)

24. Kevin, J., Hui, J.: Content based news recommendation via shortest entity distance over knowledge graphs. In: WWW 2019, San Francisco, USA, pp. 690–699. ACM (2019)
25. Chengfeng, X., Pengpeng, Z., Yanchi, L., et al.: Graph contextualized self-attention network for session-based recommendation. In: IJCAI 2019, Macao, China, pp. 3940–3946 (2019)

Knowledge Graphs

Gated Relational Graph Neural Network for Semi-supervised Learning on Knowledge Graphs

Yuyan Chen, Lei Zou$^{(\boxtimes)}$, and Zongyue Qin

Peking University, Beijing, China
zoulei@pku.edu.cn

Abstract. Entity classification is an important task for knowledge graph (KG) completion and is also crucial in many upper-level applications. Traditional methods use unsupervised representation learning to embed entities and relations into a continuous low-dimensional space, and then use the embeddings in downstream tasks. Recent years, Graph Neural Networks (GNNs) have been gaining growing interest, among which Graph Convolutional Network (GCN) is widely used in semi-supervised tasks due to its excellent capability of aggregating neighborhood features. However, GCN lacks the ability to deal with edge features, which is essential in KGs. In this paper, we propose Gated Relational Graph Neural Network (GRGNN) targeted on entity classification problem in KGs. More specifically, we apply the idea of TransE to incorporate features of entities and relations, and introduce gate mechanism to leverage hidden states of current node and its neighbors. Our method achieves state-of-the-art performance compared with other methods in FB15K and DB10K datasets.

Keywords: Knowledge graph · Entity classification · GCN

1 Introduction

Knowledge graph (KG), also known as Knowledge Base, is a multi-relational graph composed of entities (nodes) and relations (edges), where knowledge is arranged into triples in the form of (*head entity, relation, tail entity*), also denoted as (*subject, predicate, object*) ((*s, p, o*) in short). For instance, the knowledge "Washington is the capital of United States" can be represented as (*Washington, capital_of, United States*).

Ever since the concept of Knowledge Graph was proposed, it has arise more and more attention and applications, and a wide variety of problems come with it. KG-related tasks can be categorized into In-KG applications and Out-of-KG applications [27]. The former include link prediction [5], triple classification [11], entity classification [22] and entity resolution [4], and the latter mainly involve relation extraction [13], question answering [3], etc. In this paper, we aim to solve the problem of entity classification.

R. Cheng et al. (Eds.): WISE 2019, LNCS 11881, pp. 617–629, 2019.
https://doi.org/10.1007/978-3-030-34223-4_39

In most cases, entities have types in the KG, e.g., the entity *Washington* has multiple types including *location/us_country* and *location/administrative_division*. Such information can be denoted in triples with relation *isA*, e.g., (*Washington, isA, location/us_country*). In fact, the type information in a KG is usually incomplete. So it is of necessity to learn entity types using existing information. Besides, entity classification is also crucial and widely used in many natural language processing tasks [20].

To tackle this problem, some solutions have been proposed, including graph representation learning approaches and graph neural networks approaches. The former project entities and relations into continuous low-dimension vector spaces, then the embedded features are used for various downstream tasks. On the other hand, Graph Neural Networks (GNNs) are end-to-end models for specific tasks, referring to deep learning methods targeted on graph domain. Among a large number of variants of GNNs, Graph Convolutional Network (GCN) [16] is considered as a breakthrough and achieves state-of-the-art performance in node classification task due to its excellent ability to aggregate neighborhood information. However, GCN only supports processing simple graphs with labeled nodes and undirected, unlabeled edges. Although the current GCN model can be used on knowledge graphs with additional preprocessing, such procedure may involve message loss and introduce noise, as a result, the information in original KG can not be utilized adequately.

In this paper, to incorporate node and relation features of knowledge graphs as well as the interactions in between, we propose Gated Relational Graph Neural Network (GRGNN) that combines the idea of translating models in representation learning and graph neural networks. And the main contributions of our method can be summarized as follows:

- Our model is capable of dealing with diverse relations in knowledge graphs. Large-scale KGs may contain thousands of relations. To the best of our knowledge, there is no previous work that could handle such rich information of edges. GRGNN attempts to incorporate link features in a simple while effective way.
- Our model handles directed graph and applies to different features. As KG is a directed graph, the direction of edges are taken into account in our method. Besides, we adopt gate mechanism to leverage the features of current node and its neighbors. Various features can be used as inputs for our method, which makes GRGNN more flexible.
- GRGNN outperforms other methods and achieves state-of-the-art performance on the task of node classification. We evaluate the effectiveness of our method in two datasets, including FB15K and DB10K. We further analyze the influence of network depth, i.e., the number of layers. Similar to GCN, the performance of GRGNN drops as the model depth increases. The drop rate is different according to the initial features, which means some sort of features are more insensitive to model depth. And GRGNN using such features doesn't deteriorate much, as the gate mechanism could avoid over-smoothing of node embeddings. This may provide some opportunities for us to further deepen the model in future work.

The remainder of this paper is organized as follows. Section 2 gives preliminaries and annotations. Section 3 introduces related works including representation learning and graph neural networks. Section 4 describes our method in details. Section 5 shows the experiment results and Sect. 6 gives summarization and conclusion.

2   Preliminaries

A simple graph can be described by adjacency matrix $\mathbf{A} \in \mathbb{N}^{n \times n}$, where a non-zero value a_{ij} implies there is an edge from node i to node j and the value can present the weight of edge. Diagonal matrix $\mathbf{D} = \mathrm{diag}(d_1, d_2, ..., d_n)$ is the degree matrix of $\mathbf{A}$ where $d_i = \sum_j a_{ij}$. Let $\tilde{\mathbf{A}} = \mathbf{A} + \mathbf{I}_n$ and $\tilde{\mathbf{D}}_{i,i} = \sum_j \tilde{\mathbf{A}}_{i,j}$.

Given a knowledge graph $\mathcal{G} = (\mathcal{V}, \mathcal{R}, \mathcal{E})$, $\mathcal{V}$ is the set of nodes (entities), $\mathcal{R}$ is the set of relations (predicates) and $\mathcal{E} \in \mathcal{V} \times \mathcal{R} \times \mathcal{V}$ is the set of edges (triples). Each node can have zero, one or multiple labels. We use $\mathcal{N}_i$ to indicate the set of neighbor nodes of i, and $\mathcal{N}_i^r = \{j | (i, r, j) \in \mathcal{E}\}$ refers to the set of neighbor nodes of node i under relation r. Y_i denote the label set of node i, and only a subset of node labels are revealed to us. Bold lowercase letters denote vectors of features or hidden states, e.g., $\mathbf{e}$ and $\mathbf{r}$ denote the feature vector of entities and relations respectively. By putting nodes' features together, we get the feature matrices of entities $\mathbf{X_e} = [\mathbf{e}_1, \mathbf{e}_2, ..., \mathbf{e}_n]$ and relations $\mathbf{X_r} = [\mathbf{r}_1, \mathbf{r}_2, ..., \mathbf{r}_m]$, usually used as inputs of GNNs.

In a graph neural network, $\mathbf{h}_i^{(l)}$ denotes the hidden states of node i in the l-th layer, $\mathbf{h}_{ne[i]}$ denotes the embedding of i's neighbors, and $\mathbf{h}_{co[i]}$ stands for the embedding of its edges. $\mathbf{H}^{(l)} = [\mathbf{h}_1^{(l)}, \mathbf{h}_2^{(l)}, ..., \mathbf{h}_n^{(l)}]$. $\mathbf{W}$ is the weight matrix need to be trained. And we use $\mathbf{O}$ to denote the final output matrix of the network, as $\mathbf{o_i}$ refers to the output of node i. And σ is the activation function.

3   Related Work

We treat node classification as a semi-supervised problem, that is, we are aware of the whole graph structure and features of nodes as well as edges during training, but only a small amount of nodes are labeled. Our method builds upon previous work of knowledge representation learning and graph neural networks.

3.1   Representation Learning

Knowledge representation learning aims to learn low-dimensional embeddings for nodes and relations. Such methods can be roughly categorized into two groups: semantic matching models and translation-based models.

Semantic matching approaches exploit similarity-based scoring functions. Some prominent models include RESCAL [22], DistMult [30], HolE [21] and ComplEx [25].

Translation-based approaches regard the relation in a triple as a translation from head entity to tail entity. The most representative model is TransE [5], which represents entities and relations as vectors in the same space, adopting the energy function $\mathbf{s} + \mathbf{p} \approx \mathbf{o}$. In other words, the embedding of the tail entity $\mathbf{o}$ should be the nearest neighbor of $\mathbf{s} + \mathbf{p}$ when (s, p, o) holds. The loss function of TransE can be denoted as follow:

$$\mathcal{L} = \sum_{(s,p,o)\in S} \sum_{(s',p,o')\in S'_{(s,p,o)}} \max(0, \gamma + d(\mathbf{s}+\mathbf{p}, \mathbf{o}) - d(\mathbf{s}'+\mathbf{p}, \mathbf{o}')) \qquad (1)$$

where S is the training set of triples and $S'_{(s,p,o)} = \{(s', p, o)|s' \in \mathcal{V}\} \cup \{(s, p, o')|o' \in \mathcal{V}\}$ is the negative sampling set constructed by replacing head or tail entity of a positive sample.

In spite of the simplicity of TransE model, it achieves amazing performance in the task of knowledge graph completion. Extensions of TransE aim to better model 1-to-N, N-to-1 and N-N relations, such as TransH [28], TransR [19], TransD [14], etc.

3.2 Graph Neural Networks

Graph neural networks (GNNs) are deep learning methods targeted on graph domain, first introduced in [10] and [23]. The main idea of GNNs is to make use of graph topology structure as well as nodes' (and edges') features. More specifically, GNNs update a node's hidden states using its neighbors', edges' and its own hidden states in last layer [33], which can be denoted as follow:

$$\mathbf{h}_v^{(t+1)} = f(\mathbf{h}_v^{(t)}, \mathbf{h}_{ne[v]}^{(t)}, \mathbf{h}_{co[v]}^{(t)}) \qquad (2)$$

Graph representation is trained independently from target problem. In contrast, GNNs are end-to-end models targeting specific tasks. GNNs were originally introduced in [10] and [23] as extensions of recurrent neural network. Then a number of variants occurred regarding different graph types [1,24,32], diverse training methods [6,18] as well as propagation steps [12,16,26]. Among these architectures, **Graph Convolutional Network (GCN)** [16] seems to be the most appealing. It was based on a first-order approximation of spectral convolutions on graphs [8] with some simplifications and a renormalization trick, successfully improving scalability and classification performance in large-scale networks. GCN deals with simple graph with nodes features $\mathbf{X}$ and labels $\mathbf{Y}$. The propagation of GCN can be generally formulated as follow:

$$\mathbf{H}^{(t+1)} = \sigma(\mathbf{P}\mathbf{H}^{(t)}\mathbf{W}^{(t)}), \quad \text{with} \quad \mathbf{P} = \tilde{\mathbf{D}}^{-\frac{1}{2}}\tilde{\mathbf{A}}\tilde{\mathbf{D}}^{-\frac{1}{2}} \qquad (3)$$

More specifically, a two-layer GCN for multi-class classification would have the form

$$\mathbf{O} = \text{softmax}(\mathbf{P}\text{ReLU}(\mathbf{P}\mathbf{X}\mathbf{W}^{(0)})\mathbf{W}^{(1)}) \qquad (4)$$

Although GCN can only process simple graphs, where knowledge graphs are obviously not included, some preprocessing schemes can be used to tackle this problem. For instance, [31] suggests that a triple (s, p, o) can be divided into (s, p_1) and (o, p_2). In detail, each entity is treated as a node in graph, and each relation is split to two nodes (considering the direction of triples), and two edges connecting entity node and relation node are added. After the splitting, we get a heterogeneous bipartite graph that GCN is capable of computing.

Some works have been studied to enlarge GCN's capability of processing different graph types, among which R-GCN [24] aims to model relational data based on the GCN framework. Its basic idea is to train different weight matrices for different relation types, with propagation as follow:

$$\mathbf{h}_i^{(l+1)} = \sigma(\sum_{r \in \mathcal{R}} \sum_{j \in \mathcal{N}_i^r} \mathbf{W}_r^{(l)} \mathbf{h}_j^{(l)} + \mathbf{W}_0^{(l)} \mathbf{h}_i^{(l)}) \tag{5}$$

Although R-GCN introduces two strategies—basis-decomposition and block-diagonal-decomposition—for regularizing the weights so as to cope with highly multi-relational data, it can only handle limited discrete edge features, ignoring rich information behind the relations.

4   Method

In this section, we introduce the architecture of GRGNN, and then analyze its advantages through the comparison with other related work.

4.1   GRGNN

Overall, GRGNN aggregates the information of the neighbor nodes and neighbor relations of an individual node, and uses the aggregated states to update the embedding of current node.

We first describe a single layer of GRGNN as denoted by Fig. 1. Say that we are now updating node i's hidden states. Assume that i is the head entity in triples set T_1 and tail entity in triples set T_2. In formula, $T_1 = \{(s, p, o) | (s, p, o) \in \mathcal{E} \wedge s = i\}$, $T_2 = \{(s, p, o) | (s, p, o) \in \mathcal{E} \wedge o = i\}$. We want to define functions $f^{\rightarrow}(\cdot, p, o)$ and $f^{\leftarrow}(s, p, \cdot)$ to incorporate the entities states as well as the relations states. For a triple (s, p, o), based on the idea of TransE, the embedding of object is similar to the embedding of subject plus relation, which makes it reasonable to aggregate $\mathbf{s} + \mathbf{p}$ into hidden states of entity o and aggregate $\mathbf{o} - \mathbf{p}$ into states of s. So in our model, $f^{\rightarrow}(\cdot, p, o)$ is defined as $\mathbf{h}_o^{(l)} + \mathbf{h}_p^{(l)}$ and $f^{\leftarrow}(s, p, \cdot)$ as $\mathbf{h}_s^{(l)} - \mathbf{h}_p^{(l)}$. Then we sum the incorporate features of node i's neighbors. Therefore, the aggregated states of i's neighbors can be summarized as follow:

$$\mathbf{h}_{\mathcal{N}(i)} = \frac{1}{|T_1|} \sum_{(\cdot, p, o) \in T_1} (\mathbf{h}_o - \mathbf{h}_p) + \frac{1}{|T_2|} \sum_{(s, p, \cdot) \in T_2} (\mathbf{h}_s + \mathbf{h}_t) \tag{6}$$

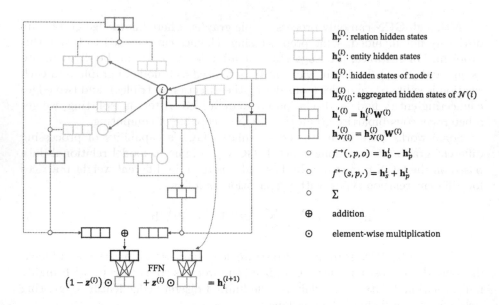

Fig. 1. A single layer of GRGNN.

Just as GCN adds self-connections to preserve the feature of current nodes, we integrate nodes' and neighbors' hidden states in a similar way. GCN assumes equal importance of self-connections and edges to neighboring nodes, while we introduce a parameter ρ to leverage self-connections and neighborhoods.

$$\mathbf{h}_i^{(l+1)} = \sigma(\rho^{(l)}\mathbf{h}_i^{(l)}\mathbf{W}^{(l)} + (1-\rho^{(l)})\mathbf{h}_{\mathcal{N}(i)}^{(l)}\mathbf{W}^{(l)}) \tag{7}$$

where $\rho^{(l)} \in (0,1)$. The larger value of $\rho^{(l)}$ is, the more weight will be put on the current node compared with its neighbors. To simplify computation, $\rho^{(l)}$ is shared among all nodes in the graph, and this parameter is learned via gradient descent by the network.

Furthermore, the self-connection weight parameter $\rho^{(l)}$ can be extended to a *gate mechanism* [7]. That is, when updating hidden states of current node, an *update gate* $\mathbf{z} \in \mathbb{R}^d$ (d is the output dimension of current layer) is used to control how much information from the previous hidden state of the current node will be carried over to next layer. The activation of node i is then computed by

$$\mathbf{h}_i^{(l+1)} = \sigma(\mathbf{z}^{(l)} \odot (\mathbf{h}_i^{(l)}\mathbf{W}^{(1)}) + (1-\mathbf{z}^{(l)}) \odot (\mathbf{h}_{N(i)}^{(l)}\mathbf{W}^{(1)})) \tag{8}$$

where $\odot$ is the element-wise multiplication.

Stacking several layers described above, we get a GRGNN. The last layer of GRGNN uses a sigmoid function as non-linearly for multi-label classification task.

It is worth mentioning that, although GRGNN borrows the idea of TransE that $\mathbf{h} + \mathbf{r} \approx \mathbf{t}$, it is not necessary to use the embedding trained by TransE as inputs. Featureless graph, where the initial features are randomized or set as one-hot encoding can also achieve a rather good performance. However, different features react differently to model depth. More detailed experiment results are shown in Sect. 5.

4.2 Comparison with Related Work

The preprocessing scheme used for GCN to deal with knowledge graphs—splitting (s, p, o) to (s, p_1) and (o, p_2)—is widely used. A significant shortcoming of such procedure is that it removes the link between s and o. Examples are demonstrated in Fig. 2.

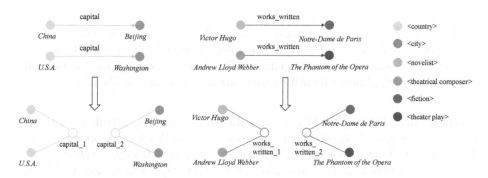

Fig. 2. Examples of the preprocessing scheme for knowledge graph from [31].

By a 2-layer GCN, *China* and *U.S.A*, *Beijing* and *Washington* convolve each other's features. There underlies the assumption that entities on the same end of one relation tend to have the same type. So it is helpful for classifying nodes of type <country> as they always present on the left end of relation *capital*, and type <city> on the right end. However, sometimes this scheme would confuse entities of different types. For example, given the triples (*Victor Hugo*, works_written, *Notre-Dame de Paris*) and (*Andrew Lloyd Webber*, works_written, *The Phantom of the Opera*), if we know that *Notre-Dame de Paris* is a fiction and *The Phantom of the Opera* is a theater play, it's not difficult to classify *Victor Hugo* to a novelist and *Andrew Lloyd Webber* to a theatrical composer. But in the split graph, *Victor Hugo* and *Andrew Lloyd Webber* would become indistinguishable. On the other hand, our GRGNN model is able to discriminate these two entities by integrating the features of *Notre-Dame de Paris* into *Victor Hugo* and *The Phantom of the Opera* into *Andrew Lloyd Webber* respectively.

As for R-GCN [24], it adopts different weight matrices for propagation on different relation types. In other words, it only separates relations regardless of the semantic information implied. What's more, the number of parameters need to train is linear to the number of relations. Although R-GCN comes with two decompositional strategies to share some parameters, it is still difficult to train such network on a knowledge graph with thousands of relations. While GRGNN only needs to train one more d-dimensional parameter per layer compared with GCN, which makes our method much more scalable than R-GCN.

5 Experiments

5.1 Dataset

In this paper, we adopt two datasets—FB15K and DB10K—to evaluate the GRGNN model on the entity multi-label classification task. Statistics of datasets are summarized in Table 1.

FB15K. FB15K [5] is a database extracted from a large-scale KG, FreeBase [2]. Following [29], we extracted all type information of entities in FB15K from Freebase, then sort them by frequencies. Top 50 types, except for the first one "common/topic" almost all entities have, are selected for classification.

DB10K. In addition to FB15K, we introduce a new dataset DB10K from DBpedia [17]—a large-scale knowledge base extracted from Wikipedia containing over 400 million facts that describe 3.7 million entities. Specifically, we start from a seed entity and use best-first-search to generate all entities to be included in the output subgraph. Then, edges between these entities are extracted. We select types that appear no less than 100 times and less than $n/2$ times, where n is the number of nodes extracted. At last, DB10K consists of 10,000 entities in 49 types, 189 relations and 54,991 triples.

Split Graph. For GCN, we did a slight change to the preprocessing scheme in [31], that is, if there are multiple edges between an entity node and a relation node, we maintain the frequency in adjacency matrix as the edge weight instead of setting them to 1.

Table 1. Dataset statistics.

	Nodes	Relations	Triples	Nodes (split)	Edges (split)	Label rate
FB15K	14,951	1,345	592,213	17,641	308,006	0.083
DB10K	10,000	189	108,473	10,378	54,991	0.166

5.2   Setup

We first compare a two-layer GRGNN with other methods, including GCN and MLP. We use the implementation provided by [16][1] for the results of GCN. As the original GCN uses a *softmax* function for classification where each node only has one type, we change the non-linearity to *logistic sigmoid* activation for multi-label classification. Other parts of the GCN code remain unchanged. Micro-F1 score is used as metrics. Then we analyze the model performance with different number of layers.

When splitting dataset for training set, we follow the guideline that positive examples for each class occur no less than 100 times. We get 1,245 training nodes in FB15K and 1,664 nodes in DB10K. We use additional 500 nodes for validation. Note that validation nodes are only used for early stopping and not used for training.

Various features can be used as GRGNN's inputs, such as the randomized features, one-hot features and embeddings trained by TransE, and these features can be concatenated to be more expressive.

We optimize the architectural hyper-parameters on FB15K dataset and then reuse them for DB10K. For GCN and MLP model, we use 0.1 for dropout rate, $1 \cdot 10^{-5}$ for L2 regularization and 128 for the number of hidden units. And for GRGNN model, parameters varies according to initial features due to the different dimensions and sparsity. For randomized and TransE features, we use the following hyper-parameters: $1 \cdot 10^{-5}$ for L2 regularization, 64 for the number of hidden units, and no dropout is adopted; for one-hot features the dropout rate is 0.5, L2 regularization weight $1 \cdot 10^{-5}$, and the number of hidden units is 256; for the concatenated features of TransE and one-hot, dropout rate of 0.3, L2 regularization of $1 \cdot 10^{-5}$ and 256 hidden units are used.

All models are initialized using Glorot initialization [9] for weight matrices and update gate. Adam SGD optimizer [15] with an initial learning rate of 0.01 is adopted to minimize cross-entropy loss on the training nodes. We use an early stopping strategy with a patience of 100 epochs to avoid over-fitting, in other words, when the micro-F1 score on validation set keeps dropping for 100 epochs, the optimization is considered finished. Then test nodes are evaluated using the parameters those perform best on validation set.

5.3   Results

Classification Performance. Results are summarized in Tables 2 and 3. Reported numbers refer to micro-F1 score in percent. We test all three models with four kinds of initial features. The TransE embeddings are 100-dimensional vectors for entities and relations, randomized features are of the same dimension as TransE features with values chosen randomly from $(-1, 1)$, one-hot encoding assigns unique features for every node in the graph, and concatenated features consist of TransE and one-hot. For GCN, the preprocessing scheme described in [31] is adopted, and since every relation is split into two nodes, we assign $\mathbf{r}$ and $-\mathbf{r}$ to every pair of relation nodes as TransE features.

---

[1] The code can be found in https://github.com/tkipf/gcn.

Table 2. Results of multi-label classification on FB15K.

Method	TransE	Randomized	One-hot	Concatenated
GCN	74.76	73.13	75.69	73.91
MLP	78.12	–	–	77.98
GRGNN	76.88	73.28	77.89	**80.10**

Table 3. Results of multi-label classification on DB10K.

Method	TransE	Randomized	One-hot	Concatenated
GCN	86.71	93.53	93.18	94.96
MLP	93.14	–	–	89.93
GRGNN	95.61	95.63	**97.07**	96.55

The results show that our method outperforms GCN and MLP significantly on both datasets. For FB15K, the best result is obtained with GRGNN model using concatenated features, as for DB10K is GRGNN using one-hot encodings.

Influence of Model Depth. It is known that in the computer vision and other traditional AI domains, neural networks can stack hundreds of layers to get better performance, as deeper structure has more parameters that can improve the expressive ability. However, GNNs have no more than three layers in most cases. [18] proves that the graph convolution is a special form of Laplacian smoothing. So as the number of graph convolution layers goes up, the features of vertices within a component tends to converge to same values, which makes nodes of different types indistinguishable. GCN achieves the best performance with two convolutional layers and suffers from converging features as well as over-fitting when adopting a deeper model.

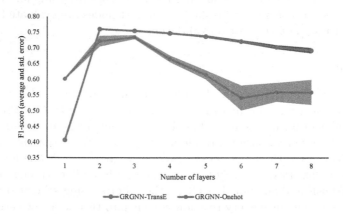

Fig. 3. Influence of model depth (number of layers) on FB15K.

Here we look into the influence of model depth (number of layers) on classification performance. We report results on models with up to 8 layers. Dropout is only adopted in the first and last layer of the network, and L2 regularization is only calculated for first layer. The number of hidden units in each layer are the same. Hyper-parameters are chosen the same as last section.

Experiments on FB15K are shown in Fig. 3, shaded area denotes standard error. The GRGNN model using TransE features and one-hot features are tested. With two or three layers, model performances do not differ much. However, using one-hot features in deeper model leads to much lower F1-score compared with TransE features. What's more, GRGNN with TransE features is insensitive to model depth, as we can see the performance doesn't drop much when the number of layers goes up. This phenomenon can be partly explained by the gate mechanism in our model. As the update gate controls the information carried over by current node and updated by neighbors, hidden states of different nodes do not necessarily converge the same values. Therefore, such mechanism and insensitive features may provide opportunities for deeper and more complex GNNs.

6   Conclusion

In this paper, we focus on the problem of entity classification in knowledge graphs and present Gated Relational Graph Neural Network (GRGNN) to incorporate features of nodes and relations, and gate mechanism is introduced to leverage hidden states of current node and its neighbors. Experiment results show that our method outperforms other methods. Besides, we analyze the influence of model depth, and find that compared with TransE features, using one-hot encodings as features leads to poorer performance with deeper model.

There are several potential extensions to GRGNN that could be addressed as future work. For instance, we can try to adopt other choice of $f^{\rightarrow}(\cdot, p, o)$ and $f^{\leftarrow}(s, p, \cdot)$. Besides, our model only uses entities and relations, while in a KG, there are a large amount of triples where the object is literal (text) carrying extra information. As literals are not used in current model, word embedding methods and language models can be adopted to incorporate literal information in future work. What's more, from the analysis of the influence of model depth on classification performance, we find that the gate mechanism may be helpful to avoid over-smoothing of nodes features, and some sort of initial features are more insensitive to model depth, which may provide opportunities for deeper and more complex GNNs.

Acknowledgements. This work was supported by The National Key Research and Development Program of China under grant 2018YFB1003504 and NSFC under grant 61932001, 61961130390, 61622201 and 61532010. This work was also supported by Beijing Academy of Artificial Intelligence (BAAI). Lei Zou is the corresponding author of this paper.

References

1. Beck, D., Haffari, G., Cohn, T.: Graph-to-sequence learning using gated graph neural networks. In: Proceedings of the 56th Annual Meeting of the Association for Computational Linguistics (Volume 1: Long Papers), pp. 273–283 (2018)
2. Bollacker, K., Evans, C., Paritosh, P., Sturge, T., Taylor, J.: Freebase: a collaboratively created graph database for structuring human knowledge. In: Proceedings of the 2008 ACM SIGMOD International Conference on Management of data, pp. 1247–1250. ACM (2008)
3. Bordes, A., Chopra, S., Weston, J.: Question answering with subgraph embeddings. In: Proceedings of the 2014 Conference on Empirical Methods in Natural Language Processing (EMNLP), pp. 615–620 (2014)
4. Bordes, A., Glorot, X., Weston, J., Bengio, Y.: A semantic matching energy function for learning with multi-relational data. Mach. Learn. **94**(2), 233–259 (2014)
5. Bordes, A., Usunier, N., Garcia-Duran, A., Weston, J., Yakhnenko, O.: Translating embeddings for modeling multi-relational data. In: Advances in Neural Information Processing Systems, pp. 2787–2795 (2013)
6. Chen, J., Zhu, J., Song, L.: Stochastic training of graph convolutional networks with variance reduction. In: International Conference on Machine Learning, pp. 941–949 (2018)
7. Cho, K., et al.: Learning phrase representations using RNN encoder-decoder for statistical machine translation. In: Proceedings of the 2014 Conference on Empirical Methods in Natural Language Processing (EMNLP), pp. 1724–1734 (2014)
8. Defferrard, M., Bresson, X., Vandergheynst, P.: Convolutional neural networks on graphs with fast localized spectral filtering. In: Advances in Neural Information Processing Systems, pp. 3844–3852 (2016)
9. Glorot, X., Bengio, Y.: Understanding the difficulty of training deep feedforward neural networks. In: Proceedings of the Thirteenth International Conference on Artificial Intelligence and Statistics, pp. 249–256 (2010)
10. Gori, M., Monfardini, G., Scarselli, F.: A new model for learning in graph domains. In: Proceedings of the 2005 IEEE International Joint Conference on Neural Networks, vol. 2, pp. 729–734. IEEE (2005)
11. Guo, S., Wang, Q., Wang, B., Wang, L., Guo, L.: Semantically smooth knowledge graph embedding. In: Proceedings of the 53rd Annual Meeting of the Association for Computational Linguistics and the 7th International Joint Conference on Natural Language Processing (Volume 1: Long Papers), vol. 1, pp. 84–94 (2015)
12. Hamilton, W., Ying, Z., Leskovec, J.: Inductive representation learning on large graphs. In: Advances in Neural Information Processing Systems, pp. 1024–1034 (2017)
13. Hoffmann, R., Zhang, C., Ling, X., Zettlemoyer, L., Weld, D.S.: Knowledge-based weak supervision for information extraction of overlapping relations. In: Proceedings of the 49th Annual Meeting of the Association for Computational Linguistics: Human Language Technologies-Volume 1, pp. 541–550. Association for Computational Linguistics (2011)
14. Ji, G., He, S., Xu, L., Liu, K., Zhao, J.: Knowledge graph embedding via dynamic mapping matrix. In: Proceedings of the 53rd Annual Meeting of the Association for Computational Linguistics and the 7th International Joint Conference on Natural Language Processing (Volume 1: Long Papers), vol. 1, pp. 687–696 (2015)
15. Kingma, D.P., Ba, J.: Adam: a method for stochastic optimization. arXiv preprint arXiv:1412.6980 (2014)

16. Kipf, T.N., Welling, M.: Semi-supervised classification with graph convolutional networks. In: International Conference on Learning Representations (ICLR) (2017)
17. Lehmann, J., et al.: Dbpedia-a large-scale, multilingual knowledge base extracted from wikipedia. Semant. Web **6**(2), 167–195 (2015)
18. Li, Q., Han, Z., Wu, X.M.: Deeper insights into graph convolutional networks for semi-supervised learning. In: Thirty-Second AAAI Conference on Artificial Intelligence (2018)
19. Lin, Y., Liu, Z., Sun, M., Liu, Y., Zhu, X.: Learning entity and relation embeddings for knowledge graph completion. In: Twenty-Ninth AAAI Conference on Artificial Intelligence (2015)
20. Neelakantan, A., Chang, M.W.: Inferring missing entity type instances for knowledge base completion: new dataset and methods. In: Proceedings of the 2015 Conference of the North American Chapter of the Association for Computational Linguistics: Human Language Technologies, pp. 515–525 (2015)
21. Nickel, M., Rosasco, L., Poggio, T.: Holographic embeddings of knowledge graphs. In: Thirtieth AAAI Conference on Artificial Intelligence (2016)
22. Nickel, M., Tresp, V., Kriegel, H.P.: A three-way model for collective learning on multi-relational data. In: Proceedings of the 28th International Conference on International Conference on Machine Learning, pp. 809–816. Omnipress (2011)
23. Scarselli, F., Gori, M., Tsoi, A.C., Hagenbuchner, M., Monfardini, G.: The graph neural network model. IEEE Trans. Neural Networks **20**(1), 61–80 (2009)
24. Schlichtkrull, M., Kipf, T.N., Bloem, P., van den Berg, R., Titov, I., Welling, M.: Modeling relational data with graph convolutional networks. In: Gangemi, A., et al. (eds.) ESWC 2018. LNCS, vol. 10843, pp. 593–607. Springer, Cham (2018). https://doi.org/10.1007/978-3-319-93417-4_38
25. Trouillon, T., Welbl, J., Riedel, S., Gaussier, É., Bouchard, G.: Complex embeddings for simple link prediction. In: International Conference on Machine Learning, pp. 2071–2080 (2016)
26. Veličković, P., Cucurull, G., Casanova, A., Romero, A., Lió, P., Bengio, Y.: Graph attention networks. In: International Conference on Learning Representations (ICLR) (2018)
27. Wang, Q., Mao, Z., Wang, B., Guo, L.: Knowledge graph embedding: a survey of approaches and applications. IEEE Trans. Knowl. Data Eng. **29**(12), 2724–2743 (2017)
28. Wang, Z., Zhang, J., Feng, J., Chen, Z.: Knowledge graph embedding by translating on hyperplanes. In: Twenty-Eighth AAAI Conference on Artificial Intelligence (2014)
29. Xie, R., Liu, Z., Jia, J., Luan, H., Sun, M.: Representation learning of knowledge graphs with entity descriptions. In: Thirtieth AAAI Conference on Artificial Intelligence (2016)
30. Yang, B., Yih, W.T., He, X., Gao, J., Deng, L.: Embedding entities and relations for learning and inference in knowledge bases (2015)
31. Yang, Z., Cohen, W.W., Salakhutdinov, R.: Revisiting semi-supervised learning with graph embeddings. arXiv preprint arXiv:1603.08861 (2016)
32. Zhang, Y., Xiong, Y., Kong, X., Li, S., Mi, J., Zhu, Y.: Deep collective classification in heterogeneous information networks. In: Proceedings of the 2018 World Wide Web Conference on World Wide Web, pp. 399–408. International World Wide Web Conferences Steering Committee (2018)
33. Zhou, J., Cui, G., Zhang, Z., Yang, C., Liu, Z., Sun, M.: Graph neural networks: a review of methods and applications. arXiv preprint arXiv:1812.08434 (2018)

Multiple Interaction Attention Model for Open-World Knowledge Graph Completion

Chenpeng Fu[1], Zhixu Li[1,2]([⊠]), Qiang Yang[4], Zhigang Chen[3], Junhua Fang[1], Pengpeng Zhao[1], and Jiajie Xu[1]

[1] School of Computer Science and Technology, Soochow University, Suzhou, China
cpfu@stu.suda.edu.cn, {zhixuli,jhfang,ppzhao,xujj}@suda.edu.cn
[2] iFLYTEK Research, Suzhou, China
[3] State Key Laboratory of Cognitive Intelligence, iFLYTEK,
Hefei, People's Republic of China
zgchen@iflytek.com
[4] King Abdullah University of Science and Technology, Jeddah, Saudi Arabia
qiang.yang@kaust.edu.sa

Abstract. Knowledge Graph Completion (KGC) aims at complementing missing relationships between entities in a Knowledge Graph (KG). While closed-world KGC approaches utilizing the knowledge within KG could only complement very limited number of missing relations, more and more approaches tend to get knowledge from open-world resources such as online encyclopedias and newswire corpus. For instance, a recent proposed open-world KGC model called ConMask learns embeddings of the entity's name and parts of its text-description to connect unseen entities to the KG. However, this model does not make full use of the rich feature information in the text descriptions, besides, the proposed relationship-dependent content masking method may easily miss to find the target-words. In this paper, we propose to use a Multiple Interaction Attention (MIA) mechanism to model the interactions between the head entity description, head entity name, the relationship name, and the candidate tail entity descriptions, to form the enriched representations. Our empirical study conducted on two real-world data collections shows that our approach achieves significant improvements comparing to state-of-the-art KGC methods.

Keywords: Knowledge Graph Completion · Attention · Open-world

1 Introduction

Knowledge Graph (KG) is a kind of large-scale structured semantic network whose nodes represent entities and edges represent relations between entities. The rise of KG in the past years has made great contributions to the success of many applications such as entity linking [6], recommendation [23] and question answering [17].

© Springer Nature Switzerland AG 2019
R. Cheng et al. (Eds.): WISE 2019, LNCS 11881, pp. 630–644, 2019.
https://doi.org/10.1007/978-3-030-34223-4_40

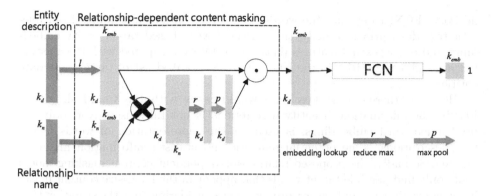

Fig. 1. Framework of Conmask model, where k_d is the length of the entity description, k_n is the length of the relationship name and k_{emb} is the word embedding size.

As more and more entities involved in a KG, a large portion of relations between entities might be missing. To deal with the case, the task of Knowledge Graph Completion (KGC) is proposed, aiming at complementing missing relation between entities in KGs. In the past years, a lot of attention has been paid on this topic, which can be roughly divided into closed-world KGC approaches and open-world KGC approaches.

The closed-world KGC approaches tend to utilize the knowledge within KG. The most active closed-world KGC methods are based on the knowledge graph embedding models such as TransE [3] and its variances [9,14,25]. By encoding the entities and relations between entities in KG into a continuous low-dimensional embedding vectors space, we could do some inference to identify some hidden relations between entities. However, closed-world KGC approaches could only complement very limited number of relation, i.e., usually lead to a low recall. On the other hand, some work tends to get knowledge from open-world resources such as online encyclopedias and newswire corpus. For instance, Description-Embodied Knowledge Representation Learning (DKRL) [26] proposes to learn the representations of entities from not only TransE, but also the description of the entities in online-encyclopedias. To achieve this, DKRL adopts to do a joint training for graph-based embeddings and description-based embeddings. They use continuous bag-of-words and deep convolutional neural network models to encode semantics of entity descriptions. However, it does not take into account that various relationships focus on different parts of the entity description, and not all information provided in its entity description is useful to predict linked entities of a given specific relationship.

A recent proposed open-world KGC model called ConMask [20] learns embeddings of an entity's name and parts of its text-description to connect unseen entities to the KG. As illustrated in Fig. 1, it first uses a so-called relationship-dependent content masking approach to select the words related to the given relationship in the relevant text description, which could effectively mitigate the presence of noisy text descriptions. Next, it trains a Fully Convolutional Neural network (FCN) to extract the word-based target entity embeddings from relevant text descriptions. Finally, the extracted word-based target entity embeddings and other textual features (entity names) are compared with the existing target candidate entities in the KG to resolve a ranked list of target candidate entities.

However, there are at least two **weaknesses** with the ConMask model. Firstly, the information of entity descriptions is not fully used. Now only the pre-trained word embeddings, is used for the representation of words in the text descriptions, some potential semantic and statistic information might be missing. Secondly, the proposed relationship-dependent content masking model could only find possible target-words that appear in the fixed-size content masking window after the indicator word, without considering that the situation that the target-words could also appear in front of the indicator word. Besides, it is difficult to determine a proper size for the content masking window.

In this paper, we propose a novel open-world KGC approach based on the same input resources with ConMask, i.e., entity names, relationship names, and entity descriptions. But different from ConMask which only uses the entity descriptions in a very simple way, we propose to use attention mechanisms to fully capture the important information generated from the multiple interactions between entity names, relationship names and entity descriptions. More specifically, the multiple interactions involved in the model include: (1) The interaction between the head entity name, the relationship name, and the head entity description. (2) The interaction between the head entity name, the relationship name, and the candidate tail entity descriptions. (3) The interaction between the description of the head entity and the candidate tail entity descriptions. In this way, our Multiple Interaction Attention (MIA) model could not only flexibly select relevant parts of the entity description according to different relationships, but also better aware of the relevant part in the head entity description and obtain the head-aware representation of the candidate tail entity description.

Besides, to make effective use of the rich information in the entity descriptions, our model encodes the head entity description, head entity name, the relationship name, and the candidate tail entity descriptions into word representations which are enhanced by additional Part-Of-Speech (POS) tags, Named-Entity-Recognition (NER) tags and handcrafted features.

To summarize, our contributions in this paper can be summarized as follows:

- We propose to use attention mechanism to simulate the interaction between the head entity name, the relationship name and the entity descriptions, such that we could dynamically select the most related information from the head entity description and the candidate tail entity descriptions according to different relations.
- We use the attention mechanism to align relevant parts between the head entity description and the candidate tail entity descriptions, such that we could enrich the representation of the candidate tail entity description.

– We also propose to make effective use of the rich information in the entity descriptions with some additional important features.

Our empirical study conducted on two real-world data collections shows that our approach achieves significant improvements on open-world KGC compared with state-of-the-art methods.

The rest of this paper is organized as follows: We cover the related work in Sect. 2, and then define our problem and introduce the framework of our approach in Sect. 3. After giving details of MIA model in Sect. 4, we report the empirical study in Sect. 5. We finally conclude the paper in Sect. 6.

2 Related Work

Knowledge graph completion (KGC) aims at completing the missing relation between entities in given KG (or KGs). So far, a lot of attention has been paid on this topic, which can be roughly divided into closed-world KGC approaches and open-world KGC approaches.

The closed-world KGC approaches tend to utilize the knowledge within KG. The most active closed-world KGC methods are based on the knowledge graph embedding models such as TransE [3] and its variances [9,14,25]. By encoding the entities and relations between entities in KG into a continuous low-dimensional embedding vectors space, we could do some inference to identify some hidden relations between entities. For a given triple $(head\ entity, relationship, tail\ entity)$, also denoted as (h, r, t), the typical embedding-based KGC model TransE [3] assumes the energy function is defined as

$$E(h, r, t) = \| \mathbf{h} + \mathbf{r} - \mathbf{t} \|, \tag{1}$$

which indicates that the tail embedding $\mathbf{t}$ should be the closeness neighbour of $\mathbf{h} + \mathbf{r}$, where $\mathbf{h}$, $\mathbf{r}$ are embeddings of head entity and relationship respectively. There are also many models that introduce more relationship-dependent parameters. TransR [14], $\mathbf{hM}_r + \mathbf{r} = \mathbf{tM}_r$ where $\mathbf{M}_r$ is a relationship-dependent entity embedding transformation. TransR [14] models entities and relations in distinct semantic space (entity space and relation spaces) and performs translation in relation space when learning embeddings. PTransE [13] maintain a simple translation function and proposes a multiple-step relation path-based representation learning model. Liu et al. [15,16] focus on the optimal social trust path selection problem in complex social networks.

Unlike topology-based models that have been extensively studied, there are several methods that use textual information for KGC. For instance, the Neural Tensor Networks (NTN) model [21] initializes the representation of the entity by using the average word embedding in entity name, and allow sharing of textual information located in similar entity names. Zhang et al. [29] represents entities with entity names or the average of word embeddings in descriptions. Jointly [27] first uses the weighted sum combination topology-embeddings and text-embeddings, and then calculates the L_n distance between the translated

Partial triple: (*Donald_Trump*, birth_place, ?)	Partial triple: (*Donald_Trump*, mother, ?)	Partial triple: (*Fred_Trump*, spouse , ?)
Head entity description: *Donald_Trump*: Donald John Trump is the 45th and current president of the United States.... Trump was born and raised in the New York City borough of Queens and received an economics degree from the Wharton School. ... His parents were Frederick Christ Trump, a real estate developer, and Scottish-born housewife Mary Anne MacLeod. ... Trump had graduated first in his class at Wharton. ...		**Head entity description:** *Fred_Trump*: Frederick Christ Trump ... and the father of Donald Trump, the 45th president of the United States,...Trump met his future wife Mary Anne MacLeod , an immigrant from Glasgow, Scotland, ...
Question: donald trump birth place ?	**Question:** donald trump mother ?	**Question:** frederick christ trump spouse ?
Candidate tail entity descriptions: *Washington,_D.C*: Washington, D.C., formally the District of Columbia and commonly referred to as Washington or D.C., is the capital of the United States. Founded after the American ... ... *Beijing*: Beijing, alternately romanized as Peking, is the capital of the People's Republic of China, the world's third most populous city proper, ... *New_York_City*: The City of New York, usually called either New York City or simply New York (NY), is the most populous city in the United States. ... five boroughs – Brooklyn, Queens, ...	**Candidate tail entity descriptions:** *Ann_Dunham*: Stanley Ann Dunham was an American anthropologist...She was the mother of Barack Obama, the 44th President of the United States. ... *Mary_Anne_MacLeod_Trump*: Mary Anne Trump was the mother of Donald Trump, the 45th president of the United States , . . . Born in the Outer Hebrides of Scotland, ... Mary Anne's husband Fred Trump died at age 93 in June 1999. ... ... *Dorothy_Howell_Rodham*: Dorothy Emma Rodham was an American homemaker ... and 2016 Democratic presidential nominee Hillary Clinton.	**Candidate tail entity descriptions:** *Michelle_Obama*: Michelle LaVaughn Obama is an American lawyer, ...She is married to the 44th U.S. president, Barack Obama, ... *Mary_Anne_MacLeod_Trump*: Mary Anne Trump was the mother of Donald Trump, the 45th president of the United States , . . . Born in the Outer Hebrides of Scotland, ... Mary Anne's husband Fred Trump died at age 93 in June 1999 .... *Hillary_Clinton*: Hillary Diane Rodham Clinton is an American politician ... she moved to Arkansas and married future president Bill Clinton in 1975; ...
Correct tail entity: *New_York_City*	**Correct tail entity:** *Mary_Anne_MacLeod_Trump*	**Correct tail entity:** *Mary_Anne_MacLeod_Trump*

Fig. 2. Open-world KGC examples with our MIA model

head and tail entities. However, closed-world KGC approaches could only complement very limited number of relations, i.e., usually lead to a low recall.

More recent work tends to get knowledge from open-world resources such as online encyclopedias and newswire corpus. For instance, DKRL [26] uses a joint training of graph-based embeddings and description-based embeddings. They use continuous bag-of-words and deep convolutional neural network models to encode semantics of entity descriptions. It can directly build representations from the description of the novel entities. A recent work proposes ConMask [20] model, which is a text-focused approach that could reduce irrelevant and noisy words by selecting words associated with relationships in the given entity description, and then fuse the relevant text through fully convolutional neural networks (FCN) to extract the word-based entity embedding and combined with background representations of other textual features (entity names) to connect unseen entities to the KG.

3 Problem Definition and the Framework

We formally define the Knowledge Graph Completion (KGC) task below:

Definition 1. Knowledge Graph Completion (KGC). *Given an incomplete Knowledge Graph $\mathcal{KG}$ with a set of incomplete relation triples in the form of $(h, r, ?)$, where h denotes the head entity, r denotes the relation, and ? is the missing tail entity t, the task of Knowledge Graph Completion (KGC) is to find t for each incomplete relation triple to consist a complete one (h, r, t).*

To illustrate how our Multiple Interaction Attention (MIA) model works on open-world KGC task, several examples are given in Fig. 2. For a given partial triple (*Donald_Trump*, mother, ?), if a human reader were asked to determine from the head entity description and some candidate tail entity descriptions, "Who is the mother of US President Donald Trump?", then human reader will first look for contextual clues such as mother, parent or family-related information from the description of the head entity "*Donald_Trump*". Here, the human reader has located the sentence "His parents were ... and Scottish-born housewife Mary Anne MacLeod" in the head entity description. So, the human reader may infer that Donald Trump's mother is a Scottish-born housewife Mary Anne MacLeod. After that, the human reader locates the description of the candidate tail entity "*Mary_Anne_MacLeod_Trump*" from the candidate tail entity descriptions. In the description of "*Mary_Anne_MacLeod_Trump*", the human reader will be pleasantly surprised to find "Mary Anne Trump was the mother of Donald Trump, the 45th president of the United States" and "Born in the Outer Hebrides of Scotland". Therefore, the human reader can more accurately reason that "*Mary_Anne_MacLeod_Trump*" is the correct tail entity of the partial triple (*Donald_Trump*, mother, ?).

We split the above reasoning process into three steps below:

1. Locating task-related information in the head entity description and the candidate tail entity descriptions, respectively;
2. Extracting the context information of the related text in the head entity description and the candidate tail entity descriptions;
3. Matching the head entity description and candidate tail entity descriptions respective relevant text context information to determine the correct tail entity.

Correspondingly, the MIA model is designed to simulate this process, which is mainly composed of three components below:

1. Multiple Interaction Attention, which highlights task-related words;
2. Text Context Encoder, which encodes context information in the relevant text;
3. Matching Prediction, which chooses a correct tail entity by matching the context information in the relevant text to calculate the similarity score between the head entity description and the candidate tail entity descriptions.

Note that we consider that the head entity, relationship, and tail entity usually appear in a snippet of the text description at the same time, so we combine the head entity name with the relationship name into a question as an input to our model to help the model locate task-related information more effectively.

4 The MIA Model

The MIA model first encodes the head entity description, question, and candidate tail entity descriptions into a word representation and enhances it by appending some other features. Then, it emphasizes and organizes relevant information by using a word-level attention mechanism to simulate the interaction

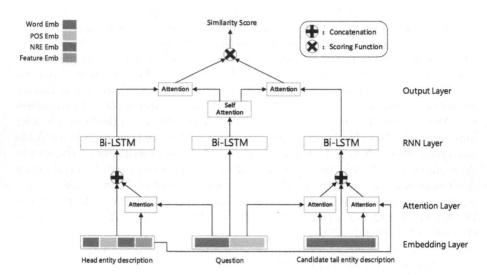

Fig. 3. Main neural architecture of the Multiple Interaction Attention (MIA) model.

between the head entity description, question and candidate tail entity descriptions. Afterwards, MIA uses Bidirectional Long Short-Term Memory network (Bi-LSTM) to encoded context information in the relevant text. Finally, through a matching prediction, it compares the representation extracted to the head entity description with the representation of each candidate tail entity description to resolve a ranked list of candidate tail entities. The architecture of MIA model is also illustrated in Fig. 3.

In the following of this section, we describe the details of the MIA model component by component. Throughout this section, we will use D_h for representing the head entity description, Q_r for representation question consisting of the head entity name and the relationship name, C_t for the candidate tail entity descriptions. Note that since the description of the operations for each candidate tail entity are the same, for the sake of simplicity, we only take one of the candidate tail entity descriptions for illustration.

4.1   Input Word Representation

We transform each word in the head entity description, question, and candidate tail entity description into continuous representations. In this paper, each training example entered during training contains a head entity description $\{w_i^{D_h}\}_{i=1}^{|D_h|}$, a question $\{w_j^{Q_r}\}_{j=1}^{|Q_r|}$, a candidate tail entity description $\{w_n^{C_t}\}_{n=1}^{|C_t|}$ and a label $y^* \in \{0, 1\}$, where $|D_h|$, $|Q_r|$, and $|C_t|$ are the length of the head entity description, question, and candidate tail entity description.

Here, we take the input representation of the i-th word $w_i^{D_h}$ in the given head entity description as an example, which is the concatenation of several components:

- **Word Embedding**: We use the publicly available pre-trained GloVe [18] embedding $\mathbf{E}_{w_i^{D_h}}^{word}$.
- **Part-Of-Speech (POS) Embedding**: We use spaCy[1] for part-of-speech tagging $\mathbf{E}_{w_i^{D_h}}^{pos}$. Similar to traditional word embeddings, we assign different trainable vectors for each part-of-speech tag.
- **Named-Entity-Recognition (NER) Embedding**: Like POS, we use spaCy for named entity recognition $\mathbf{E}_{w_i^{D_h}}^{ner}$.
- **Handcrafted Features Embedding**: We use term frequency and co-occurrence feature as handcrafted features $\mathbf{E}_{w_i^{D_h}}^{feat}$. The term frequency is calculated based on English Wikipedia. In the binary co-occurrence feature, it is true when $w_i^{D_h}$ appears in the question $\{w_j^{Q_r}\}_{j=1}^{|Q_r|}$ or candidate tail entity description $\{w_n^{C_t}\}_{n=1}^{|C_t|}$.

We concatenate four embedding components to form the final input representations for the word $w_i^{D_h}$, namely $\mathbf{E}_{w_i^{D_h}}$.

$$\mathbf{E}_{w_i^{D_h}} = [\mathbf{E}_{w_i^{D_h}}^{word}; \mathbf{E}_{w_i^{D_h}}^{pos}; \mathbf{E}_{w_i^{D_h}}^{ner}; \mathbf{E}_{w_i^{D_h}}^{feat}] \tag{2}$$

In the same way, we concatenate Word Embedding $\mathbf{E}_{w_j^{Q_r}}^{word}$ and POS Embedding $\mathbf{E}_{w_j^{Q_r}}^{pos}$ to get the input word representation $\mathbf{E}_{w_j^{Q_r}}$ of a word $w_j^{Q_r}$ in a given question.

$$\mathbf{E}_{w_j^{Q_r}} = [\mathbf{E}_{w_j^{Q_r}}^{word}; \mathbf{E}_{w_j^{Q_r}}^{pos}] \tag{3}$$

The input representation of the $w_n^{C_t}$ for a given candidate tail entity description contains only Word Embedding $\mathbf{E}_{w_n^{C_t}}^{word}$.

$$\mathbf{E}_{w_n^{C_t}} = [\mathbf{E}_{w_n^{C_t}}^{word}] \tag{4}$$

4.2 Multiple Interaction Attention

In our model, we use the interaction between the given head entity description, the question, and the candidate tail entity description to emphasize and organize relevant information accordingly. We exploit the same word-level sequence alignment attention mechanism for each interaction. In this section, we first describe the Word-level Sequence Alignment (WSA) attention mechanism in detail and then explain various interactions.

Word-Level Sequence Alignment Attention Mechanism. Following [4,11, 24] and other recent works, given two inputs $\mathbf{X}$ and $\mathbf{Y} = \{\mathbf{Y}_i\}_{i=1}^m$, let's define a attention function:

$$att(\mathbf{X}, \{\mathbf{Y}_i\}_{i=1}^m) = \sum_{i=1}^m a_{xy_i} \mathbf{Y}_i \tag{5}$$

---

[1] https://github.com/explosion/spaCy.

$$a_{xy_i} = \mathrm{softmax}(\alpha(\mathbf{WX})^T \alpha(\mathbf{WY}_i)) \tag{6}$$

where the attention score a_{xy_i} captures the similarity between $\mathbf{X}$ and each words $\mathbf{Y}_i$, $\mathbf{W}$ is a matrix, and $\alpha(\cdot)$ is a activation function with ReLU nonlinearity.

Question-Aware Head Entity Description WSA Attention. Note that words in the head entity description are not equally important, and the importance of them changes in tune with the different questions. Just like people find relevant answers from a given passage based on the question, people can always give more attention to the words that are most relevant to the question. Therefore, we can get the question-aware representation $\mathbf{E}^{qr}_{w_i^{D_h}}$ of the word $w_i^{D_h}$ in the head entity description according to the question:

$$\mathbf{E}^{qr}_{w_i^{D_h}} = att(\mathbf{E}^{word}_{w_i^{D_h}}, \{\mathbf{E}^{word}_{w_j^{Q_r}}\}_{j=1}^{|Q_r|}) \tag{7}$$

Question-Aware Candidate Tail Entity Description WSA Attention. In a similar way, we use question information as the key to extracting important information from candidate tail entity description. Then we get the question-aware representation $\mathbf{E}^{qr}_{w_n^{C_t}}$ of the word $w_n^{C_t}$ in the candidate tail entity description:

$$\mathbf{E}^{qr}_{w_n^{C_t}} = att(\mathbf{E}^{word}_{w_n^{C_t}}, \{\mathbf{E}^{word}_{w_j^{Q_r}}\}_{j=1}^{|Q_r|}) \tag{8}$$

Head-Aware Candidate Tail Entity Description WSA Attention. We find when those entities have relationships, they usually mention to each other in each other's descriptions. In order to adequately leverage the information in the head entity description, we align the candidate tail entity description with the head entity descriptions. In details, we embed the information of the head entity description into the candidate tail entity description representation so that we can better align and aware the relevant parts of the head entity description. Thereby the word $w_n^{C_t}$ in the candidate tail entity description can obtain the aware representation of the head entity description with the following equation:

$$\mathbf{E}^{d_h}_{w_n^{C_t}} = att(\mathbf{E}^{word}_{w_n^{C_t}}, \{\mathbf{E}^{word}_{w_i^{D_h}}\}_{i=1}^{|D_h|}) \tag{9}$$

4.3   Text Context Encoder

The third component of the model is the Recurrent Neural Network (RNN) layer which uses a Bidirectional Long Short-Term Memory network (Bi-LSTM) [5,19] to model the contextual information.

In order to learn long-term dependencies [1,7,8] in RNN, Long Short-Term Memory network (LSTM) was proposed by [8]. The Bi-LSTM consists of two independent LSTMs, the forward LSTM and the backward LSTM. By using three separate BI-LSTMs, we encode the head entity description, question and candidate tail entity description as follows:

$$\mathbf{B}^{D_h} = \mathrm{Bi\text{-}LSTM}(\{[\mathbf{E}_{w_i^{D_h}}; \mathbf{E}^{qr}_{w_i^{D_h}}]\}_{i=1}^{|D_h|}) \tag{10}$$

$$\mathbf{B}^{Q_r} = \text{Bi-LSTM}(\{\mathbf{E}_{w_j^{Q_r}}\}_{j=1}^{|Q_r|}) \tag{11}$$

$$\mathbf{B}^{C_t} = \text{Bi-LSTM}(\{[\mathbf{E}_{w_n^{C_t}}; \mathbf{E}_{w_n^{C_t}}^{q_r}; \mathbf{E}_{w_n^{C_t}}^{d_h}]\}_{n=1}^{|C_t|}) \tag{12}$$

4.4 Matching Prediction

We use the self-attention [28] to summarize the question sequence representation $\mathbf{B}^{Q_r}$ into the final question representation $\mathbf{R}_{Q_r}$. The definition of the self-attention function is as follows:

$$att_{self}(\{\mathbf{X}_i\}_{i=1}^m) = \sum_{i=1}^m a_i \mathbf{X}_i \tag{13}$$

$$a_i = \text{softmax}\left(\mathbf{W}_{self}^T \mathbf{X}_i\right) \tag{14}$$

where the attention score a_i indicates the importance of $\mathbf{X}_i$ in $\{\mathbf{X}_i\}_{i=1}^m$.

According to the question representation $\mathbf{R}_{Q_r} = att_{self}(\{\mathbf{B}_j^{Q_r}\}_{j=1}^{|Q_r|})$, we can get the head entity description representation $\mathbf{R}_{D_h} = att(\mathbf{R}_{Q_r}, \{\mathbf{B}_i^{D_h}\}_{i=1}^{|D_h|})$, and the candidate tail entity description representation $\mathbf{R}_{C_t} = att(\mathbf{R}_{Q_r}, \{\mathbf{B}_n^{C_t}\}_{n=1}^{|C_t|})$. Finally, we calculate the similarity score of $\mathbf{R}_{D_h}$ and $\mathbf{R}_{C_t}$ as our output y:

$$y = \text{sigmoid}\left(\mathbf{R}_{D_h} \mathbf{R}_{C_t}\right) \tag{15}$$

To train our model, we use standard cross entropy function as the loss function to minimize the gap between the prediction and the ground truth.

5 Experiments

5.1 Datasets

We use the following two public-accessed open-world datasets for evaluating the effectiveness of our approach. (1) [26] introduced the FB20k dataset, a dataset extracted from a typical large-scale KG Freebase [2]. The dataset is built upon the FB15k [3] dataset, it first removed 47 entities from FB15K which have shorter than 3 words after preprocessed or even have no descriptions, and removed all triples containing these entities in FB15K, then by adding test triples with unseen entities, which are selected to have rich descriptions. (2) [20] introduced DBPedia50k dataset for both open-world and closed-world KGC tasks, a dataset randomly sampled from a large-scale KG DBPedia [12]. We evaluate our approach on FB20k and DBPedia50k. Statistics of datasets are shown in Table 1.

Table 1. Data set statistics.

Dataset	Entities	Rel.	Triples		
			Train	Validation	Test
FB15k	14,904	1,341	472,860	48,991	57,830
FB20k	19,923	1,341	472,860	48,991	88,293
DBPedia50k	49,900	654	32,388	399	10,969

Table 2. Hyper-parameter settings.

Symbol	Descriptions	Size
$|D_h|$	Head entity description max length	512
$|C_t|$	Candidate tail entity description max length	512
k	Word embedding size	200
pos	POS-tag embedding size	12
ner	NRE-tag embedding size	8
h	Bi-LSTM hidden size	96

5.2 Experiment Setting

Due to the lack of an open-world KGC task validation set on FB20k, we randomly sampled 10% of the test triples as a validation set.

Evaluation Protocol. We use the tail entity prediction on the test set for performance evaluation. For each test triple (h, r, t) with open-world head entity $h \in \mathbf{E}'$, where $\mathbf{E}'$ is an entity superset, we rank all known entities $t \in \mathbf{E}$ by use the KGC model to calculate the actual ranking score, where $\mathbf{E}$ is an entity set. We then use three measures as our evaluation metrics: (1) Mean Rank (MR): the averaged rank of correct tail entities; (2) HITS@K: the proportion of correct tail entities ranked in top k; (3) Mean Reciprocal Rank (MRR): mean reciprocal rank of correct tail entities.

Note that there may be multiple triples in the dataset that have the same head entity and relationship but different tail entities: $(h, r, t_1), ..., (h, r, t_n)$. Following [3], when computing the Mean Reciprocal Rank (MRR), given a triple (h, r, t_i) only the reciprocal rank of t_i itself is evaluated (and not the best out of $t_1, ..., t_i, ..., t_n$, which would produce better results). This differs from Con-Mask's MRR evaluation method, which is the reason why result in Table 3 differs from [20] (see the asterisk (*) mark).

Note also that a filtering method called *target filtering* is used in ConMask: When evaluating a test triple (h, r, t), only when a triple of the form $(?, r, t')$ exist in the training set, we treat the tail entity t' as a candidate tail entity, otherwise it is skipped. Therefore, we also use *target filtering* when comparing with the Conmask model.

Table 3. Open-world tail entity prediction results on FB20k and DBPedia50k. Note that we used the same evaluation protocol with target filtering as in ConMask. The asterisk (*) indicates that the result differs from the one published, because the MRR is calculated differently.

Model	DBPedia50k					FB20k				
	HITS@1	HITS@3	HITS@10	MR	MRR	HITS@1	HITS@3	HITS@10	MR	MRR
Target filtering baseline	0.045	0.097	0.23	104	0.11*	0.17	0.32	0.41	123	0.27
DKRL (2-layer CNN)	–	–	0.40	70	0.23	–	–	–	–	–
ConMask	0.47	0.65	0.81	16	0.58*	0.38	0.49	0.63	54	0.46
MIA model	**0.64**	**0.83**	**0.93**	**5**	**0.75**	**0.45**	**0.63**	**0.80**	**21**	**0.57**

Parameter Setting. Following ConMask, we set the maximum head entity description length $|D_h| \leq 512$ and the maximum candidate tail entity description length $|C_t| \leq 512$. We apply the spaCy for tokenization, part-of-speech (POS), and named entity recognition (NER). The main hyper-parameters of our model are listed in Table 2. The word embeddings are initialized by the publicly available pre-trained 200-dimensional GloVe [18] embeddings. We use Adam [10] for parameter optimization, with initial learning rate 0.002. A mini-batch of 32 samples is used to update the model parameter per step. In order to prevent overfitting, we apply dropout [22] to input embeddings and Bi-LSTM's outputs with a drop rate of 0.4. We use PyTorch[2] to implement our model.

5.3 Open-World Tail Entity Prediction

We compare our model MIA with other open-world KGC models, the experimental results are shown in Table 3. For a fair comparison, all the results are evaluated using target filtering.

The results for Target Filtering Baseline, DKRL and ConMask were obtained by the implementation provided by [20]. The Target Filtering Baseline assigns randomly scores to all entities that pass the target filtering. DKRL uses a two-layer convolutional neural network (CNN) over the entity descriptions. ConMask uses relationship-dependent content masking and fully convolutional neural network (FCN) to extract word-level target entity embedding from entity descriptions and then combine some other text features (entity names) are compared with the candidate tail entities to resolve a ranked list of candidate tail entities.

As can be seen from the Table 3, our MIA model significantly outperforms Conmask in HITS@K, MR, and MRR by a large margin. At the same time, we also find that the MIA model performed better on the DBPedia50k dataset than on the FB20k dataset, because the entity description in the DBPedia50k dataset is more abundant than the entity description in the FB20k dataset, where DBpedia50k dataset has an average entity description length of 454 words, FB20k dataset of 147 words.

---

[2] https://pytorch.org.

5.4   Ablation Study

We carry out model ablations to further demonstrate the effectiveness of the proposed model. Firstly, we conduct an ablation analysis on the input word representation, which consists of several components: Part-Of-Speech (POS) Embedding, Named-Entity-Recognition (NER) Embedding and Handcrafted Features Embedding etc. The experimental results on DBPedia50k are shown in Table 4, we find all the input word representation components contribute to the performance of our MIA model. This suggests that it is useful to incorporate various feature into the word representation. We also remove our multiple interaction attention in the model. The results in Table 4 show a significant drop in performance by 1.5%, which indicates that the multiple interaction attention is effective in extracting the most relevant parts from the entity text description given different relationships.

Table 4. Ablations on several model components.

Model	MRR
MIA model	0.750
w/o POS	0.746 (−0.004)
w/o Handcrafted Features	0.743 (−0.007)
w/o NER	0.742 (−0.008)
w/o Attention	0.735 (−0.015)

6   Conclusions and Future Work

This paper introduces an open-world KGC model called MIA that uses a word-level attention mechanism to simulate the interaction between the head entity description, head entity name, the relationship name and candidate tail entity descriptions. Experiments on two datasets show that the MIA model has achieved significant improvement on the open-world KGC task compared to state-of-the-art models. However, MIA relies heavily on the richness of the entity descriptions, and the tail entity can be effectively predicted only when the necessary information related to the relationship is expressed in the entity description. In the future work, we consider to introduce more external knowledge into MIA to make it more robust.

Acknowledgments. This research is partially supported by National Natural Science Foundation of China (Grant No. 61632016, 61402313, 61472263), and Natural Science Research Project of Jiangsu Higher Education Institution (No. 17KJA520003), and this is a project funded by the Priority Academic Program Development of Jiangsu Higher Education Institutions.

References

1. Bengio, Y., Simard, P., Frasconi, P., et al.: Learning long-term dependencies with gradient descent is difficult. IEEE Trans. Neural Netw. **5**(2), 157–166 (1994)
2. Bollacker, K., Evans, C., Paritosh, P., Sturge, T., Taylor, J.: Freebase: a collaboratively created graph database for structuring human knowledge. In: Proceedings of the 2008 ACM SIGMOD International Conference on Management of Data, pp. 1247–1250. ACM (2008)
3. Bordes, A., Usunier, N., Garcia-Duran, A., Weston, J., Yakhnenko, O.: Translating embeddings for modeling multi-relational data. In: Advances in Neural Information Processing Systems, pp. 2787–2795 (2013)
4. Chen, D., Fisch, A., Weston, J., Bordes, A.: Reading wikipedia to answer open-domain questions. arXiv preprint arXiv:1704.00051 (2017)
5. Graves, A., Schmidhuber, J.: Framewise phoneme classification with bidirectional LSTM and other neural network architectures. Neural Netw. **18**(5–6), 602–610 (2005)
6. Hachey, B., Radford, W., Nothman, J., Honnibal, M., Curran, J.R.: Evaluating entity linking with wikipedia. AI **194**, 130–150 (2013)
7. Hochreiter, S., Bengio, Y., Frasconi, P., Schmidhuber, J., et al.: Gradient flow in recurrent nets: the difficulty of learning long-term dependencies (2001)
8. Hochreiter, S., Schmidhuber, J.: Long short-term memory. Neural Comput. **9**(8), 1735–1780 (1997)
9. Ji, G., He, S., Xu, L., Liu, K., Zhao, J.: Knowledge graph embedding via dynamic mapping matrix. In: Proceedings of the 53rd Annual Meeting of the Association for Computational Linguistics and the 7th International Joint Conference on Natural Language Processing (Volume 1: Long Papers), vol. 1, pp. 687–696 (2015)
10. Kingma, D.P., Ba, J.: Adam: a method for stochastic optimization. arXiv preprint arXiv:1412.6980 (2014)
11. Lee, K., Salant, S., Kwiatkowski, T., Parikh, A., Das, D., Berant, J.: Learning recurrent span representations for extractive question answering. arXiv preprint arXiv:1611.01436 (2016)
12. Lehmann, J., et al.: DBpedia-a large-scale, multilingual knowledge base extracted from Wikipedia. Semant. Web **6**(2), 167–195 (2015)
13. Lin, Y., Liu, Z., Luan, H., Sun, M., Rao, S., Liu, S.: Modeling relation paths for representation learning of knowledge bases. In: EMNLP, pp. 705–714 (2015)
14. Lin, Y., Liu, Z., Sun, M., Liu, Y., Zhu, X.: Learning entity and relation embeddings for knowledge graph completion. In: AAAI, vol. 15, pp. 2181–2187 (2015)
15. Liu, G., Wang, Y., Orgun, M.A.: Optimal social trust path selection in complex social networks. In: Twenty-Fourth AAAI Conference on Artificial Intelligence, pp. 1391–1398 (2010)
16. Liu, G., Wang, Y., Orgun, M.A., Lim, E.P.: Finding the optimal social trust path for the selection of trustworthy service providers in complex social networks. IEEE Trans. Serv. Comput. **6**(2), 152–167 (2013)
17. Lukovnikov, D., Fischer, A., Lehmann, J., Auer, S.: Neural network-based question answering over knowledge graphs on word and character level. In: Proceedings of the 26th international conference on World Wide Web, pp. 1211–1220. International World Wide Web Conferences Steering Committee (2017)
18. Pennington, J., Socher, R., Manning, C.: Glove: global vectors for word representation. In: Proceedings of the 2014 Conference on Empirical Methods in Natural Language Processing (EMNLP), pp. 1532–1543 (2014)

19. Schuster, M., Paliwal, K.K.: Bidirectional recurrent neural networks. IEEE Trans. Signal Process. **45**(11), 2673–2681 (1997)
20. Shi, B., Weninger, T.: Open-world knowledge graph completion. In: Thirty-Second AAAI Conference on Artificial Intelligence, pp. 1957–1964 (2018)
21. Socher, R., Chen, D., Manning, C.D., Ng, A.: Reasoning with neural tensor networks for knowledge base completion. In: Advances in Neural Information Processing Systems, pp. 926–934 (2013)
22. Srivastava, N., Hinton, G., Krizhevsky, A., Sutskever, I., Salakhutdinov, R.: Dropout: a simple way to prevent neural networks from overfitting. J. Mach. Learn. Res. **15**(1), 1929–1958 (2014)
23. Wang, H., et al.: Ripplenet: propagating user preferences on the knowledge graph for recommender systems. In: Proceedings of the 27th ACM International Conference on Information and Knowledge Management, pp. 417–426. ACM (2018)
24. Wang, L., Sun, M., Zhao, W., Shen, K., Liu, J.: Yuanfudao at semeval-2018 task 11: three-way attention and relational knowledge for commonsense machine comprehension. arXiv preprint arXiv:1803.00191 (2018)
25. Wang, Z., Zhang, J., Feng, J., Chen, Z.: Knowledge graph embedding by translating on hyperplanes. In: Twenty-Eighth AAAI Conference on Artificial Intelligence (2014)
26. Xie, R., Liu, Z., Jia, J., Luan, H., Sun, M.: Representation learning of knowledge graphs with entity descriptions. In: Thirtieth AAAI Conference on Artificial Intelligence (2016)
27. Xu, J., Chen, K., Qiu, X., Huang, X.: Knowledge graph representation with jointly structural and textual encoding. arXiv preprint arXiv:1611.08661 (2016)
28. Yang, Z., Yang, D., Dyer, C., He, X., Smola, A., Hovy, E.: Hierarchical attention networks for document classification. In: Proceedings of the 2016 Conference of the North American Chapter of the Association for Computational Linguistics: Human Language Technologies, pp. 1480–1489 (2016)
29. Zhang, D., Yuan, B., Wang, D., Liu, R.: Joint semantic relevance learning with text data and graph knowledge. In: Proceedings of the 3rd Workshop on Continuous Vector Space Models and their Compositionality, pp. 32–40 (2015)

OntoDS: An Ontology-Aware Distributed Storage Scheme for RDF Graphs

Baozhu Liu[1], Xin Wang[1,2($\boxtimes$)], Yajun Yang[1,2], and Yunpeng Chai[3]

[1] College of Intelligence and Computing, Tianjin University, Tianjin, China
{liubaozhu,wangx,yjyang}@tju.edu.cn
[2] Tianjin Key Laboratory of Cognitive Computing and Application, Tianjin, China
[3] School of Information, Renmin University of China, Beijing, China
ypchai@ruc.edu.cn

Abstract. With the development of the Semantic Web, the amount of RDF data has been increasing rapidly. It is no longer feasible to store entire data sets on a single machine, and still be able to access the data at reasonable performance. Consequently, the requirement for clustered RDF database systems is becoming more and more important. At the same time, the native storage scheme of RDF data is less mature in many aspects compared with relational storage scheme. SQL-on-Hadoop is a distributed relational database engine for big data with many factors, which uses Hadoop to improve the fault tolerance of the system and is fully transactional. However, currently, there is no SQL-on-Hadoop relational database that realizes a subsystem for RDF data storage. In this paper, we propose an Ontology-aware Distributed Storege scheme for RDF, called OntoDS, which modifies the relational RDF data storage scheme DB2RDF to build a novel scheme for RDF data and optimizes the partitioning of RDF graphs by distributing RDF triples based on ontologies to meet the need for RDF graph data storage and query load. The experimental results on the benchmark datasets show that our distributed RDF storage scheme is about 1–1.5 times faster than the state-of-the-art native storage schemes.

Keywords: RDF data storage · RDF graph · DB2RDF

1 Introduction

Among the data models of knowledge graphs, the Resource Description Framework (RDF [1]) is a model for representing Web resources, which has become a standard format for knowledge graphs and is widely used. With the development of the Semantic Web, RDF format is gaining widespread acceptance and the amount of RDF data has been dramatically increasing. The number of triples of the latest 2016-10 version of the DBpedia [2] dataset has reached 13 billion. With the rapid rise of the data volume of RDF graphs, it is no longer feasible to store entire data sets on a single machine. In order to solve the scalability

© Springer Nature Switzerland AG 2019
R. Cheng et al. (Eds.): WISE 2019, LNCS 11881, pp. 645–659, 2019.
https://doi.org/10.1007/978-3-030-34223-4_41

problem of the RDF storage scheme on a single machine, distributed RDF storage scheme has become an inevitable option. On the other hand, native storage schemes of RDF data are less mature in many aspects compared with the corresponding relational versions. Thus, we choose relational storage schemes rather than the native ones.

Although many models have been proposed to store RDF graphs [3] (e.g., triple table, horizontal table, property table, vertical partitioning [4], sextuple indexing, DB2RDF [5], and SQLGraph [6]), the existing solutions are implemented on a single machine, not in a distributed environment. SQL-on-Hadoop is a kind of data management technology based on Hadoop [7], which is a data query and storage mechanism using SQL as its query language. SQL-on-Hadoop architecture is suitable for the storage of large-scale RDF graph data due to its high degree of parallelism, robustness, reliability, and scalability while running on heterogeneous commodity hardware. Therefore, a distributed database with SQL-on-Hadoop architecture can be used to solve the storage problem of RDF graph data.

Based on SQL-on-Hadoop, some distributed RDF storage systems are proposed. The system details will be introduced in Sect. 5. Among these systems, none of them combines MPP features with ontology-aware distribution of RDF graphs, which can significantly accelerate queries.

To speed up the queries over RDF graphs in a distributed environment, it is obvious that a reasonable RDF graph distribution method needs to be first considered. Since many RDF queries depend on ontology information, it is beneficial to realize an ontology-aware data distribution for RDF graphs in a distributed cluster. Unlike the random distribution of RDF triples in the existing systems, our OntoDS storage scheme takes full advantage of ontologies associated with RDF graphs to partition and store RDF triples in a semantic-aware manner.

In this paper, we focus the distributed storage scheme of RDF graphs and propose OntoDS, which is an ontology-aware distributed RDF storage scheme. Meanwhile, based on the SQL-on-Hadoop infrastructure, we have developed a prototype system that implements the OntoDS storage scheme and supports efficient RDF query processing on top of OntoDS.

Our contributions can be summarized as follows:

(1) We propose a novel relational storage scheme for RDF data with five relations, which is flexible to handle dynamic RDF schemas, as it does not require schema changes when RDF triples being inserted.
(2) The prefix encoding used to record ontology information not only facilitates the distribution of RDF data, but also keeps the hierarchical information of ontologies. Compared with type-oriented methods, which can only provide the nearest ontology of entities, the prefix encoding can give more helpful information during queries.
(3) Extensive experiments were conducted to verify the scalability and efficiency of OntoDS. The experimental results show that OntoDS is about 1–1.5 times faster than the state-of-the-art native storage schemes.

The rest of this paper is organized as follows. Section 2 provides an overview of OntoDS. In Sect. 3, the distribution method over RDF graph of OntoDS is introduced. Section 4 shows experimental results on benchmark datasets. Section 5 briefly reviews related work. Finally, we conclude in Sect. 6.

2   RDF over Relational

There have been many attempts to shred RDF data into relational models. DB2RDF [5] is one of the entity-oriented alternatives, however it does not distribute data in a semantic-aware way, which can reduce data shuffle and further accelerate queries. Thus, we modify DB2RDF by adding ontology information for data distribution to provide a suitable scheme for distributed environment.

2.1   The OntoDS Storage Scheme

OntoDS is composed of five relations, including Direct Primary Hash (DPH), Reverse Primary Hash (RPH), Direct Secondary Hash (DS), Reverse Secondary Hash (RS), and TYPES, as is depicted in Fig. 1. The DPH and DS relations essentially encode the outgoing edges of an entity, in other words, the entities in DPH represent subjects in RDF triples. Meanwhile, the RPH and RS relations encode the incoming edges of an entity, which means the entities in RPH represent objects in RDF triples. In order to distinguish the columns of DPH and RPH, we use different subscripts on these columns, e.g., $value_{m1}$ and $value_{n1}$. RPH and RS relations are added to facilitate object-given queries. The TYPES relation records the codings of all ontologies.

In our scheme, DPH is a wide relation, in which each tuple stores a subject s in the entry column, with its ontology information stored in the *type* colomn and all its associated predicates and objects stored in the $pred_i$ and val_i columns, respectively $(0 \le i \le k)$. If subject s has more than k predicates, the extra predicates are spilled to another tuple and process continues until all the predicates for s are stored. When it comes to multi-valued predicates, a new unique identifier is assigned as the value of the predicate in DPH relation. Then, the identifier is stored in the DS relation along with its real predicate values. RPH and RS works in the same way as DPH and DS. The example of the scheme is shown in Fig. 3. The related RDF graph is shown in Fig. 2(a).

OntoDS treats the columns of a relation as flexible storage locations that are not pre-assigned to any predicate, but predicates are assigned to them dynamically, during insertion. The assignment ensures that a predicate is always assigned to the same column or more generally the same set of columns.

We refer to a query with common subject or object and its adjacent nodes as a *star query*. To execute query over OntoDS, after the triples with the same subject or object being merged, the query rewriter will construct a single SQL SELECT-statement for each star query. The ontology of entities can be generated by joining DPH (resp. RPH) with TYPES. If the subjects (resp. objects) are given in the queries, we use DPH (resp. RPH) to get the result. When star query involving

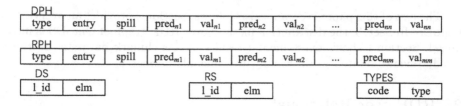

DPH									
type	entry	spill	$pred_{n1}$	val_{n1}	$pred_{n2}$	val_{n2}	...	$pred_{nn}$	val_{nn}

RPH									
type	entry	spill	$pred_{m1}$	val_{m1}	$pred_{m2}$	val_{m2}	...	$pred_{mm}$	val_{mm}

DS			RS			TYPES	
l_id	elm		l_id	elm		code	type

Fig. 1. OntoDS storage scheme.

multi-valued predicates, the SQL statement will join DPH (resp. RPH) with DS (resp. RS) together to product the actual objects (resp. subjects) of a subject (resp. object) entity.

The number of columns in DPH and RPH relations is decided by predicate inteference graph coloring, and predicates along with their corresponding objects are inserted by string hash functions. The details will be explained in next subsection.

2.2 Data Insertion

The objective of the OntoDS scheme is to dynamically assign predicates of a given dataset to columns such that:

(1) the total number of columns used across all subjects is minimized;
(2) for a subject, the probability to mapping two different predicates into the same column is minimized to reduce *spill*.

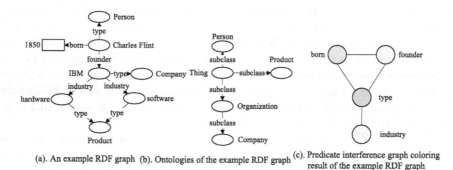

(a). An example RDF graph (b). Ontologies of the example RDF graph (c). Predicate interference graph coloring result of the example RDF graph

Fig. 2. An example typed RDF graph.

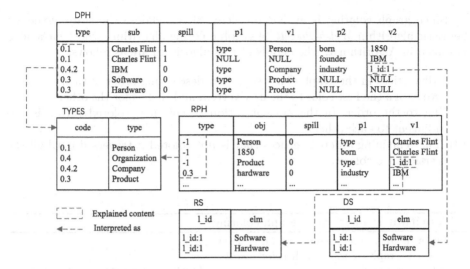

Fig. 3. Example scheme for typed RDF graph.

The same column cannot store different predicates of the same subject. Formal definition is given as follows:

Definition 1 (Predicate Mapping). *A Predicate Mapping is a function: URL → $\mathcal{N}$, and the domain of which is URIs of predicates and the range of which is natural numbers between 0 and maximum m, m is the largest allowed number on a single database row.*

$\mathcal{N}$ can be determined by predicate interference graph coloring, the definition of predicate inteference graph can be formally given as:

Definition 2 (Predicate Inteference Graph). *G_D is a Predicate inteference graph for a specific dataset D such that:*

$$V_D = \{p \mid \langle s, p, o \rangle \in D\} \tag{1}$$

$$E_D = \{\langle p_i, p_j \rangle \mid \langle s, p_i, o \rangle \in D \wedge \langle s, p_j, o \rangle \in D\} \tag{2}$$

Correspondingly, the predicate inteference graph coloring problem can be defined as Definition 3. In a predicate interference graph, where predicates with the same subject are connected, the nodes connected cannot be assigned to the same color. The coloring result of the example RDF graph's predicate interference graph Fig. 2(a) is depicted in Fig. 2(c)

Definition 3 (Predicate Inteference Graph Coloring). *For specific predicate inteference graph $G = \langle V, E \rangle$, its predicate inteference graph coloring result C is a maping from vertex v to color c, that:*

$$M(G, C) = \{\langle v, c \rangle \mid v \in V \wedge c \in C \wedge (\langle v_i, c_i \rangle \in M \wedge \langle v, v_i \rangle \in E \to c \neq c_i)\} \tag{3}$$

Since graph coloring is an NP problem [8], we choose the state-of-the-art heuristic algorithm *Welsh-Powell* [9] graph coloring algorithm, whose basic idea is shown in Algorithm 1. The details of this algorithm are as follows:

(1) All vertices in the graph G are sorted in descending order of their degrees.
(2) We assign the first color to the first vertex, and then color the others according to the order. In the same iteration of coloring, colored vertex is not adjacent to each other.
(3) The remaining ordered vertices that is not colored is traversed until all the vertices are colored.

Algorithm 1: Interference graph coloring

Data: predicate interference graph $G' = \langle V', E' \rangle$
Result: graph coloring result *color_count*

```
1  color_count := 0;                          // the counts for used colors
2  C := ∅;                                     // the set for colored vertices
3  for each vi ∈ V' do
4  │   if color(vi) = false then
   │       // this vertex is not colored
5  │       color(vi) := true;
   │       // color this vertex
6  │       color_count := color_count + 1;
   │       // the counts for used colors add one
7  │       C := C ∪ {vi};
   │       // include it into the set for colored vertices
8  │       for vj ∈ V' do
9  │       │   if not_neightbor_of(C) then
   │       │       // vj is not connected to vi
10 │       │       C := C ∪ {vj};
   │       │       // include it into the set for colored vertices
11 │       │       color(vj) := true ;
   │       │       // color the vertex
12 │       │   return color_count
```

The result of the predicate interference graph coloring guides us to build the DPH and RPH relations, and the insertion of the relations is determined by string hash functions. To minimize column collisions, eight string hash functions were selected, and the calculation method of selected ones are irrelevant. The specific workflow is shown in Algorithm 2.

RDF data insertion using eight string hash functions can be considered as the process of predicate combination composition, which is formally defined as follows:

Definition 4 (Predicate Mapping Composition). *A Predicate Mapping Composition, defines a new predicate mapping that combines the column numbers from multiple predicate mapping functions* $f_1, ... f_n$:

$$f_{m,1} \oplus f_{m,2} \oplus ... \oplus f_{m,n} \equiv \{v_1, ..., v_n \mid f_{m,i}(p) = v_i\} \qquad (4)$$

For each hash function, the random strings composed of letters and numbers are calculated. The effect of BKDRHash is the best. APHash is not as good as BKDRHash, moreover, is also worse than DJBHash, JSHash, RSHash, and SDBMHash. PJWHash and ELFHash are the worst. Except for PJWHash and ELFHash, the number of hash collisions per 10,000 strings is about 2 to 3 for each hash function, and PJWHash and ELFHash are about 30 per 10,000. It can be observed that the effects of selected eight hash functions are desirable.

Algorithm 2: String hash insertion

Data: subject s with its predicates $P(s)$ and predicates' corresponding objects $O(s)$

Result: *color_count* pairs of predicate-object array PO after hash insertion

1 **for** *each* $pred_i \in P(s)$ **do**

    // string hash results for $pred_i$

2     $v_i :=$HASH($pred_i$) % *color_count*;

    // 'HASH' refers to the eight selected string hash functions, i.e., SDBMHash, RSHash, JSHash, PJWHash, ELFash, BKDRHash, DJBHash, and APHash

3   **if** $PO[v_i] = NULL$ **then** $PO[v_i] := \{pred_i, obj_i\}$;

4   **else** split_to_another_row();

    // hash collision occurs in all functions, and the data have to be inserted into another row in DPH

The existing systems randomly distribute data by entities, so that all data with the same entity will be distributed to the same node. This distribution approach is easy to implement, but does not consider the real-world query needs. In order to accelerate type-related queries, which is common in real-world queries, OntoDS takes full advantage of ontologies associated with RDF graphs to partition and store RDF triples in a semantic-aware manner. In the next Section, we will explain the RDF graph distribution method of OntoDS in detail.

3 Ontology-Aware RDF Graph Distribution

OntoDS, which is shown in Fig. 1, records the ontology information of each entity in the corresponding type column of DPH or RPH relation, and creates a TYPES relation to store each type and its encoding. DPH and RPH are distributed by type column. Therefore, entities of the same type are assigned to the same node. As we all know, in a distributed environment, we should reduce communication between nodes as much as possible, since communication is the most time consuming process. By ontology-aware distribution, type-related queries will be greatly accelerated, since queries are processed locally, and data shuffle is significantly reduced.

Unlike type-oriented methods, OntoDS records the entity's ontology information for data distribution and querying rather than creates a separate relation for each type. This approach avoids the disadvantages of data sparseness in the type-oriented method, however still easy to obtain the ontology information, when it is needed in queries. For the query workloads provided by many benchmark datasets always first give the type of the involved entity, recording the ontology information of the entity, we can immediately limit the range of data to some nodes in the distributed environment.

3.1   RDF Ontology Information

The IRI (Internationalized Resource Identifiers) of an RDF resource contains a namespace prefix indicating the classes or attributes of the RDF entity, and RDF vocabulary indicates the meaning of these classes. Common RDF vocabularies included FOFA, Dublin Core, Schema.org etc. RDF resources are divided into various classes. Each class has its own instance, and the collection of instances is an extension of a class. Two classes may have the same set of instances but be different classes, and a class can also be its own extension. A class can have its subclasses, so RDF is actually a hierarchical structure. The example RDF ontology hierarchical structure in Fig. 4 is extracted from Lehigh University Benchmark (LUBM) [10]. The toppest ancestor of each ontology is *owl:Thing*.

As is shown in Fig. 4, *Professor* is constituted by *Dean, Chair, AssistantProfessor, FullProfessor, AssociateProfessor*, and *VisitingProfessor*. A RDF dataset could have large number of types (e.g. the DBpedia ontology contains 150K types), but not all types appear in the actual data. For example, there are 41 types in the OWL file (the file format to store ontology information of a RDF graph) of LUBM datasets, but only 13 of them are actually used. Thus, we need to record type information based on actual data, not on priori knowledge to minimize records.

The ontology semantic distribution method is easy to implement, as DB2RDF provides us with the convenience of clustered data based on entities. We only need to capture the ontology information of every entity. When extracting various predicates along with their values of an entity, we do not need to concern about the entity's ontology information, and vice versa. So, the whole process of OntoDS can be divided into two separate processes, easy to operate. The type information of the entity is determined by the triple with the predicate *type*, and the hierarchical ontology information needs to be found in the OWL file, i.e., RDFS (Resource Description Framework Scheme) and RDF often stored in different files, which in turn facilitates our separate storage process.

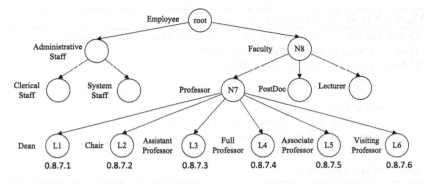

Fig. 4. RDF hierarchical structure and prefix encoding.

3.2   Type Hierarchy Coding

The type hierarchical information of RDF graph Fig. 2(a) is shown in Fig. 2(b).
All entities with no type information or strings are coded like '-1', while the
highest level type *Thing* is encoded as '0'. Every hierarchy of ontologies except
Thing will be recorded in TYPES relation. Figure 4 shows an example of type
hierarchy coding. Each time, we only focus on the ontology encoding of one leaf
node, recursively upward encode the nodes until the highest level ontology is
met. The ontology encode method is shown in Algorithms 3 and 4.

Algorithm 3: Ontology encoding

Data: the types encountered: T
Result: the codes of types: *type.code*
1 **for** $type_i \in T$ **do**
   // get the code for every type in the set
2   **return** $type_i.code :=$ getcode($type_i$)
3 **return** *type.code*

3.3   Queries on Ontologies

With the ontologies of entities recorded, we should change the query statements
to take full advantages of the storage scheme. When it comes to a specific query,
we can first point out the type of the involved entities to restrict the entities
to some node rather than the whole cluster, which will reduce data shuffle and
accelerate queries.

Algorithm 4: `getcode(`*type$_i$*`)`

Data: the type need to be encoded: *type$_i$*, the types encountered: T
Result: the code for the *type$_i$*: *type$_i$.code*

1 **if** *type$_i$* is_subclass_of *type$_j$* **then**
2   *type$_j$.key* := $T.size$;
      `// all nodes are numbered in order of appearance`
3     $T := T \cup \{type_j\}$;
      `// include this type into the set` T
4   **return** `getcode(`*type$_j$*`)` + *type$_i$.key*;
      `// '+' refers to string concatenation`
5 **else return** *type$_i$.key*;

The most typical type-related query is just like: *query the number of publications of A.* We can alter the query like: *query the number of entities whose type is Publication, and whose author is A.*

The best query order for OntoDS should be: (1) find the ontology code of the queried entity in the TYPES relation, (2) query the entity according to the ontology in the corresponding DPH or RPH relation, and (3) find the required data according to the filter information. This query order maximizes the query efficiency of type-related aggregate queries.

4 Experiments

In this section, a thorough experimental study on the RDF data benchmark dataset is conducted to evaluate the performance of OntoDS, using HAWQ [11] as our relational backend. The tested systems are deployed on a 4-node cluster, of which 3 nodes are used for segments and DataNodes, and 1 node is used for master and NameNode. Each node has 4-core, Intel(R) Core(TM) i7-6700 CPU @ 3.40 GHz system, with 16 GB of memory, running 64-bit Linux, and 50 GB hard disk. We conducted experiments on LUBM [10]. Each query was issued 4 times, the first run of which was discarded, and 3 consecutive runs after the first run were used for the average result.

4.1 Datasets

LUBM consists of a university domain ontology, along with customizable and repeatable synthetic data. As the basic idea of OntoDS is to maximize the efficiency of type-related queries, we choose some other query statements instead of using the benchmark queries LUBM provides. The chosen queries are listed in Appendix A (*lubmc* refers to the schema of RDF graph using DB2RDF, *lubmt* refers to that using OntoDS).

4.2 Experimental Results

Main Results. In general, the experimental results show that OntoDS is both efficient and scalable. Data insertion and deletion can be completed in a short time. The results show that OntoDS is suitable to store RDF data in a distributed environment. The prototype system has certain practical significance.

Data Insertion and Deletion. Although OntoDS needs more time than DB2RDF in data insertion and deletion, their time costs are on the same order of magnitude, thus are comparative, as is shown in Fig. 5. OntoDS can achieve promising insertion efficiency on small RDF data sets. As the amount of RDF data grows, the type information is more dispersed, and the insertion time on OntoDS grows faster than DB2RDF. Although OntoDS is not dominant in the data insertion and deletion, the slight overhead paid on the insertion is worthwhile compared to the gain in the efficiency of queries.

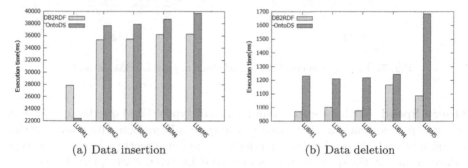

(a) Data insertion

(b) Data deletion

Fig. 5. The experimental results of data insertion and deletion on LUBM datasets.

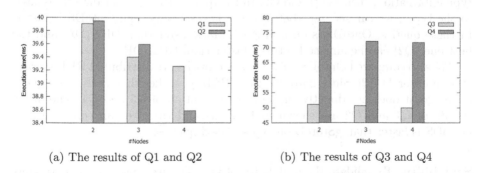

(a) The results of Q1 and Q2

(b) The results of Q3 and Q4

Fig. 6. The experimental results of scalability.

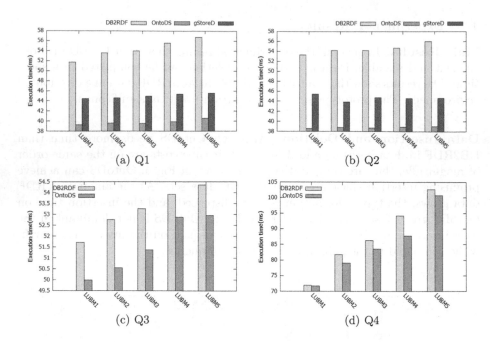

Fig. 7. The experimental results of efficiency on LUBM datasets.

Query Speed. We have selected four kinds of type-related queries: (1) directly about type information, (2) related to the type information with some filtering conditions, and (3) two kinds of queries that do not directly related to the type information but the queried data is clustered by type. Each query is executed one by one to minimize the impact on query time from external factors. Due to the dynamic changes of the network, the query time may not always follow a proportional relationship, while the overall trend remains. There are significant differences between DB2RDF and OntoDS on queries that directly related to type information, such as Q1 and Q2. In the queries that are not directly related to the types, we can see gaps between the two schemes, and OntoDS is faster. In these queries, OntoDS is on average 1 times faster than DB2RDF, and the best one (Q2) can be almost 1.5 times faster than DB2RDF.

We also compared OntoDS with the state-of-the-art distributed RDF storage system gStoreD [12]. Since gStoreD uses SPARQL rather than SQL as its query language, it does not directly support the execution of Q3 and Q4. The query results of Q1 and Q2 are shown in Fig. 7(a) and (b). We can conclude that OntoDS is faster than gStoreD on type-related queries.

Scalability. To validate the scalability of OntoDS, we conducted experiments on LUBM, varying the number of nodes from 2 to 4. The results depicted in Fig. 6 suggests that for a fixed dataset, the execution time is near-linearly decreased as the cluster size increases, in other words, OntoDS is scalable and flexible.

5 Related Work

Single Node RDF Data Storage. In the field of single node RDF data storage, there have been many attempts to shred RDF data into the relational model. The most straightforward solution is to use the characteristics of the RDF triples to store in a Triple Table. This solution takes up too much storage space. Even if it only stores small amount of RDF data, the table needs to have many rows. Another approach, Horizontal Table records all predicates and objects of a subject in one tuple. This solution does not save much space, because considering the varieties of predicates, this horizontal table can have numerous columns while each subject has fewer predicates, so the table have a lot of empty items. Except the schemes above, several storage schemes focus on the type characteristics of the RDF graph data, e.g., Property Table creates tables based on the types of the subjects; Vertical Partitioning creates tables based on the types of predicates [4]. Type-oriented approaches perform simple classifications to reduce the number of rows and empty items in the table. However, they require schema changes as new RDF types are encountered, which is unbearable. Sextuple Indexing storage scheme is created in order to facilitate various join operations, which establishes six tables by all six forms of triplets. This scheme sacrifices storage space while optimizing for queries. DB2RDF [5] is an entity-oriented alternative, which avoids both the skinny relation of the triple table, and the schema changes required by type-oriented approaches. Nevertheless, as mentioned above, DB2RDF is not suitable for a distributed environment.

Distributed RDF Data Storage. In the field of distributed RDF data storage, based on SQL-on-Hadoop, some distributed RDF storage systems are proposed. The H2RDF+ [13] system realizes Sextuple Indexing based on the HBase distributed repository. This approach trades much storage space for get a better query effect, while saving storage space as much as possible is the original intention of our scheme. Sempala [14] is an RDF graph data query engine based on the distributed SQL-on-Hadoop database Impala and Parquet columnar file format. Nevertheless, Sempala is not a relational storage scheme. Stylus [15] is a distributed RDF graph repository that uses strong type information to build optimized storage schemes and query processing. The underlying layer is based on a key-value library. gStoreD [12] is an RDF graph storage scheme that can optimize graph partitioning and store RDF graph based on query load. However, there is no consideration of ontology information in Stylus and gStoreD.

To the best of our knowledge, OntoDS is the first distributed RDF storage scheme to consider ontology information and distributed situations.

6 Conclusion

This paper presented OntoDS, an ontology-aware distributed storage scheme for RDF graphs, and implemented a prototype system of OntoDS based on HAWQ. OntoDS has additional benefits for type-related queries in distributed environment, as it reduces data shuffle between nodes. The experimental results on the benchmark datasets show that our distributed RDF storage scheme is both efficient and scalable, which is 1–1.5 time faster than the state-of-the-art schemes.

Acknowledgments. This work is supported by the National Natural Science Foundation of China (61572353, 61402323) and the Natural Science Foundation of Tianjin (17JCYBJC15400).

A   Appendix

A.1   Queries for DB2RDF

Q1: SELECT COUNT(*) FROM lubmc.dph WHERE lubmc.dph.p1='type'
AND lubmc.dph.v1='AssistantProfessor';

Q2: SELECT COUNT(*) FROM lubmc.dph WHERE lubmc.dph.p1='type'
AND lubmc.dph.v1='AssistantProfessor'
AND lubmc.dph.p2='mastersDegreeFrom'
AND lubmc.dph.v2 ='http://www.University389.edu';

Q3: SELECT COUNT(*) FROM lubmc.rph
WHERE lubmc.rph.obj LIKE '%Lecturer%' ;

Q4: SELECT COUNT(*) FROM lubmc.dph JOIN lubmc.rph
ON lubmc.dph.sub=lubmc.rph.obj;

A.2   Queries for OntoDS

Q1: SELECT COUNT(*) FROM lubmt.dph
WHERE lubmt.dph.type='0.17.16.15.2';

Q2: SELECT COUNT(*) FROM lubmt.dph
WHERE lubmt.dph.type='0.17.16.15.2'
AND lubmt.dph.p2='mastersDegreeFrom'
AND lubmt.dph.v2 ='http://www.University389.edu';

Q3: SELECT COUNT(*) FROM lubmt.rph
WHERE lubmt.rph.obj LIKE '%Lecturer%' ;

Q4: SELECT COUNT(*) FROM lubmt.dph JOIN lubmt.rph
ON lubmt.dph.sub=lubmt.rph.obj;

A.3   Queries for gStoreD

Q1: SELECT ?a WHERE {?a type AssistantProfessor.};

Q2: SELECT ?a WHERE {?a type AssistantProfessor. ?a mastersDegreeFrom
http://www.University389.edu.};

References

1. W3C: RDF 1.1 concepts and abstract syntax (2014)
2. Lehmann, J., et al.: DBpedia-a large-scale, multilingual knowledge base extracted from Wikipedia. Semant. Web **6**(2), 167–195 (2015)

3. Wang, X., Zou, L., Wang, C., Peng, P., Feng, Z.: Research on knowledge graph data management: a survey. Ruan Jian Xue Bao/J. Softw. **30**(7), 2139–2174 (2019). (in Chinese). http://www.jos.org.cn/1000-9825/5841.htm

4. Abadi, D.J., Marcus, A., Madden, S.R., Hollenbach, K.: SW-Store: a vertically partitioned DBMS for Semantic Web data management. VLDB J. **18**(2), 385–406 (2009)

5. Bornea, M.A., et al.: Building an efficient RDF store over a relational database. In: Proceedings of the 2013 ACM SIGMOD International Conference on Management of Data, pp. 121–132. ACM (2013)

6. Sun, W., Fokoue, A., Srinivas, K., Kementsietsidis, A., Hu, G., Xie, G.: SQLgraph: an efficient relational-based property graph store. In: Proceedings of the 2015 ACM SIGMOD International Conference on Management of Data, pp. 1887–1901. ACM (2015)

7. Floratou, A., Minhas, U.F., Özcan, F.: SQL-on-Hadoop: full circle back to shared-nothing database architectures. Proc. VLDB Endowment **7**(12), 1295–1306 (2014)

8. Krishnamoorthy, M.S.: A note on some simplified NP-complete graph problems. ACM Sigact News **9**(3), 24–24 (1977)

9. Welsh powell algorithm. https://iq.opengenus.org/welsh-powell-algorithm/

10. Guo, Y., Pan, Z., Heflin, J.: LUBM: a benchmark for owl knowledge base systems. Web Semant. Sci. Serv. Agents World Wide Web **3**(2–3), 158–182 (2005)

11. Chang, L., et al.: HAWQ: a massively parallel processing SQL engine in hadoop. In: Proceedings of the 2014 ACM SIGMOD International Conference on Management of Data, pp. 1223–1234. ACM (2014)

12. Peng, P., Zou, L., Chen, L., Zhao, D.: Adaptive distributed RDF graph fragmentation and allocation based on query workload. IEEE Trans. Knowl. Data Eng. **31**(4), 670–685 (2018)

13. Papailiou, N., Tsoumakos, D., Konstantinou, I., Karras, P., Koziris, N.: H 2 RDF+: an efficient data management system for big RDF graphs. In: Proceedings of the 2014 ACM SIGMOD International Conference on Management of data, pp. 909–912. ACM (2014)

14. Schätzle, A., Przyjaciel-Zablocki, M., Neu, A., Lausen, G.: Sempala: interactive SPARQL query processing on hadoop. In: Mika, P., et al. (eds.) ISWC 2014. LNCS, vol. 8796, pp. 164–179. Springer, Cham (2014). https://doi.org/10.1007/978-3-319-11964-9_11

15. He, L., et al.: Stylus: a strongly-typed store for serving massive RDF data. Proc. VLDB Endowment **11**(2), 203–216 (2017)

Learning Relational Fractals for Deep Knowledge Graph Embedding in Online Social Networks

Ji Zhang[1,2]([⊠]), Leonard Tan[1], Xiaohui Tao[1], Dianwei Wang[3],
Josh Jia-Ching Ying[4], and Xin Wang[5]

[1] University of Southern Queensland, Toowoomba, Australia
Ji.Zhang@usq.edu.au
[2] Zhejiang Lab, Zhejiang, China
[3] Xi'an University of Posts and Telecommunications, Xi'an, China
[4] National Chung Hsing University, Taichung, Taiwan ROC
[5] Southwest Jiaotong University, Chengdu, China

Abstract. Knowledge Graphs (KGs) have deep and impactful applications in a wide-array of information networks such as natural language processing, recommendation systems, predictive analysis, recognition, classification, etc. Embedding real-life relational representations in KGs is an essential process of abstracting facts for many important data mining tasks like information retrieval, privacy and control, enrichment and so on. In this paper, we investigate the embedding of the relational fractals which are learned from the Relational Turbulence profiles in the transactions of Online Social Networks (OSNs) into KGs. These relational fractals have the capability of building both compositional-depth hierarchies and shallow-wide continuous vector spaces for more efficient computations on devices with limited resources. The results from our RFT model show accurate predictions of relational turbulence patterns in OSNs which can be used to evolve facts in KGs for more accurate and timely information representations.

Keywords: Relational turbulence · Deep learning · Knowledge graph embedding · Online Social Networks · Fact evolution

1 Introduction

Knowledge Graphs (KGs) have many important real-world applications like semantic parsing, named entity disambiguation, information extraction and question answering [18]. A KG is a multi-relational structure built from real-life information which are condensed into relational tuples. These relational tuples establish ground truths between entities and objects and are also known as facts [13,18]. A fact is constructed from a group of three cardinal elements: (*subject* → *predicate* → *object*) and is also known as an atomic information construct to the parent KG superset [19]. Recent years have seen rapid growths

© Springer Nature Switzerland AG 2019
R. Cheng et al. (Eds.): WISE 2019, LNCS 11881, pp. 660–674, 2019.
https://doi.org/10.1007/978-3-030-34223-4_42

in the construction and successful practical applications of KG datasets. Examples of some prominent KG datasets include Freebase, DBpedia, YAGO, NELL, Knowledge Vault and Google Knowledge Graph [18]. KG embedding refers to the process of transplanting components of information such as entities and relations - into shallow continuous vector spaces [18,19]. The objective is to simplify computations at run time while preserving the inherent structure of the KG [19]. However, shallow structures are necessarily space consuming. In order to limit this resource consumption, deep networks which rely on the use of compositional functions to establish a hierarchy of facts remains a popular choice for reducing representional spaces required to contain these sophistication [1]. Nontheless, implementation of a compositional structure comes at high computational costs when Directed Acyclic Graph (DAG) source directed queries need to be resolved [1]. Hence, we propose the use of relational fractals which can be used to dynamically build and/or prune predicates that increases compositional hierarchy and/or extend shallow vector manifold representations as new information is continuously added to the graph. The objective function of embedding deep relational fractals in KG networks is to optimize the compositional depth computation cost to continuous vector space consumption ratio [11,17]. This approach will enable mobile and/or lightweight devices which are limited in both computational power and storage capacity to efficiently process large amounts of information efficiently, while at the same time, storing it in adequately sized KGs.

In this paper, we introduce a new model, called the Relational-Flux-Turbulence (RFT), that effectively represents the dynamism of popular key relational dimensions uncovered from previous approaches and techniques conducted on online social structure [21]. The model builds a multi-stage deep neural network from a stack of fractals with hybrid architectures of Restricted Boltzmann Machines (RBMs) and Recursive Neural Nets (RNNs) [2]. These structures are self-evolving from a meta-learning perspective. The neural network accepts as inputs, key relational feature states f_i between actors a_j and global events E_ϵ from past and present social transactions to determine the relational turbulence τ_{ij} pattern within an identified social flux F_ϵ. Turbulence may correspond to various disruptions in social communication of different environments and contexts [10]. For example, in the discussion of world events like trade wars, passive sentiments passed through public posts and comments are indicative of hostility and potential conflict which may lead to a breakdown of linked integrity between actors in many aspects like trust, influence, status, etc. and hence change the predicates of the KG in question. We develop a novel architecture from Relational Turbulence Theory and Models (RTT and RTM) to learn the relational fractal from turbulence, within a given social context describing the state of flux. Then, we evaluate our methods on Twitter, Google and Enron email datasets and demonstrate that they outperform similarity based feature and shallow unidirectional flat structural approaches in detecting social flux and turbulence.

In this study, we look at the dynamic structure of such an shallow ANN known as fractals. Fractals are the lowest principle decompositions of never ending patterns. They maintain a key property of self-similarity across different varying scales [5]. Driven by a recursive process, fractals are adaptable enough to describe highly dynamic system representations [9]. In the sections that follow, we describe the methods and experiments performed on Twitter, Google and Enron email datasets at different instances and show that structural fractals behave like cognitive super primers that can be used to decode representational information sophistication through a generative feedback loop. The main contributions of our study are:

1. Our method adaptively learns from real-time online streaming data to identify key turbulent relationships within a given OSN;
2. An innovative RFT model was developed to capture key relational features which were used to detect and profile social communication patterns of eventful states within a given OSN;
3. Experiment results show that RFT is able to offer a good modeling of relational ground truths, while FNN is able to efficiently and accurately represent evolving relational turbulence and flux profiles within a given OSN.

The remaining part of the paper is organized as follows: Sect. 2 presents a brief overview of related works and introduces key concepts drawn from social theories and relational structures. Section 3 introduces the theories and methods of our proposed model. Section 4 provides a thorough analysis of experimental design and implementation. Section 5 presents the results and discussion of this paper that leads to a conclusion and potential future directions.

2   Related Literature

Relational Turbulence, first studied in [6], was typically characterized as a resultant state in conflict of interests from competing goals between two or more actors (entities) in question [20]. Although conflict does provide the basis of stimulation for communication within a relationship that is centered in a flux, it also correlates to negative consequences in the form of detrimental event occurrences if left undetected and unchecked [12]. An important discriminator of detecting conflict and hence the resulting turbulence in any relationship model between networks of actor entities is the observation and management of relational altering events [22]. These events, if found to be in huge negative violations of expectancies between relational reciprocates of actors, can lead to instability in a relational flux [23].

   The Relational Turbulence Model (RTM) [15] builds upon the core principles of relational state shifts and conflict management to define an artificial construct. This construct enables intelligent predictions of communication behaviors during relationship transitions in an environment of continuous online social disruptions. The process of turbulent relationship development can be described as a continuous and communicative state of flux [15]. This state defines a consistent exchange of sentimental and affective information between the actor/s involved [23,24]. Each transition to another state (e.g. professional colleagues to friendship) has the probability to cause friction (conflict), which may lead to a polarization of sentiments and affective communication flux in OSNs [16]. Two key tenets of the RTM are actor interferences and relational uncertainty. These two prime relational features in OSNs enable the effective detection and prediction of conflict and event (new fact) occurrences in sentimental and affective computing of KGs.

In [8], the authors present a minimalist neural network architecture for reliably and accurately estimating emotional states based on EEG captured data. Their model uses an innovative parameter known as the reinforced gradient coefficient to tackle the vanishing gradient problem faced by deep learning architectures. Additionally, their model adopts a weighing step to extract outliers from the discrepancies between successive predictions. Although this approach may help alleviate diminishing gradients by increasing error gradients during the back propagation process, it does so at the expense of performance. Tackling larger error gradients during the forward and back propagation burn-in phases of training especially on a shallow ANN architecture means that convergence to an accurate estimation is slower with more lengthy iterations. Furthermore, the trade off in accuracy gains between the MNN and other state of the art methods (e.g. ADA, RMS, NM, etc.) included in their work does not justify the computational resource costs involved.

In [14], the authors deal with the problem of social role recognition through the use of a Conditional Random Field (CRF) layered model architecture. Their architecture is used to learn actor-environment and actor-actor behaviors from different unlabeled video streams of a given event classification. Their work derives from the motivation in the field of Role Theory in sociology. Their full model results on You-Tube social videos show a higher event-based social role classification hit-rate as compared to traditional k-Means and CRF cluster algorithms. However, for video image frames in which latent social role-based semantics exists, CRF architectures are ill-adapted to handle the complex representations of the depth of these roles in the identification process. This invariably leads to poor performance output measures of their full model method.

3   Theories and Methods

From the RTM approach, we define Relational Intensity $P(\gamma_{rl})$, Relational Interference $P(\vartheta_{rl})$ and Relational Uncertainty $P(\varphi_{rl})$ to be three key probabilistic outputs of the RFT model which represents the relational turbulence $P_{\tau_{rl}}$ of a given link in an OSN. The key element types we have identified to be contributing features between the duration of the turning point and relationship development (as an unstable/turbulent process) are the Confidence ρ_{ij}, Salience ξ_{ij} and Sentiment λ_{ij} scores in an actor-actor relationship of a social transaction in question.

Firstly, we define relational intensity during state altering events conditionally, as the integration of sentimental transactions (flux) per unit (context) area. Mathematically, this is given as:

$$\gamma_{rl} = \sum_{i,j=1}^{n} \frac{\beta_{ij}| - \frac{\nabla F_{ej}}{\nabla t}|}{L_{F_e}} \tag{1}$$

Where β_{ij} is defined as the temporal derivative of the latent topic (context) signal phase ϵ. Secondly, we define relational uncertainty as the likelihood measure of opposing sentiment mentions. Mathematically, this is given by:

$$\varphi_{rl} = \frac{\sum_{i,j=1}^{n} S_i S_j}{\sqrt{\sum_{i=1}^{n} S_i}\sqrt{\sum_{j=1}^{n} S_j}} \tag{2}$$

Where S_i and S_j are sentiments transacted from nodes i to j and from nodes j to i respectively. Finally, we define relational interference as the probability that deviations from predicted or expected outcomes of relational flux intensity and uncertainty fall outside a confidence interval centered about the mean. Mathematically, this is given as:

$$\vartheta_{rl} = E(F(\gamma_{rl}, \vartheta_{rl} : \mu_{\gamma\varphi}, \omega_{\gamma\varphi}^2))$$
$$= \frac{1}{2} + \frac{1}{\sqrt{2\pi\omega}} \sum_{\gamma_{rl},\varphi_{rl}=0}^{n} \frac{1}{2} erf(\frac{\gamma_{rl}, \varphi_{rl} - \mu}{\sqrt{2\pi}}) \exp^{\frac{-(\gamma_{rl}\cdot\varphi_{rl}-\mu)^2}{2\omega^2}} \tag{3}$$

Where,

$$F(\gamma_{rl}, \varphi_{rl} : \mu_{\gamma\varphi}, \omega_{\gamma\varphi}^2) = \frac{1}{\sqrt{2\pi\omega}} \sum_{t=-\infty}^{\gamma_{rl},\varphi_{rl}} \exp^{\frac{-(t-\mu)^2}{2\omega^2}} dt \tag{4}$$

Here, $F(\gamma_{rl}, \varphi_{rl} : \mu_{\gamma\varphi}, \omega_{\gamma\varphi}^2)$ is the cumulative distribution function, and $erf(x)$ is the error function of the predicted outcomes γ_{rl} and φ_{rl}.

The model we have chosen, with which to address the prediction problem of relational turbulence is the Fractal Neural Network (FNN) that adopts a hybrid (turing-learning based) architecture which incorporates the use of both generative and discriminative deep networked architectures. At the core of the RFT architecture is a stack of Restricted Boltzmann Machines (RBM) which constitutes the essence of a Deep Belief Network (DBN) that pretrains our Deep Neural Network (DNN) structural framework. In our architecture, the generative DBN is used to initialize the DNN weights and the fine-tuning from the backprop is carried out sequentially layer by layer. In order to tackle the problem of computational efficiency and learning scalability to large data sets, we have adopted the Deep Stacking Network (DSN) model framework for our study. Central to the concept of such an architecture is the relational use of stacking to learn complex distributions from simple core belief modules, functions and classifiers [3].

On the system level, streaming data like live tweets are first retrieved from the twitter database using the twitter firehose API matched according to our specified filters of choice (e.g. hashtag, geo-location, trending topics, etc.) in the first stage. While batch data can be easily retrieved from saved sources like Enron emails and Google datasets, live data streams are analyzed in real-time. The next stage of our system establishes a data pipe which sends data streams across to a cache filter. For batch process data, these are sent across directly. Next, Google NLP is used to determine key relational features which form independent inputs to our model in Fig. 1. The RFT architecture then predicts relational turbulence at the outputs learned from ground truths expressed in Eqs. (1), (2) and (3).

3.1   The RFT Architecture

We begin our subject of research with the definition of a soft kernel used to discover a markovian structure which we then encode into confabulations of fractal sub-structures. For a given set of data observables as inputs: $\chi \in X$ and outputs: $\Im \in \Xi$ we wish to loosely define a mapping such that the source space (X, α) maps onto a target space $(\Im, \omega)$. The conditional $P(\chi \vee \omega)$ assigns a probability from each source input χ to the final output space in ω. Each posterior state-space from in between input to output is generated and sampled through a random walk process. It is worth noting that markovian random walks are used to build a more generalized stochastic discovery process in our experiment. However for larger datasets, any one of the more sophisticated markovian sampling methods (e.g. Gibbs, Monte carlo, Metropolis-Hastings, Hamiltonian, etc.) can be used as drop-in replacements. An indicator function which we have chosen to describe the state transition rule is:

$$\Theta_{t+1} = min \begin{cases} 0 \\ \amalg_{c=1}^{n} \frac{\delta E_{t+1}^c}{\delta \chi_t^c} \end{cases} \tag{5}$$

Where δE_{t+1}^c is the error change from one hidden feature activity state $h_t \in H$ onto higher posterior confabulations. The objective function at each transition seeks to minimize error gradients to eliminate problems associated with exploding and vanishing gradients during backpropagation. This can be caused by an excessive generation of layered confabulations which leads to unnecessary increments in depth from the markovian ANN discovery mechanism. For a general finite state space markovian process, the markov kernel is thus defined as:

$$Kern(M) = \begin{cases} p : X \times \omega \rightarrow [0,1] \\ p(\chi|\omega) = \oint_{\omega} q(\chi, \Im)\nu(\delta \Im) \end{cases} \tag{6}$$

Once a unique markovian neural network has been discovered, a Single Layer Convolutional Perceptrion (SLCP) is proposed as a baseline structure to learn the fractal sub-network from pre-existing posterior confabulations. The SLCP baseline structure changes as discovered knowledge is progressively encoded during the learning process. Any one baseline model can be used to learn a morphing transposition into a fractal signature structure. In essence, methods like Progressive Neural Networks (PNNs) where activation links of neighboring DNN stacks are learned laterally across hidden layers [4] or the wide use of summarizing information from ensemble methods like distillation [7] are relevant alternatives.

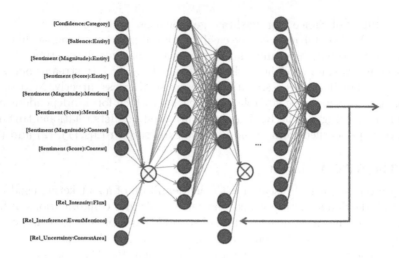

Fig. 1. The RFT architecture design

4   Experiments and Results

The experiments were conducted on three datasets using three different algorithms. The datasets are: Twitter, Google and Enron emails. The Stanford Twitter Sentiment Corpus contains APIs (http://help.sentiment140.com/api) for classifying raw tweets that allows us to integrate their classifiers into our deep learning model. Their plug-in module uses an ensemble of different learning classifiers and feature extractors to deliver the best outputs with different combinations of classifiers and feature extractors. In addition to the sentiment results obtained from their model, we cross validated the output against googles NLP API (https://cloud.google.com/natural-language/) to replicate the most accurate sentiment scores and magnitudes of context spaces and mentions. Our twitter dataset was live streamed from a twitter API account and contains a maximum of 1675882 nodes and 160799842 links.

The Google dataset was obtained from the repositories of common crawl and was sentilyzed from the stripped down WET file contents. The dataset which was used in this experiment was extracted from the April 2014 crawl data and contained 3566224 entities and 436994489 dyads. Lastly, the Enron email dataset was obtained from the David Newman website, hosted on the UCI Machine Learning Repository (https://archive.ics.uci.edu/ml/datasets/bag+of+words) and contains approximately 500000 emails generated by employees of the Enron corporation. The entire repository of email contents were extracted and sentilyzed using google's NLP model to provide the inputs we require of our training model.

Our experiments were conducted on our training model with a learning rate set to 1.1, a sliding window set to 3, an error tolerance set to 0.1 (10%), a data outlier threshold set to 1.0, with scaling set to 10, a vanishing gradient error threshold at 0 and an exploding gradient error threshold set to 100. Finally, our trust region radius parameter was set to 5 and our softmax temperature regularization parameter was staged at 1.2.

4.1 Experimental Design

Figure 1 describes the inputs into our model. Specifically, the RFT model accepts as inputs, the confidence of the detected category in every social transaction, the Salience of all detected entities in the transaction, the sentiment scores and magnitudes of entities, mentions and drifting contexts. These eight relational features form the key independent input into our RFT fractal neural network (FNN) model. Additionally, the outputs (Relational Intensity γ_{rl}, Relational Interference ϑ_{rl} and Relational Uncertainty φ_{rl}) which represent turbulence are fed back into the model as recurrent inputs into the neural network to act as memory retention for the relational turbulence profiles of previous transaction/s, and as good influential initialization points for new training sequences of extracted sentiments in later social transactions.

Relational Turbulence was calculated from conditional posteriors of γ_{rl}, ϑ_{rl} and φ_{rl} as the mathematical relation of:

$$P(\tau_{rl}) = \sum_{i=1}^{n} \frac{P(\gamma_i|\theta_i)P(\vartheta_i|\varphi_i)P(\varphi_i|\gamma_i)}{N_i P(\gamma_i)P(\vartheta_i)P(\varphi_i)} \tag{7}$$

The true values of relational turbulence in the graphs from Fig. 2(a) to (i) were obtained directly from Eqs. (1), (2) and (3). The inputs were tested across three deep architecture models and the learning results were compared using both Kendall and Spearman correlation tests to measure both strength of dependence and degree of association between input independent variables and output turbulence metrics. In addition, the different deep learning approaches were cross validated using k-fold cross validation techniques.

4.2 Experimental Performance

The Kendall (w coefficient) and Spearman (rho coefficient) tests were conducted on the results obtained from the testing procedures.

Specifically, the Kendall (tau-b coefficient) was used to measure the strength of associations between predicted and expected outputs of the learning models. The Kendall (tau-b) coefficient is given as:

$$\tau_b = \frac{N_c - N_d}{\sqrt{(N_0 - N_x)(N_0 - N_y)}} \tag{8}$$

Where,

$$N_0 = \frac{N(N-1)}{2} \tag{9}$$

And,

$$N_x = \sum_i \frac{u_i(u_i - 1)}{2} \tag{10}$$

And,

$$N_y = \sum_j \frac{v_j(v_j - 1)}{2} \tag{11}$$

Where N_c is the number of concordant paris, N_d is the number of discordant pairs, u_i is the number of tied values in the ith group of ties for the first quantity and v_j is the number of tied values in the jth group of ties for the second quantity.

The Spearman (rho coefficient) was used to measure the monotonic relationship between the independent variables (Category confidence $\mathfrak{C}_i$, Entity Sailence $\mathcal{J}_i$, Entity sentiments - magnitude and scores $(\mathfrak{S}_i, \mathbb{J}_i)$, Mention sentiments -magnitude and scores $(\mathcal{L}_i, \lambda_i)$, Context sentiments - magnitude and scores $(\mathfrak{C}_i, \mathbb{T}_i)$) and the dependent variables (Relational Intensity γ_{rl}, Relational Interference ϑ_{rl} and Relational Uncertainty φ_{rl}). Essentially, the relationship of measure is calculated as:

$$\varGamma_S = 1 - \frac{6 \sum D_i^2}{N(N^2 - 1)} \tag{12}$$

Where $D_i = rank(X_i) - rank(Y_i)$ is the difference in ranks between the observed independent variable X_i and dependent variable Y_i and N is the number of predictions to input data sets for all three sources.

The tests were run across the Single Layer Perceptron (SLP), a 45-layer Deep Convolutional Network (DCN) and a dynamically stacked Fractal Neural Network (FNN). Selected results are shown in Tables 1–11 and Fig. 5, 6–30:

4.3   Testing Results

Finally, during the experimentation, the full datasets obtained from the different sources (Twitter, Google and Enron) were partitioned into k-subsamples. One of the subsamples was retained as the validation set for each run and the validation set was chosen in a round robin fashion for subsequent experimentation runs. A noteworthy point of mention is that k-fold cross validation is used in our experimentation design to obtain a good estimate of the prediction generalization. This testing technique does not scale well to measurements of model precision. How accurately a learning model is able to predict an expected output is based on the Kendall (tau-b coefficient) results. k-fold validation was performed over all deep learning models across the Mean Absolute Percentage Error (MAPE) measurement of each run. Mathematically, MAPE can be expressed as:

$$\delta_{MAPE} = \frac{1}{N} \sum_{i=1}^{N} |\frac{E_i(x) - Y_i(t)}{E_i(x)}| \tag{13}$$

Where $E_i(x)$ is the expectation at the output of data input set i and $Y_i(t)$ is the corresponding prediction over N total subsamples. The tabulation of the k-fold cross validation used in our experimentation is given in Table 11 (Fig. 3).

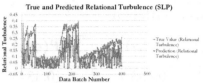

(a) Graph of True and Predicted SLP relational turbulence values for the Twitter dataset

(b) Graph of True and Predicted DCN relational turbulence values for the Twitter dataset

(c) Graph of True and Predicted RFT relational turbulence values for the Twitter dataset

(d) Graph of True and Predicted SLP relational turbulence values for the Google dataset

(e) Graph of True and Predicted DCN relational turbulence values for the Google dataset

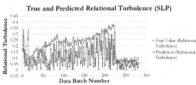

(f) Graph of True and Predicted RFT relational turbulence values for the Google dataset

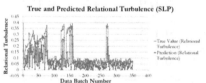

(g) Graph of True and Predicted SLP relational turbulence values for the Enron email dataset

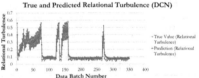

(h) Graph of True and Predicted DCN relational turbulence values for the Enron email dataset

True and Predicted Relational Turbulence (RFT)

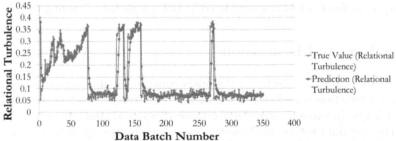

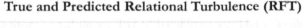

(i) Graph of True and Predicted RFT relational turbulence values for the Enron email dataset

Fig. 2. Graphs of relational turbulence predictions

Spearman (rho) coefficient			
	$P(\gamma_{rl})$	$P(\vartheta_{rl})$	$P(\varphi_{rl})$
$\mathfrak{C}_i$	0.435	0.433	0.434
$\mathfrak{J}_i$	-0.750	0.728	0.075
$\mathfrak{J}_i$	0.834	0.930	0.834
$\beth_i$	0.849	0.974	0.878
$\mathcal{L}_i$	0.784	0.818	0.811
λ_i	0.817	0.725	0.891
$\mathcal{O}_i$	-0.402	-0.322	-0.319
$,7_i$	-0.359	-0.357	-0.358

Fig. 3. Table of spearman's (rho) coefficient for Enron's email (RFT)

K	δ_{MAPE} - (SLP)	δ_{MAPE} - (DCN)	δ_{MAPE} - (RFT)
20	0.461	0.189	0.127
30	0.424	0.175	0.131
50	0.420	0.173	0.112
80	0.418	0.169	0.110
100	0.421	0.166	0.107

Fig. 4. Table of K-fold cross validated MAPE for all three learning models

5   Analysis and Discussion

As can be seen from the graphs, SLP models consistently underperforms in ranking where prediction accuracy is concerned, the Kendall (tau-b coefficient) test shows a lower (positive) correlation between expected and predicted outputs across the test data set for SLP models and much higher (positive) association for both DCN and RFT. Additionally, from the results of the Spearman (rho coefficient) test done on the independent and dependent variables, it can be seen from Tables 1–10 that the spearman coefficient indicates strongly positive

Spearman (rho) coefficient			
	$P(\gamma_{rl})$	$P(\vartheta_{rl})$	$P(\varphi_{rl})$
$\mathfrak{C}_i$	0.076	-0.077	-0.078
$\mathfrak{I}_i$	-0.805	0.764	0.002
$\mathfrak{Z}_i$	0.844	0.837	0.703
$\mathfrak{I}_i$	0.901	0.872	0.871
$\mathcal{L}_i$	0.877	0.913	0.953
λ_i	0.788	0.827	0.891
σ_i	-0.303	-0.297	-0.295
$\daleth_i$	-0.271	-0.302	-0.312

(a) Table of Spearman's (rho) coefficient for Twitter (SLP)

Spearman (rho) coefficient			
	$P(\gamma_{rl})$	$P(\vartheta_{rl})$	$P(\varphi_{rl})$
$\mathfrak{C}_i$	0.079	-0.074	-0.077
$\mathfrak{I}_i$	-0.783	0.776	0.001
$\mathfrak{Z}_i$	0.874	0.843	0.767
$\mathfrak{I}_i$	0.892	0.882	0.846
$\mathcal{L}_i$	0.864	0.891	0.921
λ_i	0.779	0.833	0.888
σ_i	-0.293	-0.289	-0.278
$\daleth_i$	-0.275	-0.276	-0.284

(b) Table of Spearman's (rho) coefficient for Twitter (DCN)

Spearman (rho) coefficient			
	$P(\gamma_{rl})$	$P(\vartheta_{rl})$	$P(\varphi_{rl})$
$\mathfrak{C}_i$	0.074	-0.070	-0.076
$\mathfrak{I}_i$	-0.738	0.825	0.007
$\mathfrak{Z}_i$	0.842	0.847	0.787
$\mathfrak{I}_i$	0.887	0.834	0.837
$\mathcal{L}_i$	0.846	0.884	0.901
λ_i	0.784	0.846	0.871
σ_i	-0.285	-0.292	-0.269
$\daleth_i$	-0.273	-0.287	-0.278

(c) Table of Spearman's (rho) coefficient for Twitter (RFT)

Spearman (rho) coefficient			
	$P(\gamma_{rl})$	$P(\vartheta_{rl})$	$P(\varphi_{rl})$
$\mathfrak{C}_i$	-0.012	-0.031	-0.048
$\mathfrak{I}_i$	-0.262	0.462	0.360
$\mathfrak{Z}_i$	0.344	0.338	0.303
$\mathfrak{I}_i$	0.401	0.472	0.371
$\mathcal{L}_i$	0.357	0.311	0.353
λ_i	0.378	0.327	0.391
σ_i	-0.576	-0.602	-0.595
$\daleth_i$	-0.514	-0.588	-0.542

(d) Table of Spearman's (rho) coefficient for Google (SLP)

Spearman (rho) coefficient			
	$P(\gamma_{rl})$	$P(\vartheta_{rl})$	$P(\varphi_{rl})$
$\mathfrak{C}_i$	0.018	0.071	0.042
$\mathfrak{I}_i$	0.270	0.301	0.289
$\mathfrak{Z}_i$	0.316	0.348	0.333
$\mathfrak{I}_i$	0.397	0.431	0.351
$\mathcal{L}_i$	0.337	0.329	0.383
λ_i	0.362	0.367	0.321
σ_i	-0.542	-0.611	-0.587
$\daleth_i$	-0.533	-0.547	-0.556

(e) Table of Spearman's (rho) coefficient for Google (DCN)

Spearman (rho) coefficient			
	$P(\gamma_{rl})$	$P(\vartheta_{rl})$	$P(\varphi_{rl})$
$\mathfrak{C}_i$	0.005	0.005	0.005
$\mathfrak{I}_i$	0.286	0.314	0.298
$\mathfrak{Z}_i$	0.414	0.337	0.345
$\mathfrak{I}_i$	0.399	0.423	0.399
$\mathcal{L}_i$	0.383	0.384	0.384
λ_i	0.392	0.379	0.377
σ_i	-0.558	-0.607	-0.597
$\daleth_i$	-0.514	-0.551	-0.548

(f) Table of Spearman's (rho) coefficient for Google (RFT)

Spearman (rho) coefficient			
	$P(\gamma_{rl})$	$P(\vartheta_{rl})$	$P(\varphi_{rl})$
$\mathfrak{C}_i$	0.581	0.435	0.437
$\mathfrak{I}_i$	-0.601	0.781	0.080
$\mathfrak{Z}_i$	0.781	0.978	0.883
$\mathfrak{I}_i$	0.842	0.901	0.857
$\mathcal{L}_i$	0.744	0.812	0.846
λ_i	0.891	0.789	0.861
σ_i	-0.481	-0.387	-0.375
$\daleth_i$	-0.361	-0.359	-0.361

(g) Table of Spearman's (rho) coefficient Enron's email (SLP)

Spearman (rho) coefficient			
	$P(\gamma_{rl})$	$P(\vartheta_{rl})$	$P(\varphi_{rl})$
$\mathfrak{C}_i$	0.446	0.443	0.444
$\mathfrak{I}_i$	-0.708	0.762	0.070
$\mathfrak{Z}_i$	0.817	0.923	0.848
$\mathfrak{I}_i$	0.831	0.947	0.837
$\mathcal{L}_i$	0.747	0.851	0.875
λ_i	0.875	0.774	0.884
σ_i	-0.451	-0.348	-0.344
$\daleth_i$	-0.377	-0.374	-0.375

(h) Table of Spearman's (rho) coefficient for Enron's email (DCN)

Fig. 5. Tables of spearman's coefficient

monotonic correlations between turbulence measures (γ_{rl}, ϑ_{rl} and φ_{rl}) and sentiment scores $[(\Im_i, \beth_i), (\mathcal{L}_i, \lambda_i), (\mathcal{O}_i, \daleth_i)]$ and moderately positive correlations between the same turbulence measures (γ_{rl}, ϑ_{rl} and φ_{rl}) to both category confidence and entity salience ($\mathfrak{C}_i, \mathcal{J}_i$).

Generally however, it can be observed from all the plots that intensity, interference and uncertainty correlates very well to expressed sentiments over entities, mentions, and (fairly well) over contexts. However, an interesting observation made from the distribution of the results is that while entity and mention sentiments are (strongly) positively correlated to the tenets of relational turbulence (i.e. higher sentiment scores expressed in these classifier manifolds are more likely to evoke a relational state altering event), context sentiments are (mediocrely) negatively correlated instead. It can also be observed that this negative correlation of contexts to turbulence is weaker in both Twitter and Enron (where communications are both specifically directed and/or semi-directed at social individuals) and stronger in Google datasets (where communications are non-specific and loosely directed at certain social groups or communities) (Fig. 4).

6   Conclusion

In conclusion, we have shown that RFT is capable of predicting relational turbulence profiles between actors within a given OSN acquired from anytime data. Furthermore, the novel FNN model which we have developed is able to rapidly scale and adaptively represent relational complexities of anytime sequenced data within a live online social scene. Our results show superior accuracies and performance of the FNN model in comparison well known baseline models like the Single Layer Perceptron (SLP) and the Deep Convolutional Network (DCN) designs. We have demonstrated the feasibility of our learning model through the implementation on three large scale networks: Twitter, Google Plus and Enron emails. Our study uncovers three pivotal long-term objectives from a relational perspective. Firstly, relational features can be used to strengthen fact evolutions in medical, cyber security and social KG applications where the constant challenges between detection, recommendation, prediction, data utility and privacy are being continually addressed. Secondly, in fintech applications, relational predicates (e.g. turbulence) are determinants to market movements - closely modeled after a system of constant shocks. Thirdly, in artificial intelligence applications like computer cognition and robotics, learning relational features between social actors enables machines to recognize and evolve. Deep learning relational graph models appear to have considerable potential, especially in the fast growing area of social networks.

References

1. Bronstein, M.M., Bruna, J., LeCun, Y., Szlam, A., Vandergheynst, P.: Geometric deep learning: going beyond euclidean data. IEEE Signal Process. Mag. **34**(4), 18–42 (2017)

2. Cai, L., Wang, W.Y.: Kbgan: Adversarial learning for knowledge graph embeddings. arXiv preprint arXiv:1711.04071 (2017)
3. Deng, L., Yu, D., et al.: Deep learning: methods and applications. Found. Trends Signal Process. **7**(3–4), 197–387 (2014)
4. Gideon, J., Khorram, S., Aldeneh, Z., Dimitriadis, D., Provost, E.M.: Progressive neural networks for transfer learning in emotion recognition. arXiv preprint arXiv:1706.03256 (2017)
5. Goh, K.I., Salvi, G., Kahng, B., Kim, D.: Skeleton and fractal scaling in complex networks. Phys. Rev. Lett. **96**(1), 018701 (2006)
6. Haunani Solomon, D., Theiss, J.: A longitudinal test of the relational turbulence model of romantic relationship development. Pers. Relat. **15**, 339–357 (2008). https://doi.org/10.1111/j.1475-6811.2008.00202.x
7. Hinton, G., Vinyals, O., Dean, J.: Distilling the knowledge in a neural network. arXiv preprint arXiv:1503.02531 (2015)
8. Keshmiri, S., Sumioka, H., Nakanishi, J., Ishiguro, H.: Emotional state estimation using a modified gradient-based neural architecture with weighted estimates. In: 2017 International Joint Conference on Neural Networks (IJCNN), pp. 4371–4378. IEEE (2017)
9. Larsson, G., Maire, M., Shakhnarovich, G.: Fractalnet: ultra-deep neural networks without residuals. arXiv preprint arXiv:1605.07648 (2016)
10. Li, X., Lou, C., Zhao, J., Wei, H., Zhao, H.: "tom" pet robot applied to urban autism. arXiv preprint arXiv:1905.05652 (2019)
11. Liu, S., Trenkler, G.: Hadamard, khatri-rao, kronecker and other matrix products. Int. J. Inform. Syst. Sci. **4**(1), 160–177 (2008)
12. McLaren, R.M., Solomon, D.H., Priem, J.S.: The effect of relationship characteristics and relational communication on experiences of hurt from romantic partners. J. Commun. **62**(6), 950–971 (2012)
13. Nickel, M., Murphy, K., Tresp, V., Gabrilovich, E.: A review of relational machine learning for knowledge graphs. Proc. IEEE **104**(1), 11–33 (2015)
14. Ramanathan, V., Yao, B., Fei-Fei, L.: Social role discovery in human events. In: Proceedings of the IEEE Conference on Computer Vision and Pattern Recognition, pp. 2475–2482 (2013)
15. Solomon, D.H., Knobloch, L.K.: Relationship uncertainty, partner interference, and intimacy within dating relationships. J. Soc. Pers. Relat. **18**(6), 804–820 (2001). https://doi.org/10.1177/0265407501186004
16. Solomon, D.H., Knobloch, L.K., Theiss, J.A., McLaren, R.M.: Relational turbulence theory: explaining variation in subjective experiences and communication within romantic relationships. Hum. Commun. Res. **42**(4), 507–532 (2016)
17. Trivedi, R., Dai, H., Wang, Y., Song, L.: Know-evolve: deep temporal reasoning for dynamic knowledge graphs. In: Proceedings of the 34th International Conference on Machine Learning, vol. 70, pp. 3462–3471. JMLR. org (2017)
18. Wang, Q., Mao, Z., Wang, B., Guo, L.: Knowledge graph embedding: a survey of approaches and applications. IEEE Trans. Knowl. Data Eng. **29**(12), 2724–2743 (2017)
19. Wang, Z., Zhang, J., Feng, J., Chen, Z.: Knowledge graph embedding by translating on hyperplanes. In: Twenty-Eighth AAAI Conference on Artificial Intelligence (2014)
20. Wilmot, W., et al.: Interpersonal Conflict, 9th edn. McGraw-Hill Higher Education, New York (2007)

21. Zhang, J., Tan, L., Tao, X.: On relational learning and discovery in social networks: a survey. Int. J. Mach. Learn. Cybern. **20**(8), 1–18 (2018). https://doi.org/10.1007/s13042-018-0823-8
22. Zhang, J., et al.: Detecting relational states in online social networks. In: 2018 5th International Conference on Behavioral, Economic, and Socio-Cultural Computing (BESC), pp. 38–43. IEEE (2018)
23. Zhang, J., Tan, L., Tao, X., Zheng, X., Luo, Y., Lin, J.C.-W.: SLIND: identifying stable links in online social networks. In: Pei, J., Manolopoulos, Y., Sadiq, S., Li, J. (eds.) DASFAA 2018. LNCS, vol. 10828, pp. 813–816. Springer, Cham (2018). https://doi.org/10.1007/978-3-319-91458-9_54
24. Zhang, J., Tao, X., Tan, L., Lin, J.C.-W., Li, H., Chang, L.: On link stability detection for online social networks. In: Hartmann, S., Ma, H., Hameurlain, A., Pernul, G., Wagner, R.R. (eds.) DEXA 2018. LNCS, vol. 11029, pp. 320–335. Springer, Cham (2018). https://doi.org/10.1007/978-3-319-98809-2_20

Graph Mining

Parameter-Free Structural Diversity Search

Jinbin Huang[1](✉), Xin Huang[1], Yuanyuan Zhu[2], and Jianliang Xu[1]

[1] Hong Kong Baptist University, Kowloon Tong, Hong Kong
{jbhuang,xinhuang,xujl}@comp.hkbu.edu.hk
[2] Wuhan University, Wuhan, China
yyzhu@whu.edu.cn

Abstract. The problem of structural diversity search is to find the top-k vertices with the largest structural diversity in a graph. However, when identifying distinct social contexts, existing structural diversity models (e.g., t-sized component, t-core, and t-brace) are sensitive to an input parameter of t. To address this drawback, we propose a parameter-free structural diversity model. Specifically, we propose a novel notation of **discriminative core**, which automatically models various kinds of social contexts without parameter t. Leveraging on **discriminative cores** and h-index, the structural diversity score for a vertex is calculated. We study the problem of parameter-free structural diversity search in this paper. An efficient top-k search algorithm with a well-designed upper bound for pruning is proposed. Extensive experiment results demonstrate the parameter sensitivity of existing t-core based model and verify the superiority of our methods.

1 Introduction

Nowadays, information spreads quickly and widely on social networks (e.g., Twitter, Facebook). Individuals are usually influenced easily by the information received from their social neighborhoods [14]. Recent studies show that social decisions made by individuals often depend on the multiplicity of social contexts inside his/her contact neighborhood, which is termed as *structural diversity* [25]. Individuals with larger structural diversity, are shown to have higher probability to be affected in the process of social contagion [25]. Structural diversity search, finding the individuals with the highest structural diversity in graphs, has many applications such as political campaigns [15], viral marketing [17], promotion of health practices [25], facebook user invitations [25], and so on.

In the literature, several structural diversity models (e.g., t-sized component, t-core and t-brace) need an input of specific parameter t to model distinct social contexts. A social context is formed by a number of connected users. The component-based structural diversity [25] regards each connected component whose size is larger than t as a social context. Another core-based structural diversity model is defined based on t-core. A t-core is the largest subgraph such that each vertex has at least t neighbors within t-core. The core-based structural

© Springer Nature Switzerland AG 2019
R. Cheng et al. (Eds.): WISE 2019, LNCS 11881, pp. 677–693, 2019.
https://doi.org/10.1007/978-3-030-34223-4_43

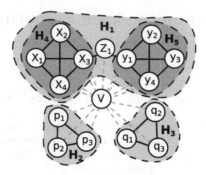

Fig. 1. The ego-network $G_{N(v)}$ of vertex v

diversity model regards each maximal connected t-core as a distinct social context. Figure 1 shows the contact neighborhood (ego-network) $G_{N(v)}$ of a user v. All vertices and edges in ego-network $G_{N(v)}$ are in solid lines. Consider the core-based structural diversity model and parameter $t = 2$. Subgraphs H_1, H_2 and H_3 are maximal connected 2-cores. H_1, H_2, and H_3 are regarded as 3 distinct social contexts. Thus, the core-based structural diversity of v is 3.

This paper proposes a new parameter-free structural diversity model based on the core-based model [12] and h-index measure [11]. Our parameter-free model does not need the input of parameter t any more. This avoids suffering from the limitations of setting parameter t. We show two major drawbacks of the t-core based model as follows.

- **Sensitivity of t-core based model.** The number of social contexts is sensitive to parameter t. On the one hand, if t is set to a large value, it may discard small and weakly-connected social contexts; On the other hand, if t is set to a small value, it may have weak ability of recognizing strongly-connected social contexts fully. Consider the contact neighborhood $G_{N(v)}$ of a user v in Fig. 1. When $t = 2$, the structural diversity of v is 3. When $t = 3$, H_2 and H_3 are 2-cores and disqualified for social contexts, due to the requirement of social contexts as 3-core. Meanwhile, H_1 is decomposed as two components of 3-core as H_4 and H_5. Thus, the structural diversity of v becomes 2. However, when $t \geq 4$, the structural diversity of v is 0. This example clearly shows the sensitivity of structural diversity w.r.t. parameter t.
- **Inflexibility of t-core based model.** Structural diversity model lacks flexibility for different vertices using the same parameter t. Generally, different social contexts should not be modeled and quantified using the same criteria of parameter t. For example, in a social network, the social contexts of a famous singer and a junior student can be dramatically different in terms of size and density. Thus, it is difficult to choose one consistent value t for different vertices in a graph. In Fig. 1, H_1 can be decomposed into two social contexts H_4 and H_5, which requires the setting of $t = 3$. However, the identification of H_2 and H_3 requires $t = 2$. This indicates the necessary of personalized parameter t for different social contexts.

To address the above two limitations, we define a novel notation of discriminative core to represent each distinct social context without inputing any parameters. Specifically, a discriminative core is a densest and maximal connected subgraph inside a user's contact neighborhood. It can be regarded as a criteria for representing unique and strong social context. However, the distribution of discriminative cores in two users' contact neighborhoods can be totally different in terms of density and quantity, which cannot be compared directly. To tackle this issue, we propose a new structural diversity model based on h-index. In the literature, the h-index is defined as the maximum number of h such that a researcher has published h papers whose citations have at least h [11]. We apply the similar idea to measure structural diversity in ego-networks. Given a vertex v, the structural diversity of v is the largest number h such that there exists at least h discriminative cores with coreness at least h. In this paper, we study the problem of top-k h-index based structural diversity search, which finds k vertices with largest h-index based structural diversity. To summarize, we make the following contributions:

- We propose a novel definition of discriminative core to provide a parameter-free scheme for identifying social contexts. To simultaneously measure the quantity and strength of social contexts in one's contact neighborhood, we propose a new h-index based structural diversity model. We formulate the problem of top-k h-index based structural diversity search in a graph (Sect. 3).
- We propose a useful approach for computing the h-index based structural diversity score $h(v)$ for a vertex v and give a baseline algorithm for solving the top-k structural diversity search problem (Sect. 4).
- Based on the analysis of the discriminative core structure and the property of h-index, we design an upper bound of $h(v)$. Equipped with the upper bound, we propose an efficient top-k search framework to improve the efficiency (Sect. 5).
- We conduct extensive experiments on four real-world large datasets to demonstrate the parameter sensitivity of the existing core-based structural diversity model and verify the effectiveness of our proposed model. Experiment results also validate the efficiency of our proposed algorithms (Sect. 6).

2   Related Work

This work is related to the studies of structural diversity search and k-core mining.

Structural Diversity Search. In [25], Ugander et al. studied the structural diversity models in the real-world applications of social contagion. The problem of top-k structural diversity search is proposed and studied by Huang et al. [12,13]. The goal of the problem is to find k vertices with the highest structural diversity scores. Two structural diversity models based on t-sized component and t-core respectively are studied w.r.t. a parameter threshold t. Recently, Chang et al. [4] proposed fast algorithms to address structural diversity search by improving the efficiency and scalability of the methods [13]. Cheng et al. [5]

propose an approach of diversity-based keyword search to solve the mashup construction problem. Different from above studies, we propose a parameter-free structural diversity model based on the novel definition of discriminative cores, which avoids suffering from the difficulties of parameter tuning.

K-Core Mining. There exist lots of studies on k-core mining in the literature. k-core is a definition of cohesive subgraph, in which each vertex has degree at least k. The task of core decomposition is finding all non-empty k-cores for all possible k's. Batagelj et al. [2] proposed an in-memory algorithm of core decomposition. Core decomposition has also been widely studied in different computing environment such as external-memory algorithms [6], streaming algorithms [23], distributed algorithms [22], and I/O efficient algorithms [26]. The study of core decomposition is also extended to different types of graphs such as dynamic graphs [1,16], uncertain graphs [3], directed graphs [19], temporal graphs [27], and multi-layer networks [9]. Recently, core maintenance in dynamic graphs has attracted significant interest in the literature [1,20,28].

3   Problem Statement

In this section, we formulate the problem of h-index based structural diversity search.

3.1   Preliminaries

We consider an undirected and unweighted simple graph $G = (V, E)$, where V is the set of vertices and E is the set of edges. We denote $n = |V|$ and $m = |E|$ as the number of vertices and edges in G respectively. W.l.o.g. we assume the input graph G is a connected graph, which implies that $m \geq n - 1$. For a given vertex v in a subgraph H of G, we define $N_H(v) = \{u$ in $H : (u, v) \in E(H)\}$ as the set of neighbors of v in H, and $d_H(v) = |N_H(v)|$ as the degree of v in H. We drop the subscript of $N_G(v)$ and $d_G(v)$ if the context is exactly G itself, i.e. $N(v)$, $d(v)$. The maximum degree of graph G is denoted by $d_{max} = \max_{v \in V} d_G(v)$.

Given a subset of vertices $S \subseteq V$, the subgraph of G induced by S is denoted by $G_S = (S, E(S))$, where the edge set $E(S) = \{(u, v) \in E : u, v \in S\}$. Based on the definition of induced subgraph, we define the ego-network [8,21] as follows.

Definition 1. *(Ego-network) Given a vertex v in graph G, the ego-network of v is the induced subgraph of G by its neighbors $N(v)$, denoted by $G_{N(v)}$.*

In the literature, the term "neighborhood induced subgraph" [12] is also used to describe the ego-network of a vertex. For example, consider the graph G in Fig. 1. The ego-network of vertex v is shown in the gray area of Fig. 1, which excludes v itself with its incident edges. The t-core of a graph G is the largest subgraph of G in which all the vertices have degree at least t. However, the t-core of a graph can be disconnected, which may not be suitable to directly depict social contexts. Hence, we define the connected t-core as follows.

Definition 2. *(Connected t-Core) Given a graph G and a positive integer t, a subgraph $H \subseteq G$ is called a connected t-Core iff H is connected and each vertex $v \in V(H)$ has degree at least t in H.*

Given a parameter t, the core-based structural diversity model treats each maximal connected t-core as a distinct social context [12,25]. To measure the structural diversity of an ego-network, one essential step is to tune a proper value for parameter t. However, such parameter setting is not easy and even critically challenging. The following example illustrates it.

Example 1. Figure 1 shows an ego-network $G_{N(v)}$ of vertex v. Given an integer $t = 2$, three maximal connected 2-core (H_1, H_2 and H_3) will be treated as distinct social contexts. The core-based structural diversity of v is 3. When we set $t = 3$, the core-based structural diversity of v will be 2, since H_4 and H_5 will be treated as two distinct social contexts. In this case, H_2 and H_3 are no longer treated as social contexts. If we set t to be some values higher than 3, no social contexts can be identified. The core-based structural diversity of v will then be 0. From this example, we can see that if the value of t is tuned too high, no social contexts can be identified. But if the value of t is set too low, some strong social contexts with denser structures cannot be captured. Thus, to choose a proper value of t for all vertices in a graph is a challenging task.

To tackle the above issue, we propose a parameter-free scheme for automatically identifying strong social contexts in one's ego-network. We firstly give a novel definition of discriminative core based on the concept of coreness as follows.

Definition 3. *(Coreness) Given a subgraph $H \subseteq G$, the coreness of H is the minimum degree of vertices in H, denoted by $\varphi(H) = \min_{v \in H}\{d_H(v)\}$. The coreness of a vertex $v \in V(G)$ is $\varphi_G(v) = \max_{H \subseteq G, v \in V(H)}\{\varphi(H)\}$.*

Definition 4. *(Discriminative Core) Given a graph G and a subgraph $H \subseteq G$, H is a discriminative core if and only if H is a maximal connected subgraph such that there exists no subgraph $H' \subseteq H$ with $\varphi(H') > \varphi(H)$.*

By Definition 4, a discriminative core H is a maximal connected component that cannot be further decomposed into smaller subgraphs with a higher coreness. It indicates that a discriminative core is the densest and most important component of a social context, which can be used as a distinct element to represent a social context. In addition, the coreness of a discriminative core reflects the strength of its representative social context. For example, H_4 is a discriminative core with $\varphi(H_4) = 3$. And H_2 is another discriminative core with $\varphi(H_2) = 2$. According to the core-based structural diversity, they cannot be identified as distinct social contexts simultaneously using the same value of parameter t. But by our discriminative core definition, they will be treated as distinct social contexts automatically without loosing the information of their strength.

For an ego-network $G_{N(v)}$, the whole network may consist of multiple discriminative cores with various corenesses, which can be depicted as a coreness distribution of discriminative cores. Moreover, to rank the structural diversity of two vertices, it is difficult to directly compare the coreness distributions of two ego-networks. Because it is not easy to measure both the number of social contexts and the strength of social contexts simultaneously.

Making use of the idea of h-index criteria, we define the diversity vector and diversity score as follows.

Definition 5. *(Diversity Vector and Diversity Score) Given a graph G and a vertex v, the diversity vector of v is the coreness distribution of discriminative cores in $G_{N(v)}$, denoted by $\mathcal{C}(v) = [c_v(1), ..., c_v(n)]$, where $c_v(r) = |\{H : \varphi(H) = r$ and H is a discriminative core in $G_{N(v)}\}|$. The h-index based structural diversity score of v, denoted by $h(v)$, is defined as $h(v) = \max\{r : \sum_r^n c_v(r) \geq r\}$. For short, diversity score is called.*

Example 2. Consider the ego-network of v shown in Fig. 1, subgraph H_1 is not a 2-core discriminative component since it can be further decomposed into two 3-cores H_4 and H_5. There is no discriminative core with the coreness of 1, so $c_v(1) = 0$. And $c_v(2) = 2$ since it has two discriminative cores H_2 and H_3 with the coreness of 2. Similarly, $c_v(3) = 2$ because H_4, H_5 are two discriminative cores with the coreness of 3. There exists no discriminative cores with coreness greater than 3. Thus, the diversity vector of v is $\mathcal{C}(v) = [0, 2, 2, 0, ..., 0]$. And the diversity score is $h(v) = 2$ by definition.

In this paper, we study the problem of h-index based structural diversity search in a graph. The problem formulation is defined as follows.

Problem Formulation. Given a graph G and an integer k, the goal of h-index based structural diversity search problem is to find an optimal answer S^* consisted of k vertices with the highest h-index based structural diversity scores, i.e.,

$$S^* = \arg\max_{S \subseteq V, |S|=k} \{\min_{v \in S} h(v)\}.$$

4   Baseline Algorithm

In this section, we introduce a baseline approach for h-index based structural diversity search over graph G. The high-level idea is to compute the diversity score for each vertex in graph G one by one. After obtaining the scores of all vertices, it sorts vertices in decreasing order of their scores and returns the first k vertices with the highest structural diversity scores. This method computes the top-k result from scratch, which is intuitive and straightforward to obtain answers.

In the following, we first introduce an existing algorithm of core decomposition [2]. Then, we present an important and useful procedure to compute h-index based structural diversity score $h(v)$ for a given vertex v.

Algorithm 1. Core Decomposition [2]

Input: a graph $G = (V, E)$
Output: the coreness $\varphi_G(v)$ for each vertex $v \in V$

1: $\mathcal{L} \leftarrow$ Sort all vertices in G in ascending order of their degree.
2: Let $t \leftarrow 1$;
3: **while** G is not empty **do**
4:      **for** each vertex $v \in \mathcal{L}$ with $d(v) < t$ **do**
5:          Remove v and its incident edges from G; Remove v from $\mathcal{L}$;
6:          $\varphi_G(v) \leftarrow t - 1$;
7:          Update the degree of the affected vertices and reorder $\mathcal{L}$;
8:      $t \leftarrow t + 1$;
9: **return** $\varphi_G(v)$ for each vertex $v \in V$;

4.1 Core Decomposition

The core decomposition of graph G computes the coreness of all vertices $v \in V$. Algorithm 1 outlines the algorithm of core decomposition [2]. The algorithm starts with an integer $t = 1$, and iteratively removes the nodes with degree less than t and their incident edges. The number of $t - 1$ is assigned to be the coreness of the removed vertices. Then, the degree of affected vertices needs to be updated, since the removal of a vertex decreases the degree of its neighbors in the remaining graph. The number t is increased by one after each iteration, until all vertices and edges are deleted from the input graph.

4.2 Computing $h(v)$

The computation of $h(v)$ includes three major steps. First, we extract from graph G and obtain an ego-network $G_{N(v)}$ for vertex v, which is the induced subgraph of G by the set of v's neighbors $N(v)$. Next, we decompose the entire ego-network $G_{N(v)}$ into several discriminative cores, and count their corenesses to derive structural diversity vector $\mathcal{C}(v)$. The detailed procedure is outlined in Algorithm 2. Finally, based on the diversity vector of $\mathcal{C}(v)$, we compute the diversity score $h(v)$ by the Definition 5 using Algorithm 3.

Discriminative Core Decomposition. Algorithm 2 outlines the detailed steps for discriminative core decomposition and diversity vector computation. For an ego-network $G_{N(v)}$ of vertex v, we firstly apply the core decomposition algorithm on it to calculate the coreness of each vertex (line 1). Then, we sort all vertices in $G_{N(v)}$ in ascending order of their coreness (line 3). For each integer t from 1 to the maximum coreness of the vertices in $G_{N(v)}$, we identify and count the number of discriminative cores with the coreness of t by using a breadth first search approach (lines 5–19). By definition, a discriminative core with the coreness of t will be only formed by the vertices with the coreness of exactly t. Thus, in each iteration, we traverse vertices with the same coreness of t to search all the discriminative cores Hs with $\varphi(H) = t$ (lines 7–19 and lines 14–15). Edges connecting the current visited vertex x to the vertices with coreness greater than t indicate that

---

Algorithm 2. Discriminative Core Decomposition

Input: an ego-network $G_{N(v)} = (N(v), \{(u, w) \in E : u, w \in N(v)\}))$
Output: the diversity vector $\mathcal{C}(v)$

1: Apply the core decomposition algorithm in Algorithm 1 on $G_{N(v)}$;
2: $t_{max} = \max_{u \in N(v)} \varphi_{G_{N(v)}}(u)$;
3: $\mathcal{L} \leftarrow$ Sort all vertices in $G_{N(v)}$ in ascending order of their coreness;
4: $Q \leftarrow \emptyset;\ visited \leftarrow \emptyset$
5: **for** $t \leftarrow 1$ to t_{max} **do**
6:     $c_v(t) \leftarrow 0$;
7:         **for** each vertex $u \in \mathcal{L}$ with the coreness of $\varphi_{G_{N(v)}}(u) = t$ **do**
8:             $Flag \leftarrow$ **true**;
9:             **if** $u \notin visited$ **then**;
10:                 $visited \leftarrow visited \cup \{u\};\ Q.push(u)$;
11:                 **while** Q is not empty **do**
12:                     $x \leftarrow Q.pop()$;
13:                     **for** each $y \in \{y : (x, y) \in E(G_{N(v)})\}$ **do**
14:                         **if** $\varphi_{G_{N(v)}}(y) = t$ **then**
15:                             Insert y to Q and $visited$ if y is unvisited;
16:                         **else if** $\varphi_{G_{N(v)}}(y) > t$ **then**
17:                             $Flag \leftarrow$ **false**;
18:             **if** $Flag =$ **true then**;
19:                 $c_v(t) \leftarrow c_v(t) + 1$;
20: **return** $\mathcal{C}(v)$;

---

the current found component can not be counted as a discriminative core and x does not belong to any discriminative cores in $G_{N(v)}$ (lines 16–17). Then the t-th element $c_v(t)$ of the diversity vector $\mathcal{C}(v)$ can be computed (lines 18–19). Finally, the diversity vector $\mathcal{C}(v)$ of v will be returned.

H-index Score Computation. The details of computing the h-index based structural diversity score are shown in Algorithm 3. After figuring out the diversity vector $\mathcal{C}(v)$ (lines 1–2), the diversity score $h(v)$ can then be calculated by Definition 5 (lines 3–6). We firstly initialize $h(v)$ as 0 (line 3). Then, for each element $c_v(t)$ in the reverse order of the diversity vector $\mathcal{C}(v)$, we keep accumulating it to $h(v)$ until the first t appears such that $h(v) \geq t$ (line 4–6). Such t is the diversity score $h(v)$ of v.

Equipped with Algorithm 3, we are able to compute the h-index based structural diversity for all the vertices in G. By sorting the diversity scores, we can obtain the top-k results for a given k.

5   Efficient Top-k Search Algorithm

The drawback of baseline method presented in the previous section is obviously inefficient and can be improved. Firstly, both the ego-network extraction and discriminative core decomposition are costly in computation. Secondly, it iteratively computes the h-index based structural diversity scores for all vertices

Algorithm 3. Compute $h(v)$

Input: a graph $G = (V, E)$; a vertex v
Output: the diversity score $h(v)$

1: Extract the ego-network $G_{N(v)}$ of v;
2: $\mathcal{C}(v) \leftarrow$ Apply the discriminative core decomposition procedure in Algorithm 2 on $G_{N(v)}$;
3: $h(v) \leftarrow 0$;
4: **for** $t \leftarrow t_{max}$ to 1 **do**
5:     $h(v) \leftarrow h(v) + c_v(t)$
6:     **if** $h(v) \geq t$ **then** $h(v) \leftarrow t$; **break**;
7: **return** $h(v)$;

---

on the entire graph G, which is expensive. Thirdly, some vertices appear to be obviously unqualified for the top-k result. And the score computations of them are reluctant and should be avoided.

In this section, we develop an efficient top-k search framework by exploiting useful pruning techniques to reduce the search space, leading to a small number of candidate vertices for score computations. Specifically, we design an upper bound $\widehat{h}(v)$ for diversity score $h(v)$, based on the analysis of the core structure.

5.1   An Upper Bound of $h(v)$

We starts with a structural property of t-core.

Lemma 1. *Given a vertex v and any vertex $u \in N(v)$, if u has $\varphi_{G_{N(v)}}(u) = r$ in ego-network $G_{N(v)}$, then u has the coreness $\varphi_G(u) \geq r + 1$ in graph G.*

Proof. We omit the proof for brevity. The detailed proof can be referred to [12]. □

Example 3. Consider vertex x_1 in Fig. 1, x_1 has coreness $\varphi_G(x_1) = 4$. However, in the ego-network $G_{N(v)}$, $\varphi_{G_{N(v)}}(x_1) = 3$. Here $\varphi_G(x_1) \geq \varphi_{G_{N(v)}}(x_1) + 1$ holds.

For a vertex v and some vertices $u \in N(v)$, the global coreness $\varphi_G(u)$ is sometimes much larger than the coreness of u in the ego-network of v, i.e. $\varphi_G(u) \gg \varphi_{G_{N(v)}}(u)$. The following lemma gives another upper bound for estimating the coreness $\varphi_{G_{N(v)}}(u)$, w.r.t. vertices v and $u \in N(v)$.

Lemma 2. *Given a vertex v and its coreness $\varphi_G(v)$, $\forall u \in N(v)$, $\varphi_{G_{N(v)}}(u) < \varphi_G(v)$.*

Proof. We prove this by contradiction. For any $u \in N(v)$, we assume $\varphi_G(v) = r$ and $\varphi_{G_{N(v)}}(u) \geq \varphi_G(v)$, which is $\varphi_{G_{N(v)}}(u) \geq r$. By the definition of coreness, there exists a subgraph $H \subseteq G_{N(v)}$ with coreness $\varphi(H) \geq r$ indicating that $\forall v^* \in V(H)$, $d_H(v^*) \geq r$. We add the vertex v and its incident edges to H to generate a new subgraph $H' \subseteq G$, where $V(H') = V(H) \cup \{v\}$ and $E(H') = E(H) \cup \{(v, u) : u \in V(H)\}$. It's easy to verify that for all v^* in H', we have $d_{H'}(v^*) \geq r + 1$. Since v is also contained in H', by definition, $\varphi_G(v) \geq r + 1$, which contradicts to the condition $\varphi_G(v) = r$. □

Combining Lemmas 1 and 2, we have the following corollary.

Corollary 1. *Given a vertex v in graph G, for any vertex $u \in N(v)$, $\widehat{\varphi}_{G_{N(v)}}(u) = \min\{\varphi_G(v), \varphi_G(u) - 1\}$ and $\widehat{\varphi}_{G_{N(v)}}(u) \geq \varphi_{G_{N(v)}}(u)$ hold.*

Based on Corollary 1, we derive an upper bound $\widehat{h}(v)$ for the h-index based structural diversity score $h(v)$ as follows.

Lemma 3. *Given a vertex v and its ego-network $G_{N(v)}$, we have an upper bound of diversity score $h(v)$, denoted by*

$$\widehat{h}(v) = \max_{x \in \mathbb{Z}_+}\{x : |\{u \in N(v) : \widehat{\varphi}_{G_{N(v)}}(u) \geq x\}| \geq x \cdot (x+1)\}.$$

Proof. Assume that $h(v) = x^*$, we prove $\widehat{h}(v) \geq x^*$. By $h(v) = x^*$, it indicates that there exists x^* discriminative cores g with $\varphi(g) \geq x^*$ in the ego-network $G_{N(v)}$. For $\varphi(g) \geq x^*$, discriminative core g has at least $x^* + 1$ nodes u with $\varphi_{G_{N(v)}}(u) \geq x^*$. Thus, the whole ego-network $G_{N(v)}$ has at least $x^* \cdot (x^* + 1)$ nodes u with $\varphi_{G_{N(v)}}(u) \geq x^*$, i.e., $h(v) = x^* \leq \max_{x \in \mathbb{Z}_+}\{x : |\{u \in N(v) : \varphi_{G_{N(v)}}(u) \geq x\}| \geq x \cdot (x+1)\}$. By Corollary 1, $\widehat{\varphi}_{G_{N(v)}}(u) \geq \varphi_{G_{N(v)}}(u)$, hence we have $\widehat{h}(v) \geq x^* = h(v)$.

According to Lemma 3, once applying the core decomposition algorithm on graph G, we can directly compute the upper bounds $\widehat{h}(v)$ for all vertices v.

Algorithm 4. Efficient Top-k Search Framework

Input: $G = (V, E)$, an integer k
Output: top-k structural diversity results

1: Apply the core decomposition on G by Algorithm 1 and obtain $\varphi_G(v)$ for all vertices $v \in V$;
2: **for** $v \in V$ **do**
3:     Compute $\widehat{h}(v)$ according to Lemma 3;
4: $\mathcal{L} \leftarrow$ Sort all vertices V in descending order of $\widehat{h}(v)$;
5: $\mathcal{S} \leftarrow \emptyset$;
6: **while** $\mathcal{L} \neq \emptyset$ **do**
7:     $v^* \leftarrow \arg\max_{v \in \mathcal{L}} \widehat{h}(v)$; Delete v^* from $\mathcal{L}$;
8:     **if** $|\mathcal{S}| = r$ and $\widehat{h}(v^*) \leq \min_{v \in \mathcal{S}} h(v)$ **then**
9:         **break**;
10:     Invoke Algorithm 3 to compute $h(v^*)$;
11:     **if** $|\mathcal{S}| < r$ **then** $\mathcal{S} \leftarrow \mathcal{S} \cup \{v^*\}$;
12:     **else if** $h(v^*) > \min_{v \in \mathcal{S}} h(v)$ **then**
13:         $u \leftarrow \arg\min_{v \in \mathcal{S}} h(v)$;
14:         $\mathcal{S} \leftarrow (\mathcal{S} - \{u\}) \cup \{v^*\}$;
15: **return** $\mathcal{S}$;

5.2   Top-K Structural Diversity Search Framework

Equipped with the upper bound $\widehat{h}(v)$, we develop an efficient top-k search framework for safely pruning the search space and avoiding the unnecessary computation of $h(v)$. The efficient top-k structural diversity search framework is presented in Algorithm 4.

Algorithm 4 starts with the initialization of the upper bound of each vertex v (lines 1–2). Then, it sorts all vertices in descending order according to their upper bounds (line 3). It maintains a list S to store the top-k result (line 4). In each iteration, the algorithm pops out a vertex v^* from the vertex list $\mathcal{L}$ with the largest upper bound $\widehat{h}(v^*)$ (line 6). Next, it checks the early stop condition: if the answer set S has k results and the minimum score in S is no less than the current upper bound, i.e. $\widehat{h}(v^*) \leq min_{v \in S} h(v)$, the current vertex v^* is safely pruned and the searching process is terminated (lines 8–9). Otherwise, the procedure of structural diversity score computation is invoked and check if v^* can be added into the result set (lines 10–14). Finally, the top-k results stored in S are returned.

5.3   Complexity Analysis

In this section, we analyze the time and space complexity of Algorithm 4.

Lemma 4. *Algorithm 3 computes $h(v)$ for each vertex v in $O(\sum_{u \in N(v)} min\{d(u), d(v)\})$ time and $O(m)$ space.*

Proof. Extracting $G_{N(v)}$ of v takes $O(\sum_{u \in N(v)} min\{d(u), d(v)\})$, since all triangles $\triangle_{vuw}$ should be listed to enumerate each edge $(u, w) \in E(G_{N(v)})$. According to [2], the core decomposition performed in $G_{N(v)}$ takes $O(|E(G_{N(v)})| + d(v))$ time. The sorting of the vertices can be finished in $O(d(v))$ time using bin sort. And the breadth first search process for identifying the discriminative cores needs $O(|E(G_{N(v)})|)$ time. In addition, the computing of the h-index based structural diversity score $h(v)$ runs in $O(\delta(G_{N(v)}))$ time, where $\delta(G_{N(v)}) = max_{u \in N(v)} \varphi_{G_{N(v)}}(u)$ is the degeneracy of $G_{N(v)}$. And $\delta(G_{N(v)})$ is bounded by the degree of v, which is $O(\delta(G_{N(v)})) \subseteq d(v)$. Overall, the time complexity of Algorithm 3 is $O(\sum_{u \in N(v)} min\{d(u), d(v)\})$.

We continue to analyze the space complexity of Algorithm 3. The storage of the ego-network of v takes $O(n + m)$ space since $G_{N(v)} \subseteq G$. And both the sorted list of vertices (line 4) and the structural diversity vector of v takes $O(n)$ space. Thus, the space complexity of Algorithm 3 is $O(n + m) \subseteq O(m)$ due to our graph connectivity assumption.

Theorem 1. *Algorithm 4 computes the top-k results in $O(\rho m)$ time and $O(m)$ space, where ρ is the arboricity of G and $\rho \leq min\{d_{max}, \sqrt{m}\}$ [7].*

Proof. Firstly, the core decomposition algorithm performed on G takes $O(m)$ time and $O(n+m)$ space. Secondly, the computation of upper bound $\widehat{h}(v)$ for all

v's takes $O(m)$ time and $O(n)space$. In the worst case, Algorithm 4 needs to compute $h(v)$ for every vertex v. This takes $O(\sum_{v \in V}\{\sum_{u \in N(v)} min\{d(u), d(v)\}\})$ time in total by Lemma 4. According to [7], we have

$$O(\sum_{v \in V}\{\sum_{u \in N(v)} min\{d(u), d(v)\}\}) \subseteq O(\sum_{(u,v) \in E} min\{d(u), d(v)\}) \subseteq O(\rho m).$$

Here ρ is the arboricity of graph G, which is defined as the minimum number of disjoint spanning forests that cover all the edges in G. In addition, the top-k results can be maintained in a list in $O(n)$ time and $O(n)$ space using bin sort. Overall, Algorithm 4 runs in $O(\rho m)$ time and $O(m)$ space.

6   Experiments

We conduct extensive experiments on real-world datasets to evaluate the effectiveness and efficiency of our proposed h-index based structural diversity model and algorithms.

Datasets: We run our experiments on four real-world datasets downloaded on the SNAP website [18]. All datasets are treated as undirected graphs. The statistics of the networks are listed in Table 1. We report the node size $|V|$, edge size $|E|$ and the maximum degree d_{max} of each network.

Table 1. Network statistics

Name	$|V|$	$|E|$	d_{max}
Gowalla	196,591	950,327	14,730
Youtube	1,134,890	2,987,624	28,754
LiveJournal	3,997,962	34,681,189	14,815
Orkut	3,072,441	117,185,083	33,313

Compared Methods: We evaluate all compared methods in terms of efficiency, effectiveness and also sensitivity to parameter setting. Specifically, we show three compared algorithms as follows.

- baseline: is the baseline method proposed in Sect. 4.
- h-core: is an improved top-k search algorithm for computing the top-k vertices with highest h-index based structural diversity in Algorithm 4.
- t-core: is to compute the top-k vertices with highest t-core based structural diversity [12]. Here, t is a parameter of coreness threshold.

Note that in the sensitivity evaluation, we test the state-of-the-art competitor t-core and compare the top-k results for different parameter t. Our h-index based structural diversity model has no input parameter, which is consistent on the top-k results.

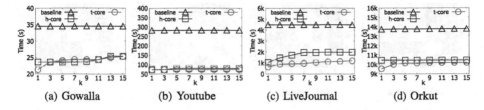

(a) Gowalla   (b) Youtube   (c) LiveJournal   (d) Orkut

Fig. 2. Comparsion of baseline, h-core and t-core in terms of running time (in seconds).

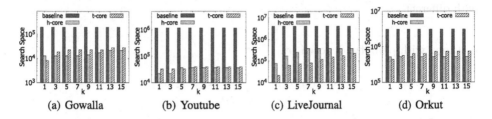

(a) Gowalla   (b) Youtube   (c) LiveJournal   (d) Orkut

Fig. 3. Comparsion of baseline, h-core and t-core in terms of search space.

6.1 Efficiency Evaluation

In this experiment, we compare the efficiency of baseline, h-core and t-core on four real-world datasets. For the t-core method, we fix parameter $t = 2$. We compare the running time and search space (i.e., the number of vertices whose structural diversity scores are computed in the search process). Figure 2 shows the running time results of three methods varied by k. It clearly shows that top-k search algorithm h-core runs much faster than baseline on all the reported datasets. Specifically, in Fig. 2(c), h-core is 5 times faster than baseline on Youtube in term of running time. Moreover, Fig. 3 further shows the search space of three methods varied on all datasets. We can observe that leveraging on the upper bound $\hat{h}(v)$, a large number of disqualified vertices is pruned during the search process by h-core. The search space significantly shrinks into less than $\frac{1}{10}$ of vertex size in graphs. It verifies the tightness of our upper bound and the superiority of h-core against baseline in efficiency. According to Figs. 2 and 3, our h-core is very comparative to the state-of-art method t-core in terms of running time and search space.

6.2 Sensitivity Evaluation

This experiment evaluates the sensitivity of t-core model. Given two different values of t, t-core model may generate two different lists of top-k ranking results. We use the Kendall rank tau distance to counts the number of pairwise disagreements between two top-k lists. The larger the distance, the more dissimilar the two lists, and also more sensitive the t-core model. We adopt the Kendall

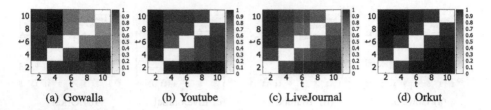

(a) Gowalla	(b) Youtube	(c) LiveJournal	(d) Orkut

Fig. 4. Sensitivity heat matrices of t-core model on all datasets. Each matrix element represents the Kendall's Tau distance between two top-100 ranking lists by t-core model with different t.

distance with penalty, denoted by,

$$K^{(p)}(\tau_1, \tau_2) = \sum_{\{i,j\} \in \mathcal{P}} \overline{K}_{i,j}^{(p)}(\tau_1, \tau_2)$$

where $\mathcal{P}$ is the set of all unordered pairs of distinct elements in two top-k list τ_1, τ_2 and p is the penalty parameter. In our setting, we set $p = 1$ and normalize the Kendall distance by the number of permutation $|\mathcal{P}|$. The values of normalized Kendall distance range from 0 to 1.

We test the sensitivity of t-core model by varying parameter t in $\{2, 4, 6, 8, 10\}$. We compute the Kendall distance of two top-100 lists by t-core model with two different t. The results of sensitivity heat matrix on four datasets are shown in Fig. 4. The darker colors reveal larger Kendall distances between two top-k lists and also more sensitive of t-core models on this pair of parameters t. Overall, sensitivity heat matrices are depicted in dark for most parameter settings on all datasets. This reflects that the top-k results computed by t-core are very sensitive to the setting of parameter t, which has a bad robustness. It strongly indicates the necessity and importance of our parameter-free structural diversity model.

6.3   Effectiveness Evaluation

In this experiment, we evaluate the effectiveness of our proposed h-index based structural diversity. We compare our method h-core with state-of-the-art t-core [12] in the task of social contagion. Specifically, we adopt the independent cascade model to simulate the influence propagation process in graphs [10]. Influential probability of each edge is set to 0.01. Then, we select 50 vertices as activated seeds by an influence maximization algorithm [24]. We perform 1000 times of Monte Carlos sampling for propagation. For comparison, we count the number of activated vertices in the top-k results by t-core and h-core methods. The method that achieves the largest number of activated vertices is regarded as the winner.

First, we report the average activated rate by h-core and t-core method on all four datasets in Fig. 5(a). Let $D = \{$ "Gowalla", "Youtube", "LiveJournal", "Orkut"$\}$. Given a dataset $d \in D$, the activated rate is defined as $f_k(d) =$

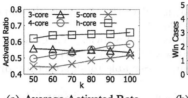
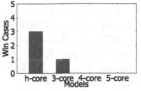

(a) Average Activated Rate   (b) Win Cases Distribution

Fig. 5. Comparison of t-core and h-core in terms of the average activated ratio and win cases on four datasets.

$\frac{ActNum_k}{k}$, where $ActNum_k$ is the number of activated vertices in the top-k result. The average activated rate is defined as $ActRate_k = \frac{\sum_{d \in D} f_k(d)}{|D|}$. Figure 5(a) shows that our method h-core achieves the highest activated rates, which significantly outperforms t-core method for all different t. It indicates that the top-k results found by h-core tend to have higher probability to be affected in social contagion.

In addition, we also report the win cases of h-core and t-core with different parameter t on all dataset. We vary $t = \{2, 3, 4\}$ and set $k = 100$ for all methods. The winner of a dataset is the method that achieves the highest number of activated vertices in this dataset. Figure 5(b) shows the win cases of t-core and h-core. As we can see, h-core wins on three datasets, which achieves the best performance. It further shows the superiority of our h-index structural diversity model. Besides, 3-core wins once, 4-core and 5-core win none, indicating that t-core performs sensitively to parameter t.

7 Conclusion

In this paper, we propose a parameter-free structural diversity model based on h-index and study the top-k structural diversity search problem. To solve the top-k structural diversity search problem, an upper bound for the diversity score and a top-k search framework for efficiently reducing the search space are proposed. Extensive experiments on real-wold datasets verify the efficiency of our pruning techniques and the effectiveness of our proposed h-index based structural diversity model.

Acknowledgments. This work is supported by the NSFC Nos. 61702435, 61972291, RGC Nos. 12200917, 12200817, CRF C6030-18GF, and the National Science Foundation of Hubei Province No. 2018CFB519.

References

1. Aridhi, S., Brugnara, M., Montresor, A., Velegrakis, Y.: Distributed k-core decomposition and maintenance in large dynamic graphs. In: DEBS, pp. 161–168. ACM (2016)

2. Batagelj, V., Zaversnik, M.: An o (m) algorithm for cores decomposition of networks. arXiv preprint cs/0310049 (2003)
3. Bonchi, F., Gullo, F., Kaltenbrunner, A., Volkovich, Y.: Core decomposition of uncertain graphs. In: KDD, pp. 1316–1325. ACM (2014)
4. Chang, L., Zhang, C., Lin, X., Qin, L.: Scalable top-k structural diversity search. In: ICDE, pp. 95–98 (2017)
5. Cheng, H., Zhong, M., Wang, J., Qian, T.: Keyword search based mashup construction with guaranteed diversity. In: Hartmann, S., Küng, J., Chakravarthy, S., Anderst-Kotsis, G., Tjoa, A.M., Khalil, I. (eds.) DEXA 2019. LNCS, vol. 11707, pp. 423–433. Springer, Cham (2019). https://doi.org/10.1007/978-3-030-27618-8_31
6. Cheng, J., Ke, Y., Chu, S., Özsu, M.T.: Efficient core decomposition in massive networks. In: ICDE, pp. 51–62 (2011)
7. Chiba, N., Nishizeki, T.: Arboricity and subgraph listing algorithms. SIAM J. Comput. **14**(1), 210–223 (1985)
8. Ding, F., Zhuang, Y.: Ego-network probabilistic graphical model for discovering on-line communities. Appl. Intell. **48**(9), 3038–3052 (2018)
9. Galimberti, E., Bonchi, F., Gullo, F.: Core decomposition and densest subgraph in multilayer networks. In: CIKM, pp. 1807–1816. ACM (2017)
10. Goyal, A., Lu, W., Lakshmanan, L.V.S.: CELF++: optimizing the greedy algorithm for influence maximization in social networks. In: WWW, pp. 47–48 (2011)
11. Hirsch, J.E.: An index to quantify an individual's scientific research output. Proc. Nat. Acad. Sci. **102**(46), 16569–16572 (2005)
12. Huang, X., Cheng, H., Li, R., Qin, L., Yu, J.X.: Top-k structural diversity search in large networks. VLDB J. **24**(3), 319–343 (2015)
13. Huang, X., Cheng, H., Li, R.-H., Qin, L., Yu, J.X.: Top-k structural diversity search in large networks. PVLDB **6**(13), 1618–1629 (2013)
14. Huang, X., Lakshmanan, L.V., Xu, J.: Community Search over Big Graphs. Morgan & Claypool Publishers, San Rafael (2019)
15. Huckfeldt, R.R., Sprague, J.: Citizens, Politics and Social Communication: Information and Influence in an Election Campaign. Cambridge University Press, Cambridge (1995)
16. Jakma, P., Orczyk, M., Perkins, C.S., Fayed, M.: Distributed k-core decomposition of dynamic graphs. In: StudentWorkshop@CoNEXT, pp. 39–40. ACM (2012)
17. Kempe, D., Kleinberg, J.M., Tardos, É.: Maximizing the spread of influence through a social network. In: KDD, pp. 137–146 (2003)
18. Leskovec, J., Krevl, A.: SNAP Datasets: Stanford large network dataset collection, June 2014. http://snap.stanford.edu/data
19. Levorato, V.: Core decomposition in directed networks: kernelization and strong connectivity. In: Contucci, P., Menezes, R., Omicini, A., Poncela-Casasnovas, J. (eds.) Complex Networks V. SCI, vol. 549, pp. 129–140. Springer, Cham (2014). https://doi.org/10.1007/978-3-319-05401-8_13
20. Li, R., Yu, J.X., Mao, R.: Efficient core maintenance in large dynamic graphs. In: TKDE, vol. 26, no. 10, pp. 2453–2465 (2014)
21. Mcauley, J., Leskovec, J.: Discovering social circles in ego networks. TKDD **8**(1), 4 (2014)
22. Montresor, A., Pellegrini, F.D., Miorandi, D.: Distributed k-core decomposition. TPDS **24**(2), 288–300 (2013)
23. Sarıyüce, A.E., Gedik, B., Jacques-Silva, G., Wu, K.-L., Çatalyürek, Ü.V.: Streaming algorithms for k-core decomposition. PVLDB **6**(6), 433–444 (2013)
24. Tang, Y., Shi, Y., Xiao, X.: Influence maximization in near-linear time: a martingale approach. In: SIGMOD Conference, pp. 1539–1554. ACM (2015)

25. Ugander, J., Backstrom, L., Marlow, C., Kleinberg, J.: Structural diversity in social contagion. PNAS **109**(16), 5962–5966 (2012)
26. Wen, D., Qin, L., Zhang, Y., Lin, X., Yu, J.X.: I/O efficient core graph decomposition: application to degeneracy ordering. TKDE **31**(1), 75–90 (2019)
27. Wu, H., et al.: Core decomposition in large temporal graphs. In: BigData, pp. 649–658 (2015)
28. Zhang, Y., Yu, J.X., Zhang, Y., Qin, L.: A fast order-based approach for core maintenance. In: ICDE, pp. 337–348 (2017)

CoreCube: Core Decomposition in Multilayer Graphs

Boge Liu[1], Fan Zhang[2]([✉]), Chen Zhang[1], Wenjie Zhang[1], and Xuemin Lin[1]

[1] University of New South Wales, Sydney, Australia
boge.liu@unsw.edu.au, {Chenz,zhangw,lxue}@cse.unsw.edu.au
[2] Guangzhou University, Guangzhou, China
fanzhang.cs@gmail.com

Abstract. Many real-life complex networks are modelled as multilayer graphs where each layer records a certain kind of interaction among entities. Despite the powerful modelling functionality, the decomposition on multilayer graphs remains unclear and inefficient. As a well-studied graph decomposition, core decomposition is efficient on a single layer graph with a variety of applications on social networks, biology, finance and so on. Nevertheless, core decomposition on multilayer graphs is much more challenging due to the various combinations of layers. In this paper, we propose efficient algorithms to compute the CoreCube which records the core decomposition on every combination of layers. We also devise a hybrid storage method that achieves a superior trade-off between the size of CoreCube and the query time. Extensive experiments on 8 real-life datasets demonstrate our algorithms are effective and efficient.

Keywords: Core decomposition · Multilayer graph · Graph processing

1 Introduction

In real-life networks, there are usually multiple types of interactions (edges) among entities (vertices), e.g., the relationship between two users in a social network can be friends, colleagues, relatives and so on. The entities and interactions are usually modelled as a multilayer graph, where each layer records a certain type of interaction among entities [9]. Because of the strong modeling paradigm to handle various interactions among a set of entities, there are significant existing studies of multilayer graphs, e.g., [6,14]. Previous works usually focus on mining dense structures from multilayer graphs according to given parameters, e.g., [29]. Nevertheless, graph decomposition, as a fundamental graph problem [22], remains largely unexplored on multilayer graphs.

Core decomposition (or k-core decomposition), as one of the most well-studied graph decomposition, is to compute the core number for every vertex in the graph [20]. It is a powerful tool in modeling the dynamic of user engagement in social networks. In practice, a user u tends to adopt a new behavior if there are a considerable number of friends (e.g., the core number of u) in the group who also adopted the same behavior [18]. Core decomposition is also

© Springer Nature Switzerland AG 2019
R. Cheng et al. (Eds.): WISE 2019, LNCS 11881, pp. 694–710, 2019.
https://doi.org/10.1007/978-3-030-34223-4_44

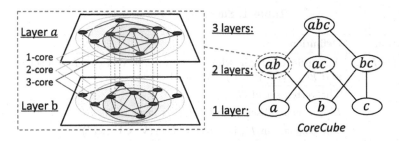

Fig. 1. Multilayer core decomposition and CoreCube of a graph

theoretically supported by Nash equilibrium in game theory [5]. It has a variety of applications, e.g., graph visualization [3], internet topology [7] and user engagement [24,26]. Extending the single-layer core decomposition to multilayer graphs is a critical task which can benefit a lot of applications considering the various real-world interactions between entities.

Given a multilayer graph, the multilayer k-core on a set of layers is defined as a set of vertices whose minimum degree in the induced subgraph of each layer is at least k. The core number of a vertex on a set of layers is the largest k such that the multilayer k-core on these layers contains the vertex. Multilayer core decomposition on a set of layers is to compute the core number for each vertex on these layers. In this paper, we propose CoreCube which records the core numbers of each vertex for every combination of layers in a multilayer graph. In the following, we show the details for some application examples.

User Engagement Evaluation. In social networks, users may participate in multiple groups with different themes, where each group forms a layer in the multilayer graph. For instance, the authors in a coauthor network have different coauthor relationship on different venues (conferences or journals). For any given user-interested combination of venues (correspond to layers), CoreCube of the coauthor network can immediately answer the engagement level for each author, i,e, the core numbers [18]. Given a degree constraint k, we can also immediately retrieve a cohesive user group from CoreCube, i.e., the multilayer k-core.

Biological Module Analysis. In biological networks, different interactions between the modules are detected with different methods due to data noise and technical limitations [11]. Analyzing module structure according to single method, i.e., on a single layer, may not be accurate. CoreCube allows us to study the connections between modules for any combination of potential methods. Thus, we can find co-expression clusters and verify the effectiveness of detection methods.

Figure 1 shows an example of CoreCube on a graph G with three layers and depicts the multilayer core decomposition on layer a and b. The 3-core on layer a and b contains 5 vertices where each vertex has a degree of at least 3 in each layer. There are 7 different combinations of layers in CoreCube of G. For each combination, we compute its multilayer core decomposition and record the core numbers in CoreCube. CoreCube can immediately answer a query for core numbers on any set of layers including the traditional single layer graph.

Table 1. Summary of notations

Notation	Definition
$G = (V, E, L)$	A multilayer graph, where V is a set of vertices, L is a set of layers, and $E \subseteq (V \times V \times L)$ is a set of edges
$V(G)$	The vertex set of G
$L'; l$	$L' \subseteq L$ is a subset of L; $l \in L$ is a layer in L
$E_{L'}$	The edge set in L', i.e., $E_{L'} = E \cap (V \times V \times L')$
u, v	A vertex in the graph
$|V|, |E|, |L|$	The number of vertices, edges, and layers in G, respectively
$N_G(v, l)$	The set of adjacent vertices of v in layer l of G
$deg_G(v, l)$	The number of adjacent vertices of v in layer l of G
d_{max}	The maximum degree, i.e., $d_{max} = \max\{deg_G(v, l) \mid v \in V \wedge l \in L\}$
$C_{L'}^k$	The multilayer k-core on a set of layers L'
$\mathcal{C}_{L'}(v)$	The core number of v on a set of layers L'
$\mathcal{C}_{L'}$	The multilayer core decomposition result on a set of layers L'
$\mathcal{C}$	The CoreCube of G, i.e., $\mathcal{C} = \{\mathcal{C}_{L'} \mid L' \subseteq L\}$

Challenges and Contributions. Although core decomposition on a single-layer graph can be computed in linear time, it becomes very challenging on a multilayer graph because the combination number of layers is exponential to the number of layers. In the general case, no polynomial-time algorithm may exist for computing the CoreCube. To the best of our knowledge, there is only one similar work [10] where the algorithms can be adapted to compute the CoreCube while it is hard to share the computation among different combination of layers. The algorithms proposed in this paper can largely speed up the computation of CoreCube. We summarize our contributions as follows:

- We propose efficient algorithms to compute the CoreCube. Several theorems reveal the inner characteristics of multilayer core decomposition.   (Sect. 3)
- We devise a hybrid storage method which has a superior trade-off between query processing time and storage size.   (Sect. 4)
- Extensive experiments demonstrate that our CoreCube computation and query processing are faster than baselines by more than one order of magnitude.   (Sect. 5)

2   Problem Definition

In this section, we give some notations and formally define CoreCube. The notations are summarized in Table 1.

We consider an unweighted and undirected multilayer graph $G = (V, E, L)$, where V represents the set of vertices in G, L represents the set of layers, and $E \subseteq (V \times V \times L)$ represents the set of edges. We use $|V|$, $|E|$, and $|L|$ to denote the number of vertices, edges, and layers, respectively. $N_G(v, l)$ is the set of adjacent vertices of v in layer l. We say a vertex u is incident to an edge, or an edge is incident to u, if u is one of the endpoints of the edge. We use $deg_G(v, l)$ to denote the number of adjacent vertices of u in layer l. When the context is clear, we omit the input graph in notations, such as $deg(v, l)$ for $deg_G(v, l)$.

Definition 1. _Multilayer k-core._ _Given a multilayer graph $G = (V, E, L)$, a set of layers $L' \subseteq L$ and an integer k, the multilayer k-core of G on L', denoted by $C_{L'}^k$, is the maximum vertex set such that every vertex v in the subgraph H induced by $C_{L'}^k$ satisfies $deg_H(v, l) \geq k$ on each $l \in L'$._

Let k_{max} be the maximum possible k such that a multilayer k-core of G on L' exists. The multilayer k-core for all $1 \leq k < k_{max}$ has the following partial containment property:

Property 1. Given a multilayer graph $G = (V, E, L)$ and a set of layers L', $C_{L'}^{k+1} \subseteq C_{L'}^k$ for all $1 \leq k < k_{max}$.

Next, we define the core number for each $v \in V$.

Definition 2. _Core Number._ _Given a multilayer graph $G = (V, E, L)$ and a set of layers $L' \subseteq L$, the core number of v on L', denoted by $C_{L'}(v)$, is the largest k such that v is contained in multilayer k-core on L', i.e., $C_{L'}(v) = \max\{k \mid v \in C_{L'}^k\}$._

Based on Property 1 and Definition 2, we can easily derive following lemma:

Lemma 1. _Given a multilayer graph $G = (V, E, L)$, a set of layers L', and an integer k, we have $C_{L'}^k = \{v \in V \mid C_{L'}(v) \geq k\}$._

Definition 3. _Multilayer Core Decomposition._ _Given a multilayer graph $G = (V, E, L)$ and a set of layers $L' \subseteq L$, the multilayer core decomposition, denoted by $C_{L'}$, computes $C_{L'}^k$ for all $1 \leq k \leq k_{max}$._

According to Lemma 1, multilayer core decomposition on L' is equivalent to computing the core number $C_{L'}(v)$ for each $v \in V$. Finally, we give the formal definition of CoreCube and the problem we tackle in this paper.

Definition 4. _CoreCube._ _Given a multilayer graph $G = (V, E, L)$, the CoreCube of G, denoted as C, computes multilayer core decomposition on all the subsets of L, i.e., $C = \{C_{L'} \mid L' \subseteq L\}$._

Problem Statement. In this paper, we study the problem of efficiently computing and compactly storing CoreCube of multilayer graphs.

3   CoreCube Computation

In this section, we present our basic CoreCube computation algorithm and then discuss how to improve the algorithm by sharing computation among multilayer core decomposition on different sets of layers.

3.1 Basic CoreCube Algorithm

Based on Property 1, given a multilayer graph $G = (V, E, L)$ and a set of layers $L' \subseteq L$, the multilayer core decomposition on L' can be computed in a bottom up manner following the paradigm used for single layer graphs [4], which increases k step by step and iteratively removing vertices whose degree are less than k. We give this algorithm Core-BU in Algorithm 1. Core-BU computes multilayer core decomposition in increasing order of k. Each time, k is selected as the minimum degree (line 3). Whenever there exists a vertex v whose degree is no larger than k in some layer $l \in L'$ (line 4), we know that the core number of v is k (line 5) and we remove v with all its incident edges from the graph (line 6). The core numbers are returned in line 7. With the help of bin sort and the efficient data structure proposed in [13] to maintain the minimum degree, Core-BU can achieve a time complexity of $O(|E_{L'}| + |V|)$.

The algorithm CoreCube-BU which computes CoreCube with Core-BU is shown in Algorithm 2. In Algorithm 2, CoreCube is computed level-by-level. Each time, we generate all the subsets of L with the same size z (line 3) and compute multilayer core decomposition on each subset (line 4–5). CoreCube is returned in line 6.

Algorithm 1: Core-BU(G, L')

> **Input**   : $G = (V, E, L)$: a multilayer graph, L' : a subset of L
> **Output** : $\mathcal{C}_{L'}$: the multilayer core decomposition on L'
> 1 $G' \leftarrow G_{L'}$;
> 2 **while** $G' \neq \emptyset$ **do**
> 3 $\quad$ $k \leftarrow \min\{deg_{G'}(v, l) \mid v \in V(G') \wedge l \in L'\}$;
> 4 $\quad$ **while** $\exists v \in V(G')$ and $l \in L' : deg_{G'}(v, l) \leq k$ **do**
> 5 $\quad\quad$ $\mathcal{C}_{L'}(v) \leftarrow k$;
> 6 $\quad\quad$ remove v and its incident edges from G';
>
> 7 **return** $\mathcal{C}_{L'}$

Algorithm 2: CoreCube-BU(G)

> **Input**   : G : a multilayer graph
> **Output** : $\mathcal{C}$: the CoreCube of G
> 1 $\mathcal{C} \leftarrow \emptyset$;
> 2 **for** $z = 1$ to $|L|$ **do**
> 3 $\quad$ $Z \leftarrow \{$all the subsets of L whose size are $z\}$;
> 4 $\quad$ **for each** $L' \in Z$ **do**
> 5 $\quad\quad$ $\mathcal{C} \leftarrow \mathcal{C} \cup \{$Core-BU($G, L'$)$\}$;
>
> 6 **return** $\mathcal{C}$

Complexity. Since there are $2^{|L|} - 1$ (expect $\emptyset$) subsets of L need to be processed and Core-BU runs in $O(|E_{L'}| + |V|)$ for any subset L', the complexity of CoreCube-BU is $O(2^{|L|} \cdot (|E| + |V|))$.

3.2   Computation-Sharing CoreCube Algorithm

Core-BU needs to remove all the edges in $E_{L'}$ when computing multilayer core decomposition on L'. This is because the core number of a vertex v is obtained only when v is removed. Therefore, CoreCube-BU computes each multilayer core decomposition independently. To improve the efficiency of CoreCube computation, we aim at devising an algorithm that shares computation among multilayer core decomposition on different sets of layers. We first extend the locality property of k-core in single layer graphs [19] to multilayer graphs.

Theorem 1. *Given a multilayer graph $G = (V, E, L)$ and a set of layers $L' \subseteq L$, we have the following recursive equations for core number $\mathcal{C}_{L'}(v)$ of a vertex $v \in V$:*

$$\forall l \in L' \quad M_l(v) = \max \; k \; s.t. \; |\{u \in N(v, l) \mid \mathcal{C}_{L'}(u) \geq k\}| \geq k \tag{1}$$

$$\mathcal{C}_{L'}(v) = \min\{M_l(v) \mid l \in L'\} \tag{2}$$

where $N(v, l)$ is the set of adjacent vertices of v in layer l.

Proof. (i) Let $k_c = \min\{M_l(v) \mid l \in L'\}$ and S be the multilayer k_c-core on L'. Firstly, S must be nonempty as there exists some vertex u satisfying $\mathcal{C}_{L'}(u) \geq k_c$. According to Eqs. 1 and 2, we have $\forall l \in L'$, $|\{u \in N(v, l) \mid \mathcal{C}_{L'}(u) \geq k_c\}| \geq k_c$. Therefore, in each layer $l \in L'$, v has at least k_c adjacent vertices in S, which means $v \in S$. Hence, $\mathcal{C}_{L'}(v) \geq k_c$. (ii) On the other hand, according to Eqs. 1 and 2, there must exist some $l_0 \in L'$ in which $|\{u \in N(v, l_0) | \mathcal{C}_{L'}(u) \geq k_c+1\}| < k_c+1$. Therefore, $\mathcal{C}_{L'}(v) < k_c + 1$. Combining the conclusion in (i) and (ii) together, it holds that $\mathcal{C}_{L'}(v) = \min\{M_l(v) \mid l \in L'\}$.

Following Theorem 1, we devise the algorithm Core-TD which computes multilayer core decomposition on L' in a top down manner. Core-TD iteratively reduces the upper bound of core number for each vertex. Initially, each vertex v is assigned an arbitrary upper bound of core number (e.g. the minimum degree of v in L'). Then Core-TD keeps updating the upper bound using Eqs. 1 and 2 until convergence. The pseudocode of Core-TD is given in Algorithm 3. Here, we use $\overline{\mathcal{C}}_{L'}(v)$ to denote the upper bound of $\mathcal{C}_{L'}(v)$. We also use $sup(v, l)$ (support of v) to denote the number of adjacent vertices of v in layer l whose upper bound is no less than $\overline{\mathcal{C}}_{L'}(v)$. That is

$$sup(v, l) = |\{u \in N(v, l) \mid \overline{\mathcal{C}}_{L'}(u) \geq \overline{\mathcal{C}}_{L'}(v)\}| \tag{3}$$

Note that if $sup(v, l) < \overline{\mathcal{C}}_{L'}(v)$, Eq. 1 does not hold for v in layer l. Therefore, we can determine whether $\overline{\mathcal{C}}_{L'}(v)$ needs to be updated by comparing $\overline{\mathcal{C}}_{L'}(v)$ with $sup(v, l)$ for each $l \in L'$ instead of scanning all the adjacent vertices of v.

Core-TD first initializes $sup(v,l)$ for every vertex based on Eq. 3 in line 1. Then it updates vertex v whose upper bound violates Eq. 1 in some layer r (line 2). c_0 records the value of $\overline{C}_{L'}(v)$ before being updated (line 3). Core-TD updates $\overline{C}_{L'}(v)$ according to Eqs. 1 and 2 (line 4–7). Then, for each layer $l \in L'$, it recomputes $sup(v,l)$ and updates $sup(u,l)$ for each adjacent vertex u of v (line 8–12). $sup(u,l)$ is decreased by 1 if v once contributed to $sup(u,l)$ but not anymore after $\overline{C}_{L'}(v)$ being updated (line 11–12). Finally, after all the upper bound converges, Core-TD sets $C_{L'}(v)$ as $\overline{C}_{L'}(v)$ for each vertex $v \in V$ in line 13 and returns $C_{L'}$ in line 14.

Complexity. In Core-TD, each time when the upper bound of some vertex v is updated, line 2–12 takes $O(\sum_{l \in L'}(deg(v,l)))$. Since $\overline{C}_{L'}(v)$ is at least decreased by 1 whenever being updated, the time complexity of Core-TD is $O(\sum_{v \in V}(\overline{C}_{L'}(v) \cdot \sum_{l \in L'} deg(v,l)))$, which is bounded by $O(d_{max} \cdot |E_{L'}|)$ as the maximum degree d_{max} can always serve as an upper bound for any vertex.

Algorithm 3: Core-TD(G, L', $\overline{C}_{L'}$)

Input : $G = (V,E,L)$: a multilayer graph, L' : a subset of L, $\overline{C}_{L'}$: upper
bound of core number on L' for each vertex in V

Output : $C_{L'}$: the multilayer core decomposition

1 $sup(v,l) \leftarrow |\{u \in N(v,l) \mid \overline{C}_{L'}(u) \ge \overline{C}_{L'}(v)\}|$ for each $v \in V$ and $l \in L'$;

2 **while** $\exists v \in V'$ and $r \in L'$: $sup(v,r) < \overline{C}_{L'}(v)$ **do**

3       $c_0 \leftarrow \overline{C}_{L'}(v)$;

4   **for each** $l \in L'$ **do**

5           $M_l(v) = \max k$ $s.t.$ $|\{u \in N(v,l) \mid \overline{C}_{L'}(u) \ge k\}| \ge k$;

6   **if** $\overline{C}_{L'}(v) > M_l(v)$ **then**

7               $\overline{C}_{L'}(v) \leftarrow M_l(v)$;

8   **for each** $l \in L'$ **do**

9           $sup(v,l) \leftarrow |\{u \in N(v,l) \mid \overline{C}_{L'}(u) \ge \overline{C}_{L'}(v)\}|$;

10   **for each** $u \in N(v,l)$ **do**

11   **if** $\overline{C}_{L'}(u) \le c_0$ and $\overline{C}_{L'}(u) > \overline{C}_{L'}(v)$ **then**

12                  $sup(u,l) \leftarrow sup(u,l) - 1$;

13 $C_{L'}(v) \leftarrow \overline{C}_{L'}(v)$ for every $v \in V$;

14 **return** $C_{L'}$

Correctness. The correctness of Core-TD is based on Theorem 1. When Core-TD terminates, Eqs. 1 and 2 are satisfied for each vertex. On the other hand, the value computed for each vertex cannot be smaller than the core number because it is always an upper bound of the core number. Hence, Core-TD correctly computes core number for each vertex.

The key issue with Core-TD is how to initialize the upper bound tight enough such that it can quickly converge. To deal with this issue, we introduce the following lemma:

Algorithm 4: CoreCube-TD(G)

Input   : G : a multilayer graph
Output : $\mathcal{C}$: the CoreCube of G

1 $\mathcal{C} \leftarrow \emptyset$;
2 **for** $z = 1$ to $|L|$ **do**
3 $\quad$ $Z \leftarrow \{$all the subsets of $|L|$ whose size are $z\}$;
4 $\quad$ **for each** $L' \in Z$ **do**
5 $\quad\quad$ **for each** $v \in V$ **do**
6 $\quad\quad\quad$ **if** $z = 1$ **then**
7 $\quad\quad\quad\quad$ $\overline{\mathcal{C}}_{L'}(v) \leftarrow deg(v, l)$ where $l \in L'$;
8 $\quad\quad\quad$ **else**
9 $\quad\quad\quad\quad$ $\overline{\mathcal{C}}_{L'}(v) \leftarrow \min\{\mathcal{C}_D(v) \mid D \subset L' \wedge |D| = |L'| + 1\}$;
10 $\quad\quad$ $\mathcal{C} \leftarrow \mathcal{C} \cup \{\texttt{Core-TD}\ (G, L', \overline{\mathcal{C}}_{L'})\}$;

11 **return** $\mathcal{C}$

Lemma 2. *Given a multilayer graph $G = (V, E, L)$ and a vertex $v \in V$, it holds that $\mathcal{C}_{L_1}(v) \geq \mathcal{C}_{L_2}(v)$ if $L_1 \subseteq L_2$.*

Proof. Let $k = \mathcal{C}_{L_2}(v)$. Based on the definition of core number, there exists a set of vertices $S \subseteq V$ such that each vertex v in the subgraph H induced by S satisfies $deg_H(v, l) \geq k$ for $l \in L_2$. Since $L_1 \subseteq L_2$, we have $\mathcal{C}_{L_1}(v) \geq k = \mathcal{C}_{L_2}(v)$.

According to Lemma 2, the core number of a vertex v on L' can serve as an upper bound of v's core number on any superset of L'. Note that if we compute CoreCube level-by-level, we will obtain core numbers on all the subsets of L' when computing multilayer core decomposition on L'. Therefore we can exploit previous computation as much as possible by initializing $\overline{\mathcal{C}}_{L'}(v)$ with the minimum core number of v on all the subsets of L', i.e., $\overline{\mathcal{C}}_L(v) = \min\{\mathcal{C}_P(v) | P \subset L'\}$. Furthermore, based on Lemma 2, we actually only need to consider the subsets whose size is only one smaller than $|L'|$ because any the subset of L' whose size is smaller than $|L'| - 1$ must be contained in some subset of L' whose size is $|L'| - 1$.

The algorithm CoreCube-TD which computes CoreCube with Core-TD is shown in Algorithm 4. Each time before it invokes Core-TD for a set of layers L', it sets the upper bound of core number for each vertex according to Lemma 2 (line 9). If $|L'|$ is 1, it sets the upper bound as the vertex degree (line 7). Finally, the CoreCube of G is returned in line 11.

Complexity. In CoreCube-TD, since the number of subsets D processed in line 9 is $|L'|$, line 5–9 takes $O(|L'| \cdot |V|)$. Considering that there are $2^{|L|} - 1$ subsets of L and Core-TD is invoked for each subset, the time complexity of CoreCube-TD is bounded by $O(2^{|L|} \cdot (|L| \cdot |V| + d_{max} \cdot |E|))$. Though the time complexity is apparently worse than that of CoreCube-BU, we find that much less vertices are visited in our experiments, especially when the number of layers is large. This is because the upper bound is initialized very close to the core number and converges quickly in Core-TD.

4   CoreCube Storage

In this section, we devise a method for compactly storing CoreCube and discuss how to process queries for core numbers on any set of layers. A straightforward method is storing core numbers on each set of layers in separate files. Given a core number query, we can directly retrieve the result from the disk. However, this method requires large disk space as we need to store each vertex in every file. To reduce space usage, we propose two optimization strategies.

Firstly, many vertices' core number on a set of layers L' can be zero when $|L'|$ is large because the core number of a vertex v is zero if $deg(v, l)$ in some layer $l \in L'$ equals to 0. Therefore, we do not record the vertex whose core number is zero. Secondly, the core number on L' can remain unchanged when a new layer l is added to L' if the core number on L' is small or the distribution of core number on l is nearly the same as that in L'. Hence, we can store the difference between core numbers on different sets of layers instead of directly storing core number for each vertex. Here, we call the file that stores nonzero core numbers as absolute storage and the file that stores the difference as relative storage. The algorithm Hybrid-Storage which uses both absolute storage and relative storage is given in Algorithm 5.

Hybrid-Storage creates a file F for each subset of L (line 3). For the subset consists of single layer, it uses absolute storage to store the nonzero core number for each vertex (line 4–6). For other subsets L', it first counts the number of

---

Algorithm 5: Hybrid-Storage(G, $\mathcal{C}$)

Input : $G = (V, E, L)$: a multilayer graph, $\mathcal{C}$: the CoreCube of G
Output : the files that stores $\mathcal{C}$

1 $Z \leftarrow \{$all the subsets of $|L|\}$;
2 **for** each $L' \in Z$ **do**
3   create a new file F;
4   **if** $|L'| = 1$ **then**
5   **for** each $v \in V$ and $\mathcal{C}_{L'}(v) \neq 0$ **do**
6   write v and $\mathcal{C}_{L'}(v)$ into F;

7   **else**
8          $n_1 \leftarrow$ the number of non zero values in $\mathcal{C}_{L'}$;
9          $P \leftarrow$ the subset of L' s.t. $|\{v \in V \mid \mathcal{C}_P(v) \neq \mathcal{C}_{L'}(v)\}|$ is minimum $\wedge |P| = |L'| - 1$;
10         $n_2 \leftarrow |\{v \in V \mid \mathcal{C}_P(v) \neq \mathcal{C}_{L'}(v)\}|$;
11   **if** $n_1 \leq n_2$ **then**
12   **for** each $v \in V$ and $\mathcal{C}_{L'}(v) \neq 0$ **do**
13   write v and $\mathcal{C}_{L'}(v)$ into F;

14   **else**
15   write P as the predecessor into F;
16   **for** each $v \in V$ and $\mathcal{C}_P(v) - \mathcal{C}_{L'}(v) \neq 0$ **do**
17   write v and $\mathcal{C}_P(v) - \mathcal{C}_{L'}(v)$ into F;

Algorithm 6: Core-Retrieve(G, L')

Input   : $G = (V, E, L)$: a multilayer graph, L': a subset of layers
Output : $\mathcal{C}_{L'}$: the multilayer core number on L'
1 $\mathcal{C}_{L'}(v) \leftarrow 0$ for each $v \in V$;
2 $flag \leftarrow true$; $P \leftarrow L'$;
3 **while** $flag$ **do**
4 | load the file F corresponding to P from disk;
5 | **if** F is relative storage **then**
6 | | $P \leftarrow$ the predecessor in F;
7 | **else**
8 | | $flag \leftarrow false$;
9 | $\mathcal{C}_{L'}(v) \leftarrow \mathcal{C}_{L'}(v) + F(v)$ for each $v \in V$;
10 **return** $\mathcal{C}_{L'}$

---

nonzero core number in $\mathcal{C}_{L'}$ as n_1 (line 8). Then, it finds the subset P of L' such that the number of different values between $\mathcal{C}_{L'}$ and $\mathcal{C}_P$ is minimum (line 9) and refers this number as n_2 (line 10). If $n_1 \leq n_2$, Hybrid-Storage uses absolute storage (line 11–13). Otherwise, it uses relative storage that stores all the difference between $\mathcal{C}_{L'}$ and $\mathcal{C}_P$ (line 16–17). It also records P as the predecessor (line 15) so that we can know from which subset the difference is made when answering queries.

The algorithm which processes queries for core numbers on a set of layers L' is shown in Algorithm 6. Core-Retrieve keeps loading files from disk according to the predecessors (line 4–6) until it meets absolute storage (line 7–8). Meanwhile, Core-Retrieve computes core numbers by summing up the difference stored in each file (line 9). Note that we use $F(v)$ to represent the value (core number or difference) associated with node v stored in file F. Finally, core numbers are returned in line 10. Note that Core-Retrieve loads at most $|L'|$ files.

Table 2. Statistics of datasets

Dataset	Vertices	Edges	Layers	Domain
Homo	$18,223$	$153,922$	7	Genetic
SacchCere	$6,571$	$247,152$	7	Genetic
Twitter	$2,281,260$	$3,827,964$	3	Social
Amazon	$410,237$	$8,132,506$	4	Co-purchasing
DBLP	$2,175,466$	$8,221,193$	10	Co-authorship
Flickr	$2,302,927$	$23,350,524$	10	Social
StackOverflow	$6,024,272$	$28,978,914$	10	Social
Wiki	$25,323,885$	$132,693,853$	10	Hyperlinks

5 Experimental Evaluation

5.1 Experimental Setting

Datasets. Eight real-life networks were deployed in our experiments. Table 2 shows the statistics of the 8 datasets, listed in increasing order of their edge numbers. Home and SacchCere are networks describing different types of genetic interactions between genes. Twitter represents different types of social interaction among Twitter users. Amazon is a co-purchasing temporal network, containing four snapshots between March and June 2003. DBLP is a co-author network. Flickr is a social network represents Flickr users and their friendship connections. StackOverflow is a temporal network represents different types of interactions on the website Stack Overflow. Wiki contains users and pages from Wikipedia, connected by edit events.

Algorithms. We test 4 algorithms for CoreCube computation. CoreCube-BU and CoreCube-TD are our algorithms, e.g., Algorithms 2 and 4.

ML-DFS and ML-Hybrid are two state-of-the-art existing solutions proposed in [10]. They compute cores for all the coreness vector k, where k is a $|L|$-dimension vector and the value k in each dimension represents that the degree of each vertex is no less than k in the corresponding layer. ML-DFS searches the space of k through depth-first search strategy. ML-Hybrid adopts both depth-first and breath-first search strategy. In our experiments, we compute CoreCube by using cores whose k has the same value in every nonzero dimension. For the sake of fairness, *we extract and report the time spent on computing these cores* in ML-DFS and ML-Hybrid instead of the total running time.

To the best of our knowledge, no existing work investigates the storage of CoreCube. We test three algorithms Naive-Storage, Nonzero-Storage and Hybrid-Storage. Naive-Storage stores core numbers without any optimization strategies. Nonzero-Storage only stores nonzero core numbers. Hybrid-Storage uses both absolute storage and relative storage, i.e., Algorithm 5.

Core-Retrieve is our algorithm for answering core number queries, i.e., Algorithm 6. CoreScratch computes core numbers from scratch for each query. We divide CoreScratch into two procedures, CoreScratch-Load and CoreScratch-Comp. CoreScratch-Load is the procedure that loads the graph from disk into main memory. CoreScratch-Comp is the procedure that computes core numbers. For CoreScratch-Comp, we test both Core-BU and Core-TD, and report the running time based on the faster one.

All algorithms are implemented in C++ with -O2 optimization level and tested on an server equipped with Intel Xeon CPU at 2.8 GHz and 128 GB main memory.

5.2 CoreCube Computation

In this set of experiments, we set the maximum running time for each test as 48 h. If an algorithm cannot stop within the time limit, we omit its running time.

Exp-1: CoreCube Computation Time on Different Datasets. We report the time cost for computing CoreCube on different datasets in Fig. 2. As shown

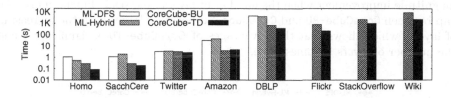

Fig. 2. CoreCube computation time in all datasets

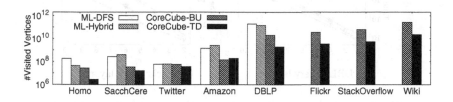

Fig. 3. The number of visited vertices in CoreCube computation in all datasets

in Fig. 2, our proposed algorithm `CoreCube-TD` is the fastest algorithm in all datasets except `Amazon` and achieves one order of magnitude improvement on average compared with existing solutions `ML-DFS` and `ML-Hybrid`. For example, in DBLP, `CoreCube-BU` and `CoreCube-TD` spend 662s and 375s respectively while `ML-DFS` and `ML-Hybrid` spend 4487 s and 3932 s respectively. In the three largest datasets, `ML-DFS` and `ML-Hybrid` cannot terminate within 48 h.

Exp-2: The Number of Visited Vertices in CoreCube Computation. To better demonstrate performance of the four CoreCube computation algorithms, we report the number of visited vertices in Fig. 3. The number of visited vertices represents how many times the value related to a vertex is modified or accessed, e.g., removing an edge or decreasing upper bound. For `ML-DFS` and `ML-Hybrid`, the number of visited vertices is collected during the computation of cores that are used for computing CoreCube. As shown in Fig. 3, the number of visited vertices in `CoreCube-TD` is smallest in all datasets except for `Amazon`. This is because the core numbers in `Amazon` vary a lot on different sets of layers, which leads to slow convergence in `Core-TD`. Compared with our algorithms, the number of visited vertices in `ML-DFS` and `ML-Hybrid` is much larger. The reason is that they need to generate a subgraph that contains some core before computing it.

Exp-3: Scalability of CoreCube Computation. In this experiment, we evaluate the performance of four CoreCube computation algorithms with varying the number of layers. We show results on DBLP and `Flickr` in Fig. 4. The trends are similar in other datasets. As shown in Fig. 4, the running time of four algorithms stably increases. The gap between existing algorithms and our proposed algorithms becomes larger as the number of layers increases. Compared with existing algorithms, our proposed algorithm `CoreCube-TD` achieves at least 1 order of magnitude improvement when the number of layers excesses 7. Furthermore, the gap between CoreCube-BU and CoreCube-TD becomes larger with the increasing of layers, which shows that the advantages of CoreCube-TD is significant when the number of layers becomes large.

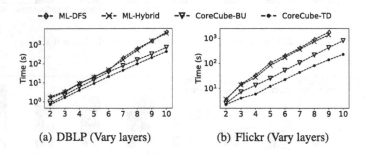

(a) DBLP (Vary layers)   (b) Flickr (Vary layers)

Fig. 4. CoreCube computation time with varying number of layers

5.3   CoreCube Storage and Query Processing

Exp-4: Disk Usage under Different Storage Methods. In this experiment, we report the disk usage of storing CoreCube of all datasets in Fig. 5. As shown in Fig. 5, the disk usage of Hybrid-Storage is smallest in all datasets. For example, in DBLP, the disk usage of Naive-Storage, Nonzero-Storage and Hybrid-Storage are 21GB, 522MB and 302MB respectively. The gap between Naive-Storage and Nonzero-Storage shows that many vertices have zero core number in CoreCube. Hybrid-Storage further reduces disk usage by storing the difference between core numbers on different subsets of layers.

Exp-5: Core Number Query Processing Time. In this experiment, we randomly generate 100 core number queries for each dataset. Each core number query asks for core numbers on a specific set of layers. The total running time of answering the 100 queries is reported in Fig. 6. As shown in Fig. 6, Core-Retrieve finishes 100 queries within 10 ms in all datasets including the time spent on loading files from disk. CoreScratch spends more than 100 s in the largest dataset even if the graph has already been loaded into memory. In real scenarios, graphs cannot always be kept in memory. The advantage of Core-Retrieve is more significant when considering the graph loading time in CoreScratch-Load.

5.4   Case Study on DBLP

In this section, we test the effectiveness of multilayer core decomposition on DBLP. Here, the multilayer graph has two layers. One layer is the coauthor network of SIGMOD conference. Another one is the coauthor network of KDD

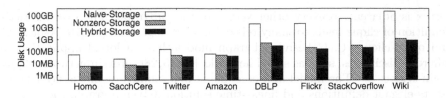

Fig. 5. CoreCube storage in all datasets

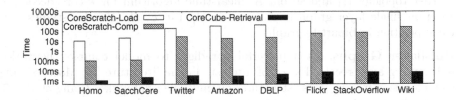

Fig. 6. Core number query processing time

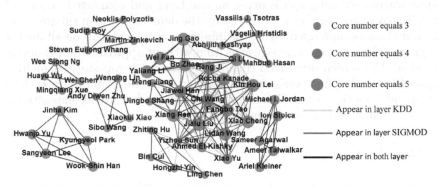

Fig. 7. Multilayer core decomposition on DBLP

conference. Two authors are connected if they collaborated on at least one paper. Both layers are extracted from data from 2013 to 2017.

Exp-6: Case Study on DBLP. We show vertices with core number no less than 3 in Fig. 7. Edges that appear exclusively in KDD and SIGMOD are colored with blue and red respectively. Edges that appear in both layers are colored with black. As shown in Fig. 7, multilayer core decomposition effectively captures authors with different engagement level in both conferences. Note that the subgraph induced by multilayer k-core are not necessarily connected.

6   Related Work

Cohesive Subgraphs. A variety of cohesive subgraph models are proposed to handle different scenarios. One of the earliest model is clique [17] where every vertex is adjacent to every other vertex in the subgraph. The over-restrictive definition of clique leads to many relaxed models, e.g., n-clique [16], k-plex [21], and quasi-clique [1]. Cohesive subgraph models have a lot of applications on different disciplines, such as social networks [18,25,28], protein networks [2] and brain science [8]. The model of k-core [20] is well-studied on single-layer graphs for its elegant definitions and linear-time solution. Given a graph, the k-core of every input k naturally forms a hierarchical graph decomposition. Core decomposition is applied to many areas of importance, e.g., graph visualization [3], internet topology [7] and so on. A linear-time algorithm for k-core decomposition on single layer graph is proposed in [4]. Liu *et al.* [15] also studies core decomposition in bipartite graphs.

Multilayer Graphs. As a powerful paradigm to model complex networks, multilayer graphs received a lot of interests in the literature [9]. Most existing works focus on mining dense structures on multilayer networks. Zhang *et al.* [27] detect cohesive subgraphs on a 2-layer graph where one layer corresponds to user engagement and the other corresponds to user similarity. Wu *et al.* [23] find subgraphs where each subgraph is dense on one layer and connected on the other layer. Jethava and Beerenwinkel [12] study the densest common subgraph problem to find a subgraph maximizing the minimum average degree on all the layers of a graph. They propose a greedy algorithm without approximation guarantees. Zhu *et al.* [29] search diversified coherent k-cores with top sizes on multilayer graphs. Li *et al.* [14] find persistent k-cores on a temporal graph where each layer corresponds to a time span. Galimberti *et al.* [10] study core decomposition and densest subgraph extraction on multilayer graphs.

7   Conclusion

In this paper, we study core decomposition on multilayer graphs and propose the CoreCube which records the multilayer core decomposition on every combination of layers. We devise algorithms for efficiently computing and compactly storing CoreCube. The experimental results validate the efficiency of our proposed algorithms and effectiveness of multilayer core decomposition.

References

1. Abello, J., Resende, M.G.C., Sudarsky, S.: Massive quasi-clique detection. In: Rajsbaum, S. (ed.) LATIN 2002. LNCS, vol. 2286, pp. 598–612. Springer, Heidelberg (2002). https://doi.org/10.1007/3-540-45995-2_51
2. Altaf-Ul-Amine, M., et al.: Prediction of protein functions based on k-cores of protein-protein interaction networks and amino acid sequences. Gen. Inf. **14**, 498–499 (2003)
3. Alvarez-Hamelin, J.I., Dall'Asta, L., Barrat, A., Vespignani, A.: Large scale networks fingerprinting and visualization using the k-core decomposition. In: NIPS, pp. 41–50 (2005)
4. Batagelj, V., Zaversnik, M.: An o(m) algorithm for cores decomposition of networks. CoRR, cs.DS/0310049 (2003)

5. Bhawalkar, K., Kleinberg, J., Lewi, K., Roughgarden, T., Sharma, A.: Preventing unraveling in social networks: the anchored k-core problem. SIAM J. Discrete Math. **29**(3), 1452–1475 (2015)
6. Boden, B., Günnemann, S., Hoffmann, H., Seidl, T.: Mining coherent subgraphs in multi-layer graphs with edge labels. In: SIGKDD, pp. 1258–1266 (2012)
7. Carmi, S., Havlin, S., Kirkpatrick, S., Shavitt, Y., Shir, E.: A model of internet topology using k-shell decomposition. PNAS **104**(27), 11150–11154 (2007)
8. Daianu, M., et al.: Breakdown of brain connectivity between normal aging and Alzheimer's disease: a structural k-core network analysis. Brain Connectivity **3**(4), 407–422 (2013)
9. Dickison, M.E., Magnani, M., Rossi, L.: Multilayer Social Networks. Cambridge University Press, New York (2016)
10. Galimberti, E., Bonchi, F., Gullo, F.: Core decomposition and densest subgraph in multilayer networks. In: CIKM, pp. 1807–1816 (2017)
11. Hu, H., Yan, X., Huang, Y., Han, J., Zhou, X.J.: Mining coherent dense subgraphs across massive biological networks for functional discovery. In: International Conference on Intelligent Systems for Molecular Biology, pp. 213–221 (2005)
12. Jethava, V., Beerenwinkel, N.: Finding dense subgraphs in relational graphs. In: Appice, A., Rodrigues, P.P., Santos Costa, V., Gama, J., Jorge, A., Soares, C. (eds.) ECML PKDD 2015. LNCS (LNAI), vol. 9285, pp. 641–654. Springer, Cham (2015). https://doi.org/10.1007/978-3-319-23525-7_39
13. Khaouid, W., Barsky, M., Srinivasan, V., Thomo, A.: K-core decomposition of large networks on a single PC. Proc. VLDB Endowment **9**(1), 13–23 (2015)
14. Li, R., Su, J., Qin, L., Yu, J.X., Dai, Q.: Persistent community search in temporal networks. In: ICDE, pp. 797–808 (2018)
15. Liu, B., Yuan, L., Lin, X., Qin, L., Zhang, W., Zhou, J.: Efficient (α, β)-core computation: an index-based approach. In: The World Wide Web Conference, WWW 2019, pp. 1130–1141. ACM, New York (2019)
16. Luce, R.D.: Connectivity and generalized cliques in sociometric group structure. Psychometrika **15**(2), 169–190 (1950)
17. Luce, R.D., Perry, A.D.: A method of matrix analysis of group structure. Psychometrika **14**(2), 95–116 (1949)
18. Malliaros, F.D., Vazirgiannis, M.: To stay or not to stay: modeling engagement dynamics in social graphs. In: CIKM, pp. 469–478 (2013)
19. Montresor, A., De Pellegrini, F., Miorandi, D.: Distributed k-core decomposition. IEEE Trans. Parallel Distrib. Syst. **24**(2), 288–300 (2013)
20. Seidman, S.B.: Network structure and minimum degree. Soc. Netw. **5**(3), 269–287 (1983)
21. Seidman, S.B., Foster, B.L.: A graph-theoretic generalization of the clique concept. J. Math. Soc. **6**(1), 139–154 (1978)
22. Wen, D., Qin, L., Zhang, Y., Lin, X., Yu, J.X.: I/O efficient core graph decomposition at web scale. In: ICDE, pp. 133–144 (2016)
23. Wu, Y., Jin, R., Zhu, X., Zhang, X.: Finding dense and connected subgraphs in dual networks. In: ICDE, pp. 915–926 (2015)
24. Zhang, F., Li, C., Zhang, Y., Qin, L., Zhang, W.: Finding critical users in social communities: the collapsed core and truss problems. In: TKDE (2018)
25. Zhang, F., Yuan, L., Zhang, Y., Qin, L., Lin, X., Zhou, A.: Discovering strong communities with user engagement and tie strength. In: Pei, J., Manolopoulos, Y., Sadiq, S., Li, J. (eds.) DASFAA 2018. LNCS, vol. 10827, pp. 425–441. Springer, Cham (2018). https://doi.org/10.1007/978-3-319-91452-7_28
26. Zhang, F., Zhang, W., Zhang, Y., Qin, L., Lin, X.: OLAK: an efficient algorithm to prevent unraveling in social networks. PVLDB **10**(6), 649–660 (2017)

27. Zhang, F., Zhang, Y., Qin, L., Zhang, W., Lin, X.: When engagement meets similarity: Efficient (k, r)-core computation on social networks. PVLDB **10**(10), 998–1009 (2017)
28. Zhang, F., Zhang, Y., Qin, L., Zhang, W., Lin, X.: Efficiently reinforcing social networks over user engagement and tie strength. In: ICDE, pp. 557–568 (2018)
29. Zhu, R., Zou, Z., Li, J.: Diversified coherent core search on multi-layer graphs. In: ICDE, pp. 701–712 (2018)

Computing Maximum Independent Sets over Large Sparse Graphs

Maram Alsahafy and Lijun Chang$^{(\boxtimes)}$

The University of Sydney, Sydney, Australia
{mals4485,lijun.chang}@sydney.edu.au

Abstract. This paper studies the fundamental problem of efficiently computing a maximum independent set (or equivalently, a minimum vertex cover) over a large sparse graph, which is receiving increasing interests from the research communities of graph algorithms and graph analytics. The state-of-the-art algorithms for both exact and heuristic computations heavily rely on kernelization techniques that use reduction rules to reduce a large input graph to a smaller graph (called its kernel) while preserving the maximum independent set. However, the existing kernelization techniques either run slow (but return a small kernel), or return a large kernel (but run fast). In this paper, we propose two techniques—aggressive incremental reduction rules and connected component checking—to speed up the kernelization process while computing a small kernel. Furthermore, for efficient maximum independent set computation, we propose to control the giant kernel connected component size, and propose to invoke maximum clique solvers for solving the kernel graph. Extensive empirical studies on large real graphs demonstrate the efficiency and effectiveness of our techniques.

Keywords: Maximum independent set · Maximum clique · Kernelization · Reduction rules · Sparse graph

1 Introduction

Large graphs with millions of vertices and edges are common to see in real world applications, *e.g.*, geographic maps, communication networks, web networks and social networks. In this paper, we study the problem of efficiently computing a maximum independent set (MIS) over a large sparse graph $G = (V, E)$, where V and E are the vertex set and edge set of G, respectively. A vertex subset I of V is an *independent set* if there is no edge in G between any two vertices of I, and an independent set is *maximum* if it has the largest cardinality among all independent sets of G. The problem of MIS computation is equivalent to the problem of *minimum vertex cover* (MVC) computation, since I is a (maximum) independent set of G if and only if $V \backslash I$ is a (minimum) vertex cover of G. Thus, solving one of the problems directly solves the other.

Computing an MIS (or equivalently, an MVC) is a fundamental problem in both the research area of experimental graph algorithms [1,7,10,12] and the research

© Springer Nature Switzerland AG 2019
R. Cheng et al. (Eds.): WISE 2019, LNCS 11881, pp. 711–727, 2019.
https://doi.org/10.1007/978-3-030-34223-4_45

area of graph analytics [6,15,21], and has many applications. For example, it has been used in map labelling for assigning the maximum number of visible non-overlapping labels on a map [18], in job/game scheduling to schedule the largest number of possible jobs/games while avoiding conflicts [4], in computing social network coverage and reach [17], and in collusion detection for voting pools [3].

The problem of MIS computation is among Karp's original 21 NP-hard problems [11], and is also NP-hard to approximate within a factor of $n^{1-\epsilon}$ for any constant $0 < \epsilon < 1$ [9] where n is the number of vertices in the input graph. Nevertheless, in view of the importance of this problem, techniques have been developed to efficiently compute an MIS in practice, either exactly (*e.g.*, VCSolver [1]) or heuristically (*e.g.*, LinearTime and NearLinear [6]). One of the most effective techniques is kernelization which uses reduction rules to reduce a large input graph to a smaller graph, called its kernel, while preserving the MIS [1,6,10]. A *reduction rule* reduces a graph by either removing some vertices from it or contracting some sets of vertices into super-vertices. In either case, the number of vertices in the graph is reduced, and moreover an MIS of the original graph can be obtained from an MIS of the reduced graph by "undoing" the reduction. A large set of reduction rules are implemented in the state-of-the-art MIS/MVC solver VCSolver to get a very small kernel graph, while LinearTime and NearLinear only use several simple reduction rules to get a near-linear running time.

Our Approaches. In this paper, we also make use of kernelization. Intuitively, the smaller the kernel and the fast the kernel computation, the better for MIS computation. However, as observed in [10], the existing kernelization techniques either run slow but compute a small kernel (*e.g.*, VCSolver), or compute a large kernel but run fast (*e.g.*, LinearTime and NearLinear). Although the recently proposed FastKer and ParFastKer [10] slightly alleviate this issue, they still compute a much larger kernel compared to that by VCSolver. In view of this, we design a kernelization algorithm MISKernel for efficiently computing a small kernel. We categorize the reduction rules into *incremental reduction rules* and *iterative reduction rules*, and aggressively apply the incremental reduction rules whenever possible by maintaining the set of all vertices that are reducible by these reduction rules. To avoid unfruitful computations, we propose to maintain the connected components of the graph and kernelize the graph in a component-by-component fashion. Thus, the connected components that are irreducible by the set of reduction rules will not be tested again and again, while other parts of the graph can be reduced.

Based on MISKernel, we then design an algorithm MISSolver for efficiently computing an MIS. Firstly, we observe that the hardness of a (connected) kernel component is generally proportional to its size, and VCSolver likely will take more than 20 h to compute the MIS over a kernel component with thousands of vertices. Thus, we propose to control the giant kernel component size, by applying the inexact reduction rule proposed in [6] whenever the giant kernel component cannot be reduced by the exact reduction rules and its size is still larger than a threshold τ. Secondly, we propose to invoke maximum clique solvers rather than VCSolver to solve the kernel components, by observing that the problem of exact maximum clique computation has been extensively studied with many advanced techniques being designed. For any graph G, the maximum clique in its complement graph $\overline{G}$ is an MIS of G; $\overline{G}$ contains exactly the same set of vertices as G, and there is an edge between u and v in $\overline{G}$ if and only if there is no such edge in G. However, note that we cannot directly invoke maximum clique solvers on the complement of the input graph due to its extremely large size, as our input graphs are large sparse graphs with millions of vertices.

Contributions. Our main contributions are summarized as follows.

- We design an efficient kernelization algorithm MISKernel for computing a small kernel, by proposing aggressive incremental reduction rules and connected component checking.
- We design an efficient algorithm MISSolver for computing an MIS, by proposing giant kernel component control and invoking maximum clique solver to solve kernel components.
- We conduct extensive empirical studies on large real graphs. The results show that our kernelization algorithm MISKernel computes a much smaller kernel than LinearTime, NearLinear, FastKer and ParFastKer, and computes a comparable kernel to VCSolver but runs much faster. Moreover, our algorithm MISSolver runs much faster than VCSolver for computing MIS.

Related Works. We categorize the related works as computing MIS/MVC, and computing maximum clique.

Computing MIS/MVC. VCSolver [1] is the only existing solver that can compute the exact MIS over large sparse graphs. It follows the branch-and-reduce framework, by heavily using kernelization techniques. Chang et al. [6] designed the reducing-peeling framework and proposed algorithms LinearTime and NearLinear, for efficiently computing an MIS. Computing large independent set in the I/O environment and for dynamic graphs are studied in [15] and [21], respectively. Iterated local search methods for improving the quality of independent set are investigated in [2,7,12]. Recently, a parallelization algorithm ParFastKer is proposed in [10] to speed up kernelization.

Computing Maximum Clique. The problem of maximum clique computation has been extensively studied in the literature. Specifically, exact algorithms have been proposed in [5,13,14,19]. They follow the *branch-and-bound* framework, where advanced bounding techniques such as approximate graph coloring [19] and incremental MaxSAT reasoning [13,14] have been designed to prune branches whose upper bounds are no larger than the size of the currently found largest clique. These bounding techniques have been shown to be very effective in pruning a large number of search branches.

2 Preliminary

In this paper, we consider an unweighted and undirected graph $G = (V, E)$, where V and E are the vertex set and edge set of G, respectively. We denote the number $|V|$ of vertices and the number $|E|$ of edges of G by n and m, respectively. An edge between $u \in V$ and $v \in V$ is denoted by (u, v), and u (resp. v) is said to be connected to and is a neighbor of v (resp. u). The set of neighbors of a vertex u is denoted by $N(u) = \{v \in V \mid (u, v) \in E\}$, and its degree is $d(u) = |N(u)|$.

For a vertex subset $S \subseteq V$, we use $N(S)$ to denote the union of the neighbors of vertices of S by excluding S itself (i.e., $N(S) \cap S = \emptyset$), and use $N[S]$ to denote $N(S) \cup S$. We use $G[S]$ to denote the subgraph of G induced by S, i.e., the subgraph of G consisting of all G's edges whose both end-points are in S. Given $S \subseteq V$, we may modify G by (1) *deleting* from G all vertices of S and their associated edges (denote the resulting graph by $G \backslash S$), or (2) *contracting* all vertices of S into a single super-vertex (denote the result graph by G/S).

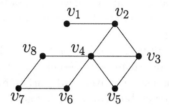

Fig. 1. An example graph

A vertex subset I of V is an *independent set* if there is no edge in G between any vertices of I. The size of an independent set is its cardinality (i.e., number of vertices). An independent set of G is a *maximum independent set* (MIS) if its size is the largest among all independent sets of G, and this size is called the *independence number* of G, denoted $\alpha(G)$. Note that, the MIS of a graph may be not unique. Consider the graph G in Fig. 1, $\{v_2, v_5, v_7\}$ is an independent set of size 3, while both $\{v_2, v_5, v_6, v_8\}$ and $\{v_1, v_3, v_6, v_8\}$ are MIS of size 4.

Problem Statement. Given a large sparse graph $G = (V, E)$ with millions of vertices, we study the problem of efficiently computing an MIS of G.

2.1   Reduction Rules

In this paper, we make use of reduction rules, and briefly describe them in below. The simplest reduction rule is that *for a degree-zero vertex u, any MIS must include u*. Thus, in the following, we assume all degree-zero vertices are handled in this way, and we do not consider degree-zero vertices.

Degree-one Reduction [8]. Given a vertex u of degree one (i.e., $d(u) = 1$) of G, there exists an MIS that contains u but not its neighbor. That is, let v be u's neighbor, then $\alpha(G) = \alpha(G \backslash \{v\})$. For example, the graph in Fig. 1 must have a MIS that contains v_1 but not v_2.

Degree-two Reduction [7,8]. Given a vertex u of degree two (i.e., $d(u) = 2$) of G and let v and v' be u's two neighbors, there are two cases.

- *Degree-two Isolation* [7]. If $(v, v') \in E$, there exists an MIS that contains u but not v or v'. Thus, $\alpha(G) = \alpha(G\backslash\{v, v'\})$. For example, as $d(v_5) = 2$ in Fig. 1, the graph must have an MIS that contains v_5 but not v_3 or v_4.
- *Degree-two Folding* [8]. Otherwise $(v, v') \notin E$, there exists an MIS that contains either u or $\{v, v'\}$. Thus, $\alpha(G) = \alpha(G\backslash\{u\}/\{v, v'\}) + 1$ and we can remove u from the graph and contract $\{v, v'\}$. Specifically, if the contracted vertex is in the MIS of the reduced graph, then both v and v' are in that of the original graph; otherwise, u is in that of the original graph.

Dominance Reduction [8]. A vertex u is *dominated* by v, if $(u, v) \in E$ and all neighbors of v, except u, are connected to u. For example, v_4 is dominated by v_3 in Fig. 1. If u is dominated by another vertex, then there exists an MIS that does not contain u; thus, $\alpha(G) = \alpha(G\backslash\{u\})$. Degree-one reduction and degree-two isolation discussed above are special cases of the dominance reduction.

Unconfined Reduction [20]. The unconfined reduction further generalizes the dominance reduction. If a vertex u is unconfined, then there exists an MIS that does not contain u. The following procedure is used to determine whether a vertex u is unconfined or not [1]. Initially $S = \{u\}$. Then, it tries to find a vertex $v \in N(S)$ such that $|N(v) \cap S| = 1$ and $|N(v)\backslash N[S]|$ is minimized. There are four cases. (1) If there is no such vertex, then u is confined. (2) If $N(v)\backslash N[S] = \emptyset$, then u is *unconfined*. (3) If $N(v)\backslash N[S]$ contains a single vertex w, then w is added to S and the algorithm continues. (4) Otherwise, u is confined.

Twin Reduction [20]. Given two vertices u_1 and u_2 of degree three that share the same neighbors (let $N(u_1) = N(u_2) = \{v_1, v_2, v_3\}$), there are two cases to reduce the graph based on whether there are edges among v_1, v_2 and v_3, in a similar spirit to the two cases of degree-two reduction.

- If there are edges among v_1, v_2 and v_3, then there exists an MIS that contains u_1 and u_2 but not v_1, v_2 or v_3 (*i.e.*, $\alpha(G) = \alpha(G\backslash\{v_1, v_2, v_3\})$). Thus, we can remove v_1, v_2 and v_3 from the graph.
- Otherwise, then there exists an MIS that contains either $\{u_1, u_2\}$ or $\{v_1, v_2, v_3\}$. Thus, $\alpha(G) = \alpha(G\backslash\{u_1, u_2\}/\{v_1, v_2, v_3\}) + 2$, and we can remove $\{u_1, u_2\}$ from the graph and contract $\{v_1, v_2, v_3\}$.

Linear Programming (LP) Reduction [16]. The MIS problem can be expressed as an 0–1 integer programming as: maximize $\sum_{u \in V} x_u$, such that $x_u + x_v \leq 1$ for all $(u, v) \in E$, and $x_u \in \{0, 1\}$ for all $u \in V$. Here, $x_u = 1$ and $x_u = 0$, respectively, means u is in and is not in the MIS. It is known that the linear programming relaxation of the above 0–1 integer programming has a half-integral optimal solution (*i.e.*, $x_u \in \{0, 1/2, 1\}$), and there is an MIS that contains all vertices with $x_u = 1$ and none of the vertices with $x_u = 0$. Moreover, the linear programming can be solved by a bipartite matching [1], which also minimizes the number of vertices with $x_u = 1/2$. Thus, we can remove all vertices with $x_u = 0$ from the graph, and include all vertices with $x_u = 1$ into the independent set.

3   Approaches

In this section, we develop techniques for efficient MIS computation. Towards that goal, we first propose an efficient kernelization algorithm MISKernel in Sect. 3.1, and then present our MISSolver algorithm in Sect. 3.2.

3.1   MISKernel Algorithm

Intuitively, the reduction rules in Sect. 2.1 can be iteratively applied one after another to the entire graph, as done in VCSolver. However, this is not efficient. We propose two techniques to speed up the kernelization process: aggressive incremental reduction rules, and connected component checking.

Aggressive Incremental Reduction Rules. We notice that all the reduction rules described in Sect. 2.1 are in the general form of "if a certain condition holds on a subset of vertices, then we reduce the graph for this vertex subset". We call the "if" part the *condition* and the "then" part the *action*. For example, for degree-one reduction, the condition is "the degree of a vertex is exactly one" and the action is "remove the neighbor of the degree-one vertex from the graph"; for unconfined reduction, the condition is "the vertex is unconfined" and the action is "remove the vertex from the graph". Generally, the actions of the reduction rules are simple and can be carried out efficiently. However, the complexities of the conditions vary dramatically for different reduction rules; some are easy to check (*e.g.*, degree-one reduction), and some are hard to check (*e.g.*, unconfined reduction). Motivated by the above, we categorize the reduction rules into *incremental reduction rules* such that all vertices satisfying their conditions can be easily and incrementally obtained, and *iterative reduction rules* otherwise.

Incremental Reduction Rules. Among the reduction rules described in Sect. 2.1, the incremental reduction rules include degree-one reduction and degree-two reduction. The set of vertices that can be reduced by these two reduction rules are, respectively, degree-one vertices and degree-two vertices, which can be easily and incrementally obtained. Thus, we propose to incrementally apply these two reduction rules. The pseudocode is shown in Algorithm 1, denoted DOTReduction. Specifically, we maintain in $V^{=1}$ and $V^{=2}$, respectively, the sets of vertices whose degrees were one and two at some point of the execution. To achieve this, we also maintain the degree for all vertices such that whenever a vertex's degree newly becomes one or two, the vertex is inserted into $V^{=1}$ or $V^{=2}$, respectively (Lines 7,14,17). Then, vertices are iteratively removed from $V^{=1}$ and $V^{=2}$ (Lines 4,9) and processed. Before processing a vertex, its degree in the current graph will be checked (Lines 5,10), because this may have changed due to modifications to its neighbors. The number of vertices that are determined, during the kernelization process, to be included into the MIS is recorded in s, which will be later used together with the independence number of the reduced graph to determine that of the original graph.

Algorithm 1: DOTReduction

Input: An undirected graph $G = (V, E)$, the set $V^{=1}$ of vertices of degree one, and the set $V^{=2}$ of vertices of degree two

Output: Size s of independent set vertices determined during the kernelization process, and the reduced graph G

1 $s \leftarrow 0$;
2 **while** $V^{=1} \neq \emptyset$ *or* $V^{=2} \neq \emptyset$ **do**
3   **while** $V^{=1} \neq \emptyset$ **do**
4   Remove a vertex u from $V^{=1}$;
5   **if** *the degree of u is not one in the current graph* **then continue**;
6          $s \leftarrow s + 1$, and add u to the independent set;
7   Remove u and its neighbor from the graph, and add the vertices whose degrees newly become one or two to $V^{=1}$ or $V^{=2}$ accordingly;

8   **while** $V^{=1} = \emptyset$ *and* $V^{=2} \neq \emptyset$ **do**
9   Remove a vertex u from $V^{=2}$;
10   **if** *the degree of u is not two in the current graph* **then continue**;
11   Let v_1 and v_2 be the two neighbors of u;
12   **if** *there is an edge between v_1 and v_2 in the graph* **then**
13             $s \leftarrow s + 1$, and add u to the independent set;
14   Remove u, v_1, v_2 from the graph, and add the vertices whose degrees newly become one or two to $V^{=1}$ or $V^{=2}$ accordingly;

15   **else**
16             $s \leftarrow s + 1$;
17   Remove u and contract v_1 and v_2, and add the vertices whose degrees newly become one or two to $V^{=1}$ or $V^{=2}$ accordingly;

18 **return** (s, G);

Iterative Reduction Rules. Among the reduction rules in Sect. 2.1, the iterative reduction rules include dominance reduction, unconfined reduction, LP reduction, and twin reduction. Note that, we do not include alternative reduction which is used in VCSolver [1]. This is because applying alternative reduction, although decreases the number of vertices, may increase the number of edges of the graph, and thus cannot be supported by our current graph representation.[1] Moreover, in this paper we also introduce a new reduction rule—we call it pyramid reduction—which handles degree-four vertices whose neighbors induce a chordless 4-cycle (see Fig. 2), as follows.

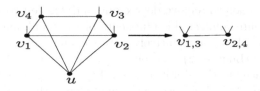

Fig. 2. Pyramid reduction

---

[1] It will be our future work to integrate alternative reduction into our algorithms.

Lemma 1. *If a vertex u of G is of degree four and its neighbors v_1, v_2, v_3, v_4 induce a chordless 4-cycle, then $\alpha(G) = \alpha(G \backslash \{u\} / \{v_1, v_3\} / \{v_2, v_4\}) + 1$. Thus, we can remove u from the graph, contract v_1 and v_3, and contract v_2 and v_4, as shown in Fig. 2.*

Due to space limits, we omit the proof of Lemma 1. The general idea is that there exists an MIS of the graph that contains either $\{u\}$, or $\{v_1, v_3\}$, or $\{v_2, v_4\}$.

For the iterative reduction rules, we need to try each of them on the entire graph. For example, for the unconfined reduction, we need to check each vertex whether it is unconfined or not; if it is, then we remove that vertex from the graph. Note that, there is a chance that after checking all vertices of the graph, none of them is unconfined and thus the graph is not reduced at all. Thus, intuitively, the iterative reduction rules are generally much harder to apply than the incremental reduction rules. Motived by this, we propose to aggressively apply the incremental reduction rules whenever possible. That is, if a modification to the graph creates degree-one or degree-two vertices, then we invoke DOTReduction to exhaustively reduce all degree-one and degree-two vertices before we check the next vertex for the iterative reduction rules. Due to limit of space, we omit the pseudocode.

Connected Component Checking. We also notice that after a few kernelization steps, we will likely get a disconnected graph, even if the input graph is connected. It is possible that some of the connected components cannot be further reduced by the reduction rules, while others can. Thus, it is a waste of time to apply all reduction rules on the entire graph. For example, consider a graph G that consists of connected components $g_1, \ldots, g_l$, and assume $g_1, \ldots, g_{l-1}$ cannot be reduced anymore while g_l can. Then all the checking on $g_1, \ldots, g_{l-1}$ for applying iterative reduction rules are wasted. Motivated by this, we propose to maintain the connected components of the graph and kernelize the graph in a component-by-component fashion. Thus, the connected components that are irreducible by the set of reduction rules will not be tested again and again, aiming for reduction, whenever other parts of the graph can be reduced.

Pseudocode of MISKernel. Based on the above ideas, our kernelization algorithm MISKernel is shown in Algorithm 2; for the current being, please ignore τ and Lines 15–17, which will be discussed in Sect. 3.2. We first reduce the graph by invoking DOTReduction to remove degree-one and degree-two vertices (Lines 1–2), and then compute connected components of the graph and put them into a queue $\mathcal{Q}$ (Line 3). Then, we iteratively pop a connected component g from $\mathcal{Q}$ (Line 6), reduce it (Line 8–12), and either put its connected components back into $\mathcal{Q}$ (if g is reduced at this iteration, Line 14) or add g into the kernel $\mathcal{K}$ (otherwise, Line 18). In Algorithm 2, we aggressively apply the incremental reduction rules during applying the iterative reduction rules at Lines 8–12.

3.2 MISSolver Algorithm

Intuitively, after obtaining the kernel graph, we can process each kernel (connected) component individually. To efficiently compute the MIS in a kernel com-

---

Algorithm 2: MISKernel

Input: An undirected graph $G = (V, E)$ and a threshold τ
Output: Size s of independent set vertices determined during the kernelization process, and the kernel $\mathcal{K}$

1   Get the degree-one vertices $V^{=1}$ and degree-two vertices $V^{=2}$ of G;
2   $(s, G) \leftarrow$ DOTReduction$(G, V^{=1}, V^{=2})$;
3   Compute the connected components of G and put them into a queue $\mathcal{Q}$;
4   $\mathcal{K} \leftarrow \emptyset$;
5   **while** $\mathcal{Q} \neq \emptyset$ **do**
6      $g \leftarrow$ pop a connected component from $\mathcal{Q}$;
7      $oldsize \leftarrow |V(g)|$;
8      $(s_1, g) \leftarrow$ apply dominance reduction together with DOTReduction on g;
9      $(s_2, g) \leftarrow$ apply unconfined reduction together with DOTReduction on g;
10     $(s_3, g) \leftarrow$ apply LP reduction together with DOTReduction on g;
11     $(s_4, g) \leftarrow$ apply twin reduction together with DOTReduction on g;
12     $(s_5, g) \leftarrow$ apply pyramid reduction together with DOTReduction on g;
13     $s \leftarrow s + \sum_{i=1}^{5} s_i$;
14   **if** $|V(g)| \neq oldsize$ **then** Compute connected components of g and put them into $\mathcal{Q}$;
15   **else if** $|V(g)| > \tau$ **then**
16        $(s', g) \leftarrow$ apply the inexact reduction together with DOTReduction on g; /* Remove the vertex with the maximum degree from g */
17        $s \leftarrow s + s'$; Compute connected components of g and put them into $\mathcal{Q}$;
18   **else** $\mathcal{K} \leftarrow \mathcal{K} \cup g$;
19 **return** $(s, \mathcal{K})$;

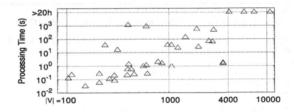

Fig. 3. Processing time of VCSolver for kernel components

ponent, we propose two techniques: giant kernel component control and invoking maximum clique solver to solve the kernel components.

Giant Kernel Component Control. Due to the NP-hardness nature of MIS computation, some of the kernel components could still be too large to be processed in a reasonable amount of time. We observe that the hardness of a kernel component is generally proportional to its size, and VCSolver likely will take more than 20 h to compute the MIS over a kernel component with thousands of vertices. For example, Fig. 3 shows the processing time of VCSolver for kernel components of the graphs in Sect. 4, where each triangle dot corresponds to one kernel component. Motivated by this, rather than waiting for hours, days, or even longer to obtain the exact MIS for these large kernel components, we propose to control the giant kernel component size. Specifically, whenever the giant kernel component cannot be reduced by the reduction rules and its size is still larger than a user-given threshold τ, then we apply the *inexact reduction rule* proposed in [6] (*i.e.*, remove the vertex with the largest degree); to contrast, we call the reduction rules in Sect. 2.1 as *exact reduction rules*. The pseudocode is shown at Lines 15–17 of Algorithm 2. As a result, all resulting kernel components will contain at most τ vertices.

Invoking Maximum Clique Solver to Solve Kernel Components. To solve each kernel component, we propose to invoke the existing exact maximum clique solvers rather than VCSolver. This is because the problem of exact maximum clique computation has been extensively studied with many advanced techniques being designed (*e.g.*, [13,14,19]), and it is expected that the maximum clique solvers are better optimized/engineered than VCSolver which is the only existing MIS solver. Note that, for any graph G, the maximum clique in its complement graph $\overline{G}$ is an MIS of G; $\overline{G}$ contains exactly the same set of vertices as G, and there is an edge between u and v in $\overline{G}$ if and only if there is no such edge in G. However, it is worth pointing out that we cannot directly invoke maximum clique solvers on the complement of the input graph which is large sparse graph with millions of vertices; note that, the complement of a sparse graph with n vertices with contain $\Omega(n^2)$ edges.

Algorithm 3: MISSolver

Input: An undirected graph $G = (V, E)$
Output: Size s of an independent set, and an upper bound s_{ub} of $\alpha(G)$

1 $(s, s_{ub}, \mathcal{K}) \leftarrow$ MISKernel$(G, 1000)$;   /* $s_{ub} = s$ + the number of times that the inexact reduction rule is applied in MISKernel */;
2 **if** $\mathcal{K} \neq \emptyset$ **then**
3 | **for each** *connected component g of* $\mathcal{K}$ **do**
4 | | $\overline{g} \leftarrow$ the complement graph of g;
5 | | $s' \leftarrow$ MoMC$(\overline{g})$;
6 | | $s \leftarrow s + s'$; $s_{ub} \leftarrow s_{ub} + s'$;

7 **return** (s, s_{ub});

Pseudocode of MISSolver. The pseudocode of our algorithm MISSolver for computing an MIS is shown in Algorithm 3. It first invokes MISKernel to compute the kernel graph such that all kernel components have at most 1,000 vertices (Line 1). Then, it processes each kernel component by invoking the state-of-the-art maximum clique solver MoMC [14] on the complement of the kernel component (Lines 3–6). Note that, an upper bound s_{ub} of the independent number $\alpha(G)$ of G is also maintained, which is the sum of the size s of the independent set obtained by the algorithm and the number of times the inexact reduction rule is applied at Line 16 of Algorithm 2. Thus, if $s = s_{ub}$, then the independent set reported by MISSolver is guaranteed to be maximum.

4   Experiments

We conduct extensive empirical studies to evaluate the efficiency and effectiveness of our techniques for computing a small kernel and for computing a maxi-

Table 1. Statistics of real graphs ($\bar{d}$: average degree)

Graphs	$|V|$	$|E|$	$\bar{d}$	Types
CA-GrQc	4,158	13,422	6.46	Collaboration network
p2p-Gnutella	8,846	31,839	7.20	Internet peer-to-peer
soc-Slashdot	82,168	50,430	1.23	Social network
CA-CondMat	21,363	91,286	8.55	Collaboration network
CA-AstroPh	18,771	198,050	21.10	Collaboration network
Email-EuAll	224,832	339,925	3.02	Communication network
soc-Epinions	75,877	405,739	10.69	Social network
Amazon0302	262,111	899,792	6.87	Product co-purchasing
DBLP	317,080	1,049,866	6.62	Online community
WikiTalk	2,388,953	4,656,682	3.90	Wikipedia
Twitter	81,306	1,342,296	33.20	Social network
web-Google	875,713	4,322,051	9.87	Web graph
as-Skitter	1,694,616	11,094,209	13.09	Autonomous system
LiveJournal	4,843,953	42,845,684	17.69	Social network
Europe	50,912,018	54,054,660	2.12	Street map
Asia	11,950,757	12,711,603	2.13	Street map
Amazon0601	403,364	2,443,311	12.11	Product co-purchasing
roadNet-PA	1,088,092	1,541,898	2.83	Road network
roadNet-TX	1,379,917	1,921,660	2.79	Road network
roadNet-CA	1,965,206	2,766,607	2.81	Road network

mum (or near-maximum) independent set. We evaluate our algorithms MISKernel and MISSolver against the following algorithms.

- VCSolver [1]: the existing exact algorithm for computing a minimum vertex cover; we also modify it to stop immediately after obtaining the kernel, and denote the kernelization algorithm as VCKernel.
- LinearTime and NearLinear [6]: the existing heuristic algorithms for efficiently computing a near-maximum independent set.
- FastKer and ParFastKer [10]: the existing efficient kernelization algorithms.

The source code of VCSolver is downloaded from https://github.com/wata-orz/vertex_cover, the source code of LinearTime and NearLinear are downloaded from https://github.com/LijunChang/Near-Maximum-Independent-Set, and the source code of FastKer and ParFastKer are obtained from the authors of [10]. We implemented our algorithms MISKernel and MISSolver in C++.

Datasets. We evaluate the algorithms on twenty real graphs that are downloaded from the Standford Network Analysis Platform[2] and the Benchmarks of

---

[2] http://snap.stanford.edu/.

Graph Clustering and Partitioning[3]. The graphs are from different domains such as social networks, road networks, collaboration networks, communication networks, product co-purchasing networks, Web graphs, autonomous systems, and online community. Statistics of these graphs are shown in Table 1.

Evaluation Metrics. We evaluate the algorithms from three aspects: kernel size, independent set size, and running time (in seconds). The smaller the kernel the better, the larger the independent set the better, and the smaller the running time the better. Experiments are conducted on a machine with an Intel Xeon 2.50 GHz CPU.

4.1   Experimental Results

Eval-I: Against VCSolver. In this testing, we evaluate the performance of MISKernel and MISSolver against the state-of-the-art exact minimum vertex cover solver VCSolver. Note that, we have modified VCSolver, as a postprocess, to report independent set instead of vertex cover.

The results of kernelization by MISKernel and VCKernel are reported in Table 2. Firstly, we can see that for all the graphs that VCKernel computes an empty kernel, our algorithm MISKernel also computes an empty kernel and runs much faster. This is due to our strategies of aggressive incremental reduction rules and connected component checking. Secondly, for graphs web-Google, as-Skitter and LiveJournal, MISKernel computes a smaller kernel and at the same time runs much faster than VCKernel. The smaller kernel is due to our newly introduced pyramid reduction. Thirdly, for the other graphs, MISKernel computes a slightly larger kernel but runs much faster than VCKernel. The slightly larger kernel is because, as noted in Sect. 3.1, MISKernel does not use the alternative reduction that are implemented in VCKernel; it will be our future work to incorporate the alternative reduction into MISKernel.

Based on the results in Table 2, we divide the twenty tested graphs into three categories according to their kernel sizes obtained by MISKernel. The first category contains the ten graphs that MISKernel computes an empty kernel, the second category contains the four graphs (Twitter, web-Google, as-Skitter, and LiveJournal) whose giant/largest kernel components obtained by MISKernel have between 1 and 1,000 vertices, and the third category contains the remaining six graphs that have larger giant kernel components than 1,000 vertices. As discussed in Sect. 3.2, in general the larger the giant kernel component, the hard the graph instance for MIS computation. Intuitively, the three categories correspond to easy, medium and hard graph instances, respectively. In particular, *for graphs in the first category,* MISKernel *already obtains the* MIS *during the kernelization process. Thus, we omit these graphs from the remaining testings.*

The results of running MISSolver and VCSolver on graphs in the second and third categories are shown in Table 3. Across all these graphs, MISSolver runs much faster than VCSolver. Specifically, for the graph as-Skitter, MISSolver

---

[3] https://www.cc.gatech.edu/dimacs10/downloads.shtml.

Table 2. Kernel size and running time of MISKernel and VCKernel (S_{GCC}: size of giant kernel component, S_{kernel}: total size of kernel)

Cat.	Graphs	MISKernel			VCKernel		
		S_{GCC}	S_{kernel}	Time	S_{GCC}	S_{kernel}	Time
1	CA-GrQc	0	0	0.014 s	0	0	0.029 s
	p2p-Gnutella	0	0	0.001 s	0	0	0.013 s
	soc-Slashdot	0	0	0.013 s	0	0	0.197 s
	CA-CondMat	0	0	0.004 s	0	0	0.070 s
	CA-AstroPh	0	0	0.008 s	0	0	0.112 s
	Email-EuAll	0	0	0.013 s	0	0	0.144 s
	soc-Epinions	0	0	0.008 s	0	0	0.108 s
	Amazon0302	0	0	0.147 s	0	0	1.773 s
	DBLP	0	0	0.047 s	0	0	0.423 s
	WikiTalk	0	0	0.115 s	0	0	0.603 s
2	Twitter	298	1,373	0.139 s	281	1,225	1.619 s
	web-Google	73	140	0.561 s	70	345	3.316 s
	as-Skitter	615	2,648	0.411 s	567	3,561	6.919 s
	LiveJournal	284	1,346	2.195 s	295	1,490	10.126 s
3	Europe	1,402	8,491	6.020 s	676	8,272	389 s
	Asia	2,438	15,789	1.248 s	1,226	15,217	158 s
	Amazon0601	8,507	8,864	14.519 s	7,455	7,750	7.877 s
	roadNet-PA	8,153	28,063	2.095 s	5,834	22,942	18.993 s
	roadNet-TX	5,122	35,529	1.714 s	4,102	31,262	20.025 s
	roadNet-CA	13,652	44,639	0.930 s	10,807	37,705	35.991 s

Table 3. Independent set size and running time of MISSolver and VCSolver (S_{IS}: size of independent set)

Graphs	MISSolver			VCSolver	
	S_{IS}	Gap to MIS upper bound	Time	S_{IS}	Time
Twitter	36,843	0	0.373 s	36,843	2.079 s
web-Google	529,138	0	0.617 s	529,138	3.329 s
as-Skitter	1,169,594	0	243 s	1,169,594	1,976 s
LiveJournal	2,629,616	0	2.377 s	2,629,616	11.113 s
Europe	25,633,431	2	20.911 s	25,633,431	398 s
Asia	5,998,332	35	425 s	5,998,335	700 s
Amazon0601	136,832	738	221 s	-	>20 h
roadNet-PA	533,620	219	245 s	-	>20 h
roadNet-TX	678,275	239	4,912 s	-	>20 h
roadNet-CA	961,831	354	7,142 s	-	>20 h

computes the exact MIS in 4 min, while VCSolver takes 33 min; this is due to our strategy of invoking maximum clique solver to solve kernel components. For graphs Amazon0601, roadNet-PA, roadNet-TX, and roadNet-CA, VCSolver cannot finish within 20 h due to the large size of giant kernel component; this validates our strategy of controlling the size of the giant kernel component. Note that, our algorithm MISSolver computes MIS for graphs in the second category, but not for graphs in the third category due to our giant kernel component control technique. Nevertheless, the gap of the size of the independent set reported by MISSolver to an upper bound of MIS size computed by MISSolver is small. Furthermore, it is worth mentioning that this gap estimation is pessimistic; for example, the gap to the MIS size for graphs Europe and Asia are 0 and 3, respectively, while the reported gap estimations are 2 and 35.

Table 4. Independent set size and running time of MISSolver, LinearTime and NearLinear (the S_{IS} gaps of LinearTime and NearLinear are with respect to the S_{IS} of MISSolver)

Graphs	MISSolver		LinearTime			NearLinear		
	S_{IS}	Time	S_{kernel}	S_{IS} gap	Time	S_{kernel}	S_{IS} gap	Time
Twitter	36,843	0.373 s	37,805	33	0.038 s	10,374	11	0.141 s
web-Google	529,138	0.617 s	321,097	140	0.451 s	1,282	7	0.433 s
as-Skitter	1,169,594	243 s	235,794	170	0.447 s	9,733	49	0.616 s
LiveJournal	2,629,616	2.377 s	271,493	378	2.529 s	10,173	33	2.620 s
Europe	25,633,431	20.911 s	1,499,996	6,606	6.795 s	695,593	3,960	11.190 s
Asia	5,998,332	425 s	626,657	4,126	1.238 s	426,195	3,249	2.285 s
Amazon0601	136,832	221 s	329,369	528	0.154 s	66,672	89	0.584 s
roadNet-PA	33,620	245 s	350,521	2,710	0.131 s	309,233	2,436	0.718 s
roadNet-TX	678,275	4,912 s	07,230	3,065	0.171 s	357,337	2,772	0.726 s
roadNet-CA	961,831	7,142 s	677,814	5,258	0.265 s	596,329	4,632	2.210 s

Eval-II: Against LinearTime and NearLinear. The results of evaluating MISSolver against the two existing heuristic MIS algorithms LinearTime and NearLinear are shown in Table 4. Specifically, the 5th column and 8th column, respectively, show the gap of the independent set reported by LinearTime and NearLinear to that by MISSolver. We can see that across all these tested graphs, MISSolver computes a much larger (*i.e.*, higher-quality) independent set than both LinearTime and NearLinear, and the gap can be several thousands for the street map graphs and road networks. This is due to the much smaller kernel computed by MISSolver; for example, the improvement of kernel size of MISSolver is up-to 2,293 times over LinearTime (*e.g.*, on web-Google), and up-to 82 times over NearLinear (*e.g.*, on Europe). Nevertheless, MISSolver as expected takes more time than LinearTime and NearLinear, due to iteratively applied more reduction rules. Thus, MISSolver is more suitable for computing a small kernel or for computing a large independent set, while LinearTime and NearLinear are suitable for a fast running time.

Table 5. Kernel size and running time of MISKernel, FastKer and ParFastKer

Graphs	MISKernel		ParFastKer		FastKer	
	S_{kernel}	Time	S_{kernel}	Time	S_{kernel}	Time
Twitter	1,373	0.151 s	1,440	0.311 s	1,345	0.173 s
web-Google	140	0.440 s	612	2.039 s	457	0.677 s
as-Skitter	2,648	0.448 s	4,620	1.178 s	4,596	0.678 s
LiveJournal	1,346	2.555 s	1,673	3.799 s	1,549	2.722 s
Europe	8,491	5.044 s	14,201	10.134 s	14,066	6.553 s
Asia	15,789	0.872 s	34,635	2.905 s	34,930	1.554 s
Amazon0601	8,864	0.853 s	14,990	3.839 s	8,888	1.241 s
roadNet-PA	28,063	0.496 s	37,924	1.04 s	38,077	1.627 s
roadNet-TX	35,529	0.487 s	44,205	1.159 s	44,313	0.754 s
roadNet-CA	44,639	0.825 s	57,847	2.033 s	57,795	1.199 s

Eval-III: Against FastKer and ParFastKer. In this testing, we evaluate our kernelization algorithm MISKernel against the two existing efficient kernelization algorithms FastKer and ParFastKer [10]. ParFastKer is a parallelized version of FastKer by utilizing multiple CPU cores, and needs to partition the input graph. Due to the parallelized computation, ParFastKer cannot be compiled and run on the machine we used in our previous experiments. As a result, we conducted this set of experiments on another machine with an Intel Core 2.6 GHz CPU, which is slightly faster. The kernel size and running time of these three algorithms are shown in Table 5. We can see that MISKernel computes a much smaller kernel and runs faster than both FastKer and ParFastKer. This is due to our strategies of aggressive incremental reduction rule applying and connected component checking. Note that, ParFastKer has an overhead of partitioning the input graph and thus could be slower than FastKer, and ParFastKer here uses only two CPU cores. Although the running time of ParFastKer could be improved by using more CPU cores, the size of the kernel will also increase with more CPU cores as shown in Table 5 and also in [10]. Thus, our kernelization algorithm MISKernel outperforms FastKer and ParFastKer in terms of computing a much smaller kernel.

5   Conclusion

In this paper, we firstly designed an efficient kernelization algorithm MISKernel by proposing two optimization strategies: aggressive incremental reduction rules and connected component checking. Then, we designed an efficient maximum independent set computation algorithm MISSolver by controlling the giant kernel component size, and invoking the existing maximum clique solvers for solving the kernel components. Experiments on large real graphs demonstrate that our kernelization algorithm computes a similar size kernel as VCSolver but runs faster, and computes smaller kernels than LinearTime, NearLinear, FastKer and ParFastKer. Our maximum independent set algorithm MISSolver runs faster than VCSolver, and computes a much larger independent set than LinearTime and NearLinear. Possible directions of future work are incorporating the alternative reduction into our algorithms, and optimize the maximum clique solver to be efficiently used as a subroute for our algorithm.

References

1. Akiba, T., Iwata, Y.: Branch-and-reduce exponential/fpt algorithms in practice: a case study of vertex cover. Theor. Comput. Sci. **609**, 211–225 (2016)
2. Andrade, D.V., Resende, M.G., Werneck, R.F.: Fast local search for the maximum independent set problem. J. Heuristics **18**(4), 525–547 (2012)
3. Araujo, F., Farinha, J., Domingues, P., Silaghi, G.C., Kondo, D.: A maximum independent set approach for collusion detection in voting pools. J. Parallel Distrib. Comput. **71**(10), 1356–1366 (2011)
4. van Bevern, R., Mnich, M., Niedermeier, R., Weller, M.: Interval scheduling and colorful independent sets. CoRR abs/1402.0851 (2014)
5. Chang, L.: Efficient maximum clique computation over large sparse graphs. In: Proceedings of KDD 2019, pp. 529–538 (2019)
6. Chang, L., Li, W., Zhang, W.: Computing a near-maximum independent set in linear time by reducing-peeling. In: Proceedings of SIGMOD 2017, pp. 1181–1196 (2017)
7. Dahlum, J., Lamm, S., Sanders, P., Schulz, C., Strash, D., Werneck, R.F.: Accelerating local search for the maximum independent set problem. In: Proceedings of SEA 2016 (2016)
8. Fomin, F.V., Grandoni, F., Kratsch, D.: A measure & conquer approach for the analysis of exact algorithms. J. ACM **56**(5), 25 (2009)
9. Håstad, J.: Clique is hard to approximate within $n^{1-epsilon}$. In: Proceedings of FOCS 1996, pp. 627–636 (1996)
10. Hespe, D., Schulz, C., Strash, D.: Scalable kernelization for maximum independent sets. In: Proceedings of ALENEX 2018, pp. 223–237 (2018)
11. Karp, R.M.: Reducibility among combinatorial problems. In: Miller, R.E., Thatcher, J.W., Bohlinger, J.D. (eds.) Complexity of Computer Computations. The IBM Research Symposia Series, pp. 85–103. Springer, Boston (1972). https://doi.org/10.1007/978-1-4684-2001-2_9
12. Lamm, S., Sanders, P., Schulz, C., Strash, D., Werneck, R.F.: Finding near-optimal independent sets at scale. In: Proceedings of ALENEX 2016, pp. 138–150 (2016)
13. Li, C., Fang, Z., Xu, K.: Combining maxsat reasoning and incremental upper bound for the maximum clique problem. In: Proceedings of ICTAI 2013, pp. 939–946 (2013)
14. Li, C.M., Jiang, H., Manyà, F.: On minimization of the number of branches in branch-and-bound algorithms for the maximum clique problem. Comput. Oper. Res. **84**, 1–15 (2017)
15. Liu, Y., Lu, J., Yang, H., Xiao, X., Wei, Z.: Towards maximum independent sets on massive graphs. PVLDB **8**(13), 2122–2133 (2015)
16. Nemhauser Jr., G. L., Trotter, L.: Vertex packings: structural properties and algorithms. Math Programm. **8**(1), 232–248 (1975)

17. Puthal, D., Nepal, S., Paris, C., Ranjan, R., Chen, J.: Efficient algorithms for social network coverage and reach. In: 2015 IEEE International Congress on Big Data, pp. 467–474 (2015)
18. Strijk, T., Verweij, A., Aardal, K.: Algorithms for maximum independent set applied to map labelling (2000)
19. Tomita, E.: Efficient algorithms for finding maximum and maximal cliques and their applications. In: Proceedings of WALCOM 2017, pp. 3–15 (2017)
20. Xiao, M., Nagamochi, H.: Confining sets and avoiding bottleneck cases: a simple maximum independent set algorithm in degree-3 graphs. Theor. Comput. Sci. **469**, 92–104 (2013)
21. Zheng, W., Wang, Q., Yu, J.X., Cheng, H., Zou, L.: Efficient computation of a near-maximum independent set over evolving graphs. In: Proceedings of ICDE 2018 (2018)

Fast Algorithms for Intimate-Core Group Search in Weighted Graphs

Longxu Sun[1(⊠)], Xin Huang[1], Rong-Hua Li[2], and Jianliang Xu[1]

[1] Hong Kong Baptist University, Hong Kong, China
sunlongxu@life.hkbu.edu.hk, {xinhuang,xujl}@comp.hkbu.edu.hk
[2] Beijing Institute of Technology, Beijing, China
lironghuabit@126.com

Abstract. Community search that finds query-dependent communities has been studied on various kinds of graphs. As one instance of community search, intimate-core group search over a weighted graph is to find a connected k-core containing all query nodes with the smallest group weight. However, existing state-of-the-art methods start from the maximal k-core to refine an answer, which is practically inefficient for large networks. In this paper, we develop an efficient framework, called local exploration k-core search (LEKS), to find intimate-core groups in graphs. We propose a small-weighted spanning tree to connect query nodes, and then expand the tree level by level to a connected k-core, which is finally refined as an intimate-core group. We also design a protection mechanism for critical nodes to avoid the collapsed k-core. Extensive experiments on real-life networks validate the effectiveness and efficiency of our methods.

Keywords: Graph mining · Weighted graphs · K-core · Community search

1 Introduction

Graphs widely exist in social networks, biomolecular structures, traffic networks, world wide web, and so on. Weighted graphs have not only the simple topological structure but also edge weights. The edge weight is often used to indicate the strength of the relationship, such as interval in social communications, traffic flow in the transportation network, carbon flow in the food chain, and so on [18–20]. Weighted graphs provide information that better describes the organization and hierarchy of the network, which is helpful for community detection [19] and community search [10, 11, 13, 26]. Community detection aims at finding all communities on the entire network, which has been studied a lot in the literature. Different from community detection, the task of community search finds only query-dependent communities, which has a wide application of disease infection control, tag recommendation, and social event organization [23, 29]. Recently, several community search models have been proposed in different dense subgraphs of k-core [2, 22] and k-truss [11, 24].

© Springer Nature Switzerland AG 2019
R. Cheng et al. (Eds.): WISE 2019, LNCS 11881, pp. 728–744, 2019.
https://doi.org/10.1007/978-3-030-34322-4_46

As a notation of dense subgraph, k-core requires that every vertex has k neighbors in the k-core. For example, Fig. 1(a) shows a graph G. Subgraphs G_1 and G_2 are both connected 3-cores, in which each vertex has at least three neighbors. K-core has been popularly used in many community search models [1, 9,16,17,23,31]. Recently, Zheng et al. [29] proposed one problem of intimate-core group search in weighted graphs as follows.

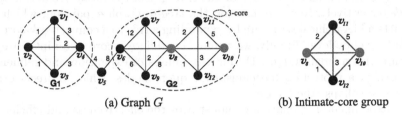

(a) Graph G (b) Intimate-core group

Fig. 1. An example of intimate-core group search in graph G for $Q = \{v_8, v_{10}\}$ and $k = 3$.

Motivating Example. Consider a social network G in Fig. 1(a). Two individuals have a closer friendship if they have a shorter interval for communication, indicating a smaller weight of the relationship edge. The problem of intimate-core group search aims at finding a densely-connected k-core containing query nodes Q with the smallest group weight as an answer. For $Q = \{v_8, v_{10}\}$ and $k = 3$, the intimate-core group is shown in Fig. 1(b) with a minimum group weight of 13.

This paper studies the problem of intimate-core group search in weighted graphs. Given an input of query nodes in a graph and a number k, the problem is to find a connected k-core containing query nodes with the smallest weight. In the literature, existing solutions proposed in [29] find the maximal connected k-core and iteratively remove a node from this subgraph for intimate-core group refinement. However, this approach may take a large number of iterations, which is inefficient for big graphs with a large component of k-core. Therefore, we propose a solution of local exploration to find a small candidate k-core, which takes a few iterations to find answers. To further speed up the efficiency, we build a k-core index, which keeps the structural information of k-core for fast identification. Based on the k-core index, we develop a local exploration algorithm LEKS for intimate-core group search. Our algorithm LEKS first generates a tree to connect all query nodes, and then expands it to a connected subgraph of k-core. Finally, LEKS keeps refining candidate graphs into an intimate-core group with small weights. We propose several well-designed strategies for LEKS to ensure the fast-efficiency and high-quality of answer generations.

Contributions. Our main contributions of this paper are summarized as follows.

- We investigate and tackle the problem of intimate-core group search in weighted graphs, which has wide applications on real-world networks. The problem is NP-hard, which bring challenges to develop efficient algorithms.
- We develop an efficient local exploration framework of LEKS based on the k-core index for intimate-core group search. LEKS consists of three phases: tree generation, tree-to-graph expansion, and intimate-core refinement.
- In the phase of tree generation, we propose to find a seed tree to connect all query nodes, based on two generated strategies of *spanning tree* and *weighted path* respectively. Next, we develop the tree-to-graph expansion, which constructs a hierarchical structure by expanding a tree to a connected k-core subgraph level by level. Finally, we refine a candidate k-core to an intimate-core group with a small weight. During the phases of expansion and refinement, we design a protection mechanism for query nodes, which protects critical nodes to collapse the k-core.
- Our experimental evaluation demonstrates the effectiveness and efficiency of our LEKS algorithm on real-world weighted graphs. We show the superiority of our methods in finding intimate groups with smaller weights, against the state-of-the-art ICG-M method [29].

Roadmap. The rest of the paper is organized as follows. Section 2 reviews the previous work related to ours. Section 3 presents the basic concepts and formally defines our problem. Section 4 introduces our index-based local exploration approach LEKS. Section 5 presents the experimental evaluation. Finally, Sect. 6 concludes the paper.

2   Related Work

In the literature, numerous studies have been investigated community search based on various kinds of dense subgraphs, such as k-core [2,22], k-truss [11,24] and clique [25,26]. Community search has been also studied on many labeled graphs, including weighted graphs [7,29,30], influential graphs [4,16], and keyword-based graphs [8,9,12]. Table 1 compares different characteristics of existing community search studies and ours.

The problem of k-core minimization [1,6,17,31] aims to find a minimal connected k-core subgraph containing query nodes. The minimum wiener connector problem is finding a small connected subgraph to minimize the sum of all pairwise shortest-path distances between the discovered vertices [21]. Different from all the above studies, our work aims at finding an intimate-core group containing multiple query nodes in weighted graphs. We propose fast algorithms for intimate-core group search, which outperform the state-of-the-art method [29] in terms of quality and efficiency.

3   Preliminaries

In this section, we formally define the problem of intimate-core group search and revisit the existing intimate-core group search approaches.

Table 1. A comparison of existing community search studies and ours

Method	Dense subgraph model	Node type	Edge type	Local search	Index-based	Multiple query nodes	NP-hard
[25]	Clique	×	×	✓	✓	×	✓
[26]	Clique	×	×	×	✓	✓	✓
[14]	k-truss	×	×	✓	✓	✓	✓
[17,31]	k-core	×	×	×	×	×	✓
[6]	k-core	×	×	✓	×	×	✓
[23]	k-core	×	×	×	×	✓	✓
[1]	k-core	×	×	✓	✓	✓	✓
[12]	k-truss	Keyword	×	✓	✓	✓	✓
[9]	k-core	Keyword	×	✓	✓	×	✓
[16]	k-core	Influential	×	×	✓	×	×
[4]	k-core	Influential	×	✓	×	×	×
[30]	k-truss	×	Weighted	✓	✓	×	×
[29]	k-core	×	Weighted	×	×	✓	✓
Ours	k-core	×	Weighted	✓	✓	✓	✓

3.1 Problem Definition

Let $G(V, E, w)$ be a weighted and undirected graph where V is the set of nodes, E is the set of edge, and w is an edge weight function. Let $w(e)$ to indicate the weight of an edge $e \in E$. The number of nodes in G is defined as $n = |V|$. The number of edges in G is defined as $m = |E|$. We denote the set of neighbors of a node v by $N_G(v) = \{u \in V : (u, v) \in E\}$, and the degree of v by $deg_G(v) = |N_G(v)|$. For example, Fig. 1(a) shows a weighted graph G. Node v_5 has two neighbors as $N_G(v_5) = \{v_4, v_6\}$, thus the degree of v_5 is $deg_G(v_5) = 2$ in graph G. Edge (v_2, v_3) has a weight of $w(v_2, v_3) = 1$. Based on the definition of degree, we can define the k-core as follows.

Definition 1 (K-Core [2]). *Given a graph G, the k-core is the largest subgraph H of G such that every node v has degree at least k in H, i.e., $deg_H(v) \geq k$.*

For a given integer k, the k-core of graph G is denoted by $C_k(G)$, which is determinative and unique by the definition of largest subgraph constraint. For example, the 3-core of G in Fig. 1(a) has two components G_1 and G_2. Every node has at least 3 neighbors in G_1 and G_2 respectively. However, the nodes are disconnected between G_1 and G_2 in the 3-core $C_3(G)$. To incorporate connectivity into k-core, we define a connected k-core.

Definition 2 (Connected K-Core). *Given graph G and number k, a connected k-core H is a connected component of G such that every node v has degree at least k in H, i.e., $deg_H(v) \geq k$.*

Intuitively, all nodes are reachable in a connected k-core, i.e., there exist paths between any pair of nodes. G_1 and G_2 are two connected 3-cores in Fig. 1(a).

Definition 3 (Group Weight). *Given a subgraph $H \subseteq G$, the group weight of H, denoted by $w(H)$, is defined as the sum of all edge weights in H, i.e.,*
$$w(H) = \sum_{e \in E(H)} w(e).$$

Example 1. For the subgraph $G_1 \subseteq G$ in Fig. 1(a), the group weight of G_1 is $w(G_1) = \sum_{e \in E(G_1)} w(e) = 1 + 3 + 5 + 2 + 1 + 3 = 15$.

On the basis of the definitions of connected k-core and group weight, we define the *intimate-core group* in a graph G as follows.

Definition 4 (Intimate-Core Group [29]). *Given a weighted graph $G = (V, E, w)$, a set of query nodes Q and a number k, the intimate-core group is a subgraph H of G if H satisfies following conditions:*

- **Participation.** *H contains all the query nodes Q, i.e., $Q \subseteq V_H$;*
- **Connected K-Core.** *H is a connected k-core with $deg_H(v) \geq k$;*
- **Smallest Group Weight.** *The group weight $w(H)$ is the smallest, that is, there exists no $H' \subseteq G$ achieving a group weight of $w(H') < w(H)$ such that H' also satisfies the above two conditions.*

Condition (1) of participation makes sure that the intimate-core group contains all query nodes. Moreover, Condition (2) of connected k-core requires that all group members are densely connected with at least k intimate neighbors. In addition, Condition (3) of minimized group weight ensures that the group has the smallest group weight, indicating the most intimate in any kinds of edge semantics. A small edge weight means a high intimacy among the group. Overall, intimate core groups have several significant advantages of small-sized group, offering personalized search for different queries, and close relationships with strong connections.

The problem of *intimate-core group search* studies in this paper is formulated in the following.

Problem Formulation: Given an undirected weighted graph $G(V, E, w)$, a number k, and a set of query nodes Q, the problem is to find the intimate-core group of Q.

Example 2. In Fig. 1(a), G is a weighted graph with 12 nodes and 20 edges. Each edge has a positive weight. Given two query nodes $Q = \{v_8, v_{10}\}$ and $k = 3$, the answer of intimate-core group for Q is the subgraph shown in Fig. 1(b). This is a connected 3-core, and also containing two query nodes $\{v_8, v_{10}\}$. Moreover, it has the minimum group weight among all connected 3-core subgraphs containing Q.

3.2   Existing Intimate-Core Group Search Algorithms

The problem of intimate-core group search has been studied in the literature [29]. Two heuristic algorithms, namely, ICG-S and ICG-M, are proposed to deal with this problem in an online manner. No optimal algorithms have been proposed yet because this problem has been proven to be NP-hard [29]. The NP-hardness is shown by reducing the NP-complete clique decision problem to the intimate-core group search problem.

Existing solutions ICG-S and ICG-M both first identify a maximal connected k-core as a candidate, and then remove the node with the largest weight of its incident edges at each iteration [29]. The difference between ICG-S and ICG-M lies on the node removal. ICG-S removes one node at each iteration, while ICG-M removes a batch of nodes at each iteration. Although ICG-M can significantly reduce the total number of removal iterations required by ICG-S, it still takes a large number of iterations for large networks. The reason is that the initial candidate subgraph connecting all query nodes is the maximal connected k-core, which may be too large to shrink. This, however, is not always necessary. In particular, if there exists a small connected k-core surrounding query nodes, then a few numbers of iterations may be enough token for finding answers. This paper proposes a local exploration algorithm to find a smaller candidate subgraph. On the other hand, both ICG-S and ICG-M apply the core decomposition to identify the k-core from scratch, which is also costly expensive. To improve efficiency, we propose to construct an index offline and retrieve k-core for queries online.

4 Index-Based Local Exploration Algorithms

In this section, we first introduce a useful core index and the index construction algorithm. Then, we present the index-based intimate-core group search algorithms using local exploration.

4.1 K-Core Index

We start with a useful definition of coreness as follows.

Definition 5 (Coreness). *The coreness of a node $v \in V$, denoted by $\delta(v)$, is the largest number k such that there exists a connected k-core containing v.*

Obviously, for a node q with the coreness $\delta(q) = l$, there exists a connected k-core containing q where $1 \leq k \leq l$; meanwhile, there is no connected k-core containing q where $k > l$. The k-core index keeps the coreness of all nodes in G.

K-Core Index Construction. We apply the existing core decomposition [2] on graph G to construct the k-core index. The algorithm is outlined in Algorithm 1. The core decomposition is to compute the coreness of each node in graph G. Note that for the self-completeness of our techniques and reproducibility, the detailed algorithm of core decomposition is also presented (lines 1–7). First, the algorithm sort all nodes in G based on their degree in ascending order. Second, it finds the minimum degree in G as d. Based on the definition of k-core, it next computes the coreness of nodes with $deg_G(v) = d$ as d and removing these nodes and their incident edges from G. With the deletion of these nodes, the degree of neighbors of these nodes will decrease. For those nodes which have a new degree at most d, they will not be in (d+1)-core while they will get $\delta(v) = d$. It continues the

---

Algorithm 1. Core Index Construction

Input: A weighted graph $G = (V, E, w)$
Output: Coreness $\delta(v)$ for each $v \in V_G$

1: Sort all nodes in G in ascending order of their degree;
2: **while** $G \neq \emptyset$
3:    Let d be the minimum degree in G;
4:    **while** there exists $deg_G(v) \leq d$
5:        $\delta(v) \leftarrow d$;
6:        Remove v and its incident edges from G;
7:        Re-order the remaining nodes in G in ascending order of their degree;
8: Store $\delta(v)$ in index for each $v \in V_G$;

---

removal of nodes until there is no node has $deg_G(v) \leq d$. Then, the algorithm back to line 2 and starts a new iteration to compute the coreness of remaining nodes. Finally, it stores the coreness of each vertex v in G as the k-core index.

4.2 Solution Overview

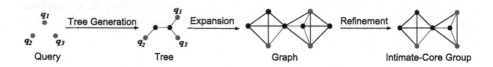

Fig. 2. LEKS framework for intimate-core group search

At a high level, our algorithm of local exploration based on k-core index for intimate-core group search (LEKS) consists of three phases:

1. *Tree Generation Phase*: This phase invokes the shortest path algorithm to find the distance between any pair of nodes, and then constructs a small-weighted tree by connecting all query nodes.
2. *Expansion Phase*: This phase expands a tree into a graph. It applies the idea of local exploration to add nodes and edges. Finally, it obtains a connected k-core containing all query nodes.
3. *Intimate-Core Refinement Phase*: This phase removes nodes with large weights, and maintains the candidate answer as a connected k-core. This refinement process stops until an intimate-core group is obtained.

Figure 2 shows the whole framework of our index-based local exploration algorithm. Note that we compute the k-core index offline and apply the above solution of online query processing for intimate-core group search. In addition, we consider $|Q| \geq 2$ for tree generation phase, and skip this phase if $|Q| = 1$. Algorithm 2 also depicts our algorithmic framework of LEKS.

Algorithm 2. LEKS Framework

Input: $G = (V, E, w)$, an integer k, a set of query vertices Q
Output: Intimate-core group H

1: Find a tree T_Q for query nodes Q using Algorithm 3 or Algorithm 4;
2: Expand the tree T_Q to a candidate graph G_Q in Algorithm 5;
3: Apply ICG-M [29] on graph G_Q;
4: Return a refined intimate-core group as answers;

4.3 Tree Generation

In this section, we present the phase of tree generation. Due to the large-scale size of k-core in practice, we propose local exploration methods to identify small-scale substructures as candidates from the k-core. The approaches produce a tree structure with small weights to connect all query nodes. We develop two algorithms, respectively based on the minimum spanning tree (MST) and minimum weighted path (MWP).

Tree Construction. The tree construction has three major steps. Specifically, the algorithm firstly generates all-pairs shortest paths for query nodes Q in the k-core C_k (lines 1–7). Given a path between nodes u and v, the path weight is the total weight of all edges along this path between u and v. It uses $\mathsf{spath}_{C_k}(u, v)$ to represent the shortest path between nodes u and v in the k-core C_k. For any pair of query nodes q_i, $q_j \in Q$, our algorithm invokes the well-known Dijkstra's algorithm [5] to find the shortest path $\mathsf{spath}_{C_k}(q_i, q_j)$ in the k-core C_k.

Second, the algorithm constructs a weighted graph G_{pw} for connecting all query nodes (lines 3–8). Based on the obtained all-pairs shortest paths, it collects and merges all these paths together to construct a weighted graph G_{pw} correspondingly.

Third, the algorithm generates a small spanning tree for Q in the weighted graph G_{pw} (lines 9–22), since not all edges are needed to keep the query nodes connected in G_{pw}. This step finds a compact spanning tree to connect all query nodes Q, which removes no useful edges to reduce weights. Specifically, the algorithm starts from one of the query nodes and does expand based on Prim's minimum spanning tree algorithm [5]. The algorithm stops when all query nodes are connected into a component in G_{pw}. Against the maximal connected k-core, our compact spanning tree has three significant features: (1) Query-centric. The tree involves all query nodes of Q. (2) Compactly connected. The tree is a connected and compact structure; (3) Small-weighted. The generation of minimum spanning tree ensures a small weight of the discovered tree.

Example 3. Figure 3(a) shows a weighted graph G with 6 nodes and 8 edges with weights. Assume that $k = 2$, the whole graph is 2-core as C_2. A set of query nodes $Q = \{v_1, v_2, v_5\}$ are colored in red in Fig. 3(a). We first find the shortest path between every pair of query nodes in Q. All edges along with

Algorithm 3. Tree Construction

Input: $G = (V, E, w)$, an integer k, a set of query vertices Q, the k-core index
Output: Tree T_Q

1: Identify the maximal connected k-core of C_k containing query nodes Q;
2: Let G_{pw} be an empty graph;
3: **for** $q_1, q_2 \in Q$
4:      **if** there is no path between q_1 and q_2 in C_k **then**
5:          **return** $\emptyset$;
6:      **else**
7:          Compute the shortest path between q_1 and q_2 in C_k;
8:          Add the $\mathsf{spath}_{C_k}(q_1, q_2)$ between q_1 and q_2 into G_{pw};
9: Tree: $T_Q \leftarrow \emptyset$;
10: Priority queue: $L \leftarrow \emptyset$;
11: **for** each node v in G_{pw}
12:      $\mathrm{dist}(v) \leftarrow \infty$;
13: $Q \leftarrow Q - \{q_0\}$; dist $(q_0) \leftarrow 0$; $L.push(q_0, \mathrm{dist}(q_0))$;
14: **while** $Q \neq \emptyset$ **do**
15:      Extract a node v and its edges with the smallest $\mathrm{dist}(v)$ from L;
16:      Insert node v and its edges into T_Q;
17:      **if** $v \in Q$ **then**
18:          $Q \leftarrow Q - \{v\}$;
19:      **for** $u \in N_{G_{pw}}(v)$ **do**
20:          **if** $\mathrm{dist}(u) > w(u, v)$ **then**
21:              $\mathrm{dist}(u) \leftarrow w(u, v)$;
22:              Update $(u, \mathrm{dist}(u))$ in L;
23: **return** T_Q;

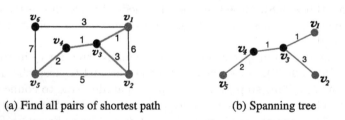

(a) Find all pairs of shortest path   (b) Spanning tree

Fig. 3. Tree construction for query nodes v_1, v_2, v_5.

these shortest path are colored in red in Fig. 3(a). For example, the shortest path between v_1 and v_2 is $\mathsf{spath}_{C_2}(v_1, v_2) = \{(v_1, v_3), (v_3, v_2)\}$. Similarly, $\mathsf{spath}_{C_2}(v_1, v_5) = \{(v_1, v_3), (v_3, v_4), (v_4, v_5)\}$, $\mathsf{spath}_{C_2}(v_2, v_5) = \{(v_2, v_5)\}$. All three paths are merged to construct a weighted graph G_{pw} in red in Fig. 3(a). A spanning tree of T_Q is shown in Fig. 3(b), which connects all query nodes $\{v_1, v_2, v_5\}$ with a small weight of 7.

Path-Based Construction. Algorithm 3 may take expensive computation for finding the shortest path between every pair of nodes that are far away from

---

Algorithm 4. Path-based Construction

Input: $G = (V, E, w)$, an integer k, a set of query vertices Q, the k-core index

Output: Tree T_Q

1: Identify the maximal connected k-core of C_k containing query nodes Q;
2: Let q_0 be the first query node of Q;
3: $Q \leftarrow Q - \{q_0\}$;
4: **while** $Q \neq \emptyset$ **do**
5:   **if** there is no path between q and q_0 in C_k **then**
6:     **return** $\emptyset$;
7:   **else**
8:     Compute the shortest path between q and q_0 in C_k;
9:     Add the $\mathsf{spath}_{C_k}(q, q_0)$ between q and q_0 into T_Q;
10:    $q_0 \leftarrow q$, $Q \leftarrow Q - \{q_0\}$;
11: **return** T_Q;

---

each other. To improve efficiency, we develop a path-based approach to connect all query nodes directly. The path-based construction is outlined in Algorithm 4. The algorithm starts from one query node q_0, and searches the shortest path to the nearest query node in Q (lines 2–8). After that, it collects and merges the weighted path $\mathsf{spath}_{C_k}(q, q_0)$ into T_Q to construct the tree (line 9). Recursively, it starts from the new query node q as q_0 to find the next nearest query node q, until all query nodes in Q are found in such a way (line 10). The algorithm returns the tree connecting all query nodes.

Example 4. We apply Algorithm 4 on graph G in Fig. 3(a) with query $Q = \{v_1, v_2, v_5\}$ and $k = 2$. We start the shortest path search from v_1. The nearest query node to v_1 is v_5, we can find the shortest path $\mathsf{spath}_{C_2}(v_1, v_5) = \{(v_1, v_3), (v_3, v_4), (v_4, v_5)\}$. Next, we start from v_5 and find the shortest path $\mathsf{spath}_{C_2}(v_5, v_2) = \{(v_5, v_2)\}$. Finally, we merge the two paths $\mathsf{spath}_{C_2}(v_1, v_5)$ and $\mathsf{spath}_{C_2}(v_5, v_2)$ to construct the tree T_Q.

Complexity Analysis. We analyze the complexity of Algorithms 3 and 4. Assume that the k-core C_k has n_k nodes and m_k edges where $n_k \leq n$ and $m_k \leq m$.

For Algorithm 3, an intuitive implementation of all-pairs-shortest-paths needs to compute the shortest path for every pair nodes in Q, which takes $O(|Q|^2 m_k \log n_k)$ time. However, a fast implementation of single-source-shortest-path algorithm can compute the shortest path from one query node $q \in Q$ to all other nodes in Q, which takes $O(m_k \log n_k)$ time. Overall, the computation of all-pairs-shortest-paths can be done in $O(|Q| m_k \log n_k)$ time. In addition, the weighted graph G_{pw} is a subgraph of C_k, thus the size of G_{pw} is $O(n_k + m_k) \subseteq O(m_k)$. Identifying the spanning tree of G_{pw} takes $O(m_k \log n_k)$ time. Overall, Algorithm 3 takes $O(|Q| m_k \log n_k)$ time and $O(m_k)$ space.

For Algorithm 4, it applies $|Q|$ times of single-source-shortest-path to identify the nearest query node. Thus, Algorithm 4 also takes $O(|Q| m_k \log n_k)$ time and

---

Algorithm 5. Tree-to-Graph Expansion

Input: $G = (V, E, w)$, a set of query vertices Q, k-core index, T_Q
Output: Candidate subgraph G_Q

1: Identify the maximal connected k-core of C_k containing query nodes Q;
2: $L_0 \leftarrow \{v | v \in V_{T_Q}\}$; $L' \leftarrow L_0$;
3: $i \leftarrow 0$; $G_Q \leftarrow \emptyset$;
4: **while** $G_Q = \emptyset$ **do**
5:     **for** each $v \in L_i$ **do**
6:         **for** each $u \in N_{C_k}(v)$ and $u \notin L' \cup L_{i+1}$ **do**
7:             $L_{i+1} \leftarrow L_{i+1} \cup \{u\}$;
8:     $L' \leftarrow L' \cup L_{i+1}$; $i \leftarrow i + 1$;
9:     Let G_L be the induced subgraph of G by the node set L';
10:    Generate a connected k-core of G_L containing query nodes Q as G_Q;
11: **return** G_Q;

$O(m_k)$ space. In practice, Algorithm 4 runs faster than Algorithm 3 on large real-world graphs, which avoids the weighted tree construction and all-pairs-shortest-paths detection.

4.4 Tree-to-Graph Expansion

In this section, we introduce the phase of tree-to-graph expansion. This method expands the obtained tree from Algorithms 3 or 4 into a connected k-core candidate subgraph G_Q. It consists of two main steps. First, it adds nodes/edges to expand the tree into a graph layer by layer. Then, it prunes disqualified nodes/edges to maintain the remaining graph as a connected k-core. The whole procedure is shown in Algorithm 5.

Algorithm 5 first gets all nodes in T_Q and puts them into L_0 (line 2). Let L_i be the vertex set at the i-th depth of expansion tree, and L_0 be the initial set of vertices. It uses L' to represent the set of candidate vertices, which is the union of all L_i set. The iterative procedure can be divided into three steps (lines 4–10). First, for each vertex v in L_i, it adds their neighbors into L_{i+1} (lines 5–7). Next, it collects and merges $\{L_0, ..., L_{i+1}\}$ into L' and constructs a candidate graph G_L as the induced subgraph of G by the node set L' (lines 8–9). Finally, we apply the core decomposition algorithm on G_L to find the connected k-core subgraph containing all query nodes, denoted as G_Q. If there exists no such G_Q, Algorithm 5 explores the $(i+1)$-th depth of expansion tree and repeats the above procedure (lines 4–10). In the worst case, G_Q is exactly the maximum connected k-core subgraph containing Q. However, G_Q in practice is always much smaller than it. The time complexity for expansion is $O(\sum_{i=0}^{l_{max}} \sum_{v \in V(G_i)} \deg(v))$, where l_{max} is the iteration number of expansion in Algorithm 5.

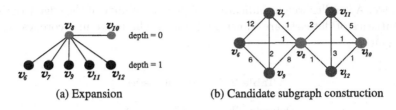

|(a) Expansion | (b) Candidate subgraph construction|

Fig. 4. Tree-to-graph expansion

Example 5. Figure 1(a) shows a weighted graph G with query $Q = \{v_8, v_{10}\}$ and $k = 3$. We first identify the maximal connected 3-core containing query nodes Q. Since there is only 2 query nodes, the spanning tree is same as the shortest path between them, such that $T_Q = \mathsf{spath}_{C_3}(v_8, v_{10})$. Next, we initialize L_0 as $L_0 = \{v_8, v_{10}\}$ and expand nodes in L_0 to their neighbors. The expansion procedure is shown in Fig. 4(a). We put all nodes in Fig. 4(a) into L' and construct a candidate subgraph G_L shown in Fig. 4(b). Since G_L is a 3-core connected subgraph containing query nodes, the expansion graph G_Q is G_L itself.

4.5 Intimate-Core Refinement

This phase refines the candidate connected k-core into an answer of the intimate-core group. We apply the existing approach ICG-M [29] by removing nodes to shrink the candidate graph obtained from Algorithm 5. This step takes $O(m' \log_\varepsilon n')$ time, where $\varepsilon > 0$ is a parameter of shrinking graph [29]. To avoid query nodes deleted by the removal processes of ICG-M, we develop a mechanism to protect important query nodes.

Protection Mechanism for Query Nodes. As pointed by [3,27,28], the k-core structure may collapse when critical nodes are removed. Thus, we precompute such critical nodes for query nodes in k-core and ensure that they are not deleted in any situations. We use an example to illustrate our ideas. For a query node q with an exact degree of k, it means that if any neighbor is deleted, there exists no feasible k-core containing q any more. Thus, q and all q's neighbors are needed to protect. For example, in Fig. 4(b), assume that $k = 3$, there exists $deg_G(v_{10}) = k$. The removal of each node in $N_G(v_{10})$ will cause core decomposition and the deletion of v_{10}. This protection mechanism for query nodes can also be used for k-core maintenance in the phrase of tree-to-graph expansion.

5   Experiments

In this section, we experimentally evaluate the performance of our proposed algorithms. All algorithms are implemented in Java and performed on a Linux server with Xeon E5-2630 (2.2 GHz) and 256 GB RAM.

Datasets. We use three real-world datasets in experiments. All datasets are publicly available from [15]. The edge weight represents the existence probability of an edge. A smaller weight indicates a higher possibility of the edge to existing. The statistics of datasets are shown in Table 2. The maximum coreness $\delta_{max} = \max_{v \in V} \delta(v)$.

Table 2. Network statistics

Datasets	$|V|$	$|E|$	δ_{max}
wiki-vote	7,115	103,689	56
Flickr	24,125	300,836	225
DBLP	684,911	2,284,991	114

Algorithms. We compare 3 algorithms as follows.

- ICG-M: is the state-of-the-art approach for finding intimate-core group using bulk deletion [29].
- LEKS-tree: is our index-based search framework in Algorithm 2 using Algorithm 3 for tree generation.
- LEKS-path: is our index-based search framework in Algorithm 2 using Algorithm 4 for tree generation.

We evaluate all algorithms by comparing the running time and the intimate-core group weight. The less running time costs, the more efficient the algorithm is. Smaller the group weight of the answer, better effectiveness is.

Queries and Parameters. We evaluate all competitive approaches by varying parameters k and $|Q|$. The range of k is $\{2, 4, 6, 8\}$. The number of query nodes $|Q|$ falls in $\{1, 2, 3, 4, 5, 6, 7\}$. We randomly generate 100 sets of queries by different k and $|Q|$.

Exp-1: Varying k. Figure 5 shows the group weight of three algorithms by varying parameter k on all datasets. The results show that our local search methods LEKS-tree and LEKS-path can find intimate groups with lower group weights than ICG-M, for different k. The performance of LEKS-tree and LEKS-path are similar. Figure 6 shows that LEKS-path performs the best for most cases, and runs significantly faster than ICG-M. Interestingly, ICG-M can find answers quickly for $k = 4$, which achieves similar performance with LEKS methods.

Exp-2: Varying $|Q|$. Figure 7 reports the group weight results of three algorithms for different queries by varying $|Q|$. With the increasing $|Q|$, LEKS-tree and LEKS-path methods can always find intimate groups with smaller weights than ICG-M. LEKS-tree and LEKS-path perform similarly. Figure 8 reports the results of running time. It shows that our methods are always faster than ICG-M.

Exp-3: Quality Evaluation of Candidate Intimate-Core Groups. This experiment evaluates the subgraphs of candidate intimate-core groups by all methods, in terms of vertex size and group weight. ICG-M takes the maximal

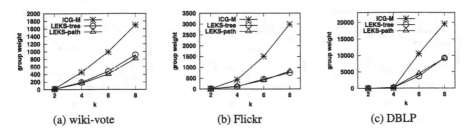

Fig. 5. Effectiveness evaluation by varying k

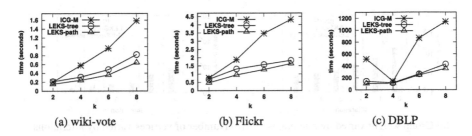

Fig. 6. Efficiency evaluation by varying k

connected k-core subgraph containing query nodes as an initial candidate, and iteratively shrinks it. LEKS-tree and LEKS-path both generate an initial candidate subgraph locally expanded from a tree, and then iteratively shrink the candidate by removing nodes. We use $k = 6$ and $|Q| = 5$. We report the results of the first 5 removal iterations and the initial candidate at the #iteration of 0. Figure 9(a) shows that the group weight of candidates by our methods is much smaller than ICG-M. Figure 9(b) reports the vertex size of all candidates at each iteration. The number of vertices in the candidate group by LEKS-tree and LEKS-path at the #iteration of 0, is even less than the vertex size of candidate group by ICG-M at the #iteration of 5.

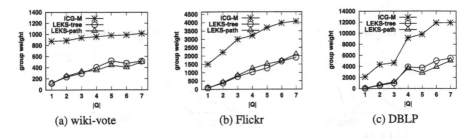

Fig. 7. Effectiveness evaluation by varying $|Q|$

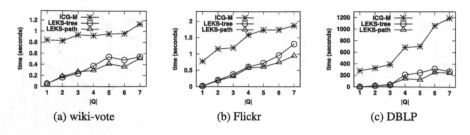

Fig. 8. Efficiency evaluation by varying $|Q|$

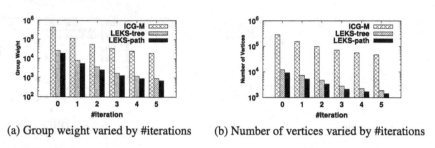

(a) Group weight varied by #iterations   (b) Number of vertices varied by #iterations

Fig. 9. The size and weight of intimate-groups varied by #iterations

Exp-4: Case Study on the DBLP Network. We conduct a case study of intimate-core group search on the collaboration DBLP network [29]. Each node represents an author, and an edge is added between two authors if they have co-authored papers. The weight of an edge (u, v) is the reciprocal of the number of papers they have co-authored. The smaller weight of (u, v), the closer intimacy between authors u and v. We use the query $Q = \{$"Huan Liu", "Xia Hu", "Jiliang Tang"$\}$ and $k = 4$. We apply LEKS-path and ICG-M to find 4-core intimate groups for Q. The results of LEKS-path and ICG-M are shown in Fig. 10(a) and Fig. 10(b) respectively. The bolder lines of an edge represent a smaller weight, indicating closer intimate relationships. Our LEKS method discovers a compact 4-core with 5 nodes and 10 edges in Fig. 10(a), which has the group

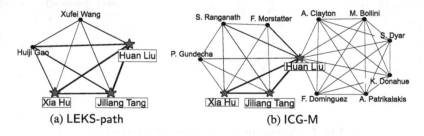

(a) LEKS-path   (b) ICG-M

Fig. 10. Case study of intimate-core group search on the DBLP network. Here, query $Q = \{$"Huan Liu", "Xia Hu", "Jiliang Tang"$\}$ and $k = 4$.

weight of 1.6, while ICG-M finds a subgraph with 12 nodes, which has a larger group weight of 16.7 in Fig. 10(b). We can see that nodes on the right side of Fig. 10(b) has no co-author connections with two query nodes "Xia Hu" and "Jiliang Tang" at all. This case study verifies that our LEKS-path can successfully find a better intimate-core group than ICG-M.

6 Conclusion

This paper presents a local exploration k-core search (LEKS) framework for efficient intimate-core group search. LEKS generates a spanning tree to connect query nodes in a compact structure, and locally expands it for intimate-core group refinement. Extensive experiments on real datasets show that our approach achieves a higher quality of answers using less running time, in comparison with the existing ICG-M method.

Acknowledgments. This work is supported by the NSFC Nos. 61702435, 61772346, U1809206, RGC Nos. 12200917, 12200817, and CRF C6030-18GF.

References

1. Barbieri, N., Bonchi, F., Galimberti, E., Gullo, F.: Efficient and effective community search. DMKD **29**(5), 1406–1433 (2015)
2. Batagelj, V., Zaversnik, M.: An O(m) algorithm for cores decomposition of networks. arXiv preprint arXiv:cs/0310049 (2003)
3. Bhawalkar, K., Kleinberg, J., Lewi, K., Roughgarden, T., Sharma, A.: Preventing unraveling in social networks: the anchored k-core problem. SIAM J. Discrete Math. **29**(3), 1452–1475 (2015)
4. Bi, F., Chang, L., Lin, X., Zhang, W.: An optimal and progressive approach to online search of top-k influential communities. PVLDB **11**(9), 1056–1068 (2018)
5. Cormen, T.H., Leiserson, C.E., Rivest, R.L., Stein, C.: Introduction to algorithms (2009)
6. Cui, W., Xiao, Y., Wang, H., Wang, W.: Local search of communities in large graphs. In: SIGMOD, pp. 991–1002 (2014)
7. Duan, D., Li, Y., Jin, Y., Lu, Z.: Community mining on dynamic weighted directed graphs. In: ACM International Workshop on Complex Networks Meet Information & Knowledge Management, pp. 11–18 (2009)
8. Fang, Y., Cheng, R., Chen, Y., Luo, S., Hu, J.: Effective and efficient attributed community search. VLDBJ **26**(6), 803–828 (2017)
9. Fang, Y., Cheng, R., Luo, S., Hu, J.: Effective community search for large attributed graphs. PVLDB **9**(12), 1233–1244 (2016)
10. Fang, Y., et al.: A survey of community search over big graphs. arXiv preprint arXiv:1904.12539 (2019)
11. Huang, X., Cheng, H., Qin, L., Tian, W., Yu, J.X.: Querying k-truss community in large and dynamic graphs. In: SIGMOD, pp. 1311–1322 (2014)
12. Huang, X., Lakshmanan, L.V.: Attribute-driven community search. PVLDB **10**(9), 949–960 (2017)

13. Huang, X., Lakshmanan, L.V., Xu, J.: Community Search over Big Graphs. Morgan & Claypool Publishers, San Rafael (2019)
14. Huang, X., Lakshmanan, L.V., Yu, J.X., Cheng, H.: Approximate closest community search in networks. PVLDB **9**(4), 276–287 (2015)
15. Huang, X., Lu, W., Lakshmanan, L.V.: Truss decomposition of probabilistic graphs: semantics and algorithms. In: SIGMOD, pp. 77–90 (2016)
16. Li, R.-H., Qin, L., Yu, J.X., Mao, R.: Influential community search in large networks. PVLDB **8**(5), 509–520 (2015)
17. Medya, S., Ma, T., Silva, A., Singh, A.: K-core minimization: a game theoretic approach. arXiv preprint arXiv:1901.02166 (2019)
18. Newman, M.E.: Scientific collaboration networks. II. Shortest paths, weighted networks, and centrality. Phys. Rev. E **64**(1), 016132 (2001)
19. Newman, M.E.: Analysis of weighted networks. Phys. Rev. E **70**(5), 056131 (2004)
20. Opsahl, T., Agneessens, F., Skvoretz, J.: Node centrality in weighted networks: generalizing degree and shortest paths. Soc. Netw. **32**(3), 245–251 (2010)
21. Ruchansky, N., Bonchi, F., García-Soriano, D., Gullo, F., Kourtellis, N.: The minimum wiener connector problem. In: SIGMOD, pp. 1587–1602 (2015)
22. Saríyüce, A.E., Gedik, B., Jacques-Silva, G., Wu, K.-L., Çatalyürek, Ü.V.: Streaming algorithms for k-core decomposition. PVLDB **6**(6), 433–444 (2013)
23. Sozio, M., Gionis, A.: The community-search problem and how to plan a successful cocktail party. In: KDD, pp. 939–948 (2010)
24. Wang, J., Cheng, J.: Truss decomposition in massive networks. PVLDB **5**(9), 812–823 (2012)
25. Yuan, L., Qin, L., Lin, X., Chang, L., Zhang, W.: Diversified top-k clique search. VLDBJ **25**(2), 171–196 (2016)
26. Yuan, L., Qin, L., Zhang, W., Chang, L., Yang, J.: Index-based densest clique percolation community search in networks. ICDE **30**(5), 922–935 (2017)
27. Zhang, F., Zhang, W., Zhang, Y., Qin, L., Lin, X.: OLAK: an efficient algorithm to prevent unraveling in social networks. PVLDB **10**(6), 649–660 (2017)
28. Zhang, F., Zhang, Y., Qin, L., Zhang, W., Lin, X.: Finding critical users for social network engagement: the collapsed k-core problem. In: AAAI (2017)
29. Zheng, D., Liu, J., Li, R.-H., Aslay, C., Chen, Y.-C., Huang, X.: Querying intimate-core groups in weighted graphs. In: IEEE International Conference on Semantic Computing, pp. 156–163 (2017)
30. Zheng, Z., Ye, F., Li, R.-H., Ling, G., Jin, T.: Finding weighted k-truss communities in large networks. Inf. Sci. **417**, 344–360 (2017)
31. Zhu, W., Chen, C., Wang, X., Lin, X.: K-core minimization: an edge manipulation approach. In: CIKM, pp. 1667–1670 (2018)

Text Mining

Unsupervised Ontology- and Sentiment-Aware Review Summarization

Nhat X. T. Le[✉] [ID], Neal Young [ID], and Vagelis Hristidis [ID]

Department of Computer Science and Engineering, University of California,
900 University Avenue, Riverside, CA 92521, USA
{nle020,neal.young}@ucr.edu, vagelis@cs.ucr.edu

Abstract. In this Web 2.0 era, there is an ever increasing number of customer reviews, which must be summarized to help consumers effortlessly make informed decisions. Previous work on reviews summarization has simplified the problem by assuming that aspects (e.g., "display") are independent of each other and that the opinion for each aspect in a review is Boolean: positive or negative. However, in reality aspects may be interrelated – e.g., "display" and "display color" – and the sentiment takes values in a continuous range – e.g., somewhat vs very positive. We present a novel, unsupervised review summarization framework that advances the state-of-the-art by leveraging a domain hierarchy of concepts to handle the semantic overlap among the aspects, and by accounting for different sentiment levels. We show that the problem is NP-hard and present bounded approximate algorithms to compute the most representative set of sentences or reviews, based on a principled opinion coverage framework. We experimentally evaluate the proposed algorithms on real datasets in terms of their efficiency and effectiveness compared to the optimal algorithms. We also show that our methods generate summaries of superior quality than several baselines in short execution times.

Keywords: Review summarization · Unsupervised extractive summarization · Online customer review · Aspect based sentiment analysis

1 Introduction

Online users are increasingly relying on user reviews to make decisions on shopping (e.g., Amazon, Newegg), seeking doctors (e.g., Vitals.com, zocdoc.com) and many others. However, as the number of reviews per item grows, especially for popular products, it is infeasible for customers to read all of them, and discern the useful information from them. Therefore, many methods have been proposed to summarize customer opinions from the reviews [5,9,13,17]. They generally either choose important text segments [13], or extract product concepts (also referred as aspects or attributes in other works), such as "display" of a phone, and customer's opinion (positive or negative) and aggregate them [5,9,17].

© Springer Nature Switzerland AG 2019
R. Cheng et al. (Eds.): WISE 2019, LNCS 11881, pp. 747–762, 2019.
https://doi.org/10.1007/978-3-030-34223-4_47

However, neither of these approaches takes into account the relationship among product's concepts. For example, assuming that we need the opinion summary of a smartphone, showing that the opinions for both *display* and *display color* are very positive is redundant, especially given that we would have to hide other concepts' opinion (e.g., "battery"), given the limited summary size. What makes the problem more challenging is that the opinion of a user for a concept is not Boolean (positive or negative) but can take values from a linear scale, e.g., "very positive", "positive", "somewhat positive", "neutral", and so on. Hence, if "display" has a positive opinion, but "display color" has neutral, the one does not subsume the other, and both should be part of the summary. Further, a more general concept may cover a more specific but not vice versa.

Our key contribution is a novel review summarization framework that accounts for the relationships among the concepts (product aspects), while at the same time supporting various sentiment levels. Specifically, we model our problem as a pairs coverage problem, where each pair consists of a concept and a sentiment value, and coverage is jointly defined on both of them. We show that the problem of selecting the best concepts and opinions to display is NP-hard even when the relationships among the concepts are represented by a Directed-Acyclic-Graph (DAG). For that, we propose bounded approximation algorithms inspired by well-studied graph coverage algorithms.

To summarize, the review summarization framework consists of the following tasks: *(a) Concept Extraction*: we build upon existing work for extracting hierarchical concepts (aspects) from reviews. *(b) Sentiment Estimation*: estimate the sentiment of each mentioned concept on a linear scale. *(c) Select k representatives*: depending on the problem variant, a representative is a concept-sentiment pair (e.g., "display" = 0.3), or a sentence from a review (e.g., "this phone has pretty sharp display") or a whole review. Our proposed selection algorithms can be used to select representatives at any of these granularities. Note that our summarization approach is unsupervised, thus does not require any labeled dataset which is expensive to create in a new domain.

Our contributions can be summarized as below:

- We propose a fresh perspective for the review summarization problem that exploits available concept hierarchies and a novel opinion coverage definition. We model the problem as a coverage optimization problem (Sect. 2) and show how to map a set of reviews to our model (Sect. 5.1).
- We prove that the problem is NP-hard and propose several efficient approximation algorithms with guaranteed bounds (Sect. 4).
- We carry out a thorough evaluation on the cost and time of our proposed algorithms. We experimentally evaluate our methods on real collections of online doctor patient reviews, using popular medical concept hierarchies [10], and corresponding concept medical extraction tools [1].
- We perform qualitative experiments on both online doctor patient reviews and online cell phone buyer reviews. Using various intuitive summary quality measures, we show that our method outperforms state-of-the-art review summarization methods (Sect. 5.3).

2   Problem Framework

Define an item (for example, a doctor or a camera) as a set of reviews, where each review is a set of *concept-sentiment* pairs $\{(c_1, s_1), (c_2, s_2), \ldots, (c_n, s_n)\}$, and $s_j \in \mathbb{R}$ is the sentiment for concept c_j in the review. We shows how to extract the concepts and their sentiments from the text of the reviews in Sect. 4.1, and Related Work (Sect. 6). The set of concepts are related based on a hierarchical *ontology* such as WordNet [19] and ConceptNet [23]. For instance, the "part-whole" relation in those ontologies can be utilized to create the hierarchy of aspects suitable for our framework. Alternatively, Kim et al. [12] automatically extract an aspect-sentiment hierarchy using a Bayesian non-parametric model.

We define the (directed) *distance* $d(p_1, p_2)$ between two concept-sentiment pairs $p_1 = (c_1, s_1)$ and $p_2 = (c_2, s_2)$, based on the concepts' relationship in the hierarchy, as follows.

Definition 1. *The distance $d(p_1, p_2)$ is:*

$$d(p_1, p_2) = \begin{cases} d(r, c_2) & \text{if } c_1 \text{ is the root } r, \text{ or} \\ d(c_1, c_2) & \text{if } c_1 \text{ is the ancestor of } c_2 \text{ and } |s_1 - s_2| \leq \epsilon, \text{ or} \\ \infty & \text{otherwise} \end{cases}$$

where the concept distance $d(c_1, c_2)$ is the shortest-path length from c_1 to c_2 in the hierarchy, r is the root of the hierarchy, and $\epsilon > 0$ is the sentiment threshold.

If pair p_1 has finite distance to p_2, we say that p_1 *covers* p_2. Pair p_1 covers p_2 iff p_1's concept c_1 is an ancestor of p_2's concept c_2, and either c_1 is the root concept or the sentiments of p_1 and p_2 differ by at most ϵ. Figure 1 shows an example of how the concept-sentiment pairs of an item's reviews are mapped on the concept hierarchy, where the dashed line is the path from the root, and concept c_6 doesn't have any pairs. For instance, pair $(c_1, 0.7)$ represents an occurrence of concept c_1 in a review with sentiment 0.7. The same pair is also represented by the circled 0.7 value inside the c_1 tree node.

Given a set $P = \{p_1, p_2, \ldots, p_q\}$ of concept-sentiment pairs for the reviews of an item, and an integer k, our goal is to compute a set $F = \{f_1, f_2, \ldots, f_k\} \subseteq P$ of k pairs that best summarize P. To measure the quality of such a summary F, we define its cost $C(F, P)$ as the distance from F to P, defined as follows.

Definition 2. *The distance from F to a pair p is the distance of the closest pair in $F \bigcup \{r\}$ to p: $d(F, p) = \min_{f \in F \bigcup \{r\}} d(f, p)$. The cost of F is the sum of its distances to pairs in P: $C(F, P) = \sum_{p \in P} d(F, p)$.*

We introduce two summarization problems as following:

1. **k-*Pairs Coverage:*** given a set P of concept-sentiment pairs (coming from a given set of reviews for an item) and integer $k \leq |P|$, find a subset $F \subseteq P$ with $|F| = k$ that summarizes P with minimum cost: $\min_{F \subseteq P, |F| = k} C(F, P)$
2. **k-*Reviews/Sentences Coverage:*** given a set R of reviews (or sentences) and integer $k \leq |R|$, find a subset $X \subseteq R$ with $|X| = k$ that summarizes

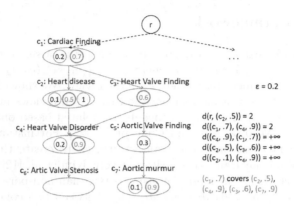

Fig. 1. Representation of concept-sentiment pairs on SNOMED-CT concept hierarchy

R with minimum cost: $\min_{X \subseteq R, |X|=k} C(P(X), P(R))$, where $P(R)$ is the set of concept-sentiment pairs derived from the set R of reviews/sentences, and $P(X)$ is the set of concept-sentiment pairs derived from the subset X of R.

Intuitively, the first problem is appropriate when the summaries consist of concise concept-sentiment pairs, e.g. "good Heart Disease management", extracted from the reviews, and may be more suitable for mobile phone-sized screens. The second problem is appropriate if the summaries consist of whole sentences of reviews, which better preserves the meaning of the review, but may require more space to display.

The k-Pairs Coverage problem can be viewed as a special case of the k-Reviews/Sentences Coverage problem, when each review/sentence has just one pair. For presentation simplicity, we first present our NP-hard proof and algorithms for k-Pairs Coverage in Sect. 4, then describe how they can be applied to the k-Reviews/Sentences Coverage in Sect. 4.5.

3   Both Problems are NP-Hard

This section proves both proposed problems NP-hard.

Theorem 1. *The k-Pairs Coverage problem is NP-hard.*

Proof. The decision problem is, given a set P of concept-sentiment pairs, an integer $k \leq |P|$, and a target $t \geq 0$, to determine whether there exists a subset $F \subseteq P$ of size k with cost $C(F, P)$ at most t. We reduce Set Cover to it. Fix any Set-Cover instance (S, U, k) where U is the universe $\{u_1, u_2, \ldots, u_n\}$, and $S = \{S_1, S_2, \ldots, S_m\}$ is a collection of subsets of U, and $k \leq |S|$. Given (S, U, k), first construct a concept-hierarchy (DAG) with root r, concepts c_i and e_i for each subset S_i, and a concept d_j for each element u_j. For each set S_i, make c_i a child of r and e_i a child of c_i. For each element u_j, make d_j a child of c_i for

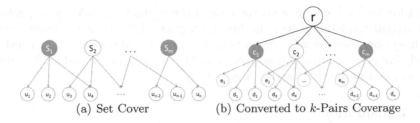

(a) Set Cover   (b) Converted to k-Pairs Coverage

Fig. 2. Reduction from set cover

each set S_i containing u_j. (See Fig. 2.) Next, construct $2m+n$ concept-sentiment pairs $P = \{p_1, \ldots, p_{2m+n}\}$, one containing each node in the DAG other than the root r, and all with the same sentiment, say 0. Take target $t = 3m + n - 2k$. This completes the reduction. It is clearly polynomial time. Next we verify that it is correct. For brevity, identify each pair with its node.

Suppose S has a set cover of size k. For the summary $F \subseteq P$ of size k, take the k concepts in P that correspond to the sets in the cover. Then each d_i has distance 1 to F, contributing n to the cost. For each set in the cover, the corresponding c_i and e_i have distance 0 and 1 to F, contributing k to the cost. For each set not in the cover, the corresponding c_i and e_i have distance 1 and 2 to F, contributing $3(m - k)$ to the cost, for a total cost of $n + 3m - 2k = t$.

Conversely, suppose P has a summary of size k and cost $t = n + 3m - 2k$. Among size-k summaries of cost at most t, let F be one with a maximum number of c_i nodes. We show that the sets corresponding to the (at most k) c_i nodes in F form a set cover. Assume some $c_{i'}$ is missing from F (otherwise $k \geq m$ so we are done). For every e_i in F, its parent c_i is also in F. (Otherwise adding c_i to F and removing e_i would give a better summary F', i.e., a size-k summary of cost at most t, but with more c_i nodes than F, contradicting the choice of F). No e_i is in F (otherwise removing e_i and adding the missing node $c_{i'}$ would give a better summary F'). No d_j is in F (otherwise, since neither $e_{i'}$ nor $c_{i'}$ are in F, removing d_j from F and adding $c_{i'}$ would give a better summary F'). Since no e_i or d_j is in F, only c_i nodes are in F. Since the cost is at most $t = n+3m-2k$, by calculation as in the preceding paragraph, the sets S_i corresponding to the nodes c_i in F must form a set cover.   □

When we already have k-Pairs Coverage as a NP-hard problem, it's natural to prove the following theorem.

Theorem 2. *The k-Reviews/Sentences Coverage problem is NP-hard.*

Proof. K-Reviews/Sentences Coverage is a generalization of k-Pairs Coverage, so the theorem follows from the previous theorem.

4   Algorithms

We implement three algorithms for k-Pairs Coverage. The first, which is the only one generates an optimal solution, solves the standard integer-linear program

(ILP) for the problem, as a special case of the well-known k-Medians problem. The second randomly solves the linear program (LP), then randomly rounds the fractional solution achieving a bounded approximation error. The third is a greedy bounded approximation algorithm. The three algorithms share a common initialization phase that we describe first.

4.1   Initialization

The initialization phase computes the underlying edge-weighted bipartite graph $G = (U, W, E)$ where vertex sets U and W are the concept-sentiment pairs in the given set P, edge set E is $\{(p, p') \in U \times W : d(p, p') < \infty\}$, and edge (p, p') has weight equal to the pair distance $d(p, p')$. The initialization phase builds G in two passes over P. The first pass puts the pairs $p = (c, s)$ into buckets by category c. The second pass, for each pair $p = (c, s)$, iterates over the ancestors of c in the DAG (using depth-first-search from c). For each ancestor c', it checks the pairs $p' = (c', s')$ in the bucket for c'. For those with finite distance $d(p, p')$, it adds the corresponding edge to G.

For our problems, the time for the initialization phase and the size of the resulting graph G are roughly linear in $|P|$, because the average number of ancestors for each node in the DAG is small.

4.2   ILP for Optimal Solution

Given the graph $G = (U, W, E)$, we adapt the standard k-Medians ILP for our non-standard cost function as below.

minimize $\quad\quad \sum_{(p,q) \in E} y_{pq} \times d(p, q)$

subject to $\quad\quad x_r = 1; \quad \sum_{p \in P \setminus \{r\}} x_p = k; \quad \sum_{\forall q \in W, p:(p,q) \in E} y_{pq} = 1$

$\quad\quad\quad\quad\quad (\forall (p, q) \in E \quad 0 \le y_{pq} \le x_p; \quad (\forall p \in U) \quad x_p \in \{0, 1\}$

Our first algorithm solves the ILP using the Gurobi solver. Of course, no worst-case polynomial-time bounds are known for solving this NP-hard ILP, but on our instances the algorithm finishes in reasonable time (Details are in Sect. 5).

4.3   Randomized Rounding

The second algorithm computes an optimal fractional solution (x, y) to the LP relaxation of the ILP (using Gurobi, details in Sect. 5), then randomly rounds it as shown in Algorithm 1: it chooses the summary F by sampling k pairs p at random from the distribution $x / \|x\|_1$. No good worst-case bounds are known on the time to solve the LP, but on our instances the solver solves it in reasonable time. The randomized-rounding phase can easily be implemented to run in linear time, $O(n)$ where $n = |P|$. This randomized-rounding algorithm is due to [27] (see also [4]). The following worst-case approximation guarantee holds for this algorithm, as a direct corollary of the analysis in [4]. Let $\text{OPT}_k(P)$ denote the minimum cost of any size-k summary of P.

Algorithm 1. Randomized Rounding Algorithm

Input: fractional solution x, y

Output: summary F

1: **procedure** RANDOMIZED ROUNDING
2:     Define probability distribution q on $P' = P \setminus \{r\}$ such that $q(p) = \frac{x_p}{\sum_{p \in P'} x_p}$.
3:     $F = \emptyset$
4:     **while** $|F| < k$ **do**
5:         Sample one pair p without replacement from q.
6:         Add p to F.
7:     Return F.

Theorem 3. *The expected cost of the size-k summary returned by the randomized-rounding algorithm is $O(\text{OPT}_{k'}(P))$ for some $k' = O(k/\log n)$.*

In our experiments it gives near-optimal summary costs.

4.4 Greedy Algorithm

The greedy algorithm is Algorithm 2. It starts with a set $F = \{r\}$ containing just the root. It then iterates k times, in each iteration adding a pair $p \in P$ to F chosen to minimize the resulting cost $C(F \cup \{p\}, P)$. Finally, it returns summary $F \setminus \{r\}$. This is essentially a standard greedy algorithm for k-medians. Since the cost is a submodular function of P, the algorithm is a special case of Wolsey's generalization of the greedy set-cover algorithm [26].

After the initialization phase, which computes the graph $G = (U, W, E)$, the algorithm further initializes a max-heap for selecting p in each iteration. The max-heap stores each pair p, keyed by $\delta(p, F) = C(F \cup \{p\}, P) - C(F, P)$. The max-heap is initialized naively, in time $O(m + n \log n)$ (where $m = |E|$, $n = |P|$). (This could be reduced to $O(m + n)$ with the linear-time build-heap operation.) Each iteration deletes the pair p with maximum key from the heap (in $O(\log n)$ time), adds p to F, and then updates the changed keys. The pairs q whose keys change are those that are neighbors of neighbors of p in G. The number of these updates is typically $O(d^2)$, where d is the typical degree of a node in G. The cost of each update is $O(\log n)$ time. After initialization, the algorithm typically takes $O(kd^2 \log n)$ time. In our experiments, our graphs are sparse (a typical node p has only hundreds of such pairs q), and k is a small constant, so the time after initialization is dominated by the time for initialization. The following worst-case approximation guarantee is a direct corollary of Wolsey's analysis [26]. Let $H(i) = 1 + 1/2 + \cdots + 1/i \approx 1 + \log i$ be the ith harmonic number. Let Δ be the maximum depth of the concept DAG.

Theorem 4. *The greedy algorithm produces a size-k summary of cost at most $\text{OPT}_{k'}(P)$, where $k' = \lfloor k/H(\Delta n) \rfloor$.*

In our experiments, the algorithm returns near-optimal size-k summaries.

Algorithm 2. Greedy Algorithm

Input: $G = (U, W, E)$ from initialization, computed from P.
Output: Size-k summary F

1: **procedure** GREEDY
2:     Define $\delta(p, F) = C(F \cup \{p\}, P) - C(F, P)$.
3:     Initialize $F = \{r\}$, and max-heap holding $p \in U$ keyed by $\delta(p, F)$.
4:     **while** $|F| < k + 1$ **do**
5:         Delete p with highest key from max-heap.
6:         Add p to F.
7:         **for** w such that $(p, w) \in E$ **do**
8:             **for** q such that $(q, w) \in E$ **do**
9:                 Update max-heap key $\delta(q, F)$ for q.
10:     **return** $F \setminus \{r\}$

4.5   Adaptation for k-Reviews/Sentences Coverage Problem

When whole reviews or sentences (each containing a set of concept-sentiment pairs) must be selected, the above algorithms can still be applied with only a modification of the initialization stage. In particular, we modify the construction of bipartite graph $G = (U, W, E)$, so instead of having both U and W be concept-sentiment pairs in P, U represents the set of candidate reviews or sentences R, and W represents concept-sentiment pairs as before. Therefore the edge set E becomes $\{(r, p) \in U \times W : d(r, p) < \infty\}$, and edge (r, p) has weight equal to the distance $d(r, p)$ from review/sentence r to pair p. After this initialization, the algorithms work as usual.

5   Experimental Evaluation

In this section we conduct both quantitative and qualitative evaluations. The quantitative evaluation measures the time and accuracy trade-offs of the proposed approximate summarization algorithms compared to the optimal solution. The qualitative evaluation evaluates the quality of the summaries generated by the proposed methods, compared to baseline state-of-the-art unsupervised summarization methods using several intuitive measures.

5.1   Experiment Setup

Datasets: We utilize two real-world datasets: health care and online consumer reviews. Our first dataset consists of 68,686 patient reviews of the 1000 most reviewed doctors from vitals.com, which is a popular doctor rating website. As the second dataset, we crawled customer reviews of 60 unlocked cell phones, which are featured in the first five pages on Amazon and have at least 100 distinct reviews each. Table 1 presents basic statistics of the two datasets.

Table 1. Dataset characteristics.

	Doctor reviews	Cell phone reviews
#Items (doctor/product)	1000	60
#Reviews	68686	33578
Min #reviews per item	43	102
Max #reviews per item	354	3200
Average #sentences per review	4.87	3.81

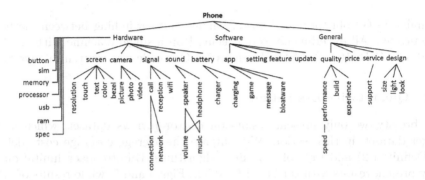

Fig. 3. Cell phone aspect hierarchy

Concepts and Sentences Extraction: To extract medical concepts in doctor reviews we use automated tool MetaMap [1] and SNOMED CT [10] ontology, which has more than 300,000 concepts and is suitable for our problem given its focus on describing medical conditions. For example, for sentence *"Dr Robert did an awesome job with my tummy tuck and liposuction"*, concepts *"tummy tuck"* (UMLS ID = C0749734) and *"liposuction"* (ID = C0038640) are extracted. In cell phone reviews dataset, we employ Double Propagation method [22] to extract aspects such as screen and battery. We only focus on the 100 most popular extracted aspects. Since there is no available hierarchy of cell phone aspects, we manually built a hierarchy from the extracted aspects as shown in Fig. 3.

Sentiment Computation: To compute the sentiment around a concept, we compute the sentiment of the containing sentence and assign this sentiment to the concept. We adopt a neural network based representation learning approach *doc2vec* to represent sentences by fixed-size vectors [15]. Then, sentence's sentiment estimation is formulated as a standard regression problem using the sentence vector representation.

Configuration: We evaluate the three methods proposed in Sect. 4: Integer Linear Programming - ILP, Randomized Rounding - RR, and Greedy algorithm. For ILP and RR, we use the Gurobi optimization library version 6.0.5 [8] with Dual-Simplex as the default method. This method is chosen because it shows the best performance in our case after experimental trials on different options

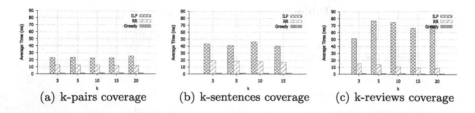

Fig. 4. Time evaluation with threshold 0.5

available in Gurobi (primal simplex, barrier, auto-switching between methods, concurrent). All experiments were executed on a single machine with Intel i7-4790 3.60 GHz, 16 GB RAM, Windows 10. Our code was written in Java 8.

5.2 Quantitative Evaluation

For brevity, we only present results on doctor reviews dataset, which is the larger dataset, in this section. We compare the average coverage cost (defined in Definition 2) and time of our three algorithms. Due to space limitation we only present results with threshold (ϵ) 0.5 in Figs. 4 and 5, while results of other thresholds show similar trends.

A key observation from these experiments is that Greedy is always the fastest algorithm while maintaining reasonable costs compared to ILP and RR. Of course, ILP gives optimal solution, thus offers the cheapest cost. The Greedy algorithm has the worst cost but never more than 8% higher than the optimal ($\leq$ 5% most of time). In terms of time, the Greedy outperforms ILP by a factor up to 19×, 32× and 63× in the top pairs, top sentences and top reviews problems, respectively. Similarly, Greedy runs faster than RR, at most 14 times, and usually takes only 1–2 ms per doctor. RR algorithm works similarly to ILP regarding cost, specifically, the difference is about 1–2%. The speedup of RR over ILP is about 2–5×. This is because RR only solves a Linear Program system and then randomizes the solution instead of finding an optimal integer solution.

We also notice that with the same threshold, the cost decreases from top pairs to top sentences, and then to top reviews problem. The reason is that a sentence or review can have multiple pairs, so they typically cover more pairs than a single pair can cover. Therefore, k sentences or reviews usually cover more pairs than k pairs can, which leads to smaller costs. Similarly, the elapsed time of all algorithms for top sentences/reviews problem are larger than for top pairs problem. It's because for top sentences/reviews, there are more connections (edges) between selecting candidates and pairs to consider.

In general, the results suggest that our problem has latent structures friendly to Greedy algorithm. Therefore, the optimal solution from ILP algorithm seem to be close to the one of Greedy algorithm which can be achieved much faster. Therefore, we choose Greedy algorithm for the next qualitative experiments.

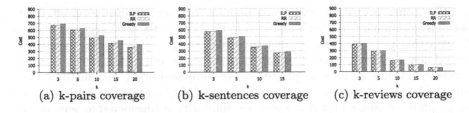

(a) k-pairs coverage   (b) k-sentences coverage   (c) k-reviews coverage

Fig. 5. Cost evaluation with threshold 0.5

5.3 Qualitative Evaluation

The goal of this section is to study the quality of the summarization achieved by the proposed algorithms, compared to several state-of-the-art unsupervised baselines. We focus on the sentence selection problem variant, which offers a balance between conciseness and semantic completeness.

Selecting Sentiment Threshold ϵ Used by Greedy Algorithm: We select the threshold value ϵ for which the rate of covered sentences significantly drops if we further increase ϵ. For that, we employ the elbow method, which shows that the sentiment threshold's elbow is at 0.5 most of time (details removed for brevity). Intuitively, this sentiment threshold is also reasonable in the sense that a very positive sentiment of value 1.0 can cover a positive sentiment of value 0.5. Therefore, we choose sentiment threshold 0.5 for our greedy summarizer.

Baseline Summarization Methods: Our baselines come from two areas: one from opinion summarization approach, and the other from multi-document summarization. Specifically, the first baseline method to select top k sentences is adapted from Hu et al. [9]. This algorithm was designed to summarize customer reviews of online shopping products. It first extracts product aspects (attributes like "picture quality" for product "digital camera"), then classifies review sentences that mention these aspects as positive or negative, and finally sums up the number of positive and negative sentences for each aspect. To have a fair comparison, we adapt their method to select top k sentences into summaries. We first count the number of pair (concept, positive) or (concept, negative), for example: aspect "picture quality" with sentiment "positive" occurs in 200 sentences. Then, we select k most popular pairs and return one containing sentence for each selected pair. Note that the aspect extraction task is common in both the baseline and our methods. We refer to this baseline as *"most_popular"* since their summarizer favors the most popular aspects.

The second baseline from the opinion summarization area is adapted from a review summarizer [3] of local services (such as hotels, restaurants). This method selects the (aspect, positive/negative) pairs proportionally to the pair's frequency instead of selecting the most popular pair as in *"most_popular"* method. Then, it pick the new, most extremely polarized sentence to represent each selected pair (concept, positive/negative). We name this summarizer as *"proportional"*.

The other set of baselines are popular extractive, unsupervised multi-document summarizers that are agnostic to a concept's sentiment orientation. Contrasting to abstractive summarizers that compose summaries by creating brand-new sentences, extractive summarizers make use of original documents' sentences, hence it is appropriate to be compared with our method. TextRank [18] summarizer applies PageRank algorithm on text by modelling text as graph of sentences in which sentences' similarity is considered as sentence-to-sentence edge weight. LexRank [6] is another document summarizer relying on a sentence graph for detecting the most important sentences. The last baseline in this line is Latent Topic Modelling (LSA) based summarizer [24], which utilizes the sentence's vector representation calculated using Singular Value Decomposition (SVD) on a term-sentence matrix. In our experiments, We utilize Sumy [2] library for these three methods. We summarize all baselines with brief descriptions in Table 2.

Table 2. Baseline unsupervised summarizers

Most_popular	[9]	Pick representative sentences of popular aspect-polarity pairs
Proportional	[3]	Pick representative sentences with extreme sentiments after selecting aspects proportionally
TextRank	[18]	No sentiment, use sentence graph with word overlap for sentence similarity
LexRank	[6]	No sentiment, use sentence graph with cosine-based sentence similarity
LSA-based	[24]	No sentiment, utilize SVD on term-sentence matrix

Summary Quality Measures and Results: We evaluate all methods on a new measure, named *"sentiment error"* (or *"sent-err"*), to avoid giving an unfair advantage to our method. Note that typical multi-document summarization measures such as ROGUE are not applicable in our context since they do not consider sentiment and concept relationship. The key idea is to look at the difference between every concept's sentiment in the original reviews and that concept's sentiment (extrapolated if concept not in summary) in the summary. That is, for each pair in the original reviews, we find the closest concept in the summary and measure the *sentiment distance* between them. In contrast, in Definition 2, we measure the *concept distance* (in hierarchy edges) between a review concept and its nearest covering summary concept.

Recall that we summarize a set of concept-sentiment pairs P by a subset F contained in k sentences. We define *"sent-err"* of F with respect to P in a root-mean-square error manner: $sent\text{-}err(P, F) = \sqrt{\frac{1}{|P|} \sum_{p \in P} err_{p,F}^2}$, where p is a pair of (concept c_p, sentiment s_p). $err_{p,F}$ (Eq. 1) is the smallest difference between s_p and that concept's sentiments in a pair in F. When concept c_p does not appear in F, we use the sentiments of c_p's lowest ancestor in F if available.

When neither c_p nor its ancestors appear in F, we consider a neutral sentiment 0. The intuition is that the error models the difference of every concept's sentiment and the closest sentiment of that concept or its ancestors in summary.

$$err_{p,F} = \begin{cases} \min_{f \in F, c_f = c_p} |s_f - s_p| & : c_p \in F \\ \min_{f \in F, c_f = c_p\text{'s ancestor}} |s_f - s_p| & : c_p \notin F \wedge c_p\text{'s ancestor} \in F \\ |0 - s_p| = |s_p| & : \text{otherwise} \end{cases} \quad (1)$$

Another version of this measure penalizes the case of missing concept c_p and its ancestor in summary F by considering the largest possible error of c_p's sentiment. In another words, the third branch of Eq. (1) becomes $err_{p,F} = \max(|1 - s_p|, |-1 - s_p|)$. Note that $+1$ and -1 are the extreme sentiments in our model. We name this measure version as *"sent-err-penalized"*.

Results: Figure 6 compare the errors of our method and the baselines on cell phone review dataset (similar results on doctor reviews dataset). On the first measure, *sent-err* (Fig. 6(a)), we find that our method always leads to the smallest sentiment error, i.e. highest-quality summaries. It can reduce the error of the second best performance method ("most_popular") by 4.1% on average, and other methods by 14.6% on average. The multi-document summarization methods generally perform poorly since they ignore the sentiment. Our method reduces those multi-document summarizers' error by up to 23.7%. The errors of all methods drop when the number of summary sentences increases, as expected.

On *sent-err-penalized* measure (Fig. 6(b)), our method beats all baselines with larger margins. Specifically, our method improves the error of second best performance method ("most_popular") 14.9%, and other methods by 19.8% on average. This result indicates that missing concepts in summary problem is more severe in baseline methods, and our method is smarter in choosing sentences.

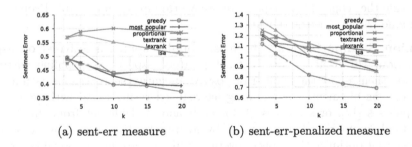

(a) sent-err measure   (b) sent-err-penalized measure

Fig. 6. Sentiment error on cell phone reviews dataset (lower error is better)

6   Related Work

Multi-document Summarization: This is a traditional problem with the most well-known applications in summarizing online news articles.

Goldstein et al. presented a typical method [7], which extend single-document summarization techniques. A key difference is that there is more redundancy across documents of a similar topic than within a single document. This is an observation we also adopt in our work. TexRank [18] and LexRank [6] are two popular, similar methods based on building weighted graph of document sentences, which are rated by Pagerank algorithm to pick the important ones. Steinberger et al. [24] proposed an LSA-based summarizer that utilizes sentence's vector representation in their latent index space. Recent deep learning based approaches [11,16] are supervised and/or abstractive summarizers while our summarization method is unsupervised and extractive. However, none of above methods consider the sentiment in input documents. We incorporate some of these methods (TextRank, LexRank and LSA-based) as baselines in our evaluations (Sect. 5).

Sentiment Analysis: The methods fall into two categories, using unsupervised or supervised learning. The unsupervised methods [25] focus on building a comprehensive opinion word dictionary, or use linguistic rules to find opinion phrases containing adjectives or adverbs in a document. An early supervised method [21] applies a Bag-Of-Word model to classify movie reviews as positive or negative. Recently, a common approach [15] is to use neural network model to extract the better review's vectors, thus get the better results on sentiment classification task. Any of these methods can be plugged into our framework.

Aspect Extraction: A common review analysis task is to extract the product aspects. Traditional methods [9,22] use association mining to find frequent aspects, then apply pruning rule to remove meaningless, redundant ones; later they also have a rule to discover additional infrequent aspects based on both frequent ones and opinion words. A semi-supervised approach based on topic modelling extract product aspects as multi-grain topics [20]. Extracting aspects is outside the scope of this paper. We use Metamap [1] and Double Propagation technique [22] in our experiments.

Opinion Summarization: The most popular approach is based on aspect extraction. Hu et al. [9] first extract product aspects from online customer reviews, then report the number of positive/negative sentences for each aspect. This can be augmented by showing aggregated rating along with representative phrases [17], or sentences [3] for each aspect. Different from this kind of statistical summaries, Lappas et al. [13] formulates the problem as selecting k reviews that optimize the aspect coverage while rewarding high-quality reviews, or maintaining their proportion of aspect opinions.

A key difference of this paper from all the above works is that they do not consider the relationships between the aspects nor a continuous sentiment scale. A preliminary version of this work, which focuses on coverage measure of the greedy algorithm on the doctor reviews dataset, was published as a poster [14].

7 Conclusions

We introduced a novel review summarization problem that considers both the ontological relationships between the review concepts and their sentiments. We described methods for extracting concepts and estimating their sentiment. We proved that the summarization problem is NP-hard even when the concept ontology is a DAG, and for that we presented efficient approximation algorithms We evaluated the proposed methods extensively with both quantitative and qualitative experiments. We found that the Greedy algorithm can achieve quality comparable to the optimal is much shorter time, comparing to other algorithms. Moreover, using various coverage measures and sentiment error measures, we show that the Greedy outperforms a baseline method on selecting k sentences to summarize real reviews.

Acknowledgment. This work was partially supported by NSF grants IIS-1838222, IIS-1619463, IIS-1901379 and IIS-1447826

References

1. Aronson, A.R.: Effective mapping of biomedical text to the UMLS metathesaurus: the metamap program. In: AMIA (2001)
2. Belica, M.: Sumy: Module for automatic summarization of text documents (2017)
3. Blair-Goldensohn, S., Hannan, K., McDonald, R., Neylon, T., Reis, G.A., Reynar, J.: Building a sentiment summarizer for local service reviews. In: WWW Workshop on NLP in the Information Explosion Era (2008)
4. Chrobak, M., Kenyon, C., Noga, J., Young, N.E.: Oblivious medians via online bidding. In: Correa, J.R., Hevia, A., Kiwi, M. (eds.) LATIN 2006. LNCS, vol. 3887, pp. 311–322. Springer, Heidelberg (2006). https://doi.org/10.1007/11682462_31
5. Ding, X., Liu, B., Yu, P.S.: A holistic lexicon-based approach to opinion mining. In: WSDM 2008. ACM (2008)
6. Erkan, G., Radev, D.R.: LexRank: graph-based lexical centrality as salience in text summarization. J. Artif. Intell. Res. **22**, 457–479 (2004)
7. Goldstein, J., Mittal, V., Carbonell, J., Kantrowitz, M.: Multi-document summarization by sentence extraction. In: NAACL-ANLP (2000)
8. Gurobi Optimization: Gurobi optimizer manual (2015). http://gurobi.com
9. Hu, M., Liu, B.: Mining and summarizing customer reviews. In: KDD (2004)
10. International: SNOMED CT (2016). https://www.snomed.org/snomed-ct
11. Jadhav, A., Rajan, V.: Extractive summarization with swap-net: sentences and words from alternating pointer networks. In: ACL (2018)
12. Kim, S., Zhang, J., Chen, Z., Oh, A.H., Liu, S.: A hierarchical aspect-sentiment model for online reviews. In: AAAI (2013)
13. Lappas, T., Crovella, M., Terzi, E.: Selecting a characteristic set of reviews. In: KDD (2012)
14. Le, N.X., Hristidis, V., Young, N.: Ontology- and sentiment-aware review summarization. In: ICDE, pp. 171–174 (2017)
15. Le, Q.V., Mikolov, T.: Distributed representations of sentences and documents. In: ICML (2014)

16. Liu, L., Lu, Y., Yang, M., Qu, Q., Zhu, J., Li, H.: Generative adversarial network for abstractive text summarization. In: AAAI (2017)
17. Lu, Y., Zhai, C., Sundaresan, N.: Rated aspect summarization of short comments. In: World Wide Web 2009 (2009)
18. Mihalcea, R., Tarau, P.: TextRank: bringing order into text. In: EMNLP (2004)
19. Miller, G.A.: WordNet: a lexical database for English. Commun. ACM **38**(11), 39–41 (1995)
20. Mukherjee, A., Liu, B.: Aspect extraction through semi-supervised modeling. In: 50th Annual Meeting of ACL (2012)
21. Pang, B., Lee, L., Vaithyanathan, S.: Thumbs up?: Sentiment classification using machine learning techniques. In: ACL-2002 Conference on EMNLP (2002)
22. Qiu, G., Liu, B., Bu, J., Chen, C.: Opinion word expansion and target extraction through double propagation. Comput. Linguist. **37**(1), 9–27 (2011)
23. Speer, R., Havasi, C.: Representing general relational knowledge in conceptNet 5. In: LREC (2012)
24. Steinberger, J., Jezek, K.: Using latent semantic analysis in text summarization and summary evaluation. In: Proceedings of ISIM 2004 (2004)
25. Taboada, M., Brooke, J., Tofiloski, M., Voll, K., Stede, M.: Lexicon-based methods for sentiment analysis. Comput. Linguist. **37**(2), 267–307 (2011)
26. Wolsey, L.A.: An analysis of the greedy algorithm for the submodular set covering problem. Combinatorica **2**(4), 385–393 (1982)
27. Young, N.E.: K-medians, facility location, and the Chernoff-Wald bound. arXiv preprint arXiv:cs/0205047 (2002)

Enriching the Context: Methods of Improving the Non-contextual Assessment of Sentence Credibility

Aleksandra Nabożny[1]([⊠])[iD], Bartłomiej Balcerzak[2], and Danijel Koržinek[2]

[1] Gdańsk University of Technology, 80233 Gdańsk, Poland
alenaboz@pg.edu.pl
[2] Polish-Japanese Academy of Information Technology, 02008 Warsaw, Poland

Abstract. This paper presents several methods of automatic context enrichment of sentences that need to be evaluated, tagged or fact-checked by human judges. We have created a corpus of medical Web articles. Sentences from this corpus have been fact-checked by medical experts in two modes: contextually (reading the entire article and evaluating sentence by sentence) and without context (evaluating sentences from all articles in random order). It is known that non-contextual evaluation is faster, but some sentences are impossible to evaluate without context. We have designed and evaluated several methods of summarizing context that we hypothesized were suitable for supporting evaluation of sentences without reading the entire text. Then, we collected new assessments from medical experts for the sentences with enriched context. The context enrichment methods have been evaluated using two measures: conversion, which calculates how frequently a method allows experts to evaluate sentences that were impossible to evaluate without context, and agreement, which depends on how frequently the new expert evaluations match with evaluations from experts who had read the whole text before rating a sentence. Our results show that the best method achieves a high conversion rate, while providing experts with a condensed context summary. Moreover, the method significantly reduces the time needed to evaluate one sentence, compared to the baseline method (which provides the expert with the entire paragraph surrounding the target sentence). The problem of automatically enhancing the context of a sentence for fast fact-checking or tagging has not appeared in other studies before. We present preliminary results of the research in this area and a framework for testing potential new methods.

Keywords: Information credibility · Fact-checking · Text summarization · Context enrichment

1 Introduction

People often search the Web for medical or health-related information ("medical Web content" for short). As a matter of fact, eight in ten people browse the Web

© Springer Nature Switzerland AG 2019
R. Cheng et al. (Eds.): WISE 2019, LNCS 11881, pp. 763–778, 2019.
https://doi.org/10.1007/978-3-030-34223-4_48

for health-related content, which makes a consultation with "Dr Google" one of the most common Internet activities. Unfortunately, what the users ultimately find is often misleading, incomplete, and non-credible. The Web is filled with a myriad of humbug therapies, mysterious superdrugs and pseudo-doctors. As it can be easily guessed it may, and often does, lead to grave consequences, as can be seen by the example of the anti-vaccine movement, a global community that is largely present on the Web.

On the other hand, recent findings show that the trend of "googling" the diagnoses indeed permanently changes patient-doctor dialogue [2], with positive results to both sides. According to [20] it is likely that most people will experience at least one diagnostic error in their lifetime. Research suggests that the traditional diagnosis process is error-prone and it might be improved with some supporting methods, e.g. patient assistance in gathering knowledge about their own health condition.

Ordinary Web users' credibility evaluation of medical Web content can be supported in several ways. Existing fact-checking sites, such as Snopes.com, have separate categories of non-credible medical Web content. Other sites, such as hon.ch, offer specialized search engines of medical Web content and run a certification process for medical Websites. Classifiers of Webpage credibility can be applied to augment output of Google search (using a browser plugin) with indications of search result's credibility [26]. A more detailed method is to mark sentences contained in a medical Webpage with credibility indicators - research has shown that evaluations of statements can impact overall Webpage credibility evaluations [7,8]. This method also works for social media posts [19]. To obtain such information, evaluations of these sentences must be available. Thus, in this article we shall focus on supporting the process of acquiring sentence credibility evaluations from medical experts.

Consider a situation when a medical expert is asked to rate credibility of medical Web content. The most accurate, but also most time-consuming method would be for the expert to read the entire Webpage and to mark credibility of selected sentences. However, this method has a low output, because experts' time is limited. Presenting experts with sentences selected for evaluation by Web users (or automatically) is another method. In this case, the medical expert only needs to read short sentences and can immediately give an evaluation. While this method has an advantage of speed, it has a drawback: contextual information that may be necessary for evaluation is missing.

It should be emphasized that in this study we limited ourselves to the medical field as a domain of research to reduce the amount and variety of necessary experts, keywords, etc. However, credibility assessment can be supported using the presented methods in any domain that requires an expert for the proper content evaluation. Other domains may include climate sciences, psychology, history, etc. Credibility evaluation is also important on Wikipedia, where teams of editors oversee quality and veracity of statements [22].

In order to cope with this challenge, we have designed and evaluated methods of context enrichment that help experts in assessing credibility of sentences retrieved from medical Webpages. We also performed qualitative analysis of the sentences marked by experts as impossible to assess without context. We identified different types of such sentences and adapted the appropriate context-filling methods to each type. They are described in Sect. 4.

We asked medical experts to rate the credibility of sentences that have been found previously as impossible to evaluate without context, using each method separately. Results of the experimental evaluation are shown in Sect. 5. Our work bases on a dataset obtained in a previous experiment [14] (see Sect. 3), that is made available to interested researchers on request. The problem of automatically enriching the context in order to assess the credibility of a sentence has not appeared in any studies before. Note that this problem occurs whenever sentences need to be fact-checked quickly by experts, which means that it is significant for most kinds of fake news verification. Fast fact-checking is also crucial in the era of deep-learning models that need large sets of training data. Data that most oftenly needs to be tagged by human judges. We propose the first approach to the problem defined in this way, while opening the field for further research.

2   Related Work

Modelling of context is present in many solutions for downstream natural language processing tasks, however in most cases it serves only as an intermediate tool to enhance performance of these algorithms. These tasks include e.g. semantic text similarity, sentiment analysis and machine translation. The context is usually coded in a way that is not possible to interpret by a human, but only by a neural network that processes this context. In [12] surrounding sentences are used to better learn vector representations of the input sentence, similarly to the way that word2vec algorithm learns representation of the word. In [24], on the other hand, context summarized in a hierarchical way is integrated with neural machine translation model as a source for updating decoder states. In [18] the authors take advantage of contextual relations among sentences so as to improve the performance of sentence regression for text summarization.

In our study we aim to retrieve the context directly, which can be later accessed in a human readable format. A variety of methods exist that include direct extraction of context. Cloze-style reading comprehension problem has recently became a well known baseline NLP task. It is a task where the level of text understanding of a system is tested by asking it questions, the answer for which can be inferred from the document. In [6] the query is designed in a form of a short sentence that summarizes some statement which appears in a text, but lacks one named entity. Predicting the missing component requires a deep understanding of the context. The authors takes advantage of the popular deep-learning architectures with recurrent neural networks and attention to solve this problem. Their approach was to review and simplify the existing solutions, such as Pointer Networks [23] or Memory Neural Networks for text comprehension [5], which resulted as a new state of the art.

Aside from the aforementioned NLP methods, there is a whole other branch of methods which utilize rule-based algorithms for context extraction. These methods are used to support decision-making by retrieving context from electronic medical reports. For example ConText algorithm [4] derives information such as negation, experiencer and temporality of the medical condition. One of the methods presented in this study is also rule-based, but as an addition to the more general keyword-based approach.

Some of the previous works, designed to support credibility assessment of the query, take advantage of the automatically retrieved context. [27] uses global context (derived from the whole set of documents retrieved by the search engine) to prompt the user with sentences that may indicate controversy related to the given query, whereas [21] uses context to provide the information whether given article supports or rejects the statement contained in the query. Unlike in our approach, both studies are focused on the regular internet user (not an expert) and treat the query as a whole, not as part of a larger content.

3   Datasets

In this study we examine two datasets. One has been previously collected, analysed, and described in detail in [14] and we will focus only on one part of this dataset, which we will refer to as the first dataset. The second dataset has been collected especially for this study, using results from the first dataset.

The first dataset contains credibility assessments of articles retrieved from medical Webpages, as well as individual credibility evaluations of all sentences from those articles. The assessments were made by medical experts (doctors, Ph.D. students in medicine) on medical textual Web content (popular science articles addressed to a lay recipient). All articles were in Polish and have been assessed by medical experts who were Polish native speakers. This dataset was available for researchers upon request.

Experts evaluated individual sentences in one of the two possible procedures:

1. sentences in a given evaluation round were put one after another and formed the whole article (contextual mode),
2. sentences in a given round were taken at random from the whole corpus of articles (non-contextual mode).

Experts evaluated credibility of sentences marking them with one of the following labels:

0 - non-credible sentence,
1 - neutral sentence,
2 - credible sentence,
−1 - impossible to assess due to the missing context (only in non-contextual mode).

This article focuses on the last case. In the reference study, during the first round, whole articles were additionally evaluated as either credible (2) or noncredible (0) regardless of the evaluation of individual sentences in the article. The full dataset summary is as follows:

- 247 evaluations of the whole articles (with only 0 or 2 label) collected,
- 11034 contextual evaluations of sentences collected, of which 3035 sentences were labelled as impossible to evaluate without context.

Interestingly, the percentage of sentences that were impossible to assess without context was 27.5% of the corpus of sentences. 24% of sentences marked as credible in the contextual mode were labelled as '−1' in the non-contextual mode. Similarly, 24% of those marked as non-credible in the contextual mode were labelled as '−1' in the non-contextual mode. Lastly, 36% of sentences marked as neutral in the contextual mode turned out to be impossible to assess without context in the non-contextual mode.

In this study we further investigate the last subset of sentences. We have performed an experiment in which 5 experts evaluated 500 randomly selected sentences from the subset of 3035 sentences that were impossible to assess in a non contextual mode. All those 500 sentences were grouped into the evaluation groups of 100 sentences. Each round consisted of 20 sentences with a pronoun, and 80 without pronouns, to reflect the distribution from the full set. 5 groups multiplied by 4 methods of context enrichment resulted in 2000 evaluations in total. The groups were sent to the respondents so that they would not encounter the same sentence twice (Fig. 1).

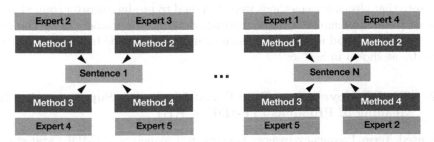

Fig. 1. Evaluation procedure: every sentence has its context enriched by all four methods designed throughout the study and is evaluated by different experts (every expert evaluates one sentence only once, but ultimately all experts use all methods on different sets of sentences).

Basing on the previous dataset we re-investigated the results for the purpose of this more focused study and we have found that experts had difficulty with assessing sentences that:

1. contained an "aggregate of meaning", that is, a word being a hypernym that refers to a category of specific medical terms (for example, "This virus is dangerous" contains the word "virus", which is considered a hypernym for the words "HIV", "HPV" and "measles"),

2. contained an anaphoric, cataphoric or dialectical pronoun, eg. "They have serious consequences.",
3. do not contain a subject (or subject is hard to identify),
4. that lacked a general context.

This analysis demonstrates the generality of the problem of missing context. This problem cannot be reduced to a single, well-known NLP problem.

4   Context Summarizing and Evaluation Methods

We have designed context summarizing methods described below, which we hypothesized were suitable for solving most problems with non-contextual sentence credibility evaluation. Experts were confronted with the sentences with added context summaries to compare (basing on the contextual evaluations from the first dataset) how the automatically extracted context changes their perception of the sentence.

4.1   Context Window (CW)

The first method consisted in retrieving the context window from the surrounding sentences. Two preceding sentences and one following sentence formed the context summary in this method. We treat this method as a baseline approach, in order to see to what extent the full, unprocessed information about the context is needed to correctly assess the credibility of a sentence.

Note that alternative methods were designed to produce shorter context summaries than the context window method. Sentence evaluation using the context window method would require significantly longer time from the evaluating experts, as shown in Sect. 5.

4.2   TF-IDF Keywords + Rule-Based Method of Supplementing the Meaning of Pronouns (TF-IDF + RB)

We used Term Frequency-Inverse Document Frequency (TF-IDF) statistic to retrieve 5 most relevant words from each article. We then attached them as keywords to the sentence that was to be evaluated. Then, for all sentences that contained pronouns, we applied the rule-based method of supplementing the meaning of pronouns, as described in Algorithm 1.

4.3   TextRank Keywords + Rule-Based Method of Supplementing the Meaning of Pronouns (TextRank + RB)

This method is a modification of the method described in Sect. 4.2. However, instead of calculating the Tf-Idf scores for words from the entire document, only 3 most relevant sentences were used. This method was used in order to focus on the most relevant parts of a document. This is important due to the fact that some documents may be long. These sentences are selected based on the TextRank algorithm as described by [13], 3 sentences with the highest TextRank scores are selected. Next, the same rule-based method as described in Sect. 4.2 is used to complement the meaning of pronouns.

4.4 Rule-Based Method of Supplementing the Meaning of Pronouns

After we have identified sentences that consisted of anaphoric, cataphoric or dialectic pronouns, we used the Algorithm 1 to retrieve related noun phrase to the given pronoun. The longest considered noun phrase consists of a noun and a corresponding adjective. We used the Concraft [25] tool from Multiservice [16] web service for morphosyntactic tagging.

Result: noun [adjective]
INPUT: contextWindow (two preceding sentences) and a targetSentence (each sentence is tokenized; each token has morphosyntactic tags attached);
for *pronoun p in a targetSentence* **do**
    candidate_target_nouns = all nouns in a target sentence that has a matching subset of {number, case, gender} with *p* ;
    if *length of candidateTargetNouns* $\neq$ 0 **then**
        return first noun from the list ;
    else
        candidate_context_nouns = all nouns in a context sentences that has min. of 2 overlapping values of {number, case, gender} with p ;
        if *length of candidate_context_nouns* $\neq$ 0 **then**
            result = (if exist) first noun from the closer sentence (else) last noun from the further sentence ;
            if exists adjective in a 2-word context window for the resulting noun **then** return *result + adjective* ;
            else return *result* ;
        else
            return empty string ;
        end
    end
end

Algorithm 1: Rule-based algorithm for supplementing the meaning of pronouns

---

4.5 Coreference Resolution (COREF)

The purpose of coreference resolution is to find words or groups of words that are linked to the same concept. These links can span over multiple sentences and serve as an important tool to explain the meaning, especially when analyzing fragments out of context.

In order to perform coreference resolution, we used the Multiservice [16] web service developed for the Polish language by researchers from the Clarin project (https://clarin-pl.eu). The coreference resolution pipeline consists of several tools activated in sequence:

1. Concraft [25] - initial segmentation and morphosyntactic tagging
2. Spejd [1] - morphosyntactic disambiguation and segment grouping
3. Nerf - named entity recognition

4. MentionStat [9] - detection of potential coreference candidate, so-called
   mentions
5. Bartek3 [10] - coreference resolution engine

The result of the whole process is a list of mention clusters. A mention is a
segment or a sequence of segments representing the basic unit that can participate in a coreference. It can be a named entity or a term with an important
semantic role, but it can also be a pronoun or another word that can easily be
linked in the coreference. The mentions are grouped in clusters that form an
equivalence relation between all the mentions in a given cluster. In theory, it
would be possible to make the relations more meaningful by providing actual
roles to the mentions, but the current version of the tool supports only equivalence relations.

The purpose behind using the tool was to enrich the context of sentences. In
order to perform the coreference analysis, a set of short text fragments containing
the mentioned sentences was prepared, with a context of two previous and one
following sentence, giving four sentence per fragment. After collecting the list
of mention clusters only those were kept that contained at least one mention
within the analyzed sentence and at least one mention outside that sentence.

After collecting the coreferences for the whole data set, about 31% of the
sentences didn't contain any mentions.

4.6   Performance Evaluation Measures

In this section, we present performance evaluation measures that will be used to
evaluate the proposed methods of improving non-contextual sentence credibility
assessment.

Conversion rate: This measure represents the percentage of sentences that
had their credibility assessment changed from undetermined (in non-contextual
evaluation) to either credible or not credible (after context enrichment). This
measure shows the general efficiency of the method. The exact definition of the
measure is provided by the following formula:

$$C = N_d/N_{n,d}$$

where N_d is the number of sentences which were given the evaluation (in place of
the previous indeterminate assessment), while $N_{n,d}$ is the number of all sentences
which had previously an indeterminate assessment.

Strong agreement rate: This measure represents the quality of the method
for assessment improvement. It is a ratio of agreement between a non-contextual
sentence evaluation (after the enhancement was applied), and the contextual
assessment. A high ratio would indicate that the context transferring method
was successful in recreating the context of the document in which the sentence

ORIGINAL SENTENCE:
(pl) Pojawiają się one dopiero w momencie, gdy organizm zakażonej osoby zaczyna walczyćz wirusem i wytwarza przeciw niemu przeciwciała.
(eng) They appear only when the body of the infected person begins to fight the virus and makes antibodies against it.

TFIDF + RB
(pl) Słowa kluczowe artykuću: HIV zakazić wirus zakażenie test . Wybrane zdanie z artykułu: Pojawiają się one [Objawy] dopiero w momencie , gdy organizm zakaż onej osoby zaczyna walczyć z wirusem i wytwarza przeciw niemu przeciwciała.
(eng) Article keywords: HIV infect virus infection test. Selected sentence from the article: They appear [Symptoms] only when the body of an infected person begins to fight the virus and makes antibodies against it.

TextRank + RB
(pl) Słowa kluczowe: zakazić HIV wirus test kobieta . Wybrane zdanie z artykułu:,Pojawiają się one [Objawy] dopiero w momencie , gdy organizm zakażonej osoby zaczyna walczyć z wirusem i wytwarza przeciw niemu przeciwciała.
(eng) Keywords: infect HIV virus test woman. Selected sentence from the article: They appear [Symptoms] only when the body of the infected person begins to fight the virus and makes antibodies against it.

CW
(pl) Często może być tego faktu zupełnie nieświadoma, ponieważ infekcje mogą przebiegać przez długi czas bezobjawowo. Objawy HIV We wstępnej fazie zakażenia, kiedy HIV wnika do organizmu, żadne objawy nie są zauważalne ani odczuwalne. Pojawiają sią one dopiero w momencie, gdy organizm zakażonej osoby zaczyna walczyć z wirusem i wytwarza przeciw niemu przeciwciała. Zakażony może wówczas czuć się tak, jak podczas grypy: będzie miał bęle głowy, mięśni i lekkie nabrzmienie węzów chłonnych.
(eng) She can often be completely unaware of this, because infections can be asymptomatic for a long time. In the initial stages of HIV infection, when HIV enters the body, no symptoms are noticeable or felt. They appear only when the body of an infected person begins to fight the virus and makes antibodies against it. The infected person may then feel as if they have the flu: she will have headaches, muscle aches and a slight swollen lymph nodes.

COREF
(pl) Pojawiają się one dopiero w [długi czas]momencie, gdy organizm zakażonej osoby zaczyna walczyć z wirusem i wytwarza przeciw niemu przeciwciała.
(eng) They appear only at a [long time] when the body of an infected person begins to fight the virus and makes antibodies against it.

Example 1.1. Exemplary sentence enriched with the presented context enrichment methods

was placed. Thus, it is a measure complementary to the conversion rate. The agreement rate is expressed by the following formula:

$$A = \frac{s_{agr}}{k} \tag{1}$$

where s_{agr} stands for the sum of all consistent pairs

$$s_{agr} = \sum_{n=1}^{S} 1\left[s_{c,n} == s_{nce,n}\right] \tag{2}$$

s_c stands for contextual evaluation of the sentence, s_{nce} - enriched non-contextual evaluation, and k - number of pairs where s_{nce} does not equal -1

$$k = \sum_{n=1}^{S} 1\left[s_{nce,n} \neq -1\right] \tag{3}$$

Weighted Agreement Rate: This measure is a modification of the strong agreement rate. We weigh the outcome of the assessments comparison so that:

1. the largest weight w_1 is assigned to the strong agreement (credible-credible and noncredible-noncredible assessment pairs between the contextual assessment mode and the non-contextual mode with context enrichment);
2. much smaller weight w_2 is assigned to all the pairs that contained "neutral" label on one side (either on C or NCE) (either on contextual or non-contextual with context enrichment). We justify this modification based on the assumption that misinterpretation of informative non-credible sentence with neutral, as well as informative credible with neutral, is potentially less harmful than misinterpretation of informative non-credible with informative credible. Moreover, there is much more randomness in assigning "neutral" labels to sentences by human judges than any other label type;
3. smaller weight w_3, but closer to w_1 than to w_2, is given to the neutral-neutral pairs, for the same reason related to the randomness of the assessments.
4. zero (weight w_4) is given in the strong disagreement scenario (credible-noncredible).

Ultimately, the weights are assigned as follows:

$$w = \begin{cases} 1, & \text{if } \{s_c, s_{nce}\} == \{2,2\} \text{ or } \{0,0\} \\ 0.33, & \text{if } \{s_c, s_{nce}\} == \{1,2\} \text{ or } \{2,1\} \text{ or } \{1,0\} \text{ or } \{0,1\} \\ 0.8, & \text{if } \{s_c, s_{nce}\} == \{1,1\} \\ 0, & \text{otherwise} \end{cases}$$

And the formula is:

$$A_w = \frac{\sum_{n=1}^{S} w_n}{k} \tag{4}$$

where k, s_c, s_{nce} are defined as in Eqs. 2 and 3.

Standardized Length Factor. In the evaluation process we took into consideration the length factor of the retrieved context. It is expressed as a mean retrieved context length per method, divided by the mean length of the article. All lengths are represented as numbers of tokens.

$$LF = m_{cl}/m_{al} \tag{5}$$

where m_{cl} is the mean retrieved context length and m_{al} is the mean article length. We invert the LF because this is the factor that we want to minimize in the aggregate measure

$$ILF = 1 - LF \tag{6}$$

We standardize obtained values with min-max normalization function to the interval (0.5, 1) in order to make the outputs appear on the same scale as the other measures

$$SLF = MinMax(0.5, 1, min_{LF}, max_{LF}, v_{LF}) \tag{7}$$

$MinMax$ is a linear transformation that takes as arguments: minimum and maximum value of the considered set, minimum and maximum value of the new interval, and the variable itself

$$MinMax(min_{new}, max_{new}, min_{LF}, max_{LF}, v_{LF})$$
$$= \frac{v_{LF} - min_{LF}}{max_{LF} - min_{LF}} * (max_{new} - min_{new}) + min_{new} \tag{8}$$

Weighted Harmonic Mean. In this paper, we claim that an optimal context enrichment method ought to maximize conversion rate and agreement rate, while at the same time minimize the length of the added context. Besides, the conversion rate is slightly less important than the agreement rate. That is why we introduced the summarizing measure: weighted harmonic mean. We consider this measure for both strong and weighted agreement rate. Eventually, the formula looks like follows:

$$WHM = \frac{3}{\frac{0.8}{C} + \frac{1.2}{A} + \frac{1}{SLF}} \tag{9}$$

5   Results and Discussion

Table 1 contains quality measures for all considered methods on the full dataset, while Table 2 limits the results to sentences that contained pronouns.

The Context Window method performs best when taking into consideration conversion, strong agreement and weighted agreement rates (as seen in Table 1). Surprisingly, the agreement rate of the Context Window method is only a few percent higher than the results for methods $TFIDF+RB$ and $TextRank+RB$. Application of the length factor makes the Context Window method suboptimal, according to our criteria (maximizing C and A, and minimizing LF). TextRank

Table 1. Performance measures calculated for the full dataset of sentences with enriched context. Evaluations obtained with each method are compared to the fully contextual evaluations from the first dataset. C stands for Conversion rate, A - strong agreement rate, A_w - weighted agreement rate, SLF - standarized length factor, WHM_{SA} - weighted harmonic mean with strong agreement rate and WHM_{WA} - weighted harmonic mean with weighted agreement rate. All measures take values from 0 to 1, except SLF that takes values from 0.5 to 1.

	TFIDF + RB	TextRank + RB	CW	COREF
C	0.796	0.814	**0.912**	0.394
A	0.5	0.499	**0.564**	0.563
A_w	0.593	0.605	**0.655**	0.642
SLF	0.918	0.918	0.5	1
WHM_{SA}	0.67	**0.677**	0.599	0.581
WHM_{WA}	0.73	**0.74**	0.637	0.612

Keywords + rule-based algorithm for supplementing the meaning of pronouns appears to be optimal, taking into consideration both strong and weighted agreement rate. The results are close to those obtained by $TFIDF + RB$ method. It may indicate that shorter context, but collected from the full content of the article (as opposed to the context collected only from the surrounding sentences), proves to be sufficient for the expert to correctly assess the credibility of the sentence.

Table 2. Performance measures calculated only for the sentences that contained pronouns

	TFIDF + RB	TRK + RB	CW	COREF
C	0.84	0.83	**0.93**	0.54
A	0.44	0.506	**0.548**	0.352
A_w	0.569	0.591	**0.665**	0.491
SLF	0.918	0.918	0.5	1
WHM_{SA}	0.629	**0.678**	0.594	0.509
WHM_{WA}	0.723	**0.735**	0.643	0.609

We have also checked to what extent the methods for complementing the meaning of pronouns (both rule-based and the method resulting from finding coreferences) affect the overall score. We have selected and calculated measures only for the subset of sentences that contained a pronoun. While the conversion rate is significantly higher for all methods (especially for coreference resolution), it does not affect aggregate measures (as seen in Table 2).

We have performed chi-squared tests to check whether the evaluated context enrichment methods improve accuracy of credibility evaluation when compared to random evaluations. In case of all methods, for both strong and weighted agreement rate, we can reject the null hypothesis that context enrichment does not improve accuracy at a 99% confidence level.

In general, the method of applying co-referent mentions to the sentences, at the current stage of model development, proved to be sub-optimal to the task of tagging the credibility of sentences. From studying the individual steps of the co-reference resolution pipeline, we did not notice any large-scale issues within the initial steps of the processing. We suspect that the main issue lies in the actual co-reference resolution engine Bartek. From what we can gather, the system is trained on the Polish Co-reference Corpus [15] which is hand-annotated co-reference corpus based largely on the Polish National Corpus [17] and other sources of news articles. The corpus does mention a very small percentage of scientific texts, but it is very unlikely it would contain any significant amount of medical texts.

In the course of the study we also experimented with the WordNet based method for completing the meaning of hyponymous expressions (eg. we tried to complement a word [virus] with [HIV], or [cancer] with [breast], [malignant]). We have collected the terminology from the article and compared it to the set of units that were linked in WordNet to the given word as hyperonyms. We faced the problem of too much generality in a WordNet structure and we decided to abandon this approach in favor of other methods. However, experiments with more domain-specific knowledge networks will be subject of our future work.

Reduction of Assessment Time. In the reference study [14] the Authors report the average article assessment time as approximatelly 10 min. Articles in the corpus have an average of 771 tokens, which should give about 3.5 min per article (given the average reading speed of 200 words per minute, according to [11]). This discrepancy points to the fact that credibility assessment is a longer and more complicated process than reading comprehension. In case where an expert has many sentences to evaluate, factors such as monotony and monotype may influence time of the assessment as well.

In the current study we used standarized length factor as a measure to approximate the amount of time and attention needed to evaluate a sentence. It is however possible that other factors might affect the final time as well, such as complexity of the extracted text, difficulty with relating tags to the actual content or connecting pronouns with suggested nouns and adjectives. We suspect that some methods might be more time-consuming than the others, considering different types of thinking they impose on a reader. A detailed study of time required to evaluate the text obtained by the presented methods is yet to be performed and should be addressed in the future work.

We therefore propose a simplification, which allows to estimate a proportional shortening of the expert's time in relation to the situation in which he or she is forced to get acquainted with the full context before assessing the sentence. Basing on our LF measure, our best method $TextRank + RB$ allows the expert to assess the credibility of the sentence approximately in 4% of the time needed for credibility assessment of a full article, and in 25% of the time needed for assessment using our baseline Context Window method.

Based on our experience from the previous study [14] that showed the overall consistency of the credible and non-credible articles (credible articles contain mostly credible sentences and vice versa, ordering of credible/non-credible sentences is not important), we were not surprised by these results. We can hypothesize that non-credible medical content is built upon some finite set of key phrases and words that can be easily detected by an expert. The keyword-based context enrichment methods therefore proved to be suitable to extract the most important features from the context that allow fast and accurate credibility assessment.

6 Conclusions and Future Work

In this article, we have studied the problem of context enrichment for fast fact-checking (credibility evaluation) of sentences. Individual sentences - being part of some larger content - can be evaluated by experts in full context (experts read the entire text to which the sentence belongs) or with partial or no context. We have studied the problem of automatic context enrichment empirically on the case of medical Web content in the Polish language. Our main findings and statements can be summarized as follows:

- In our study the best-performing methods (TextRank + RB, TFIDF + RB) rely on simple NLP tools (such as taggers) and we are confident that our results can be generalized to other languages.
- Evaluation of individual sentences with partial context provided by our methods is faster than evaluation of the same sentences with full context.
- The context enrichment methods presented in this article are not domain-specific and could be applied to fact-checking tasks apart from the medical domain.
- Short textual information acquired in the process of context enrichment could also prove useful in information retrieval tasks.
- As shown by our analysis of types of sentences that cannot be evaluated due to missing context, the problem of context enrichment is a general NLP problem that differs significantly from other NLP problems such as pronoun matching or coreference resolution.

References

1. Buczyński, A., Przepiórkowski, A.: Spejd demo: An open source tool for partial parsing and morphosyntactic disambiguation (2008)
2. Chen, Y.Y., Li, C.M., Liang, J.C., Tsai, C.C.: Health information obtained from the internet and changes in medical decision making: questionnaire development and cross-sectional survey. J. Med. Internet Res. **20**(2), e47 (2018)
3. ELRA: Proceedings of the Eighth International Conference on Language Resources and Evaluation, LREC 2012 (2012)

4. Harkema, H., Dowling, J.N., Thornblade, T., Chapman, W.W.: Context: an algorithm for determining negation, experiencer, and temporal status from clinical reports. J. Biomed. Inform. **42**(5), 839–851 (2009)
5. Hill, F., Bordes, A., Chopra, S., Weston, J.: The goldilocks principle: Reading children's books with explicit memory representations. arXiv preprint arXiv:1511.02301 (2015)
6. Kadlec, R., Schmid, M., Bajgar, O., Kleindienst, J.: Text understanding with the attention sum reader network. arXiv preprint arXiv:1603.01547 (2016)
7. Kąkol, M., Jankowski-Lorek, M., Abramczuk, K., Wierzbicki, A., Catasta, M.: On the subjectivity and bias of web content credibility evaluations. In: Proceedings of the 22nd International Conference on World Wide Web, pp. 1131–1136. ACM (2013)
8. Kakol, M., Nielek, R., Wierzbicki, A.: Understanding and predicting web content credibility using the content credibility corpus. Inf. Process. Manage. **53**(5), 1043–1061 (2017)
9. Kopeć, M.: Zero subject detection for Polish. In: Proceedings of the 14th Conference of the European Chapter of the Association for Computational Linguistics, Volume 2: Short Papers, pp. 221–225. Association for Computational Linguistics, Gothenburg, Sweden (2014)
10. Kopeć, M., Ogrodniczuk, M.: Creating a coreference resolution system for Polish. In: LREC [3], pp. 192–195
11. Lewandowski, L.J., Codding, R.S., Kleinmann, A.E., Tucker, K.L.: Assessment of reading rate in postsecondary students. J. Psychoeducational Assess. **21**(2), 134–144 (2003)
12. Logeswaran, L., Lee, H.: An efficient framework for learning sentence representations. arXiv preprint arXiv:1803.02893 (2018)
13. Mihalcea, R., Tarau, P.: TextRank: bringing order into text. In: Proceedings of the 2004 Conference on Empirical Methods in Natural Language Processing (2004)
14. Nabożny, A., Balcerzak, B., Wierzbicki, A.: Automatic credibility assessment of popular medical articles available online. In: Staab, S., Koltsova, O., Ignatov, D.I. (eds.) SocInfo 2018. LNCS, vol. 11186, pp. 215–223. Springer, Cham (2018). https://doi.org/10.1007/978-3-030-01159-8_20
15. Ogrodniczuk, M., Głowińska, K., Kopeć, M., Savary, A., Zawisławska, M.: Coreference in Polish: Annotation, Resolution and Evaluation. Walter De Gruyter (2015). http://www.degruyter.com/view/product/428667
16. Ogrodniczuk, M., Lenart, M.: Web Service integration platform for Polish linguistic resources. In: LREC [3], pp. 1164–1168
17. Przepiórkowski, A., Górski, R.L., Łazinski, M., Pezik, P.: Recent developments in the national corpus of Polish. NLP, Corpus Linguistics, Corpus Based Grammar Research, p. 302 (2010)
18. Ren, P., Chen, Z., Ren, Z., Wei, F., Ma, J., de Rijke, M.: Leveraging contextual sentence relations for extractive summarization using a neural attention model. In: Proceedings of the 40th International ACM SIGIR Conference on Research and Development in Information Retrieval, pp. 95–104. ACM (2017)
19. Samuel, H., Zaïane, O.: MedFact: towards improving veracity of medical information in social media using applied machine learning. In: Bagheri, E., Cheung, J.C.K. (eds.) Canadian AI 2018. LNCS (LNAI), vol. 10832, pp. 108–120. Springer, Cham (2018). https://doi.org/10.1007/978-3-319-89656-4_9
20. National Academies of Sciences, Engineering and Medicine: Improving diagnosis in health care. National Academies Press (2016)

21. Shibuki, H., Nagai, T., Nakano, M., Miyazaki, R., Ishioroshi, M., Mori, T.: A method for automatically generating a mediatory summary to verify credibility of information on the web. In: Proceedings of the 23rd International Conference on Computational Linguistics: Posters, pp. 1140–1148. Association for Computational Linguistics (2010)
22. Turek, P., Wierzbicki, A., Nielek, R., Datta, A.: WikiTeams: how do they achieve success? IEEE Potentials **30**(5), 15–20 (2011)
23. Vinyals, O., Fortunato, M., Jaitly, N.: Pointer networks. In: Advances in Neural Information Processing Systems, pp. 2692–2700 (2015)
24. Wang, L., Tu, Z., Way, A., Liu, Q.: Exploiting cross-sentence context for neural machine translation. arXiv preprint arXiv:1704.04347 (2017)
25. Waszczuk, J.: Harnessing the CRF complexity with domain-specific constraints. the case of morphosyntactic tagging of a highly inflected language. In: Proceedings of COLING 2012, pp. 2789–2804 (2012)
26. Wierzbicki, A.: Web Content Credibility. Springer, Heidelberg (2018). https://doi.org/10.1007/978-3-319-77794-8
27. Yamamoto, Y.: Disputed sentence suggestion towards credibility-oriented web search. In: Sheng, Q.Z., Wang, G., Jensen, C.S., Xu, G. (eds.) APWeb 2012. LNCS, vol. 7235, pp. 34–45. Springer, Heidelberg (2012). https://doi.org/10.1007/978-3-642-29253-8_4

ConceptMap: A Conceptual Approach for Formulating User Preferences in Large Information Spaces

Alireza Tabebordbar[1(✉)], Amin Beheshti[2], and Boualem Benatallah[1]

[1] University of New South Wales, Sydney, Australia
{alirezat,boualem}@cse.unsw.edu.au
[2] Macquarie University, Sydney, Australia
amin.beheshti@mq.edu.au

Abstract. In a large information space a user needs to iteratively investigate the data to formulate her preferences for IR systems. In recent years several visualization techniques have been proposed to help a user to better formulate her preferences. However, using these solutions a user needs to explicitly specify her preferences for IR systems in forms of keywords or phrases. In this paper we present ConceptMap, a system that takes the advantage of deep learning and a knowledge lake to provide a conceptual summary of the information space. ConceptMap allows a user to specify her preferences implicitly as a set of concepts without the need to iteratively investigate the information space. It provides a 2D Radial Map of concepts where a user can rank items relevant to her preferences through dragging and dropping. Our experiment results shows that ConceptMap can help users to better formulate their preferences when they need to retrieve varied and comprehensive list of information across a large amount of data.

Keywords: Formulating user preferences · Conceptual visual summary · Conceptual information retrieval

1 Introduction

Information Retrieval (IR) systems have been extensively used to extract and locate users information. These systems retrieve a ranked list of items ordered by their relevancy and allow a user to skim and pick items from the list. Exploratory search is a part of information exploration process in which a user is unsure about the way to retrieve her information needs, and often becomes familiar with the information space over time. Normally, in an exploratory search, a user relies on text based queries for formulating her preferences. Text queries are made up of a few keywords or phrases [12] and allow a user to explore and retrieve the information. However, formulating queries has been proven to be painstakingly difficult as a user needs to iteratively read and synthesize a large amount of information. This problem is exacerbated as humans have a limited memory

© Springer Nature Switzerland AG 2019
R. Cheng et al. (Eds.): WISE 2019, LNCS 11881, pp. 779–794, 2019.
https://doi.org/10.1007/978-3-030-34223-4_49

capacity in absorbing information, which can lead to information overload or attention management [24]. In past years, several studies [13, 27, 28] have been conducted to formulate user's preferences through rules, e.g., boolean operators. However, these studies have concluded that comprehension of the information space is needed in order a user formulates her preferences properly.

In recent years several pioneering solutions [11, 26, 29] have been proposed to couple Human-Computer-Interaction (HCI) techniques with IR systems to aid users to develop insight and absorb greater amounts of information. These solutions fuse the traditional text based queries with various visualization elements, such as bar-graph [15], table [29], and relevance map [23]. Although visual encoding lowers user's cognitive load [12, 15], still the user needs to iteratively explore the information space to identify the relation between attributes, e.g., keywords, phrases, named entities, in documents to formulate her preferences. This process is challenging for several reasons: (1) in many cases an exploratory search scenario contains too many topical subspaces[1] and is difficult for a user to formulate her preferences in forms of keywords or phrases; (2) Sensemaking of the information space is incomplete as text queries only retrieve a small part of the information space and the rest remains invisible; and (3) Relying on text queries are time-consuming as the user is not familiar with the information space and needs to create a large number of queries to retrieve her information.

For example, consider a user intends to analyse citizens' opinions on social media platforms, e.g., Twitter and Facebook, to identify issues in *Health Care Services* that need improvement. Currently, the user needs to read and scan the information space to identify the query terms that properly retrieve items relevant to a large number of topical subspaces. Such search scenario needs the user to spend a long period of time to identify the content bearing terms associated to each subtopic. Alternatively, Carterette et al. showed users are more willing to express their preferences relatively [10], instead of precisely specifying attributes associated to it. In this context, we follow a similar trend by generating a conceptual summary of the information space and helping a user to formulate her preferences *implicitly* as a set of concepts.

In this paper, we present the *ConceptMap*, a system for lowering user's cognitive load in ranking and exploring the information space. While previous systems allow a user to formulate her preferences explicitly, e.g., keywords and phrases, and observing changes in rankings to understand the data, we provide a different ranking and data presentation approach. Our work focuses on creating a conceptual summary of the information space to help a user to understand the data and relate it to her preferences. Hence, we focus on boosting a user's cognitive skill in understanding the data and formulating that understanding to extract information relevant to her topic of interest. We do this by interacting with user to explore her preferences in a $2D$ Radial Map. A user can refine her preferences

---

[1] Topical subspaces are subtopics relevant to the topic a user seeking for, e.g., "medical centres", "aged care services", and "mental health" are subtopics relevant to health care.

through dragging and dropping concepts into a Query Box to update document rankings, representing the relevancy of concepts and documents.

ConceptMap is made up of two main technical achievements: *Knowledge Lake* [2–4], which is a centralized repository containing several knowledge bases, providing a contextualization layer for annotating attributes within the information space with a set of facts and information. *Summarization* [6,14], takes the advantage of a deep learning skip gram embedding network [22] to learn the associations between attributes and groups the similar ones. We discuss how conceptual summary lowers user's cognitive load to better formulate her preferences. We also discuss how the insight developed from conceptual summary aids a user to formulate her preferences through more advanced IR systems features, such as rules.

The rest of this paper is organized as follows: in Sect. 2, we discuss the previous works regarding formulating users preferences and comprehending the information space. Section 3 describes the interface of the ConceptMap and it's components. Section 4, we discuss our proposed approach for generating a conceptual summary of the information space. Then, in Sect. 5, we discuss the experiment results, before concluding the paper in Sect. 6.

2   Related Work

Formulating User Preferences. The relevance judgement for IR systems has been made on binary scale, where a document is considered relevant to a query or not [10]. Such judgement requires a user to precisely formulate her preferences to locate and retrieve the relevant documents. Typically, for formulating preferences a user needs to conduct exploratory search by iteratively investigating the information space to develop insight and create a mental structure of it [21]. Exploratory search is beyond the basic information seeking task of looking for a few relevant documents. In an exploratory search, a user learns to formulate her preferences by investigating the context over time [19]. Previous works have focused on augmenting users' comprehension of the information space with visual encoding to formulate their preferences more precisely [15]. However, recently some solutions [10,29] have shown that it is easier for a user to make relative judgement of her preferences rather than explicitly specifying attributes associated to it. For example, a user may formulate "mental health" as a "disorder", but unable to specifying the attributes associated to it. In this context, we allow a user to formulate her preferences *implicitly*, as a set of abstract concepts, such as topics and named entities. Our approach automatically identifies the relation among attributes within the information space and groups them based on their semantical relations. In the next sections we will accentuate approaches focused on augmenting users' comprehension of the information space.

Comprehension and Sensemaking of the Information Space. Sensemaking of the information space is defined as processes and activities a user undertakes to frame the information space in an understandable schema [24]. Sensemaking has been identified as a quintessential task of information retrieval [21], especially when a user has varied information needs across a large amount of data [23]. During the past years several solutions have been proposed to enhance user comprehension and sensemaking of the data. One category of these solutions focused on augmenting the ranked lists of search result with different visualization elements. For example, TileBar [17] represents the relevancy of ranked documents to query terms with shaded blocks. LineUp [15], used bar charts to visualize the ranking of multi-attributes data, while other approaches highlight ranked lists with stacked bar and metaphor-/snippet-based layout.

Another types of works have coupled visualization with HCI techniques for augmenting a user to gain a better understanding of the information space. For example, Wall et al. [29], proposed table layout to present a holistic view of the information space for multi-attribute ranking systems. Di-Sciascio et al. [11] boosts user comprehension of the information space by contributing previous users search terms. Peltonen et al. [23] provides a topical overview of the information space by interacting with the user to visualize the association between keywords on a relevance map. However, these approaches focus on enhancing user capacity to better identify the relations between attributes in documents. But, in a large information space many of these relations remain invisible to user, either due to user inability in identifying them or visual clutter [26]. Instead, ConceptMap provides a conceptual summary of the data and offloads user to iteratively investigate the information space to discover associations between attributes in documents.

Topic Modeling Techniques. In addition to interactive methods for augmenting user's comprehension, we consider it appropriate to include approaches that provide a topical overview of the information space. Topic modeling is a generative approach, which aims at discovering groups of words that frequently co-occur in documents [8]. Latent Dirichlet Allocation (LDA) is the most common topic modeling algorithm, which has been used extensively for providing a topical overview of the information space. For example, TIARA [20] is one of the early works on topic based text summarization, which creates a visual summary of the information space through visualizing the result of LDA algorithm with a stacked graph. Serendip [1] is a topic modeling system focused on structuring exploration of information for supporting multi-level discovery. TopicNets [16] is an interactive topic modeling system that visualizes documents on a graph of connected network. Although topical modeling can provide an overview of the information space, these algorithms cannot identify the semantical relation between attributes in information space [18]. Moreover, topical modeling algorithms have a high performance requirements and is computationally expensive to rely on for dynamic and real time search scenarios.

3   ConceptMap Interface

In this section, we describe ConceptMap interface, a system that provides a conceptual summary of the information space. ConceptMap discovers the semantical relation between attributes and lets a user to formulate her preferences as a set

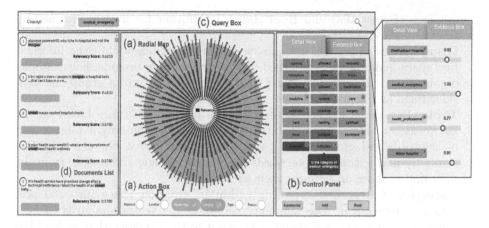

Fig. 1. ConceptMap interface and its component.ConceptMap interface and its component.

of abstract concepts. The ConceptMap interface is made up of four components: the Radial Map (Fig. 1a), the Control Panel (Fig. 1b), the Query Box (Fig. 1c) and the Documents List (Fig. 1d). Followings explain the character of each component in detail.

Radial Map. The Radial Map is the main data view in the ConceptMap, and shows a summary of the most frequent concepts within the information space (Fig. 1a). A user can select her preferred summaries through interacting with the *Action Box*. It has six toggles for visual encoding of the Radial Map: *Persons, Organizations, Locations, Categories, Topics,* and *Keywords*. Each toggle represents a specific summary and colors the summary if its checked by a user. A user can observe concepts associated with summaries through the Radial Map by pressing the *Summarize* button. The coloring of concepts within the Radial Map corresponds to the Action Box, where concepts associated to a summary mirror that color. By default, the Radial Map is divided into 50 wedges, where each wedge represents a concept. Each wedge augmented with a grid line and shows the relevancy of the concept to the information space. The value of the grid line is between 0 to 1, where zero represents the least relevancy and one represents the highest relevancy. Augmenting wedges with grid lines let a user to grasp an overview of the information space along with their relevancy as a whole at a glance. The following explains the types of summaries a user can choose.

- **Location:** Provides a summary of places within the information space based on their geographical distances. For example, location summary may represent a concept such as *Suburbs in Sydney, Australia* through grouping suburbs located within it, e.g., Five Dock, Canada Bay, Kensington.
- **Person:** Identifies person names within the information space and groups people based on their title. For example, person summary may create a concept like "Health Ministers" by grouping persons, such as Greg Hunt (Health

minister of Australia) and Brad Hazzard (Minister for health and medical research, NSW, Australia).

- **Organization:** Identifies companies and organizations within the information space and groups them based on the services they provide. For example, organization summary may place financial companies into a group and consider them as a concept.
- **Topic:** Provides a topical summary of the information space based on the keywords semantical relationships. It examines the hypernym relationship of keywords and groups them based on their similarity. For example, this summary may extract the keywords pigeon, crow, eagle and seagull from the information space and groups them as *bird*.
- **Category:** Categorizes keywords within information space into 200 prevalidated topics [14]. This summary computes the vector similarity of keywords and categories and assign a keyword to the category that yielded the highest similarity score.
- **Keyword:** Presents the most frequent keywords and their co-occurrence within information space and lets a user manually create her topic of interest.

Control Panel. The second component of the ConceptMap is the Control Panel (Fig. 1b), which allows a user to modify concepts based on her preferences. The Control Panel is made up of two components: Details View and Evidence Box. We will now turn to explain the character of components.

Details View. The Details View provides a detailed representation of concepts to a user. It shows attributes, e.g., keywords, named entities, associated to a concept and allows a user to modify the concept based on her preferences. Attributes within the Details View are colored to represent their relevancy to a concept. The opacity of the color shows the measure of relevancy between a concept and its attributes. Darker color means higher relevancy, while lighter color shows lower relevancy. A detailed description of attributes can be seen in a tooltip by hovering over them. The tooltip provides a small textual description, and lets a user better judge the relevancy of attributes to the concept. In cases where a user identifies an attribute as irrelevant, the user may remove the attribute by pressing the ($\times$) button located on top right side of it. A user can organize the potential concepts relevant to her preferences into the Evidence Box by pressing the Add button, located below the Control Panel.

Evidence Box: The Evidence Box acts as a central repository and aids a user to gather potential concepts relevant to her preferences. Collecting concepts altogether in a place, allows a user to create a mental structure of the information space. It is particularly important in large information spaces as humans have a limited memory capacity in absorbing information. The concepts within the Evidence Box follows the same coloring scheme applied to the Radial Map (Fig. 1a). Making it easier for a user to identify the type of summaries stored in the Evidence Box. A slider placed horizontally below concepts to visually encode the weight of concepts. Initially, the slider shows a pre-computed weight for each concept, which is the average TF-IDF score of attributes associated to it. A user may change the weight of a concept by moving the slider indicator to left/right.

Fig. 2. The Query Box provides two interfaces for formulating a user preferences: (a) the Concept Only, which allows a user to formulate her preferences as a set of concepts; (b) the Concept + Rule, which aids a user to formulate her preferences through rules.

Query Box. In this section, we explain the Query Box a component for examining a document-concept relevancy. A user can interact with the Query Box through two interfaces: *Concept Only* (Fig. 2a) and *Concept + Rule* (Fig. 2b). The *Concept Only* interface allows a user to specify her preferences implicitly by dragging and dropping concepts from the Evidence Box. A user can drag and drop several types of concepts, e.g., Topic, Category, Person, into the Query Box, allowing to rank documents from different perspectives.

The second interface *Concept + Rule*, aids a user to formulate her preferences through rules. Rule based techniques have been coupled with IR systems for an extended period of time. Often, however, users are making mistake in using rules to formulate their preferences [13]. The goal of this interface is to study whether providing conceptual summary of the information space augments users ability to understand and utilize rules more effectively or not. The *Concept + Rule* interface provides four operators for formulating user's preferences through rules '[', 'AND', 'OR', ']'. '[', indicates the start of a rule clause, ']', indicates the end of a rule clause. 'AND' operator implies a document must contain a specified concept to be ranked by the ConceptMap. 'OR' operator implies at least one of the user specified concepts must appear in a document to be ranked by the ConceptMap.product of concepts as below: Next, we will explain how the ConceptMap formulates user's selected concepts for IR systems to rank documents.

As each concept represents a set of attributes rather than a single keyword, we need to transform concepts into forms of text queries for ranking through IR systems. We do this by computing the cartesian product of concepts. Formally, we denote a concept C, as a set of attributes $C = \{c_1, c_2, c_3, ..., c_n\}$, where each attribute $c \in C$ represents a keyword or named entity associated to it. We also, denote a query Q as a set of concepts $Q = \{C_1, C_2, C_3, ..., C_n\}$, where each concept $C \in Q$ represents a preference dropped into the Query Box through the user. To score documents based on the user preference, we compute the cartesian product of attributes associated to concepts $Q = C_1 \times C_2 \times C_3 \times ... \times C_n$. The resulting set $Q = \{q_1, q_2, q_3, ..., q_n\}$ are the queries—ConceptMap computes their relevancy to documents. More clearly, suppose a user intends to analyse citizen's opinion about government budget in "Health Care System". Assume, the user selects the

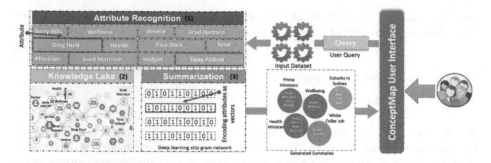

Fig. 3. Overview of the proposed summarization technique. (1) Extracts the potential attributes from the information space. (2) Annotates the attribute using the Knowledge Lake. (3) performs analogous reasoning through mapping attributes on a vector space.

"Health Ministers" concept from the *Persons* summary $Health\ Ministers = \{Greg\ Hunt, Brad\ Hazzard\}$, and the "budget" concept from the *Category* summary $Budget = \{fund, money, budget\}$. To generate the queries that represent the given concepts, ConceptMap computes the cartesian product of concepts as below: $Q = \{(Greg\ Hunt, fund), (Greg\ Hunt, money), (Greg\ Hunt, budget),$ $(BradHazzard, fund), (Brad\ Hazzard, money), (Brad\ Hazzard, budget)\}$. The following explains how ConceptMap computes the relevancy of queries and documents. ConceptMap arranges documents based on their relevancy to queries. To compute the relevancy of a query and a document, we implemented a Vector Space Model (VSM) model. The model transforms documents d and queries $q \in Q$ into vectors and computed their relevancy using a cosine similarity.

$$S(d, q) = \sum \frac{tfidf(c, d).W_C}{||d||.||q||}$$

Where $tfidf(c, d)$ is the TF-IDF score for the attribute c in document d. W_C is the weight a user specified for the concept C. Also, $||d||$ and $||q||$, are the Euclidean norms for vectors d and q. ConceptMap arranges documents in descending order based on their cosine similarity scores to queries.

Documents List. The Documents List (Fig. 1d) provides a list of documents ranked based on the user's preferences. Documents List relies on stacked bar charts for visual encoding of documents. It shows a barchart below each document, illuminating the relevancy of documents to user preferences. To aid a user to better comprehend documents-concepts relevancy, Document List applies the same coloring scheme as the Radial Map. The coloring allows a user to identify the contribution of each concept to a document, and provides an explanation of why one document ranked higher than another. The ranking of documents are updated as a user modifies her preferences through adding/removing concepts.

4 Solution Overview

In this section, we explain how the ConceptMap generates a conceptual summary of the information space. To generate a summary, ConceptMap utilizes a *Knowledge Lake* [3,4] and a deep learning skip gram embedding network [22]. Knowledge Lake, is a central repository made up of several knowledge bases and provides a contextualization layer for transforming the raw data into contextualized knowledge. It annotates attributes within the information space with a set of facts and information. Deep learning network measures the conceptual commonality existing between attributes and groups attributes with similar characteristics. Overall, our approach is made up of three stages, *Attributes Recognition*, *Knowledge Lake*, and *Summarization*. Followings explain each stage in detail.

Attributes Recognition. The initial step in generating the summary is identification of content bearing attributes exists within the information space. These attributes allows ConceptMap to discover the aboutness of data. Today, ConceptMap extracts two types of attributes: *Keyword* and *Named Entity*. To extract the attribute type of keyword, ConceptMap performs a preprocessing task by removing the stopwords, keeping the proper names capitalized, and filtering out of the ungrammatical and irrelevant tokens, e.g., URLs' or imoji. Also, It applies the WordNet lemmatizer over the remaining tokens to increase the probability of matching between words with the common base, e.g., "playing", "playful", "plays" all reduce to the base form "play". The second attribute type is named entity. Named entities are the span of words in a text which refer to real-world objects, such as person and company names, or gene and protein names. Today, ConceptMap extracts three types of named entities: persons, locations, and organizations. For recognizing named entities, we used the system proposed by Beheshti et al. [5,7]. It provides a pipeline for various data curation tasks, including named entity recognition, information linking, similarity computation, indexing. After extracting attributes, we annotate them through a Knowledge Lake to identify the concepts existing within the information space. In the next section, we explain how Knowledge Lake contributes to our work.

Knowledge Lake. Knowledge Lake allows a user to understand attributes within the information space and provides a foundation to measure the commonality between them. We utilize several readily available knowledge bases and taxonomies to create the Knowledge Lake: (1) *Geoname*, is a geographical database and contains information over 25 millions geographical places around the world (geonames.org/), (2) *Wikidata*, is a central storage of several Wikimedia data, including Wikipedia, Wikivoyage, Wikisource (wikidata.org/), (3) *WordNet*, is a semantic lexicon and grouped English words into sets of synonyms called synsets (wordnet.princeton.edu/citing-wordnet), (4) Empath, is a deep learning skip-gram network, which categorizes text over 200 built-in categories [14], and (5) Google Knowledge Graph, is a knowledge base based on a graph database, and provides information about real-world entities, including persons, locations, business (developers.google.com/knowledge-graph/). These knowledge bases annotate the attributes within the information space in a more generalized or understandable form. Knowledge Lake K acts as a function $K : c_i \longmapsto \ell(c_i)$, that receives an attribute c_i as the input and returns an annotation $\ell(c_i)$ to describe the attribute.

Summarization. In this section, we explain how ConceptMap generates a conceptual summary of the information space. For example, we explain how it identifies two persons, e.g., Greg Hunt and Brad Hazzard can be similar and forms a concept.

As we discussed, ConceptMap annotates attributes[2] within the information space through the Knowledge Lake. ConceptMap uses these annotations to generate the summaries. For generating the *Topic* summary, ConceptMap utilizes the WordNet and groups attributes based on their hypernym relations. For example, it groups attributes, such as $\{doctor, physician, dentist\}$ as "medical_practitioner", while attributes like $\{health \ and \ wellness\}$ as "wellbeing".

For generating the *Category* summary, ConceptMap relies on the EMPATH, which categorizes text into 200 built-in human-validated categories. For example, it may categorize $\{doctor \ and \ physician\}$ attributes as relevant to "medical_emergency", "occupation" and "white_collar_job" categories, while $\{health\}$ as "medical_emergency", but not "occupation" and "white_collar_job". ConceptMap visualizes categories with the highest frequency within the information space. To generate the *Person, Organization, and Location* summaries, ConceptMap takes the advantage of a deep learning skip gram network [22] to predict the semantic similarity between attributes within the information space. By training the skip gram network, it learns a representation of words within the information space, which known as neural embeddings. The neural embeddings construct a vector space model and allows to measure the similarity between attributes in an unsupervised fashion. We used word2vec neural embeddings model[3] to map attributes onto a vector space.

For attributes annotated with the Knowledge Lake, ConceptMap encodes attributes as vectors by querying the vector space model trained on the word2vec. Then, it performs "analogous reasoning"[4] [14] by conducting the vector arithmetic on generated attributes vectors, e.g., the vector arithmetic for words "Women + King - Man" generates a vector similar to "Queen". The following explains how ConceptMap performs analogous reasoning to measure the similarity between attributes.

For each attribute c, ConceptMap tokenizes its annotation $\ell(c)$ into a set of words $\ell(c) = \{\ell'_1(c), \ell'_2(c), \ell'_3(c), ..., \ell'_n(c)\}$. Then, for each $\ell'(c) \in \ell(c)$, ConceptMap queries the vector space model and extracts the vector $V(\ell'(c))$ corresponds to it. It performs analogous reasoning by computing the vector sum of all $V(\ell'(c))$. The resulting vector $V(c)$ represents the attribute c in vector space.

$$V(c) = \sum\nolimits_{i=1}^{n} \ell'_i(c)$$

---

[2] ConceptMap generates *topic* and *category* summaries from the attribute type of keyword, while *person, organization* and *location* summaries from the named entities.

[3] https://drive.google.com/file/d/0B7XkCwpI5KDYNlNUTTlSS21pQmM/edit.

[4] A form of comparison to highlight respects in which two attributes can be similar.

ConceptMap groups similar attributes based on their vectors similarity using the cosine measure. For example to compute the vector similarity of two attributes, the cosine similarity is

$$cos(\theta) = \frac{V(c_1) \cdot V(c_2)}{||V(c_1)|| \cdot ||V(c_2)||}$$

Where $V(c_1)$ and $V(c_2)$ representing vectors of attributes c_1 and c_2, and $||V(c_1)||$ and $||V(c_2)||$ are their lengths. More clearly, consider the attribute $c = \{Greg\ Hunt\}$, which annotated with the Knowledge Lake as the $\ell(c) = \{Health\ Minister\ of\ Australia\}$. To represent this attribute as a vector ConceptMap tokenizes $\ell(Greg\ Hunt) = \{Health,\ Minister,\ Australia\}$ and query a VSM to extract their corresponding vectors $\ell(Greg\ Hunt) = \{V(Health),\ V(Minister),\ V(Australia)\}$. Then, it computes the vector sum of all attributes $V(Greg\ Hunt) = V(Health) + V(Minister) + V(Australia)$. The resulting vector $V(Greg\ Hunt)$, represents the attribute $c = \{Greg\ Hunt\}$ in vector space, and allows ConceptMap to compute its similarity with other attributes vector using the cosine similarity. ConceptMap groups attributes that their cosine similarity is above a pre-defined threshold[5]. Figure 3 shows an overview of the summarization technique.

5 Experiments

In this section, we study the performance of ConceptMap in formulating users preferences with respect to a traditional keyword-based UI. The study followed a repeated measures design ANOVA with two independent variables: *tool*: ConceptMap, which consists of Concept-Only and Concept-Rule interfaces, and a traditional keyword-based UI—and *items*. The Keyword-Based UI allows users to investigate the information space by entering their keywords and observing the resulting set ordered by their relevancy in the Documents List. Also, to aid users to better identify the relation among keywords we visualize the most frequent keywords co-occurred with users selected keywords. To counterbalance the experiment, we conducted the study on two different topics relevant to social issues: *Health Care* and *Budget*, where topic treated as random variable.

The study simulates an exploratory search scenario, where users need to write queries to retrieve Tweets relevant to the given topics. We divided users task into two subtasks: a focused exploratory search scenario and a broad exploratory search scenario. The focused search scenario, requires users to investigate the information space to retrieve items for a limited number of topical subspaces, e.g., retrieve a list of Tweets contain information relevant to medical centres within Australia. The broad search scenario simulates cases where users need to retrieve items for a larger number of topical subspaces, e.g., retrieve citizens' opinions using the Twitter about people who involved in health care system; and Tweets about the quality of the services provided by health care centres.

---

[5] Currently, we consider attributes over 0.7 % cosine measure as similar.

We invited five post-graduate students from a lab to take part in experiments. None of the participants were knowledgable in the topics selected for the study. For evaluating the performance of tools participants selected a topic and performed the search scenarios. The goal of these experiments were to reflect the behaviour of users in formulating their preferences, while they investigating the information space with different information needs. For each task, participants filled a 7-point likert scale NASA TLX questionnaire. The questionnaire is a multidimensional assessment tool that rates perceived workload in order to assess a task or system. We limited the duration of focused search scenario to 5 min and the broad search scenario to 10 min. During the experiment we reminded participants when their allotted time was almost over, but we didn't force them to stop using the tools.

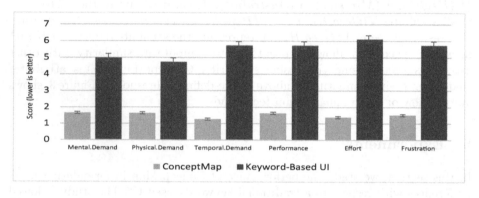

Fig. 4. The bar chart shows ConceptMap imposes lower workload across different dimensions. Error bars shows the standard error.

Results: Workload and Performance Analysis. A repeated measures ANOVA revealed the impact of ConceptMap in lowering participants overall workload $F(1,5) = 58.803$, $p = 0.05$. This tendency can be observed in detail in Fig. 4, which shows participants were more relaxed while interacting with ConceptMap for formulating their preferences. The results showed ConceptMap could significantly lowers participants temporal demand $F(1,5) = 162.00$, $p = 0.001$. We also observed a similar impact on improving participants efforts $F(1,5) = 83.308, p = 0.003$, and performance $F(1,5) = 43.560$, $p = 0.007$. Based on the obtained results we concluded that providing several summaries of information space could boost participants comprehension and sensemaking of data: this impact is more significant on improving users performance and time.

We also analyzed the effectiveness of ConceptMap in aiding participants in formulating their preferences through rules. A repeated measure ANOVA showed that ConceptMap could lowers participants overall workload $F(1,5) = 12.60, p = 0.05$. The results revealed that ConceptMap reduces participants effort in crafting rules $F(1,5) = 18.00$, $p = 0.005$. We also observed a similar impact on participants performance $F(1,5) = 22.04, p = 0.003$.

In addition to previous experiments, we analyzed the performance of tools by aggregating top 20 items collected by participants and verifying their relevancy. Thus, we created two datasets from the aggregated items. The first dataset represents items collected by participants through interacting with the ConceptMap, and the second dataset contains items collected through interacting with the Keyword-Based UI. Then, we verified whether a retrieved item is relevant to the topics assigned by participants or not. The results showed participants could retrieve items more precisely using the ConceptMap compared to the Keyword-Based UI. Figure 5a shows the average precisions obtained using the tools. Based on the observed results, allowing users to examine the relevancy of data from different perspectives has a positive impact on retrieving items.

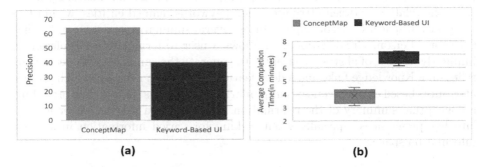

(a)   (b)

Fig. 5. (a) The average precision of participants obtained using ConceptMap and Keyword-Based UI. (b) The time participants spent for retrieving their information.

Results: Completion Time and Usability Analysis. In the second study we analysed the completion time and the usability aspect of ConceptMap. We assigned a task to participants and let them to accomplish the task using the tools in a more natural settings, e.g., without times-up. The task was a broad exploratory search scenario for retrieving Tweets relevant to issues and people involved in budget planning. Then, we asked participants to fill a standard Software Usability Scale (SUS) questionnaire [9]. SUS provides subjective assessments of a software usability, where a statement is made and respondents can indicate the degree of agreement and disagreement on a five point scale. We averaged over all participants questions, the mean score amounted to 87.5 out of 100, which falls ConceptMap in the 90–95 percentile range in the curve grading scale interpretation of SUS scores [25].

We also calculated the time participants spent to accomplish their task. We observed using the ConceptMap participants could accomplish their task in a shorter timer compared to the Keyword-Based UI (Fig. 5b). The results confirms the impact of ConceptMap on lowering participants temporal demand.

At the end of the study, we asked participants to share their impressions on strengths and weakness of the ConceptMap. All participants were agree that

ConceptMap could enhance users comprehension and sensemaking of the information space. One of the participants noted to the Evidence Box, allowing her to store potentially relevant concepts altogether in one place and shortening the time needs to examine the relevancy of concepts and documents. Another participant mentioned that providing several summaries of the information space could help him to formulate his preferences from different perspectives. Another participant mentioned to the potential of ConceptMap to support users in exploratory search scenarios where there is no well-defined goal.

6 Conclusion

In this paper, we introduced the ConceptMap, a system that automatically identifies the relation between attributes, e.g., keywords, named entities, within the information space. ConceptMap produces several summaries of data, e.g., Topic, Category, Person, Organization, and allows a user to formulate her preferences implicitly as a set of abstract concepts. To generate the summaries, ConceptMap relies on a Knowledge Lake and a deep learning skip gram network, which groups attributes based on their conceptual similarity. Our results showed that providing a conceptual summary of the information space allows a user to better formulate her preferences especially when seeking for varied information in a large information space.

Acknowledgements. We Acknowledge the AI-enabled Processes (AIP) Research Centre for funding part of this research.

We Acknowledge the Data to Decisions CRC (D2D CRC) and the Cooperative Research Centres Program for funding part of this research.

References

1. Alexander, E., Kohlmann, J., Valenza, R., Witmore, M., Gleicher, M.: Serendip: topic model-driven visual exploration of text corpora. In: 2014 IEEE Conference on Visual Analytics Science and Technology (VAST), pp. 173–182. IEEE (2014)
2. Beheshti, A., Benatallah, B., Nouri, R., Chhieng, V.M., Xiong, H., Zhao, X.: CoreDB: a data lake service. In: Proceedings of the 2017 ACM on Conference on Information and Knowledge Management, CIKM 2017, Singapore, 06–10 November 2017, pp. 2451–2454 (2017)
3. Beheshti, A., Benatallah, B., Nouri, R., Tabebordbar, A.: CoreKG: a knowledge lake service. PVLDB **11**(12), 1942–1945 (2018)
4. Beheshti, A., Benatallah, B., Tabebordbar, A., Motahari-Nezhad, H.R., Barukh, M.C., Nouri, R.: DataSynapse: a social data curation foundry. Distrib. Parallel Databases **37**(3), 351–384 (2019)
5. Beheshti, A., Vaghani, K., Benatallah, B., Tabebordbar, A.: CrowdCorrect: a curation pipeline for social data cleansing and curation. In: Mendling, J., Mouratidis, H. (eds.) CAiSE 2018. LNBIP, vol. 317, pp. 24–38. Springer, Cham (2018). https://doi.org/10.1007/978-3-319-92901-9_3

6. Beheshti, S., Benatallah, B., Venugopal, S., Ryu, S.H., Motahari-Nezhad, H.R., Wang, W.: A systematic review and comparative analysis of cross-document coreference resolution methods and tools. Computing **99**(4), 313–349 (2017)

7. Beheshti, S., Tabebordbar, A., Benatallah, B., Nouri, R.: On automating basic data curation tasks. In: Proceedings of the 26th International Conference on World Wide Web Companion, Perth, Australia, 3–7 April 2017, pp. 165–169 (2017)

8. Blei, D.M., Ng, A.Y., Jordan, M.I.: Latent Dirichlet allocation. J. Mach. Learn. Res. **3**, 993–1022 (2003)

9. Brooke, J., et al.: SUS-a quick and dirty usability scale. Usability Eval. Ind. **189**(194), 4–7 (1996)

10. Carterette, B., Bennett, P.N., Chickering, D.M., Dumais, S.T.: Here or there preference judgments for relevance. In: Macdonald, C., Ounis, I., Plachouras, V., Ruthven, I., White, R.W. (eds.) ECIR 2008. LNCS, vol. 4956, pp. 16–27. Springer, Heidelberg (2008). https://doi.org/10.1007/978-3-540-78646-7_5

11. di Sciascio, C., Brusilovsky, P., Veas, E.: A study on user-controllable social exploratory search. In: 23rd International Conference on Intelligent User Interfaces. ACM (2018)

12. di Sciascio, C., Sabol, V., Veas, E.E.: Rank as you go: user-driven exploration of search results. In: Proceedings of the 21st International Conference on Intelligent User Interfaces, pp. 118–129. ACM (2016)

13. Dinet, J., Favart, M., Passerault, J.-M.: Searching for information in an online public access catalogue (OPAC): the impacts of information search expertise on the use of boolean operators. J. Comput. Assist. Learn. **20**(5), 338–346 (2004)

14. Fast, E., Chen, B., Bernstein, M.S.: Empath: understanding topic signals in large-scale text. In: Proceedings of the 2016 CHI Conference on Human Factors in Computing Systems, pp. 4647–4657. ACM (2016)

15. Gratzl, S., Lex, A., Gehlenborg, N., Pfister, H., Streit, M.: LineUp: visual analysis of multi-attribute rankings. IEEE Trans. Vis. Comput. Graph. **19**(12), 2277–2286 (2013)

16. Gretarsson, B., et al.: TopicNets: visual analysis of large text corpora with topic modeling. ACM Trans. Intell. Syst. Technol. (TIST) **3**(2), 23 (2012)

17. Hearst, M.A.: TileBars: visualization of term distribution information in full text information access. In: CHI, vol. 95, pp. 59–66 (1995)

18. Hu, Y., Boyd-Graber, J., Satinoff, B., Smith, A.: Interactive topic modeling. Mach. Learn. **95**(3), 423–469 (2014)

19. Klouche, K., Ruotsalo, T., Cabral, D., Andolina, S., Bellucci, A., Jacucci, G.: Designing for exploratory search on touch devices. In: 33rd Annual ACM Conference on Human Factors in Computing Systems, pp. 4189–4198. ACM (2015)

20. Liu, S., Zhou, M.X., Pan, S., Qian, W., Cai, W., Lian, X.: Interactive, topic-based visual text summarization and analysis. In: Proceedings of the 18th ACM Conference on Information and Knowledge Management, pp. 543–552. ACM (2009)

21. Marchionini, G.: Exploratory search: from finding to understanding. Commun. ACM **49**(4), 41–46 (2006)

22. Mikolov, T., Sutskever, I., Chen, K., Corrado, G.S., Dean, J.: Distributed representations of words and phrases and their compositionality. In: Advances in Neural Information Processing Systems, pp. 3111–3119 (2013)

23. Peltonen, J., Belorustceva, K., Ruotsalo, T.: Topic-relevance map: visualization for improving search result comprehension. In: Proceedings of the 22nd International Conference on Intelligent User Interfaces, pp. 611–622. ACM (2017)

24. Pirolli, P., Card, C.: The sensemaking process and leverage points for analyst technology as identified through cognitive task analysis. In: Proceedings of International Conference on Intelligence Analysis, pp. 2–4 (2005)
25. Sauro, J., Lewis, J.R.: Standardized usability questionnaires. In: Quantifying the User Experience, pp. 185–240 (2012)
26. Sultanum, N., Singh, D., Brudno, M., Chevalier, F.: Doccurate: a curation-based approach for clinical text visualization. IEEE Trans. Vis. Comput. Graph. **25**(1), 142–151 (2019)
27. Sun, C., Rampalli, N., Yang, F., Doan, A.: Chimera: large-scale classification using machine learning, rules, and crowdsourcing. Proc. VLDB Endowment **7**(13), 1529–1540 (2014)
28. Tabebordbar, A., Beheshti, A.: Adaptive rule monitoring system. In: Proceedings of the 1st International Workshop on Software Engineering for Cognitive Services, SE4COG@ICSE 2018, Gothenburg, Sweden, 28–29 May 2018, pp. 45–51 (2018)
29. Wall, E., Das, S., Chawla, R., Kalidindi, B., Brown, E.T., Endert, A.: Podium: ranking data using mixed-initiative visual analytics. IEEE Trans. Vis. Comput. Graph. **24**(1), 288–297 (2018)

Helpfulness Prediction for Online Reviews with Explicit Content-Rating Interaction

Jiahua Du[1], Jia Rong[1,2(✉)], Hua Wang[1], and Yanchun Zhang[1]

[1] Institute of Sustainable Industries and Liveable Cities, Victoria University, Melbourne, VIC, Australia
jiarong@acm.org
[2] Faculty of Information Technology, Monash University, Clayton, VIC, Australia

Abstract. Automatic helpfulness prediction aims to prioritize online product reviews by quality. Existing methods have combined review content and star ratings for automatic helpfulness prediction. However, the relationship between review content and star ratings is not explicitly captured, which limits the capability of rating information in influencing review content. This paper proposes a deep neural architecture to learn the explicit content-rating interaction (ECRI) for automatic helpfulness prediction. Specifically, ECRI explores two methods to interact review content with star ratings and adaptively specify the amount of rating information needed by review content. ECRI is evaluated against state-of-the-art methods on six real-world domains of the Amazon 5-core dataset. Experimental results demonstrate that exploiting the explicit content-rating interaction improves automatic helpfulness prediction. The source code of ECRI can be obtained from https://github.com/tokawah/ECRI.

Keywords: E-commerce · Review helpfulness · Explicit content-rating interaction · Deep learning

1 Introduction

Automatic helpfulness prediction for online product reviews plays a significant role in current e-commerce fields. With the proliferation of online product reviews, many customers rely on collective wisdom to make informed purchase decisions. Nonetheless, the quality of online product reviews is diverse and unpredictable [25], depending on reviewers' education backgrounds, social statuses, and moods. With the number of reviews continuing to increase, locating useful information becomes challenging. Although e-commerce platforms gather users' opinions on review helpfulness for quality assessment, in practice, the voting data is scarce or even missing in less popular products. The proposed automatic helpfulness prediction aims to learn from the voting data to identify and recommend high-quality reviews to customers.

Previous literature on automatic helpfulness prediction [1,11,30] highly relies on review content and star ratings that respectively represent quantitative and

© Springer Nature Switzerland AG 2019
R. Cheng et al. (Eds.): WISE 2019, LNCS 11881, pp. 795–809, 2019.
https://doi.org/10.1007/978-3-030-34223-4_50

qualitative aspects [48] of product reviews. Review content describes reviewer opinions toward product properties [44,45] where the majority of helpful information is located. The accompanying star rating employs another more straightforward form to summarize the views, and its extremity affects customers in perceiving helpfulness [39]. More importantly, the interaction between review content and star ratings [42] also affects a consumer's attitude/trustworthiness towards a review. Since star ratings are subjective reflecting on self needs [20,38], the valence (positive or negative) of review content can differ from that of star ratings. The content-rating interaction imitates customers measuring the consistency between text valence and rating valence on aspects of product reviews, which is essential to helpfulness prediction.

Nonetheless, existing methods have yet to capture the explicit content-rating interaction when combining both features. In most studies [9,36,41], review content is transformed into a feature vector where the constituent elements encode latent aspects of a product. The content vector is then directly concatenated with raw rating values. Such a simple method ignores that star ratings result from customers summarizing different opinions toward a product. The significant dimensional imbalance also limits the capacity of star ratings. In [34], the authors decompose star ratings by similarly mapping each rating value into a feature vector. For interaction, the learning of a rating vector is coupled with the encoding of a content vector, assuming that a star rating is entailed in review content. Although enlarged, the capacity of rating information is still locally affected by review content. Review content and star ratings are two measures for the same customer opinion. Capturing the element-wise relationship between content vectors and rating vectors (i.e., explicit content-rating interaction) can hopefully facilitate rating information contributing to helpfulness modeling.

Inspired by [48], this paper proposes a deep neural architecture to learn the explicit content-rating interaction (ECRI) for automatic helpfulness prediction. To this end, both review content and star ratings are first embedded into feature vectors of the same dimensionality. Different from previous work, the learning of rating vectors is decoupled from the encoding of review content to maintain the global influence of rating information. The explicit content-rating relationship is then established by aligning the content and rating vectors, forcing each element pair to encode the same latent aspect of product reviews. ECRI explores two methods to combine the aligned content and rating vectors to obtain representations used for helpfulness prediction. The valence difference between review content and star ratings may affect the perceived helpfulness variously. ECRI further adopts gating mechanisms to adaptively learn the amount of rating information needed by content vectors during combination.

In a nutshell, the main contribution is the first work explicitly modeling the interaction between review content and star ratings in predicting review helpfulness. Additionally, the amount of rating information needed for each content is learned adaptively. Extensive experiments on six real-world domains of online product reviews show that ECRI can exploit the explicit content-rating interaction to improve automatic helpfulness prediction for online product reviews and outperforms state-of-the-art methods.

The remaining of the paper is organized as follows. Section 2 presents related work. Section 3 introduces the overall framework of ECRI and its learning components. Section 4 describes the datasets and experiment settings. Section 5 demonstrates the effectiveness of ECRI, along with ablation studies and discussions on the use of ECRI components. Finally, Sect. 6 concludes the paper.

2 Related Work

Recent studies have shown the feasibility of using deep learning for automatic helpfulness prediction. Currently, review content is largely used for feature learning since it contains rich information of reviews. The work that combines review content and star ratings is also gaining increasing attention. This section introduces current methods using sole review content (Sect. 2.1) and the conjunction of review content and star ratings (Sect. 2.2), respectively.

2.1 Automatic Helpfulness Prediction

The emergence of deep learning [22, 27, 28, 31] brings new paradigms into automatic helpfulness prediction. With the help of neural architectures, features used for model training can be extracted automatically, bypassing the procedure of laborious feature engineering [30] used in traditional machine learning methods [12–15, 23, 26, 32, 43, 46].

In particular, recent studies based on Convolution Neural Networks (CNNs) [17, 18, 24] have shown the feasibility in modeling helpfulness in an end-to-end manner. The main idea behind these methods is to learn automatically continuous representations that encode context-aware semantics from review content. Chen et al. [2] propose Embedding-gated CNN (EG-CNN) for review helpfulness prediction. This work adopts a CNN framework to learn multi-granularity text features from review content. Based on the assumption that words in a review may contribute diversely to its helpfulness, the authors suggest using gates to control word embeddings to be fed into the CNN model. To alleviate the out-of-vocabulary problem commonly occurred in small datasets, character-level embeddings [3] learned via another CNN framework are incorporated into the word embeddings before gating.

Following previous works, this paper adopts CNNs [17] as the base model to learn features from review content. Similar to [2], gating mechanisms are taken to control the flow of intermediate features, but the gates are calculated differently and instead used to combine word embeddings and the convoluted features for the final content representations.

2.2 Interaction Between Review Content and Star Ratings

In addition to review content, emotions expressed by reviewers such as star ratings, are employed to enhance helpfulness encoding. In [34], Qu et al. propose two CNN frameworks to combine rating information and review content. The first combination method (CM1) performs feature fusion by simply concatenating raw star ratings and the learned content representations. The second combination method (CM2) treats star ratings as new vocabulary and trains rating embeddings as individual words of a review along with other word embeddings to learn the content representations.

Fan et al. [8] formulate review helpfulness prediction as a multi-task neural learning (MTNL) problem. Specifically, a CNN framework is first employed to learn continuous features from review content. In this work, instead of treating review star ratings as input data, the learned content representations are used as shared features to predict both review helpfulness and raw star ratings.

As shown in Fig. 1, although combining review content and star ratings has shown promise in predicting helpfulness, the aforementioned methods either limit the capacity of rating information or fail to explicitly capture the interaction between the two features, which constrains rating information from providing more direct and accurate details to the learned content representations.

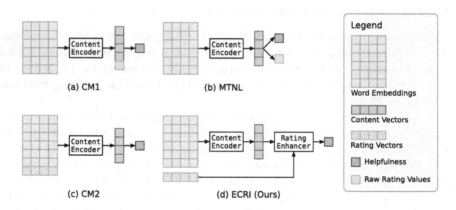

Fig. 1. ECRI captures the explicit content-rating interaction.

This paper embeds ratings separately from the encoding of review content to allows for the global influence of rating information on the learned content representations. Inspired by [48], rating embeddings are set to the same dimensionality as the learned content representations to perform the element-wise aligning between the two features. Two rating enhancement methods are then proposed to fulfill the content-rating interaction, where rating embeddings are gated to adaptively adjusted the amount of rating information for latent aspects of the learned content representations.

To the best of our knowledge, our work is the first deep learning approach to combine rating information and review content while modeling their explicit relationship for automatic helpfulness prediction for product reviews.

3   Explicit Content-Rating Interaction Networks

The prediction of review helpfulness is formulated as a binary classification task. Given a review, ECRI aims to learn from its content s and accompanying star rating r to predict a value $\hat{y}$ that approximates review helpfulness $y \in \{0, 1\}$, where 0 is unhelpful and 1 helpful. The helpfulness labels $\mathbf{y}$ depend on human assessment, which will be discussed in Sect. 4.

As Fig. 2 illustrates, ECRI involves two learning phases: (1) the content encoder learns hidden content representations $\mathbf{h}$ from review content via CNNs; and (2) the rating enhancer establishes the explicit content-rating relationship and incorporates rating information $\mathbf{e}_r'$ into the learned content representations. Figure 2 illustrates the ECRI architecture.

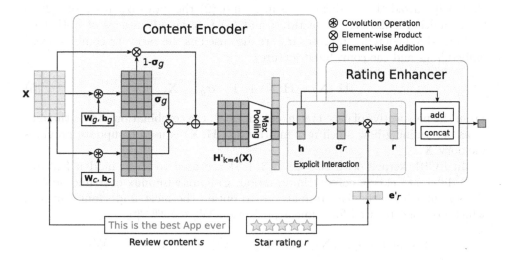

Fig. 2. The ECRI architecture.

3.1   Content Encoder

Given a collection of raw online product reviews $\mathcal{S}$, each review $s \in \mathcal{S}$ is tokenized into a sequence of n words $s = (x_1, x_2, \ldots, x_n)$. The vocabulary V is constructed by indexing all unique words in $\mathcal{S}$. The content encoder begins by associating each word $x \in s$ in the review with a d-dimensional word vector $\mathbf{e}_x$. Given an embedding lookup table $\mathbf{E} \in \mathbb{R}^{|V| \times d}$ where each row is a word vector corresponding to the vocabulary, x is encoded using the one-hot encoding scheme into $\mathbf{x} \in \mathbb{R}^{|V|}$ to select the corresponding word vector $\mathbf{e}_x$. Therefore, s can be represented by an embedding matrix $\mathbf{X} \in \mathbb{R}^{n \times d}$:

$$\mathbf{X} = [\mathbf{e}_{x_1}, \mathbf{e}_{x_2}, \ldots, \mathbf{e}_{x_n}], \tag{1}$$

$$\mathbf{e}_x = \mathbf{E}^\top \mathbf{x}. \tag{2}$$

The content encoder extracts hidden features from each input sample via Gated Linear Units (GLUs) [7]. Specifically, two sets of CNNs are applied to **X** for separate convoluted matrices. The hidden features $\mathbf{H}(\mathbf{X})$ are obtained by gating one matrix with the other whose values are squashed between 0 and 1. More formally,

$$\mathbf{H}(\mathbf{X}) = (\mathbf{X} * \mathbf{W}_c + \mathbf{b}_c) \otimes \boldsymbol{\sigma}_g, \tag{3}$$

$$\boldsymbol{\sigma}_g = \sigma(\mathbf{X} * \mathbf{W}_g + \mathbf{b}_g), \tag{4}$$

where kernels $\{\mathbf{W}_c, \mathbf{W}_g\} \in \mathbb{R}^{k \times d \times m}$ and biases $\{\mathbf{b}_c, \mathbf{b}_g\} \in \mathbb{R}^m$ are parameters to be estimated, k and m are the patch size and number of kernels for convolution, σ is the sigmoid function, and $\otimes$ is the element-wise product between matrices.

Using GLUs is beneficial: (1) GLUs provide a multiplicative skip connection that helps avoid the gradient vanishing; and (2) the gates $\boldsymbol{\sigma}_g$ resemble [2] to learn multi-granularity text features; and (3) to take advantages of both low- and high-level context, the gates $\boldsymbol{\sigma}_g$ are also used as the ratios to combine word embeddings and the convoluted features.

$$\mathbf{H}'(\mathbf{X}) = \mathbf{H}(\mathbf{X}) + (1 - \boldsymbol{\sigma}_g) \otimes \mathbf{X}. \tag{5}$$

From the perspective of gated recurrent units [4], the combination can be thought of as determining how much new information $\mathbf{H}(\mathbf{X})$ is used to update previous memory **X**.

In ECRI, kernels with size $k = \{3, 4, 5\}$ are used to learn hidden features with different fields of context information. Column-wise max-overtime pooling [5] is applied over the feature matrices to obtain the most important features, which are concatenated for a linear projection $\mathbf{W}_h \in \mathbb{R}^{3m \times m}$:

$$\mathbf{h} = [\max\{\mathbf{H}'_{k=3}(\mathbf{X})\}; \max\{\mathbf{H}'_{k=4}(\mathbf{X})\}; \max\{\mathbf{H}'_{k=5}(\mathbf{X})\}]\mathbf{W}_h, \tag{6}$$

where $[;]$ is the concatenation among the pooled feature vectors, $\mathbf{h} \in \mathbb{R}^m$ is the continuous representation of review content.

3.2   Rating Enhancer

To incorporate rating information into the learned content representations, star ratings are first embedded into individual feature vectors. Let $r = \{1, 2, 3, 4, 5\}$ be a raw star rating value. Similar to the word embedding process, each rating value is associated with a m-dimensional feature vector $\mathbf{e}'_r \in \mathbb{R}^m$ via another lookup table $\mathbf{E}' \in \mathbb{R}^{5 \times m}$. Rating embeddings not only allow for larger capacity for encoding rating information, but also are more robust to noise since raw rating values are distributed.

The rating enhancer establishes the explicit content-star connection via two constraints: (1) the rating embedding $\mathbf{e}'_r$ is set to have the same dimensionality as the learned content representation $\mathbf{h}$ for element alignment, reinforcing the correspondence between both feature vectors on a dimension level; and (2) unlike previous work, $\mathbf{e}_r'$ is defined separately from the encoding of review content to maintain the global influence of rating information. Consequently, a star rating can interact with review content for more direct and accurate rating details.

The rating enhancer steps further to adjust the rating embedding $\mathbf{e}_r'$ before incorporation. In practice, the influence of reviewers' ratings on customers in perceiving helpfulness varies depending on the mentions in a review. In other words, each part of the review content has diverse requirements for rating information. To enable such flexibility, a fully-connected layer parameterized by the weight $\mathbf{W}_r \in \mathbb{R}^{m \times m}$ and bias $\mathbf{b}_r \in \mathbb{R}^m$ is built upon the learned content representations $\mathbf{h}$ to estimate the amount of rating information needed by each review.

$$\boldsymbol{\sigma}_r = \sigma(\mathbf{W}_r^\top \mathbf{h} + \mathbf{b}_r), \tag{7}$$

$$\mathbf{r} = \mathbf{e}_r' \otimes \boldsymbol{\sigma}_r. \tag{8}$$

The gates $\boldsymbol{\sigma}_r$ are determined adaptively based on the learned content representation $\mathbf{h}$ to amplify or reduce the influence of star ratings on individual content dimensions. The adjusted rating embedding $\mathbf{r}$ imitates a more realistic situation that review content may have sway over customers' perception of review emotions. It is worth noting that setting the gates $\boldsymbol{\sigma}_r$ to all ones will allow full flow of rating information, whereas all zeros switch off rating enhancement.

Two rating enhancement methods are explored to combine the learned content representations $\mathbf{h}$ and adjusted rating embeddings $\mathbf{r}$. Since review content often contains emotional words, the learned content representations are already encoded with certain forms of internal emotions, for example, text valance. Given that the source and amount of internal emotions differ from rating information, one type of emotions can be incompatible with another. ECRI incorporates rating information based on the compatibility of the two emotion types.

Element-Wise Addition. The first method assumes that the internal emotions and rating information tend to be more homogeneous. As such, the adjusted rating embedding $\mathbf{r}$ is used to enrich the learned content representation $\mathbf{h}$ in an element-wise manner.

$$\hat{\mathbf{h}} = \mathbf{h} + \mathbf{r}. \tag{9}$$

Concatenation. On the contrary, the second method assumes the less homogeneity between the internal emotions and rating information. Thus, the adjusted rating embedding $\mathbf{r}$ is used as new information to supply the learned content representation h with additional dimensions.

$$\hat{\mathbf{h}} = [\mathbf{h}; \mathbf{r}]. \tag{10}$$

The feature vector $\hat{\mathbf{h}}$ represents the rating-enhanced review content. For simplicity, the two combination methods are henceforth called ECRI_{add} and $\text{ECRI}_{\text{concat}}$, respectively.

3.3   Training

The rating-enhanced content representation $\hat{\mathbf{h}}$ is forwarded into a dropout layer, followed by logistic regression to predict the helpfulness of the current review.

$$\hat{y} = \sigma(\mathbf{W}_o^\top \hat{\mathbf{h}} + b_o). \tag{11}$$

ECRI is trained via cross entropy minimization over N training samples:

$$\mathcal{L} = -\frac{1}{N}[\mathbf{y}^\top \log(\hat{\mathbf{y}}) + (1 - \mathbf{y})^\top \log(1 - \hat{\mathbf{y}})], \tag{12}$$

where $\hat{\mathbf{y}}$ are the predicted helpfulness labels and $\mathbf{y}$ actual helpfulness labels.

4   Experiment Setup

ECRI is evaluated through a series of experiments. Section 4.1 introduces datasets used throughout the experiments. Section 4.2 describes baseline methods for performance comparison. Section 4.3 presents hyperparameters for training ECRI and the baseline models.

4.1   Datasets

Experiments are conducted on the publicly available Amazon 5-core dataset [10] covering 142.8 million online product reviews collected between May 1996 and July 2014. Six largest domains of the dataset are selected.

  The following pre-processing steps are applied to all domains: (1) remove non-English reviews, empty reviews, and nearly identical reviews [6] in line with [16]; and (2) discard reviews with less than 10 votes [35,37,40] to alleviate the effect of words of few mouths [47], i.e., voting biases; and (3) lowercase, tokenize, and remove articles from the remaining reviews; and (4) keep the most frequent 30k terms as vocabulary, and replace the others with the <UNK> token; and (5) normalize raw star ratings into $r \in \{0.2, 0.4, 0.6, 0.8, 1\}$.

  Following [9,21,28], a product review is labeled as helpful if the percentage of received helpful votes is larger than 0.6 and unhelpful otherwise. An equal amount of helpful and unhelpful reviews are randomly selected to construct datasets with balanced labels. Each dataset is randomly split into 80%, 10%, and 10% for training, validation, and testing, in a stratified manner. Descriptive statistics of the datasets are in Table 1.

4.2   Baseline Methods

ECRI is benchmarked against seven baselines, including unigram TFIDF representations, average domain-specific word embeddings, and five state-of-the-art deep learning methods for helpfulness prediction.

- **TFIDF+SVM**: Linear SVM trained on unigram TFIDF of reviews, with document frequency no less than 1% of the training samples.

Table 1. Descriptive statistics of the balanced datasets.

	Domain	#Reviews	#Tokens	#Sentences	$\frac{\#\text{Tokens}}{\#\text{Reviews}}$	$\frac{\#\text{Tokens}}{\#\text{Sentences}}$	$\frac{\#\text{Sentences}}{\#\text{Reviews}}$
D1	Apps for Android	20,416	1,184,650	107,702	58.03	11.00	5.28
D2	Video Games	23,100	7,522,835	469,856	325.66	16.01	20.34
D3	Electronics	33,962	8,255,411	537,996	243.08	15.34	15.84
D4	CDs and Vinyl	105,934	23,096,933	1,468,718	218.03	15.73	13.86
D5	Movies and TV	164,052	40,549,434	2,510,899	247.17	16.15	15.31
D6	Books	306,430	71,632,822	4,405,047	233.77	16.26	14.38

- **EMB+SVM**: Linear SVM trained on the average of pre-trained domain-specific word embeddings of reviews, with out-of-vocabulary words ignored.
- **CNN** [17]: The vanilla CNN architecture for sentence classification.
- **EG-CNN** [2]: A variant of the vanilla CNN architecture where character embeddings and word-level embedding gates are used before convolution.
- **CM1** [34]: A variant of the vanilla CNN architecture where raw rating values and the learned content representations are concatenated.
- **CM2** [34]: A variant of the vanilla CNN architecture where rating vectors and word embeddings are concatenated to learn content representations.
- **MTNL** [8]: A variant of the vanilla CNN architecture for multi-task learning, with character and word embeddings as inputs, attention on the convoluted feature maps, and raw rating regressing as the secondary task.

4.3 Hyperparameters

The lookup table **E** in neural architectures is initialized with domain-specific word embeddings trained using the continuous Skip-gram model [29] on the whole product review collection. During training, word vectors are kept non-static in CNN and static in other neural architectures, which is determined by the validation set of each domain. The lookup table **E**$'$ for mapping raw rating values is randomly initialized from a uniform distribution in range $[-0.05, 0.05]$.

ECRI is trained using: $d = m = 200$, rectified linear units, dropout rate of 0.5, mini-batch size of 64, and early stopping when the validation loss has no improvement for 10 epochs. Neural weights are updated through stochastic gradient descent over shuffled mini-batches using the Adam [19] update rule.

The neural baselines are re-implemented following the original hyperparameter setting in the papers except word vector initialization. For SVM classifiers, the penalty term C is chosen via a grid search of $\{0.01, 0.1, 1\}$.

5 Result Analysis

The empirical evaluation is conducted in three steps: (1) Sect. 5.1 demonstrates the effectiveness of ECRI; and (2) Sect. 5.2 performs ablation studies to validate the ECRI components; and (3) Sect. 5.3 discusses the behavior and choice between the two rating enhancement methods ECRI$_{\text{add}}$ and ECRI$_{\text{concat}}$.

5.1 Comparison with Baseline Methods

Table 2 reports the results of ECRI against the baseline models in helpfulness prediction. The performance on the six balanced datasets is evaluated by classification accuracy. The highest results are in bold, whereas results higher than the baselines are in italics. For result reliability, all neural models are trained and evaluated five times on each domain to report the average accuracy. SVM-based models are run once since the results are deterministic.

Table 2. Results of ECRI against other methods.

Model	D1	D2	D3	D4	D5	D6
TFIDF+SVM	67.68	76.71	75.66	82.52	78.58	75.03
EMB+SVM	68.76	75.54	74.72	81.97	77.92	74.32
CNN	70.38	77.60	77.50	84.04	80.76	77.81
EG-CNN	70.60	78.21	78.63	85.01	81.50	78.38
CM1	71.09	77.82	78.58	84.85	81.37	78.26
CM2	71.00	77.99	79.37	85.39	81.49	78.52
MTNL	67.79	75.60	75.21	82.45	78.42	75.72
ECRI$_{add}$	**72.24****	**79.00****	_80.06_*	_87.01_**	**83.58****	_80.45_**
ECRI$_{concat}$	_72.04_**	_78.37_	**80.22****	**87.22****	_83.50_**	**80.57****

$^{*}p < 0.05$, $^{**}p < 0.01$.

In brief, both ECRI$_{add}$ and ECRI$_{concat}$ outperform the baseline models by approximately 1%–5% in accuracy across all datasets. Among the baselines, CM2 beats CM1 and achieves the closest performance to ECRI due to the vectorized encoding of rating information, which can be thought of as an implicit form of content-rating interaction. In particular, CM2 gives similar performance to ECRI on D3, showing that rating information has larger local influence on the encoding of review content; on the other domains, however, ECRI is about 2% higher in accuracy than CM2.

5.2 Ablation Studies

To better understand the effectiveness of ECRI, four model variants are considered to alter the adaptively-learned gates σ_r in Equation (8) to validate: (i) the content encoder; (ii) the explicit content-rating interaction; and (iii) the gating mechanisms for adaptive rating information assignment. The first two variants, called ECRI$_{add+full\ rating}$ and ECRI$_{concat+full\ rating}$ respectively, set $\sigma_r = 1$ for both combination methods ECRI$_{add}$ and ECRI$_{concat}$, allowing all rating information to flow into the learned content representations $\mathbf{h}$. The third variant, called ECRI$_{plain}$, sets $\sigma_r = 0$ to exclude all rating interaction from ECRI, using the learned content representations $\mathbf{h}$ to predict review helpfulness directly. The fourth variant, called ECRI$_{plain+raw\ ratings}$, concatenates raw rating values and the learned content representations $\mathbf{h}$, which is similar to CM1. Table 3 illustrates the accuracy of four ECRI variants.

Table 3. The performance of ECRI with different gate settings.

Variants	D1	D2	D3	D4	D5	D6
1 $ECRI_{add+full\ rating}$	71.97	78.23	80.04	86.80	82.90	80.00
2 $ECRI_{concat+full\ rating}$	72.33	78.83	79.86	86.89	82.93	79.92
3 $ECRI_{plain}$	70.35	77.89	78.81	85.09	81.55	78.55
4 $ECRI_{plain+raw\ ratings}$	70.36	77.38	78.79	85.12	81.43	78.75

Effectiveness of the Review Content Encoder. $ECRI_{plain}$ is more capable of helpfulness prediction than other baseline encoders. This mainly relies on the gated combination utilizing both high- and low-level contextual text features. As shown in the table: (i) $ECRI_{plain}$ outperforms CNN over all domains except D1 and EG-CNN except D1 and D2; and (ii) $ECRI_{plain}$ achieves comparable results on D1 and superior results in the remaining domains to CM1; and (iii) $ECRI_{plain}$ even outperforms CM2 on D5 and D6. Compared with EG-CNN, $ECRI_{plain}$ is less effective on D1 and D2 because EG-CNN adopts character-level information to tackle the out-of-vocabulary issue in small datasets.

Effectiveness of the Explicit Interaction. Models ($ECRI_{add}$, $ECRI_{concat}$, $ECRI_{add+full\ rating}$, and $ECRI_{concat+full\ rating}$) capturing the explicit content-rating interaction significantly improve $ECRI_{plain}$ by about 1%–2% in accuracy, whereas the improvement of $ECRI_{plain+raw\ ratings}$ using raw ratings is trivial. This proves that embedding star ratings allows for larger capacity of encoding rating information. Decoupling the learning of rating embeddings from the encoding of review content helps maintain the influence of rating information. The alignment between content and rating vectors further provides more accurate and direct information flow.

Effectiveness of the Gating Mechanisms. Using gating mechanisms improves helpfulness prediction in most cases. The comparison between $ECRI_{add}$ and $ECRI_{add+full\ rating}$ and between $ECRI_{concat}$ and $ECRI_{concat+full\ rating}$ confirm that controlling the amount of rating information flowing into review content improves the explicit content-rating interaction. The reason for the improvement is that n-grams encoded into the learned content representations require different extent of rating information, for example, "the best movie" and "the movie is". The gating mechanisms handle the requirement by assigning adaptive weights to each rating dimension.

5.3   Comparison of Rating Enhancement Methods

As Table 2 shows, both rating enhancement methods used in ECRI show comparable performance, with $ECRI_{add}$ slightly outperforming $ECRI_{concat}$ on D1, D2, and D5. Such a phenomenon can be related to the text valence of reviews in a corpus. As Table 4 reports, the three domains where $ECRI_{add}$ outperforms also possess higher ratios of emotional components, allowing the learned content representation h to encode more internal emotions and thus be more homogeneous with the rating embedding r. Given that the ratios are not proportional to the performance gains, in practice, the choice between $ECRI_{add}$ and $ECRI_{concat}$ on new domains may require further domain-specific analysis.

Table 4. Average ratio of emotional words across domains via LIWC analysis [33].

	D1	D2	D3	D4	D5	D6
Positive emotion (%)	7.17	4.78	3.57	4.66	4.41	4.03
Negative emotion (%)	2.37	2.45	1.49	1.96	2.60	2.23
Sum	9.54	7.22	5.05	6.62	7.01	6.25

6   Conclusion and Future Work

The automatic prediction and recommendation of helpful online product reviews can assist customers in making informed purchase decisions. This paper has presented ECRI, a deep neural architecture for review helpfulness prediction that learns rating-enhanced content representations. In contrast to previous work using rating information only, ECRI focuses on modeling the explicit interaction between review content and star ratings as two individual elements. ECRI also assigns review content with different amount of rating information learned adaptively on a dimension level. Extensive experiments on six real-world datasets against the state-of-the-art methods show promise of ECRI in learning text features and utilizing rating information. Further analysis of the ECRI components demonstrates that both establishing the explicit content-rating interaction and using adaptive rating assignment are critical in improving prediction performance.

There are several future directions to be addressed. Extensive experiments will be conducted on hyperparameter tuning to investigate model performance further. The interaction between review content and star ratings can be constructed using attention mechanisms or sentence-level rating information. Developing a method that can cope with different levels of homogeneity between internal emotions and rating information is desired. Frequent interaction patterns can be summarized from the trained models for diverse helpfulness prediction. Further behavior analysis on the content-rating interaction and domain-specific characteristics is necessary.

References

1. Charrada, E.B.: Which one to read? Factors influencing the usefulness of online reviews for re. In: 2016 IEEE 24th International Requirements Engineering Conference Workshops (REW), pp. 46–52, September 2016
2. Chen, C., et al.: Review helpfulness prediction with embedding-gated CNN. CoRR abs/1808.09896 (2018)
3. Chen, C., Yang, Y., Zhou, J., Li, X., Bao, F.S.: Cross-domain review helpfulness prediction based on convolutional neural networks with auxiliary domain discriminators. In: Proceedings of the 2018 Conference of the North American Chapter of the Association for Computational Linguistics: Human Language Technologies, Volume 2 (Short Papers), pp. 602–607. Association for Computational Linguistics, New Orleans, Louisiana, June 2018
4. Chung, J., Gülçehre, Ç., Cho, K., Bengio, Y.: Empirical evaluation of gated recurrent neural networks on sequence modeling. CoRR abs/1412.3555 (2014)
5. Collobert, R., Weston, J., Bottou, L., Karlen, M., Kavukcuoglu, K., Kuksa, P.: Natural language processing (almost) from scratch. J. Mach. Learn. Res. **12**, 2493–2537 (2011)
6. Danescu-Niculescu-Mizil, C., Kossinets, G., Kleinberg, J., Lee, L.: How opinions are received by online communities: a case study on amazon.com helpfulness votes. In: Proceedings of the 18th International Conference on World Wide Web WWW 2009, pp. 141–150. ACM, New York (2009). https://doi.org/10.1145/1526709.1526729
7. Dauphin, Y.N., Fan, A., Auli, M., Grangier, D.: Language modeling with gated convolutional networks. In: Proceedings of the 34th International Conference on Machine Learning, vol. 70, pp. 933–941. JMLR.org (2017)
8. Fan, M., Feng, Y., Sun, M., Li, P., Wang, H., Wang, J.: Multi-task neural learning architecture for end-to-end identification of helpful reviews. In: 2018 IEEE/ACM International Conference on Advances in Social Networks Analysis and Mining (ASONAM), pp. 343–350, August 2018
9. Ghose, A., Ipeirotis, P.G.: Estimating the helpfulness and economic impact of product reviews: mining text and reviewer characteristics. IEEE Trans. Knowl. Data Eng. **23**(10), 1498–1512 (2011)
10. He, R., McAuley, J.: Ups and downs: modeling the visual evolution of fashion trends with one-class collaborative filtering. In: Proceedings of the 25th International Conference on World Wide Web. WWW 2016, International World Wide Web Conferences Steering Committee, Republic and Canton of Geneva, Switzerland, pp. 507–517 (2016). https://doi.org/10.1145/2872427.2883037
11. Hoffait, A.S., Ittoo, A., Schyns, M.: Assessing and predicting review helpfulness: Critical review, open challenges and research agenda. In: 29ème conférence européenne sur la recherche opérationnelle (EURO2018) (2018)
12. Hu, H., Li, J., Wang, H., Daggard, G.: Combined gene selection methods for microarray data analysis. In: Gabrys, B., Howlett, R.J., Jain, L.C. (eds.) KES 2006. LNCS (LNAI), vol. 4251, pp. 976–983. Springer, Heidelberg (2006). https://doi.org/10.1007/11892960_117
13. Huang, J., Peng, M., Wang, H., Cao, J., Gao, W., Zhang, X.: A probabilistic method for emerging topic tracking in microblog stream. World Wide Web **20**(2), 325–350 (2017)
14. Khalil, F., Li, J., Wang, H.: An integrated model for next page access prediction. IJ Knowl. Web Intell. **1**(1/2), 48–80 (2009)

15. Khalil, F., Wang, H., Li, J.: Integrating markov model with clustering for predicting web page accesses. In: Proceeding of the 13th Australasian World Wide Web Conference (AusWeb 2007), pp. 63–74. AusWeb (2007)
16. Kim, S.M., Pantel, P., Chklovski, T., Pennacchiotti, M.: Automatically assessing review helpfulness. In: Proceedings of the 2006 Conference on Empirical Methods in Natural Language Processing EMNLP 2006, pp. 423–430. Association for Computational Linguistics, Stroudsburg, PA, USA (2006)
17. Kim, Y.: Convolutional neural networks for sentence classification. In: Proceedings of the 2014 Conference on Empirical Methods in Natural Language Processing (EMNLP), pp. 1746–1751. Association for Computational Linguistics, Doha, Qatar, October 2014
18. Kim, Y., Jernite, Y., Sontag, D., Rush, A.M.: Character-aware neural language models. In: Proceedings of the Thirtieth AAAI Conference on Artificial Intelligence AAAI 2016, pp. 2741–2749. AAAI Press (2016)
19. Kingma, D.P., Ba, J.: Adam: A method for stochastic optimization. CoRR abs/1412.6980 (2014)
20. Kozinets, R.V.: Amazonian forests and trees: multiplicity and objectivity in studies of online consumer-generated ratings and reviews, a commentary on de Langhe, Fernbach, and Lichtenstein. J. Consum. Res. **42**(6), 834–839 (2016). https://doi.org/10.1093/jcr/ucv090
21. Krishnamoorthy, S.: Linguistic features for review helpfulness prediction. Expert Syst. Appl. **42**(7), 3751–3759 (2015)
22. Lee, S., Choeh, J.Y.: Predicting the helpfulness of online reviews using multilayer perceptron neural networks. Expert Syst. Appl. **41**(6), 3041–3046 (2014)
23. Li, H., Wang, Y., Wang, H., Zhou, B.: Multi-window based ensemble learning for classification of imbalanced streaming data. World Wide Web **20**(6), 1507–1525 (2017)
24. Li, L., Situ, R., Gao, J., Yang, Z., Liu, W.: A hybrid model combining convolutional neural network with XGBoost for predicting social media popularity. In: Proceedings of the 25th ACM International Conference on Multimedia, pp. 1912–1917. ACM (2017)
25. Liu, J., Cao, Y., Lin, C.Y., Huang, Y., Zhou, M.: Low-quality product review detection in opinion summarization. In: Proceedings of the 2007 Joint Conference on Empirical Methods in Natural Language Processing and Computational Natural Language Learning (EMNLP-CoNLL), pp. 334–342. Association for Computational Linguistics, Prague, Czech Republic, June 2007
26. Ma, J., Sun, L., Wang, H., Zhang, Y., Aickelin, U.: Supervised anomaly detection in uncertain pseudoperiodic data streams. ACM Trans. Internet Technol. (TOIT) **16**(1), 4 (2016)
27. Ma, Y., Xiang, Z., Du, Q., Fan, W.: Effects of user-provided photos on hotel review helpfulness: an analytical approach with deep leaning. Int. J. Hospitality Manage. **71**, 120–131 (2018)
28. Malik, M., Hussain, A.: Helpfulness of product reviews as a function of discrete positive and negative emotions. Comput. Hum. Behav. **73**, 290–302 (2017)
29. Mikolov, T., Sutskever, I., Chen, K., Corrado, G.S., Dean, J.: Distributed representations of words and phrases and their compositionality. In: Advances in Neural Information Processing Systems, pp. 3111–3119 (2013)
30. Ocampo Diaz, G., Ng, V.: Modeling and prediction of online product review helpfulness: a survey. In: Proceedings of the 56th Annual Meeting of the Association for Computational Linguistics (Volume 1: Long Papers), pp. 698–708. Association for Computational Linguistics, Melbourne, Australia, July 2018

31. Paul, D., Sarkar, S., Chelliah, M., Kalyan, C., Sinai Nadkarni, P.P.: Recommendation of high quality representative reviews in e-commerce. In: Proceedings of the Eleventh ACM Conference on Recommender Systems RecSys 2017, pp. 311–315. ACM, New York (2017). https://doi.org/10.1145/3109859.3109901
32. Peng, M., Zeng, G., Sun, Z., Huang, J., Wang, H., Tian, G.: Personalized app recommendation based on app permissions. World Wide Web **21**(1), 89–104 (2018)
33. Pennebaker, J.W., Boyd, R.L., Jordan, K., Blackburn, K.: The development and psychometric properties of LIWC2015. Technical report. The University of Texas at Austin (2015)
34. Qu, X., Li, X., Rose, J.R.: Review helpfulness assessment based on convolutional neural network. CoRR abs/1808.09016 (2018)
35. Roy, G., Datta, B., Mukherjee, S.: Role of electronic word-of-mouth content and valence in influencing online purchase behavior. J. Market. Commun. **0**(0), 1–24 (2018)
36. Salehan, M., Kim, D.J.: Predicting the performance of online consumer reviews: a sentiment mining approach to big data analytics. Decis. Support Syst. **81**, 30–40 (2016)
37. Scholz, M., Dorner, V.: The recipe for the perfect review? Bus. Inf. Syst. Eng. **5**(3), 141–151 (2013)
38. Schuckert, M., Liu, X., Law, R.: Insights into suspicious online ratings: direct evidence from tripadvisor. Asia Pac. J. Tourism Res. **21**(3), 259–272 (2016)
39. Shin, S., Chung, N., Xiang, Z., Koo, C.: Assessing the impact of textual content concreteness on helpfulness in online travel reviews. J. Travel Res. **58**(4), 579–593 (2019)
40. Siering, M., Muntermann, J.: What drives the helpfulness of online product reviews? from stars to facts and emotions. Wirtschaftsinformatik **7**, 103–118 (2013)
41. Siering, M., Muntermann, J., Rajagopalan, B.: Explaining and predicting online review helpfulness: the role of content and reviewer-related signals. Decis. Support Syst. **108**, 1–12 (2018)
42. Tsang, A.S., Prendergast, G.: Is a "star" worth a thousand words?: The interplay between product-review texts and rating valences. Eur. J. Market. **43**(11/12), 1269–1280 (2009)
43. Wang, K.N., Bell, J.S., Chen, E.Y.H., Gilmartin-Thomas, J.F.M., Ilomäki, J.: Medications and prescribing patterns as factors associated with hospitalizations from long-term care facilities: a systematic review. Drugs Aging **35**(5), 423–457 (2018)
44. Yang, Y., Chen, C., Bao, F.S.: Aspect-based helpfulness prediction for online product reviews. In: 2016 IEEE 28th International Conference on Tools with Artificial Intelligence (ICTAI), pp. 836–843, November 2016
45. Yang, Y., Yan, Y., Qiu, M., Bao, F.: Semantic analysis and helpfulness prediction of text for online product reviews. In: Proceedings of the 53rd Annual Meeting of the Association for Computational Linguistics and the 7th International Joint Conference on Natural Language Processing (Volume 2: Short Papers), pp. 38–44. Association for Computational Linguistics, Beijing, China, July 2015
46. Zhang, J., Tao, X., Wang, H.: Outlier detection from large distributed databases. World Wide Web **17**(4), 539–568 (2014)
47. Zhang, R., Tran, T., Mao, Y.: Opinion helpfulness prediction in the presence of "words of few mouths". World Wide Web **15**(2), 117–138 (2012)
48. Zhou, S., Guo, B.: The interactive effect of review rating and text sentiment on review helpfulness. In: Stuckenschmidt, H., Jannach, D. (eds.) EC-Web 2015. LNBIP, vol. 239, pp. 100–111. Springer, Cham (2015). https://doi.org/10.1007/978-3-319-27729-5_8

Author Index

Printed in the United States
By Bookmasters